The **Sustainable Development Goals Series** is Springer Nature's inaugural cross-imprint book series that addresses and supports the United Nations' seventeen Sustainable Development Goals. The series fosters comprehensive research focused on these global targets and endeavours to address some of society's greatest grand challenges. The SDGs are inherently multidisciplinary, and they bring people working across different fields together and working towards a common goal. In this spirit, the Sustainable Development Goals series is the first at Springer Nature to publish books under both the Springer and Palgrave Macmillan imprints, bringing the strengths of our imprints together.

The Sustainable Development Goals Series is organized into eighteen subseries: one subseries based around each of the seventeen respective Sustainable Development Goals, and an eighteenth subseries, "Connecting the Goals," which serves as a home for volumes addressing multiple goals or studying the SDGs as a whole. Each subseries is guided by an expert Subseries Advisor with years or decades of experience studying and addressing core components of their respective Goal.

The SDG Series has a remit as broad as the SDGs themselves, and contributions are welcome from scientists, academics, policymakers, and researchers working in fields related to any of the seventeen goals. If you are interested in contributing a monograph or curated volume to the series, please contact the Publishers: Zachary Romano [Springer; zachary.romano@springer.com] and Rachael Ballard [Palgrave Macmillan; rachael.ballard@palgrave.com].

Sustainable Development Goals Series

Mette Ramsgaard Thomsen ·
Carlo Ratti · Martin Tamke
Editors

Design for Rethinking Resources

Proceedings of the UIA World Congress of Architects Copenhagen 2023

Springer

Editors
Mette Ramsgaard Thomsen
CITA (Centre for Information Technology and Architecture)
The Royal Danish Academy—Architecture, Design, Conservation
Copenhagen, Denmark

Carlo Ratti
Senseable City Lab
Massachusetts Institute of Technology
Cambridge, MA, USA

Martin Tamke
CITA (Centre for Information Technology and Architecture)
The Royal Danish Academy—Architecture, Design, Conservation
Copenhagen, Denmark

ISSN 2523-3084 ISSN 2523-3092 (electronic)
Sustainable Development Goals Series
ISBN 978-3-031-36553-9 ISBN 978-3-031-36554-6 (eBook)
https://doi.org/10.1007/978-3-031-36554-6

This Springer imprint is published by the registered company Springer Nature Switzerland AG
The registered company address is: Gewerbestrasse 11, 6330 Cham, Switzerland

Scientific Committee

General Reporter, Alternate General Reporter

Mette Ramsgaard Thomsen, Professor and Head of CITA (Centre for Information Technology and Architecture), The Royal Danish Academy—Architecture, Design, Conservation.
Martin Tamke, Associate Professor, CITA (Centre for Information Technology and Architecture), The Royal Danish Academy—Architecture, Design, Conservation.

Panel Chairs

Panel 1: Design for Climate Adaptation.
Billie Faircloth, Research Director KieranTimberlake, Adjunct Professor University of Pennsylvania.
Maibritt Pedersen Zari, Educator and Researcher, Associate Professor, Auckland University of Technology.

Panel 2: Design for Rethinking Resources.
Carlo Ratti, Professor and Director of the Senseably City Lab, MIT, Founding Partner of CRA—Carlo Ratti Associati.
Mette Ramsgaard Thomsen, Professor and Head of CITA (Centre for Information Technology and Architecture), The Royal Danish Academy—Architecture, Design, Conservation.

Panel 3: Design for Resilient Communities.
Juan Du, Professor and Dean of the John H. Daniels Faculty of Architecture, Landscape and Design, University of Toronto.
Anna Rubbo, Senior Researcher, Center for Sustainable Urban Development, The Climate School, Columbia University.

Panel 4: Design for Health.
Arif Hasan, Chairperson Urban Resource Center Karachi, former Visiting Professor NED University Karachi, former member of UNs Advisory Group on Forced Evictions (PA)

Christian Benimana, Co-Executive Director and Senior Principal MASS Design Group

Panel 5: Design for Inclusivity.
Magda Mostafa, Autism Design Principal Progressive Architects, Professor of Design at the Department of Architecture, The American University in Cairo.
Ruth Baumeister, Associate Professor of Theory and History of Architecture, Aarhus School of Architecture.

Panel 6: Design for Partnerships for Change.
Sandi Hilal, Co-director DAAR (Decolonizing Architecture Art Research), Lise Meinert Visiting Professor, Lund University.
Merve Bedir, Fellow at BAK (Basis voor Actuele Kunste) Utrecht.

Special Advisors

Katherine Richardson, Professor in Biological Oceanography and Leader of Sustainability Science Centre, University of Copenhagen.
Chris Luebkeman, Leader of the Strategic Foresight Hub, Office of the President, ETH Zürich.
Thomas Bo Jensen, Head of Research, Aarhus School of Architecture.
Camilla Ryhl, Research Director, Bevica Fonden.

Peer Review Committee

We would like to thank all members of the Peer-Review Committee for this volume for their enduring effort and valuable advice.
Luca Breseghello,
Nadja Gaudilliere,
Niels Martin Larsen,
Fujing Ma,
Dagmar Reinhardt,
Diego Alejandro Velandia Rayo,
Adrien Rigobello,
Dana Saez,
Hugo Mulder,
Hao Wu,
Xan Browne,
Tomas Vivanco,
Tom Svilans,
Mia Münster,
Jonas Runberger,
Mathilde Marengo,
Mokolade Johnson,
Tobias Schwinn,
Saqib Aziz,
Paul Nicholas,
Aswini Balashanmugam,
Lotte M. Bjerregaard Jensen,
Shafique Rahman,
Tom Buckland,
Ingrid Halland,
David Jenny,
Deepika Raghu,
Anders Kruse Aagaard,
Oliver Tessmann,
Catherine De Wolf,
Kathy Velikov,
Thomas Wortmann,
Axel Kilian,
Aleksandra Anna Apolinarska,
Giovanni Betti,

Rodrigo Velasco,
Olga Popovic Larsen,
Kathrin Dörfler,
Philipp Eversmann,
Philippe Block,
Anita Ollár,
Brady Peters,
Markus Hudert,
Anja Kunic,
Hanaa Dahy,
Svenja Keune,
Anke Pasold,
Elise Elsacker,
Matthias Haeusler,
Vladimir Simon Montoya Torres,
David Andreen,
Stefanie Weidner,
Mariana Popescu,
Urszula Kozminska,
Timo Carl,
Nikol Kirova,
Marina Diamantidou,
Mohammad Habib Reza,
Jana Revedin,
Alexandre Monnin,
Henriette Bier,
Tim Schork,
Jan Wurm,
Felix Heisel,
Daniel Lee,
Yu Ye.

Preface

Introduction

In the autumn of 2022, as part of the preparations for the UIA World Congress of Architects 2023 Copenhagen we invited Panel Chair and MASS Design Group architect Christian Benimana to Copenhagen to speak to our collegiate and students. In his introduction he outlined the dramatic land use change in Rwanda following the country's population growth over the last 50 years. Pointing to the maps he argued that we have passed a tipping point and that our given societal infrastructures cannot simply be extended or optimized to support this new situation. We cannot build 500 universities or 600 hospitals, he said, instead we need systemic change to rethink what a university is, what a hospital can be. We need to question how our institutions, infrastructures and communities can change in the way they address those in need and what access can be, and therefore also how architecture, its practices, embedded knowledge and products can be methods of instigating change.

The present proceedings present six volumes examining the knowledge foundation for such change. As proceedings for the Science Track of the UIA World Congress of Architects 2023 Copenhagen Sustainable Futures—Leave no one behind, they contain a total of 296 papers investigating, show casing and arguing for how change can be imagined across the built environment. By asking how architecture can help achieving the UN Sustainable Development Goals (SDGs), the presented papers collect the research- and practice-based results of a global community. Together they ask what the future of the built environment can be and how design as action and as knowledge can create new roles for architecture and the communities it serves.

This preface starts with the articulation of our profound gratitude to the Scientific Committee and the community of submitting authors and peer reviewers that have been part of this effort. During the last two and half years we have worked together with the Scientific Committee's Panel Chairs and Special Advisors to form a vision for the Science Track. The process has been an education. Not only in our understanding of the SDGs, the transformative power of design creation or the wider societal role of the built environment, but also in keeping our minds open to the many positions that architecture can be thought through and its critical role in engaging—interfacing, informing and developing—different knowledge cultures and perspectives. We therefore start by thanking the 17 members of the Scientific Committee, the contributing 656 authors of the 296 accepted papers, the 1486 authors of the more than 750 paper submissions and the 536 peer reviewers that have all made this project possible.

Platform

The UIA World Congress 2023 Copenhagen starts with an ambition. Pitched in 2017, only 1 year after the launch of the UN Sustainable Development Goals, the central nerve is the articulation of the profound agency of architecture and how it plays an acute role in achieving the SDGs. In the congress, the Science Track is given a particular role. Initiated early in the planning process, the aim has been to place the Science Track at the heart of the congress in order to collect its underpinning knowledge foundation and shape its criticality through a broad outreach to a global community. Sustainability, like architecture, is a wicked problem. Its solutions are dependent on the way we ask, the methods we use and the contexts in which we work. To ask how architecture can be part of the dynamic fulfilment of the UN SDGs is to ask: who are the communities we design with and for, what is the knowledge we draw upon and how can its sharing change how we think about what our built environment can be.

One of the central drivers in our preparatory work for the Science Track has been the realizations of the blindness of the UN SDGs to the agency of architecture. The SDGs seek to steer behaviour both through impacting legislation and wider societal value sets. They establish priorities and galvanize efforts across communities by identifying targets and providing shared yard sticks in the form of indicators. In doing so they inscribe a world view of its defining actors; the governmental bodies, industries and communities that can be leveraged upon to instigate change. And in this world view architecture is strangely absent. At present, none of the UN SDGs declare targets that directly articulate architecture as a driver for change nor are there any indicators that evaluate its role. The built environment is only mentioned as a driver for resilient communities but without real value setting of the role of planning and design. This despite the extensive and complex impact architecture holds on human and non-human well-being; the way we live our lives, shape equity and use our resources.

For us this realization has led to the overarching aim of using the congress to build awareness. To argue for and demonstrate how architecture has the ability to afford change in the way we understand and construct the world around us and therefore how it as a situated practice engaging directly with both legislation, industry and the communities in which architecture *takes place* can become a direct way of effecting change.

Vision

The Science Track is formed around six panels of which this volume is one. The vision of the six panels is to articulate six differentiated perspectives onto how architecture can be part of achieving the SDGs while reinforcing their interconnectedness. The panels are in part mapped to existing fields while at the same time suggesting new. By bringing together otherwise fragmented knowledge across the breadth of architecture's research and practices, the aim is to bring together knowledge across research, practice and education to provoke new perspectives, new alliances and concrete action. In articulating the panels, the Scientific Committee asks pertinent and provocative questions that challenge the field and position the SDGs as active goal posts. These questions form the chapters of each volume asking how architectural knowledge creation can innovate the thinking, design and making of architecture.

- ***Design for Climate Adaptation*** With profound urgency, global communities are acting and adapting to the earth's changing climate. Our built environment, the most common habitat of humans, should interact with the earth's ecosystems and climates in a sustainable and regenerative way. 'Design for Climate Adaptation' emphasizes people, multiple forms of research, knowledges and action for high- and low-tech solutions that make buildings, neighbourhoods, landscapes, cities, and regions regenerative, resilient and adaptive to climate change impacts.
- ***Design for Rethinking Resources*** Design shapes our world, from the places we live in to objects we use every day. As we grow more aware of the limits of our planet's resources shifting from an exploitative to a restorative, regenerative and circular design ideology becomes fundamental. 'Design for Rethinking Resources' examines approaches to resourcefulness in architecture; how sustainability challenges the foundations of our material practices and how they can change with it.
- ***Design for Resilient Communities*** A resilient community anticipates, adapts to and recovers from adversity. Climate change, the global pandemic and political upheavals in many countries have revealed social, economic and environmental inequalities that threaten communities worldwide. These fault lines disproportionately impact the poor, people of colour, the racially or ethnically marginalized and women. 'Design for Resilient Communities' encourages innovative solutions and facilitates the development of knowledge and skills necessary for adaptation and recovery.
- ***Design for Health*** Architecture and health are inseparable. From the direct design of hospitals and places for healing to the strategic design of infrastructures and city planning, architecture affects physical and mental health of individuals and communities. 'Design for Health' asks how architecture can reconceive health as a design issue. How land rights impact healthy living; how legislation, planning and building impact inequality and access to water and how single buildings and the civic construction of hospitals, health clinics and community buildings can

operate in unison with local environments and ecologies to create a safe and healthy space for all.

- ***Design for Inclusivity*** No individual deserves to experience space in a manner that is less safe, less comfortable or less accessible as a result of their identity or challenges. Sustainability, in its most holistic definition, cannot be achieved without a collective act. 'Design for Inclusivity' aims to critically define the constructs and categories of who exactly we are excluding, and why, in order to mindfully develop strategies to mitigate this exclusion.
- ***Design for Partnerships for Change*** 'Design for Partnerships' is about recognizing the asymmetrical relationships between states, public spaces, civil societies and private domains to find new balances for the existing power structures. By challenging the ontology of universalism, it examines how architecture and the built environment can play an essential role in creating a ground for care through local governance, space making practices, imaginaries and scenarios of plural(istic) political, socially and ecologically sustainable futures.

Critical Positions

The two and half years of preparation has been an inspiring experience through which we have witnessed the power of architectural thinking in action—its interweaving of the critical and the creative ideation as well as its inherent inventiveness orientation towards the future. As part of the curation of this work we have defined a series of critical positions by which to understand the correlation between architectural thinking and the UN SDGs. A first position has been to challenge the inherent anthropocentrism and perceived lack of hierarchy between the goals; the Tabula Rasa effect as Johan Rockström names it [Rockström 2016]. The SDGs have been criticized for failing to recognize that planetary, people and prosperity concerns are interconnected [Kotzé 2022]. In forming the six panels of the Science Track we seek to position a rupture to the modernist axiom that the environment is situated outside of us. Instead, we understand the SDGs as a balancing between planetary and human needs which needs to be holistically addressed.

A second position is the critical appreciation that the SDGs retain an adherence to an underlying model of growth. The Science Track asks what the future practices of architecture can be, what the ethical roles of architectural design are and how architecture knowledge can create change in how architecture is produced both within and without of models of growth. It seeks to identify who the partners of architecture practice can be both through grassroot community action and through industry-based models.

A third position is the challenge of the embedded universalism within the SDGs. The SDGs maintain a universalism that is common to the UN system and underlies much of UN's work. However, this fundamentally modernist position of understanding sustainability as 'a problem to be solved' and placing agency with legislation leaves questions of agency, voice and power unchallenged. The Science Track seeks to incorporate this criticism through the panel calls and their associated sub-questions by provoking reflection on the perceived neutrality of architecture's own humanist traditions and insists on the question of how architecture is produced, by people and for people.

The challenge to universalism has also led to a review of the scientific practice of knowledge dissemination. The call for papers deliberately encourages exchanges and learnings across different knowledge and practice silos. This is effected through differentiated publication formats that include scientific knowledge production as well as design-based knowledge production, narrative formats such as oral history, visual essays, as well as dialogue-based exchanges and argumentative essays. The aim of these formats is to expand the possibility of transdisciplinary knowledge exchange and include voices that are not commonly part of academic and professional discourse.

The fourth and final position is to understand the SDGs as part of a changing world. The SDGs set out a 14-year long project. Any project of that length needs to build in methods of reviewing its own fundamental value sets and core conceptual foundation. The intensifying and accumulating effects of climate change, the aftermath of the COVID-19 pandemic, the continued stress on the world's resources and the escalating multi-partisan war in Ukraine have deep and unequal repercussions on global communities. To engage with the SDGs is to correlate the goals to a changeable understanding of both needs and means. It is to commit to a continual address of both the contexts and instruments of change-making. In the Science Track, our focus on the concrete and the actionable through presentations of cutting-edge research, real-world case studies and near-future focused arguments call for a situated understanding of the SDGs. This emphasis contextualizes the SDGs within the multiple and diverse practices of architecture as well as the disparate places in which architecture takes place. The perspectives, methods and means are purposefully broad. They seek to represent the breadth of the solution space needed for the systemic change needed. They also purposefully include different voices and different styles to make present the different actors, different knowledge streams and different institutions that create this change.

Perspective

The result is a six volume proceedings tracking a wide and multifarious interpretation on how architecture can be part of achieving the SDGs. Across their individual chapters we see a breadth of enquiries asking who the communities are, who the actors are and what the means of architectural production are. They ask how we can shape the methods of architectural thinking we well as their associated technologies, how they can be distributed and what is the consequence of their sharing.

The proceedings instantiate a moment in time. As research strands, they are part of larger trajectories of knowledge creation. Where our aim for the World Congress is to facilitate new discussions and exchange enabling synergy across silos and geographies, it is clear that the full potential of this conversation is only just beginning. The World Congress coincides with the half-way mark of the SDGs. Launched in 2016 and with a projected completion date of 2030, we need to transition from a place of planning and speculating to one of action. The work of the Science Track is therefore marked by a sense of urgency. The desire is to define the effort of this work not in terms of their individual results, but more as a launch pad for future exchange and collaboration. We hope that what is created here is a community of dedicated actors all with a shared stake in the well-being of future generations. Our hope is that the legacy of this project will be that we can retain this commitment and grow its stakeholders to mature these propositions into actionable change.

We profoundly thank the Scientific Committee for their immense effort and profound engagement in shaping the Science Track. Thank you to: Billie Faircloth, Maibritt Pedersen Zari, Carlo Ratti, Anna Rubbo, Juan Du, Arif Hasan, Christian Benimana, Magda Mostafa, Ruth Baumeister, Sandi Hilal, Merve Bedir, Katherine Richardson, Chris Luebkeman, Thomas Bo Jensen and Camilla Ryhl.

Mette Ramsgaard Thomsen
General Reporter
Copenhagen, Denmark

Martin Tamke
Alternate General Reporter
Copenhagen, Denmark

References

Kotzé LJ, Kim RE, Burdon P, du Toit L, Glass L-M, Kashwan P, Liverman D et al (2022) Chapter 6: planetary integrity. In: Sénit C-A, Biermann F, Hickmann T (eds) The political impact of the sustainable development goals: transforming governance through global goals? Cambridge University Press, Cambridge, pp 140–171

Rockström J, Sukhdev P (2016) The SDGs wedding cake—Stockholm resilience centre. Retrieved from https://www.stockholmresilience.org/research/research-news/2016-06-14-the-sdgs-wedding-cake.html. Accessed on 04 Apr 2023

Editorial

The Weight of Cities

"How much does your house weigh?" Buckminster Fuller famously asked this question in the 1920s, while promoting his lightweight, mass-produced Dymaxion house that he hoped would democratize homeownership in the same way that the Model T had done for cars. The 1929 version provided an incredible ratio of living space to mass: 150 m^2 for only 2700 kilos (Reed 2002).

The question of weight was effective because it focussed people's attention to an oft-overlooked concept: the weight of a building. Such a measure may seem nonsensical at first, as most people never consider it. However, today, as the climate crisis threatens cities everywhere, Fuller's question is becoming a heavier one. Much of a building's weight does not press into the ground, but instead floats in the atmosphere and traps sunlight. In other words, mass often indicates embodied energy and CO_2 emissions. We may not feel the weight of the built environment, but the planet does. We need to lighten the load.

Towards a Metabolistic Architecture

Conceptualizing the weight of cities is at the heart of the field of industrial urbanism and the subject of the UN Resource Panel's reports (IRP 2018). Here, weight is understood not only as a measure of resource consumption but—more generally—a means of managing environmental impacts over the full life cycle of production. By asking us to consider cities as *a circular metabolism* through which resources flow, this highly multidisciplinary approach challenges twentieth-century perceptions of resources as fundamentally neutral, apolitical and endlessly available at market cost. Currently, cities have little control over their environmental footprints. By provoking us to imagine the weight of the urban organism to which this metabolism belongs, researchers such as Mark Swilling and Marina Fischer-Kowalski warn us of the hidden costs of *resource ignorance*. They challenge city leaders to radically rethink how resources are strategized and deployed across our society—envisaging circular design strategies that cascade materials through different use-scenarios to most fully exhaust the extracted, avoiding waste and its detrimental impact on the environment.

But how do we measure the weight of the city? Much of a city's weight does not press into the ground, but instead floats in the atmosphere, choking our lungs and trapping our sunlight. Of course, the relationship between mass and emissions is not a perfect correlation. As designers we can consider how every pound of material we employ falls differently on the scales of sustainability. If we use a measure like CO_2 emissions, we could even say that there are forms of 'positive weight' and 'negative weight'. Producing a tonne of concrete generates around 600 kilos of CO_2 (Nature 2021). When we compare this with the emerging field of advanced timber construction, which has far greater potential for carbon storage, more weight might mean more environmental integration. It is estimated that producing one tonne of dry timber removes 1.8 tonnes of CO_2 from the atmosphere (Cambridge 2019).

To achieve such changes, we need to improve our ability to measure the problem. It may seem daunting to manage the complex accounting of carbon, land, energy, water, raw materials and cost, but the rise of digital technologies is giving us a twenty-first-century toolkit with which to approach this task. As we craft the buildings of the future, the advanced computing of 'digital twins' and Life-Cycle Assessment (LCA) allows us to track the emissions implications of a given design choice. Once built, sensors can help us to verify and validate our models. Advanced sensing technology and artificial intelligence analysis allow us to track—precisely and in real time—how our schemes for sustainability hold up against the tests of reality. These sophisticated simulations and their real-life counterparts in sensor data enable critical assessments of design in context of how resources are deployed, their intensity and their performance.

Further expansions of these methods push us to integrate end-of-life scenarios into early-stage design. Our agency as designers is being expanded by our ability to conceptualize not only the building programme and lifetime, but also calculate and calibrate the further propagation of building components and materials in future uses. Designing for adaptation, for disassembly or for re-materialization challenges the conceptualization that design is an end-point, instead positioning it as an intermediate position along the cascading flow of materials where resources are not only extracted, transported, used and disposed, but also filtered for multiple reuses as they cascade into more and more intricate life cycle paths.

Could weight become a means of accounting for resource flow through our larger built environment? If *lightness* directed the postmodern movement both through high-tech visions and through economy-driven axioms of resource optimization, *weight* here implies both conceptual and direct ways of broadening designers' vision to consider the impact they make on the wider world.

Consequences of Weight

But what is weight in context? Returning to Swilling and the UN Resource Panel's report, the core message is that a city's well-being, its sustainable performance and the ability to design and manage its material flow are closely tied with its ability to afford social equity (ibid). Resources are inevitably deeply entangled in the socio-ecological networks of their extraction and deployment.

Resource scarcity tends to be understood as a quantitative problem. But resource extraction, and the implied results of over-extraction, such as illegal mining or the felling of protected forests, have deep impacts on the communities and ecologies in which they take part. As such, resource deployment and design are fundamentally ethical decisions. Existing design methods have little or no ability to inform designers on the ethics of resource choice. At a coarse level we might be conscious of particular endangered species or build awareness of specific unethical fabrication processes, but we lack consolidated conceptual as well as practical models to steer the impact of design decisions on their socio-ecological contexts.

The present emergence of such models straddles a high- and low-tech divide. On the one hand, systems design and resilience modelling enable the analysis of socio-ecological systems and the way extraction impacts on their embedded balances (Tamberg 2020). On the other hand, new models of grassroots-driven, participatory and community-focussed resource thinking challenge the globalization of industrial building practice, instead insisting on the localization of resource and building technology. Local crafts, traditional knowledge and a return to grown resources allow alternative ways of accounting for and sustainably ensuring resource longevity.

What results is a challenge to the embedded universality of the concept of weight. As if we were occupying varying gravities, weight is not the same from place to place. While this insistence on context and place supports the rupture with resources as neutral and apolitical, it complicates the transferability and value of its results.

Building Vision on a Global Research Stage

It is through this complexity we enter the conceptualization of the UIA World Congress of Architects 2023 CPH Science Track panel *Design for Rethinking Resources*. It draws upon the UN Sustainable Development Goals and 2030 Agenda, whose central, transformative process is 'leave no one behind'. In adopting and situating this agenda for architecture, this promise is a guide for the UIA World Congress of Architects 2023 CPH. These Proceedings collect the contributions of a global collegiate of designers, researchers and educators. Together, our global collegiate of designers, researchers and educators report on a wide array of projects that challenge, rethink and propose new ways for architecture to understand its material practices. *Design for Rethinking Resources* challenges us to question the origins and entangled belongings of the materials we use and how their extraction, displacement and deployment disrupt socio-ecological belongings while creating new possibilities.

In this new paradigm, architecture, the design and production of our built environment, becomes not only the creation of the new or the response to changing societal needs, but also a strategization of resources; an ethical practice shaped around the concerns of how materials flow through our societies impacting our environment, its ecologies and the people it engages. The 2030 Agenda and SDGs ask us to consider our *material footprints*. In the central indicator 8.4.1 (UN2022), they challenge us to measure the attribution of global material extraction per capita and per GDP to be able to understand the impact and equity of our consumption.

The work of conceptualizing and curating this UIA World Congress Science Track panel centres around these questions. In one of our pre-events leading up to the congress, resource management theorist Helmut Rechberger asked us to address not just resource scarcity, but also the impact that business-as-usual models have on our landfills. Challenging us to imagine the projected weight of the city of Vienna and the fact that it will double by 2050, Rechberger reminds us that cities act as unwanted end-points as resources meet their end-of-life.

Design for Rethinking Resources journeys through these questions. Through the six chapters collecting 46 papers, authors present novel scientific research questioning what resource streams, practices and associated technologies can amount to a sustainable architecture. They ask us to rethink our material practices by (re)introducing new and old crafts, conceptualizing the city as a resource bank, creating new computational practices to steer the cascading of resources through the built environment as a material chain, ideating what bio-based renewables can be, and probing the post-extractive practices of tomorrow.

Our process started with a mapping. By reading through the SDGs and their associated indicators and targets we searched for the way questions of resource impact the Sustainable Design Goals (SDG). What we found was an overwhelming sense that the conceptualization of resources—their origin, ownership and rights for deployment act as a cornerstone to all 17 goals. The SDGs ask: Who has the right to resources and by what means? Who extracts? What pollutes? What land, what water, what energy and what body is implicated in extraction? Who has access and at what cost? What are the balances of our material footprint, from country to country and from Global South to Global North? What is the impact on CO_2 emissions and how can technology ensure resource efficiency? What are the needs of the environment and what model of social equity must they be balanced with?

In response, we have defined *Design for Rethinking Resources* across six questions defining six sub-panels together challenging the way resources are thought of across architecture. The questions follow an unravelling of how resources flow through the built environment, moving from questions of extraction, through production, assembly and management to models of recycling. In places they consolidate known stepping stones and in others they expand the dialogue to bring in further dimensions to what it means to understand design as the strategization of resources.

Each question has attracted papers from a global community across disparate sub-fields creating new interfaces between knowledge fields. In line with the wider UIA World Congress 2023 CPH Scientific Committee focus on fostering broad inter-cultural exchanges of independent, coherent and authoritative scientific papers, our aim is to create a global conversation enabling knowledge sharing across localities and knowledge cultures. As such, each sub-panel brings in culturally specific contributions that support each other and bring perspective to the breadth of conditions in which architecture is produced.

Six Questions for Design for Rethinking Resources

The six chapters that structure these proceedings move along the flow of materials from sources through use to waste. They question how resources are employed, advance the techniques and technologies by which they are used, track their cascading and ideate methods for understanding the city as a resource map. They ask how bio-based material practices can present alternate replenishable sources for the material of the built environment and challenge us to reimagine the role of heritage and the traditions of craft.

Post-extractive Visions
The proceedings start with a series of provocations. Arriving from different corners of the field, the papers bring together theory and case studies to imagine architecture beyond extraction. The papers offer a broad critical assessment of past and present extractive practices and their impact on local, regional and national ecologies. They ask us to question what post-extractive technologies might be and how biological technologies that employ living organisms for material production can present new horizons for the way we understand the built environment.

Localizing Resources
Resource management challenges the way we understand equality across multiple social, economic and political divides. The second chapter prompts us to rethink the globalization of resource deployment. Localizing Resources asks how participatory and community-focussed local resource thinking can challenge the way we work with materials. From eggshells to recycled timber, from earthen bricks to coconut fibres, the chapter presents a series of cases demonstrating how 'thinking local' can foster socio-ecological networks that facilitate meaningful and contextualized resource cascading.

Heritage to High-Tech
Local resources entail local craft. Heritage to High-Tech positions craftsmanship in the context of resources. It reminds us how local traditions and vernaculars are situated in sensitively balanced socio-ecological networks that already incorporate circular principles of cascading and renewable strategies. In developing this position for a sustainable future, it asks both to

carefully examine what these practices entail and what their interfacing with transposed building knowledge and new technologies can be.

Fabrication Futures

Digital modelling, analysis and fabrication allow us to innovate design practice. Advanced modelling for material optimization, cyber-physical augmentation along with LCA, digital twins, material passports and data basing for resource stock suggests new territories for data-driven material management. Fabrication Futures asks us to rethink how materials are deployed. How can their intensity and performance allow for a lighter and less impactful material practice? Designing for disassembly can help us to imagine a more flexible and reconfigurable building stock.

Restarting from Renewables

Bio-based materials present a particular perspective on circular thinking. They are renewable, potentially abundant, carbon-neutral and recyclable. Restarting from Renewables asks us to imagine our material world as one we are co-producing through practices of growing and harvesting, transforming the perception of our surroundings from something inert to something alive. By spanning different technological perspectives, from the traditions of bamboo construction to the fermentation of mycelium reinforced weaves to the implementation of biochar cementitious composites, the papers present a series of cutting-edge case studies.

The Value of Waste

Circular principles ask us to reconsider the waste streams of production. How can models of cascade thinking and recycling reallocate material resources? What models of circularity can transform production and mitigate waste streams such as pollution, water overuse and energy efficiency? The Value of Waste asks what are the necessary infrastructures for cascading and how can buildings, cities and the wider built environment be understood not only as shelters for the present, but as resources for the future.

Conclusions

With our new research, we could address problems around cities and sustainability that have bedevilled us for decades. In 1995, Richard Rogers called for architects to 'define an ethical stance', and pursue the creation of sustainable buildings that would 'work into the cycle of nature'. He noted that 'cities are consuming three-quarters of the world's energy and causing three-quarters of the world's pollution', and warned that these environmental consequences would only worsen because 'over the next 30 years, a further two billion people are expected to be added to the cities of the developing world' (Rogers 1998).

Those decades have gone by, but the condition Rogers described has scarcely changed but for growing more urgent. Today the effects of climate change are here, and 30 years from now the urban population is projected to double.

Now, finally, a new shared research foundation in our field has a chance to truly change our trajectory. The linear economy has condemned our resources and planet to the same one-way path from extraction to waste. Adopting high-tech and low-tech methods, and more importantly embracing a fundamental change in mindset about the costs our built environment incurs, can take us from one track to the other. We must develop the consciousness—of cities, of the environment, and of ourselves—which Buckminster Fuller exhorted us to acquire when he first asked what a building weighs. Instead of weighing us down, cities, architecture and the resource strategizing they embody can lift us up.

Mette Ramsgaard Thomsen
Professor and Head of
CITA—Centre for Information Technology and Architecture
The Royal Danish Academy—Architecture, Design, Conservation
Copenhagen, Denmark

Carlo Ratti
Professor and Director
Senseable City Lab
Massachusetts Institute of Technology
Cambridge, USA

Founding Partner
CRA—Carlo Ratti Associati
Turin, Italy

References

Nature Board (2021) Concrete needs to lose its colossal carbon footprint. Nature 597:594–597. https://www.nature.com/articles/d41586-021-02612-5

Reed P, McQuaid M (eds) (2002) Envisioning architecture: drawings from the museum of modern art. The Museum of Modern Art, New York, pp 64–65

Rogers R (1998) Cities for a small planet. Westview Press, Boulder Colorado, pp 27

SDG indicator metadata (2022). https://unstats.un.org/sdgs/metadata/files/Metadata-08-04-01.pdf

Sowing seeds for timber skyscrapers can rewind the carbon footprint of the concrete industry. University of Cambridge (2019). https://www.cam.ac.uk/research/news/sowing-seeds-for-timber-skyscrapers-can-rewind-the-carbon-footprint-of-the-concrete-industry

Swilling M, Hajer M, Baynes T, Bergesen J, Labbé F, Musango JK, Ramaswami A, Robinson B, Salat S, Suh S, Currie P, Fang A, Hanson A, Kruit K, Reiner M, Smit S, Tabory S (2018) The weight of cities: resource requirements of future urbanization. A Report by the International Resource Panel (IRP). United Nations Environment Programme, Nairobi, Kenya

Tamberg L, Heitzig J, Donges J (2020) A guideline to modelling resilience of complex systems

Contents

Post-extractive Visions

Upcycled Regionalism: The Aesthetics of Geopolymer Concrete

Ingrid Halland and Stian Rossi

Abstract

Sometimes the more powerful act is not to make new, but to make anew with what we already have. Since the built environment accounts for around 50% of all raw material extracted from the earth (European Commission 2018), the transition to a circular economy for this industry will require extensive restructuring in terms of resource extraction and production patterns. This paper argues, however, that aesthetics also plays a significant role for rethinking resources. The paper examines the aesthetics of Geopolymer concrete (GPC)—a low carbon sustainable building material based on circular principles—recently developed by the Norwegian start-up company SafeRock. GPC answers the demand for circular economy in the building industry through upcycling of waste. SafeRock announced their first pilot project in 2021, a GPC made with mining residues from the mine Titania in Norway. SafeRock's GPC is based on on-site production, eliminating both transport emissions and time. Importantly, the site-specificity of the mine becomes an *aesthetic quality* of GPC surfaces; different waste products from different mines produce different textures and colors. Like a "spolia" for our times, site-specific traces of a modernist paradigm of mass-extraction are aesthetically visible in the upcycled GPC surface. The paper aims to develop a conceptual framework for critical reflection on aesthetics and upcycling of materials. By combining theory from the humanities with examples from architectural practice, the paper introduces the term "upcycled regionalism" to unpack the aesthetic potential of GPC.

Keywords

Upcycling · Aesthetics · Architecture theory · Ethics · Deep ecology

In 1864, the Norwegian Titanic Iron Ore Company bought the mining rights to a geologically complex area close to Jøssingfjorden in Rogaland, Norway, which was rich in ilmenite—a black titanium-iron mineral. The iron mine was relatively unprofitable until the company Elektrokemisk Industri (today called Elkem) became interested in the *mining waste*, namely, titanium. Elektrokemisk was founded in 1904 with the goal of creating an international industrial company based on mass-utilization of Norwegian natural resources. In the early 1900s, an ethos of technological optimism dominated the young

I. Halland (✉)
The Oslo School of Architecture and Design, University of Bergen, Bergen, Norway
e-mail: ingrid.halland@aho.no

S. Rossi
Saferock As, Chief Commercial Officer, Sola, Norway

M. R. Thomsen et al. (eds.), *Design for Rethinking Resources*, Sustainable Development Goals Series,
https://doi.org/10.1007/978-3-031-36554-6_1

Fig. 1 Nedre Helleren hydropower plant from 1907. *Photo* JanOve Grastveit

nation-state (Norway gained independence from Sweden in 1905). The optimism was largely caused by path-breaking technological innovations in hydroelectric power that had boosted the rather poor country's capability for industrial production (Fig. 1). The Norwegian Ministry of Trade joined forces with Elkem in 1907, and a special committee for electrometallurgy was appointed to evaluate the unprofitable ilmenite ore in Jøssingfjorden, the aim being: (1) to utilize the surplus energy from the area's recently-built dams and hydropower plants (Figs. 1 and 2) to find a new future for the unprofitable mining waste titanium.

After years of experiments, reports, mistakes, and patents, a team of chemists, entrepreneurs, and geologists found a solution that was to revolutionize the modern world. The patent for the industrial production of titanium dioxide—filed in 1910 by Norwegian chemist Peder Farup—marked the start of a world industry: the international pigment industry based on the sulfate process for mass-producing a non-toxic, absolute white paint that resisted miscoloring due to dirt

Fig. 2 The Titania land deposit will reach its limit in 2024. *Photo* Johnslien (2019)

and rust. Production for the global market began in the mine Titania AS in Sokndal, Norway in 1916—and the company is still a key player in the global production of titanium dioxide. Throughout the twentieth century, titanium dioxide was increasingly used in mass-production of surfaces (as a coating for concrete, glazing for ceramics, and additive in plastic), thereby titanium dioxide changed the aesthetics of surfaces in architecture and design, making surfaces—all over the globe—smoother, brighter, flatter, and more opaque (see Halland and Johnslien 2023). Titanium dioxide can be considered a material signature of modernism; an almost invisible hermetic techno-natural skin that materialized values of technological progress, homogenization, and durability into and into objects and environments worldwide.

After a hundred years of mining, the extraction of ilmenite has left an irreversible change in the local landscape. The environmental trace of mass-extraction of natural resources consists of a vast cut through the surface of the earth and a gray artificial desert of mining waste (Fig. 2). The desert of waste designates a regressive and outdated approach to resources—an approach that considers raw materials as limitless resources. As we have known for over fifty years,[1] this approach cannot be sustained.

1 Challenge I: What can Architecture Do with Modernism's Waste Products?

Resource scarcity, natural disasters, and wars have throughout global history caused creative architectural reconstructions based on available resources. The use of found objects and the reuse of waste building materials—so-called *spolia*—is an ancient tradition that is shared globally through all societies and ages. The word "spolia" derives from Latin "spolium," meaning stripping, depriving, or robbing of something that covers—or the act of stripping off an animal's skin. Another meaning is to strip, deprive, or rob of arms or equipment, in other words, to disarm. The ancient use of spolia was not only functional, but profoundly symbolic; to use waste products as an aesthetic strategy to rewrite the past on a metaphorical level, through disregarding, violating, and cannibalizing the achievements of predecessors. The time has come to use modernism's waste products as spolia. The time has come to make use of the numerous deserts of mining waste—that is putting land under pressure and disrupting ecosystems. Yet how can architecture combine such critical and aesthetic design strategies as "upcycled spolia" with real-life economic and functional renewal?

2 Challenge II: How to Combine Critical Design with Real-Life Solutions?

Perhaps a more well-known material signature of 20th Century architecture is Portland cement (OPC) (Forty 2012; Langdalen et al. 2018). During the post-war period, OPC became the most widely used material in the construction industry. It also developed into one of the most lethal on the planet, closely associated with toxic substances like asbestos, cadmium, and silica. For every ton of Portland cement produced, 4000–7500 MJ of energy are consumed, releasing approximately 1 ton of CO_2 in the process (Mehta and Monteiro 2014). This accounts for a

[1] The environmental movement owes much of its legacy to the 1972 book *The Limits to Growth.* The book has been claimed to be the most successful publication in the field of environmental studies, mainly due to the book's wide-ranging influence, selling 12 million copies in more than 30 languages. The argument of the book was that five factors—pollution, population, agricultural production, natural recourses, and industrial production—were fully entangled and dependent on each other, so no single action addressing one, or two, of the factors would have any effect on the whole. According to the authors, the most likely future scenario would be a total world collapse by mid-21st Century. Donella H. Meadows, Dennis L Meadows, and Jørgen Randers, *The Limits to Growth: A Report for the Club of Rome's Project on the Predicament of Mankind* (London: Earth Island Ltd., 1972).

significant portion of global CO_2 emissions. Clearly, the hegemony of OPC has to be challenged in the name of sustainability. It is for that reason that scientists, industry players, and governments are actively seeking low carbon alternatives to OPC concretes.

However, recent research on Danish architectural practice has proven the value of material upcycling: "Results show that the recycling/upcycling strategy is the most effective in reducing the embodied carbon" (Birkved et al. 2020). The Norwegian start-up company Saferock aims to produce fully CO_2 neutral concrete by upcycling byproducts from the mining industry. Geopolymer concrete (GPC)—recently developed by Saferock—is a low carbon sustainable building material based on circular principles. Made from mining waste, which represents 27% of the total waste generated from economic activities and households in the EU (Eurostat 2016), GPC provides a more stable and sustainable long-term supply, in addition to offering the potential of circular economy value propositions through upcycling of waste (Fig. 3).

Geopolymers, also known as inorganic polymers, are mainly composed of aluminosilicate obtained from waste streams which react in presence of alkali silicate solution (Khalifeh et al. 2018). On a laboratory scale, geopolymers developed for zonal isolation have shown superior properties compared to OPC. The raw materials used in the traditional geopolymers are susceptible to change or have high CO_2 emissions as they are usually fly ash-based or metakaolin (clay) based products. Due to a lack of consistency in the waste streams obtained from coal power plants or heavy industry, there is either a technical limitation to deliver a unique solid phase to be mixed with the hardener, or the material simply does not have the potential to significantly reduce the carbon footprint. In response, Saferock has patented true geopolymeric materials in which its hardener phase is a potassium silicate solution (Fig. 4). The rock-based formulations are based on (i) norite, a solid waste generated during the production of ilmenite, and (ii) granite particles below 63 μm, treated as a waste when producing granite as an aggregate to concrete and asphalt. Saferock's ambition is to offer unique rock-based geopolymer binders as a sustainable substitute to OPC in an increasing number of applications for the building and construction industry. As a breakthrough innovation in sustainable building materials, the technology will cater to large-scale technical requirements, health and environmental conditions, and market needs. Based on mining

Fig. 3 *Photo* OIOIOI (2021)

Fig. 4 From ilmenite to mining waste to geopolymer. Displaying material circularity in Saferock's material lab. *Photo* Halland (2021)

residues, raw materials are abundantly available for widespread international use. Research grants from the Research Council of Norway and the University of Stavanger, and financial support from Innovation Norway, Enova, Equinor Ventures, and private sponsors, have secured USD 5 million to finance Saferock's first pilot factory. The new site-specific factory will be operative before the end of 2023 and will use waste materials from the Titania mine in Sokndal (see Fig. 2) By using modernism's waste products as a spolia for our times, Saferock's GPC is a concrete example of a critical design project with profound symbolic value, that at the same time also offers a functional real-life solution for the industry.

As an almost ironic twist of circularity, Saferock now reconfigures history by reusing the unprofitable waste from the Titania mine—that was originally initiated by the ambition of making the waste product titanium into a profitable product. Today, 100 years after the titanium dioxide innovation, an interdisciplinary team of entrepreneurs, geologists, and engineers again visit the mine in order to innovate a new future. After years of experiments, reports, mistakes, and patents, Saferock is now ready for its first pilot project with GPC from the Titania mine. However, there is a major difference between these two approaches to rethinking waste: 100 years ago, the rethinking of waste was entirely economically driven—governed by an attitude to resources that considered raw materials as a limitless supply. Today, Saferock aims to rethink waste conditioned by principles of *systemic circularity*—an attitude to resources that considers raw materials as a limited supply. Moreover, today designers play a key role in rethinking waste.

3 Challenge III: How to Make Systemic Change Desirable?

Culture theorist Mark Fisher (2009, 2012) has argued that in order to find a realistic way of making a sustainable future, we need to reclaim capitalism's most essential drive, namely *desire*. Fisher challenges the notion that capitalism is the only driver associated with desire. In agreement with Fisher, the authors of this paper believe that a key gap for future sustainability is to rethink aesthetic desire. Architecture needs to demonstrate that aesthetic desire can exist outside the capitalist drive for mass-extraction, production, and consumption. We must use aesthetics as a driving force for change. New solutions in

Fig. 5 Future circular design in 1969: From the exhibition "Og etter oss..." [And After Us...], made by students from The Oslo School of Architecture and Design (1969). Illustration is taken from p. 56–57 in the exhibition catalog

architecture must be aesthetically desirable; they must display the systemic circularity in the very aesthetics of the surface (Fig. 5).

Saferock's GPC is based on on-site production using site-specific waste products. Importantly, the site-specificity of the mine becomes an *aesthetic quality* of GPC surfaces; different waste products from different mines will produce different textures and colors. GPC is not only a functional low carbon sustainable building material based on circular principles, like the spolia, the material is profoundly symbolic. GPC disarms and reclaims the past by "stripping down" the endless gray desert of mining waste that now covers a whole valley. In the Titania mine, circular principles now cannibalize modernism's achievements, and up from the desert arise a new aesthetics, the aesthetics of a *deep circularity* that considers natural resources to be a limited supply.

In 1983, architectural critic Kenneth Frampton formulated his call to counter modernism's drive to universalization. His theoretical concept "Critical Regionalism" argues for an architectural resistance to modernism that is situated in the topographic, climatological conditions of a site and that highlights the tectonics of architectural construction and the tactile sensibility of the locality. What we see in Saferock's GPC is a reconfigured design resistance for our time. As described by Frampton, the visual and tactile in GPC—the aesthetic desire—is a key design component (Fig. 6).

4 Conclusion

Today, there is a consensus that sustainability cannot be achieved by quantitative methods alone: There is a need for actionable change, innovative partnerships, and, arguably, a cognitive shift in regard to human beings' modes of thinking and patterns of behavior—all in which design is uniquely positioned to encourage. The two authors of this paper come from two distinctly different fields; one is from architecture history and theory and one is from the architectural industry. By writing this paper, we have attempted to bridge history, criticism, theory, and real-life solutions. We have attempted an innovative partnership aiming to change patterns of behavior and attitudes toward waste.

In 1802, the German poet Friedrich Hölderlin suggested that "where danger is, grows the saving power also."[2] Perhaps he was right. If locating "dangerous" attitudes toward natural resources in mass-extraction of natural resources (Fig. 7), architecture must find strategies to forcefully disarm and *un-skin* these outdated attitudes. Yet importantly, by drawing on Frampton's argument, we suggest that the act of disarming modernism's outdated attitudes to mass-extraction of resources, must be both tactile, site-specific, and aesthetically desirable. By

[2] Friedrich Hölderlin's poem "Patmos" quoted in Martin Heidegger, *The Question Concerning Technology, and Other Essays* (New York: HarperCollins, 2013), 42.

Fig. 6 Samples of Saferock's new geopolymer concrete (GPC) photographed in the desert of mining waste in Titania, Sokndal. *Photo* OIOIOI (2021)

Fig. 7 The Titania open pit mine in Sokndal Norway. *Photo* OIOIOI (2021)

using modernism's waste as spolia, we can reach an upcycled regionalism—making systemic change a bit more realistic.

References

Birkved M, Birgisdottir H, Rasmussen FN (2020) Low-carbon design strategies for new residential buildings—Lessons from architectural practice. Taylor & Francis

European Commission (2018) Report on critical raw materials and the circular economy. P. O. o. t. E. Union

Eurostat (2016) Energy, transport and environment indicators. Publications Office of the European Union

Favier A, Habert G, Roussel N, d'Espinose de Lacaillerie J-B (2015) A multinuclear static NMR study of geopolymerisation. Cement Concr Res 75:104–109

Fisher M (2009) Capitalist Realism: Is there no alternative? Zero Books

Fisher M (2012) Postcapitalist desire. In what we are fighting for: a radical collective manifesto, edited by Federico Campagna and Emanuele Campiglio. Pluto Press

Forty A (2012) Concrete and culture: a material history. Reaktion Books

Halland I (ed) (2021) Ung uro: unsettling climates in nordic art, architecture & design. Cappelen Damm Akademisk

Halland I, Johnslien M (2023) 'With-On' White: inconspicuous modernity with and on aesthetic surfaces, 1910–1950. The Aggregate Architectural History Collaborative

Johnslien M (2019) Circumstantial sculpture. Reflection. Ph.D. dissertation. Oslo National Academy of the Arts

Khalifeh M, Saasen A, Larsen HB, Hodne H (2017) Development and characterization of norite-based cementitious binder from an ilmenite mine waste stream. In: Advances in materials science and engineering, pp 6849139

Labuhn B (Forthcoming) (Dis)entanglements of architecture and environmentalism in Norway, 1965–1975. Ph.D. dissertation. The Oslo School of Architecture and Design

Langdalen EF, Pinochet A, Szacka L-C (eds) (2018) Concrete Oslo. Torpedo Press

Mehta PK, Monteiro PJM (2014) Concrete: microstructure, properties, and materials, 4th edn. McGraw-Hill Education

Post-extractive Material Practice: The Case of Quarried Stone

Jonathan Foote, Urszula Kozminska,
and Nikola Gjorgjievski

Abstract

Humans have extracted materials from the earth for millennia, but only recently have our extraction practices had planetary consequences. The scale, intensity, and violence of current practices are theorized as 'extractivism', and as a number of recent publications and exhibitions have made clear, a post-extractive future is urgently needed. How and under what conditions can we address such an imperative? Is it possible to aim toward a pre-modern notion of extraction as a custodial practice of care in relation to earth's resources? Following recent projects to re-introduce load-bearing stone, known loosely as 'The New Stone Age', this argumentative essay explores a post-extractive future through the case of quarried stone. We see it as a poignant case of how we might move from extractivism to a more custodial form of extraction, and show the potential to reduce our need for materials such as steel, concrete, and wood, industries that are driven in large part by extractivist principles. Based on Timothy Morton's notion of ecological thought, we argue for a new way to understand stone extraction as an ecology of interconnections. On the one hand, great potential emerges in the circularity and reversibility of stone construction, along with its extremely long life cycle, and on the other we see a strong potential to recover the historical connections between architecture and the anthropogenic landscapes of stone extraction. In the end, we position the issue of material extraction within its multifaceted entanglements of landscape, construction, socio-cultural and economic contexts.

Keywords

Ecological thought · Extractivism · Post-extraction · Quarried stone

1 Introduction

The Earth is the very quintessence of the human condition and earthly nature.—Hannah Arendt (1958)

When Hannah Arendt bound the human condition to the Earth, in her seminal book, *The Human Condition* (1958), NASA had not yet released the iconic image of the "Blue Marble", taken from the Apollo 17 mission in 1972. From then onward, humans were no longer an earth-bound entity, creating a classical Cartesian separation between object and subject and challenging both Lovelock's Gaia hypothesis and

J. Foote (✉) · U. Kozminska · N. Gjorgjievski
Aarhus School of Architecture, Aarhus, Denmark
e-mail: jofo@aarch.dk

M. R. Thomsen et al. (eds.), *Design for Rethinking Resources*, Sustainable Development Goals Series,
https://doi.org/10.1007/978-3-031-36554-6_2

Næss' concept of Deep ecology that Earth is a living system of which we are part, not apart (Lovelock and Whitfield 1982; Naess 1973). By the turn of the century, thousands of satellites circling the earth defined the landscape as a digital field observed and controlled from afar, disabling and reducing tactile human interaction.

Our perspective of the earth as a finite entity should encourage architects to be stewards, but the opposite seems increasingly apparent. As Luke Jones points out in his essay, "The Carbon Tectonic", we have multiplied our rate of material extraction from the earth by more than twenty times during the last century (Jones 2021). Most buildings include elements and materials from all continents, and construction activities are sustained by vast global supply chains (Wigley 2021). This hyper-industrialization of construction materials reduces our understanding of materials' inherent and cultural properties, their whereabouts, and their connection to the environment. Extraction was seen as a technical issue that machinery could solve, and the matter beneath the soil turned into a disposable space at the mercy of human demand. "Seemingly static buildings are actually pieces of mining equipment, actively devouring the planet," writes Mark Wigley, "as buildings rise in one place, a deadly net of holes, gaps, cracks, collapses, deficiencies, floods, and famines appear elsewhere" (Wigley 2021).

Humans have extracted materials from the earth for millennia, but only recently have our extraction practices had planetary consequences. Following the establishment of colonial capitalism in the eighteenth century and its subsequent industrialization throughout the nineteenth century, the scale and exploitative practices have been on a one-way, accelerated expansion. The intensified reliance on large-scale, mineral and cultural exploitation started to receive systematic critique in Latin America in the 1970s under the term *extractivísmo* (Durante et al. 2021; Acosta 2017). More recent scholarship has identified some of the core logic of extractivism: capital intensive, foreign investment aimed primarily toward exportation; little or minimal processing at the site of extraction; maximizing economic benefit at the expense of cultural violence, loss of biodiversity, and ecological degradation. The direct impact of extractivism is rarely seen in the developed world, where we have become accustomed to "cheap architectures" (Grima 2021), as the underlying displacement logic purposely conceals these consequences under the guise of globalized processing and supply chains (Hutton 2020). For architects and consumers, however, materials arrive from 'somewhere else,' or they simply are not considered at all, what Joseph Grima recently described as the problem of "externalities" (Grima 2021).

As a number of recent publications and exhibitions have made clear, a post-extractive future is urgently needed. How and under what conditions can we address such an imperative? Architecture can no longer overlook the impact of where and how materials are exploited (Moe 2020), and the discipline must do more to propose creative alternatives to normative practices of building material extraction (Grima 2021). Is it possible to aim toward a pre-modern notion of extraction as a custodial practice of care in relation to earth's resources? A deeper sensibility is necessary, one that goes beyond simple solutions and focuses on extraction as an ecology of architectural actions and consequences (Morton 2007). This interconnectedness operates within multiple scales and various time perspectives, embedding a diversity of actors and tools to address entangled complexities of extractive processes and materials. The post-extractive approach contributes to UN SDGs, especially goal 12 "Responsible Consumption and Production" by questioning the reliance of the construction sector on natural resources and by advocating for more ecological patterns of architectural production, which have the potential to act against climate change, biodiversity loss, and pollution. Indirectly, it addresses goals 5, 8, 10, and 11 by protecting terrestrial ecosystems and accounting for a broader understanding of socio-economic entanglements.

2 Extraction and Natural Stone

Exploring the ecology of extraction requires a material test case in order to unfold these interconnections. For this, we examine one of architecture's most exploited materials: natural stone. Today, stone is typically quarried in an export economy and used for surface applications such as pavings, building facades, and interior finishes, an ineffective use of resources that leads to a significant amount of generated waste. This a-tectonic use of materials seems highly inadequate in our epoch of finite resources and multiplying crises. However, a number of architects and engineers have begun experimenting with load-bearing stone, an area of research that has not been seriously pursued since the late nineteenth century. The stone exoskeleton building at 15 Clerkenwell Close in London, by Groupwork, stands as a notable example, as do a number of social housing projects built in and around Paris in recent years (Le Dréan and Kuratli 2022). The latest generation of practitioners often cites the environmental advantages of structural stone: nearly half the embodied energy and one-fourth the carbon footprint of reinforced concrete elements (Webb 2022), the ability to be easily disassembled and reused in perpetuity (Brilliant and Kinney 2016), and its capacity to support biodiversity (Chartier-Dalix 2019). This shift toward exploiting the modular and structural properties of stone, also combined with wood and other materials, is what has recently been dubbed "The New Stone Age," in *Architectural Review* from April 2022 (Webb 2022). These developments point toward a "tectonically simpler architecture based on monolithic and 'literal' materials" (Jones 2021), one that reduces the reach of a building's material supply chains.

The emerging potential of massive stone construction offers a poignant test case for re-imagining architecture's relation with material extraction. There are several reasons for this. To begin, more than most contemporary materials, stone retains its connection with the site of extraction. Larvikite from the Lundhs company, for example, is advertised as "Natural Stone from Norway," accompanied by images depicting the rugged, timeless beauty of the Norwegian landscape. This present-day marketing tool follows a long tradition where the qualities and aesthetic values of stone have been associated with the culture and geology of the extraction site. The architecture of the National Romantic period, for example, relied heavily on the link between geology, culture, and national identity (Ringbom 1987). This also included religious significance, as in the Inca builders, for whom sacred architecture was only achievable through stone quarried from a specific, venerated quarry site (Ogburn 2013). Similarly, the use of the so-called 'Jerusalem Stone,' a limestone quarried today primarily in the West Bank, is throughout the world considered a politically-charged symbol of Jewish identity (Abusaada 2022). Thus, when imagining a post-extractive future, the link between stone and locality seems hard-wired into our collective memory.

The sheer difficulty of extraction and transport meant that architecture often had profound connections with the local landscape and geology. In pre-modern, lithic societies, the local geology was a vehicle for a symbiotic relationship between building culture and landscape (Abu-Jaber et al. 2008). In the Balearic Islands, for instance, a wide-spread geological formation in soft sandstone, known locally as *marès*, supported the development of both monumental and vernacular building culture. At one time, the island of Mallorca hosted over 1500 *marès* quarries, each one supporting an infra-local economic and social community (Mata 2018). The same could be observed in a more contemporary context with the case of Lutetian limestone from the Paris Basin, known also as the 'Paris Stone'. Its pervasive use in the monuments and streets of Paris created a unity between the colors and textures of the city, its regional geology, and its landscape (Barrault et al. 2018).

3 Post-extraction Ecology

These promising trajectories connect to an ecological agenda that considers the broader entanglements of material use, from geology to quarry, from quarry to stone, and the quarry again as a landscape of healing or reuse. The alternatives to extractivism are investigated through an approach to architectural design inspired by principles of ecosystemic plurality and interconnectedness (Tsing 2015; Latour 2018; Ait-Touati et al. 2022) and ecological thought (Morton 2007, 2010) by positioning extractive practices and design actions within the current multi-crisis and engaging with existing environmental, socio-cultural, and economic realities indicating evolutionary and post-extractive material strategies.

Those strategies employ five attributes of hyperobjects, defined as "things that are massively distributed in time and space relative to humans" (Morton 2013). They include *viscosity, nonlocality, temporal undulation, phasing, and interobjectivity* (Morton 2013), which act as lenses to explore post-extractive material practices. Viscosity and nonlocality, by questioning the perception of distance and proximity, reveal the limitations of approaches that focus only on immediate contexts and local markets. Temporal undulation argues that surrounding objects are also distributed in time which shows the necessity to explore different temporal frames for extraction and design practices. Phasing, by stating that we are experiencing only fractions of the entities, directs us to questions of scale and tools to address ungraspable material complexities. Finally, interobjectivity leads beyond anthropocentric and local approaches that indicate fundamental repositioning of extraction and reversible design processes that favor multispecies coexistence and community building.

Everything known and unknown is interconnected (Morton 2018). Thus, a post-extractive building and design approach values *multiplying viewpoints* over a *single vision* (Latour 2018) and employment of *the art of noticing*, understood as simultaneous noticing, observing, documenting, and tracing of multiple phenomena that manifest *the patchiness of the world* (Tsing 2015). This viewpoint means that design choices are embedded in multi-scalar and multi-local entanglements (Ait-Touati et al. 2022). The ambiguity of distances and scales in times of globalization influences the invisibility of environmental and socio-economic associations embedded in material substance and construction processes. Currently, available advanced tools and technologies operate with more and more data enabling optimization of environmental and economic impacts, although this optimization still operates within the system where resources are perceived as commodified entities.

The post-extractive approach investigates the consequences of design actions on sites of material extraction, production, supply, and construction by asking "how can we fundamentally transform supply chains that are built on the exploitation of cheap work and cheap nature …?" (Malterre-Barthes 2021). This approach necessitates multidimensional perspectives, transdisciplinary collaboration, diverse scales of exploration, and multifaceted architectural responses that consider ecosystemic entanglements. The scope of analysis needs to be a multifocal perspective that includes the changes in the extractive landscapes (quarry), the reservoir of finite resources (stone), embedded environmental impacts (embodied carbon), and modifications of socio-cultural structure (local communities).

How to track those impacts when the materials are extracted all over the world, shipped, and constructed in distant locations? It is easy to advocate for using locally extracted resources as it happens, for example, with *marès* stone in Mallorca (Fig. 1). For years, this sandstone, which rarely if ever leaves the island, has been extracted from adjacent quarries in standardized blocks with dimensions that mirror the modular system of architectural elements that construct Mallorcan villages and cities (Inyesta and Sunyer 1997). The case of *marès* shows that stone can be used effectively, avoiding unnecessary waste during extraction and at the end-of-life as material durability and repetitive modularity enables reuse (Fig. 2). Moreover, Mallorcan stone shows

Fig.1. The scale of extraction of marès in quarries in Mallorca: 1–2., 4 Petra, 3. Bank of marés in Petra, 5. Can Picafort. *Source* U. Kozminska

Fig. 2. Diagram of post-extractive approach in marès staone, illustrating the main points of transdisciplinarity, circularity, and multi-focal/mult-scalar perspectives. *Source* N. Gjorgjievski

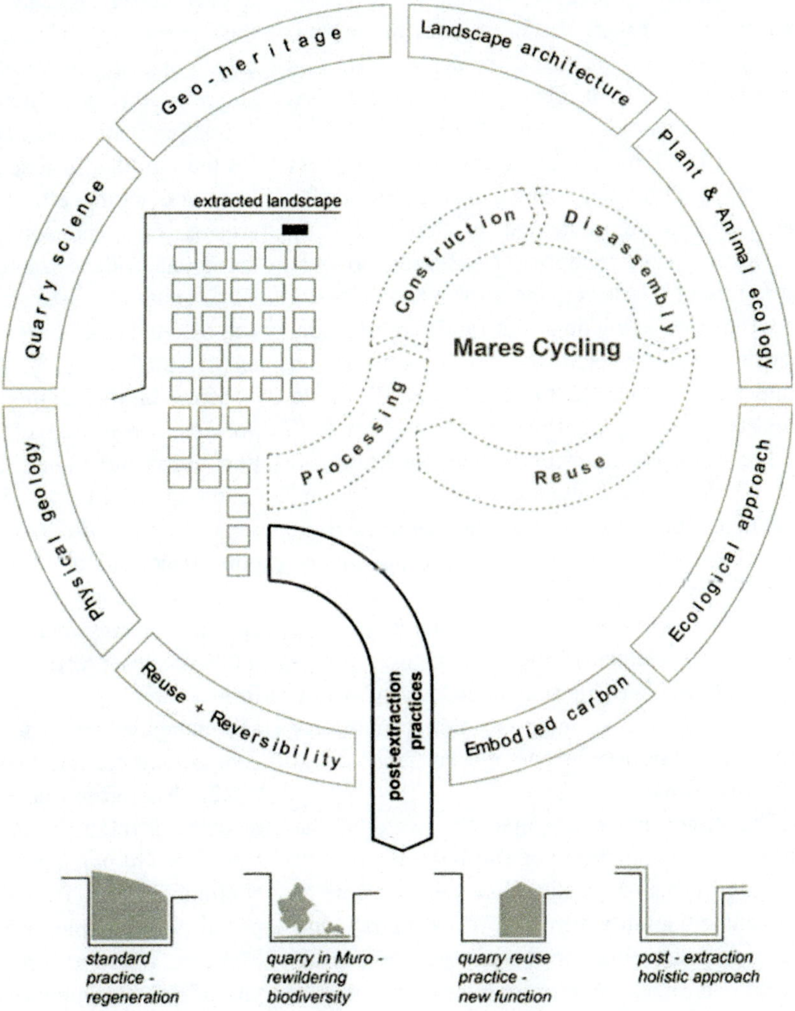

that the scale of production limited to local markets makes a positive, reciprocal impact on the local environment, landscape, and communities (Matas 2018).

A contrary situation can be observed in the quarry in Larvik, Norway, which exemplifies most extractive practices in stone today. Here, the local stone, known as larvikite, is extracted in large blocks of 15–20 tons to be shipped to processing facilities elsewhere in Europe, where they are usually turned into finished elements. Due to the high reliance on marketable and aesthetic standards, the waste generated during the extraction process in Larvik constitutes approximately 50% of extracted resources. This is actually quite low in relation to the global standard, which can yield up to 70% of waste (Carredu 2019). Optimal carbon trajectories are less obvious as Portuguese, Spanish or Italian facilities score better than the ones located in Norway due to the heavy reliance on sea transport over truck freight. However, the apparent disconnect between the extraction site and architectural element results in relatively inefficient use of the finite resource, not to mention other displaced impacts.

These two examples of diverse extraction and use of stone not only show certain inspiring directions, but also ambiguities embedded in current extractive practices. They bring up the questions of the sustainable scale of extraction, the relation between material production and architectural elements, the contemporary, perhaps carbon-related definition of locality within the global contexts, and tools that can inform ecological design decisions and sustainable use of natural stone.

The discussion concerning the scale of the extraction process does not end here. Ecological thinking makes us "realize that there are lots of different temporality formats" (Morton 2018) and that the consequences of our material choices are spread over time. Therefore, a post-extractive use of resources, stone included, requires that we redefine the temporal frame of an architectural project that equally cares for what is taken as well as left behind. The discussion concerning resources cannot be limited to calculating embodied carbon, but it needs to address the entire life cycle of the material, starting with sites of extraction, processing, production, and construction, followed by anticipation of diverse usage and performance patterns to plan for maintenance, repair, and reuse.

The post-extractive approach considers the changes within the landscape. In the case of natural stone, this means that designers enter a geological time perspective and are to reflect on the life cycle of the quarry that material is sourced from and related changes to its ecosystem. Moreover, it is not only about understanding the current environmental and social impacts, but also foreseeing the future ones. It is about considering the afterlife of the quarry while extracting materials but also taking responsibility for the modified landscape and maybe extracting less and more efficiently. As Mark Wigley argues, there is a need for architecture less complicit with extractive economies, and one that: "must at least return a gift of the architect and take the risk of seeing what might come after architecture" (Wigley 2021).

How to return this doubtful gift? Existing policies concerning quarry landscape regeneration (Jorba and Vallejo 2010) are often limited to filling in the excavated territories that look the same as they did before extraction. Although quarry pits are often seen as deep scars in the landscape today (Palmer 2014), this visual approach seems to forget that terraforming is a multi-species practice and that extraction transforms chemical and physical soil composition and embedded ecosystems that unfold around the zone of direct interference (Ait-Touati et al. 2022). Therefore, instead of filling up closed quarries, it might be more ecological to employ *techniques of patching* to reestablish ecosystemic continuities (Ait-Touati et al. 2022). This happened, for example, in a closed quarry in Muro, Mallorca where a thin layer of soil covering the pit and some orange seeds created a new natural habitat in place of previous *marés* extraction. Therefore, it may be crucial to look for less anthropocentric approaches which treat extractive ruins as new topographies and allow the territories to create their own "feral dynamics" that flourish "on infrastructures of human disturbance" (Bubandt and Tsing 2018). Maybe the way forward is about doing less—extracting, controlling and covering up less, and embracing instabilities, including more than human perspectives and developing over time.

Hopefully, the attitude of more mindful and efficient extraction would reduce the amount of generated waste during material extraction and production. Current statistics concerning quarry waste demonstrate that around 50% of extracted material is classified as waste due to irregular shapes. Moreover, present circular approaches to reduce the amount of quarry waste include turning it into aggregates, pebble stone, rubble, sawdust, and stone sludge for agriculture, breakwater, and paving. Thus, due to legislation issues and waste definitions, large, defective blocks are crushed and downcycled despite the fact that their physical–chemical and mechanical properties are often defined as equal to raw material (Carredu 2019). A shift toward more use of structural stone opens the door to significantly reducing waste even further (Webb 2022).

The issue of waste is not limited to quarry waste. The post-extractive approach requires that we reflect on future waste streams to be generated at the end of the life of a designed building. Ideally, we should avoid generating it at all by creating long-lasting buildings that are maintained and repaired to function undisturbedly due to the suitable layered organization of their elements (Brand 1994). However, considering the fact that, on average, a building lasts around 60 years, it is crucial to consider its future disassembly and reuse scenarios, where "in the most optimistic ecological scenario, architecture would be so generous that it would disappear" (Wigley 2021).

Natural stone has the potential to fulfill this ecological agenda by creating new reversible building systems that rely on circular construction (Guldager Jensen 2016) and design for disassembly principles (Crowther 2001) while working with such reversible building actions as, for example, *stacking, resting, overlapping, consoling, spanning, lifting or splitting* (Nielsen 2012). However, more reversible strategies for stone construction require that we embrace its load-bearing capacity and use elements of larger dimensions instead of employing the material only as finishings. This approach starts to appear in architecture, for example, in the aforementioned modular stone system proposed by Groupwork or the biodiversity wall systems by Chartier-Dalix.

Lastly, it is crucial to explore reversible tectonics and landscape approaches in broader systemic contexts investigating the environmental and socio-cultural impacts while rethinking existing economies, business models, and supply chains that construct the new building systems. It is advisable to follow the discourse of Jane Hutton, who advocates for more *solidarity* to avoid colonizing practices and *reciprocity* to create mutual, inter-species obligations for land and resources (Hutton 2021).

4 Conclusion

As the practice of architecture continues to dictate material extraction, it is prudent to question the new potentials of buildings, no longer perceived as disconnected finalities but as catalysts for change. Yet, even the staunchest opponents of resource extraction acknowledge that a moratorium on mining and quarrying would be practically unrealizable. For this reason, we need to look closer at the materials we extract for architecture and take "an approach to the designed environment that takes complete responsibility for itself" (Grima 2021). Such an approach requires that we flatly reject our widespread reliance on site-intensive and extractivist development. To do so, we must look critically at extraction and find ways not only to reduce the demand on earth's resources, but also to use what we take in the best way and to encourage responsible approaches to the disturbed landscapes that employ more plural perspectives and operate within the diversity of species, localities, temporalities, scales, and disciplines. We encourage the visualization of the link between the building and its extracted landscape, highlighting that the design challenge of the future must not solely address the spatial problem of the building, but the opened landscape as well.

Although only in a nascent state, new research into natural stone offers a poignant case of how we might move from extractivism to a more custodial form of extraction. It shows the potential of reducing our need for materials such as steel, concrete, and wood; industries that are driven in large part by extractivist principles. Furthermore, it is not so challenging to imagine how to reinstate historical connections between architecture and the anthropogenic landscapes of stone extraction. In this way, we position the issue of material extraction within its multifaceted entanglements of landscape and sociocultural and economic contexts. Only when those contextual associations are acknowledged and understood, can it be possible to develop post-extractive, reversible, and more circular modes of building with natural resources.

References

Abu-Jaber N, Bloxam EG, Degryse P, Heldal T (2008) Quarryscapes: ancient stone quarry landscapes in the eastern mediterranean. NGU Geological Survey of Norway, Norway

Abusaada N (2022) Jerusalem stone: the history and identity of Palestinian stereotomy. Architectural Review: Stone

Acosta A (2017) Posextractivismo: del discurso a la práctica—Reflexiones para la acción. International Development Policy, Revue internationale de politique de développement. https://journals.openedition.org/poldev/2496#quotation

Ait-Touati F, Arenes A, Gregoire A (2022) Terra forma. MIT Press, Cambridge, A Book of Speculative Maps

Arendt H (1958) The human condition. The University of Chicago Press, Chicago

Barrault T, Pressacco C, Petkova N (2018) Pierre: Révéler la resource, Explorer le matériau. Pavillon de l'Arsenal, Paris

Brand S (1994) How buildings learn: what happens after they're built. Penguin Books, London

Brilliant R, Kinney D (2016) Reuse value: Spolia and appropriation in art and architecture from Constantine to Sherrie Levine. Routledge, New York

Bubandt N, Tsing A (2018) Feral dynamics of post-industrial ruin: an introduction. J Ethnobiol 38:1–7

Carredu N (2019) Dimension stones in the circular economy world. Resour Policy 60:243–245

Chartier-Dalix (2019) Mur biodiversitaire/ChartierDalix en partenariat avec le Museum d'Histoire Naturelle (laboratoire CESCO) et l'école d'architecture de Paris Malaquais (laboratoire GSA). http://chartier-dalix.com/fr/ressources/prototypes-rue-buffon-paris-5

Crowther P (2001) Developing an inclusive model for design for deconstruction, Deconstruction and material reuse: technology, economic and policy, CIB publication 266

Durante F, Kröger M, LaFleur W (2021) Extraction and Extractivisms: definitions and concepts. In: Shapiro J, McNeish SJ-A (eds) Our extractive age : expressions of violence and resistance. Routledge, Abingdon Oxon

Grima J (2021) Design without depletion: On the need for a new paradigm in architecture. In: Space caviar (ed) non-extractive architecture. On designing without depletion, vol 1. Sternberg Press, Berlin, pp 7–26

Guldager Jensen K (2016) Building a circular future. GXN Innovation, Denmark

Hutton J (2020) Reciprocal landscapes: stories of material movements. Routledge, Abingdon Oxon

Hutton J (2021) Notes on reciprocity and solidarity. In: Space Caviar (ed) Non-extractive architecture. On designing without depletion, vol 1. Sternberg Press, Berlin, pp 279–289

Inyesta G, Sunyer O (1997) Construir en Marès. Col.legi Oficial D'Arquitectes de Balears, Palma de Mallorca

Jones L (2021) Carbon tectonic. In: Space Caviar (ed) Non-extractive architecture. On designing without depletion, vol 1. Sternberg Press, Berlin, pp 113–125

Jorba M, Vallejo V (2010) Manual para la restauración de canteras de roca caliza en clima mediterráneo. Generalitat de Catalunya

Palmer L (2014) In the Aura of a Hole: exploring sites of material extraction. Black Dog Publishing, London UK

Latour B (2018) Down to earth. Politics in the new climatic regime. Polity Press, Cambridge

Le Dréan M, Kuratli J (2022) La Pierre banale: Logements collectifs en pierre massive, régione parisienne, 1948–1973. EPFL Press

Lovelock JE, Whitfield M (1982) Life span of the biosphere. Nature 296

Malterre-Barthes C (2021) The Devil is in the Details: 'Who is it that the Earth belongs to?' In: Space Caviar (ed) Non-extractive architecture. On designing without depletion, vol 1. Sternberg Press, Berlin, pp 85–96

Mata CS (2018) Les pedreres de marès: Identitat oblidada del paisatge de Mallorca. Lleonard Muntaner, Palma de Mallorca

Moe K (2020) Unless. Actar, New York

Morton T (2007) Ecology without nature: rethinking environmental aesthetics. Harvard University Press, Cambridge

Morton T (2010) The ecological thought. Harvard University Press, Cambridge

Morton T (2013) Hyperobjects. In: Philosophy and ecology after the end of the world. University of Minnesota Press, Minneapolis
Morton T (2018) All art is ecological. Penguin Random House, UK, London
Naess A (1973) The shallow and the deep, long-range ecology movement: a summary. Interdiscip J Philos 16:95–100
Nielsen S (2012) The tectonic potential of design for deconstruction (DfD). In: CIB W115 green design conference proceedings, vol 366, pp 21–26
Ogburn D (2013) Variation in Inca building stone quarry operations in Peru and ecuador. In: Tripcevich N, Vaughn K (eds) Mining and Quarrying in the ancient andes : sociopolitical economic and symbolic dimensions. Springer, New York
Ringbom S (1987) Stone style and truth: the vogue for natural stone in Nordic architecture 1880–1910. Helsinki
Tsing A (2015) The mushroom at the end of the world. In: On the possibility of life in capitalist ruins. Princeton University Press, Princeton
Webb S (2022) Stone age: a new architecture from an old material. Architectural Review: Stone
Wigley M (2021) Returning the gift: running architecture in reverse. In: Space Caviar (ed) Non-extractive architecture. On Designing without depletion, vol 1. Sternberg Press, Berlin, pp 41–57

Towards a Nature-Inspired Bio-digital Platform Powered by Microbes as a Circular Economy Infrastructure in the Practice of the Built Environment

Rachel Armstrong

Abstract

This paper examines the role of microbial technologies in fundamentally rethinking the relationship between waste, energy, human inhabitation, and microbial "life." Based on the Microbial Fuel Cell (an ecologically "just" platform that provides bioelectrical energy, data, and chemical transformation from human waste streams), a series of transactional systems between humans and microbes are outlined and exemplified by the Living Architecture project, the 999 years 13 m^2 (the future belongs to ghosts) installation, and the Active Living Infrastructure: Controlled Environment (ALICE) prototype. These nature-inspired "living" infrastructures comprise a combined utility system capable of forming a domestic circular economy based on the microbial commons—a regenerative, transactional system for ecological regeneration founded on microbial metabolisms—that is accessed, understood, and engaged using a bio-digital interface. These simultaneously "sustainable" and "smart" demonstrators substantiate the future trajectory of the wider uptake of microbial technologies in an urban context, ultimately resulting in "Living" cities that are based on microbial (micro)economics and are fundamentally bioremediating.

Keywords

Microbes · Microbial fuel cells · Living bricks · Circular economy · Bio-digital interface · Bioelectricity

1 Introduction

Life, when you boil it right down, is a flow of electrons. (Brahic 2014)

The hardest part of altering our building impacts is changing our thinking, our habits, and our concepts of what a *good life* entails. Modernity's machinic living spaces are at odds with the increasingly urgent need for ecological lifestyles. To change how we live and work requires a different technological platform that aligns with the natural world—and the only technology that enables truly circular design is "life" itself[1] (de Lorenzo 2015).

[1] The turn of the millennium saw the rise of "living" technology, which directly engaged the characteristics of life moving away from the machine as the dominant metaphor for innovating new ways to perform useful work.

R. Armstrong (✉)
Ku Leuven, Architecture, Ghent, Belgium
e-mail: rachel.armstrong@kuleuven.be

M. R. Thomsen et al. (eds.), *Design for Rethinking Resources*, Sustainable Development Goals Series,
https://doi.org/10.1007/978-3-031-36554-6_3

2 Introducing the Microbial Commons

"Microbial technologies" are a potential new *life-based platform* for circular design comprising a variety of emerging biotechnology applications that use electrode–bacteria interactions to generate material change including the production of electricity, wastewater treatment, bioremediation, and the production of valuable products (Kracke et al. 2015). Although the average lifespan of a bacterium is around twelve hours or so, collectively, microbes (bacteria, archaea, fungi, viruses, protists) have sustained the living world for billions of years through their critical contributions to biogeochemical functions, e.g., decomposition and nutrient cycling (Fenchel et al. 2012). Most microbes (80% of bacteria, archaea) reside in biofilms which are organized heterogeneous assemblages of microbial cells that are encased within a self-produced matrix—and within this essay will be considered as *microbial cities* (Penesyan et al. 2021). Producing biomolecules from resources in the environments, biofilms, and planktonic microbes (which do not live in a biofilm) collectively form the "microbial commons" (Dedeurwaerdere 2010) where the free and open exchange of microbial materials takes place, forming a foundational biomolecular currency (metabolism) that performs critical "ecosystem services," which generate ecosystem benefits at little, or no resource cost, within the urban and global context (Bell et al. 2005; Balvanera et al. 2006). For example, nitrification and denitrification are microbial processes that are extensively used in urban wastewater treatment (Bitton 2011), providing food for plants, and removing toxins respectively, with substantial "health" benefits to the local environment. The term *microbial commons* originates from the biotechnology revolution where the generous exchanges of microorganisms among culture collections, laboratories, and researchers worldwide, formed an Open Source facility for research activities to advance the scientific understanding of microbes. Commercial pressures from biotechnology firms in the late twentieth century, however, altered this situation by restricting this access through Intellectual Property (IP) protections and gene patenting. In this essay, the term microbial commons is deanthropocentrized, so that rather than referring to a laboratory commodity, instead it refers to the global exchange of microbial goods in the living world and introduces *microbial technologies as an economic system for microbial/human interactions*.

3 Principles of Microbial Economics

Over the eons, microbes have used their commons to invent all major forms of metabolism, multicellularity, nanotechnology, organic metallurgy, sensory systems, locomotive apparatuses, reproductive strategies, community organization, metabolism, and mineral conversion platforms, creating the basis for versatile and nature-based microbial technologies (Margulis 1981). The incorporation of these agents into our living spaces within specific contexts creates a transactional platform for sustainable interventions, potentially replacing modern plumbing with bioreactor systems that transform a variety of household wastes into usable resources (Lahiji and Friedman 1997). Such technologically mediated access to the *microbial commons* operates according to mutualistic principles—*if you give me your waste, then I shall give you [something useful]*—escaping the capitalist logic of resource inequality (extreme stockpiling leading to resource depletion and gross inequalities), as the same things are valued differently by different actors, facilitating equitable transactions. Mutualistic relationships, however, break down in the presence of pathogens, which account for less than one percent of all microbes, and are an exception to the principle rather than the rule—so much so, that biofilms actively destroy them (Pasternak et al. 2019).[2]

[2] The exceptional and important behavior of pathogens is documented in detail in many medical papers. It is not my aim to downplay this important aspect of microbial behavior but, rather to emphasize most of all microbes are beneficial to all life and can, therefore, be engaged within our living spaces using appropriate principles of hygiene.

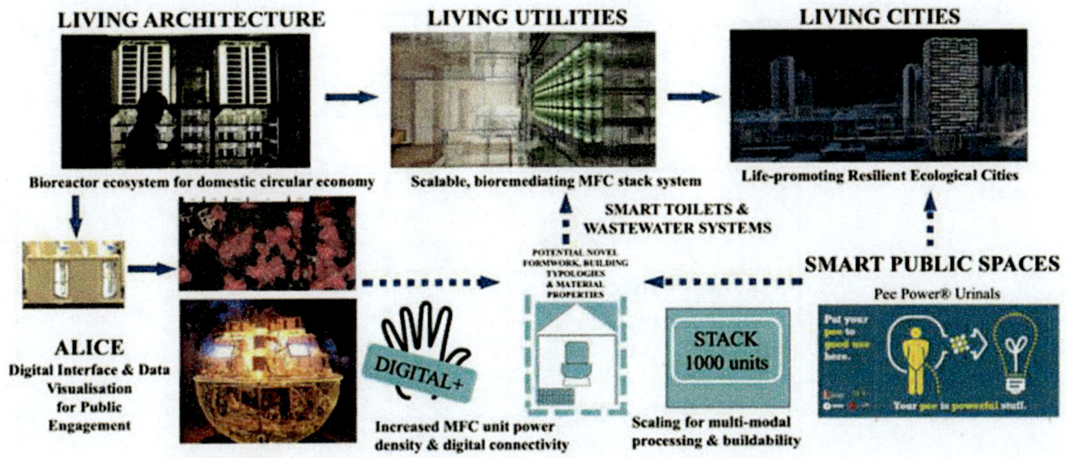

Fig. 1 Environmentally regenerative "Living" cities are enabled by microbial technologies that perform *building operations* based on an ecosystem of MFC-based technologies—PeePower®, Living Architecture (LIAR), Active Living Infrastructure: Controlled Environment (ALICE), which provide a range of ecosystem services and energy for low-powered electronic devices in a fundamentally bioremediating manner. The integration of MFCs systems into building structures enables the further exploration of novel formwork, aesthetics, and user experiences

While a regenerative technological platform that enables meaningful resource exchanges between humans and microbes may sound like science fiction, the functional integration of microbes into the built environment is feasible through the designed incorporation of biofilms and specific types of microbial consortia[3] into bioreactors, where environmental parameters can be adjusted to encourage the microbes to perform different kinds of work. Collectively, these microbial populations turn waste streams into low-cost, low-power energy that can be made available to citizens across multiple social, economic, and political divides. Forming the operational basis for community transactions through access to the microbial commons, such microbial technologies establish a platform for a circular resource economy enabling new kinds of urban exchanges (Timmis et al. 2019). A vision outlining the principles of transforming urban impacts by incorporating microbial technologies into our buildings is shown (Fig. 1).

[3] A microbial consortium or microbial community is when two or more bacterial or microbial groups live symbiotically but have not formed a biofilm.

4 Introducing the Microbial Fuel Cell

The microbial fuel cell (MFC) is a specific type of microbial technology—a kind of organic battery, which was first reported by Potter in 1911 (Potter 1911) (Fig. 2). Each cell has an anode and cathode, separated by a proton-exchange membrane that divides the liquids into the two chambers while allowing protons to pass from the anode to the cathode. Organic waste flows into the anode as feedstock for an anaerobic biofilm, which excretes electrodes that are captured by electrodes, generating an electrical current sufficient to power electronic devices, while protons pass through the membrane where they combine with oxygen to produce water.

MFCs, however, produce more than just bioelectricity as the biofilms also recover nutrients, synthesize biofertilizer (in the stabilized sludge they produce, which is a rich source of nitrogen and phosphate), make disinfectant, treat wastewater, and kill pathogens while they excrete electrons—much like soil biofilms in natural ecosystems. MFCs also act as biosensors by generating voltage, which linearly correlates

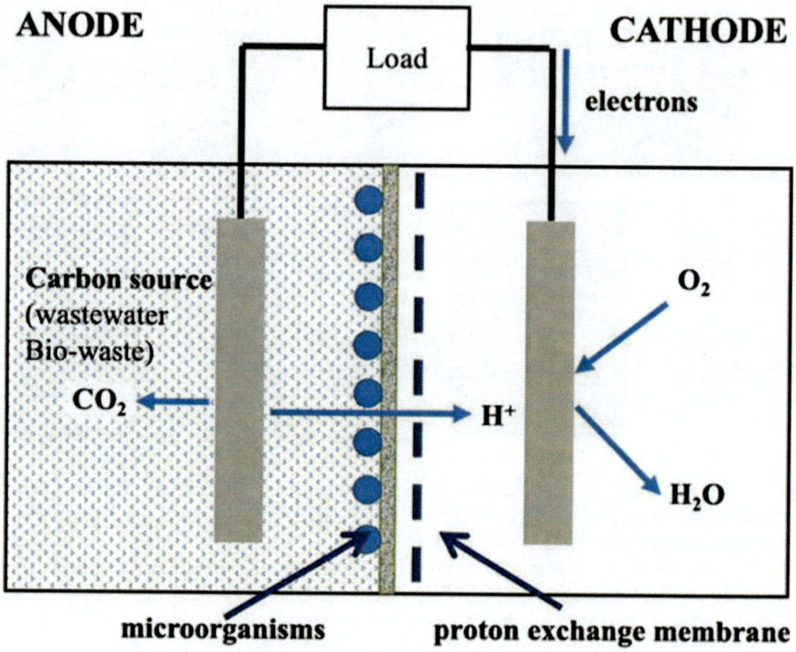

Fig. 2 Technical diagram showing electrons produced from an anaerobic biofilm that are captured by electrodes to produce electricity while performing ecosystems services (bioremediation of wastewater)

with specific quantities of toxins like heavy metals (copper, chromium, and zinc), and organic compounds (*p*-nitrophenol (PNP), formaldehyde, and levofloxacin) (Zhou et al. 2017). In this sense, MFCs are completely unlike batteries or other types of modern utilities, which are designed to process one type of resource at a time. Providing a foundational platform that can change the impacts of human settlement across a range of parameters, MFCs catalyze the transformation of a range of organic wastes (e.g., urine, greywater, and black water which are feedstock for microbes) into bioelectricity while simultaneously providing environmental services (bioremediation, detoxification, and water purification) generating an overall net-positive impact on ecosystem health.

5 State of the Art

MFCs are not the only microbial technological platform to process waste into a range of new resources. Anaerobic digestion in *biodigesters* also produces fuel (*biogas*), removes biochemical oxygen demand (BOD) from sewage, conserves nutrients (especially nitrogen compounds), and reduces pathogens. MFCs, however, have some advantages over biodigesters as they produce electricity without combustion, act as sensors, and can be used for the treatment of low-concentration substrates at temperatures below 20 °C, where anaerobic digestion generally fails to function. This creates specific application niches for MFCs that do not compete with, but complement, anaerobic digestion technology. MFCs still face important limitations in terms of large-scale application including investment costs, upscale technical issues, and the factors limiting the performance, both in terms of anodic and cathodic electron transfer (Pham et al. 2006). To date, MFCs have not been used widely as they are neither competitive with fossil fuels nor renewables as a single source of electrical energy. Over the last two decades, MFCs have largely gained traction as a source of bioremediating green energy owing to breakthrough advancements, which are predominantly confined to the lab. Field/pilot trials for the treatment of wastewater streams such as urine, greywater, and blackwater, by MFC stacks have increased in number, e.g., Pee Power® urinals

(Walter et al. 2018) and Urine-tricity (Oxfam International 2015), but further research and development is still needed to advance the technology's commercial readiness (TRL) for novel applications such as smart toilets that ultimately, will be installed in ecohomes. At scale, pilots engaging MFCs (such as in the brewing industry) are at relatively immature stages of development, so their potential impacts are cautiously reported (Singh and Sharma 2010).

6 Bioelectricity for Building Operations

Although a vast range of microbially mediated processes is of value to households and industry, to date, the electrical outputs of bioreactors have generally been too weak to drive conventional electronic hardware (Koffi and Okabe 2020). While MFCs cannot compete with the sheer power provided by other electricity generating systems (renewables, fossil fuels), their (material) circularity is unsurpassed, providing a circular flow of resources within a household or building, which are metabolically constrained by the carrying capacity of the site. Part of the challenge for MFCs becoming a household system is that innovation in electrical appliances for the last 150 years has effectively operated within a conceptual frame where energy is unlimited and can be used to solve all challenges from refrigeration to climbing stairs, resulting in the voltage outputs for modern homes being standardized for 230 V. Despite current innovation outputs being benchmarked against industrial expectations, significant advances in biotechnology, material sciences, and hardware design are creating the context for the installation of smaller, lighter, and more efficient MFCs that are currently being trialled in wastewater processing and enabling their installation in buildings. This new generation of MFCs benefits greatly from their compatibility with low-powered digital technologies, enabling the juxtaposition of organic and "smart" (electronic) platforms at the bio-digital interface.

7 Introducing the Bio-digital Realm

The bio-digital interface is located at the MFC electrodes where the organic and electronic domains meet. Making an intrinsic connection between microbial metabolisms and digital systems (screens, LEDs, USB ports, etc.) the bio-digital interface mediates the direct relay of MFC outputs (electrons, data, and chemical transformation) from an organic environmental platform (microbes) to low-power electronic devices. As the production of electrons by the biofilm are based on metabolic reactions, alterations in electron transport and carbon metabolism can also influence bioelectricity production, establishing the potential for two-way electro-biochemical exchanges between human and microbe, which are based on an electron economy to establish the principles of a human/microbial trading system. The following precedents offer a series of case studies based on operational MFC prototypes that are designed for implementation in building systems. Each prototype is spatially organized in a unique configuration, which operate at low-power thresholds of around 2–3 V.

8 Living Architecture

The *Living Architecture project* is a "living" combined utility infrastructure that uses fifteen MFC complexes consisting of four chambers[4] to turn liquid household waste such as urine and grey water into valuable resources (electricity, biomass, water, reclaiming phosphate from washing-up liquids and removing poisonous gases from the air) that can be re-used in the household (Armstrong et al. 2017) (Fig. 3). The performance of the biofilms is optimized using an AI, that is also powered by the MFCs, which operates simple mechanical controls that strategically deliver feedstock within the arrays.

[4] These chambers comprise two MFCs separated by two photobioreactors separated by ceramic plates that enable the different resident microbial systems to exchange nutrients between their respective populations.

Fig. 3 Detail of the fully inoculated *Living Architecture* "wall" and apparatus installed at the University of the West of England, Bristol, the *Living Architecture* project, 2019. Photography by Rolf Hughes © Rolf Hughes

Enabling the smarter use of electrons, multiple tasks are performed within the apparatus from generating power to providing data and enabling a range of metabolic transformations as the basis for a circular resource economy established between humans and microbes. *Living Architecture's* circular economy is enabled by its integrated infrastructure, where people exchange their waste for clean water, bioelectricity, and a range of useful biomolecules, which are compatible with everyday activities of daily living. To facilitate this exchange, the MFC design was optimized by creating geometries that combined structure and flow to generate electricity-producing building blocks, or "living bricks" (You et al. 2019). The electrochemical characteristics of three different kinds of conventional house bricks were tested from two source locations. When fed with human urine, European standard off-the-shelf house bricks generated a maximum power of 1.2 mW (13.5 mW/m^2), whereas Ugandan house air bricks[5] produced a maximum power of 2.7 mW (32.8 mW/m^2). The integration of MFCs into buildings activates the microbial commons to enable a domestic economy that frees the household from its obligate consumption of resources, reducing electricity and utilities bills, while mitigating the amount of untreated waste discharged into the environment. Interrupting modern technological silos *Living Architecture's living bricks* are suitable for different uptake communities within different architectural typologies for the construction of ecohomes, where the contributions by all who carry out the work-of-life are valued through their ecological transactions, turning domestic spaces into sites of wealth-generation. Inhabitants now have choices to make about how they use their ecological resources to reduce their dependency on centralized utilities or help others.

9 999 Years 13 m^2 (The Future Belongs to Ghosts)

The original version of *Living Architecture* could not be exposed directly to the public owing to the presence of genetically modified organisms, so an alternative wild-type microbial experience entitled *999 years* 13 m^2 *(the future belongs to ghosts)*, was developed for the *Is This Tomorrow* exhibition at the Whitechapel Gallery in collaboration with artist Cecile B Evans (Bevan, 2019) (Fig. 4). An installation was prefigured as a

[5] The choice of Ugandan bricks was based on using building blocks from a Pee Power® field site.

Fig. 4 The installation 999 years 13 m^2 (the future belongs to ghosts) is an art installation embodying a posthuman apartment comprised of a bank of MFCs and digital screens. The installation is by Cecile B. Evans and Rachel Armstrong for the *Is This Tomorrow?* exhibition at the Whitechapel Gallery, London. Photograph by Rolf Hughes, 2019 © Rolf Hughes

minimal housing space (13 m^2) conferred with the longest possible lease (999 years) and powered using the natural biofilms in an array of fifteen MFC complexes from the *Living Architecture* project producing ~200 mW/L urine. The only observable traces of humans were digital manifestations of the human past, present, and future (ghosts), powered by microbes, highlighting the interdependencies between the species through rituals of cleaning, feeding, and care in exchange for a range of services from cleaning wastewater to generating power for the installation. Emphasizing how poorly modern society values the microbial realm and discards nutrient-rich resource streams as "waste," the installation invited audiences to adopt an ethical position in relation to resource recycling and consider what kinds of transactions are acceptable for activating the *microbial commons*.

10 Active Living Infrastructure: Controlled Environment (ALICE)

The *Active Living Infrastructure: Controlled Environment* (ALICE) prototype (2019–2021), advances the ambitions of the *Living Architecture* project by establishing the first bio-digital interface using its fifteen MFC complexes to establish the foundations for collaboration with microbes using electron flow as a real-time language (Armstrong and Hughes 2021). Electrical activity from the biofilms provided a source for both power and data, generating ~200 mW/L urine, which was translated by software into animations that conveyed the overall status of the biofilm in relatable terms. Audiences could, therefore, respond to the microbial behaviour, not by looking at unpleasant "slime" (the natural "face" of microbial colonies), but instead by interacting with appealing forms on a familiar screen-based interface. This world of "Mobes," a characterful term coined for the data-based representations of microbes, offers a simple, probiotic approach to interspecies communication within the highly situated realm of microbes in a relatable manner that could even become part of our everyday routines establishing new value systems that invite different kinds of (house) work, as well as stimulating alternative domestic routines for our living spaces (Fig. 5). Since MFCs are *living*, possessing a force and agency of their own, they require our appropriate care and attention if they are to engage with us in a productive and co-constitutive manner (Bellacasa 2017).

Presently, ALICE exists as a permanent online exhibit (ALICE 2019) that can be accessed under the section Bio-Digital Interface by clicking the Launch Artwork button, which leads to an animated set of "Mobes". On selecting

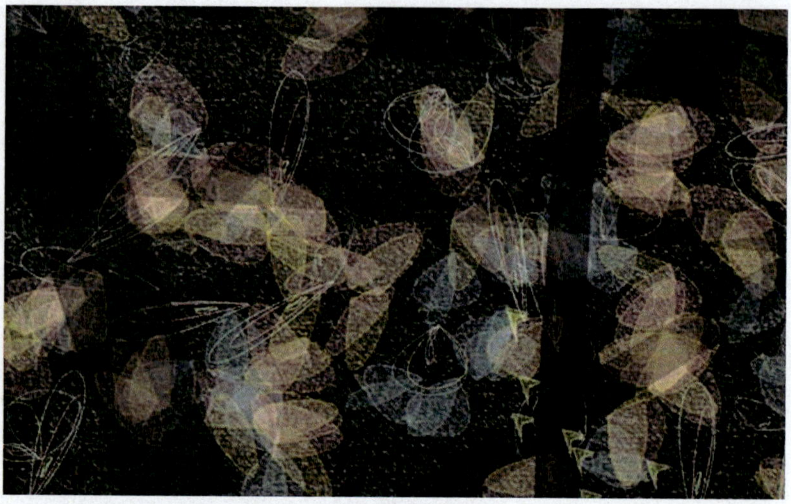

Fig. 5 "Mobes," from the ALICE website (http://alice-interface.eu) showing dynamic, interactive, and graphical representations of microbes, courtesy the ALICE consortium, 2021. © ALICE consortium, screenshot from the website

different options from drop-down menus, the environmental parameters (temperature, pH) and performance (power output) of a real-world microbial community can be interrogated that inhabits a permanent MFC array installed in a scientific laboratory. Depending on how the visitor reads the health of the microbes, they can respond to the "Mobes" by feeding them using a remotely operated valve system, or by speeding up their metabolisms by activating an LED to warm them gently. The graphical symbols provide a language where factual propositions (environmental parameters) are represented, and further truths can be inferred directly, or by means of a calculating system, which can be overridden by human intervention.

ALICE also existed as an embodied, real-world installation that premiered at the Digital Design Weekend, V&A, London, UK, as part of the London Design Festival from 24 to 26 September 2021 (Barto 2021) and was installed at the Electromagnetic Field Festival, from 2 to 5 June 2022 (Electromagnetic Field Camp 2022). Inviting meaningful human/microbial transactions, ALICE demonstrated the potential for microbially powered technologies with bio-digital interfaces to become creature-like, becoming relatable entities that do not ooze, stink, or repulse, as an engaging a way of organizing our daily routines differently, where the microbial/human transactions are more than functional but have meaning, value, and can be culturally adopted in life-promoting ways. Since these microbial systems are not based on simple substitutions for modern building systems, a set of design principles and protocols for their uptake and implementation is needed through the recognition of novel elements like bio-digital interfaces and an ethics of care that can assist the designer/architect in developing, innovating, and scaling new applications that are appropriately situated in an increasing variety of sites. By incorporating life-bearing microbial technologies into buildings, the architect's role is to connect the home with the biosphere through economic transactions at the bio-digital interface, which ultimately enhances the overall life-bearing potential of a site (Fig. 6).

11 Critical Reflections on Near-Future Developments

The three MFC-based installations presented provide demonstrators that each take different steps towards the implementation of a platform for a circular design that is made accessible via the installation of a bio-digital interface. As the MFC platform is further refined, living bricks will become increasingly stackable, lightweight, and versatile in their design; enabling new kinds of formwork and ultimately novel building

Fig. 6 The ALICE installation, a transparent orb powered by microbes that animate LEDs and iPads was installed during the Digital Design Week at the Victoria and Albert Museum, September 2021, embodying the bio-digital platform through the integration of microbial and artificial intelligence with biological and technical bodies. Courtesy of the ALICE consortium: Ioannis Ieropoulos, Julie Freeman, and Rachel Armstrong. Photograph by Julie Freeman. © ALICE consortium

typologies. Currently, a 1000-unit MFC stack comprises 46 modules of 5 L geometrical volume each, which is 230 L total volume. With a 50 L header tank and peripherals, this volume becomes 300 L. With smaller MFC modules, 2 tiers of 22 MFCs/module can be produced reducing the total number of modules to 23, which would be 115 L total volume ($\sim$190 L with tanks and peripherals). The power density of MFC modules is also set to reach 1 mW/mL feedstock (Gadja et al. 2018). 1,000 MFC units can be reasonably expected to operate at power levels of 1–2 mW/MFC, 1–2 W, 24–48 Wh for 1 day. This means that 1,000 MFC units can deliver 168–336 Wh per week, 720 Wh–1.4 kWh for 1 month, and 8.7–17.5 kWh for 1 year. Such a scale-up takes a significant step towards resource circularity for human settlements, raising the possibility of 12 V domestic lifestyles that go beyond survival and bioremediate our surroundings while meeting—and perhaps one day surpassing—the basic expectations of modern existence. Importantly, the *bio-digital platform* substantially increases the resilience of the built environment by providing low-power energy autonomy from waste, enabling material resource circularity (turning effluents to electricity, cleaned water, and biomass), and creating a digitally compatible "hub" that bioremediates its surroundings in ways that can be monitored (smart) and are biologically sustainable (nature-compatible); ultimately establishing the building blocks for resource circular cities. **Heralding an era of change** microbial technologies that are cultured and grown in our homes and communities alongside us comprise a paradigm shifting building technology being able to fundamentally change the impact of the built environment on our living world by reaching new levels of sustainability and resilience against climate change. Integrated microbial technologies provide new sources of bioenergy, sanitation, circularity for natural resources, increased soil and water health, thereby making significant contributions to SDGs 6, 7, 11, 4 and 17.[6] Neither optional extras nor an architectural fashion, installing microbial technologies in our homes, buildings, and cities

[6] MFCs contribute to the following SDGs: **6** via clean water and sanitation; **7** providing affordable and clean energy; **11** by creating a platform for sustainable cities and communities; **4** enabling quality education via bioremediating digital services; **17** in creating partnerships for the goals by strengthening the means of implementation and revitalizing the global partnership for sustainable development Technology.

will, literally, save lives. At the time of climate emergency, escalating fuel prices, and the displacement of people from war, having access to basic utilities as a combined processing system can provide clean water, shelter, power, and sanitation, which maintains a basic livability made possible through the microbial commons to generate freely available materials, even in extremis.

References

ALICE (2019) Active living infrastructure: controlled environment. https://alice-interface.eu. Accessed 10 June 2022

Armstrong R, Hughes R (2021) Cyberneticisation as a theory and practice of matter. Footprint 28 15 (1):29–44

Armstrong R, Ferracina S, Caldwell G, Ieropoulos I, Rimbu G, Adamatzky A, Phillips N, De Lucrezia D, Imhof B, Hanczyc MM, Nogales J, Garcia J (2017) Living architecture (LIAR): metabolically engineered building units. In: Hebel DE, Heisel F (eds) Cultivated building materials: industrialized natural resources for architecture and construction. Birkhauser, Berlin, Germany, pp 170–177

Balvanera P, Pfisterer AB, Buchmann N, He JS, Nakashizuka T, Raffaelli D, Schmid B (2006) Quantifying the evidence for biodiversity effects on ecosystem functioning and services. Ecol Lett 9:1146–1156

Barto M (2021) Digital design weekend. https://www.vam.ac.uk/blog/design-and-society/digital-design-weekend-2021. Accessed 3 Sep 2022

Bellacasa MP (2017) Matters of care: speculative ethics in more than human worlds. University of Minnesota Press, Minneapolis

Bell T, Newman JA, Silverman BW, Turner SL, Lilley AK (2005) The contribution of species richness and composition to bacterial services. Nature 436:1157–1160

Bevan R (2019) Is this tomorrow? Review: installations show a troubled mood of the future. Evening Standard. https://www.standard.co.uk/culture/is-this-tomorrow-review-installations-show-a-troubled-mood-of-the-future-a4066551.html Accessed 1 Sep 2022

Bitton G (2011) Wastewater microbiology. Wiley, Hoboken

Brahic C (2014) Meet the electric life forms that live on pure energy, New Scientist. https://www.newscientist.com/article/dn25894-meet-the-electric-life-forms-that-live-on-pure-energy/. Accessed 2 Sep 2022

de Lorenzo V (2015) It's the metabolism, stupid! Environ Microbiol Rep 7:18–19

Dedeurwaerdere T (2010) Self-governance and international regulation of the global microbial commons: introduction to the special issue on the microbial commons. Int J Commons 4(1):390–403

Electromagnetic Field Camp (2022) Pee is Powerful! From artwork to new world infrastructures with ALICE. https://www.emfcamp.org/schedule/2022/307-pee-is-powerful-from-artwork-to-new-world-infrastructures. Accessed 4 June 2022

Fenchel T, King G, Blackburn TH (2012) Biogeochemical cycling in soils: bacterial biogeochemistry. Academic Press, London, pp 89–120

Gajda I, Stinchombe A, Merinao-Jimenez I, Pasternak G, Sanchez-Herranz D, Greenman J, Ieropoulos IA (2018) Miniaturized ceramic-based microbial fuel cell for efficient power generation from urine and stack development. Front Energy Res 6(84). https://doi.org/10.3389/fenrg.2018.00084

Koffi N'DJ, Okabe S (2020) High voltage generation from wastewater by microbial fuel cells equipped with a newly designed low voltage booster multiplier (LVBM). Nat Sci Rep 10. https://doi.org/10.1038/s41598-020-75916-7

Kracke F, Vassilev I, Krömer JO (2015) Microbial electron transport and energy conservations—The foundation for optimizing bioelectrochemical systems. Front Microbiol 6. https://doi.org/10.3389/fmicb.2015.00575

Lahiji N, Friedman V (1997) Plumbing: sounding modern architecture. Princeton Academic Press, New York

Margulis L (1981) Symbiosis in cell evolution: life and its environment on the early earth. W H Freeman, San Francisco, p 347

Oxfam International (2015) Pee power to light camps in disaster zones. https://www.oxfam.org/en/press-releases/pee-power-light-camps-disaster-zones

Pasternak G, Greenman J, Ieropoulos I (2019) Removal of Hepatitis B virus surface HBsAg and core HBcAg antigens using microbial fuel cells producing electricity from human urine. Sci Rep 9(11787). https://doi.org/10.1038/s41598-019-48128-x

Penesyan A, Paulsen IT, Kjelleberg S, Gillings M (2021) Three faces of biofilms: a microbial lifestyle, a nascent multicellular organism, and an incubator for diversity. Nat Biofilms Microbio 7(80). https://doi.org/10.1038/s41522-021-00251-2

Pham TH, Rabaey K, Aelterman P, Clauwert P, De Schamphelaire L, Boon N, Verstraete W (2006) Microbial fuel cells in relation to conventional anaerobic digestion technology. Eng Life Sci 6(3):285–292

Potter MC (1911) Electrical effects accompanying the decomposition of organic compounds. Proc R Soc B 571(84):260–276

Singh AM, Sharma VN (2010) Treatment of brewery wastewater and production of electricity through microbial fuel cell technology. Int J Biotechnol Biochem 6(1):71–80

Timmis K, Cavicchioli R, Garcia J-L, Nogales B, Chavarría M, Stein L, McGenity TJ, Webster N, Singh BK, Handelsman J, de Lorenzo V, Pruzzo C, Timmis J, Martín JLR, Verstraete W, Jetten M, Danchin A, Huang W, Gilbert J, Lal R, Santos H,

Lee SY, Sessitsch A, Bonfante P, Gram L, Lin RTP, Ron E, Karahan ZC, van der Meer JR, Artunkal S, Jahn D, Harper L (2019) The urgent need for microbiology literacy in society. Environ Microbiol 21(5):1513–1528

Walter XA, Merino-Jimenez I, Greenman J, Ieropoulos I (2018) Pee power Urinal II—Urinal scale-up with microbial fuel cell scale-down for improved lighting. J Power Sour 392:150–158

You J, Rimbu GA, Wallis L, Greenman J, Ieropoulos I (2019) Living architecture: towards energy generating buildings powered by microbial fuel cells. Front Energy Res. https://doi.org/10.3389/fenrg.2019.00094

Zhou T, Han H, Liu P, Xiong J, Tian F, Li X (2017) Microbial fuels cell-based biosensor for toxicity detection: a review. Sensors 17(10). https://doi.org/10.3390/s17102230

Place-Based Processing: Industrial Process Architecture for Sumptuous Convivialities

Sarah Kantrowitz

Abstract

Today, technologies of the emerging bioeconomy present one focused opportunity to unwind and transmute industrial operations into sustainable, regenerative work. Relying on standing industrial process design methods, however, may tether this hope to the same consequences and disparities industrialization has already suffered us the last few centuries. Instead, process design methods reconsidered to begin from specificity of place and reverence for relationship may be a helpful balance to methods that begin from abstract process operations or intended product outcomes, alone. This essay posits that drawing architects deeper into the pragmatics of designing and delivering factories, refineries, waste/energy plants, or other industrial infrastructure might support the above re-tooling of process design methods toward kind, non-modern practice. Process architecture is an established professional capacity for working spatially and relationally on plant design in collaboration with process engineering, though it is often practiced more by former plant operators than by architects. When included, the role of Process Architect can also be technically well-positioned within standing project delivery structures to help identify and implement better designs for the substantial footprint this building type has in ecological relationship, carbon emissions, material and waste flows, worker's rights, concentration of capital, and patterns of land use and urbanization. To invite more architects to this seat at the table, this essay offers a very basic introduction (for architects) to the roles of process architecture and process engineering in industrial operations today and to how these disciplines might support a deeper sustainability in the near future. With a stronger coalition of architects working in skillful partnership with the vast webs of human and non-human assemblages shaping today's industrial landscape, perhaps together we may help re-weave some of today's most barren and extractive industrial practices into thriving, mutually nourishing convivialities.

Keywords

Bioeconomy · Industrial bioprocessing · Process engineering · Process architecture · Regenerative design · Degrowth · Bathing

S. Kantrowitz (✉)
Water, Cambridge, MA, USA
e-mail: sarah@bathe.stream

M. R. Thomsen et al. (eds.), *Design for Rethinking Resources*, Sustainable Development Goals Series,
https://doi.org/10.1007/978-3-031-36554-6_4

1 Design Need: Let Go of Industrial Emissions

Industrialization has been complex. We make a lot of things we don't need, but also a lot of things we do. Organizing together to work a lot something can be joyful for those involved and generous for others relieved of the responsibility, but it can also enclose commons, impose distance, concentrate power, and exploit brutally. Across the world, consumption of energy and goods or proximity to extraction and pollution have never been distributed evenly. Future generations living cooperatively may want physical infrastructures for shared resource and waste work, but perhaps far less or very different infrastructures than those driving climate crisis today. One aspect of industrialization at least is relatively unequivocal: it has become on the whole a wildly unsustainable source of carbon emissions, serving a dwindlingly small population at the increasing precarity of a growing global majority. This is much more than a design problem, but within it, there are ways architects might be in greater service. To invite more architects into the pragmatics of this work, this essay offers an introduction to the roles that process engineering and process architecture play in the physical design of industrial operations today, and to how these disciplines might be practiced differently with more architects involved in the future.

Between energy consumption and process reactions, industrial operations today account for roughly 30% of current carbon emission rates. Industrial processes are also the fastest-growing source of emissions, jumping by 203% since 1990, according to data from the World Resources Institute (Ge et al. 2020). To help focus collective attention on addressing this problem, No. 09 of the 17 United Nations Sustainable Development Goals (UN SDGs) states the objective to "Build resilient infrastructure, promote inclusive and sustainable industrialization and foster innovation". In 2023, the UIA World Congress of Architects asked how and where architects can better advance the UN SDGs. This essay is one response. Of the eight official targets established within SDG 09, three seem particularly salient to the built form of physical process design. These three targets are stated as follows:

> **Target 9.3**: Increase the access of small-scale industrial and other enterprises, in particular in developing countries, to financial services, including affordable credit, and their integration into value chains and markets
> **Target 9.4**: By 2030, upgrade infrastructure and retrofit industries to make them sustainable, with increased resource-use efficiency and greater adoption of clean and environmentally sound technologies and industrial processes, with all countries taking action in accordance with their respective capabilities
> **Target 9.5**: Enhance scientific research, upgrade the technological capabilities of industrial sectors in all countries, in particular developing countries, including, by 2030, encouraging innovation and substantially increasing the number of research and development workers per 1 million people and public and private research and development spending

Given the scale, scope, and diversity of industrialization's reach, there are many parallel and interconnected pathways to move toward these targets. Today, mainstream political and economic activity around technologies of the emerging bioeconomy present one focused, substantive design opportunity to do more with less, plan for resilience against uncertain conditions, and embrace a regenerative industrial culture after modernity (Tan and Lamers 2021; Heinzle et al. 2007; D'Adamo et al. 2020; Aguilar et al. 2019). Many points in this essay may be applicable to other areas of manufacturing, energy, or waste management infrastructure. For the sake of brevity and to share from personal experience, I will focus here on the design and delivery of bioprocessing facilities.

The bioeconomy is a new term for the broad portion of industry based on the processes of presently living biological organisms and ecosystems, like fuel brewed from freshly grown algae to replace petroleum's long-deceased hydro-carbon sources. Beyond just fuel, the International Energy Agency estimates that petrochemical feedstocks account for roughly

12% of global crude oil demand (IEA 2018) and are at least partially constituent to 96% of all products manufactured in the US and Europe. As such, reducing the embodied emissions in a product like petroleum-derived synthetic rubber requires reworking not just the fuel source that powered the factory that made the synthetic rubber, but also reworking the process design recipe for making the material itself, including the embodied emissions in every individual piece of equipment at the factory stacked together into the process flow that produces the rubber. Standing physically inertial against change, every operation is a web of interdependent and interlocking, heavily capitalized design decisions, all dripping in petroleum.

Bioprocessing might offer pathways to release both petroleum and its associated input-intensive, control-based manufacturing technologies, and instead embrace lively, renewable biogenic carbon sources and biotic, partnership-based self-assembling production and decomposition processes. The use of biological processes to serve industrial, agricultural, biomedical, energy, waste management and environmental remediation applications is of course nothing new. Evidence of the first use of bioprocessing goes back at least 8,000 years to the practice of making leavened bread. Beginning about 5,000 years ago, medical practitioners in China used moldy soybean curd to treat skin infections, and about 4,500 years ago, Egyptians began malting and fermenting barley into beer (Ladisch 2002). Bioprocessing is also regularly practiced by a wide range of more-than-human beings, such as the leaf-cutter ants that ferment plant matter into fungal biomass to maintain a domestic food source inside their colonies.

Over the last few years, a range of new biotechnologies have been coming online and dropping in cost that might help to soften the walls between industrial and living processes. As techniques advance, we may find new old ways to steward evolution in cells and consortia to give rise to more of the metabolic pathways and congenial transformations of matter and energy we desire, sans gadgetry. Bioprocessing also has a history of being an enjoyable activity, suited to serving the social and economic equity also tasked by SDG No. 09. As with traditional fermentation and other large-volume biotechnologies like farming or forestry, the regenerativity and resiliency of bioprocessing may best be leveraged by operations that are adaptive, decentralized, small or medium-sized, slow, and site-specific. When organized accordingly, these operations can be well-suited to management by collectivities across a range of scales, levels of capitalization, irregularity, and mode of cultural integration. Good design requires being part of these socio-political and economic organizing processes in concert with organizing towards physical form (Shapira et al. 2022; Bookchin 1971).

Overall, as late enlightenment re-learns how to observe and negotiate with the genetics and population dynamics of microorganisms, funguses, parts of living cells, plants, animals, and microbial or other vital consortia (Bennett 2010), many opportunities open for industry to un-learn the distinction between producing technologies and participating in living systems, and to hold space for passing the shame and grief wound into that complex legacy of alienation. According to a report released by the McKinsey Global Institute, "as much as 60 percent of the physical inputs to the global economy could, in principle, be produced biologically," (Chui et al. 2020; Adom et al. 2014). A recent report issued by the United States Congressional Research Service to shape federal investment in the emerging bioeconomy offers that, "some experts estimate the direct economic impact of bio-based products, services, and processes at up to $4 trillion per year globally over the next 10 years" (Gallo 2022).

While dreams of self-repairing personal electronics grown from waste sludge still flicker out on the horizon, other novel bioprocesses are ready to go. Basic science has demonstrated viability in both broad applications like bioplastics and biofuels as well as in niches like enzymes that improve efficiency in pulp and paper bleaching, soil microbes that reduce fertilizer use and sink greater atmospheric carbon loads, polyurethane foams made from algae oil waste streams in omega-3 fatty acid production, and sustainable fish meal culled from methane that was previously being vented to the atmosphere (Hodgson et al. 2022).

2 Design Need: Professional Roles and Competencies

We have so much knowledge in sustainable practices, but dismantling unsustainable practices remains a challenge. It is often speculated that this is due more to siloed hoarding of wealth and power than to any limits in our socio-technical design capabilities, but I am enough of a materialist to feel that these things go together. Billions of dollars and as many research hours have been funneled toward prototyping and proving scientific possibility across many verticals at benchtop volume, but the development of reliable processes to grow complex, dynamic, and highly variable living systems into appropriately sized commercial operations is still its own challenge. To quote Schmidt Futures on the gap between research and practice in this space, "an analogy would be turning a home-based, one-carboy beer fermenter into a [network of] full-fledged brewer[ies] capable of producing enough beer to stock every liquor store, bar, and restaurant," (Hodgson et al. 2022). Or to use a fun historical example, it was 1928 when Alexander Fleming made the incidental observation that colonies of *Penicillium notatum* could inhibit the growth of *Staphylococcus* culture in a petri dish, but more than a decade passed before it was worked out how to produce the volume of penicillin needed for commercial distribution. Finally in 1941, Mary Hunt devised a method working at the U.S. Department of Agriculture to batch growth of pencillin on moldy cantaloupes set bobbing in submerged-culture tanks of corn steep liquor, a byproduct of the cornstarch industry whose sugars were found to accelerate the targeted mold growth (Shuler and Kargi 1991), and that is all before even beginning to mobilize the resources to crystallize this process into a manufactory.

On Page 02 of a recent report to US President Joe Biden entitled "Biomanufacturing to Advance the Bioeconomy", the President's Council of Advisors on Science and Technology (2022) state that they have "identified three key gaps that are slowing the country's progress and must be addressed if we are to realize this enormous potential…: insufficient manufacturing capacity, regulatory uncertainty, and an outdated national strategy". Likewise, the 2022 Schmidt Futures Report "The U.S. Bioeconomy: Charting a Course for a Resilient and Competitive Future" (Hodgson et al. 2022) offers "building a national infrastructure for bioproduction scale-up capacity" as number two of its four core investment recommendations.

Given the scale, complexity, and uncertainty of the design and construction task ahead in delivering retrofit or new bio-industrial infrastructure, it is beyond the scope of this paper to make specific facility design recommendations. More critically, it is the concern of this paper to make recommendations about the type of design roles and competencies that can be developed in order to have design professionals able to support specification of those recommendations in-situ as this need matures.

As things currently stand, architects are not on the list. Taking the above mentioned 2022 Schmidt Futures Report as a litmus test of current influential views in industry, government, and academic research, it is generally perceived that "a further constraint on developing bioproduction capabilities… is that there is a severe shortage of bioprocess engineering talent… one that raises the need for education in bioprocess engineering at all levels, from community college to graduate school (as well as) process development research and training programs". Besides bioprocess engineers, the other professionals presented in the report as necessary to design, build, and run bioproduction processes are "automation engineers, manufacturing science and technology staff, downstream processing staff, and commissioning, qualification, and validation engineers".

While it may feel obvious to many that process engineering is the design competency in greatest demand to support bioprocess scaling and implementation, it is my sense that many of process engineering's core design precepts have also gotten us into this climate pickle in the first place. A chemical engineer either currently occupies or has previously occupied the CEO position for 3M, DuPont, Union Carbide, Texaco, General Electric, Dow Chemical, and Exxon. To quote the great Audre Lorde, "the master's tools will never dismantle the master's house" (Lorde 1981).

Based on my light preliminary experience as an (unlicensed) architect working with process engineers to plan roughly 600,000 SF of commercial processing operations across the sectors of medical biologics, precision fermentation, and relationship-based food systems, it seems that current dominant approaches in process development may risk recapitulating in biological process many of the same downsides to currently dominant petro-chemical activity. In their article "Building a Bottom-Up Bioeconomy", Shapira et al. (2022) frame this challenge aptly: "rather than trying to industrialize biology, the real task is biologizing industry." Where this work touches the physical design of industrial processes at scale, I am humbly of the view that architecture might offer a set of design tools quite useful in helping 'biologize industry'. I am also of the view that architects can contribute to this work more effectively by entering a standing role (like process architect) on project delivery teams and by working in close partnership with process engineering.

3 Design Methods: Process Engineering

Process engineering is the work of designing, controlling, and operating process and instrumentation lines to transmute raw matter into desired material at industrial scale. While process engineers can work with all manner of matter in all manner of industries from mining to pharmaceutical production to wastewater treatment, a consistent material theme is that process engineers tend to deal with the manipulation of batched or continuous flows of substance (e.g., aluminum alloy, latex, cheese powder, etc.), as opposed to the manufacturing engineers who build assembly lines for discrete products (e.g., cars, dental dams, boxed mac & cheese, etc.) (Brodkey and Hershey 1987). As discussed above, this essay takes interest in modes of bioprocessing that are themselves biological, using life to manage life (Lorimer 2020) and allowing complexity to self-assemble, but to get there, we are reviewing the dominant shape of industry tools as they stand today and the history on which they have been built. Process engineering emerged as a branch of chemical engineering (Walker et al. 1937; Underwood 1965). Even if the source material being handled by a given process is biological, current modes of engagement are often largely chemical and mechanical: running mass balances and material flows through equipment like plastic tanks, rigid metal piping, pumps, valves, disposable filters, centrifuges, and automation controls to monitor and adjust properties like phase, density, viscosity, particle-size distribution, pressure, or temperature in order to optimize for throughput rate, process yield, and product purity.

It is generally agreed upon that the first consultant to begin calling himself a chemical engineer was George Davis, working in Britain in the late nineteenth century just as the manufacture of a handful of chemicals were growing to industrial proportions. Davis had studied chemistry and begun his career at Bealey's Bleach Works in Manchester, followed by roles at several other chemical firms, culminating in his appointment to the Alkali Inspectorate in 1881 (Cohen 1996). His responsibility was to inspect Leblanc alkali works of the Midland regions to enforce the Alkali Act of 1863. Industrialization of the (then) new Leblanc process had enabled production of sodium carbonate (soda ash) from salt, coal, and limestone as demand from the growing soap, glass, and textile industries outpaced harvests of naturally-occurring soda from kelp and barilla. This process had also begun spewing dangerous quantities of gaseous hydrochloric acid into the air. One of the first pollution control laws ever passed, the

Alkali Act sought to curb this unregulated discharge from the burgeoning soda industry.

As the potency of the Leblanc process culled together what had been a stew of craft operations working on a theoretical basis akin to witchcraft (Federici 2004; Stanley 1995) and transfigured them into the modern chemical industry, Davis' job as an Alkali Inspector also gave him a front-row seat to scrutinize and compare each different factory's attempt at rationalizing their operations. Perhaps even more so, he became acutely concerned with the rapidly growing industry's failures of rationalization. Frustrated by the abundance of wasteful, polluting, and poorly managed plants with little understanding of their own operations, Davis wrote a letter to the editor of the Chemical News in June 1880 proposing the creation of a Society of Chemical Engineering to crystallize this new class of professional expertise: "Many processes can be carried on very successfully by chemists in the laboratory, but few are able to make chemical processes go on the large scale, and simply because they have a lack of physical and mechanical knowledge combined with their chemistry… Chemists with a thorough knowledge of physics combined with a fair knowledge of mechanics…Such men are not very plenty we know, but they would, by forming such a society, help to diffuse that knowledge through the next generation, if not in the present." (Flavell-While 2012).

While his petition was at the time unsuccessful (it would take another 40 years for the first professional society of chemical engineers to be founded in 1922), he continued working to spread the understanding that addressing the problems of a given chemical plant could be abstracted into general principals, systemized, and compared to other plants or even to other process types, as problems of engineering. While other industrial chemistry books written at the time were siloed to a given type of chemical process like brewing or acid production, in 1887, Davis gave a series of 12 lectures at the Manchester School of Technology presenting an approach that instead broke complex, historically entangled processes down into a relatively small number of universal subset operations, each of which could then be independently refined and applied back to any other industrial chemical process (Davis 1901).

This foundational concept of the newborn discipline of Chemical Engineering was further developed and given a name by Arthur D Little around 1915, while he was helping shape the chemical engineering curriculum at the Massachusetts Institute of Technology: Little called it the Unit Operation. A 1922 report he chaired for the American Institute of Chemical Engineers' Committee on Education lays out the framework clearly:

> Chemical engineering as a science, as distinguished from the aggregate number of subjects comprised in courses of that name, is not a composite of chemistry and mechanical and civil engineering, but a science of itself, the basis of which is those unit operations which in their proper sequence and coordination constitute a chemical process as conducted on the industrial scale. These operations, as grinding, extracting, roasting, crystallizing, distilling, air-drying, separating, and so on, are not the subject matter of chemistry as such nor of mechanical engineering. Their treatment in the quantitative way with proper exposition of the laws controlling them and of the materials and equipment concerned in them is the province of chemical engineering. It is this selective emphasis on the unit operations themselves in their quantitative aspects that differentiates the chemical engineer from industrial chemistry, which is concerned primarily with general processes and products (Van Antwerpen 1980)

As with all modernist, enlightenment epistemologies, the abstraction and atomization of the Unit Operation is both its profound potency and its liability. The inevitable cost of systematic isolation and operable control is alienation from felt sense of relational meaning or purpose and falling deaf to signals and feedback loops, checks and balances in the lost intimacy of mutual knowing between a process and its role in an environment. This abstraction of processes away from place and path-dependent interrelationships has granted process engineering an impressive capacity to re-plumb the physical transformation processes of industrialization hard and fast past any sentimentality in millennia of accrued metabolic pathways, interwoven ecological assemblages, or socio-environmental rhythms, but we have also done just that (Tsing 2021; Haraway 2016). Now that we find ourselves living in the abiotic tailing ponds of petrochemical industrialization's fever dream, will the disassociated abstraction of unit operations find the wholism to build more sustainable, resilient, and equitable shared processing infrastructures for basic food, fuels, materials and medicines, while generally also building less?

Sustainability-minded engineers have offered that the methods of resource accounting in a mass and energy balance can just scale from plant to planet, re-situating a given process design within its finite or circular context (Woinaroschy 2016; Zhang et al. 2017). In this view, process engineering, "can/must define its subject area as the whole socio-metabolic system and its target as sustainability transition. Through philosophical and sociological arguments, concepts of human existence, society and social system, production/consumption, exchange and economy, politics and policies, information, science and technology are clarified with a hope that these issues can be incorporated into chemical engineering principles in the future" (Horio 2019). Is this approach holistic or subjugating in its reach for totality?

Even in the most advanced realms of biological engineering, working in hushed proximity with the most intricate details of living, breathing beings, it is easy for industrial biotechnology to default to the metaphor of casting life as a machine to be controlled: "With biotechnology, we can direct the nanoscale machinery of living cells to produce self-contained factories that perform on a characteristic scale of one micron… The challenge facing bioengineers is to redirect genetic and cellular machinery to make economically important molecules when the cells are placed in controlled environments. Engineers must design, build, and operate hardware and integrated systems that can multiply a cell's output by a factor of one trillion, as well as recover and purify the products in a cost-effective manner. Bioprocess engineering is the next frontier." (Ladisch 2004).

While this latter view may be a singularly impoverished imagination of the connective, relational expanse of linkage and attunement available through process engineering, it is helpful to read lucidly this historically-entrenched chemical-mechanical ideology while here in attendance to the forces powerfully reworking the dis- or a-biotic ever on towards the biotic, with or without us. By holding understanding of these path-dependencies neutrally and candidly alongside respect and admiration for the rigor, potency and accomplishments of process engineering, it may be possible to build more richly integral and balanced creative partnerships between engineering and architecture in ecological design of industrial operations (Gibbs 2008).

4 Design Methods: Process Architecture

Process architecture today is generally practiced as a supporting or translational role between process engineering and conventional architecture in the delivery of an industrial facility. Despite the name, it is often practiced by people who are not trained in architecture. Like lab planning for R&D environments or medical planning for hospitals, many process architects enter the work through having been users or operators of facilities themselves, intimate with the spaces needed to run operations via first-hand experience. After a process engineer has developed diagrammatic process flows and equipment sizing for an industrial project, today's process architect can help to spatialize the process and to manage coordination between process requirements and overall facility planning. For example, if the process engineering team has specified an equipment line with clean classification requirements, process architecture might then draw up a set of rooms sized and sequenced to organize adjacencies, separations, clearances and circulation between equipment placement, service connections, material flows, operator and maintenance access, and hazard management. Process architecture can coordinate with the facility MEP engineering team to layer in process utility point-of-use sizing and placement, various destinations for differentiated waste drainage, back-up power requirements, or directional air flow. Coordination with the structural engineering team can layer in accommodation for equipment loads, specialized overhead support requirements, or metrological isolation against vibration. Coordination with the architecture team can then layer in requirements for envelope penetrations, ceiling access, emergency containment curbing, a specific thickness of epoxy floor finish, interlocking of door control hardware for entry/exit airlocks gating clean flow of personnel and materials, or a comfortably sized janitor's closet. As in all building projects, coordination with construction planning, project financing, regulatory compliance, and permitting can all also feed back iteratively on the design process, incrementally aligning towards a resolved, built whole.

Process architecture is by no means a part of every industrial project team. When needed, it can play a critical and central role in developing a facility, but is also generally contracted onto a project long after the most core process design commitments have already been made. Alternatively, there are many ways that the goal of sustainable industry could be well served by developing both a more robust professional design competency for process architecture and by increasing integration of process architecture into a broader scope and earlier stages of bioprocess development. For example, I think this new process architecture would be able to take the lead on much needed work in better integrating worker rights and quality of life into the core of operations planning, de-intensifying process design to decarbonize and welcome more ecological complexity and resiliency into material/energy flows, or working through affordably scaled-down mid-size facility standards that better enable decentralized operations, local or cooperative ownership, degrowth, onshoring and regional cultural integration. To simplify, this set of design priorities for process architecture might all fall together under the mantle of a place-based or a relationship-based approach to process design. In their essence, sustainability and place might both be about dense ecologies of relationship that work well together. As things hum and life invites more life, we dip from the sustainable to the regenerative to the sumptuously convivial.

Three major rock formations run beneath Dallas, Texas. In a rainstorm, the expansive clay soils of the Eagle Ford shale can swell enough to move the foundation of a building up to one inch. Somewhere on the northeastern fringes of the city, the proposed design for a 300,000 SF cleanroom blood protein extraction plant recoils at the possibility of this subtle external perturbation. The reams of hygienic stainless steel process piping with rigid validated-weld connections are only able to tolerate one-quarter inch of movement in their run from the 5,000L buffer tanks up overhead to the building's steel structure and over and down to the column chromatography skids. The evolving design of the plant reacts, injecting a few million dollars of rebar and labor into hardening its concrete foundation slab against the surrounding earth. Executing this work sucks in more global byproducts of the petrochemical industry and belches more carbon emissions out into the atmosphere, but succeeds in the goal of safeguarding further isolation, fending off the unwelcome variability of the plant's surroundings, and exerting further control over the conditions of its interior. A glimpse of a brilliantly white, smooth, and sterile chamber glistens out of view as the doors of three nested airlocks snap shut (Fig. 1).

Meanwhile, across the Atlantic Ocean and deep within the bowels of the Alps, thousands of open-format bioreactors come alive to the daily moisture-wick of porous stone walls, the seasonal cycle of humidifying glacial melt, ventilating winds gifted by fissures, and the vast thermal mass holding cavernous cells cool but not frozen. Erosion from subterranean rivers of glacial melt collapse adjacent rock formations and grant the cheese aging cave new real estate, spaces tuned to thrive on the geological upheaval of the facility's broader host environment and eager to be populated with biofilms of companion microflora. As millions of kilos of cheese blossom in these mountain caves, one almost

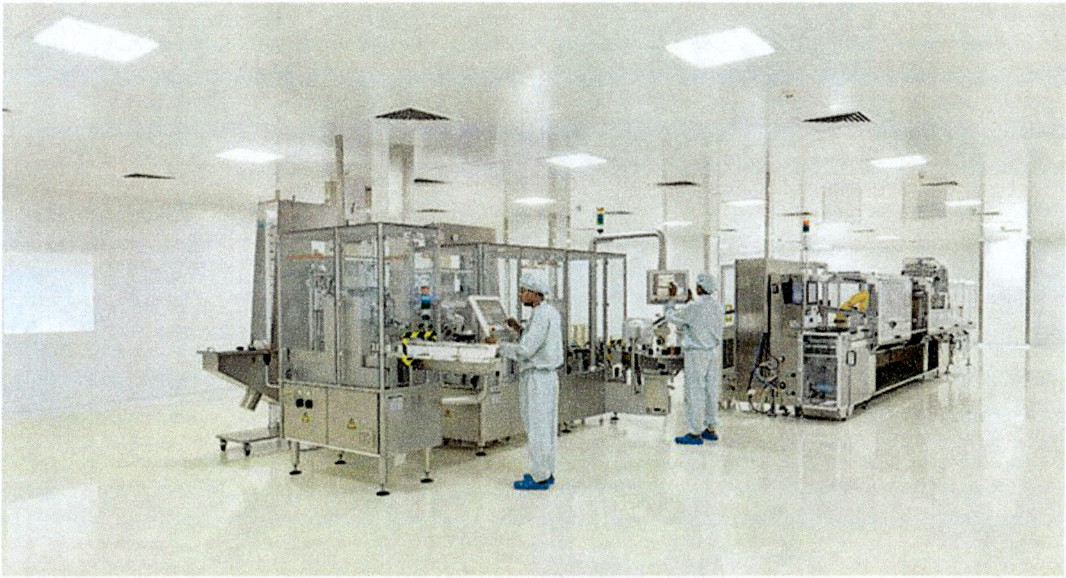

Fig. 1 Inside an oral solid dosage and injectable pharmaceutical manufacturing facility in Nagpur, India. It looks very clean. © Kunal Kampani

gets the idea that this lively family of fermented foods co-evolved with or even emerged from the microbially- and hydro-/geologically-entangled process architectures used to manufacture them (Fig. 2).

These two tales of bioprocess architecture present a very different relationship to place. In the first scenario, process design performance is premised on maximizing control of interior conditions by maximizing isolation from the surrounding environment. In the second scenario, the performance of this bioprocessing facility is premised on interdependence with and making use of existing environmental conditions, cultivating processes that are able to adapt fluidly to and benefit from the variability in these conditions, precisely because of their intimacy with and embodied knowledge of their place.

While helping design the blood fractionation plant in the first scenario as a member of the cGMP process architecture team at a large corporate architecture firm, I experienced the role of process architecture more or less as it stands in the industry today. The process engineering firm on the project had been brought on by the client roughly five years prior to the beginning of our architectural engagement. In that time, the process engineers had helped take the client's bench-top volume discoveries from the laboratory (roughly 0–50 L per batch), test scaling them through pilot volumes (roughly 50–500 L per batch), and prepare plans for the full commercial volume demo plant my team was then brought on to help deliver (10,500L per batch or roughly 1,500,000 L per year). Given the complexity of current bioprocessing facilities, path dependencies bake so many practice ideologies and embodied or operating emissions sources into thousands of interlocking decisions on project financing, building scale and siting, material supply chains, building HVAC and other environmental controls, building and equipment construction materials, maintenance, and waste. Under typical project delivery workflows, the energy demands and dependencies of many bioprocessing operations can also be fixed in place by early-stage process development decisions without the option to revise as full scaling impacts or opportunities become legible to all parties. By the time the process architecture team was involved in the above project, the vast majority of these decisions were already locked in.

Fig. 2 The Cooperativa Produttori Latte e Fontina near Aosta, Italy receives fresh cheese weekly from its 200 or so member cheese makers and then manages the 2–3 month aging process of the wheels in a cave complex with a capacity for roughly 150,000 wheels at a time. Dug out into the rock, the cavernous spaces of the aging cave were originally cleared as the access tunnel to a former copper mine and then later as military storage depots during World War II. The warehouses passively maintain a humidity of >90% and a temperature of around 10 °C year-round, further stabilized by cool streams of glacial melt plumbed through open trenches at ground level along the perimeters of all the aging chambers. When I asked the operators how they handle any regulatory concerns over the mold or algae found growing everywhere on their cave walls, they shrugged and responded that the mold was there when they arrived. I took this photo visiting in July 2019

Where process engineering might be a method of building industrial process design down from the abstract non-location of unit operations and laboratory environments, process architecture might become the design method of growing process design up from the specificity of place, working with communities back from affordances culled out of the particular ecological assemblages, flows, and socio-cultural practices of a unique geographic location engaged in the development of a given piece of industrial process infrastructure. Together these two approaches might meet in the middle and iterate to align on a shared vision for a given facility and its associated material/energy cascades, circles, or industrial symbionts (Grann 1997; Chertow 2008; Bezama 2016). Following the case of the cheese aging caves, and with humility to be lead and not reinvent the waterwheel, the Nashtifan windmills, or millennia of indigenous land stewardship practices (Whitewashed Hope 2022; Watson 2021), bioprocessing infrastructure can begin by feeling a place closely and learning what emerges there.

After conducting field research to survey a dozen or so northern Italian cheese aging caves for found process design strategies in modulating environmental temperature, humidity, ventilation, and microbial growth, my feeling is that aged cheese did indeed emerge from the affordances of its processing environment. The design of the product follows from the conditions of the facility rather than the the facility following from the product. As the cheese ages in the cave, losing water at a controlled rate to the moisture capacitance of porous lithic walls abundant in the processing environment, the solidifying biotic mass becomes the vessel of its own batched unit. The cheese rind is a self-assembling bioreactor whose functionality arises in concert between its living contents, the conditions of its processing environment, and gentle periodic human intervention (typical practices include hand pats, brushing, or brining at different time intervals depending on the type of cheese). When it is no longer needed, the rind bioreactor decomposes or becomes its own snack.

5 Next Steps: Growing Design Community in Bioprocess Infrastructure

In my experience, many builders, operators, and engineers are open to more holistic collaboration in the development of ambitiously sustainable or regenerative industrial projects. Equally, to move from idea to reality in any large, organizationally complex project, it is necessary for a team to all be able to speak the same language, to have clear methodologies, and to build trust. It can sometimes feel like a common reprise from architects that we ought to be consulted on all manner of things on which we are currently not consulted. This is not my proposal. I propose that architects might enter a (more or less) standing industry role by gaining the shared subject matter expertise and understanding of existing project delivery structures needed to add legible value to this type of project, and from there, within client relationships and project teams, weave in expanded design considerations and steward design scope and phasing strategies into a more balanced and reciprocal collaboration with other disciplines. Industrial operations often appear heavily technical today but they are not impossibly complex systems to learn about at the foundational level needed to collaborate.

Developing this more holistic professional role for sustainable process architecture requires building a lot of relationships. Developing this role requires listening, relearning, and growing new understandings of place-based technical expertise in how to participate generatively in biological processes, and the decolonial complement of this work in unlearning current paradigms of domination in industrial operations. On balance, architects are generally already well enough trained in many of the skills needed for a place-based approach to process architecture: reading and leveraging siting, softness, sensitivity to materials, doing more with less, space and flow planning, core principals of passive design, mediating between technical and somatic valences, trusting intuition, feeling, and spotting and cultivating nested socioecological relationships. It seems that young architects today also have a growing appetite to plug into the substantive physical design challenges of our time with both social sensitivity and technical competency. In this regard, I believe the disciplinary partnership proposed in this essay would both support architects in bringing value to process design and would support process design in bringing value to architects.

This essay is just preliminary work, and surely in need of further revision, exposition, and alignment with other like efforts. If this framing of the potential to grow process architecture as a place- and relationship-based bioprocess design competency strikes a chord, it may be helpful to next codify this proposal against more systematic case studies or to more thoroughly identify gaps in literature or training resources. For myself, I have fallen into what little I know about process design largely through experience so for now I have presented primarily first-person impressions and speculations, accordingly.

If these opportunities resonate with you, I hope we can work together to cull out more practices in bioprocessing that are themselves biological and life affirming (Lorde 1984). By partnering more deeply and more skillfully with each other and with the more-than-human world, perhaps together we may more fully bear witness to the wounds of the anthropocene and help re-weave some of today's most barren and extractive industrial practices into thriving, mutually-nourishing, and sumptuous convivialities

Acknowledgments I am immensely grateful to the many colleagues, mentors, friends, and lovers who have shared their work in industrial infrastructure, biorefining, waste management, fermentation, hologenomics, process design over the biorelational continuum, and other modes of hydrofeminist practice. Thank you.

References

Adom F, Dunn JB, Han J, Sather N (2014) Life-cycle fossil energy consumption and greenhouse gas emissions of bioderived chemicals and their conventional counterparts. Environ Sci Technol 48:14624–14631

Aguilar A, Twardowski T, Wohlgemuth R (2019) Bioeconomy for sustainable development. Biotechnol J 14:1800638

Bennett J (2010) Vibrant matter: a political ecology of things. Duke University Press, Durham NC

Bezama A (2016) Let us discuss how cascading can help implement the circular economy and the bio-economy strategies. Waste Manag Res 34:593–594

Bookchin M (1971) Post-scarcity anarchism. Ramparts Press, Menlo Park

Brodkey RS, Hershey HC (1987) Transport phenomena: a unified approach. Mc Graw-Hill, New York

Chertow M (2008) 'Uncovering' industrial symbiosis. J Ind Ecol 11(1):11–30

Chui M, Evers M, and Manyika J et al (2020) The bio revolution: innovations transforming economies, societies, and our lives. McKinsey Global Institute, p vi

Cohen C (1996) The early history of chemical engineering: a reassessment. Br J History Sci 29(2):171–194

D'Adamo I, Falcone PM, and Morone P (2020) A new socio-economic indicator to measure the performance of bioeconomy sectors in Europe. Ecol Econ 176

Davis G (1901) A handbook of chemical engineering. Davis Brothers, Manchester

Federici S (2004) Caliban and the witch: women, the body and primitive accumulation. Autonomedia, New York

Flavell-While C (2012) George E Davis–Meet the Daddy. The Chemical Engineer. https://www.thechemicalengineer.com/features/cewctw-george-e-davis-meet-the-daddy/. Accessed 02 Oct 2022

Gallo M (2022) The bioeconomy: a primer (Version 03). Congressional Research Service, Washington DC. https://sgp.fas.org/crs/misc/R46881.pdf

Ge M, Friedrich J, Vigna L (2020) 4 Charts explain greenhouse gas emissions by countries and sectors. World Resources Institute. https://www.wri.org/insights/4-charts-explain-greenhouse-gas-emissions-countries-and-sectors Accessed 04 Oct 2022

Gibbs D (2008) Industrial symbiosis and eco-industrial development: an introduction. Geogr Compass 2 (4):1138–1154

Grann H (1997) The industrial symbiosis at Kalundborg, Denmark. In: DJ Richards (ed) The industrial green game. National Academy Press, Washington

Haraway DJ (2016) Staying with the trouble: making kin in the chthulucene. Duke University Press, Durham NC

Heinzle E, Biwer A, Cooney C (2007) Development of sustainable bioprocesses: modeling and assessment. Wiley, Hoboken NJ

Hodgson A, Alper J, Maxon ME (2022) The U.S. bioeconomy: charting a course for a resilient and competitive future. Schmidt Futures, New York

Horio M (2019) On the historical role of chemical engineers in sustainability transition. Malays J Chem Eng Technol 2:41–56

International Energy Agency (2018) The future of petrochemicals. IEA, Paris. https://www.iea.org/reports/the-future-of-petrochemicals. Accessed 23 Dec 2022

Ladisch M (2002) Bioprocess Engineering (Biotechnology), pp. 434–459 in Vol. 1 of Scientific Encyclopedia, 9th ed. Van Nostrand, New York

Ladisch M (2004) The role of bioprocess engineering in biotechnology. Bridge Natl Acad Eng 34(3)

Lorde A (1981) The master's tools will never dismantle the master's house. In: Moraga C, Anzaldua G (ed) This bridge called my back: writings by radical women of color. Persephone Press, Watertown MA, p 98

Lorde A (1984) The uses of the erotic: the erotic as power. In: Sister outsider: essays and speeches. Crossing Press, Trumansburg, NY

Lorimer J (2020) The probiotic planet: using life to manage life. University of Minnesota Press, Minneapolis

President's Council of Advisors on Science and Technology (2022) Report to the president: biomanufacturing to advance the bioeconomy. Executive Office of the President. https://www.whitehouse.gov/wp-content/uploads/2022/12/PCAST_Biomanufacturing-Report_Dec2022.pdf. Accessed 06 Jan 2023

Shapira P, Matthews N, Cizauskas C, Aurand E, Friedman D et al (2022) Building a bottom-up bioeconomy. Issues Sci Technol 38(3). https://issues.org/building-bioeconomy-engineering-biology-shapira/. Accessed 14 July 2022

Shuler M L and Kargi K (1991) Bioprocess engineering: basic concepts. Prentice Hall, Englewood Cliffs, NJ

Stanley A (1995) Mothers and Daughters of Invention, Notes for a revised history of technology. Rutgers University Press, New Brunswick NJ

Tan ECD, Lamers P (2021) Circular bioeconomy concepts—A perspective. Front Sustain 2:701509

Tsing A (2021) The Mushroom at the end of the world: on the possibility of life in capitalist ruins. Princeton University Press, Princeton NJ

Underwood AJV (1965) Chemical engineering—Reflections and recollections. Trans Inst Chem Eng 43

Van Antwerpen FJ (1980) Origins of chemical engineering, In: Furter W (ed) Advances in chemistry. American Chemical Society, Washington, DC, p 190

Walker WH, Lewis WK, McAdams WH, Gilliland ER (1937) Principles of chemical engineering. McGraw-Hill, New York

Watson J (2021) A new mythology of technology: interview with Julia Watson. Next Nat Mag. https://nextnature.net/magazine/visual/2021/a-new-mythology-of-technology. Accessed 24 Aug 2022

Whitewashed Hope: A message from 10+ Indigenous leaders and organizations regenerative agriculture & permaculture offer narrow solutions to the climate crisis https://docs.google.com/document/d/1er9ixhlZWmwNgywzKPNPuGVfrM5KjeRBdVMiIsjtLUM/edit. Accessed 08 Aug 2022

Woinaroschy A (2016) A paradigm-based evolution of chemical engineering. Chin J Chem Eng 24(5):553–555

Zhang YHP, Sun J, Ma Y (2017) Biomanufacturing: history and perspective. J Ind Microbiol Biotechnol 44 (4–5):773–784

Towards a Bacterially-Induced Textile Architecture

Aurélie Mosse, Daniel Suárez Zamora, and Bastian Beyer

Abstract

Bacteria are rarely considered as belonging to the architectural realm. Yet they are part of the built environment as they are part of our own bodies. In the emerging field of biodesigned architecture, bacteria are involved in the visual expression as well as the structural performance of buildings. In this visual essay, we reflect on the results of a curriculum-based workshop focused on the biocalcification of textiles, a rigidification process informed by the bacterially induced precipitation of calcite. Building on the pioneering work of architects Bastian Beyer and Daniel Suarez in this area, the results point at the rich textile vocabulary through which biocalcification can be expressed and the value of interdisciplinary collaborations sitting at the intersection of textile design, architecture and microbiology.

A. Mosse (✉)
Ensadlab, Ecole Nationale Supérieure des Arts Décoratifs, Université Paris Sciences et Lettres, Paris, France
e-mail: aurelie.mosse@ensad.fr

D. S. Zamora
Matters of Activity, Humboldt Universität Zum Berlin, Berlin, Germany

B. Beyer
Matters of Activity, Humboldt University, Berlin, Germany

Keywords

Biocalcification · Textile design

Bacteria are rarely considered as belonging to the architectural realm. Yet they are part of the built environment as they are part of our own bodies (Gilbert and Stephens 2018). Meanwhile, bacteria became an expanding field of investigation for biotechnology-driven architecture where bacteria interact with the visual expression as well as by structural performance of construction materials. Biocalcification recently gained a lot of attention for its potential to solidify structures. However, it is mainly used by engineers for the consolidation and repair of sand, brick and concrete structures (Esnault Filet et al. 2012; Rahman et al. 2020). In the design community, biocalcification is similarly mostly applied to sand and other granular substances (Myers 2012; Arnadottir et al. 2021). Most recently, however, 3D printed polymer structures or knitted textiles expanded the material scope of biocalcification (Xin et al. 2021; Beyer et al. 2019). In this visual essay, we build upon the work of Beyer et al. (2019) to report on the "Biocalcified Textile Architecture" workshop undertaken in the context of the 1st year MA programme in Textile Design at Ecole des Arts Décoratifs in spring 2022. The workshop aimed to explore how an interdisciplinary exchange can contribute to the research on biocalcified textile architecture. The process and results informed discourse and

M. R. Thomsen et al. (eds.), *Design for Rethinking Resources*, Sustainable Development Goals Series,
https://doi.org/10.1007/978-3-031-36554-6_5

speculation about potential applications of the materials within the architectural realm. The achievements of the workshop are, firstly, an expansion of the vocabulary of possible textile structures and expressions and secondly to test a novel bioreactor setup (Figs. 1, 2, 3, 4, 5, 6, 7, 8, 9, 10, 11, 12, 13, 14, 15, 16, 17 and 18).

1 Discussion

The general aim of this workshop was the open exploration of various textile techniques and their subsequent biocalcification. Informed by design proposals such as bioreceptive structures and flexible architectural membranes, biomimetic

Fig. 1 Close-up of a biocalcified textile. Biocalcification or microbioally-induced calcite precipitation (MICP) is a form of biologically-induced mineralisation relying on the formation of calcium carbonate which results in solidification of the medium in which the process occurs

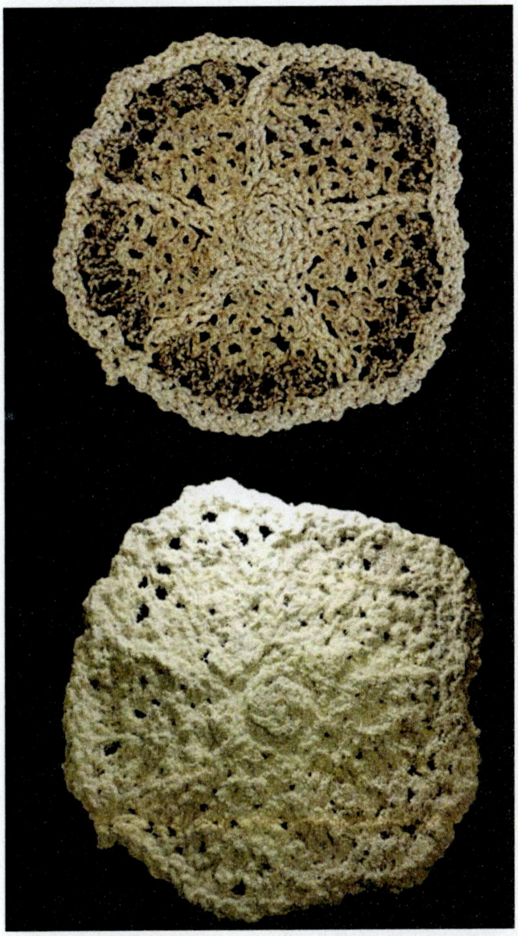

Fig. 2 Circular textile structure made with a crochet technique. Sample before (top) and after biocalcification (bottom) (diameter approx. 1 m). Crochet allowed for quick iterations and upscaling while providing great opportunities for structural gradients such as openings, material accumulations as well as inlays. Sample by: Violette Husson and Arnaud Mazel

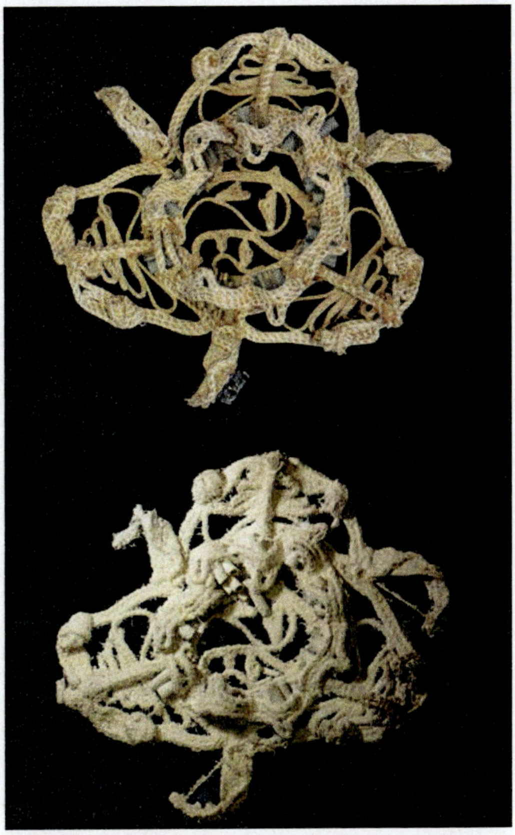

Fig. 3 This piece, reminiscent of gothic architecture is inspired by Radiolaria, a form of zooplankton producing intricate mineral skeletons. Sample by: Jeanne Guillet and Chloé Ménard

ornaments, urban furniture or modular shading elements, these experiments demonstrate the rich textile vocabulary through which biocalcification can be expressed. While Beyer and Suarez explored in the Column (2019) an open-work jute-based vertical knitted structure, here the production offers variations of patterns and textures, not only induced by different contextures and fibre's natures, but also by the intensity of the precipitation induced by the bacteria, leading to more or less dense, figurative and intricated expressions. The specificity of this language lies in its whiteness, granularity and rigidity evoking frozen times.

Further in-depth exploration of the structural behaviour of the materials in relation to the technique applied and optimisation of the bio-treatment are still to be assessed. The novel custom-built bioreactor proved to be working very well; however, it was found that the samples in the centre of the bioreactor accumulated more calcite than the ones on the periphery (see Fig. 11). For a further iteration, this distribution system inside the reactor could be improved to deliver more homogeneous results.

Fig. 4 Detail of Fig. 3 after treatment: the textile scaffold is transformed through MICP while the intricate textile structure remains perceivable

Fig. 5 Close-up of Fig. 3 and 4 demonstrating the ability of MICP to sustain textile curvature after removal of temporary substructure

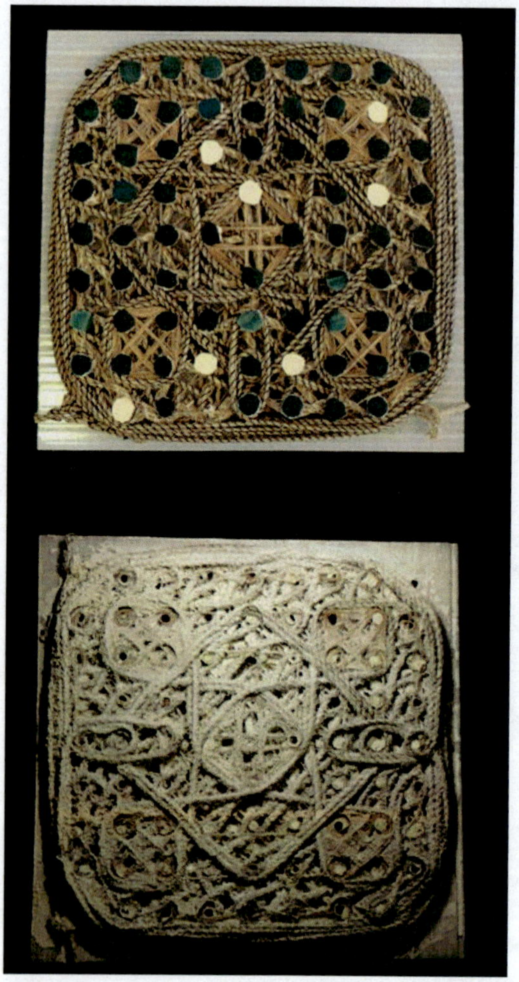

Fig. 6 Biocalcification also allows for sustaining an open weblike pattern like this mashrabyia-inspired module by Victor Manceau and Oscar Nahan. Similar to the lacing technique, yarns are pre-tensed in a specific configuration; thanks to pin-like supports which are later removed once the mineralisation has occurred

Furthermore, it could be observed that pieces incorporating more jute material such as sample 1 responded much better to the treatment as delicate and dense samples (see Fig. 14) This points towards the importance of the permeability of the material for the treatment as well as the necessity to balance the relationship between fibre and calcite matrix.

Fig. 7 This sample by Juliette Clapson and Nina Dos Santos was made by stitching jute and cotton-based woven fabric together. The large number of plies within one side of the fabric was intentionally designed to explore interstitial calcification

In general, the project provided a multidisciplinary learning experience for the students engaging with unconventional textile-based composites while they reflected on the implications of addressing the architectural scale. Furthermore, bacterially generated calcite is a biogenic material that can be found in various ecosystems. Together with the use of natural fibres, it forms a sustainable composite material. Therefore, it provided a valuable learning experience on how sustainable materials can be generated through microbiology.

They were able to engage in all the steps of the process while applying their own textile

Fig. 8 Close-up detail of Fig. 7 shows the dependency between structure and calcification

Fig. 9 Here biocalcification is tested on woven samples based on a honeycomb pattern and two distinct qualities of yarns. The differentiated response of the textile due to their qualities leads to a graded material with stiff and flexible areas. This allows the material to hinge due to its structure. Samples by Mélanie Dhubert Riollet and Maude Guirault

Fig. 10 (*Top*) Macroscopic images of biocalcification on jute-based textile yarns showing calcite crystal formations. (*Bottom*) From fibre to yarn production. Previous experimentation showed that jute fibres present a favourable substrate for the bacteria colony to adhere to. Three different yarn thicknesses (5, 8 and 12 mm) were produced by bundling regular raw twisted jute cords of approx. 1.5 mm with the help of Kemafil® (KEM) technology (Arnold et al. 1994)

Fig. 11 One of the workshop's goals was to make the students familiar with the concept of hierarchical materials. Normally, textiles operate at three different levels fibre-yarn-fabric (Eadie and Gosh 2011). Biocalcification asks for considering the bacterial microscopic level too

Fig. 12 Crafting textile samples: The framework of the workshop allowed the students to explore various textile structures and inform the project with their own textile experiences. Multiple small-scale tests were executed and then scaled up. The final pieces were crafted with the tailored yarn

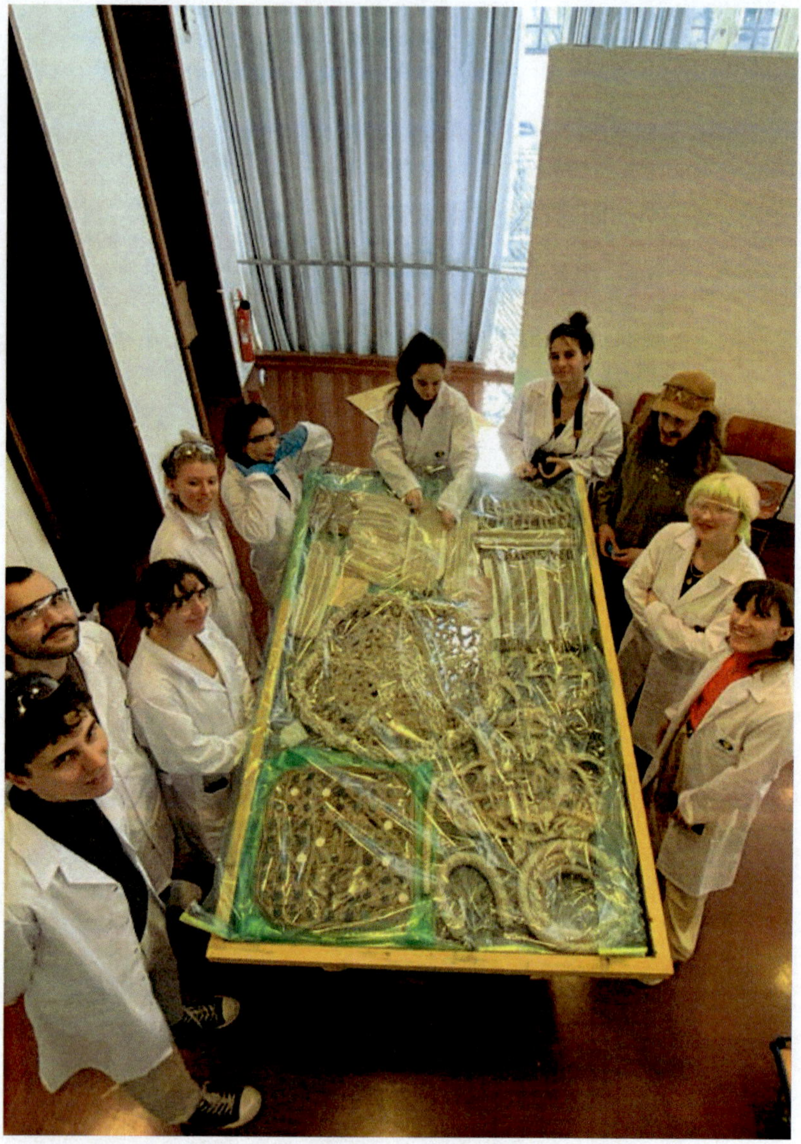

Fig. 13 Bioreactor: The device allows the bacterial solution to permeate the textile samples in a controlled way. A series of repeated treatment cycles gradually builds up bacterially induced calcite deposition on the fibres and ultimately solidifies the textile substrate

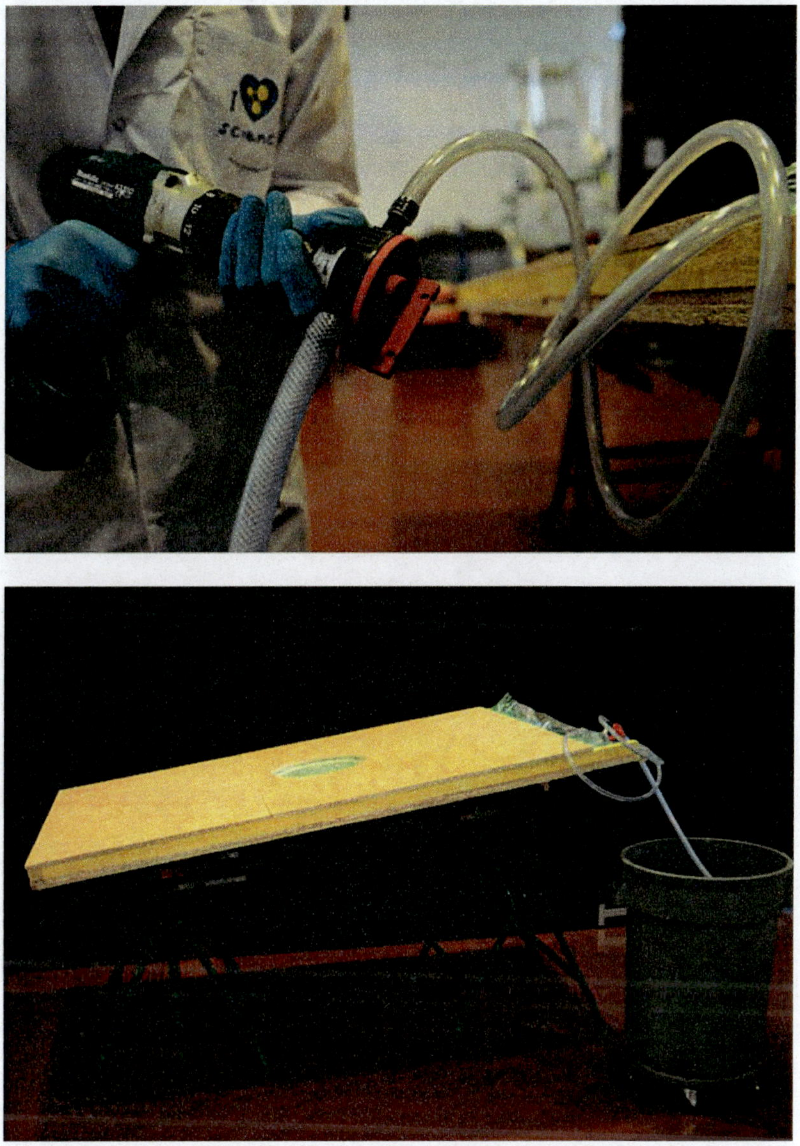

Fig. 14 The biocalcification process lies in imbuing the textiles with the chosen bacterial colony and a solution of urea and calcium carbonate. A liquid distribution system inside the bioreactor allows spreading of the bacteria solution efficiently to the textile media. The bacterial solution is circulated in an enclosed envelope in which the calcification on the textile occurs

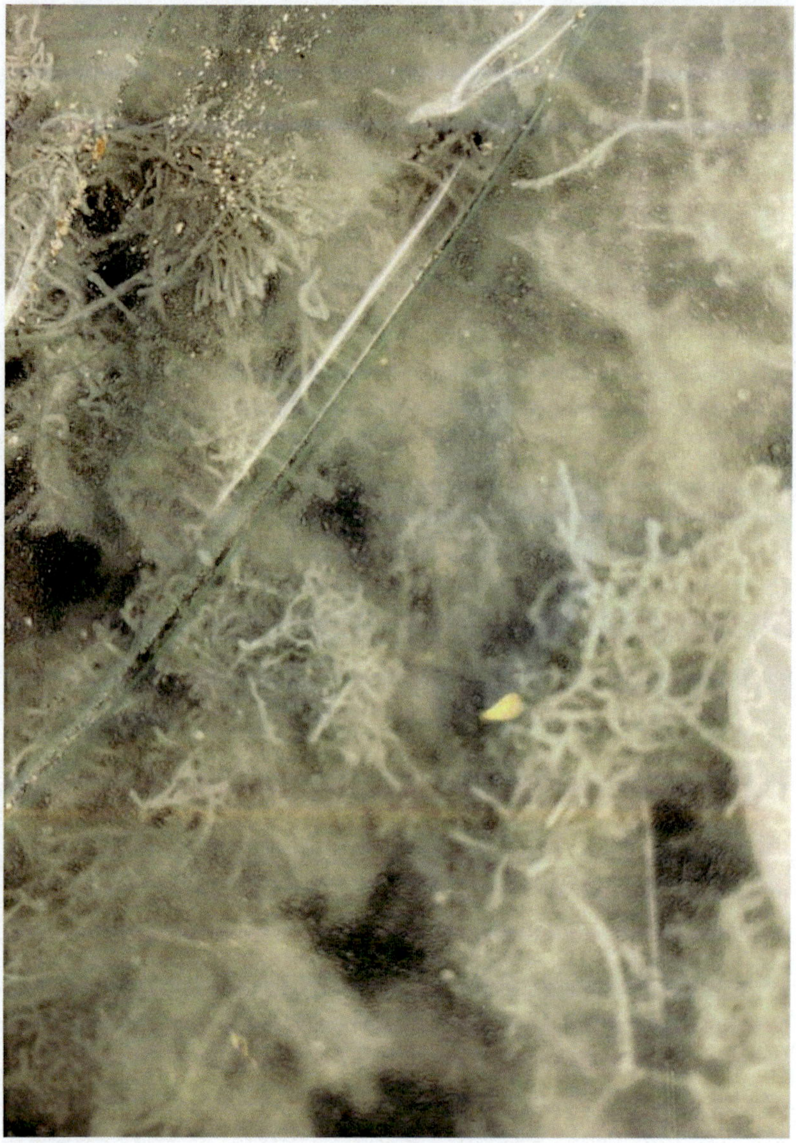

Fig. 15 Close-up of one sample inside the bioreactor. The students could observe how the biocalcification took place through a transparent area in the bioreactor

Fig. 16 First look inside the opened bioreactor after 3 days of biocalcification

Fig. 17 Close-up detail of Fig. 9. While not all those different fibres react well to the treatment, however, on a macroscopic view, it is surprising to observe how calcite grows in a very organised manner on top of the plastic monofilament

Fig. 18 Participants of the workshop. 1st MA in Textile & Material Design students and workshop leaders (Aurélie Mosse, Bastian Beyer, Daniel Suárez)

knowledge. Especially the engagement with microbiology bacterially-informed materials delivered a new experience that is not common in the conventional textile design curriculum, critically engaging with sustainable and ethical considerations of working with living organisms.

All in all, the course provided a testbed for a textile-based multidisciplinary framework where students gained a hands-on experience of the potential and the challenges of working between the disciplines.

Acknowledgements The Biocalcified textile architecture workshop benefited from the respective support of I-Portunus Houses pilot scheme2 I-*Portunus,* Soletanche Bachy, the textile design department of Ecole des Arts Décoratifs, the Cluster of Excellence "Matters of Activity. Image Space Material" funded by the Deutsche Forschungsgemeinschaft (DFG, German Research Foundation) under Germany's Excellence Strategy—EXC 2025-390648296 and the STFI Textile Institute. Picture credits: Daniel Suárez (8–9, 11 *bottom* 10, 17), Bastian Beyer (11–13, 15) Beryl Libault (1, 17, 19) Copyright SB/GL Angers University & Soletanche Bachy © (*top* 10).

References

Arnadottir TH, Dade-Robertson M, Mitrani H, Zhang M, Christgen B (2021) Turbulent casting: bacterial expression in mineralized structures. In: Slocum B, Ago V, Marcus A (eds) ACADIA 2020 distributed proximities: 40th annual conference of the association for computer aided design in architecture: Volume I: Technical Papers, Keynote Conversations, vol 1. Association for Computer Aided Design in Architecture, pp 300–309

Arnold R, Bartl A M, Hufnagl E (1994) Production of cord and narrow fabric products with Kemafil technology, Band und Flechtindustrie, vol 31, pp 48–52

Beyer B, Suarez D, Palz N (2019). Microbiologically activated knitted composites: reimagining a column for the 21st century. In: Sousa JP, Henriques GC, Xavier JP (eds) 37th eCAADe and XXIII SIGraDi joint conference: architecture in the age of the 4th industrial revolution, Porto 2019. Blucher, São Paulo, pp 541–552

Dade-Robertson M, Keren-Paz A, Zhang M, Kolodkin-Gal I (2017) Architects of nature: growing buildings with bacterial biofilms. Microb Biotechnol 10(5).1157–1167

Eadie L, Ghosh T (2011) Biomimicry in textiles: past, present and potential. An overview. J R Soc Interface 8 (59):761–775. https://doi.org/10.1098/rsif.2010.0487

Esnault Filet A et al (2012) Biocalcis and its applications for the consolidation of sands. In: Johnsen LF, Bruce DA, Byle MJ (eds) Grouting and deep mixing 2012

Fourth international conference on grouting and deep mixing, New Orleans. https://doi.org/10.1061/9780784412350.0152

Gilbert JA, Stephens B (2018) Microbiology of the built environment. Nat Rev Microbiol 16(11):661–670. https://doi.org/10.1038/s41579-018-0065-5

Myers W (2012) Biodesign: nature, science, creativity. Thames & Hudson, UK

Rahman MM, Hora RN, Ahenkorah I, Beecham S, Karim MR, Iqbal A (2020) State-of-the-art review of microbial-induced calcite precipitation and its sustainability in engineering applications. Sustainability 12 (15). https://doi.org/10.3390/su12156281

Xin A, Su Y, Feng S, Yan M, Yu K, Feng Z, Hoon Lee K, Sun L, Wang Q (2021) Growing living composites with ordered microstructures and exceptional mechanical properties. Adv Mater 33(13):2006946. https://doi.org/10.1002/adma.202006946

Synthetic Natures

Winka Dubbeldam

Abstract

In the Anthropocene the distinction between what is natural and what is human-made, which has informed the distinction between the 'wild' and the cultured for centuries, has become blurred. So, what is natural and what is artificial in the Anthropocene? The human struggle to contain systems running out of control, species going extinct, and catastrophic effects impacting our safety, has thrown our idea of 'home' into a fundamental crisis. According to the latest World Health Organization (WHO) report, 8 million people die every year globally because of air pollution alone. However, we also find strange beauty in the Anthropocene, where unexpected side effects and alternate symbiotic relationships in nature occur. Synthetic natures, a term used here to look at the purposeful symbiotic merging of nature's intelligence with building structures, studies how these new life forms (species) can alter buildings and material ecologies, and drastically reduce the buildings' environmental impact, altogether re-defining architecture as being an instrumental part of the solution rather than solely creating the problem. Architecture needs to learn/adapt and hybridize. Rather than simply applying vegetation on otherwise inert structures, buildings themselves need to learn from nature's intelligence, by adopting plant intelligence, hybridizating and initiating symbiotic relationships with natural systems. In the near future we can 'grow' new construction materials with ever more intelligent behaviors. These buildings will generate oxygen, absorb carbon, and are self-healing, creating continuous biomorphic surfaces. In the future city where building structures grow and lower the carbon footprint rather than adding to it, the human race might have a chance of not only surviving, but symbiotically living with these synthetic natures as 'home'.

Keywords

Synthetic natures · Hybrid · Anthropocene · Climate change · Architecture

1 Synthetic Nature

"Imagine an architect looking at the subnatures of a city: a cloud of smoke, a pool of mud, a pile of debris. He or she might see such things as emblematic of mismanagement, abandonment, or catastrophe….." (Gissen 2009).

W. Dubbeldam (✉)
Weitzman Architecture, University of Pennsylvania, Philadelphia, USA
e-mail: winka@design.upenn.edu

M. R. Thomsen et al. (eds.), *Design for Rethinking Resources*, Sustainable Development Goals Series,
https://doi.org/10.1007/978-3-031-36554-6_6

The distinction between what is natural and what is human-made, which has informed the distinction between the 'wild' and the cultured for centuries, has become blurred. So, what is natural and what is artificial in the era of the Anthropocene? In the Anthropocene, an era defined by human activity as the dominant influence on climate and the environment, 'nature' or the wild has become increasingly scarce and the man-made has taken over. This same Anthropocene has an ever-increasing devastating influence on the climate and the environment. The human struggle to contain systems running out of control, species going extinct, and afford environmental catastrophic events, has thrown our idea of a safe 'home' into a fundamental crisis.

Luis Barragan describes in his 1980 Laureate his notion of beauty and the "wild": "I suddenly discovered, to my astonishment, small secret green valleys the shepherds call them "jewels" surrounded and enclosed by the most fantastic, capricious rock formations wrought on soft, melted rock by the onslaught of powerful prehistoric winds. The unexpected discovery of these "jewels" gave me a sensation like the one experienced when, having walked through a dark and narrow tunnel of the Alhambra, I suddenly emerged into the serene, silent and solitary "Patio of the Myrtles" hidden in the entrails of that ancient palace. Somehow, I had the feeling that it enclosed what a perfect garden no matter its size should enclose: nothing less than the entire Universe".

This kind of "sublime" and awe for these rough rock formations holding the idyllic tufts of green, as untouched, idyllic beauty is becoming increasingly rare. At this time, nearly 50% of the US population, an estimated 150 million, live in areas that don't meet federal air quality standards. Cars and heavy-duty trucks are a major source of overall pollution, which includes ozone, particulate matter, and other smog-forming emissions. Particulate matter is singlehandedly responsible for up to 30,000 premature deaths each year. In 2013, transportation contributed more than half of the carbon monoxide and nitrogen oxides, and almost a quarter of the hydrocarbons emitted into our air (Unknown, Vehicles, Air Pollution, and Human Health 2014). According to the latest World Health Organization (WHO) report, 8 million people globally die every year because of air pollution.

A shocking and devastating realization, seemingly without a positive outlook or future solution. However, David Gissen, a New York author and designer who works in the fields of architecture, landscape, urbanism, and Professor of Architecture and Urban History at Parsons School of Design/New York University, suggests otherwise, rather than trying to "fix" the pollution mishaps, he suggests absorbing them, incorporating the dirt and discarded in our built environment. He states in "Subnatures, Architectures' Other Environments": "to take on the less favorable natures (subnatures) such as weeds, smoke, mud, debris and other pollution': and proposes that perhaps the architects consider these strange forms of nature as a material endemic to architecture and cities, as opposed to an aberration that must be consolidated, removed, or dismissed (Gissen 2009)." He goes on to suggest that buildings incorporate these subnatures, rather than only dealing with the "proper natures" that merely produce more technologically orientated structures, rather than the subversive more critical approach. Gissen suggests an alternative view of natural processes and ecosystems and how they relate to human society and architecture, a view that allows for the discarded to become productive and human debris to be absorbed in new material ecologies.

These two realities, the annihilation of the wild and mutuality, which on the surface appear to be antithetical, will be further investigated here. It could also be said that the Anthropocene's environmental crises have forced and accelerated the rate of architectural innovation. Until recently architecture's development has been slow in its technological development, and lacking innovation. Where we know we can expect our cars to be fully AI-operated and automated, building technology, on the other hand, is still at a nineteenth century level. Could these subnatures somehow be informing architecture, creating strange objects and synthetic hybrids, coalescing in a symbiotic, opportunistic relationship that benefits both organic and inorganic beings?

2 Anthropocene

The start of the Anthropocene was estimated by the Stratigraphy Commission of the Geological Society of London to be in 1780 when the steam engine was invented. Originally used for mining, steam engines as early as 1800 were also driving steamboats and trains instigating mass transport as we know it. Steam locomotives were gradually superseded by electric and diesel locomotives, with railways fully converting to electric and diesel power beginning in the late 1930s, marking the start of fossil fuel depletion, major pollution, and climate change.

The United States was long considered the world's biggest polluter in terms of carbon dioxide and other greenhouse gases, but China has moved into the top spot, according to the Global Carbon Project 2020. The consequences of pollution and the prolonged ignoring of its consequences are astronomically expensive: some economists and climate scientists have calculated that climate change could cost the United States the equivalent of nearly 4% of its gross domestic product a year by 2100, which is likely a conservative estimate; as it leaves out consequential damages from drought, floods, and climate migration. The annual US cost of flooding today is over $32 billion nationwide, flooding has cost US taxpayers more than $850 billion since 2000. Hurricane Sandy alone, which hit New Jersey in 2012 as a potent cyclone, caused major flooding, and caused over $60 billion in damage, claiming at least 125 lives in the United States and knocking out power to more than 7 million customers. New York was off the grid for weeks. Would it not be wise for the human race to pay more attention, and come up with pro-active measures rather than passively observing climate change and its consequences?

3 Mutants and Monsters

However, we also find a strange beauty in the Anthropocene, where unexpected side effects and alternate symbiotic relationships occur. The contamination of our soil–water–plant system with heavy metals causes a chemical environmental overload, with dire consequences for human and non-human health, and economic and ecological impact. Research in plant intelligence and adaptation shows that plants are capable of phytoremediation of heavy metal contamination, plants not only potentially clean the environment, but also adapt to the "new normal" in different ways, such as their altered growth and mutation.

The 2011 Earthquake in the Fukushima Prefecture, Japan, is one of these examples where reactors at the Fukushima Daiichi Nuclear Power Plant melted down after the earthquake, and new plant forms and /or mutations were found. Fields of flowers suddenly appeared to have elaborately contorted shapes, often elongated and doubling in shape (Fig. 1).

The reading of these phenomena varies. Evolution generally proceeds at a slow and steady pace, the Anthropocene however has caused a total reset, as illustrated here by the radical mutants that upset the norm. The geneticist Richard Goldschmidt sees these new forms of evolution as radical mutants, and calls them "Hopeful Monsters", in reference to a concept from evolutionary biology. These flowers, due to Fukushima's raised radiation levels gradually changed over time, and went through a form of fasciation, also known as cresting. The mutant vascular plants exhibit a relatively rare condition of abnormal growth in which the growing tip, which normally is concentrated around a single point and produces approximately cylindrical tissue, instead becomes elongated perpendicularly to the direction of growth, producing flattened, ribbon-like, elaborately contorted tissue, called phyllotaxy. Fascination starts in the apical meristem (similar to stem cells in humans), which plays a vital function in plant

Fig. 1 Flowers mutate and show cresting due to environmental stress

organogenesis [regenerative attributes] and pattern formation. Fascination may also cause plant parts to increase in weight and volume, especially the rose, legume, sunflower, dandelion, snapdragons, and cactus families are sensitive to fasciation. Goldschmidt's position is that these 'Hopeful Monsters' might be incredibly relevant in the adaptation and creation of new species, and exhibit features that are perhaps more resistant to environmental change. Surface elongation [mutant flowers], or fractalization (coast of England, medieval cities) has historically been a method of defense. Others consider these mutations "mistakes" and advise to weed them out, but more interestingly, the alternative remains closer to David Gissen's "Subnatures" argument we instead celebrate these flower's strange beauty, and see it as a lesson in resilience and beauty.

4 Synthetic Natures/New Hybrids

Going back to the initial premise that the Anthropocene's technological advances have compelled and accelerated architectural innovation, could the fasciation of these mutant flowers, and their capacity of phytoremediation be an indicator of what future architecture holds? Could these Hopeful Monsters thus exemplify a useful paradigm for architectural innovation, incorporating the 'mutant', resilient, and self-healing rather than obsessing over an "Idyllic Form"? History teaches us that major technological inventions, such as the 1930s introduction of the car, the sixties invention of mass media and travel to the moon, and of course the nineties invention of the home computer and the internet, accelerated architecture's change This had consequences. Architects today find themselves in a huge dilemma; the buildings designed based on these techno-centric paradigms are no longer the safe havens for human comfort they were intended to be, but actually generate a large share of global pollution, as close to 40% of human-made carbon emissions come from construction and building materials alone. A drastic design change in material ecologies with embedded intelligence is required, with the architecture itself gaining agency on-, and recreating a more holistic environment. An urgent recalibration of natural and synthetic material properties and what they mean for human and non-human occupancy and the surrounding city.

In this next wave of the Anthropocene where climate change and environmental challenges threaten our and other species' existence, just building 'green' no longer suffices.

Green and sustainable movements have been prevalent as early as the 1970s. At that time, the devastating impacts of the Santa Barbara oil spill had shocked the nation, resulting in the largest civil movement to that day, when 20 million people, 10% of the nation, rallied in the United States to protest environmental destruction (Yeo 2020). This resulted in the first Earth Day, where concerns for clean water and air initiated an environmental movement that expressed worries not only locally, but also for the planet as a whole. As a consequence, republicans and democrats went together and passed 28 environmental federal laws over the next 10 years to protect our nature, food supply, and endangered species. While this sounds hopeful, it was not enough. Environmental systems had to run 50 more years further out of control, for humanity to realize that it is our collective responsibility to resolve what we initiated and ignored for so many years. Architects and construction have an important role in this as households alone are responsible for 5.43 gigatons of carbon dioxide equivalent emissions every year. About 82.3% of those emissions are produced domestically.

For the first two decades of our office research has been focused on designing and implementing of new more efficient building systems, often in collaboration with advanced manufacturers. This resulted in custom and optimized solutions creating new facade systems, material innovations, and passive solar applications, resulting in well-insulated buildings with reduced energy costs. We realized that custom prefabrication with all its benefits [fast, affordable, technologically advanced, and clean] has replaced the standardized mass-produced production of the seventies. However, humanity is quickly coming to the realization that recent climate laws and measures will not suffice. Rapid adaptation is needed to adjust the course of environmental decline. Our research focus hence has been on the building's future symbiotic adaptation of nature's intelligence.

5 What Does that Mean?

Architecture itself needs to learn/adapt and hybridize. Rather than simply applying vegetation on otherwise inert structures, symbiotic relationships with natural systems and materials can be developed. This adaptation of plant intelligence then can generate mutually beneficent, self-sustaining, and even self-repairing habitats and environments. Structures that create positive feedback to their environment by absorbing carbon and emitting oxygen, while cleaning and filtering air and water. As Gissen writes: "By actively reflecting on the alienating material of the social natural environment, we might as a profession arrive at a truly radical and alternative concept of the environment for the contemporary architect (Gissen, 2009)." The Anthropocene's sudden human & environmental impact has caused a total reset, sometimes resulting in strange beauty where unexpected mutant side effects and alternate symbiotic relationships occur. Here we look at an Architectural reset as response to the Anthropocene in the form of potential new hybrids, buildings as Synthetic Nature that have the capacity to transform into a mutually beneficial coexistence of nature and architecture, resulting in symbiotic performance and behavior rather than bio-aesthetic analogies. The aim is to build clever, efficiently, minimize waste, and to develop new prototypes.

The shapes as a result of the prototyping process can be described as Protobodies or Thing-Shapes as described in Edmund Husserl's text the "Origin of Geometry" from 1937, that touch on the idea of becoming, approximating, and therefore, occupying a territory between idea and space, exactly where boundaries are not yet sharply differentiated or defined. This thing–shape is not yet decipherable by habit or felt "at home", which causes a sense of uncanniness (see also "Origin of our Ideas of the Sublime and Beautiful" by Edmund Burke, published in 1958) (Burke 1958). It accepts being other, combines awe and beauty, the "almost ugly", from which identity and character ensue.

6 Green Lung

A recent example is our recently completed masterplan for the 47 hectares Eco park for the Asian Games in Hangzhou China, with 7 buildings integrated into the landscape. This 1.6 km long park functions as a green lung, which absorbs carbon and creates oxygen while also absorbing and filtering stormwater as a run off from the otherwise hard surfaces of the fast-growing city around it. With a strict government stipulation for an 85% park and a "sponge city" concept, it was quite the challenge to position the required 185,000 m^2 of buildings. Only the 2 stadiums could be visible, the other 5 buildings had to be morphed into landscape formations with slivers of glass providing natural ventilation and daylight. A green valley with glass pavilions functioning as a shopping artery, dips under the road and river bisecting the park, reconnecting the 2 park sections in one fluid sweep, while promoting ongoing cultural, retail and hospitality activities.

This originally flat site is transformed into an Eco park by implementing a "zero-earth strategy", a strategy that re-distributes excavated earth from the newly restored wetlands and the shopping Valley to create a hilly landscape covering the submerged buildings. These hills not only dampen city sounds, filter, and retain stormwater, but also create a calm serene park landscape. Their 64,000 m^2 combined green roofs act as an agent of environmental change by releasing 83,408 kg of O_2. And absorbing 114,850 kg of CO per year. The masterplan, designed in collaboration with landscape architects! Melk, also implements a Sponge City model by recreating the wetlands, inserting islands in the river to filter and oxygenate the water, and placing porous pavement enhancing hydrology. The project achieved Green building evaluation 3 star, the Chinese equivalent of LEED platinum. The existing river as it snakes through the site, passes over the Village Valley as an aqueduct, forms a crucial part of the wetland system, mitigating stormwater runoff and offsetting the impact of adjacent newly built areas. Solar paneled awnings flank the Shopping Valley, providing solar-powered electricity, and shade, and their wing-shaped awnings also guide the fresh air to the shopping below. Local vegetation is re-introduced in the Eco park, bringing back local birds and insects to help restore the natural biome.

7 Mutants and Hybrids

To create a legacy for this Eco Park, the two stadiums are designed to function as attractors and generators of new sociocultural activities and hence are proposed as Hybrids; merging sports events with a concert hall/event space for Hangzhou after the games. This future reuse stabilizes the carbon footprint and transforms the park into a constantly evolving active urban participant, rather than creating a static, representative object that would become a vacant white elephant after the 2-week Asian Games. Inspired by the contorted and doubling shapes of the cresting of vascular plants as they adapt to their changed and challenged environment, the stadium's hybrid character was developed in a similar manner. Its shape moved away from a singular form and used into the more complex intersection of two double-curved ellipsoids; one clad in brass shingles, one created from a steel and glass oversized diagrid. The resulting shape—two elongated wobbly disk—are offset, creating an optimization of adaptability, allowing for the hybridity to be fully expressed. The slippage between the two solids creates overlapping spaces fluidly connecting the interior and exterior. The inner bowl features a hybrid seating configuration adapting the central viewing layout preferred for sporting events into an asymmetric, stage-centered configuration for performances. The result is a highly adaptable set of volumes that can seamlessly convert to an event space after the games. The mutant's adaptability is further expressed through the introduction of a self-spanning suspendome roof, which is supported on the inner bowl to cantilever out over the public lobbies. The double-curved glass diagrid façade is then simply suspended from the suspendome's edge, essentially rendering the entire building column-free. Through 3D digital design and integration in BIM coordinated workflows, we were able to minimize material use (saving 1130 tons in steel), shorten construction time by 20%, and save significant funds for the client in this ambitious project. Air Travel was drastically reduced, by performing site visits by drone (by the local architects) and sending the drone footage to the USA. By reviewing these short movies details could be reviewed up close and coordinated with the GC in Hangzhou over zoom (Fig. 2).

Strange intersections and bulges occur along the mutant's double-curved shapes, where brass shingled surfaces intersect with steel and glass

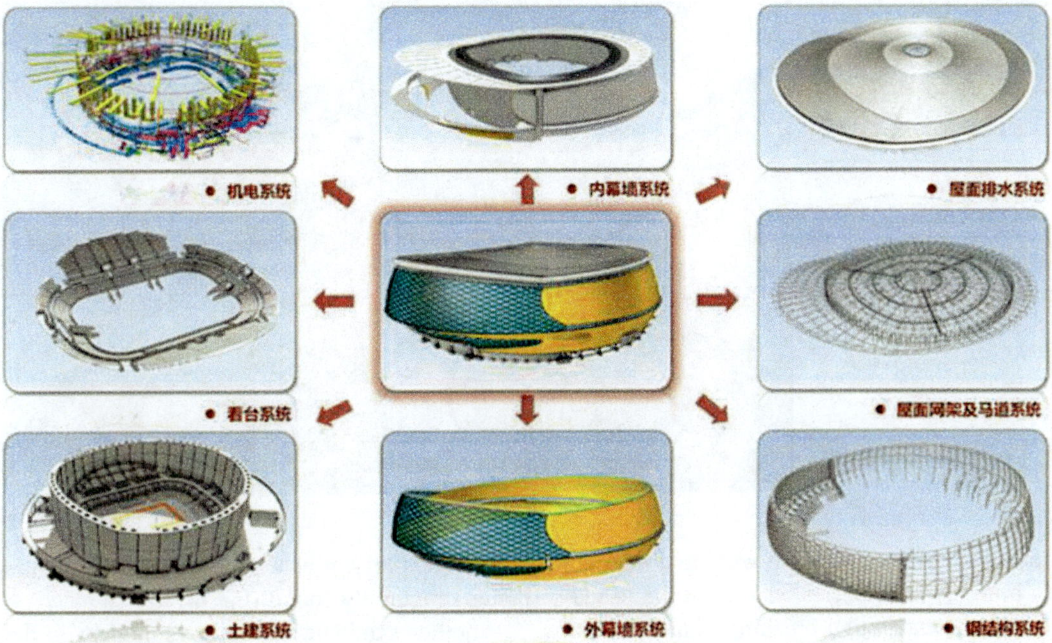

Fig. 2 BIM coordination of the Hybrid Stadium's integration of structures and passive climate control systems

diagrid rings. These bulges not only express the mutant character of the building, but also add identity and character to this strange object, making it recognizable that this is a neighborhood participatory social condenser. The double-curved inner bowl is entirely clad in locally sourced bamboo with recessed LED lighting. A central skylight with a light diffuser below creates indirect daylighting in the arena and blends technologically generated forms with renewable materials that naturally adjust to their immediate environment. The building also ties into the surrounding nature; cooling water is extracted from newly reconstituted wetlands, and energy is preserved by only cooling the seats of the viewers. The building literally 'breaths", as cool air is pulled in at night, and automatically openable skylight and ring of windows under the suspendome roof allow hot air out by day. The building is literally breathing in and out (Fig. 3).

8 Future Mutants—Materials

Not only is 'synthetic natures' a term used here to look at the purposeful symbiotic merging of nature's intelligence with building structures, but also how these new life forms (species) can alter buildings and material ecologies, drastically reducing the buildings' environmental impact and altogether re-defining architecture as being an instrumental part of the solution rather than solely creating the problem. Kent H. Redford and William M. Adams in their book Strange Natures, look at what is natural and what is artificial in the Anthropocene, they state:" The tools of synthetic biology are changing the way we answer that question. Gene editing technology is already transforming the agriculture and biotechnology industries. What happens if synthetic biology is also used in conservation to

Fig. 3 The Hybrid's intersecting bulging disks allow for hybrid use and the building's legacy for the city

control invasive species, fight wildlife disease, or even bring extinct species back from the dead?" They postulate a ground-breaking examination of the implications of synthetic biology for biodiversity conservation and the protection of natural species.

However, the focus here is not solely on the conservation of nature, but also on the creation of a synthetic nature, building symbiotic alignments between organisms and matter, where matter takes on organic qualities and therefore has a more significant agency on its environment and can start to lower the large percentage of carbon emissions due to construction and building materials.

One of these synthetic materials is the mushroom 'brick,' where mycelium, the fibrous roots of a mushroom, is combined with straw or some other agricultural waste and allowed to grow. Typically, it grows for up to two weeks, when it is halted by either firing in an oven or treating it to remove the fungus to stabilize the brick. The benefit of mycelium is that it is carbon neutral and durable, added benefits are that the material can be self-repairing and self-healing; the living fungus can grow together to form a stable structure, and if incurred damage can self-repair. We, however, were more interested in a hybrid version, creating a possibility for larger structures and greater climate impact. A hybrid of Mycelium with precast concrete creates a bio-wall that affords natural growth and air filtering while being structurally sound (Fig. 4).

Another synthetic nature development is the recently developed algae and glass façade. In their article on microalgae bio-reactive facades (Talaei et al. 2020), researchers Maryam Talaei, Mohammad-Javad Mahdavinejad, and Rahman Azari state: "Algae convert sunlight and carbon dioxide into oxygen, heat, and biomass. The integration of microalgae, as a photobioreactor-based source of biofuel, with buildings, has the potential to transform high-performance architecture". Instead of vegetation growing up the façade here, algae facades create natural shading and insulation, converting light into heat and biomass through photosynthesis, thus producing energy for the building, while lowering its carbon footprint. The Algae façade also creates ambient light, insulates the space, and shimmers in the dark.

We also find other productive, air-cleaning properties in nature, the Echeverria species has, through phytoremediation, huge air-cleaning qualities. These qualities can be further mutated to accelerate their productive symbiotic aspect, their air-cleaning capacity. Our indoor air is typically 2–5 times more contaminated than outdoor air, substances such as chemical-based cleaners and air fresheners, furniture coated with glues, paints, and fire retardants, can pollute and contribute to poor health. Succulents such as the

Fig. 4 The caption organic matter integrated into concrete, allowing for organic growth

Fig. 5 Echeverria are being mutated to enhance their air-cleaning properties

Echeveria can improve air quality and absorb toxic chemicals from common household products in astounding amounts (Fig. 5).

NASA's Clean Air Study found that Echeverria is able of removing 87% of volatile organic compounds (VOC). They do this through photosynthesizing via the Crassulacean acid metabolism, allowing a higher uptake of CO_2. in their fatty leaves. Their storage capacity is based on the structure of their tissue; succulents provide morphological and phylogenetic diversity, which can be accelerated through mutation, further increasing the air-purifying quality. A proto-hyper body trained in toxic absorption. The cultivars have such unique, highly effective traits that Nasa is looking into creating plant-based air filtering in their spaceships.

Not only Echeveria have air phytoremediation capabilities; other plant structures are capable of water remediation and remove, transfer, stabilize, and/or destroy contaminants in our water and soil. It's well-known that sunflowers are "hyperaccumulators", plants that are capable of taking up high concentrations of toxic materials, removing harbor pollution and nuclear waste as they absorb zinc, copper, and other common pollutants across a variety of their genome. Our urban wastelands can be cleaned, purified, and beautified by introducing sunflower islands in harbors and other polluted areas in greenhouses. Once absorbed, they must be eliminated similar to any toxic waste system.

9 Conclusion

Our research points at the possibilities to accelerate the development of these mutant organic materials, enhancing phytoremediation aspects, as we 'grow' new hybrid construction materials with ever more intelligent behaviors. In the future city, where buildings are designed to absorb carbon, transmit oxygen, and produce energy, the human race might have a chance of not only surviving, but also symbiotically living with these mutants and synthetic natures as 'home'. Ultimately replacing the question of what is the 'wild' and the cultured with a symbiotic system; a synthetic nature, where mutant hybrid structures perform not unlike a wide variety of intelligent plant behaviors, as they are implanted in our construction materials and built environments. These Synthetic Natures and strange new objects grow and adapt, absorb toxins, generate energy, filter, and retain water, or promote biodiversity.

Our research is currently focused on creating these hybrids; where biomatter is added to concrete mixtures allowing for plant life to become an integral part of the building's façade structure, allowing for the façade to generate oxygen, absorb carbon, and filter pollution out of the surroundings. This embedded biomatter will also have the added benefit of further insulating the interior spaces and creating maintenance-free, self repairing buildings (Fig. 6).

Fig. 6 The caption for the hybrid embedded in the eco park for the Asian games 2023

References

Burke E (1958) A philosophical enquiry into the origin of our ideas of the sublime and beautiful. In: A philosophical enquiry into the origin of our ideas of the sublime and beautiful. Columbia University Press

Gissen D (2009) Subnature: architecture's other environments. Princeton Architectural Press

Talaei M, Mahdavinejad M, Azari R (2020) Thermal and energy performance of algae bioreactive façades: a review. J Build Eng 28:101011

THE PLAN 140 (2022) An earlier version of synthetic natures by Winka Dubbeldam was originally published in THE PLAN 140, pp 13–18. https://www.theplan.it/eng/magazine/2022/the-plan-140-09-2022

Unknown, Vehicles, Air Pollution, and Human Health (2014). https://www.ucsusa.org/resources/vehicles-air-pollution-human-health

Yeo S (2020) How the largest environmental movement in history was born. https://www.bbc.com/future/article/20200420-earth-day-2020-how-an-environmental-movement-was-born

Hardware Stories. DIY Practices as More-than-Human Material Activism

Antonio Bernacchi and Alicia Lazzaroni

Abstract

In relation to an understanding of the symbiotic nature of living organisms, the project "Hardware Stories" is presented as a counternarrative to modernist hygienist standardization. It is a set of diegetic prototypes of construction elements, with relative manufacturing and maintenance tools, and is showcased through a floor component. This work questions how much the mechanism of production of architectural construction materials, with its quality control mechanisms and the economic scale that its implementation requires, actually prevents any form of alternative approaches. The project proposes to consider DIY practices as a form of mundane resistance that can empower multiple actors as a practice of material activism. Through the description of material sourcing, properties, manufacturing processes and installation, this visual essay aims to explain how these enable to conceive different relationships of cohabitation that are critically framed through the concept of "domestication-as-rewilding", introduced by anthropologist Anna Tsing.

Keywords

DIY · Material activism · Diegetic prototype · More-than-human · Domestication-as-rewilding

1 A Counternarrative to Sterilization

Almost every account of modern architecture history relates some of its origins to the so-called "hygienist movements" in the second half of the nineteenth and beginning of the twentieth century. On a macro-scale, infrastructures of sanitation and sewage were built to improve the poor health conditions of major cities, but on a small and micro-scale, the built environment has undergone a more specific process of "sterilization" and systematic elimination of dust and bacteria, in line with the modernist praise of cleanliness on both a metaphorical and literal level.

A long time after the structural conditions that originated the hygienist movement were no longer present, the scientific community's positioning had substantially evolved contributing to questioning the need and desire for cleanliness that never really left the architectural imaginary inherited from modernism, which obviously did not include any space for non-human life. Consider, for instance, the acknowledgement of the role of symbiosis in the evolution of cells,

A. Bernacchi (✉) · A. Lazzaroni
Aarhus School of Architecture, Aarhus, Denmark
e-mail: abe@aarch.dk

A. Lazzaroni
e-mail: al@aarch.dk

M. R. Thomsen et al. (eds.), *Design for Rethinking Resources*, Sustainable Development Goals Series,
https://doi.org/10.1007/978-3-031-36554-6_7

introduced by American biologist Lynn Margulis in the 1980s (Margulis 1991).

The artistic project “Hardware Stories” aims to suggest a counternarrative to the imperative for pristine sterilized surfaces in the built environment through speculative crafting, proposing a reflection on the value of entangled cohabitation with other species, as sustained by thinkers like Donna Haraway (Haraway 2016).

“Hardware Stories”, co-authored by Animali Domestici (A. Lazzaroni & A. Bernacchi), Jakob Sieder-Semlitsch, and Lynn Hyun Kieffer, is constituted by a set of prototypes of architectural components together with some hacked and redesigned tools of manufacturing and maintenance, which are used to empower independent processes of material experimentation. The floor prototype presented in this visual essay was included in the exhibition “Architectures of Cohabitation” held at SchauFenster gallery in Berlin in May 2022 and curated by Arch+.

These artefacts were conceived within a perspective of design fiction, intended as the “deliberate use of diegetic prototypes to suspend disbelief about change” (Sterling 2013). To understand this mode of operation, Bruce Sterling explains that “design fiction doesn't tell stories—instead, it designs prototypes that imply a changed world”, therefore “a design fiction is a creative act that puts the viewer into a different conceptual space—for a while”.

In this sense these prototypes deal with Dunne’s concept of para-functionality (Dunne 1999), meaning that “leveraging the seductive language of design, design artefacts are crafted to provoke people to imagine it in use and the possible future that would manifest around it” (Wakkary et al. 2015).

2 Technification and Standardization

The current relevance and influence of a hygienist approach to the design of interiors are put in question by a different understanding of symbiotic relationships at large, within a broader framework related to the crucial importance of biodiversity. Nevertheless, the disciplinary infrastructure of standards and protocols of quality control developed to enable it is still in place and is geared into the industrial machinery of production, ensuring a process of overall systematic sterilization.

This system supports the “expectation, fuelled by the insurance industry, that risk and performance can be easily and rapidly quantified, which has pushed construction further towards the use of standardised composite building products” (Material Cultures 2022). In terms of costs, time and manpower, the demands for a product to be approved within the current sophisticated systems of certification, that can be found in Western European countries and in North America, can be sustained, however, only for large-scale or mass production. This, therefore, further increases the resulting process of standardization as “any technologies which deviate from this pattern are harder to contract, harder to insure and harder to justify” (Material Cultures 2022).

In turn, it is becoming quite evident that “a culture obsessed with measurement can easily ignore or even forget the things it finds harder to measure. For example, carbon, cost and energy are all relatively easily quantified, whereas biodiversity loss and quality of life for workers are harder to measure” (Material Cultures 2022).

3 “Hardware Stories”: DIY as a Space for Alternatives

The floor prototype is a patchwork of multiple tiles made of very different materials, like, for instance, ceramic and seeds conglomerates, or dog fur on a mycelium base block, held by a timber substructure of irregular geometry and variable heights, which ensures the flush alignment of the top surface of the floor aggregation. The project articulates through material experiments an inquiry into the potential and value of porosity for bio-receptivity, as a vehicle to support the cohabitation of non-human beings in everyday architectural spaces, considering how cavities' properties like scale, shape and connectivity might influence the effectiveness in creating a hosting environment.

Do-It-Yourself (DIY) is defined in the Oxford English Dictionary as "the activity of making, repairing or decorating things in the home yourself, instead of paying somebody to do it", where the usual assumption is that this activity is done at an amateur level, instead of engaging a skilled professional, who would get paid for the service, which poses the accent on the difference in skillset or experience. Nevertheless, there are various crucial distinctions between professional and DIY activities, like the relation with instructions and protocols, the alternative networks of resources, the absence of systematic quality control and preconceived target parameters, but also, and perhaps more importantly, the dimension of time and a very different notion of efficiency. All these factors might, in fact, be much more relevant than the actual manufacturing expertise in determining the innovative and transformative potential of such practices to explore alternative modes of operation, alternative spaces and materialities, and subsequently alternative attitudes to cohabitation.

Analysing various aspects of the development of Hardware Stories, like resources, material properties, manufacturing and installation, in close connection to the visual documentation of both process and outcome, could shed a light on how they have been conceived to explore and redefine various of the core elements of distinction of DIY practices, to investigate their potential in first hand, and to narrate them as "storied matter" (Iovino and Oppermann 2014). This intends to position the work within a perspective of material ecocriticism, as defined by S. Iovino and S. Oppermann, which emphasizes the crucial role of narrative in a critical discourse on environmental questions (Iovino and Oppermann 2014).

4 DIY as Material Activism

Within this framework "Hardware Stories" aims to raise the question of whether DIY practices could represent a form of "mundane resistance".

Within the education and research in design, there is an increasing interest in including material manipulation within the design process and its fields of experimental investigation, a phenomenon that resonates with the concept of "Material Activism", coined by designer Miriam Ribul, referring to a "low-tech approach to the democratization of production" (Ribul 2014). The inspirations and methods that enable it could come from very different everyday practices, like cooking, as "common tools and ingredients are adapted in pursuit to create alternative aesthetics and processes for materials and making" (Ribul 2014). As the term "activism" suggests, this reclamation of material experimentation is often aimed at pushing for a change in the conditions and mechanism of the production industry; nevertheless, it is usually still conceived within the disciplinary domain of designers identified as such. In this sense, the project "Hardware Stories" suggests a possible step further, creating the conditions and toolkits to empower non-designers from different age groups and backgrounds, considering the demographics that are typically involved in DIY practices, to engage in the process directly (Figs. 1, 2, 3 and 4).

Looking more in detail at the manufacturing of the prototype, it is important to first describe the search for alternative sources, partly related to the constrained availability of materials. These have been mainly reused, found in fabrication workshops from previously discarded mock-ups and prototypes, or picked from leftover stacks in workshops or around fabrication facilities, like in the case of timber, bricks or stone tiles. Some of them have been sourced freely from public forests, like timber twigs and branches, or moss, and others were even harvested directly from organic sources, like the poodle fur, trimmed by the

Fig. 1 Floor prototype, overall

Fig. 2 Details

authors as part of a regular grooming cycle. A few others of smaller volume have, instead, been bought specifically for the project, like paper porcelain and the relative ceramic glazing, or like some "ingredients" from other fields like baking, where we can find edible food colourings, grains, seeds and fibre mats.

The different sourcing had, however, an obvious impact on the unevenness of the materials, like in the case of chipped stone or cracked timber, or the natural irregularity, which has been accommodated through the patchwork pattern arrangement, turning into an incentive to evolve the formal resolution of the entire prototype (Figs. 5 and 6).

The specificities of the materials, in some way their "specifications", were highly related both to their potential for manipulation and crafting, and to their targeted properties and formal features at a "micro-scale", especially in terms of porosity.

Looking in detail into a few of the tiles, the different natures of the base fluid material, from plaster to rough clay to paper clay or to finer paper porcelain, have a very different softness and drying time, which determines a varied degree of manipulability and, through a diverse

Fig. 3 Details

Fig. 4 Maintenance practice

set of crafting experiments, achieved multiple surface treatments and cavities. For instance, plaster, which can be made quite fluid for casting, allowed floating inside it a number of expansion jelly beads that left a variably connected network of spherical voids on the top side of the plaster tile, when removed after a very slow drying.

Less malleable materials, instead, demanded other types of treatments, many of which also involved the temporary insertion of foreign bodies within the mass of the tile to be later eliminated towards the end of the process. Cooking provided again a variety of inspirations and opportunities, as different grains, like quinoa or chia seeds, but also wheat products, like bulgur or couscous, present varied dimensions and shapes, which produced a number of different effects, from unsmoothed micro-roughness to a spread of mixed-diameter holes. The grains were also distributed differently depending on the substrate properties, some were lightly sprinkled and gently pressed inside, while others were manually micro-punched and naturally followed

Fig. 5 Exploring alternative materialities: sources

Fig. 6 Exploring alternative materialities: sources

Fig. 7 Exploring alternative materialities: material properties

Fig. 8 Exploring alternative materialities: material properties

a more intentional pattern, as in the case of porcelain butterfly feeders (Figs. 7, 8, 9, 10, 11 and 12).

These examples introduce some considerations on the manufacturing processes and their agency. Here an important role has been played by the customized adaptation and modification of crafting tools through hacking and home 3d printing, like in the case of ceramic extruders, or even just in the fitting of chiselled timber sticks on roller handles, to adapt them for even pressure spread in the crafting of patterned ceramic tiles. The creation of new tools, whose blueprints, 3d models and specifications are shared in open source for collective use and development, represents one of the instruments to empower non-standard processes also for users outside of design disciplines. The production of tiles composed of multiple sub-parts required the design and adaptation of multiple joints and support elements, from dowels to 3d printed water trays, which all ultimately contributed to increment as much as possible the acceptance of tolerances at a tile level, in order to reduce the non-fitting

Fig. 9 Exploring alternative materialities: material properties

Fig. 10 Exploring alternative materialities: material properties

materials and potential wastage to the minimum (Figs. 13, 14, 15 and 16).

A similar adaptive logic informed the design of the substructure. Being a raised floor with a cavity, similar to an old floating parquet, it could accommodate very different types and shapes of tiles but also house complementary elements, like water trays, maintaining their top surface level more or less consistent. Made of leftover timber profiles of very different sizes stacked in three layers. First, a thick layer at the bottom could accommodate unevenness or protruding elements in the surface below. Then, a layer of mid-sized profiles, whose thicknesses are all different, is grouped in clusters with the top layer that is totally adaptive to the specific positions, but also to their imperfections, requiring a meticulous process of individual adjustment, which became an empathic act of careful adaptation, where all the micro-properties of each element had to be re-apprehended and observed with attention (Figs. 17, 18, 19 and 20).

All of the explorations carried out throughout the development of Hardware Stories were aimed at supporting enjoyment and appropriation by non-human species in a number of different

Fig. 11 Exploring alternative materialities: material properties

Fig. 12 Exploring alternative materialities: material properties

Fig. 13 Exploring alternative craft: manufacturing processes

ways, like providing different microclimatic conditions, or embedding nutrients, for instance.

Many of the tile experiments took inspiration from existing mundane devices that are developed informally and often documented in blogs and online videos, like butterfly feeders, which are usually constituted by a tray of butterfly nectar (water with approximately 10% of sugar and some fruit juice), where the liquid should be accessible, but not fully to avoid interference from rodents or similar, while using colour to attract, as butterflies target red, blue and purple, for instance.

The butterfly feeders in Hardware Stories have a ceramic mass, which is an aggregate of porcelain compressed spheres with a surface dotted by scattered holes, sitting on top of a plastic tray 3d printed in shiny purple PLA and including small pieces of natural sponges in between the spheres that carry the nectar to the top surface by capillarity, allowing it to spill a bit in and around the surface cavities (Fig. 21 and 22).

Other ceramic tiles, instead, were aimed at creating a hosting environment for small-scale cryptogams, like mosses or lichens, which reproduce through spores that need to find a suitable balance of wet and humid conditions with a specific pH range to thrive. In these cases, the heterogeneity of form and the microporosity are crucial in providing a wide range of degrees of moisture and exposure to air of small pockets and wrinkles within the material. Therefore, a custom-made pneumatic extruder has been utilized to pour highly hydrated paper porcelain mixed with chia seeds, giving a highly jagged texture and avoiding spoiling that through manual manipulation. The chia seeds are later burned in the kiln firing process, leaving micro-cavities that enhance the natural absorption properties of the material.

Some tiles' bioreceptive properties were instead related to their chemical composition, and the manufacturing process and surface treatments allowed certain substances to be more

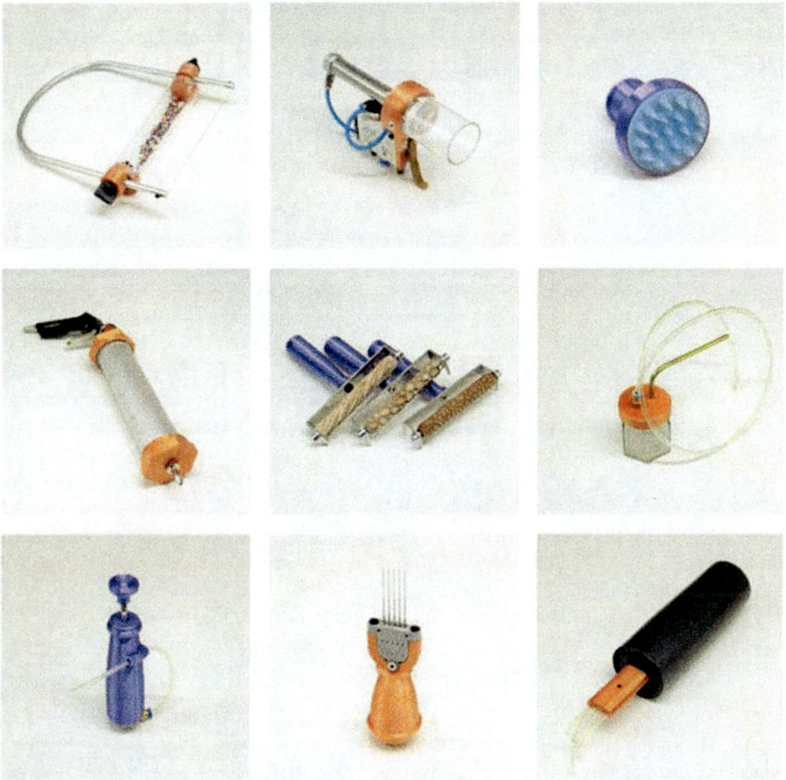

Fig. 14 Exploring alternative craft: manufacturing processes

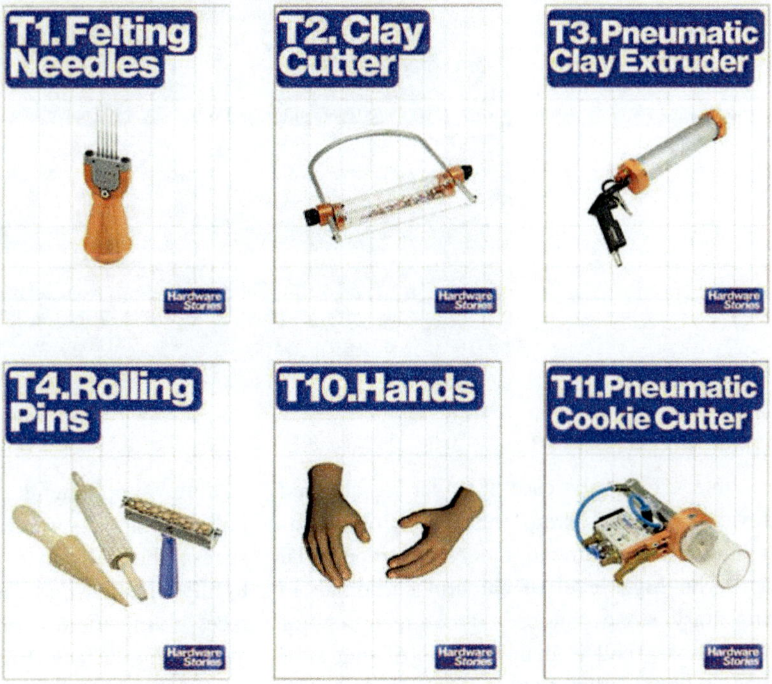

Fig. 15 Exploring alternative craft: manufacturing processes

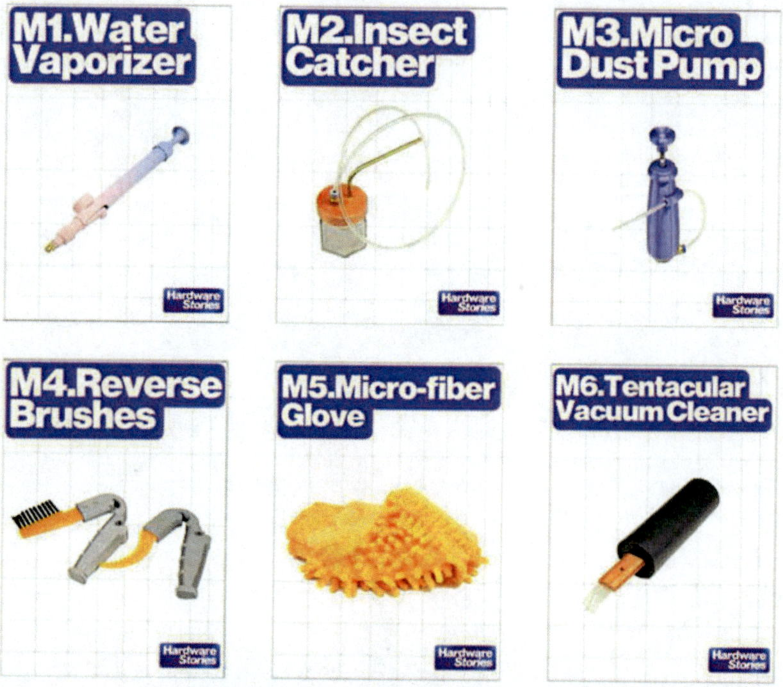

Fig. 16 Exploring alternative craft: manufacturing processes

Fig. 17 Exploring alternative adaptation: Installation

accessible. This is the case of travertine chipped leftover tiles, which were reused by slowly cutting them with diamond blades in trapezoidal shape. The high level of calcium carbonate in sedimentary stone can be a great resource for snails, whose shell is mainly made of that substance and that often need to be provided with calcium supplements; nevertheless, a honed or polished stone surface does not allow to easily access the substance, therefore the tiles were lightly brushed and sandblasted on the top surface, giving them a more rough texture and an undulating surface (Fig. 23).

Fig. 18 Exploring alternative adaptation: Installation

Fig. 19 Exploring alternative adaptation: Installation

5 Closing Paragraphs

5.1 Narrating Alternative Forms of Coexistence: Domestication as "Rewilding"

Hardware Stories proposes a reflection on different forms of cohabitation with other species within the domestic environment, not by staging or depicting them explicitly, but by opening the imagination about them through the physical manifestation of a space that is conceived for a different set of relationships and hierarchies that are less linear and straightforward than the typical notion of domestication.

In her text "nine provocations for the study of domestication", anthropologist Anna Tsing aims to put into question the fundamental assumption of a binary distinction between what pertains to domestication and what doesn't. Referring to multiple examples of coexistence determined by mutual help, rather than coercion or captivity, as for "wild chickens" in the Meratus Forest in

Fig. 20 Exploring alternative adaptation: Installation

Fig. 21 Micro-environments of cohabitation

Fig. 22 Micro-environments of cohabitation

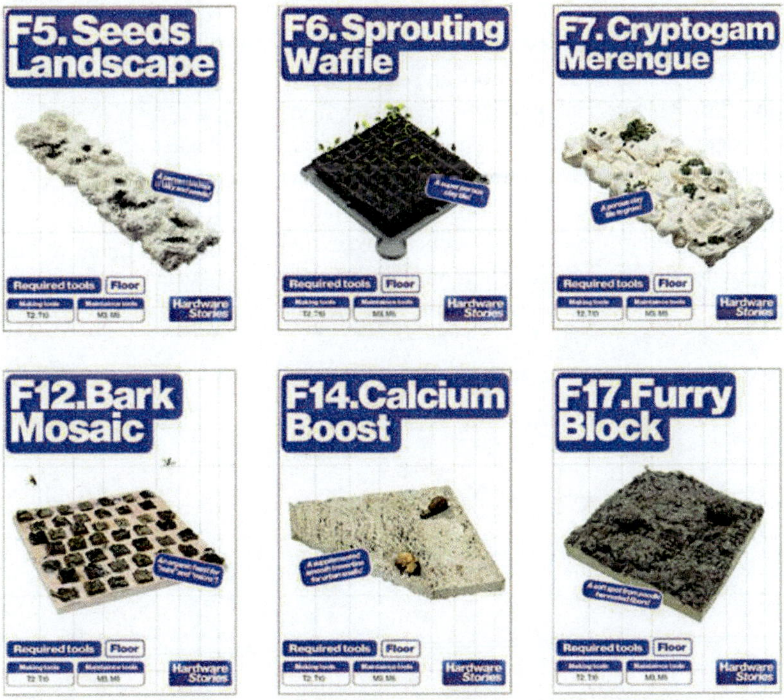

Fig. 23 Micro-environments of cohabitation

Indonesia, where having "no fences and no prohibitions" she asks: "Is this domestication?" (Tsing 2018). The reductive understanding of multispecies coexistence where humans are always in a position of control is also because "we don't count the many species with which we have other kinds of intimate relations", but there are not even terms for those, that "point to cospecies landscapes in which no species can be said to be in charge" (Tsing 2018). She later elaborates on how different "arrangements", like "domestication-as-rewilding", "offer hopeful alternatives in imagining multispecies life with humans as a component", and even further flipping the question in reverse, noting that "a majority of the species left on earth are here because they have figured out a way to live with humans- a domestication of humans for their purposes, if you will" (Tsing 2018).

5.2 Debating About Alternative Worlds Through Material Activism

"Hardware Stories" investigates the potential of alternative forms of innovation and experimentation through material activism developed from the user side, which can provide counter-narratives in order to promote different conceptions of multispecies relationships and the relative frameworks that they imply, which take bio-receptivity and more broadly a more-than-human perspective into crucial consideration. The project instigated debates on multiple levels, directly, within the exhibition space and in an open symposium connected to it, but also indirectly through the dissemination in publications, where the prototype was chosen as the cover image of the week's art section of the most diffused national newspaper, somehow underlining the instrumental role of its formal qualities.

References

Dunne A (1999) Hertzian tales: electronic products, aesthetic experience, and critical design. MIT Press, Cambridge, MA

Haraway DJ (2016) Staying with the trouble: making kin in the Chthulucene. Duke University Press, Durham, NC

Iovino S, Oppermann S (2014) Material ecocriticism. Indiana University Press

Margulis L, Fester R (eds) (1991) Symbiosis as a source of evolutionary innovation: speciation and morphogenesis. MIT Press, London, England

Material Cultures (2022) Material cultures material reform. Mack books

Ribul M (2014) Recipes for materials activism. https://issuu.com/miriamribul/docs/miriam_ribul_recipes_for_material_a. Accessed Oct 2022

Sterling B (2013) Patently untrue: fleshy defibrillators and synchronised baseball are changing the Future. Wired UK

Tsing A (2018) Nine provocations for the study on domestication. In: Lien ME, Ween GB (eds) Swanson HA. Politics and practices of multispecies relations. Duke University Press, Domestication Gone Wild, pp 231–251

Wakkary R, Odom W, Hauser S, Hertz G, Lin H (2015) Material speculation: actual artifacts for critical inquiry. Aarhus series on human centered computing, vol 1

Localising Resource

The Soil of New Culture Studios: A Spring for African Architecture

Jite Brume, Alvaro Velasco Perez, and Demas Nwoko

Abstract

This paper reflects on the contemporary need for reconsidering circularity in architecture. For this, it investigates a vernacular building system rooted in the Niger Delta that supports an alternative form of understanding materiality. Framed within current research on re-animating soils, the practices of Demas Nwoko (Nigeria, 1935)—whose work has questioned circularity since the 60s—have resisted anachronism through challenges of contemporary architecture. From a physical material basis, his perception of 'local' is established by using the site's soil applying techniques originating in ancient Nok ceramics. His understanding is that architecture is created within contemporary culture through a perpetual process of collaboration. From a temporal basis, he establishes a dynamism where the past and the future converge—digging up the past and building up foundations for the future are one and the same act; a cycle of life and birth. We argue that ours is a time of urgent reconsideration of human-soil interdependency; within it, ecological engagement with soil signifies hope. Focusing on Nwoko's New Culture Studios project (Ibadan, 1967–), the article uses critical commentary to examine the processes placing them at the intersections of modernisation, post-colonialism and material resilience. We argue that this examination extracts important principles on how appropriate technology can evolve with use of local resources. This creates a possibility of thinking design circularity as an evolutionary spiral leading to the emergence of new sustainable cultures for architectural practice.

Keywords

Soil · Earth construction · Mud · Nigerian architecture · Demas Nwoko · New culture studios · Local resources

J. Brume (✉)
Mappamundi Architects, London, UK
e-mail: jite@mappamundi-arc.com

A. V. Perez
Architectural Association School of Architecture, London, UK

D. Nwoko
New Culture Centre, Idumuje-Ugboko, Delta State, Nigeria

1 Introduction

I have always found something alluring and timeless about earth and mud. The mud cities of Africa were created as far back as the turn of the first millennium and were part of a typical process worldwide. These cities were spread all over

M. R. Thomsen et al. (eds.), *Design for Rethinking Resources*, Sustainable Development Goals Series,
https://doi.org/10.1007/978-3-031-36554-6_8

Fig. 1 New culture studios under construction c.1970/Demas Nwoko personal archive

the Sudan belt of West Africa. Deep in the tropical forest region were kingdoms that rose west of the river Niger, including the Benin Empire and the Yoruba cities of Ife, Owo, Oyo and Ibadan. The mud citadels grew expansively and have all lasted to this day, surviving the ravages of climate and time and averaging over 500 years. Apart from the communal edifices where successive kings continue to live and worshippers have not stopped using, this readily available local material has ceased being used to build domestic buildings in these regions. At the advent of colonialism, derivatives of Portland cement replaced the earth, even in rural buildings. While these changes in architectural materials gain currency, they sacrifice comfort and sustainability due to the cost implications inherent in imported materials.

On the other hand, nostalgia for experiences of the African past could be a rich resource to learn from when situated within some contemporary debates. In this paper, I would like to share my memories of material production and life cycles of soil construction. This methodology of drawing from my personal experience but situating it within the contemporary discussion aims to cast light on the efforts to achieve Sustainable Development. Looking back at the photos of the groundworks for my New Culture Studios in Ibadan taken around 1970 (Fig. 1), there is something primordial about them. The site had a tension between being simultaneously broken and shaped, a traction that holds even now because the project is still a work in progress. Nowadays, the Nigerian construction industry makes the photo look like prehistory; perhaps, that is precisely the point. Today, West African architecture has gained the interest of an international audience—we see it in museums, *Pritzker prizes*, and alternative histories. However, these works often lose their historical perspective. They are pre-historical in that they do not belong in historical conditions but in a sort of primaeval hour.

2 Soils Today

This is particularly the case in some of the discussions I read today on soils. The shift in soil thinking came some 25 years ago, with William Bryant Logan's classic *Dirt, The Ecstatic Skin of the Earth* (1995), which encouraged us to conceive soil as a living entity. Though parts of his thinking were formulated almost a century ago in the writings of Lady Eve Balfour (*The Living Soil*, 1943), both coincided in recovering 'life' as a critical characteristic of soil. However, the ecological condition of the 1990s made Logan see it as a matter of life and death. More recently, authors like Tim Ingold and Maria Puig de la Bellacasa argue that 'life' is not enough; we need to see the 'spirit' of the soil (Ingold 2006, Puig de la Bellacasa 2019). Something called 'animated soils' is emerging. It follows 'a renewed captivation for the life in soils [that] has become a common leitmotif animating imaginaries of soils across the sciences, global institutional initiatives, community groups, policy bodies, creative arts and popular media representations', as Puig de la Bellacasa argues (Puig de la Bellacasa 2019, 392). I agree that 'animated soils' might introduce new forms of understanding the ground, which the contemporary ecological context needs very much. However, I fear the risk that while trying to give a soul to the soil, we are burying the historical contexts in which animism emerged. Isabelle Stengers already drew our attention to the risk of the current interest in animism being an appropriation of Western thought and academia (Stengers 2021). We must not illicitly appropriate animism from its cultures of origin while trying to deal with a planetary emergency by overpassing the downfalls of Western thought.

For architects, thoroughly dealing with soil is an urgent and intriguing challenge. There is a beautiful essay by Heinrich Jennes in which he reflects on how European architecture has traditionally disregarded the ground. Though a slightly generic claim, he argues that as architects, we need to move beyond the idea that soil 'must be dug out and disposed of elsewhere' (Jennes 2009, 133) and engage in the material processes in which the ground is involved in a new creative way. One of these potential approaches is derived from the discipline of archaeology, which sees soil as a deposit of history through excavation (Jennes 2009, 135–139). This angle resonates with Sekulic, Tomic and Sattler in their article for the Architectural Review 'Digging up the past: Soil as Archive' where the soil transforms to 'an ecofact and artefact (…), [transcending] the divide between nature and culture' (Sekulic et al. 2020).

In my Nigerian context, I have witnessed a transition away from the cooperative social activity of building accessible housing within a building industry that has lost its mud-building technologies. This loss is due to the dominance of industrially processed materials that emerged due to colonisation and the industrial revolution, which created the global market that has unified the world's economy. Consequently, the same finished industrial products are consumed universally, irrespective of the climate and cost to individuals and the planet. With mud losing its pride of place, it has become unfeasible to build economically and sustainably. Therefore, adequate housing becomes problematic for most of the world's population to attain. There is a pressing need to develop appropriate new building materials to produce affordable housing for people in both urban and rural environments. One answer to the problem lies between the two extremes of traditional and conventional materials. New materials should be capable of mass production. The search for such a new material led me to investigate the production of reconstituted soil in 1967.

3 Local Resources

When working with earth in the 1960s, my concerns differed from those that led to 'animated soils'. Groundworks for the New Culture Studios began in 1967 with a programme that consisted of an experimental art school, including teaching, production and exhibition, along with the residence for my family. Digging the soil was not an inaugural act of removal for us but a way of resourcing material for the works. The principle that architecture should come from onsite or local sources guided us. I was experimenting with earth materials based on empirical tests on site.

I eventually developed a stabilised earth block from the reformed laterite soil, which I called 'latcrete'. The *latcrete* is a hollow building block produced in conventional sandcrete block-making moulds, both manual and mechanical. It uses Portland cement mixed with the site's earth in varying ratios, depending on load-bearing requirements. Variations in the colour and hue of the blocks could be achieved depending on the soil excavation depth, giving the contemporary facade the sense of a stratified archaeological site. The blocks were neither plastered nor painted, avoiding expenditure on finishes. Beyond this initial saving is the economy on future redecorations because the *latcrete* wall is permanently self-coloured. It also accommodates dust pollution aesthetically as it has the same pigmentation.

At the time, I was just a lecturer who had purchased a piece of land in one of the high-density areas of Ibadan with all my savings, leaving very little to build with. As I cleared the land, I noticed the composition of the earth and guessed that if I mixed cement with this type of soil, the result could be a type of concrete since it contained sand and gravel. The resultant concrete could be stronger than the typical sandcrete building blocks. So I built a mould and cast a sample of my mix into a block. The result confirmed my expectations, as the block turned out to be stronger than the local sandcrete blocks. Moreover, the material cost was just cement and nothing else, amounting to half the cost of standard blocks in 1967. Although the invention of *latcrete* began as the search for a cheaper way to build, its use in the New Culture Studios at that time transcended the pure economy of means in several ways. Beyond the logic of material scarcity, my resource for the local additionally integrated the political conditions of construction by finding a loophole in contemporary byelaws that prohibited the use of mud for building.

Ever since this experimental success, I have built all my architectural designs with walls of this material, from private houses to public buildings. I met the critical structural demands of the buildings with reinforced concrete pillars and beams and used sandcrete blocks and stone boulders for improved foundations. In addition, I improved the waterproof coating of pure mud walls by developing a render, coating it 10 millimetres thick for durable wall finishes or for casting facade tiles by substituting the sand component of cement mortars with laterite mud. I used this *latcrete* mortar to cast the traditional fluted tiles of Edo architecture, which I used as a finish for the entrance wall of the Benin Cultural Centre.

4 Collaboration and Technologies

Another guiding principle I followed was that all new technology invented should be familiar to the local context and support techniques shaped by the native culture, climate and environment. By using the same skills and equipment available in indigenous construction, I could better guarantee the adoption of the technology. Encouraged by the initial success of *latcrete*, I invited a local block maker and labourers to mass-produce the blocks onsite. For the production of *latcrete*, we substituted washed sand used in conventional sandcrete blocks with the site's soil. The block maker heaped up the earth wherever he placed his metal block mould, poured cement on the mound, mixed as usual and moulded the blocks. The only machine I invented for the earth construction at the New Culture Studios was a simple wooden jig that enabled the local labourers to lay the blocks perfectly. The training required for each worker could be completed in a day. During the excavation, I discovered that the gravel earth composition gave way to orange-coloured laterite, decomposed quartzite, mixed sand crystals and mica. I mixed this with cement and cast it, producing a colourful brick-like block of less strength and great beauty. I used this to build up a pictorial mosaic with the darker-coloured top gravel earth blocks for the walls of my studio.

Perched atop one of Ibadan's commanding hills at Oremeji and glowing in the rising and setting sun, my studios built with *latcrete* blocks attracted great attention. More satisfying was that everybody who subsequently built in that vicinity moulded their blocks the same way because the processing needed before the application was minimal in its technological demand. Following the African model of traditional building, the *latcrete* technology was no secret, and I did not implement it alone. People came to the building site, observed how we did it, and went ahead and did the same thing. However, they did not dare to leave their walls fair-faced. Following contemporary trends, they plastered and painted their *latcrete* blocks, gaining only the initial saving on their block production. The studio was never plastered or painted, yet it still maintains its unique aesthetic appeal. After over five decades of exposure, the fully load-bearing *latcrete* blocks have not shown any sign of deterioration. The case of *latcrete* is not one of outright invention if invention means bringing forth what did not exist before. It borrowed from sandcrete blocks, a conventional material, and the mud bricks of earlier town buildings, the new factor being the site earth mix. My intention was not to create something unique but to solve a problem using local techniques.

Over time, I discovered that merely substituting the cement and sand mix with pure earth did not significantly reduce the adverse conductivity of the wall. That means that my *latcrete* innovation needed to go further in reproducing the conducive thermal comfort of pure mud walls, which possess complete insulating properties against the external elements. It becomes evident that the conductive element in the wall is the industrially processed material. So to achieve any further improvement, Portland cement has to be eliminated entirely from the constitution of the core wall.

5 Nok Culture and Post-Colonial Culture

In the years prior to the New Culture Studios, I had experimented with local techniques to counter-colonial culture. My interest in the local art heritage started in the late 1950s during my student years at the Art Department of the Nigerian College of Art, Science and Technology in Zaria. Together with my colleagues Uche Okeke and Simon Okeke, we formed the Nigerian Art Society—better known in literature as Zaria Art Society. The society advocated for a national identity for contemporary art, articulated by what Uche Okeke subsequently termed 'Natural Synthesis', essentially a study of local traditions to formulate a type of modernism. In other words, our interest for the local was not a counter to the modern, but rather an attempt to rescue the indigenous art forms moving away from the idea of 'local subject matter' (folklorism)—that was dominating the arts in the country—into 'experimentation with culturally familiar *form* or aesthetic qualities' (Okeke-Agul 2015, 140).

In 1964, this interest led me to research the terracotta sculptures of Nok culture (Okeke-Agul 2015, 201–208, Williams 1966, 4–13). Materially, the clay sculptures directly connected with the soil, the territory and the claim for a national identity against a colonial power, which was interesting for the Zaria Art Society. However, my focus was more than just a revival of formal aesthetics. I was also interested in the technical way in which the Nok fired their pottery. I was trying to parallel their durability, so I had to go beyond the firing techniques available in Nigeria in the 1960s. I devised a kiln where the fuel burnt in contact with the pieces, giving them strength and nuanced hues. The research intended the resurrection of local traditions; however, the logic was in the technique of producing the ceramics rather than in the aesthetic revival of a local language only. Constructing the New Culture Studios, I incorporated these same explorations of technique and interplay between traditional and contemporary. Updating the local to generate a modern language did not remain in the invention of *latcrete* only. Entering the studios, you descend the sunken entrance to find a sudden drop of auditorium seats. A large performance space, Greek in its adapting the steps to the hill, but very much Igbo in the proximity between actors and audience. The stage was fully operational for some years and instrumental in producing some of my most relevant artworks, like the *Children of Paradise* play for FESTAC'77 (Nwoko 2016). The archival images show the void excavated during the works with an empty background—the onsite laterite quarry that later became the theatre space. The half-built skeleton of the grid of reinforced concrete structure of the compound occupies the midground (Fig. 2, left part photograph). The building is a syntax of the two languages. While the former indicates a vernacular logic, the latter shows up-to-date construction systems, a combination that continues in the villa attached to the complex where the project originated; a sand-cast

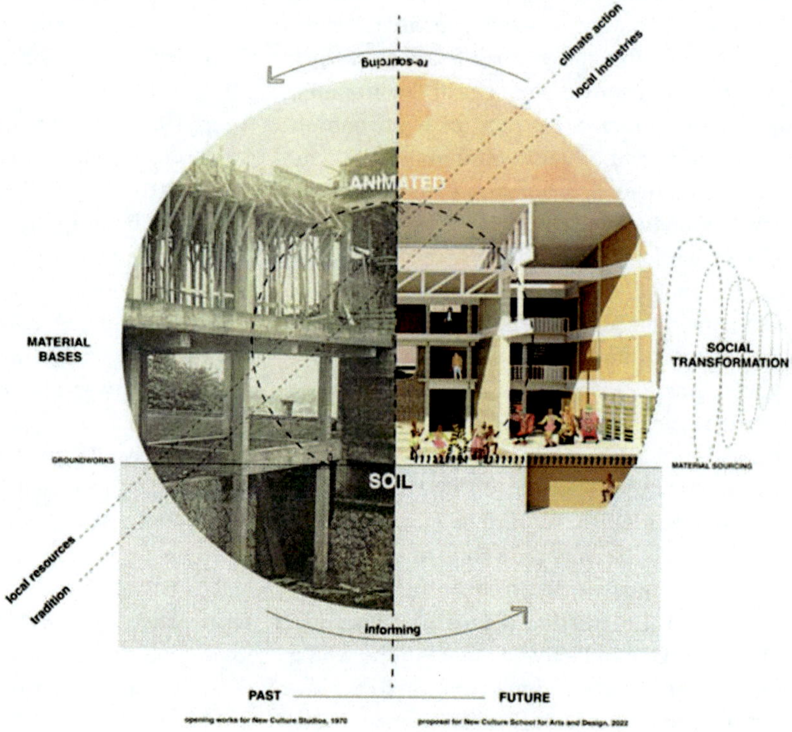

Fig. 2 Diagram showing the connections between the original project for the new culture studios and the proposal for its future. Illustration by the authors and Lloyd Martin

concrete screen, a form of brise-soleil crowning the façade; the concrete columns holding cantilevered upper floors, carved in local motifs; or a symbolic finishing for a solution that improves ventilation in the house.

6 Strange Tongues

Nigeria's independence in 1960 initiated a threshold for everyone working in cultural production in the country. Despite abandoning their colonial past, the future did not offer a culture detached from the metropolis. This modernisation process left an unfinished project, and a group of artists challenged the notion that modernity should perpetuate colonial ways. Creating a 'New Culture' was necessary, and I attempted it through a series of practices we developed during the post-independence years. The studios were only a first step aimed at complementing academic art education with explorations of African artistic expressions. In the '70s, we initiated a publishing house, the main outcome being *New Culture, A Review of Contemporary African Arts*. The journal continued the legacy of Ulli Beier's pioneer review *Black Orpheus*, where my work figured significantly in the early years of the periodical (Moore 1961, 28)—including an article on the construction of the New Culture Studios themselves (Barrett 1968, 40–41). We did not only inherit *Black Orpheus*' interest in African artistic production but Beier's unusual attitude, what John Thompson described as a 'border operator'. '[Ulli Beier] was a border operator—on the border between the European and the local, the traditional. And he could cross back and forth from one border to another and find things that they had in common. (…) Ulli was able to go back like a smuggler (…) from *avant garde* European art (…) to modern and traditional African art and literature' (John Thompson, in an interview at the Rockefeller Foundation, July 29,

1980, as quoted in Benson 1983, 431). Our journal published articles on contemporary African art production with reviews of current and historical Nigerian architecture, a section dedicated to the American scene and another to the European scene, and even a monthly segment specifically for children. However, from inception, our 'smuggler' position was not about cultural imports. My quest for a new culture was a question of language. In the editorial statement of the first issue that outlined the project ambitions, I wrote,

> This is another point in the planned development of the New Culture Studios organisation. (…) It is my considered belief that only one art stylistic idiom is valid to the African and the Blacks of African descent the world over, its origin being the too well-known form of traditional African arts; a form that was created and nurtured to maturity by the African peoples themselves, with a history that dates beyond two thousand years. (…) To this basic idiom… can be added new forms. The idiom is their language, and artists have to speak this stylistic language in order to make themselves understood by their kind. (Nwoko 1978, 1)

I proposed that an African form of modernism could be articulated, unearthing its ruins to the language of African arts. However, the trouble I found—and what the journal aimed to remedy—was that 'most contemporary African artists have found themselves strangers among their own people, because they have learnt to speak in strange tongues, articulating foreign cultural values and aesthetics which they little comprehend. These artists are the casualties of time' (Nwoko 1978, 1). I tried to offer another approach to modernity in Africa, diverging from the established 'tropical modernism'. Since independence, Maxwell Fry's 'Tropical Modern' was presented as a valid adaptation to the local postcolonial climate. However, this was arguably another question of language. Fry's was a translation exercise of the narrative of modernism into the vernacular language. But in his case the narrative was generated in the metropolis and adjusted in the Commonwealth. *Tropical Architecture* (Fry 1956) was more a kind of manual for Western architects working in the colonies (Godson 2019, 111), or the dictionary with which to translate Western modernism to African dialects. It was partly how the locals 'learnt to speak in strange tongues'. I proposed an alternative, a marriage between the native and the modern that works in the opposite direction. We departed from the local in order to argue the possibility of a new narrative, a different form of modernism to replace the top-down approach being applied that suffused the education system and the development of the environment post-independence. With members of the Zaria Art Society, we attempted to integrate expressions of local culture in the creative process, pioneering innovations of new tools and materials that could be made locally. Ours was a methodology that could be applied across the creative disciplines as well as some technical forms of production. As art historian Chika Okeke-Agulu argues, the 'Natural Synthesis' that emerged from the Zaria Art Society 'is a form of postcolonial modernism that defeats the idea that the modernism in Africa was a mimicry of Western modernism' (Okeke-Agul 2015, 2).

Writing this I remember the recent tragedy of Bruno Latour's passing. In the latter part of his career, the French philosopher argued that an urgent call for solving the environmental crisis was to be able to depart from the local. His logic was that, in the last 30 years, the discourse of global warming, climate change and the destruction of environments was built up in the scientific realms. However, Latour argued, the political spheres had not been able to communicate the due urgency to their corresponding polities. This, he ascertained, was because of the environmental science's epistemological point of view—observing the world from without. I was glad to see that Latour advocated for a shift, trying to provide solutions to the global crisis departing from the local. What we need, he titled his book, was bringing the environmental crisis 'down to earth' (Latour 2018). Taking the metaphor literally, I believe architects dealing creatively with local soils should have the political will and audacity to offer renewed possibilities of action within the environmental crisis.

7 Conclusion: A New Culture Philosophy

To conclude, at this time of urgent reconsideration of human-earth interdependency, New Culture Studios and its underlying philosophy offer a template for the cultural rebirth of African architecture and global environmental consciousness through local engagement. Using soil as a building material is the departure point for arguing the case for adopting an intrinsically local logic. Multiple possibilities of the soil emerge, encompassing the past, the present and the future, and going beyond nature and culture.

The reflections I developed while writing this paper gave me several insights towards achieving the UN Sustainable Development Goals. Firstly, working with soil is vital to a specialist's contribution to Climate Action and Sustainability in Architecture [SDGs 11 and 13]. The carbon footprint of working with earth, and its availability as a local resource, is more promising than that of Portland cement and has the potential to play a significant part in ensuring Responsible Consumption and Production [SDG 12]. However, I find particular value in the knock-on effect of working with soil as a contribution to Social Equality [SDGs 10 and 16]. Although building materials have changed with technical development, their evolution has not been sensitive to cultural values, local aesthetics and global environmental concerns. Appropriate new materials capable of mass production and between the two extremes of traditional and conventional materials are needed for sustainable industrialisation to occur [SDG 9]. The New Culture philosophy localises responsibility for architectural development, creating regional variations that conform to culture and geography. This approach questions the appropriateness of global styles propagated by multinational standards and conventions. It demonstrates a marriage between the native and the modern can work in the opposite direction, departing from the local to argue the possibility of a new narrative. The forthcoming New Culture School for Arts and Design will be a place where architects, artists and artisans can learn, explore ideas and develop a collective body of knowledge generated from the best practices of traditional African arts and architecture. It will interpret the idioms of vernacular logic and design to the contemporary scene, drawing on elements that include physical forms for localised identity, cultural demands of the people, and locally available materials and technology.

Whether we bring the environmental crisis down to earth through active participation in a grounding politics, or take a more literal approach by working creatively with local soils to address the emergency, there is great potential for actors in the industry to provide renewed possibilities in achieving Sustainable Development Goals. We need a multi-faceted approach, working on aspects that converge through education, collaboration and culture. I hope that the New Culture's emergent new grammar, forming part of a new language, will enable the industry to mature along appropriate lines by generating new sustainable cultures for architectural practice.

References

Balfour E (1943) The living soil. Faber and Faber, London

Barrett L (1968) A studio for Ibadan. Black Orpheus 2 (1):40–41

Benson P (1983) Border operators: black orpheus and the genesis of modern African art and literature. Res Afr Literat 14(4):431–473

Denyer S (1978) African traditional architecture. Heinemann Educational Books, London

Dmochowski ZR (1990) An introduction to Nigerian traditional architecture. Ethnographica, London; National Commission for Museums and Monuments, Lagos

Fry M, Drew J (1956) Tropical architecture in the humid zone. Batsford, London

Godson L (2019) Tropical tropes. Archit Rev 1466:111

Godwin J, Hopwood G (2007) The architecture of Demas Nwoko. Farafina, Lagos

Ingold T (2006) Rethinking the animate, re-animating thought. Ethnos (routledge) 7(1):9–20

Jenner HF (2009) Soil, subsoil, priming and architectural design. In: Landa ER, Feller C (eds) Soil and culture. Springer Netherlands, Dordrecht (pp 133–145)

Latour B (2018) Down to earth: politics in the new climatic regime. Polity Press, Cambridge, MA

Logan WB (1995) Dirt. The Ecstatic Skin of the Earth. W. W. Norton & Co, New York

Moore G (1961) The achievement of 'black orpheus'. Transition (1):28

Nwoko D (1978) New culture. New culture, a review of contemporary African arts, vol 1, no 1, pp 1–2

Nwoko D (1979) New culture: review of contemporary African arts, vol 1, no 6

Nwoko D, in an interview with Ajeluorou A (2016) Nwoko: progress is not copying (...). Guardian Arts, May 1. https://guardian.ng/art/nwoko-progress-is-not-in-copying-other-cultures-be-it-technology/. Accessed 11 Mar 2021

Nwoko D (2022a) Concrete thinking by Demas Nwoko. New Culture Publications, Idumuje Ugboko, Nigeria

Nwoko D (2022b) The happy little African prince. New Culture Publications, Idumuje Ugboko, Nigeria

Okeke-Agulu C (2006) Nationalism and the rhetoric of modernism in Nigeria: the art of Uche Okeke and Demas Nwoko, 1960–68. Afr Arts 39(1):26–37, 92–93

Okeke-Agulu C (2015) Postcolonial modernism: art and decolonization in twentieth-century Nigeria. Duke University Press, London

de la Bellacasa P, Maria, (2019) Re-animaiting soils: transforming human-soil affections through science, culture and community. Sociol Rev Monogr 67 (2):391–407

Sekulić D, Tomić M, Sattler P (2020) Digging up the past: soil as archive. Archit Rev 1468(February):16–23

Stanley Okoye I (2002) Architecture, history, and the debate of identity in Ethiopia, Ghana, Nigeria, and South Africa. J Soc Archit Hist 61(3):381–396

Stengers I (2021) Reclaiming animism. E-flux J 36. https://www.e-flux.com/journal/36/61245/reclaiming-animism/. Accessed 11 Mar 2021

Williams D (1966) A revival of terra-cotta at Ibadan. Nigeria Mag 88:4–13

From Traditional (Vernacular) to Contemporary (New) Architecture: A Lesson from Southwest Nigeria

Dorcas Oluwaseyi Adeoye, Babajide Agboluaje,
Olubukola Abosede Akindele,
and Samuel Bolaji Oladimeji

Abstract

Every cultural group in the globe has its own peculiarity of its own traditional architecture, though the approach may differ from places to places. This paper looks at the vernacular architecture of the Southwestern Nigeria as practices by the indigenous Yorubas who have a peculiar climatic and socio-cultural and religious practice. The approach to architecture by this ethnic group was looked into with the intention of finding positive values in the traditional way which can be applied to the contemporary one. It was deduced that local building materials are cheap, easily to come by and are able to meet housing needs without having negative effects on the local climate and the natural environment. It was recommended that more emphasis should be given to the use of locally sourced building materials in the building industry. The paper concluded that traditional values which apply to the cultural and climatic needs of an area should be given adequate priority when contemporary architecture is the choice in order to create an eco-friendly environment to meet the needs of man.

Keywords

Traditional architecture · Southwest Nigeria · Contemporary · Building · Environment

1 Introduction

Traditional architecture is not a blind repetition of the past but increases the possibilities for artistic expression making it as relevant as contemporary architecture. Le-Corbusier, the pioneer of modern architecture, quoted that ***'the past was his only teacher'***. Sadly, little or fewer measures had been done in the documentation of traditional architecture by African Architects. There is a need to retain the settlements patterns, forms and techniques of the traditional architecture of Africa and incorporate useful characteristic into contemporary architecture. Adeyemi (2008, 8) describes traditional architecture by citing Professor Zibgniew Dmochowski, a great scholar, visionary and a researcher, to have perceived it as 'technical activity by poets'. Poetry according to him is an attribute of a true architecture patterning to a culture must grow out of its own root and analyses in its own language, if not, it loses in

D. O. Adeoye (✉)
Department of Architecture, Ladoke Akintola University of Technology, Ogbomoso, Nigeria
e-mail: odadeoye@lautech.edu.ng

B. Agboluaje · S. B. Oladimeji
Department of Architecture, University of Ilorin, Ilorin, Nigeria

O. A. Akindele
Department of Architecture, Caleb University, Lagos, Nigeria

M. R. Thomsen et al. (eds.), *Design for Rethinking Resources*, Sustainable Development Goals Series,
https://doi.org/10.1007/978-3-031-36554-6_9

worth and significance. Vernacular architecture is an expression of the people's life and traditional values (culture).

According to Elleh (1997, 19), the word traditional architecture evolves several images to different professionals regardless of their nationality as mere huts made of mud and roof with grass but there is more to it, because it reaches back to the monumental building of ancient times and construction activities of the earlier time, similarly, Umar (2008, 72) stated that African traditional architecture is an ingenuity method of construction that dates back to earlier times. An architecture that has survived through knowledge passed down from generation to generation refers to as the world's oldest and richest tradition. Similarly, Amole (2000, 17–18) termed vernacular or post-traditional architecture as a style that evolves gradually from selective borrowing from external source embraced by the community, resulting in changes in the traditional forms, morphology and material but the pre-cursor traditional dwellings are sustained. Vernacular architecture, according to Osasona and Ewemade (2009), is a mixture of traditional architecture and external influences that emanate from the political and economic forces. Although there is no evidence of trained professional or apprentice but there still exist usage of local materials and traditional design. However, it is rather hard to discuss the diversity of African architecture due to conflicting publications, African traditional architecture is a collective idea or techniques defined by social status and religions and has been passed down through oral instructions, perfected over many generations to suit the need of an individual or group, and as a result, no trained builder or architect is evidence in the building process and feast is the only remuneration for the assistant renders by the community.

Traditional houses don't only house its owner but express the stage and status. It is a direct invocation of the physical environment and conforms to his or her societal needs, accommodates different traditions orientations and climatic conditions. In the same way, Opaluwa et al. (2007, 98) asserted that African architecture construction approach not scientifically documented does not mean it fails to please situation such as usage, suitability and luxury. Hence, the traditional building cannot be under rate to be short of architecture, it has culturally evolved to fulfill aesthetics, comfort and sustainability. The vernacular architecture of Nigeria can be given a description of the building material, forms and techniques leading to traditional forms of architecture especially with respect to the architecture of the three major ethic groups. The vast ethnic groups in Nigeria include southwest zone (Yoruba), northern zone (Hausa) and the eastern zone (Igbo). Therefore, architectural forms within this context are tied to different ethnic cultural practices.

Over the years, many occurrences have led to the changes in the traditional structures of regions in Nigeria. These include: British colonization and the return of repatriated slaves from Brazil and North America (Afro-Brazilian/ Brazilian architecture) in the southwest regions, Islamic influence in the northern region (the mosque, palaces and city gates, ornamentation and house form) and colonialization or westerners which abolished the existing traditional beliefs and establishing principle of individualism in the eastern region. Today, indigenous Nigeria's traditional architecture is becoming outdated; Non-Africans that have produced measures of its revival are mostly from non-architectural perspectives, their background made it agitated for them to capture the socio-cultural and socio-spatial importance of African built-form. The need for a revival of traditional architectural value to be blend with contemporary architecture is crucial and of great importance, according to William Morris (1889), that historic building represents the true concept of history and that 'our past is part of our present'.

This paper is focussed at understanding the rationale behind the traditional dwellings and significant motives behind the construction of the '*Yoruba*' southwest traditional architecture, in order to explore the possibility of incorporating useful traditional architecture characteristics in the contemporary architecture. The aim is to examine and analyse distinct characteristic features and factors (physical, environmental and cultural aspects) adopted in the formation of geometric forms of traditional Yoruba architecture in southwest, Nigeria. In that sense, this research will help to reduce substitution of Nigeria's creative traditions and heritage cultures with external styles and explores the possibility of incorporating its characteristics into the contemporary architecture to evolve new construction techniques that does not neglect various considerations on culture, climate, security, privacy and communal living.

2 Study Area

2.1 Nigeria

Nigeria lies between Latitudes 4° 16′ and 13° 53′ north of the Equator and Longitudes 2° 40′ and 14° 41′ east of the Greenwich Meridian Line. It is bordered by Niger Republic in the north, Chad in the northeast, Cameroon in the east, Benin Republic in the west and to the south; it is bordered by approximately 850 km of the Atlantic Ocean, stretching from Badagry in the west to the Rio del Rey in the east. Its total land area is 923,768 km^2. Nigeria is the fourteenth largest country in Africa. The country came into existence as a nation-state in 1914 through the amalgamation of the Northern and Southern protectorates. Nigeria became fully independent in October 1960. It is made up of 36 administrative states and a Federal Capital Territory (FCT) (Fig. 1).

The country is entirely in the tropics. The climate is semi-arid in the north, gradually becoming humid in the south. Overall relief is gentle, but there is a gradual loss of height from the north to the coast. The climate of Nigeria is classified into two. These are the Southern part of the country, where the Igbos and Yorubas are found is mostly hot and humid and has a high annual rainfall of between 1500 and 2000 mm a year. The Northern part of the country, where the Hausa predominate, is characterized by hot, dry climate and extremes of temperature between day and night. Rainfall is minimal and often less than 500 m per year. The vegetation is made up of mangrove swamp forest in the Niger Delta and Sahel grassland in the north resulted from the varying climatic condition. The three major ethnic groups in the country are Hausa, Yoruba and Igbo.

Meanwhile, Nigeria is the most populous country in Africa and the tenth in the world. The 2006 Population and Housing Census puts Nigeria's population at 140,431,790, with a national growth rate estimated at 3.28% per annum. With this the current population of Nigeria is estimated to be over 200,000,000 million people. Nigeria is a multi-religious country made up of two predominant religions, i.e. Christianity and Islam, while the population is divided roughly in half between these two major religions.

2.2 The Southwest Zone, Nigeria

The Yoruba occupies the southwestern part of Nigeria comprising of six states, namely, Ekiti, Lagos, Ogun, Ondo, Osun and Oyo. They practiced Islam, Christianity and traditional worship. According to Dmochowski (1990, 1–5), demographically, Yoruba lies within the tropics with high temperature and humidity and remains the second largest ethnic groups in Nigeria. They are subsistence farmers as similar to the Hausas, they cultivate the land only for family consumption and the need for more hands-on farm lead to them having many wives to have more children, as farming to them is considered a way of life and not a means of earning (Fig. 2).

2.3 Describing Tradition Architecture in Nigeria

2.3.1 Nigerian Traditional Architecture

The history and development of traditional architecture have been opined as a way of life that has given way to the introduction of new building techniques and expressions through

Fig. 1 The three major zones of Nigeria. *Source* Authors' compilation (2021)

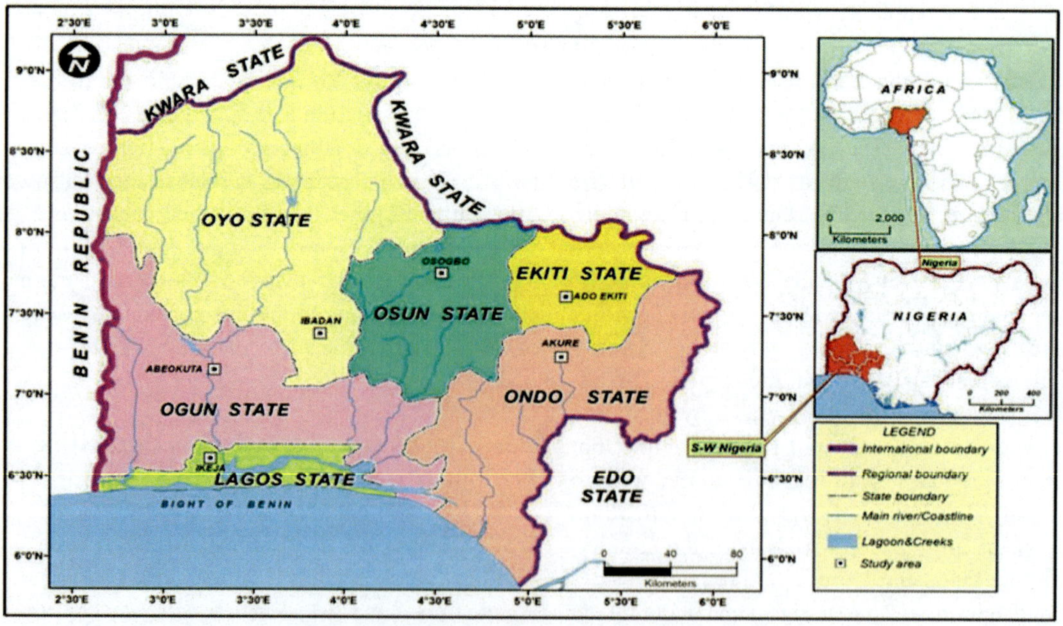

Fig. 2 Southwest Nigeria within the national contest. Adeoye (2018)

adaptation and rigorous study of the environment. Nigerian traditional architecture is similar in many ways, reflecting the different culture and traditions of the people, as well as the climatic factors. This diversity ranges from Hausa traditional earth architecture of wattle and daub walls in the North, to the thatch roof houses of the South, as well as the eighteenth-century royal houses or palace in the eastern parts of Nigeria.

2.3.2 Yoruba Settlement Patterns

The factors influencing the construction of dwellings of Yoruba include socio-cultural factors, spatial characteristic of the building, extended family, materials and previous laid down construction methods. In the year 1990, Ralph assumed the Yoruba Town as a 'wheel or circular in shape. He observed that the Yoruba Town radiated from the Oba's palace. This could be found in the Palace of Oyo located at the centre or hub (core) of the town and links other parts (see Fig. 3).

3 Methodology

The aim of carrying out a research is to add to the existing body of knowledge in a particular area of research. This can be achieved through the processes of identifying problems, defining them and refining it, develop hypothesis and theories formulation. By these, new approaches are developed and possible solutions are suggested. However, the process can involve the use of strategic and systematic methods to arrive at a logical solution and develop a problem-solving problem (Rajasekar et al. 2006).

This study adopted the exploratory research method to collect data from existing publications on the research from journal articles, books of related topics, reports, personal observations and case studies. The case study was carried out by focussing on the selected traditional dwellings in cities of Ogbomosho, Ile-Ife and Ibadan among other towns located in Southwest zone of Nigeria. Hence, a lot of site appraisals and descriptive

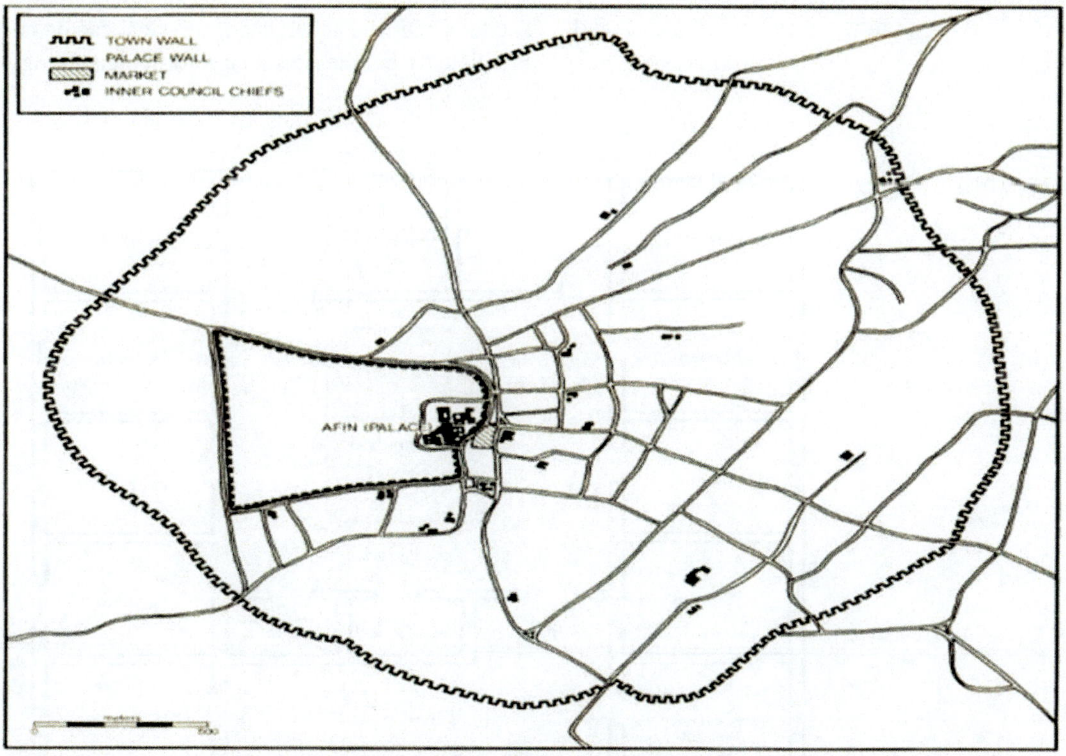

Fig. 3 Typical circular plan of Yoruba City settlement pattern. *Source* Ojo (1996)

survey were done to select houses within the study area. In this case, qualitative analysis was used, in which sketches of the floor plan and elevation were produced with a means to describe the built-forms cutting across the physical and cultural attributes of the selected buildings. The fieldwork focuses on traditional compound houses and Afro-Brazilian/Brazilian. This was considered in order to study the values of traditional architecture and gives the influence of Afro-Brazilian/Brazilian style on the contemporary architecture of Southwest Nigeria with a view to integrated into the proposed design concepts.

4 Case Studies

4.1 Yoruba House Form—Creating Factors

Yoruba house form can be classified into three (3) categories on the basis of the complexity, namely, commoners compound, Quarters for the Chief and Afin (traditional palace) (Plates 1 and 2).

Commers compound is the genesis of others and is usually headed by 'Baale' (compound head). While quarter for the chief is of the same pattern as those of the ordinary extended families with fairly large courtyard depending on the chief's rank in the traditional government. 'Afin', traditional palace, is the most complex among the three classes. It is usually designed to conform to the status of Oba (King) of a town as the arch-priest of the living members of the society. He needs an utmost security as he is regarded as unseen, unheard and untouched except on few occasions when he performs his spiritual and political duties.

In line with the above 'Afin' design has to take into consideration enhancement of cultural values, norms and effective social interaction among the members of the lineage. In support of these, Afolabi Ojo wrote in one of his books that traditional Yoruba 'Afin' housing unit is a compound type (akodi), the architecture of which fully takes into consideration of circumstances of life. 'Afin' is also influenced by the role of Oba as the social head and his spiritual leadership functions. On the other hand, physical factor is classified to be the secondary form generating

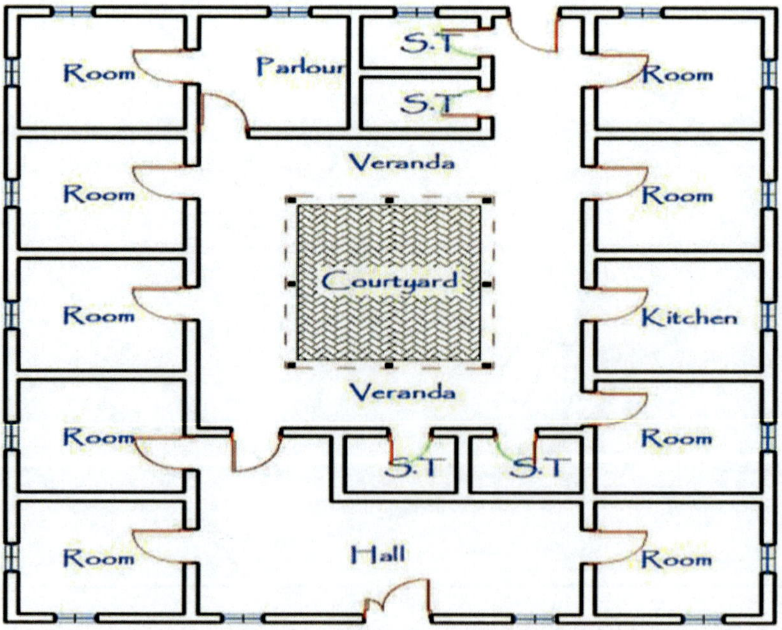

Plate 1 A commoner's compound. *Source* www.pinterest.com

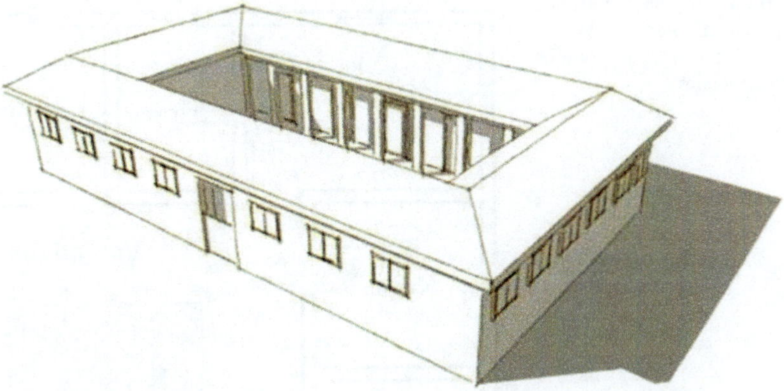

Plate 2 Quarters for the Chief. *Source* www.pinterest.com

forces to which the house responds through the use of certain materials, building structure and its orientation (Plates 3 and 4).

4.2 Materials for Construction in Yoruba House Form of Southwest Nigeria

Yoruba houses in southwest Nigeria are constructed using the limited available resources of the environment at their disposals; the soil, forest trees and the grasses of the savannah are used; the adoption of these materials is postulated by their traditions, religious activities, durability and location. Generally, materials selection is characteristic with belief and norms attributed by religious and lineage taboos, some materials are considered unwise to use due to believed that it brings bad luck to any household that uses them. Yoruba traditional architecture is constructed locally using sourced materials in abundant, this material includes adobe or clays of laterite soil used in walls and floors, wood used in making doors, windows and support, juice extract from locus beans used for wall plaster, thatch is use on roof and mud plastered with lattice palms are used as ceiling materials to prevent fire. The local materials for roofing consist of wooden poles of varying sizes, commonly used as girders, beams, rafters and joists.

Furthermore, the choice of mud is due to the high resistance to erosion, weathering process, climatic effect of mud that makes the interior space to be cool and dark all day, while the choice of roofing depends on the hardness, durability and high-resistance insects. The materials used in traditional Yoruba house and its properties are listed as follows:

1. Earth: This is the most popular traditional building material used in walls and sometimes in roofs, as stated by Srinivas (2007), earth has great values and regards as the most sustainable and ecologically balanced materials. Its greater advantages is the completely recycled properties that don't leave any environmental footprint in comparison to other construction materials.

Likewise, Dobson (2000, 3) earth has a good thermal, sound insulating and fire-resistant properties, this property necessitates its use in the tropics. They moderate between the interior and the exterior humidity and temperature and have been regarded as having the ability to breath. Table 1 summarizes the attribute of earth materials (Plate 5).

The common example or method of the usage in walls includes adobe blocks (mud bricks), rammed earth, pressed bricks and cob. Definition of each method listed above is explained below.

i. **Adobe Blocks:**: It's also known as mud bricks, an air-dried brick made from foot kneaded earth mix formed in a mould; its constituent includes clay and silt and sometimes contains stabilizer and straw.
ii. **Rammed Earth**: It's made from damp or moist soil, with or without stabilizer that is

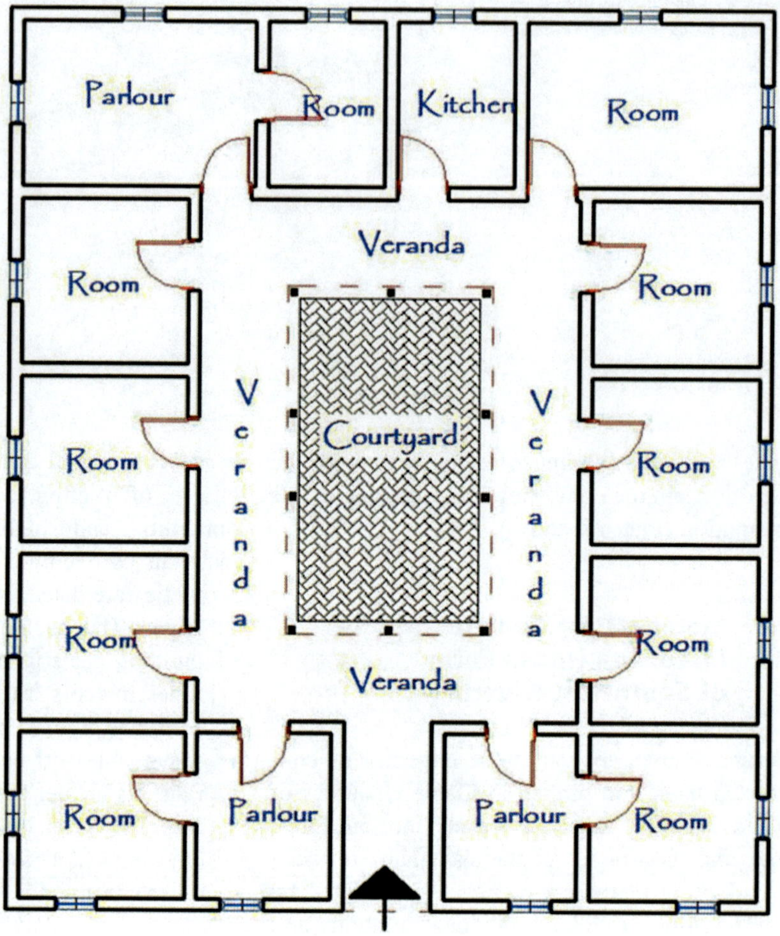

Plate 3 Yoruba architecture with Islamic Arches (Afin, Oyo, Nigeria). *Source* www.pinterest.com

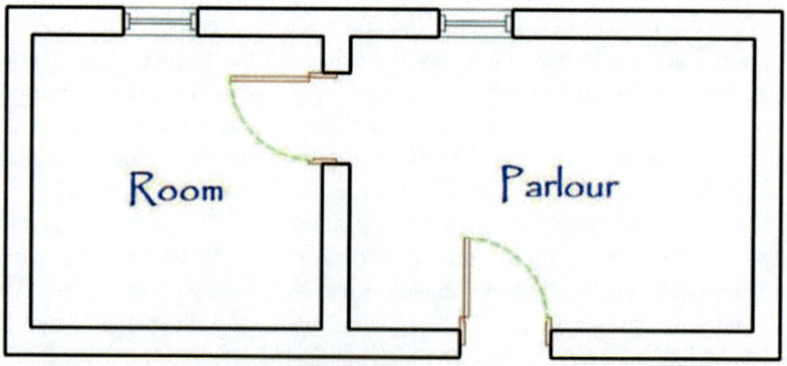

Plate 4 The Palace of Ooni of Ife, Nigeria. *Source* www.pinterest.com

Table 1 Attribute of earth materials

Attributes of earth	Evaluation
Authenticity	Traditional
Compressive strength	Good
Heat retention	Good
Water resistance	Poor
Torsional strength	Poor
Cost	High
Design flexibility	Good
Insulation	Poor
Acoustic separation	Good

Source Authors' compilation (2021)

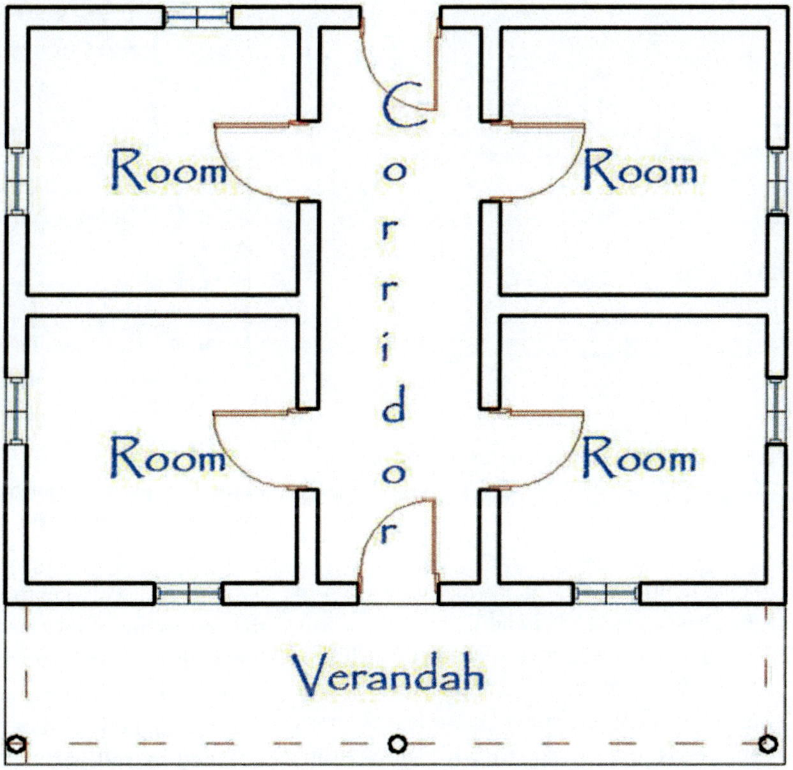

Plate 5 Bricks made with earth materials. *Source* www.desertphile.org

kneaded using temporary movable formwork, it has good thermal properties as compared to other examples.

iii. **Pressed Earth Bricks**: It's an earth brick formed using a machine-operated or hand-operated mechanics.

iv. **Cob**: It's an ancient technique used in the erection of monolithic walls, this involves using of moist earth mix with straw and has similar thermal properties with adobe and rammed earth bricks. The processing involves mixing the subsoil with clay and straw or other fibrous materials to stiffing the mud formed into small loaves. This is thrown to the builder during construction and mashed together to form a monolithic wall that is built on a solid foundation of either stone or concrete, each layer during

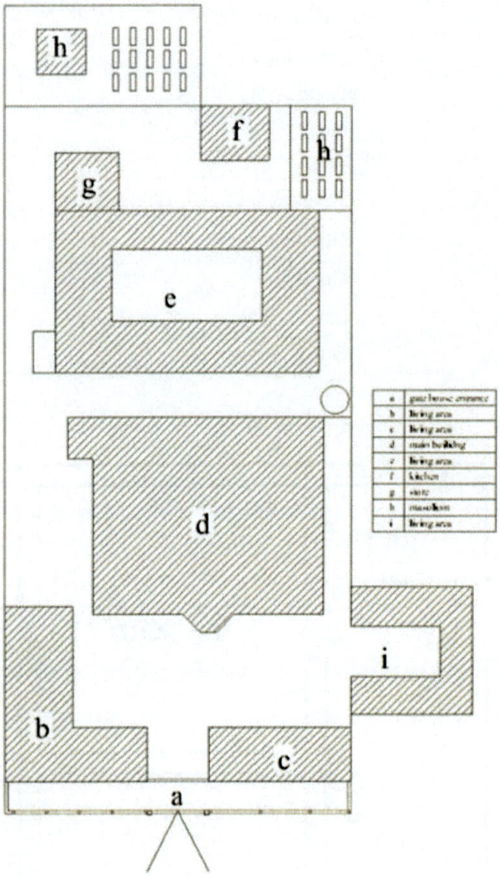

Plate 6 The example of thatch materials used in roofing. *Source* www.mastergardenproducts.com/thatchroofingand-more

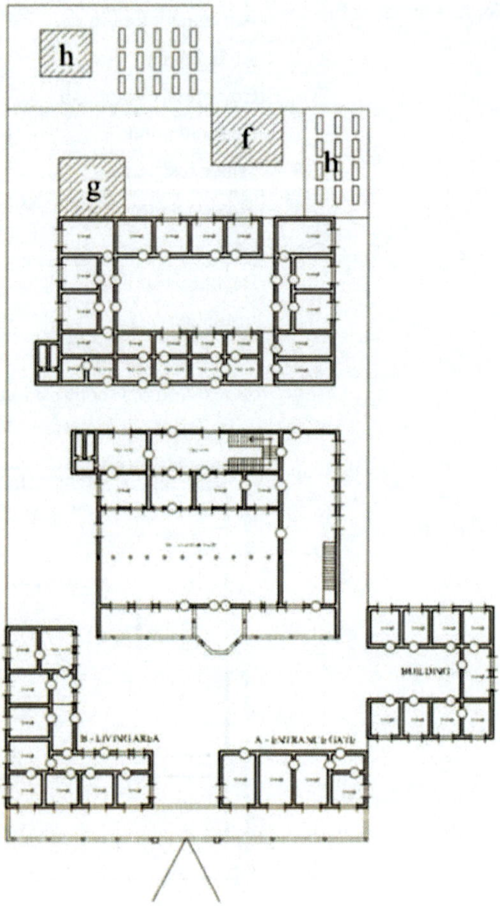

Plate 7 A thatched roof building. *Source* www.bbc.co.uk/blogs/africahaveyoursay

construction is left to be sun-dried before the successive layer is laid.

2. **Thatch**: According to Kennedy (2008), hatch is a generic name for vegetative material, including straw, palms leaves, grass and reeds, used in roofing. It's an environmentally friendly material and can be breath; this reduces its susceptibility to rotting. As seen in Plates 6 and 7, respectively.

Normally, an accurately constructed thatch roof can withstand up to 60 years. Thatch roofs are weather resistant if well maintained doesn't absorb a large amount of water; it is a natural insulator with pocket of air on grasses found on their stems. This characteristic gives a building the insulating ability needed to insulate a building in both warm and cold weather. It's a versatile and flexible material and an organic material and this makes it to get decomposed and attacked by rodents.

Wood: Wood is a strong, easily processed and beautiful building material. It has been regarded as a non-maintainable material in its purest state unless on few cases when treated. Timber is biodegradable and by no means harmful to the environment, it is an elastic material and incredibly strong while bending and flexible under load.

4.3 Procedures for Construction of Yoruba House

As stated by Osasona (2009, 11–13) construction of Yoruba dwellings is mostly a community issue, the best season selected for construction is November to March of each year. Ojo (1996, 147) indicated that during this period the rains began to cease and coincide with when there is reduced farm work resulting into more hands available for construction of shelter than food. The water needed for the construction are gotten from streams and puddles and the continuous prospect of sunshine helps the mud bricks to be used for construction in this period to dry easily as they are piled up on each other. The construction of shelter begins by the clearing of the proposed site and digging of huge amount of laterite from a deep pit located close to the site.

The first approach to construction regarded as slow is explained by Osasona (2009, 11–13) that after an adequate quantity of laterite earth has been excavated, further processing is puddling done by adding pressure using the underfoot as water is added simultaneously, this done until a malleable and homogenous state of soil is achieved. After puddling, the processed earth is covered with leaves and left to cure for few days, before additional water is added and massaged once more, the next stage is the construction of the shelter, creepers and vines, branches are used for the setting out and building configuration is marked out, the walls are built in about seven layers of swish-mud, each layer is roughly one foot and one-foot thick as found mostly in traditional Yoruba houses. The earlier layers are left to dry before the successive course is added in a means for it to be adequately hardened to bear the weight of the next layer due to wet earth not able to support its own weight. The window and door positioning are reserved as the construction is done.

The second approach is messaging the laterite earth with foot after excavation and it's moulded into bricks using rectangular moulds and consistently left to sun-dried for 2 weeks. The sun-dried bricks are stacked together to form a kiln and set on fire to burn for a couple of days, this is done for increasing the strength and compactness.

Furthermore, the bricks are later laid in course and joined with mortar until the desired height is achieved. The roofing and thatching materials to be used are readily available due to the period of the construction when the trees are adequately vegetated. The common materials used for roofing includes the coconut or fan palms or other termite-resistant trees are used as a rafter, purlin and cross beams. The roofing is completed mostly within the intervals of 2 or 3 days. Ojo (2005, 147) stated the most important part of their building is the wide verandah, supported with carved post spaced at intervals to each other used for numerous form of activities and shaded by the cantilevered eaves from the sun. It is devised for activities such as sleeping and access to rooms, privacy and storage of goods this afforded by the saddled back construction easing the use of ceiling as storage. The saddle-back roof in most rectangular compounds of the Yoruba's house which helps to connect or collect rain water into the courtyard that is collected in jars and surplus water is drained out through a hole beneath the hallway of the entrance door.

4.4 Building Types

The building types found in different geographical regions in Nigeria are caused by factors such as regional difference, culture, owners' status, ethnic diversity, socio-cultural activities, western influence, and indigenous traditional beliefs and newly introduced one, such as Islam and Christianity. However, the beginning of technology, slave trade by the British colonials, the establishment of Islam and Christianity beliefs and improved skills of workmanship lead to more diverse building types found in geographical locations in Nigeria. Examples of indigenous traditional house found among the Yoruba are Farmhouses, compound house and rooming house, custom built due to their polygamous nature and large family.

4.5 Compound House

This can also be referred to as traditional courtyard house characterized with one or multiple courtyards that defer on the ethnic built-form approach, the building knowledge is passed down from generations and practiced due to the need to accommodate large family and multi-habitation. This building type is devoid of ornamentation except few pattern found on timber window and doors, evidence of materials used includes mud bricks, thatch and corrugated iron sheets. An example of the compound house is shown in Figs. 4 and 5.

The Use of Courtyard

This is an important aspect of regional traditional architecture in Nigeria and exists among the three ethnic groups.

The courtyard has been devised culturally by traditional pre-cursor for unity, communal living, security, socio-cultural activities, privacy, seclusion and a means to curb crimes and insecurity. Furthermore, Adedokun (2014, 42) stated that the courtyard to the Yoruba is (Agbala) an open space within the compound, devised for entertaining visitors, rearing of domestic animals, cooking and collection of rainwater, sleeping, night story tales area for children and ease of extension for newly married men (see Fig. 6).

4.5.1 Farm and Tent House

This is a temporary house form found among various ethnic groups in Nigeria, the nomadic Hausa-Fulani device the tent house made with shrubs or grass and tree trenches, while Yoruba made theirs with mud and roof with shrubs, as shown in Fig. 7 and Plate 8.

4.5.2 Brazilian House

This can also be referred to as the rooming house, it is an influence that dominates southwest part of Nigeria, shaping the construction technique and high ornate features of building in Lagos Island and some parts in Ogbomoso Township at the turn of the twentieth century. Major changes occur after the repatriation of African slave from Bahia in Brazil, Cuba and other Latin American countries.

According to Osasona and Ewemade (2009, 60–61), the two groups of repatriated slaves that settled in Lagos Island are the Agudas (slaves repatriated from Brazil and other Latin America) and the Saros (returnees from England who first settled in Freetown, Sierra Leone and later moved to Lagos Island). The Agudas were

Fig. 4 A typical Yoruba compound house layout. *Source* Author's compilation (2021)

Fig. 5 A model of a traditional compound house type. *Source* Authors' compilation (2021)

Fig. 6 A typical Yoruba compound house showing the courtyard location. *Source* Authors' compilation (2021)

trained craftsmen who designed specialized Hispano-Portuguese style of building, commonly referred to as Brazilian Architecture, characterized with a complex roof, and highly ornate craftsmanship portraits on the column, staircase, balustrade and dormer windows. They also feature elaborate forms and large space verandahs covered by projected roofing supported with the aid of timber post placed at a respective distance apart. While the 'Saros' were smart businessmen

Fig. 7 A typical Yoruba farm house plan. *Source* Authors' compilation (2021)

Plate 8 A typical Yoruba farm house. *Source* Osasona (2007)

that popularized the construction of two-storey building archetype, their adopted style is the British Colonial building practice, distinguished by the use of timber framed and boarded houses frequently with en-framing verandah with features like, carved fascia, attics, dormer windows and broad timber framework at the eaves (Plate 9).

This building style was favoured and widely accepted because of its embellish decoration which has common characteristics with art that was primarily practiced by the Yoruba builders. Another reason for its acceptance is the non-imposing lifestyle and the introduction of a central hallway in the middle that can be likened to a compressed courtyard with the possibility of using local materials for its construction. Studies into resemblances in mythical representation between the Yoruba and Brazilians showed that Yoruba language is spoken in some part of Cuba and Brazil. Marafatto (1983) describes five examples of Brazilian House. These include

Plate 9 Brazilian house. *Source* www.pinterest.com

The Detached House or Bungalow: This idea *is copied from early Portuguese* bungalow house. The space arrangement is often rooms planned *along with a central hallway* (see Fig. 8 and Plate 10).

1. ***The Upper-Floored House***: They are mostly two floors or more and the staircase made of timber is placed at the end of the central hallway as mostly found in Brazilian houses. The spaces on the ground floor *are sometimes used for commercial and residential use, while the rooms on the upper floor spaces are routine for sleeping and relaxing at night (see Plate* 11*).*
2. ***The Modified Upper-Floored House*** : The significant qualities of this type are the lavish ornamentation and decorations done on its exterior, the external positioning of the staircase and inclusion of galleries further differentiate it. This house type portico *is always levelled with the road (see Plate* 12*).*
3. ***The Typical Upper-Floored House***: The ground floor plan and the first floor have typical floor plans and this is leading to the presence of

Fig. 8 A typical Yoruba rooming house (typical of Portuguese house) layout. *Source* Authors' compilation (2021)

Plate 10 An Afro-Brazilian (rooming) house at Ipetumodu, Nigeria. *Source* Authors' compilation (2021)

Plate 11 A Brazilian upper-floored house in Osogbo, Nigeria. *Source* www.pinterest.com

Plate 12 A modified upper-floored house, Nigeria. *Source* www.pinterest.com

external balconies above, the staircase is located at the hallway end close to the back door of the building as seen in Plate 13.

4. ***The Complex Upper-Floored House***: This technique is recent of all and mostly linked to the eighteenth and nineteenth-century Brazilian house, the internal space organization is complex and the use of rich ornate on windows and ribbed pillars and stuccoes are found under this example, as shown in Plate 14.

5. ***Afro-Brazilian House***: Osasona (2007, 61) described Afro-Brazilian house as a progressive form of Brazilian style or a technique derived from the marriage of cultural heritage

Plate 13 A typical upper-floored house, Nigeria. *Source* www.pinterest.com

Plate 14 A complex upper-floored house in Ogbomoso, Nigeria. *Source* Authors' compilation (2021)

with Brazilian style. It is less expensive and easily replicated due to fewer ornamentation and simplicity of model similar to the precursor buildings. The style becomes a noticeable building typology in Lagos Island and progressively to the hinterland of other towns.

According to Vlach (1984, 12), the major foremost builders of the Afro-Brazilian House are poor relatives of rich repatriated 'Saro' that have become wealthy from trading in Brazil (see Plate 15).

4.5.3 The House of Adebisi Giwa of Idikan, in Ibadan Nigeria

The mansion of Adebisi has been portrayed in the Ibadan worldview as a structure like the Mapo Hall in grandeur, elegance and splendour (Adeoye et al. 2018). The prevalent belief at the time the structure was built was that materials for the construction of the building could only have come from Europe. The building demonstrates the importance attached to the brilliance of human creativity (Plate 16). The building

Plate 15 An Afro-Brazilian house, Nigeria. *Source* www.pinterest.com

Plate 16 The Approach View of the house of Adebisi Giwa of Idikan, Ibadan, Nigeria. Adeoye et al. (2018)

symbolizes the wealth of Adebisi as cultural metaphor and a significant historical connotation for the Oyo Yoruba groups in Southwest Nigeria. The building takes on a dual role of residence and court in the manner that contest with the loftiest abodes of Ibadan royals (Figs. 9 and 10).

The annotations as shown in Figs. 9 and 10 are: a. Entrance façade; b. Living area 1; c. Living area 2; d. Main Building; e. Living area 3; f. Kitchen; g. Store; h. Cemetery/mausoleum.

4.6 Transformation in Yoruba House Form

There are a lot of transformation that had existed over the years due to intervention of colonialism, technology improvement, new means of transportation, the evolution of new styles by immigrants, new materials and a boost in money economy leading to changes in the social status of individuals and taste in the choice of house form. These listed factors contribute hostile result on the Yoruba kinship and family authority.

Fig. 9 Site layout of Sanusi Adebisi house. *Source* Adeoye et al. (2018)

Fig. 10 Ground floor plans of Sanusi Adebisi house. *Source* Adeoye et al. (2018)

1. Influence and Effect of Colonialism: This comprises missionaries, colonial administration and repatriated slaves from Brazil and Latin America. The Yoruba house form has been weakened as a result of socio-economic contact with the western world. This includes usage of imported building material, new architectural styles and religion as it collectively weakens the socio-cultural factors. The Christianity idea of one man and one wife suggested the building of a new individual house in distance to the lineage compounds. Other factors include western education, intra-ethnic war and migration that emerges by new means of transportation.

2. Boost in Money Economy: In the year 1990, Nigeria experienced a boost in economy due to the exportation of cocoa, palm oil and palm kernel, this created good incomes to farmers and many individuals and initiated the sprinkling of new (foreign) designs and use of new materials. The indigenous materials are termed inferior while the new incomes give rise to distinctive taste among members of the lineage willing to improve their portion of the compounds to taste. This contributes unpleasantly result on the kinship and family pedigree.

3. The Invention of New Building Materials: According to Mabogunje (1969, 118), the American Baptist missionary replaced the mud wall with baked clay. The use of cement to plaster was used on walls and floors. The mid-nineteenth century witnessed the introduction of corrugated iron sheets used for roofing. The introduction of these new materials although weakened the cultural values but assisted the buildings to withstand tropical climate and making it more durable.

5 Conclusion

Traditional (vernacular) architecture has undergone numerous changes over the years as a result of modernity which has led to the emergence of contemporary approaches to building construction in Nigeria. However, the values of vernacular nowadays are still very relevant to our contemporary buildings. The advantages of vernacular architecture are numerous and are not limited to the use of eco-friendly material to accommodate local climatic conditions. The use of adobe as a traditional building material characterized with its natural thermal regulating qualifies and no detrimental environmental effects is a good example here. Most of the modern building materials are not eco-friendly compared with the local building materials causing a lot of threats to the natural environment especially with the issue of climate change and global warning. Thus, there is need to revise the use of our locally available building materials. These materials could be improved upon and made to be functioning in the same capacity just as the modern building materials and with no harm to the environment. Creative means of expressing culture needs to be encouraged since cultural practices and values define the practices of traditional architecture in Nigeria which is absent in the contemporary architecture. The use of yet attractive traditional wall finishes, for instance, can be applied in the contemporary buildings rather than the expensive ones with no cultural expression being used. The introduction of courtyard system helps to resolve ventilation problem in buildings. Preserving these forms and styles will promote cultural heritage as well as be an economical way of saving cost of building construction.

References

Adedokun A (2014) Environmental and adaptation in architecture planning and building design: lesson from the forest region of West Africa. Brit J Environ Sci 2 (1):9–20

Adeyemi EA (2008) Meaning and relevance in Nigerian traditional architecture: the dialectics of growth and change. Covenant Univ Publ Lect Ser 1(21):9–20

Adeoye DO, Akande A, Oladiti AA (2018) Heritage architecture in Ibadan, Nigeria: THE HOUSe of Adebisi Giwa of Idikan. J Art Archit Stud (JAAS) 7 (1):11–20

Amole SA (2000) Yoruba vernacular architecture as an open system. Legacy 2(2):17–18

Culture of Nigeria—history, people, clothing, traditions, women, beliefs, food, customs, family (2017) Everyculture.com. http://www.everyculture.com/Ma-Ni/Nigeria.html. Accessed 3 Jan 2017

Dmochowski ZR (1990) An introduction to nigerian architecture—South-West Nigeria. Ethnographical Ltd. London 1:1–5

Dobson S (2000) Continuity of traditional: new earth building. Terra, University of Technology, Sydney, Earth research Forum, p 3

Elleh N (1997) African architecture; evolution and transformation. The McGraw Hill Companies, New York, pp 19–22

Justine MC (1983) The art and aesthetics of the Yoruba: African arts. UCLA James S. Coleman African Studies Center, vol 16(2), pp 56–59+93–94+100. http://www.jstor.org/stable/333585. Accessed: 29 Nov 2016

Kennedy JF (2016) An overview of natural building techniques. http://www.strawhomes.com. Accessed Nov 2016

Mabogunje AL (1969) Urbanization in Nigeria. Univ, Of London Press, London

Marafatto M (1983) Nigerian Brazilian houses. Istituto Italiano Cultura, Lagos

Ojo GJA (1996) Yoruba culture: a geographical analysis, the university of Ife and University of London Press Ltd, pp 147–149

Opaluwa E, Obi P, Osasona OC (2007) Sustainability in traditional African architecture: a springboard for sustainable urban cities, sustainable futures. Archit Urban Glob South Kampala, Uganda 201:27–30

Osasona CO, Ewemade FO (2009) Upgrading Ile-Ife's vernacular architecture heritage. WIT Trans Built Environ 109:60–62 © 2009 WIT Press. www.witpress.com. ISSN 1743-3509

Rajasekar S, Philominathan P, Chinnathambi V (2006) Research methodology,Ar XIV Physics. Retrieved from the web 14th March 2012, from https://arvix.org/abs/physics/0601009

Srinivas N (2007) Most dependable building material. Deccan Herald e-paper

Umar KG (2008) Transformation in Hausa traditional residential architecture: a case study of some selected parts of Kano metropolis between 1950–2005. PhD thesis, architecture, ABU. Zaria

Vlach J (1984) The Brazilian house in Nigeria: the emergence of a 20th-century vernacular house type. J Ame Folklore 97(383):3. https://doi.org/10.2307/540393

India's Informal Reuse Ecosystem Towards Circular Construction

Deepika Raghu and Catherine De Wolf

Abstract

As countries continue to develop, the amount of construction and demolition waste generated is exponentially increasing. There is an urgent need to move from linear, take-make-waste systems to more circular systems that extend materials lifespans. India is known for its material reuse and recycling businesses. Currently, most of these businesses are run locally and do not involve any government registration but they significantly contribute to the economic growth of the country. Despite the importance of this informal sector, there is still not enough understanding of how they operate. This paper examines the processes and people involved in reusing construction elements in Mumbai and Bangalore, and their current models of resource procurement and trade. Field visits and interviews were conducted to understand through whom materials are transferred in the informal ecosystem, what types of materials and quantities can be currently procured in informal supply chains, what the storage practices are, and how reusable construction elements are retrieved and processed. A qualitative content analysis method helped to understand the informal material reclamation processes in India. Findings show that although a robust informal material reuse ecosystem exists across India, organisation and governmental policies are needed for effective contribution towards sustainable development goals and a circular economy.

Keywords

Circular economy · Informal ecosystem · Material reuse · Resource consumption

1 Introduction

A circular economy (as opposed to a linear one) has no net impact on the environment and avoids the generation of waste in the production process. Its descriptive meaning relates to the concept of a cycle: the biogeochemical cycle and the idea of recycling products (Ellen McArthur Foundation 2010). Although the terminology may be relatively recent, the phenomenon of a circular economy has roots in ancient history. Romans sorted trash in Pompeii (Alberge 2020), glass was recycled in the Byzantine times in the city of Sagalassos, Turkey (Degryse et al. 2006) and copper, bronze, and iron tools were refashioned from broken ceramics in the Persian Gulf (Hoffman 2014). For millennia, a robust reuse

D. Raghu (✉) · C. De Wolf
Department of Civil, Environmental and Geomatic Engineering, ETH Zurich, Zurich, Switzerland
e-mail: draghu@ethz.ch

M. R. Thomsen et al. (eds.), *Design for Rethinking Resources*, Sustainable Development Goals Series,
https://doi.org/10.1007/978-3-031-36554-6_10

culture has also existed in India. *Juggad*, the Indian practice of using limited resources with ingenuity, brings forth creative innovations in an economy of scarcity. This can be seen in local markets, where old cotton saris are upcycled into handbags; in food establishments where items are only delivered in reusable plastic cartons, or even in repair cafés, where busted appliances, worn-out clothes, damaged jewellery, run-down garden equipment or broken furniture are given a new life.

In the context of the built environment, circular methods of construction have long been prevalent. Despite this historical phenomenon, documentation of reuse in South Asia remains limited (Patel 2009). Salvaged materials today include doors and windows, glass panels, partitions, furniture, bathroom fixtures, pipes, electric equipment, and wires. These reuse and recycling businesses are run locally by the informal sector and do not involve any government registration or legal authorization.[1] Such small- and mid-size enterprises (SMEs) significantly contribute to India's economic growth (Bagale et al. 2021), yet there is still insufficient information on how they operate and their potential impact on India's overarching circular economy goals. The informal workers effectively contribute to the realisation of the Sustainable Development Goals (SDGs) (UN 2016). Yet, policy-makers have ignored the knowledge and capacities of informal workers. The value of their work is often overlooked because informality challenges mainstream assumptions of what work should look like. A redistribution of resources and power is necessary to address social and environmental inequities. As SMEs quickly adapt to digitalization, India's digital boom also needs to be leveraged to address issues such as the absence of technology adoption by the informal material reuse economy.

The circular economy is a holistic concept, with its practices extending far beyond material use, and its impacts extending beyond the environment (SDG Knowledge Hub 2022). This paper examines the processes and people involved in the reuse of construction elements in two Indian cities, Mumbai and Bangalore, and their current models of resource procurement and trade. It explains through whom materials are transferred in the informal ecosystem, material types and quantities that can be currently procured in informal supply chains, what the storage practices are, and how reusable construction elements are retrieved and processed. The contribution of the informal reuse economy towards the SDGs is also outlined. This could aid in policy-making and research towards sustainable cities, where informal material reuse economy entrepreneurs are at the forefront.

2 Background

In recent years, India has become a pivotal voice for positive climate change negotiations. The Indian Government ratified the Paris Agreement in 2015 to ensure resource efficiency and a circular economy (Ratha 2020). Other initiatives such as Atmanirbhar Bharat Abhiyaan, a campaign to lower dependence on imports and foster local production, also promote efficient resource management ("Aatmanirbharbharat", 2020). More recently, India's Prime Minister, Narendra Modi also launched "Mission Circular Economy 2022–23" which recognises that both developmental and environmental goals need to be aligned and cannot be dealt with exclusively (Modi 2022). India's first construction and demolition waste management regulation was passed in 2016 (Central Pollution Control Board 2017), which outlined measures to be taken by contractors, recycling facilities, waste generators, and state governments. It does not, however,

[1] According to the International Labour Organization (ILO 1993, 2003, 2013), workers are considered part of the informal sector if they are (a) self-employed people in business activities that are not registered with any national authority and maintain only partial or no accounting records; (b) employees whose employer does not make social insurance contributions for them or provide them access to paid annual leave and paid sick leave; or (c) contributing family workers: people who work without pay in a business or farm owned by a family member or have informal jobs by default due to the nature of their employment.

mention the informal sector—a key agent responsible for managing waste.

With the advancement in digitalisation, these standards are evolving. The Digital India initiative (Government of India 2015) has enabled Internet access to over five hundred million Indians. Over the last decade, increased availability of bandwidth, cheap data plans, and increased awareness have caused massive growth in Internet users, especially among lower-income communities. A proposed revision to the statistical standards on informality may be implemented in 2023 to understand the ways in which informal workers can adopt digitalisation and e-formalisation (OECD 2021).

The construction industry in most developing countries comprises a regulated formal part and an unregulated informal part (Mlinga and Wells 2002). In India, numerous links between the formal and informal sectors exist. These links are based on politics, economics, and technology. Since the informal sector operates outside the legal system, the political link reflects how the informal sector is accepted by the formal sector. Economic links involve direct transactions between the formal and informal sectors, such as when the informal sector provides construction materials and components to the formal sector or if the formal sector employs the informal sector through subcontracting. Technological links involve the transfer of technology and skills between the sectors during movement of skilled workers or subcontracting.

Most construction waste management systems for developing countries are based on models developed by the West, which do not include the informal sector (Hande 2019). The following section explores a qualitative content analysis method to understand the informal material reclamation processes in India.

3 Methods

Resources belong to a complex geo-located marketplace. The assessment of the circular economy in informal construction requires context-specific interpretations. For this study, two Indian cities, Mumbai and Bangalore, are selected to carry out field studies and interviews. Both Mumbai and Bangalore are metropolitan cities. Mumbai is 603 square kilometres with a population of 2.8 crores (28 million), while Bangalore is 709 square kilometres with a population of 1.23 crores (12.3 million). These cities were selected due to their existing large-scale reuse and recycling informal infrastructure. Across the Mithi River in Mumbai, the scrap market of Dharavi employs over 250,000 people in the reuse and recycling industry, who work in about 15,000 single-room factories. Further south, in the city of Bangalore, the reuse industry employs about 11,000 people (Biyani and Anantharaman 2017), a much smaller proportion compared to Mumbai.

For identifying the relevant stakeholders to interview, a list of stakeholders was identified in collaboration with a local NGO. Three stakeholders for each category were contacted. Each respondent was individually interviewed in a face-to-face interview. All interviews were recorded and transcribed with the permission of the participants. The interviews were conducted in April 2022 and ranged from 20 to 90 min. Data for this exploratory study was first collected through semi-structured interviews (Bryman 2016). This was followed by a qualitative content analysis (Bengtsson 2016) using both deductive and inductive coding techniques (Drisko and Maschi 2016). The codes were then organised into categories and enumerated through presence/absence criteria.

4 Findings

4.1 Activities Carried Out in Informal Reuse Ecosystems

Once a building is marked for demolition, demolition/dismantling experts inspect buildings to see what materials can be reclaimed. Generally, doors, windows, glass, furniture, and bathroom fixtures are retrieved. Once a building is demolished, the experts even collect steel rebar, wires, pipes, and other viable materials from the demolition site. On the other hand, waste pickers collect components from landfills or other dump sites. They collect about 60–90 kg of waste per day while working for about 8–10 h. Both the demolition experts and waste pickers do not have sophisticated transportation of their own. The demolition experts hire a van for large quantities of materials. For smaller quantities, bullock carts are used. Waste pickers generally use push carts or sacks on their backs to transport the materials (Fig. 1). These mobile devices help them cover large distances to find waste materials.

Since these actors don't have storage facilities of their own, reclaimed materials are transported to small scrap aggregators, colloquially known as *kabadiwalas*. Here, materials are collected, stored, and minimally processed. *Kabadiwalas* are generally located in residential areas, near industries or landfills, to ensure a constant supply of post-use materials. Waste segregators sort all the materials into reusables, recyclables, and non-recyclables. The items are also sorted by colour and quality. Waste processors are then employed to clean the items and remove connective hardware or any other artefacts from the reclaimed materials (Fig. 2).

After the materials are processed, the materials are sold to larger bulk aggregators (Fig. 3) on a weekly or biweekly basis. These aggregators are either remanufacturers or reuse material suppliers who can store larger volumes of materials than the *kabadiwalas*. The reuse material suppliers have greater specialisation in terms of segregation and/or processing. They prefer to set up their units close to industries or on the periphery of the city.

Clients buy reclaimed materials from these suppliers for usage in new construction projects. The materials that are not sold or those of low quality are sent to local carpenters, who remanufacture them for further use (Fig. 4). The completely unusable wastes generated by the carpenters are put through a powder sander. The sawdust is combined with water to create a natural mixture that is used to patch blemishes on the products made by the carpenters.

The industry structure of the reuse businesses is often vertically disintegrated and highly specialised. The chain of production is broken up into pieces. The proximity between the different units allows them to collaborate and contributes to the productivity of the place, as transport costs

Fig. 1 Transport of metal panels using bullock cart (left); Reclaimed wood panels secured to pushcart (right)

Fig. 2 Connective hardware to be removed from reclaimed materials

Fig. 3 Reclaimed materials stored by local suppliers

Fig. 4 Carpenter warehouse (left); Secondary goods made from wood waste (right)

needed for the production chain are minimised. In this way, a complete informal decentralised ecosystem is run in the cities of Bangalore and Mumbai, thereby ensuring a circular economy (Fig. 5). However, this ecosystem is not limited to the cities studied. Respondents mentioned that material waste was received from across India, such as from the cities of Hyderabad, Pune, Chennai, and Delhi. Depending on the quantity of waste to be processed, different locations were chosen for selling the reclaimed materials. Generally, the waste was processed and treated locally if in small volumes, but higher volumes are sent to Mumbai as they have the largest and most robust system for reuse.

4.2 Interactions Between Formal and Informal Reuse Sectors

The construction industry supply chain is extremely complex and competitive. Hence, large tensions exist between the informal sector and formal sector, especially in terms of procurement and pricing. Negative connotations are often attached to working with informal reuse actors, which can hinder their profits. Increased community engagement can increase their efficiency and establish transparent and stable pricing. Smaller aggregator respondents are open to some sort of organisational changes but lack of knowledge and competencies necessary to

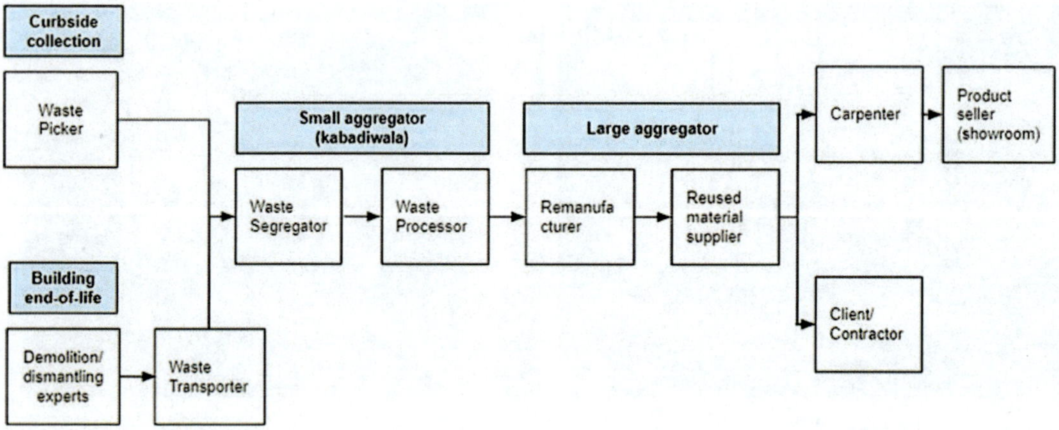

Fig. 5 Informal reuse ecosystem

complete registrations. Respondents also emphasised the lack of trust in formal enterprises. Larger aggregator respondents mentioned the difficulties in storing materials when volumes are more than their storage capacity. Sometimes items are kept on the footpath in front of their establishments due to lack of space. Larger aggregators are more receptive to formalisation and new models of trade and procurement to increase their sales.

4.3 Social and Economic Challenges

The networked reuse organisations are heavily reliant on strong social ties. The geographic concentration of reuse activities allows for a minimisation of costs, and maximisation of positive externalities, such as the sharing of expertise, innovation, and machinery. The income of demolition experts or waste pickers varies according to the region in which they are located, as a higher-income areas have higher quality materials. On average, the respondents without vehicles earned Rs. 500 per day, while the ones with push carts earned Rs. 1500 per day. The *kabadiwalas* take the material from the waste pickers and after initial cleaning and processing sell the materials to the larger dealers (Table 2). The informal reuse ecosystem comprises a community that runs on the entrepreneurial spirit of the people to support themselves. Yet, due to the stigmatisation of these businesses, their products and services are considered low value leading to low income and limited opportunities. While the wages and living conditions of different strata of informal waste workers differ greatly, most of them (especially, waste pickers) work and live in hazardous conditions. Respondents from local authorities and governmental units mentioned that waste pickers are not covered under any labour legislation. As a result, they do not benefit from social security and medical insurance schemes. Thus, there is a dire need to initiate policy action for their social and economic upliftment.

4.4 Health and Safety for the Informal Workers

Several respondents who are waste pickers mentioned physical injuries while procuring materials. The injuries were primarily because of lacerations caused by shards of glass, followed by muscle sprains. Frequent incidences of severe injuries on demolition sites by demolition/dismantling experts were also stated. Similarly, waste segregators, processors, and carpenters said exposure to various hazardous fumes results in respiratory problems. On the field, it was seen that protective clothing such as gumboots,

Table 1 List of relevant stakeholders in informal reuse ecosystem

A. Stakeholders who affect local reuse ecosystems	Stakeholders actively participating in reuse ecosystem	A.1. Waste picker
		A.2. Demolition expert
		A.3. Waste transporter
		A.4. Waste segregator
		A.5. Waste processor
		A.6. Reused material supplier
		A.7. Remanufacturer
		A.8. Contractor
		A.9. Client
		A.10. Developer
		A.11. Planning and design team
	Stakeholders involved in determining the context and policies	A.12. Local authority: Department of Planning; Department of Labour
		A.13. Regional government departments: Solid Waste Management Department, Municipal corporation, Ministry of Social Justice and Special Assistance
		A.14. Non-departmental public bodies: Housing corporations
B. Stakeholders who are affected by local reuse ecosystems		B.1. Users of building
		B.2. Local community groups and associations
C. Potentially interested entities		C.1. Non-governmental organisations
		C.2. Researchers/Academics

Table 2 Selling price of reclaimed construction materials

Material	Price (Rs/kg)	
	Kabadiwalas (small aggregators) buy material from waste pickers	Large aggregators buy material from *kabadiwalas*
Wood	3	7
PVC pipes	3	20
Glass	1.5	5
Iron rods	15	18

gloves, and masks were not used by any of the respondents, which increased their vulnerabilities and health risks.

4.5 Technology Integration

Inclusive digital transformation can ensure that digital technologies are universally available, accessible and address the needs of the most vulnerable. In the era of accelerated online consumerism, digitalization is no longer an option but a necessity for businesses to thrive. For many small-scale businesses, digitalization is about building resiliency, collaboration, and reaching a broad spectrum of consumers (URBZ 2021). However, for many informal reuse actors, this exacerbated digital divide is making it difficult for them to survive. Respondents between the ages of 30 and 40 mentioned that they have non-smart mobile phones, so their awareness and knowledge of using digital applications are still limited. However, they all mentioned that the next generation that would take over their businesses was keen to integrate digital technologies. Younger respondents between the ages of 20 and 30 communicated that they were smartphone users with an interest to digitalise aspects of their work. Technologies can strengthen social capital and de-centralise production networks by fostering a transparent, collaborative economy. Digital technology integration can decrease the exploitation of the informal reuse workers and ensure the correct flow of credit. Digital platforms and marketplaces for the informal economy can provide these businesses with an online presence which not only makes them competitive but also empowers them to professionally connect and share their skills. Such applications can empower the workers in not only structural, psychological, and resource dimensions but can also help achieve political, social, and economic inclusion.

4.6 Policies

In Mumbai, respondents from local and regional government units mentioned that the Government of Maharashtra has shown slight receptiveness towards the informal sector. In 1999, the state government issued an order to municipalities to provide identity cards to all waste pickers. Another order followed in 2002, directing municipalities to register cooperatives that collect waste from homes, shops, marketplaces, and other organisations. However, most respondents in both Mumbai and Bangalore were unaware of any laws and policies that protect their rights. It is of paramount importance to implement informal worker welfare laws that recognise and integrate informality into the waste management supply chain. The law must include basic provisions for waste collection, segregation, and sorting; access to personal protective equipment to minimise occupational hazards; right to basic necessities like water, sanitation, and facilities for clean living; and health insurance.

4.7 Discussion

Decision-makers play an important role today: the way governments, employers, and workers jointly decide to implement circular strategies will shape the future of people and the planet. The following section highlights important aspects to be considered by policy-makers and sustainability researchers to aid informal reuse ecosystems.

4.8 Contribution to Sustainable Development Goals (SDGs)

If organised and supported by public policies, informal reuse ecosystems can effectively contribute to the SDGs (Table 3). At the global level, reusing is one of the cheapest methods to reduce GHG emissions (Singh 2021). Using the WARM model developed by the USEPA, it is estimated that the informal recycling sector in Delhi avoided 962,133 tonnes of CO_2 equivalent emissions annually, thereby referring to waste pickers as 'cooling agents' (Municipal Solid Waste Knowledge Platform 2009). The informal reuse sector's logistics and processes emit significantly less GHG than formal sector practices (Vergara et al. 2016). Exact calculations of the informal sector's emission reductions citywide remain elusive due to poor data collection by municipal and national authorities. To capture the full carbon benefits that the informal reuse ecosystem provides, the Government of India must develop material-specific emissions factors customised to the Indian context and methodological tools for life cycle analyses of emissions from waste.

However, the informal reuse ecosystem also negatively influences SDGs in some respects. The workers do not have adequate working

Table 3 Contribution of informal reuse ecosystem to SDGs and a circular economy

SDG	Contribution from informal reuse ecosystem	SDG	Contribution from informal reuse ecosystem
1 NO POVERTY	Informal reuse workers are given the opportunity to generate an income	10 REDUCED INEQUALITIES	A majority of the workforce in the informal reuse sector comes from disadvantaged communities
2 ZERO HUNGER	Informal workers recover organic material waste which is used as compost or animal feed	11 SUSTAINABLE CITIES AND COMMUNITIES	Cooperatives promote social inclusion, de-stigmatization and community building
3 GOOD HEALTH AND WELL-BEING	Reusing materials limits carbon emissions generated from landfill incineration	12 RESPONSIBLE CONSUMPTION AND PRODUCTION	Raw material extraction and depletion is reduced by reuse
5 GENDER EQUALITY	A majority of waste pickers are women; thus, they are given equal opportunities to participate	13 CLIMATE ACTION	Reusing materials reduces GHG emissions by reducing energy consumption and material extraction
6 CLEAN WATER AND SANITATION	Management of waste materials avoids disposal of construction waste in water bodies, thereby reducing water pollution	14 LIFE BELOW WATER	The informal reuse sector's contribution helps avoid waste being dumped in our oceans
7 AFFORDABLE AND CLEAN ENERGY	Informal workers recover organic material waste to produce biogas for energy usage	15 LIFE ON LAND	Land degradation and release of hazardous chemicals due to waste is being avoided by informal sector's interventions
8 DECENT WORK AND ECONOMIC GROWTH	The revaluation of waste into resources boosts the economy		

conditions, they are exposed to hazardous materials and are compelled to carry out manual labour. The uncertain legal status of the informal sector can also negatively impact communities that do not receive taxes from businesses that are not registered entities.

4.9 Legal Protection and Financial Inclusion

Informal workers need to be provided identification cards that can legalise their services. Formalisation can also help individual informal suppliers against undue harassment from public officers and town residents. Their inclusion in social welfare organisations and schemes should be mandatory. Member-based organisations (MBOs) should also be endorsed. MBOs allow informal workers to advocate for their rights and create political traction as a collective agency. MBOs can help in securing jobs and incomes, facilitating access to basic services and social protection. The government also needs to provide the workers access to pension and healthcare schemes. The informal workers could be compensated for their reuse and recycling services through the provisions of adequate working spaces and machinery.

4.9.1 Training and Awareness

The work carried out by the informal sector involves occupational as well as health hazards. To ensure the health and safety of the workers, government or NGO training is required on how to use protective equipment and why safe work practices are important. Municipalities should also ensure that the workers have access to appropriate safety gear. Additionally, connectivity, access to adequate technology, and digital literacy must be ensured so that the informal workers can reap the benefits of these new opportunities. Digital skills training initiatives for empowerment and inclusion, regardless of education, age or gender should also be implemented.

5 Conclusion

To tackle climate change, we must move beyond 'carbon tunnel vision' (Konietzko 2021), and not only analyse carbon emissions but also equally examine the environmental, economic, and social burdens that it presents. Informal reuse workers in India and developing countries across the globe are climate entrepreneurs who contribute real and measurable reductions in greenhouse gas emissions. The informal workers who are greatly contributing to the fight against climate change need to be identified, assisted, and rewarded for their work. Due to a lack of compensation and widespread resistance, they risk missing out on important opportunities that could create long-term value, not only for themselves and their community but the planet at large. Engaging with the informal sector can help transition towards a circular economy as well as contribute to dignified livelihoods for those in need. Furthermore, technologies should be used when they draw from practices that are compatible with local cultures and contexts, or they run the risk of failure. Across levels of government, policy-makers should seize opportunities for ad hoc negotiations and specific action plans must be developed in collaboration with grassroots organisations. As informal waste workers are key allies in the climate transition, such supportive regulatory environments are the need of the hour.

Acknowledgements This work was supported by the ETH for Development (ETH4D) Research-to-Action Grant and the Swiss National Science Foundation under grant 200021E_215311.

References

Aatmanirbharbharat [WWW Document] (2020) https://aatmanirbharbharat.mygov.in/. Accessed 10 Oct 22

Alberge D (2020) Pompeii ruins show that the Romans invented recycling. The Observer

Bagale GS, Vandadi VR, Singh D, Sharma DK, Garlapati DVK, Bommisetti RK, Gupta RK, Setsiawan R, Subramaniyaswamy V, Sengan S (2021) Small and medium-sized enterprises' contribution in digital technology. Ann Oper Res 1–24. https://doi.org/10.1007/s10479-021-04235-5

Bengtsson M (2016) How to plan and perform a qualitative study using content analysis. NursingPlus Open 2:8–14. https://doi.org/10.1016/j.npls.2016.01.001

Biyani N, Anantharaman M (2017) Aligning stakeholder frames for transition management in solid waste: a case study of Bangalore. Revue internationale de politique de développement 8. https://doi.org/10.4000/poldev.2483

Bryman A (2016) Social research methods. Oxford University Press

Central Pollution Control Board (2017) Guidelines on environmental management of construction and demolition waste

Degryse P, Schneider J, Haack U, Lauwers V, Poblome J, Waelkens M, Muchez P (2006) Evidence for glass "recycling" using Pb and Sr isotopic ratios and Sr-mixing lines: the case of early Byzantine Sagalassos. J Archaeol Sci 33:494–501. https://doi.org/10.1016/j.jas.2005.09.003

Drisko JW, Maschi T (2016) Content analysis. Oxford University Press

Ellen McArthur Foundation (2010) The biological cycle of the butterfly diagram [WWW Document]. https://ellenmacarthurfoundation.org/articles/the-biological-cycle-of-the-butterfly-diagram. Accessed 10 Oct 2022

Government of India (2015) Digital India Initiative [WWW Document]. URL https://www.digitalindia.gov.in/di-initiatives. Accessed 10 Nov 2022

Hande S (2019) The informal waste sector: a solution to the recycling problem in developing countries. Field Actions Science Reports. J Field Actions 28–35

Hoffman B (2014) Production and consumption of copper-base metals in the Indus civilization. Archaeometallurgy in Global Perspective

ILO (1993) Resolution concerning statistics of employment in the informal sector [WWW Document]. http://www.ilo.org/global/statistics-and-databases/standards-and-guidelines/resolutions-adopted-by-international-conferences-of-labour-statisticians/WCMS_087484/lang--en/index.htm. Accessed 10 Oct 2022.

ILO (2003) Guidelines concerning a statistical definition of informal employment [WWW Document]. http://www.ilo.org/global/statistics-and-databases/standards-and-guidelines/guidelines-adopted-by-international-conferences-of-labour-statisticians/WCMS_087622/lang--en/index.htm. Accessed 10 Oct 2022

ILO (2013) Resolution concerning statistics of work, employment and labour underutilization [WWW Document]. http://www.ilo.org/global/statistics-and-databases/standards-and-guidelines/resolutions-adopted-by-international-conferences-of-labour-statisticians/WCMS_230304/lang--en/index.htm. Accessed 10 Oct 2022

Konietzko J (2021) Moving beyond carbon tunnel vision with a sustainability data strategy [WWW Document]. Digitally Cognizant. https://digitally.cognizant.com/moving-beyond-carbon-tunnel-vision-with-a-sustainability-data-strategy-codex7121. Accessed 10 Dec 2022

Mlinga RS, Wells J (2002) Collaboration between formal and informal enterprises in the construction sector in Tanzania. Habitat Int 26:269–280. https://doi.org/10.1016/S0197-3975(01)00048-0

Modi N (2022) Circular economy is need of the hour: PM Modi [WWW Document]. www.narendramodi.in. https://www.narendramodi.in/text-of-prime-minister-narendra-modi-s-address-at-a-post-budget-webinar-on-energy-for-sustainable-growth--560459. Accessed 10 Oct 2022

Municipal Solid Waste Knowledge Platform (2009) Cooling agents—an examination of the role of the informal recycling sector in mitigating climate change. Municipal Solid Waste Knowledge Platform

OECD (2021) Development co-operation report 2021: shaping a just digital transformation. Organisation for Economic Co-operation and Development, Paris

Patel A (2009) The historiography of reuse in South Asia. Arch Asian Art 59:1–5

Ratha KC (2020) India at the paris summit: a potential game changer. Indian J Asian Aff 33:112–119

SDG Knowledge Hub (2022) Guest article: three ways sustainability policy isn't serving the SDGs—and how it could|SDG knowledge hub | IISD. https://sdg.iisd.org:443/commentary/guest-articles/three-ways-sustainability-policy-isnt-serving-the-sdgs-and-how-it-could/. Accessed 10 Jan 2023

Singh DR (2021) Integration of informal sector in solid waste management. strategies and approaches. Centre for Science and Environment, New Delhi 18

UN (2016) Take action for the sustainable development goals. United Nations Sustainable Development. https://www.un.org/sustainabledevelopment/sustainable-development-goals/. Accessed 10 Dec 2022

URBZ (2021) Made in Dharavi : a digital scheme for SME's [WWW Document]. https://urbz.net/articles/made-dharavi-digital-scheme-smes. Accessed 10 Dec 2022

Vergara SE, Damgaard A, Gomez D (2016) The efficiency of informality: quantifying greenhouse gas reductions from informal recycling in Bogotá, Colombia. J Ind Ecol 20:107–119. https://doi.org/10.1111/jiec.12257

Making a Beam Social—In Search of a Localised Production Paradigm

Xan Browne, Olga Popovic Larsen, and Will Bradley

Abstract

Making a Beam Social investigates how communities can gather heterogenous wood waste streams and steer them into load bearing building components. In response to the current wasteful wood practice being single use in the construction sector, there is a need to reconsider discarded wood's potential. This research project seeks an alternative to relying on uniform traits and economies of scale, proposing novel, high value applications for waste wood. A full-scale loadbearing beam is constructed as an exemplar of alternative interactions between human and non-human agency, forming new social relations between locally salvaged timber artefacts, and community. Through a hands-on workshop, the project considers three aspects of facilitating local production with waste wood resources: • **Archiving** the existing traits of found timber pieces by utilising drawing as a means to represent the information of found materials. This situates the salvaged timber as evolving independent artefacts, and not only material resources. • **Grading** individual timber elements using established indicating properties as a means to predict material performance. • **Fabricating** the beam parts whilst retaining as much as possible their elemental identities. The physical outcome is a loadbearing member developed through interlacing the social and the technical. As an architectural element, it works towards a culture of local production.

Keywords

Timber construction · Wood cascading · Local production · Upcycling

1 Introduction

Timber construction is gaining momentum as a future alternative to more energy intensive materials such as steel and concrete; a notion that is prompted by a need for the building sector to re-evaluate its carbon impact. However, the viability of transitioning to increasingly biobased construction must also consider the pressure this may impose on the future availability of timber, and as such motivate studies into the efficient use of wood in construction. This is coupled with uncertainties regarding future forest health, as a consequence of a changing climate (Seidl et al. 2017).

X. Browne (✉) · O. P. Larsen
Royal Danish Academy: Architecture, Design, Conservation, Institute of Architecture and Technology, Copenhagen, Denmark
e-mail: xbro@kglakademi.dk

W. Bradley
Independent, London, UK

M. R. Thomsen et al. (eds.), *Design for Rethinking Resources*, Sustainable Development Goals Series,
https://doi.org/10.1007/978-3-031-36554-6_11

Currently, timber utilisation in Europe is wasteful, with enormous quantities meeting energy recovery after first use (Vis et al. 2016). Circumstances where timber is utilised for more than one application is limited to homogenising timber, such as in the production of particle-based products like OSB (Oriented Strand Board) and MDF (Medium Density Fibre Board). These two applications are a common practice in countries such as Finland, who have a well-established timber product industry (Husgafvel et al. 2018). Timber cascading offers a framework for extending the lifetime of wood by keeping it as high in the value chain as possible, and for multiple applications (Vis et al. 2016). Often, material is immediately downcycled after its first use into fibre-based products, chemicals and eventually energy, due to limited strategies for reusing timber in its solid form (Hughes et al. 2021).

During the recent period of timber's development as a building material, numerous engineered wood products (EWPs) have evolved, enabling the large-scale production of structural building components. Typically, these rely on highly reliable timber feedstock in standard dimensions with consistent performance. These requirements have minimised the diversity of timber species and formats that are used in construction, as well as forming large scale, centralised component production facilities (Ramage et al. 2017). Some studies, investigating the potential of reclaimed timber as feedstock for building components meet incompatibility between the demands of current industry and material qualities. This includes short material lengths, lack of favourable legislation, and heterogeneous and inconsistent material supplies (Sakaguchi et al. 2016; Jarre et al. 2020).

Research into utilising reclaimed timber for common EWP production has for the most part involved re-standardising waste streams to uniform cross sections, relying on significant processing, and subsequently reduced material yields (Rose et al. 2018; Risse et al. 2019). These endeavours place significant focus on material properties and processes, as opposed to incorporating spatial dimensions such as regionality and infrastructure. These latter aspects are highlighted by Bezama as fundamentals for implementing biomass cascading[1] (Bezama 2016). One example that focuses more specifically on local material availability and production is the Stavne Block, developed in Trondheim as a loadbearing block system assembled by laminating short lengths of salvaged timber (Nordby et al.). As individual components, they successfully assimilate an inconsistent supply of timber, later functioning as a modular building system of blocks slotting into one another. As well as this, the Stavne Block investigates upcycling of timber for structural use as a local, social initiative, involving members of the community in the development, production, and use of the system.

The need to rethink modes of production and industrialisation is highlighted by the United Nation's Sustainable Development Goals, that aim to define 17 explicit courses of action to work towards a more sustainable future (United Nations). Particularly relevant to this study are goal 9: Industry, Innovation and Infrastructure, and goal 12: Responsible Consumption and Production. Indicators for meeting sub targets within these goals include reduced CO_2 emissions, reduced domestic material consumption, and improved national recycling rates (United Nations). Each of these, although not quantified in this study at a national scale, are a part of the investigation's context and motivation.

This study seeks to challenge current modes of production, by avoiding dependency on the standardised lumber expected by current industry. Instead, it proposes a dialogue with waste streams, the unintended consequences of local decision making. As salvaged timber is characterised by unpredictability and heterogeneity, the course of action entails a negotiation process between the previous and new functions of timber artefacts. In turn, this strives to expose the ingrained relationship between social lives and timber, which overall aims to work towards a

[1] Biomass cascading is a framework that aims to extend the availability of biobased material over a series of applications.

culture of effective and long-term utilisation of wood in construction.

Making a beam social investigates hyperlocal production of structural building components, utilising discarded timber elements as feedstock. Recognising the cultural value of waste streams, the project investigates a continued correspondence with found timber artefacts, with an aim to maintain their enrolment in the social lives of human beings.

2 Concept and Method

Making a beam social was investigated through a student workshop at Central Saint Martins in London during August 2022. The local material sourcing and production premise was explored through a reconstellation of typical EWP requirements. As such, local reclaimed timber was gathered, archived, processed, and finally assembled into a beam (Fig. 1).

2.1 Beam Concept

Making a beam social retains reclaimed wood in its solid form for extending the material lifespan, as opposed to homogenising it for later cascaded uses. This approach capitalises on timber's strength as a strong fibrous material, a property that is lost after crushing for use in man-made fibre composites, such as MDF.

The project departed from the idea of a beam, an archetypal structural building component, with specific functional demands. Beams conjure images of rectilinear, recognisable components, that are immediately part of our existing pan-cultural architectural fabric. A concept was developed (Fig. 3) as an alternative approach to applying discarded timber to existing commonplace EPWs. The design is inspired by the structural system patented by Hilding Brosenius, which was used to create large span timber components in the middle of the twentieth century (Brosenius 1990). Recent research has

Fig. 1 Completed beam made during the workshop

demonstrated its potential for utilising waste wood, as a large portion of the beam can make use of short, unplaned lengths common characteristics of salvaged timber (Browne et al. 2022). The system is an I-beam, comprising of two main parts, the flanges which form the top and bottom cords of the beam, and the web, a series of arrayed elements at 45°. In contrast to a glulam which requires elements to be processed to the same standard cross section, the 45° elements can vary in cross-sectional geometry. This challenges the low yield experienced when applying reclaimed wood to existing EWP technologies. Furthermore, the concept retains the elemental identities of the individual salvaged elements, which is expanded on further in later sections. The advantage of applying an alternative material stream to the structural principles of an existing technology is that the in-use performance is precedented, and in this case across a number of demanding building typologies (Brosenius 1990).

3 Gathering Artefacts: In Theory and Practice

Vernacular architectures are for the most part steered by the local accessibility of natural matter, entwining communities with the resources used in their construction. Whilst today's spatial scale of extracting resources is incomparable, streams of material that we regard as waste have high context specificity, despite potentially distant provenance. Species of wood that might not grow locally, may occur in abundance in post-consumer form, as part of a waste stream in highly urbanised forestless contexts.

At present in Europe, the data available to describe timber waste is limited to weight, with little information relating to species, geometry, treatment, or previous function. Knowing a quantity of timber by weight provides a rough estimate of its fuel value, the most common second, and end of life application for timber in much of Europe today (Husgafvel et al. 2018). This mode of representation sees heterogeneous accumulations of timber elements as ambiguous formless material lumps. A scenario affecting how we value wood, as well as pointing to one of the main cited barriers previously mentioned for applying waste timber as construction material feedstock, its unpredictable heterogeneity. Therefore, as a point of departure, this project looks at how gathered heterogeneous timber elements can both be appropriately described, as well as inform the outcome of a structural building component.

The anthropologist Tim Ingold explores the relational states of materials, objects, and artefacts. In his view, whether a timber waste stream is a lump of material or pile of objects is a question of perspective, as the energy industry may align with the former, and a craftsperson with the latter. Is it then simply a case of acknowledging waste timber as material that has been formed?

Ingold writes:

> We are accustomed to thinking of making as a project. This is to start with an idea in mind, of what we want to achieve, and with a supply of the raw material needed to achieve it. And it is to finish at the moment when the material has taken on the intended form. At this point, we say, we have produced an artefact (Ingold 2012).

Or is the waste timber element, no longer the 'intended form', returned to the status of material? If artefacts embody a combination of material and idea, it would be difficult to suggest that a discarded timber element has entirely removed itself of the latter.

Our distinction between matter and form can be traced back to Aristotle's *hylomorphism*, that sees the combination of these two entities, matter and form, as defining the foundation of all things. This notion has been criticised for viewing matter as inert, and without any form giving capacity of its own. The artist and philosopher Manuel de Landa refers to the morphological, and even computational capacity of materials (De Landa 2000). This challenges the notion that objects are in fact finite and resolved, suggesting that they can take on future forms in continued response to varying environmental conditions.

Recent scholars, including Ingold, have proposed an alternative to the hylomorphic model, where instead of imposing form upon materials, practitioners seek instead for a correspondence between entities (Ingold 2012). Marcin Wojcik, a researcher into alternative timber construction applies this notion from a design perspective as a *dance of agency* (between designer and material), in which favouring the demands of one or the other would invalidate the (design) process (Wójcik 2015).

In the context of this project, the distinction between matter and form, or material and form giver is a little less clear. The prerequisite for attending *Making a Beam Social,* is to salvage a timber element from the local context, with a minimum length of 600 mm, and ideally a rectangular cross section. All these objects, in one way or another, evidence other lives. As found things, they speak of a context they were once in, an integral part of a building, a temporary function, or merely an offcut from the fabrication of something else.

What we are inheriting then is in a way a hybrid, a post-object, that's formed by a combination of a previous application, and wood's capacity to morph in response to its environment. This morphing leads to common 'defects' present in wood, such as cupping and bowing. These are geometric alternations that are a consequence of an element's once desired form, the rectilinear cross section, responding to fluctuating humidity over time. Adopting the form of found timber artefacts intends to contribute to effective utilisation increased lifetimes of timber, a core aim of wood cascading.

If today's industrialised production of building components can be defined as enabling a broad diversity of outcomes (forms), from a minimal inventory of parts (standardised elements). This project explores the opposite- how a diverse inventory can be assimilated into standardised elements, whilst retaining an indication of what the material has been, what it is now, and offering anticipation of what it might become.

4 Archiving

During the workshop, drawing was explored as a method for archiving the found timber feedstock, as well as documenting the overall yield from timber elements to beam. Wood is a heterogenous material at a scale perceivable by humans (Dinwoodie 2000). This heterogeneity is typically not represented at a lumber yard, and material descriptions are limited to cross section, length, as well as strength and or appearance grade. These explicit descriptions, such as the timber grade C24, categorise a group of material traits into a common language across professions.

In its simplest form, the drawing collects basic geometric descriptions of the found elements, as well as cut web members, offcuts, and feedstock cross sections. The drawing asks, what and how much information is necessary? Aside from the basic geometric information, it continues as an analogue tool for collecting quantitative and qualitative data, and thus it is open to workshop participants to represent many layers of information in 2D. The drawing may be utilised to record and represent many of the broader consequences of gathering artefacts, such as provenance and previous function, contributing to a nuanced spatial narrative around the representation of gathered materials. Today, more advanced methods for registering material traits exist that enable the accurate representation of complex geometries and heterogeneous materials. This study is contextualised with these methods in the discussion section of the paper.

Two artistic works that are relevant to include in context of this project's material archiving strategy. The first is 'How to lay out a croissant', a how to guide conceived by architects (Miralles and Prats 2006). The guide offers instructive advice on steps for measuring the geometry of a croissant, a familiarly complex form, that is not typically regarded as necessary to record. This questions the resolution and formats by which we need to measure objects, after all, as mentioned in the guide, croissants are merely 'meant to be eaten'.

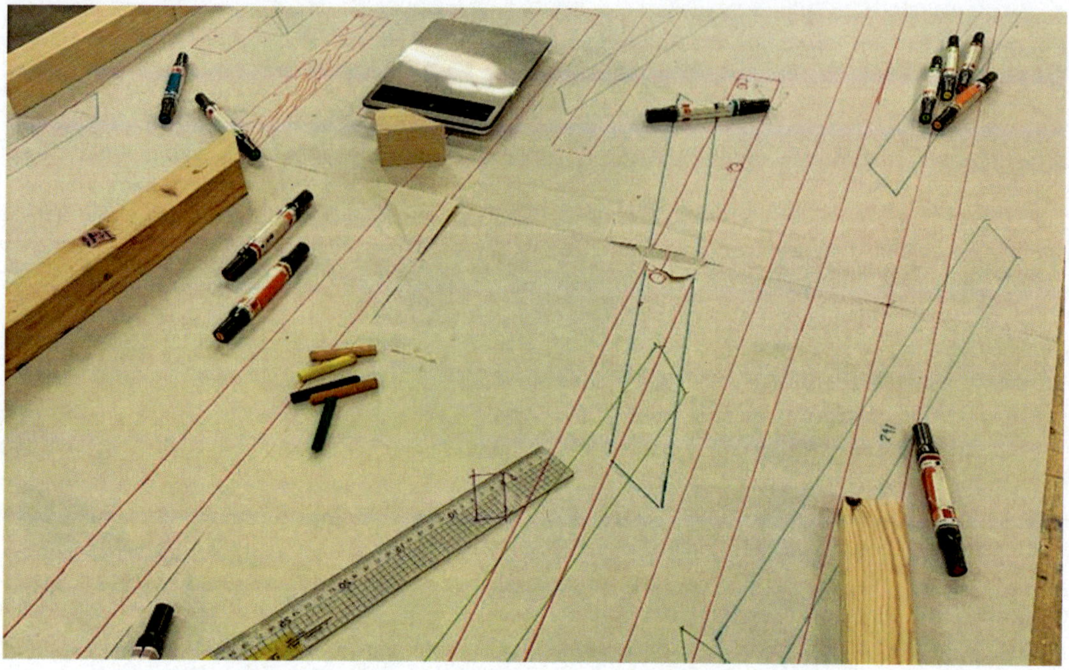

Fig. 2 Snapshot of the archive drawing during production

The other work that relates to how simple instructions can conceive context-specific descriptions of objects is that of the American abstract artist Sol Lewitt. Lewitt outlines simple instructions for creating large wall drawings that can be completed in his absence (Lewitt 1970). Some of these can be executed similarly in each context they are drawn, whereas others are formed in relation to existing elements in the wall. For example, *Wall Drawing #51* connects all perceivable points on a wall with one another, corners of door frames, light switches, fire exit signs, air vents, etc. Thus, a rule-based drawing emerges in correspondence with its context, as well as being a means to represent a specific contextual condition. In the same way that *Wall Drawing #51* will differ each time it is executed, in the same way the drawing used in the workshop will differ depending on the waste stream and design application.

Furthermore, the archiving methodology is inspired by the *situated drawing* method, a representational tool for engaging multiple actors as authors within a given scenario (Lang 2018). It was developed as a means to involve non-expert members of the public during design development, where more traditional drawing methods inhibited the participation of community members. In this project, the drawing's authorship extends to the found timber elements, and subsequently also their previous engagement in social lives. This challenges traditional methods for representing wood that typically neglect origin and provenance (Fig. 2).

5 Grading

As our arboreal ancestors would swing from tree to tree, they would assess the elasticity of branches before risking their lives on their strength (Ennos 2020). In more recent times, obtaining societal trust in building materials has fallen into the hands of craftspeople, and today it is limited to established standards and norms. Strength grading has emerged as a common method for authenticating trust in timber's performance across professions. Wood is a fibrous, natural material that exhibits varying properties depending on species, age, growing conditions,

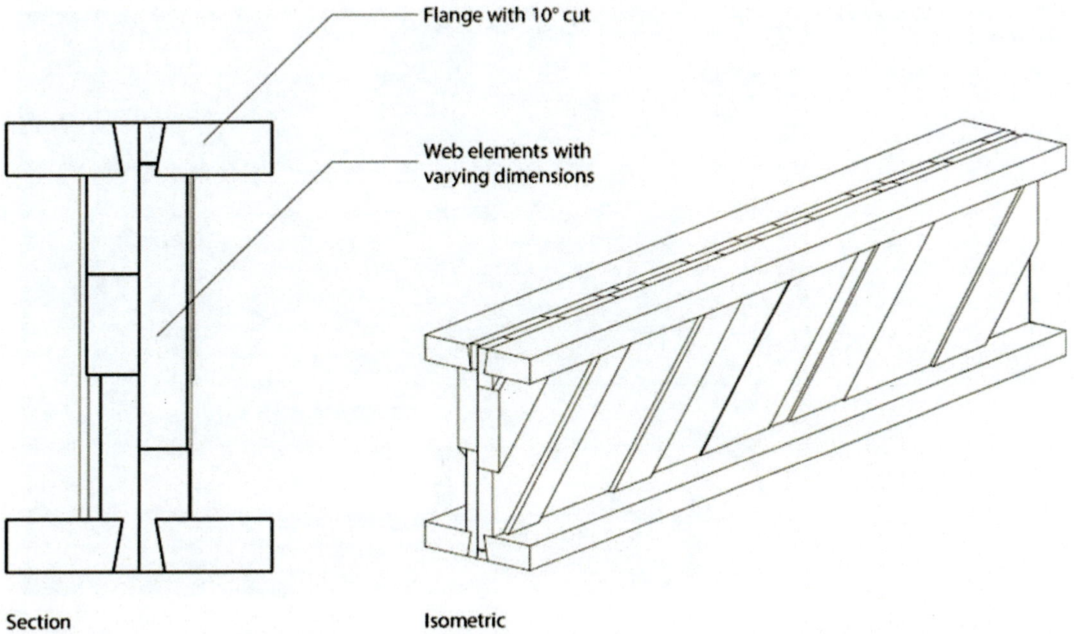

Fig. 3 Drawing of the beam concept. Note that each of the web elements is only connected to the flange, and not to one another

and part of the tree (Dinwoodie 2000). It is thus challenging to predict the strength performance of timber non-destructively, compared to more homogeneous materials such as steel and concrete. Grading timber typically relies on indicating properties (IP) to predict the performance of timber, enabling it to be taxonomized and subsequently applied appropriately to building applications (Dinwoodie 2000).

Timber's density varies significantly, also within the same species. This variation correlates broadly with its strength performance and can be utilised as an appropriate IP for grading (Dinwoodie 2000). During the workshop, the density of individual web elements is recorded for assigning grading values. These can be utilised to inform the arrangement of elements within the beam, in response to the given functional requirements/load case. It should be mentioned that density is not a suitable metric alone for predicting all performance parameters in timber and should be combined with other IPs, such as young's modulus in future studies (Ridley-ellis 2020).

The topic of the structural strength of reclaimed timber is of high importance in studies for timber reuse. The beam concept, at the component scale, has design advantages over more common, glued EWPs. A virtue of the notched connection between the web and flange is that each of the web elements performs as individuals. This enables beam components to be designed with redundancy, as a new load path is secured upon the failure of one, or more web elements (Fig. 3).

6 Fabrication

Fabricating EWPs, such as glulam, typically takes place in large production facilities that rely on high capital investment and economies of scale, a mode of production typically requiring uniform material feedstock of accurate standard dimensions. Consequently, this leads to timber requiring significant processing, large transportation distances, as well as significantly lower yields compared to some examples of vernacular timber construction.

Fig. 4 Image of the routing jig positioned at 10°

For *making a beam social,* a fabrication method was developed to enable heterogeneous elements to be a part of a single beam system, whilst retaining as much of the original geometry as possible. In practice, this requires fabrication techniques that are agnostic to a variety of forms and properties, as opposed to being limited to a single cross-sectional dimension, species, or sensitive to hazards such as fastenings or paints.

Non-standard elements were semi-standardised to each have the same length and notch geometry. This ensured that the parts could properly interface with one another, but only in specific places, and as necessary. Furthermore, the fabrication allowed for common defects in wood, such as twist, cupping, and bowing, to remain present in individual elements, avoiding processing procedures such as planing. In a sense this is a process of negotiation, between wood's resistance to standardisation, and our current building culture of standardised parts.

The routing jig (Fig. 4), used in the beam fabrication, was designed to accommodate a maximum cross section of 200 × 60 mm, but could also be configured to work with larger dimensions. The jig also functioned as a flat reference plane, as the potentially non-planar (due to twist and cupping) found elements will need to sit on a planar surface within the beam. For the beam design, the jig was positioned to mill each element with a repeatable notch of 10° of the same depth, regardless of the element's cross-sectional dimension.

7 Assembly

With two prepared flanges, cut with a profile of 10°, the beam elements can be assembled as a kit of parts. The oblique 10° angle in both the web elements and flanges omits the need for fastenings in each of the individual elements. As such, the entire beam could be assembled using few fastenings to keep the two flanges from separating. It is possible that each web element will have unique geometry; however, their interfacing relationship to one another and the flange is always standard. The advantages of this specific relation between standard and heterogeneity, and how it is contextualised amongst more contemporary assembly methods are explored further in the discussion section.

Allocating material with varying properties is a well-precedented approach in research contexts. Often, this is investigated by strategising performance parameters, such as strength and

Fig. 5 Workshop participants assembling the beam

stiffness to areas of high or low demand in a design case. Nadir et al. make interesting links between optimisation and reconfigurability by arranging uniform length web members of different sections based on a specific truss load case (Nadir et al. 2004). This can also be viewed as a discrete approach that has similarities to the balance between variation and uniformity in this study's beam concept.

Other studies take this notion further, deriving direct links between material properties and morphology. The 'Form Follows Availability' investigation by Bruttning et al. parametrically designs a series of trusses based on a constrained stock of reused steel (Brütting et al. 2019). Returning to wood, investigations by Svilans et al. utilise high resolution CT scans of timber as design data for complex spatial outcomes (Svilans et al.).

However other aspects that relate to the applied context of the beam are just as relevant, such as the aesthetic expression and relationship to other building functions. Timber's diversity is expressed visually, through different tones, textures, and patterns, all of which can be curated to form the overall expression of the beam. During the workshop, the aesthetic preferences of participants are formalised by the arrangement of web elements within the beam. The process of negotiation, as introduced earlier, is further tested through contrasting desires. It can be argued that the entangled relations of material choices and how they are arranged to make buildings have traditionally been more socially informed, whilst still recognising the constraints of contextual availability (Love 2013). This notion is closer to vernacular methods that saw people significantly involved with materials and building. A greater human involvement may enable the stark variation of qualities in waste streams to be processed and evaluated, constructing refined preferences from an unquantifiable array of values. This works towards production that engages inhabitants as participants, rather than only spectators of architecture (Groba 2022) (Fig. 5).

8 Results

Figure 6 depicts the core components of Making a Beam Social's methodology and outcome, with the beam situated in the context it has emerged from. The waste wood elements, seen here as the web in the beam, are also represented by the

Fig. 6 Photograph of the archive drawing, fabrication jigs, and assembled beam

archive drawing above. The means for processing them are shown as the two routing jigs on the left side of the image.

Timber Artefacts and Process

During the workshop, participants developed a drawing (Fig. 7) measuring 2.5 × 1.1 m. As the pile of reclaimed elements depleted, the drawing became populated with information about the waste stream and beam parts. The five key pieces of information recorded at full scale are the original outline, the cut element outline, the offcuts outline, the section outline, and the weight in grammes of the cut elements. Furthermore, other information layers relating to timber were added, such as the density of growth rings and surface representations using wax rubbing. At a first glance, the drawing appears chaotic depicting the heterogeneity of the waste stream. Upon closer inspection, one can pick out singular outlines with alarming accuracy. This combination of chaos and precision documents both the heterogeneity of the overall waste stream, as well as unique identifiers (representation of the timber's geometry) of each individual element, and how it has transformed from waste stream to beam part. Including multiple layers of information on the same drawing kept relationships between different geometries intact. Over time, the drawing evolved to represent at 1:1 a simplified form of previous artefacts, and a part of an artefact they are about to become. This drawing's representation of the timber elements enabled a low-resolution digital twin to be developed for further analysis.

9 Material Yield

Figure 8 shows the standardised timber lengths mapped onto the waste stock, as well as the range of timber widths. In this study, two ways for calculating the material yield from waste stock to the beam web elements were used. The first is by area, and the second is by volume. In the former, the material stock represented a total area of 2.2

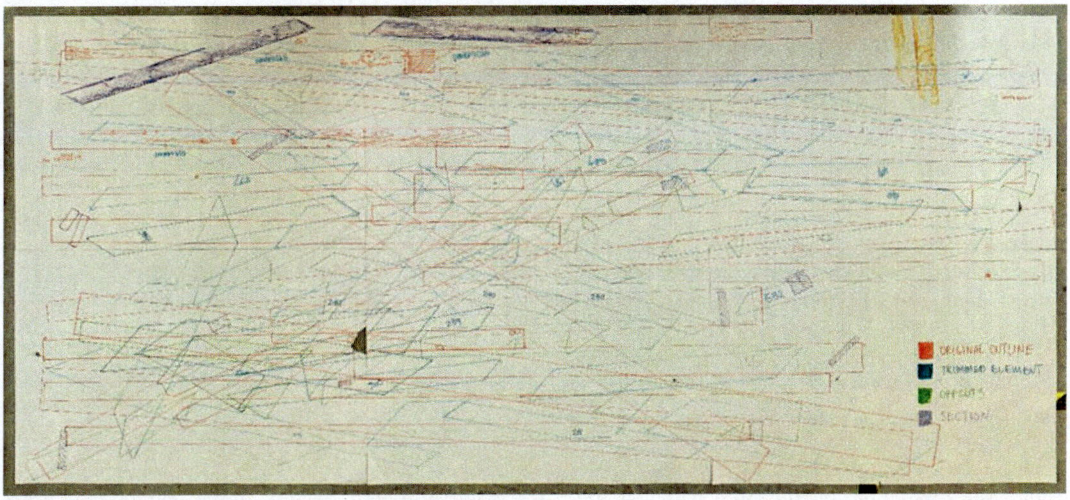

Fig. 7 Archiving drawing: chaotic at first glance, yet at closer inspection—a representation at high precision

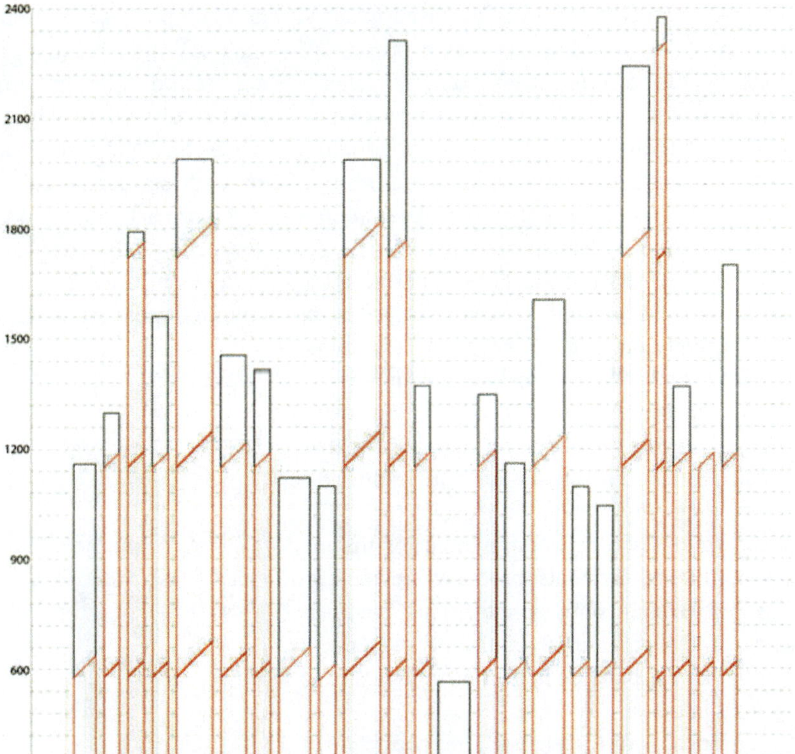

Fig. 8 Timber elements mapped onto timber feedstock, all dimensions in mm. One stock element was too short for the required web element length

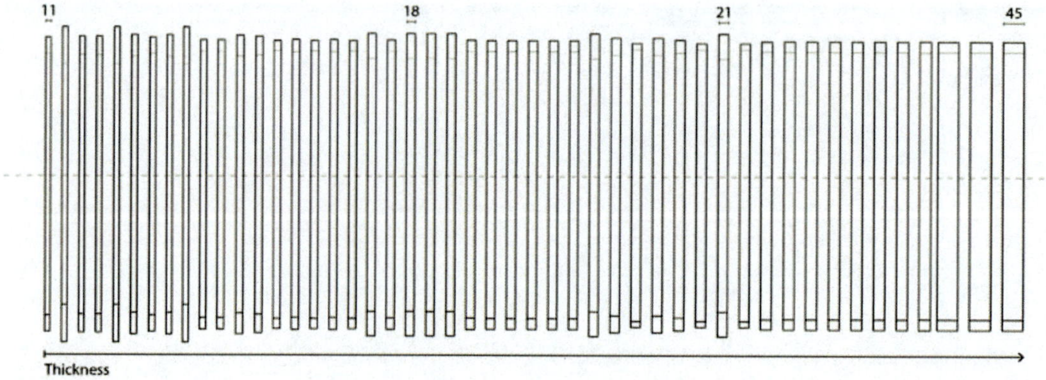

Fig. 9 Elements in order or thickness, dimensions in mm

m^2, with the elements consuming 1.6 m^2 of the stock area, concluding a material yield of 72%. In general, shortening the required lengths of the web elements would lead to higher material yields, although coupled with increased material processing. This could be seen as a process of quantising the discrete web element lengths according to the available lengths of waste wood stock to determine an optimum negotiation between available material and beam design. Naturally, this will also impact the I-beam depth and its structural performance.

By volume, the yield was 63%. Measuring the yield by volume is more arbitrary, in terms of how effectively the beam concept can assimilate the waste stock. This is due to the depth and width of the elements having no relation to the length. However, volume is an important metric for calculating utilisation, and therefore the potential longevity of wood in its solid form. The reduced yield value by volume indicates that on average, thicker elements produced fewer elements per overall stock length. This factor is outside of the parameters of the beam concept currently, as the purpose is to remain agnostic to varying thickness.

The thickness of the elements ranged from 11 to 45 mm (Fig. 9). Note that the element lengths appear different in the figure due to the varying width and 45° angle cut in each end. Arguably, some of the elements are too thin to perform optimally in the beam concept, whilst others are unnecessarily over dimensioned. This relates to the beam concept design, and the fact that the notch only engages a part of the element. Therefore, longitudinal shear would be expected in the individual web elements, both in tension and compression (Fig. 10). The added benefit of material not directly engaged with the notch connection requires further verification and would ultimately be an important parameter when arranging the elements based on their strength performance.

Figure 10 shows a section of the I-beam, vertically loaded, resulting with one web member in tension, and the other in compression. The coloured areas show the parts of the web that are

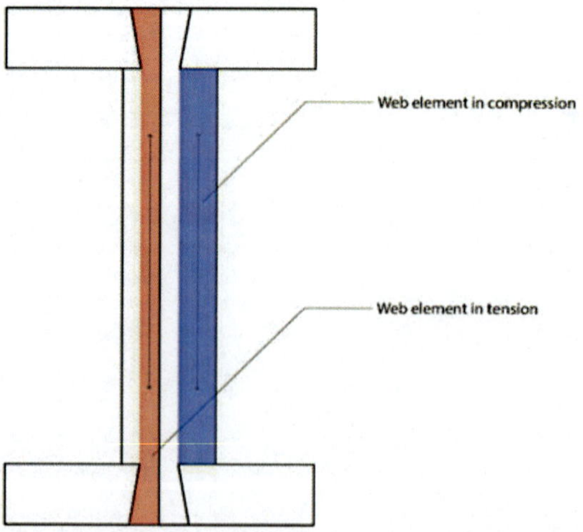

Fig. 10 Drawing describing areas of compression and tension in the web elements, and subsequent longitudinal shear

directly engaged with the flange, whereas the non-coloured areas are not optimally loaded, as the stresses are transferred laterally in longitudinal shear. One could argue that as these non-coloured areas contribute less to the structural performance, they are unnecessary. However, their inclusion in the beam enables an extended material life, therefore offering optimised effective material utilisation.

10 Discussion

Making a beam social exercised an alternative assembly of social entities, both human and non-human, to investigate a new form of structural component production. By demonstrating hyperlocal material sourcing, the varying identities of timber elements can be examined and assimilated as part of a beam system. This engages not only with the aspects relating to timber's technical concerns, but also relational factors of working with the inherited, such as the social lives embodied by found timber artefacts.

The project succeeded in collecting data points to accurately describe some aspects of heterogeneous material in varied forms, as well as documenting significantly improved yield in comparison to other studies applying reclaimed timber to EWPs. The dimensions of the beam exemplified here were determined prior to the workshop, otherwise, future versions could investigate calibrating a beam design, or even multiple beams based on an available material stock.

Other, more advanced techniques are evolving for obtaining material descriptions. For example, LiDAR scanning provides surface descriptions, and CT (computed tomography) scanning provides volumetric descriptions. Both these have recently demonstrated potential in designing specific relations between material traits and design (Svilans et al. 2019; Svilans et al. 2020). However, their workflows remain dependent on costly infrastructure and specialised knowledge for the management and strategisation of data sets (Svilans et al. 2019). In this study, the focus is to demonstrate that one can work with highly varied resources with a minimum complexity of data.

Furthermore, the methodology explored in this study is not alienated from current approaches to automated assembly. The interchangeability of the elements within the system, despite their variability, presents some advantages. The system can be seen as 'discrete' as defined by the architectural researcher Gille Retsin in the publication 'Robotic Building' (Retsin et al. 2019). His combination of speculative and built works investigate the necessity of repeated elements to work towards more 'programmable', 'reconfigurable' and 'modular' building systems, subsequently making them more favourable for automated assembly (Retsin 2020). In future version of this study with an alternative focus, the beam concept and fabrication approach could be readily transferred to automated assembly processes. However, this may challenge the authorship of individuals in the development of new production paradigms.

The gathering of offcut information is important for situating the production of timber building components amongst the other sectors that utilise wood. The trade-offs of removing wood from energy production and into another sector are important considerations, as well as the reduced demand for virgin wood for building when applying waste streams. Furthermore, the geometry information of offcuts is also available for assessing the cascading use of the material. This data may enable offcut material to be utilised for applications other than fuel, contributing further to extended material lifetimes.

Assimilating material in the form it is found leads to latent applications of timber material. This makes an interesting scenario regarding optimisation criteria, as some timber elements may be overstructured for the required function, but nevertheless contributes to overall material longevity. Material longevity of timber returns to the background of this project, in working towards effective cascading use of wood through high utilisation over many applications. Furthermore, recognising that timber is a subtractive material in production requires us to think of 'latent applications of material' as 'effectively stored material', which have positive implications regarding longevity and carbon storage, respectively.

11 Conclusions

Working with the inherited form of timber artefacts not only entwines the previous social existences of timber applications, but also contributes to higher utilisation of timber during reuse. Through embodying the context of the elements it contains, the beam differs from typical EWPs such as glulam. In that sense, the beam component is situated amongst the context required for its production.

To extend this idea further and to bring it closer to application, research building on this work is needed. That includes:

- Establishing simplified data collection of unique timber elements and reliable methods for matching them with their physical counterpart. This includes information relating to geometry, as well as indicating properties for establishing grading values. Furthermore, the qualitative qualities of varied timber cannot be ignored, and methods for including them in production processes should be considered in less 'hands-on' production environments.
- Extending the arrangement of actors outside of an academic context to real-world examples of material gathering, fabrication, and assembly. This would further investigate the necessary negotiation that takes place between material artefacts, production, and design.

Overall, this project and the concept it tested intends to work towards a more sustainable utilisation of timber resources in buildings, as well as production flows that are more resilient, through being more local. As a context-specific approach, it intends to foster regional architectures, whilst incorporating the ecology that emerges around waste streams.

Acknowledgements The research leading to these results received funding from Innovation Fund Denmark, Realdania, and Lendager as part of an industrial Ph.D. project by the first author under Grant Agreement No. 0153-00111B. The authors would also like to express their gratitude to the Spatial Practices Programme at Central Saint Martins of University of the Arts London, lead by Andreas Lang as well as workshop technicians Savvas Papasavva and Billy Dickinson.Workshop participants: Connie Beauchamp, Mike Parish, Will Hayter, Flynn Williams, and Jack Sweet.

References

Bezama A (2016) Let us discuss how cascading can help implement the circular economy and the bio-economy strategies. Waste Manag Res 34:593–594. https://doi.org/10.1177/0734242X16657973

Brosenius H (1990) HB-Balken: Projektering, Beräkning, provn ing och tillverkning. Statens råd för byggnadsforskning

Browne X, Larsen OP, Friis NC, Kühn MS (2022) Material Value(s): motivating the architectural application of waste wood. Archit Struct Constr 2:575–584. https://doi.org/10.1007/s44150-022-00065-6

Brütting J, Senatore G, Fivet C (2019) Form follows availability—designing structures through reuse. J Int Assoc Shell Spat Struct 60:257–265. https://doi.org/10.20898/j.iass.2019.202.033

De Landa M (2000) A thousand years of nonlinear history. Zone Books

Dinwoodie JM (2000) Timber: Its nature and behaviour, 2nd edn. E & FN Spon, London

Ennos R (2020) The wood age. William Collins, London

Groba UC (2022) Eloquent timber: Tacit qualities, telling materiality, and the inhabitants' voice. Archit Struct Constr 2:545–552. https://doi.org/10.1007/s44150-022-00029-w

Hughes M, Niu Y, Rasi K, Hughes M, Halme M, Fink G (2021) Prolonging life cycles of construction materials and combating climate change by cascading: the case of reusing timber in Finland Resources. Conserv Recycl Prolong Life Cycles Constr Mater Combat Clim Chang Cascad. https://doi.org/10.1016/j.resconrec.2021.105555

Husgafvel R, Linkosalmi L, Hughes M, Kanerva J, Dahl O (2018) Forest sector circular economy development in Finland: a regional study on sustainability driven competitive advantage and an assessment of the potential for cascading recovered solid wood. J Clean Prod 181:483–497. https://doi.org/10.1016/j.jclepro.2017.12.176

Ingold T (2012) Toward an ecology of materials. Annu Rev Anthropol 41:427–442. https://doi.org/10.1146/annurev-anthro-081309-145920

Jarre M, Petit-Boix A, Priefer C, Meyer R, Leipold S (2020) Transforming the bio-based sector towards a circular economy - What can we learn from wood cascading? For Policy Econ. https://doi.org/10.1016/j.forpol.2019.01.017

Lang A (2018) Situated drawing-pedagogical tools for civic practice

Lewitt S (1970) Wall drawing #51

Love S (2013) Architecture as material culture: building form and materiality in the pre-Pottery neolithic of anatolia and levant. J Anthropol Archaeol 32:746–758. https://doi.org/10.1016/j.jaa.2013.05.002

Miralles E, Prats E (2006) How to lay out a croissant. El Croquis Miralles 1983–2000

Nadir W, Kim IY, Hauser D, de Weck O (2004) Multidisciplinary structural truss topology optimization for reconfigurability. In: 10th AIAA/ISSMO multidisciplinary analysis and optimization conference. American Institute of Aeronautics and Astronautics, Albany, New York

Nordby AS, Wigum KS, Berge B, Developing the Stavne timber block ; Life cycle design in practice, pp 35–40

Ramage MH, Burridge H, Busse-Wicher M, Fereday G, Reynolds T, Shah DU, Wu G, Yu L, Fleming P, Densley-Tingley D, Allwood J, Dupree P, Linden PF, Scherman O (2017) The wood from the trees: the use of timber in construction. Renew Sustain Energy Rev 68:333–359. https://doi.org/10.1016/j.rser.2016.09.107

Retsin G (2020) Discrete timber assembly. Fabricate

Retsin G, Jimenez M, Claypool M, Soler V (2019) Robotic building: architecture in the age of automation. Detail

Ridley-ellis D (2020) Sorting, assessment & grading of timber in existing buildings

Risse M, Weber-Blaschke G, Richter K (2019) Eco-efficiency analysis of recycling recovered solid wood from construction into laminated timber products. Sci Total Environ 661:107–119. https://doi.org/10.1016/j.scitotenv.2019.01.117

Rose CM, Bergsagel D, Dufresne T, Unubreme E, Lyu T (2018) Cross-laminated secondary timber : experimental testing and modelling the effect of defects and reduced feedstock properties sustainability cross-laminated secondary timber : experimental testing and modelling the effect of defects and reduced feedstock P. https://doi.org/10.3390/su10114118

Sakaguchi D, Takano A, Hughes M (2016) The potential for cascading wood from demolished buildings: the condition of recovered wood through a case study in Finland. Int Wood Prod J 7:137–143. https://doi.org/10.1080/20426445.2016.1180495

Seidl R, Thom D, Kautz M, Martin-Benito D, Peltoniemi M, Vacchiano G, Wild J, Ascoli D, Petr M, Honkaniemi J, Lexer MJ, Trotsiuk V, Mairota P, Svoboda M, Fabrika M, Nagel TA, Reyer CPO (2017) Forest disturbances under climate change. Nature Clim Change 7:395–402. https://doi.org/10.1038/nclimate3303

Svilans T, Gatz S, Ramsgaard Thompsen M, Ayres P, Tamke M, Deep sight—a toolkit for design-focused analysis of volumetric datasets

Svilans T, Tamke M, Thomsen MR, Runberger J (2019) New workflows for digital timber

United Nations The 17 Goals. https://sdgs.un.org/goals

Vis M, Mantau U, Allen B (2016) CASCADES. Study on the optimised cascading use of wood

Wójcik M (2015) ARROW @ TU Dublin Counterculture, Ju-jitsu and emancipation of wood

Babassu Coconut Fibre-Reinforced Natural Rubber Biocomposite for Fabrication and Use in Remote Locations

German Nieva

Abstract

The work concentrates on how the largely disposed of by-products from the babassu palm tree industry and the liquid natural rubber latex extracted from the rubber trees can help produce local construction material on demand. Natural rubber and babassu natural occurrences overlap in Mapia and several other regions in Brazil; therefore, a locally fabricated biocomposite using these natural resources could be replicable in several regions across Brazil, establishing a regional collaboration between communities. Four samples with different babassu-natural rubber loading were produced following the steps commonly used in the production of biocomposites: fibre extraction, chopping of the fibres, mixing and post-processing. The mechanical properties of specimens cut out from samples with 0, 5, 7.5 and 10 pph fibre content were tested to ASTM standards. The material testing outcome follows broadly the trends highlighted in previous studies on natural rubber composites, with the addition of fibre increasing the stiffness of the natural rubber material and its tensile strength at lower elongations. The 7.5 pph sample shows a more significant increase in strength by 40% (at a 600% elongation) than non-reinforced rubber. However, the overall performance of the biocomposite is well under the expected and observed in previous studies. Even though in the early stages and with much refining required, the outcome of this work is promising. It allows foreseeing strategies for maximising the babassu–natural rubber biocomposite mechanical performance to respond to more specific applications in building projects.

Keywords

Babassu · Biowaste · Natural rubber · Biocomposite · Low-tech · Material

---

Tropisms – https://www.tropisms.co.uk/

G. Nieva (✉)
University College London, London, UK
e-mail: german@tropisms.co.uk

M. R. Thomsen et al. (eds.), *Design for Rethinking Resources*, Sustainable Development Goals Series,
https://doi.org/10.1007/978-3-031-36554-6_12

1 Introduction

One of the most difficult questions brought by the Anthropocene is: how to deal with the global inequality? (Lewis and Maslin 2020)

The most significant direct driver of terrestrial biodiversity loss is land-use change (WWF 2020). In the Brazilian Amazonas rainforest, the expansion of largely illegal logging, cattle ranching and agriculture is the leading cause of deforestation. These often-illegal practices limit access to land and natural resources by indigenous and traditional people who live off the extraction of non-timber forest products (Vadjunec et al. 2009).

The traditional Amazonian agroecological knowledge has the potential to contribute to sustainable agricultural development at the time when the modern market economy is increasingly and actively looking for new crops and products, acknowledging the need for more biodiversity-friendly production practices as identified by the WWF (World Wide Fund for Nature) (Clement 2016; Abramovay 2019; United Nations 2019; WWF 2020).

In recent years, I have been involved in the design and construction of building facilities within the framework of socio-economic projects in line with the UN's Sustainable Development Goals and explicitly concerning deforestation and social justice (Figs. 1, 2, 3, 4 and 5). Access to rural communities can be problematic, the transport of materials and skilled builders is costly and often not viable, traditional architecture is rarely suitable for specialist infrastructure and the alternative materials available are often limited to uncertified timber, cement, polycarbonate and steel sheets. Hence, this work focuses on identifying local and regional opportunities to develop and produce in situ sustainable construction materials on demand. This decentralised strategy, opposite to the traditional industrial process, could pave the way for a new approach to construction in rural and remote locations, minimising transportation, enhancing interregional cooperation and adding socio-environmental value to the products.

This study concentrates on developing a babassu palm tree coconut fibre-reinforced natural rubber composite for remote locations. Babassu oil extraction generates a large amount of lignocellulosic waste, hence the interest in

Fig. 1 Location of built projects and at the design stage within Brazil. **a** Boa Vista, Marahnao & Installation at the XII International Architecture Biennale in Sao Paulo. **b** Serra, Maranhao. Babassu mesocarp flour production facility. **c** Jaguanum Island, Rio. Beach kiosk and boat refuge. **d** Mapia, Amazonas. Building module for rural constructions (Design stage)

developing a material that can integrate this. Furthermore, the ruderal babassu palm plays a central socio-economical role within the traditional and indigenous populations, and its ecological importance is bound to increase considering the current rates of deforestation within the Cerrado biosphere in the north and north-east of Brazil (Gehring et al. 2011; Carrazza et al. 2012).

The use of natural rubber is linked to a project in Mapia, a community located within the State of Amazonas and close to the State of Acre, known for producing natural rubber latex from the Seringueira trees (Hevea Brasiliensis). Natural rubber is a versatile polymer, locally available within Mapia, inexpensive, does not create health or environmental hazards and is a natural, renewable and sustainable product (Masłowski et al. 2017).

Natural rubber tappers (seringueiros) and the quebradeiras rely on the extraction of non-timber forest products (NTFPs), and it is within their interest to protect the rainforest against deforestation. Therefore, adding value to the natural rubber and babassu by-products can help these communities continue to have dignified life stability in their communities and positively impact the protection of the forests (Peres and Pastore 2019).

Rubber and babassu natural occurrences overlap in Mapia and several other regions in Brazil, especially in the transition areas between the tropical rainforest and the Cerrado (Figs. 6 and 7). The Brazilian Institute of Geography and

Fig. 2 Quebradeiras in Boa Vista. Documenting their workday for the installation at the Architecture Biennale in Sao Paulo

Fig. 3 Babassu mesocarp flour production. Polycarbonate sheets (**a**). Concrete and steel (**b**)

Fig. 4 Beach kiosk and fishermen boat refuge. Polycarbonate roofing (**a**). Concrete and PVC

Statistics (IBGE) identifies the babassu and natural rubber as one of the most collected and profitable non-timber vegetable products behind the better-known acai fruit (Fig. 8). In addition to this, the collection's economic value does not concentrate on a few producers but benefits many extrativist communities (Fig. 8). Therefore, a locally fabricated biocomposite using natural rubber as polymer and babassu as reinforcement has the potential to be replicable in several regions across Brazil, establishing a regional collaboration between communities.

Fig. 5 Mapia's building prototype. Polystyrene mesh screens (**b**)

Fig. 6 Occurrence of babassu palm and natural rubber in Brazil (Adapted from Lorenzi, 2010 & Silva de Souza, 2018). **a** Attalea phalerata (Babassu palm tree). **b** Athalea speciosa (Babassu palm tree). **c** Hevea brasiliensis (Natural Rubber). **d** Built projects (see Fig. 1). **e** Mapia, State of Amazonas

Fig. 7 States with production of both Babassu and Natural Rubber: Collection (left). Number of producers (right). IBGE—Censo Agropecuário (2006)

Babassu (Coconut and Kernel) (t)	Natural Rubber (Liquid and Coagulum) (t)
1	2433
2	267
6142	5691
313665	165
171	4013
167	1131
28	279
4469	303

State	Babassu (Coconut and Kernel)	Natural Rubber (Liquid and Coagulum)
Acre	11	1275
Amazonas	9	534
Bahia	100	885
Maranhão	62097	7
Minas Gerais	122	76
Pará	119	102
Rondônia	83	135
Tocantins	542	9

Fig. 8 Non-timber vegetable products extracted in Brazil (Excludes timber related sub-products). *Natural Rubber includes liquid and coagulated latex. IBGE Censo Agropecuário (2006)

Quantity collected (Tons)	Acai	Babassu coco	Babassu kernel	Natural Rubber*
	267499	170235	163420	17986

Value of collection ($R x 1000)	Acai	Babassu kernel	Babassu coco	Natural Rubber*
	176380	109060	65200	25696

Value of sells ($R x 1000)	Acai	Babassu kernel	Natural Rubber*	Babassu coco
	133792	98609	25589	14368

Number of producers	Babassu kernel	Acai	Babassu coco	Natural Rubber*
	68741	41479	5356	3062

2 Background

2.1 Babassu

The babassu (Attalea speciosa Mart.) is a palm tree that occurs naturally within the Cerrado biome in Brazil (Fig. 9). Often, it is the primary source of income and nutrition in this region, with more than 60 different known uses for its trunk, leaves and fruit: human and animal food, handicraft objects, biofuel, construction material, cosmetics and medicines (Carrazza et al. 2012; Almeida Campos et al. 2015). However, the babassu is mainly used to extract lauric oil from the coconut's almonds (kernels) and, to a lesser degree, to produce flour from the coconut mesocarp layer (Figs. 10 and 11).

Approximately 70% of the coconut shell, composed of the epicarp, mesocarp and endocarp (Fig. 12), is disposed of or used to make charcoal with minimal added value (Ferrari and Soler 2015; Amaral et al. 2019). This lignocellulose-rich by-product is increasingly seen as an alternative energy source (Paula et al. 2014), water pollutant adsorbent material (Gomes et al. 2016) and as biopolymers films (Alves Lopes et al. 2020), among other applications. The epicarp and endocarp fibres and particles have been tested as reinforcement and filling in particleboards

Fig. 9 Babassu palm trees in Boa Vista

(Lima et al. 2006; Machado et al. 2017); thermoset composites (Fonteles et al. 2016); green thermoplastic biocomposites (Beber et al. 2018; dos Santos Silva et al. 2020); and in biocomposites produced with polypropylene via extrusion and mould injection (de Lemos et al. 2017). However, there is currently minimal research on using babassu fibres as reinforcement in biocomposite materials for construction.

Most of the research on biocomposites using babassu as filling have relied on micronised mesocarp powder with particle sizes of approx. 0,074 mm (Teixeira et al. 2018; dos Santos Silva et al. 2020); between 0,044 and 0,149 mm (Nunes et al. 2019); and non-specified grain sizes (Beber et al. 2018; Hoffmann et al. 2019; Reul et al. 2019). The particle–matrix ratio when used as filling varied from 1 to 30% max. Fewer studies have looked at the use of epicarp and endocarp fibres as reinforcement in composites with particle sizes ranging from approx. 0,8 to 4 mm (Lima et al. 2006; Machado et al. 2017); and the size of particles used with Polypropylene (PP) ranging between 0.5 and 0.045 mm having a more significant proportion of the fibres between 0.5 and 0.25 mm (de Lemos et al. 2017). The load in these tests varied between 85% of fibre (particleboard) and 40% of fibre

Fig. 10 Breaking coconuts

Fig. 11 Break of the coconut to obtain the kernel (left) and mesocarp (right)

(when tested with PP) (Fig. 13). Figure 14 compares the lignin, hemicellulose and cellulose content (%) between babassu fibres; more commonly used natural fibres in the fabrication of composites, such as wood, sugarcane, flax and bamboo and other natural fibres already used and tested as reinforcement in natural rubber composites.

2.2 Natural Rubber Composites

Natural rubber is a renewable, inexpensive and environmentally friendly natural polymer obtained from the Amazonas rainforest-native rubber tree (Hevea brasiliensis) in the form of field latex. Field latex is composed of rubber particles (33%), water (60%), proteins, fatty

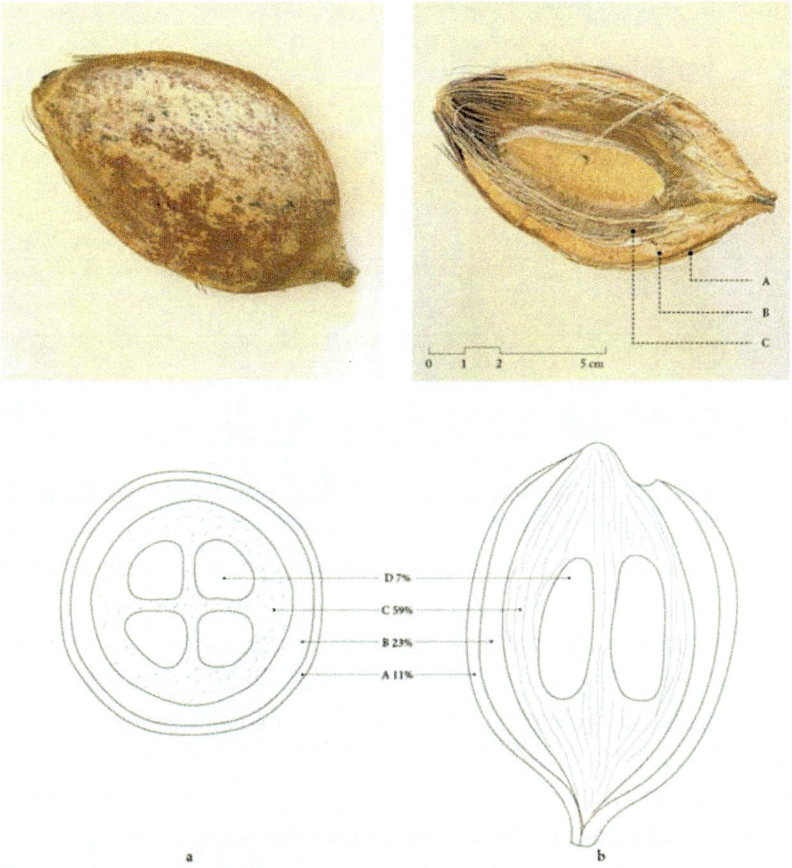

Fig. 12 Babassu coconut shell (top) and Babassu coconut transversal and longitudinal sections (bottom) and yield of fractions (%): A. Epicarp (11%); B. Mesocarp (23%); C. Endocarp (59%); D. Kernel (7%). Adapted from Teixeira et al. (2018)

acids, sugars, ash, trace elements and other impurities (Fig. 15a). This raw latex can be chemically, physically or biologically treated to produce natural rubber (NR) (Nair and Joseph 2014; Peres and Pastore 2019).

NR—long-chain polymer—is one of the most versatile elastomers, with high strength, elasticity, flexibility, resilience and abrasion resistance, characteristics conferred by its high molecular mass (>106 daltons) and its stereoregular microstructure formed almost entirely (approximately 94%) by cis-1,4 polyisoprene (Fig. 15b) (Sreekumar et al. 2013; Peres and Pastore 2019). NR does, however, have poor resistance to crack initiation, ozone and oils, overcome partially with methods such as vulcanisation, a chemical process that can improve its elasticity, tensile strength, hardness and weather resistance, properties that can be further enhanced by the addition of reinforcing fillers such as natural fibres (Nair and Joseph 2014; Masłowski et al. 2017).

Several studies have looked at physically improving the mechanical properties of natural rubber, particularly its strength, by incorporating natural fibres. Variables that affect a fibre-reinforced natural rubber composite performance are fibre length and aspect; fibre concentration; dispersion and orientation of the fibres; void content; and rubber matrix–fibre adhesion (de Lemos et al. 2017).

The need to improve the compatibility between the hydrophobic rubber matrix and hydrophilic cellulose fibres has been identified, by the majority of the precedents considered in this work, as a critical parameter in the enhancement of the mechanical properties of the NR composites (Geethamma et al. 2004;

Year	Authors	Filler / Reinforcement	Matrix	Load	Material Type / Use
2002	**Almeida et al., 2002**	Mesocarp, epicarp and endocarp	Portland cement	1:2.75 of babassu to cement ratio	Cement-bonded Particleboard
2006	**Lima, Ademi Moraes et al., 2006**	Epicarp fibres. Hammer milled with mesh of 12mm followed by a mesh of **4,37 mm.**	Urea formaldehyde	10%, 20%, 30%	Particleboard
2016	**Lemos et al., 2016**	Particles of babassu (not specific shell layer - **Sizes 354 um)**	Homopolymer Polypropylene hPP	40% babassu	Polypropylene (PP) composite
2017	**Machado et al., 2017**	Endocarp / Epicarp Particles of **0,8 > 4,0 mm & > 4 mm**	Oleo de mamona - resin	12% External & 15% Internal layers.	Particleboard
2018	**Nunes, Mário A.B.S et al., 2018**	Micronised powder mesocarp (particle size ranging between 44 - 149 um)	Poly(butylene adipate-co-terephthalate) & Thermoplastic starch (PBAT/TPS Blend)	17% Mesocarp	PBAT/TPS/Babassu Mesocarp biocomposite.
2018	**Reul, Lizzia T. A et al., 2019**	Micronised mesocarp, Fibrous epicarp	Poly (e-caprolactone) PCL	10%, 20%, 30%	PLC Compounds
2018	**Beber, Vinicius C et al., 2018**	Micronised powder mesocarp.	PBAT / PHB (two polymers)	20% Mesocarp	PBAT / PHB Biodegradable blends
2019	**Hoffmann, R et al., 2019**	Micronised babassu mesocarp	Poly(butylene adipate-co-terephthalate) & poly(3-hydroxybutyrate) (PBAT/PHB Blend)	1% - 3%	PBAT / PHB Films
2020	**dos Santos Silva, I. D. et al., 2020**	Mesocarp (sieved through a 0.074 mm mesh)	Polylactic Acid (PLA) Biopolymer 3251	1, 3 & 5%	PLA Composites

Fig. 13 Previous research on biocomposites made with babassu by-products

Year	Authors	Fibre	Lignin (%)	Hemicellulose (%)	Cellulose (%)
2016	**De Lemos, Mauss and Santana, 2016**	**Babassu Shell**	**14.73**	22.21	38.65
		Wood	15.23	23.1	40.95
		Sugarcane	14.82	28.67	39.69
2017	**Machado et al., 2017**	**Babssu Epicarp**	**35.64**	20.82	32.28
		Babassu Endocarp	**36.57**	25.72	29.8
2007	**Lovely and Rani, 2007**	**Isora**	**23**		74.8
2015	**Mohammed et al., 2015**	**Coir**	**40-45**	0.15-0.25	32-43
		Flax	2.2	18.6-20.6	71
		Bamboo	21-31	30	26-43
2017	**Maslowski, Miedzianowska and Strzelec, 2017**	**Barley**	**24.9 ±4.08**		39.1 ±4.34
		Corn	**25.2 ±3.49**		42.2 ±2.69
		Wheat	**23.1 ±4.29**		31.2 ±4.63
2017	**Kanoth et al., 2017**	**Coir**	**40-45**	0.15-0.25	32-43

Fig. 14 Babassu fibre compared to other natural fibres

(a)

Rubber hydrocarbon	Protein	Acetone Extract	Moisture	Ash
93%	3%	2.9%	0.6%	0.4%

(b)

Rubber	Water	Protein	Fatty Acids	Sugars	Ash Content	Trace Elements (Cu, Mn)	Impurities (dirt, sand)
33%	60%	2-3%	1-3%	1%	1%	2-3 ppm	8-10 ppm

Fig. 15 **a** Field Latex Composition (Adapted from Nair and Joseph 2014). **b** Natural Rubber Composition (Adapted from Sreekumar et al. 2013)

Lopattananon et al. 2006; Masłowski et al. 2017; Stelescu et al. 2017; Hoffmann et al. 2019). A better fibre–matrix adhesion can be achieved by modifying the polymer or the surface of the fibre. Even though several methods exist to modify the surface of the fibres most of the precedents studied rely on mercerisation (Sreekumar et al. 2013).

Mercerisation is an alkali treatment process that modifies the fibre's surface by partially removing cemented materials like hemicellulose and lignin, and other impurities. The effectiveness of mercerisation depends on the concentration of sodium hydroxyl (NaOH) in the solution, the temperature and the time used to treat the fibres. Alkali-treated fibre composites exhibit higher elongation at break; however, NaOH in concentrations lower than 5% does not show tensile strength improvements (Lopattananon et al. 2006). Figure 38 shows the percentage of NaOH, temperature and time different types of fibres have been treated in the previous studies. Geethamma et al. (2004), De et al. (2006), Lopattananon et al. (2006), Parambath Kanoth et al. (2019).

Generally, the increase in fibre reinforcement reduces its tensile and tear strength properties though it improves the hardness and toughness of the composite (Sreekumar et al. 2013). The maximum loading tested within the precedents studied was 50% fibre though the composite achieved the best mechanical performance at a 20% fibre content (Masłowski et al. 2017). Other precedents tested loadings ranging from 2.5 to 40% though the most effective load varied considerably between the different composites. The table in Fig. 38 only includes the loadings with the best performance in each study.

Fibre length, orientation and distribution are also variables that can significantly affect the performance of the natural rubber composite. The extent of stress transfer depends on the fibre length; therefore, it is vital to prevent the fibres from breaking excessively during the composite fabrication. A study with isora fibre testing lengths of 6, 10 and 14 mm, found that the 'critical' fibre length (this is when the transmittance of stress from the matrix to the fibre is at a maximum) was 10 mm. Hence, the samples showed the maximum value for tensile and tear strength for fibres having an original length of 10 mm before mixing (Mathew and Joseph 2007). The critical fibre length varies depending on the loading, type of fibre and treatment of the fibre (Fig. 38); therefore, the most efficient length for the babassu fibre-reinforced natural rubber composite will need to be tested.

One key challenge in preparing rubber composites is to achieve a uniform dispersion of the fibre in the matrix. A low-tech method successfully tested in NR composites is the hand lay-up technique alternating latex and fibres layers up to

	Tensile strength (MPa)	Tear strength (MPa)	Elongation at break (%)	Hardness (Shore A)	Flexural Load (N)	Flexural Strength (MPa)
Market / Natural Rubber Sheets						
Delta Rubber Black Natural rubber 60° Shore A	17		500	60		
Delta Rubber Black Natural rubber 40° Shore A	18		600	40		
Silex Silicones LTD Silicone Rubber Sheeting	10	22 (N/mm)	400	60		
Market / Synthetic Roof Sheets						
Jieli Roof (1 mm thick) Polypropylene Roof Tile	30				800	72
CorraPOL (2 mm thick) PVC Sheet	65					93

Fig. 16 Market products with potential similar uses to the natural rubber composite. *Source* products data sheets from manufacturers

the thickness required (Parambath Kanoth et al. 2019).

This work aims to enhance the strength of the natural rubber using babassu fibres obtained from the epicarp and endocarp (by-products) as reinforcement (Fig. 17). Potential uses of the composite can then be assessed based on the mechanical properties at different fibre loading, comparing its performance to natural rubber biocomposite precedents and market products used locally that it could replace (Figs. 16 and 38). The author foresees the new biocomposite in the form of sheets or tiles providing an alternative to the polycarbonate currently use on roofs.

3 Babassu Fibre-Reinforced Natural Rubber Composite

3.1 Materials

The babassu coconut shells composed of epicarp, mesocarp and endocarp were obtained from the Associação de Emprendedores Rurais de Boa Vista, Alto Alegre Pindare, Maranhao, Brazil. A total of 1.16 kg of shells were delivered to London.

For the extraction and treatment of the fibres: sodium hydroxide Pearl 99% (NaOH) was purchased from Source Chemicals Ltd, UK, and Hydrogen Peroxide solution 3% 10 Vols (H_2O_2) brand Care+, UK. Deionised water.

Low ammonia, pre-vulcanised natural rubber latex (60% dry rubber content) was purchased from Liquid Latex Direct, UK.

3.2 Methods

The composite fabrication followed the steps commonly used in the production of biocomposites: fibre extraction, chopping of the fibres, mixing and post-processing (Fig. 18). The mechanical properties of specimens cut out from four samples with 0, 5, 7.5 and 10 pph fibre content were tested to ASTM. The cutting of the samples and testing took place at the Biocomposite Centre—Bangor University in Wales.

Fibres extraction and alkali treatment

The fibres were extracted via a chemical retting process. The babassu coconut shells (Fig. 19) were first washed in water to remove dirt and then immersed in a 5% w/v NaOH solution at room temperature for 12 hs to loosen the fibres and partially remove lignin and hemicellulose from the surface of the fibres (Fig. 20a). The loosened fibres were separated manually off the shells (Fig. 20b) and washed free of alkali with water (Fig. 20c). Mostly fibres from the shell's

Fig. 17 Babassu by-products (top) and Quebradeira breaking coconuts in Boa Vista (bottom)

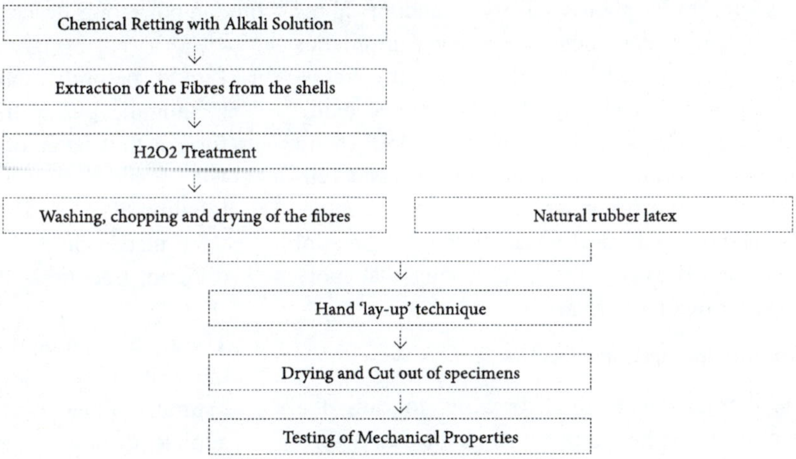

Fig. 18 Diagram showing the key stages of the composite preparation

outer layer—epicarp—were possible to extract following the first alkali treatment.

The endocarp layer was immersed for a second time in a 5% w/v NaOH solution for an additional 72 h until the fibres were sufficiently loosened to partially isolate them from the shell. Only the fibres within the layer immediate below the mesocarp were possible to extract from the endocarp.

Both the epicarp and endocarp fibres were washed with water and immersed in a 3% hydrogen peroxide (H_2O_2) solution for 8 hs (Fig. 20d). Following the H_2O_2 treatment, the fibres were washed with deionised water

Fig. 19 Babassu coconut shells

(Fig. 20e), chopped (Fig. 21a) and dried in an air oven at 45 °C for 6.5 hs until the weight loss stabilised to eliminate any moisture (Fig. 21b). It was not possible to isolate chopped fibres with the same length; these varied between 2 and 12 mm approximately. The total weight of the chopped and dried fibres obtained was 39.7 g.

Microscopic photos of five random fibres were taken to document the physical properties and condition of fibre's surface after the NaOH and H_2O_2 treatment (Fig. 22). The fibres' diameter varies significantly from about 350 μm to 1300 μm—this may be linked to the layer (endocarp or epicarp) where they were extracted. The photos also show residual matrix material still attached to the fibres, indicating that more chemical treatment is needed.

Composites preparation

The fabrication method broadly follows the preparation of macrofibre composites as described by Parambath Kanoth et al. (2019). As already mentioned, a key challenge in preparing rubber composites is to achieve a uniform dispersion of the fibre in the matrix; therefore, the low-tech hand layering approach has been seen as a possible way to avoid undesired concentrations of the fibre fibres.

Four samples were prepared with the following fibre content expressed in parts per hundred rubber (pph): sample A (0 pph): sample B (5 pph); sample C (7.5 pph); and sample D (10 pph). The formulation of the mixes and the number of specimens are given in Fig. 23.

The steps followed for the preparation of the biocomposite are described below:

1. Liquid natural rubber was poured into a rectangular plastic tray of 138 × 193 mm to form a thin layer of coagulated latex. There were 5 layers of 16 g of liquid natural rubber totalling 80 g of matrix per sample. The trays were placed on a levelled surface to avoid a different thickness throughout the composite (Fig. 24a).
2. The fibres were spread evenly by hand, with the help of a sieve, over the first layer of natural rubber and gentle pressure was applied to flatten the fibres against the latex (Fig. 24b).
3. A second layer of 16 g was poured over the fibres and coagulated (Fig. 24c).
4. Alternative layers of fibre (total of 4) and natural rubber (total of 5) were built up to an approximate 2 mm thickness as required under the ASTM standards for testing of

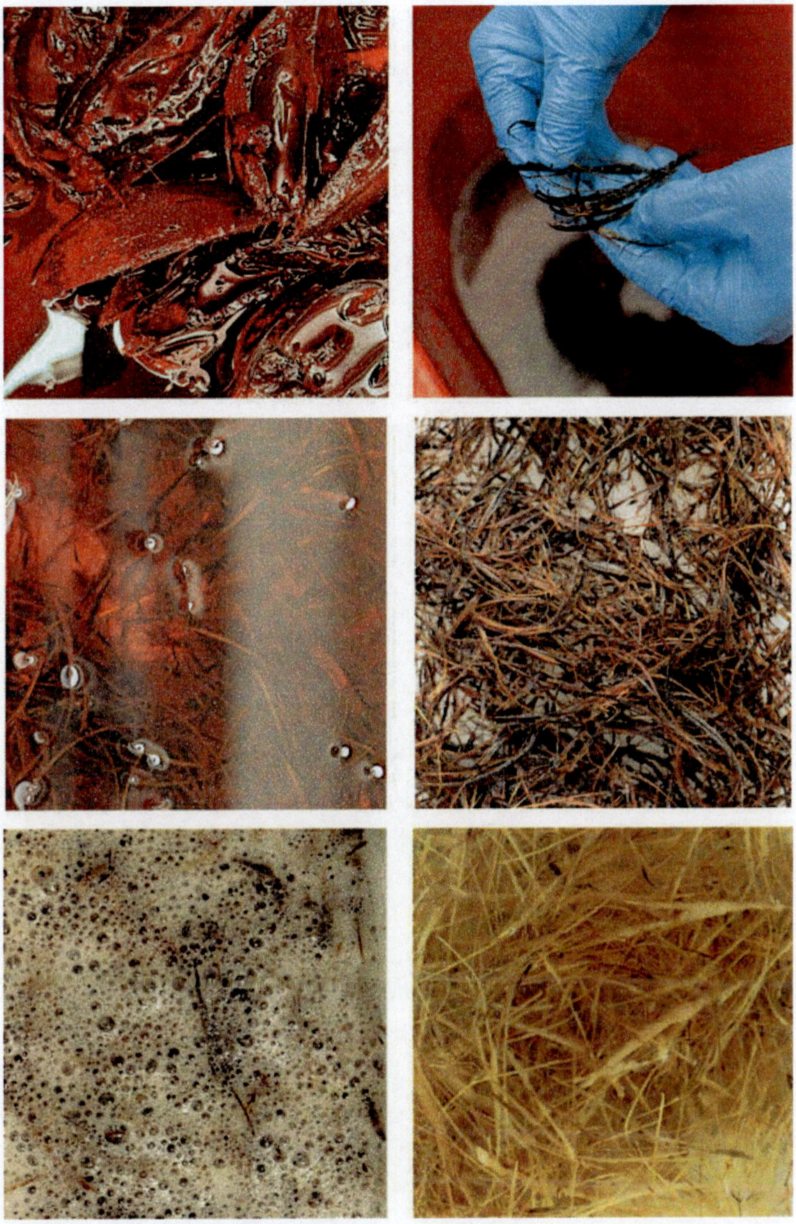

Fig. 20 **a** Alkali treatment. **b** Fibre extraction. **c** Washing of the fibres. **d** N_2O_2 treatment. **e** Wash in deionised water

mechanical properties of composite materials and elastomers (Fig. 25)

5. After the coagulation of the final layer, the sheets were removed from the mould and left to air dry at room temperature for at least 72 hs (Figs. 26 and 27).

Even though care was taken during the preparation of the composites to guarantee a uniform spread of fibres, the resulting samples show fibre concentration in many areas. An accurately homogenous dispersion of the fibres is not envisaged to be possible with the lay-up technique, but it may be appropriate if specific material regions require more or less reinforcement. Air bubbles trapped between the fibres were also observed, in particular within the 10 pph sample. Low-tech options for the preparation of the composite need to be explored further. From previous papers (Lopattananon et al. 2006)

Fig. 21 **a** Chopping. **b** Dried fibres

Fig. 22 Microscopic photos

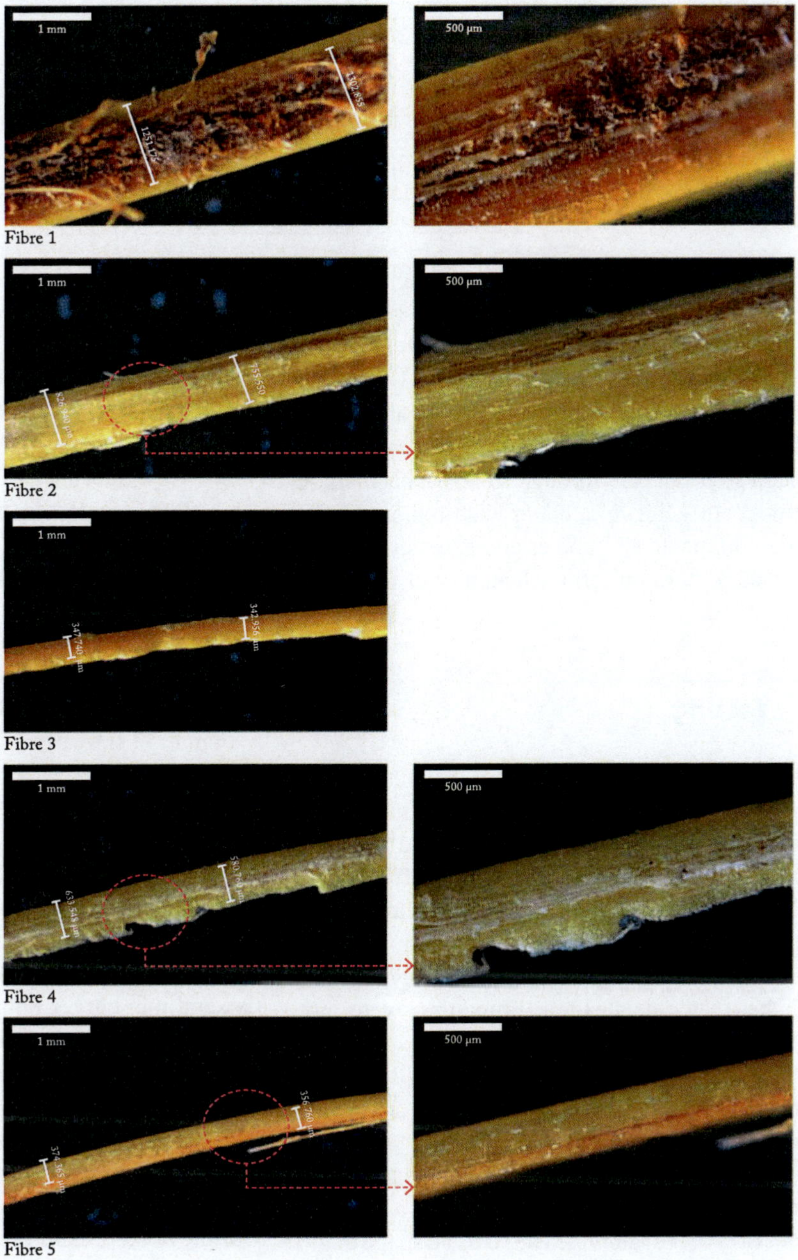

Fig. 23 Formulation of the rubber–fibre composites

Samples	A		B		C		D	
	pph	g	pph	g	pph	g	pph	g
Natural Rubber	100	80	100	80	100	80	100	80
Fibre	0	0	5	4	7.5	6	10	8

and relying on local machinery already used in the manufacture of rubber sheets, an alternative could be mixing the latex with the fibres into a dough so it can pass through rollers. This method could help achieve a better distribution of the fibres, eliminate air bubbles and excess water and provide a more uniform orientation of the fibres if required.

4 Testing

Tensile tests were carried out to the ASTM D638-14 'Standard Test Method for Tensile Properties of Plastics' and reinforced plastics. Tear tests were carried out to the ASTM D624-00 'Standard Test Method for Tear Strength of Conventional Vulcanised Rubber and Thermoplastic Elastomers'. Due to the non-rigid nature of the material, flexural properties were not tested.

Three dumbbell specimens for tensile testing were cut out from each sample with a CEAST Specimen Preparation Punching Machine (Fig. 29). Only two specimens for tear testing were possible to hand-cut with the aid of a template from each of the samples. The size and shape of the specimens are shown in Fig. 28. Tensile and tear tests of the composites were done at the Biocomposite Centre—Bangor University, utilising an Instron 3345 universal material testing machine (Fig. 30) and associated software Bluehill Universal. The tensile speed of testing was 20 mm/ min, and for the tear testing, 500 mm/min. Figure 31 shows all the specimens post-testing.

5 Results and Discussions

None of the specimens fractured during the tensile testing due to the limitation of the Instron 3345 machine to extend the specimens beyond a strain of 600%. Hence, the results obtained from the tensile testing show a partial outcome without being able to identify max strain at break or max strength at break. Smaller samples or a different machine will be required in future testing.

Figure 32 shows the data measured during the testing. Tensile stress–strain max/min curves for samples A, B, C and D are shown in Fig. 33. Sample A (without fibres) exhibit a curve, up to 600% strain, similar to the ones observed in the previous research papers and noted as the point where stress-induced crystallisation start to manifest (Parambath Kanoth et al. 2019). The max and min curves obtained for sample A vary significantly compared to the other samples. This could be linked to a potential variation of thickness between specimens.

The fibre-reinforced samples show an increase in the stress required to achieve a 600% tensile strain compared to sample A. This is consistent with previous research in fibre-reinforced rubber composites, where even though the ultimate strength and strain tend to decrease in comparison with non-reinforced natural rubber, the impact of the fibres does improve the stress–strain behaviour in the first section of the curve up to approximately 400–600% strain (Lopattananon et al. 2006; Parambath Kanoth et al. 2019) (Fig. 34).

Fig. 24 **a** First layer of liquid latex and coagulation. **b** First layer of fibres. **c** Final layer

Fig. 25 Layering up

Fig. 26 Dried samples. Fibre content **a** 0 pph. **b** 5 pph. **c** 7.5 pph. **d** 10 pph

a b

c d

Fig. 27 Dried biocomposites

Unlike previous research on natural rubber composites, following a sharp increase of strength up to an approx. 30% strain, the stress–strain curve of the composites then follows an almost parallel curve to sample A. From 450 to 550%, the composites start to show a second increase in strength compared to the non-reinforced sample. This performance could be linked to the distribution of the fibres in the composite, leaving areas free of fibre that behave as the non-reinforced sample. These clusters were observed and documented during the tests (Fig. 35a–b). The incipient second increase of tensile strength may be due to the more homogenous distribution of the fibres and rubber once the sample is stretched further.

Across the composites, sample C with fibre at 7.5 pph shows the maximum tensile strength at 600% strain, an increase of about 40% over sample A. This may be due to sample B (at 5 pph) having a large number of areas not covered by fibres, and in sample D (at 10 pph) agglomeration of fibres, generating fibre-to-fibre only areas and air bubbles between the fibres that were not filled in with rubber compromising the fibre–matrix interface (Fig. 36).

The efficiency of the fibre-matrix ratio, in this case, is closely linked to the method used to manufacture the composite, but it could also be a result of poor matrix–fibre adhesion (Fig. 39c). The pouring of the liquid latex and layering of the fibres is very challenging at a ratio over the 7.5 pph and start to require more substantial pressure to guarantee the matrix covers all the fibres.

It is noticeable the low strength (at 600% elongation) and modulus (at 300% elongation) achieved by all the samples in comparison to previous studies. The max tensile strength of non-reinforced sample A was 1.09 MPa, well

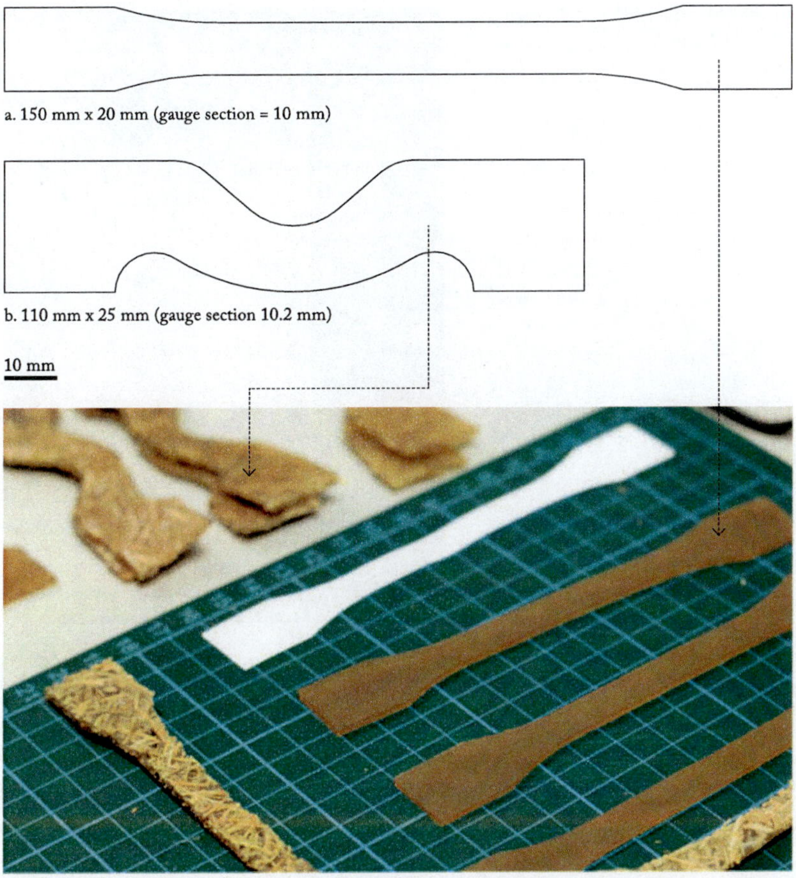

Fig. 28 Specimens for tensile (**a**) and tear testing (**b**) and Cut-out specimens

Fig. 29 CEAST Punching Machine

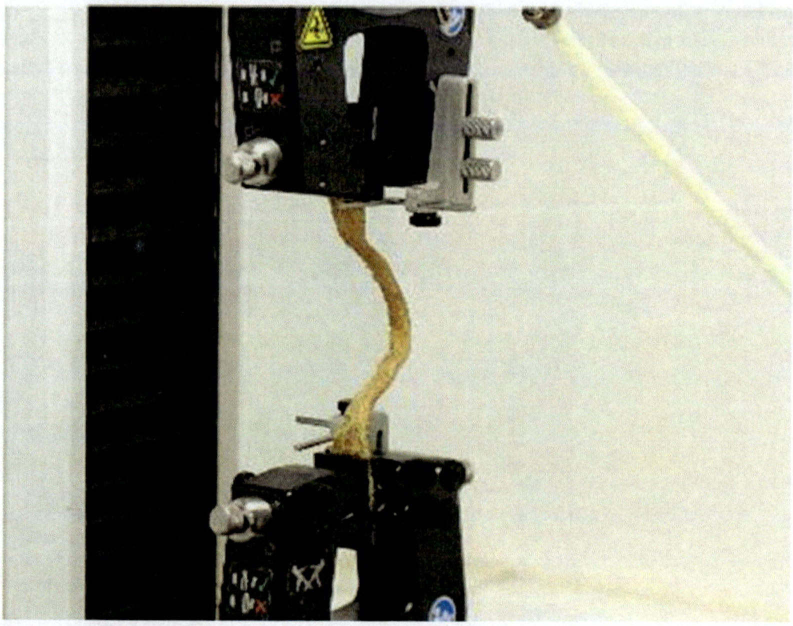

Fig. 30 Instron 3345—Universal Material Testing Machine

below the approx. 3 and 12 MPa measured in previous papers for natural rubber-only specimens (Fig. 34 and 36). The same applies to the module, where the 0.217 MPa achieved by sample A represents only 9% of what has been noted in previous studies (Mathew and Joseph 2007) (Figs. 37 and 38). This may be due to the type of liquid natural rubber used or post-processing of the samples following coagulation.

5.1 Tear Strength

The effect of fibre content on tear strength could not be concluded. The data collected shown in Fig. 39 can be used as a precedent for future testing, though there is a need for additional specimens and the use of a different testing machine/smaller specimens to determine a trend. It is also critical to also use a standard die cutter rather than hand-cut the samples.

Only one specimen cut-out from sample B and two specimens from sample C fractured during the tear testing (Fig. 42). Even though the tear strength at approx. 800% strain (max achieved by the machine) tends to increase with the addition of fibre, there is not a progressive increase of tear strength with an increase of fibre content as shown in identified in the previous studies (Parambath Kanoth et al. 2019) (Figs. 40 and 41). Since the dispersion of the fibre was not uniform, this outcome may be due to the nick in the specimen having cut areas with more or less fibre density or air bubbles that could impact the overall thickness of the specimen at the cross section.

6 Conclusions and Next Steps

This study demonstrates that babassu, a vital socio-economic, natural resource, already widely consumed by traditional and indigenous communities across the rainforest and the Brazilian savanna, can provide an alternative to the less sustainable materials regularly used in the region for non-traditional constructions. Research on the babassu by-products, rich in lignocellulose, is very minimal for its use in building materials; therefore, this research represents a departure point to explore innovative ways to use this low-cost renewable resource. Furthermore, this work also explores new uses in the construction field for natural rubber. It proposes a new approach to

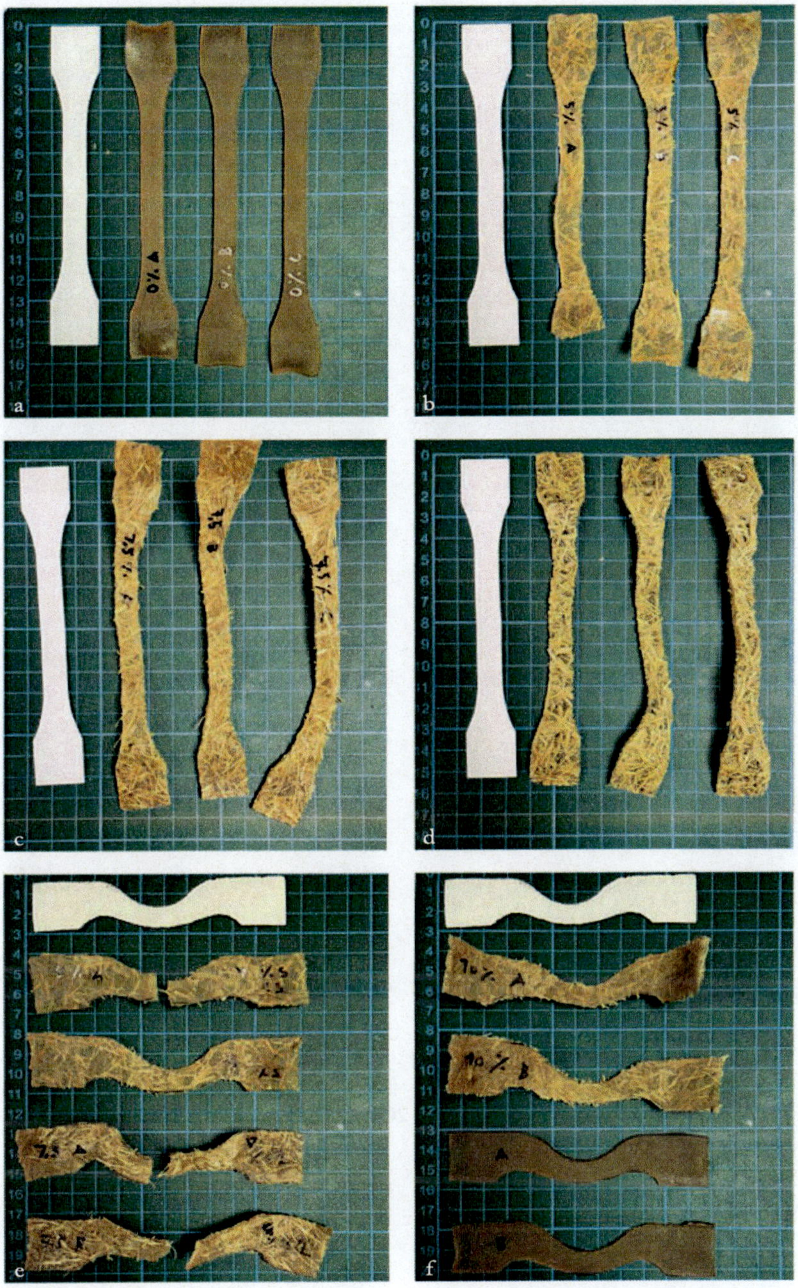

Fig. 31 Specimens post-testing and its relation to the die cut template **a–d** Tensile. **e–f** Tear

Samples	Maximum Load (N)	Tensile Extension at Maximum Load (mm)	Nominal Strain of Fracture (%)	Strain at Max Extension	Strain at Max Extension (%)	Tensile Stress at Max Extension (MPa)	Young's Modulus (stress at 300% elongation) (MPa)
A (0 pph)	21.91433	400.11136	1600.44543	5.16	515.56	1.0957	0.217
B (5 pph)	26.95452	400.23016	1600.920637	5.16	515.74	1.3477	0.255
C (7.5 pph)	31.13435	400.08882	1600.355307	5.16	515.52	1.5567	0.300
D (10 pph)	27.88163	400.08904	1600.35618	5.16	515.52	1.3941	0.273

Fig. 32 Tensile testing data

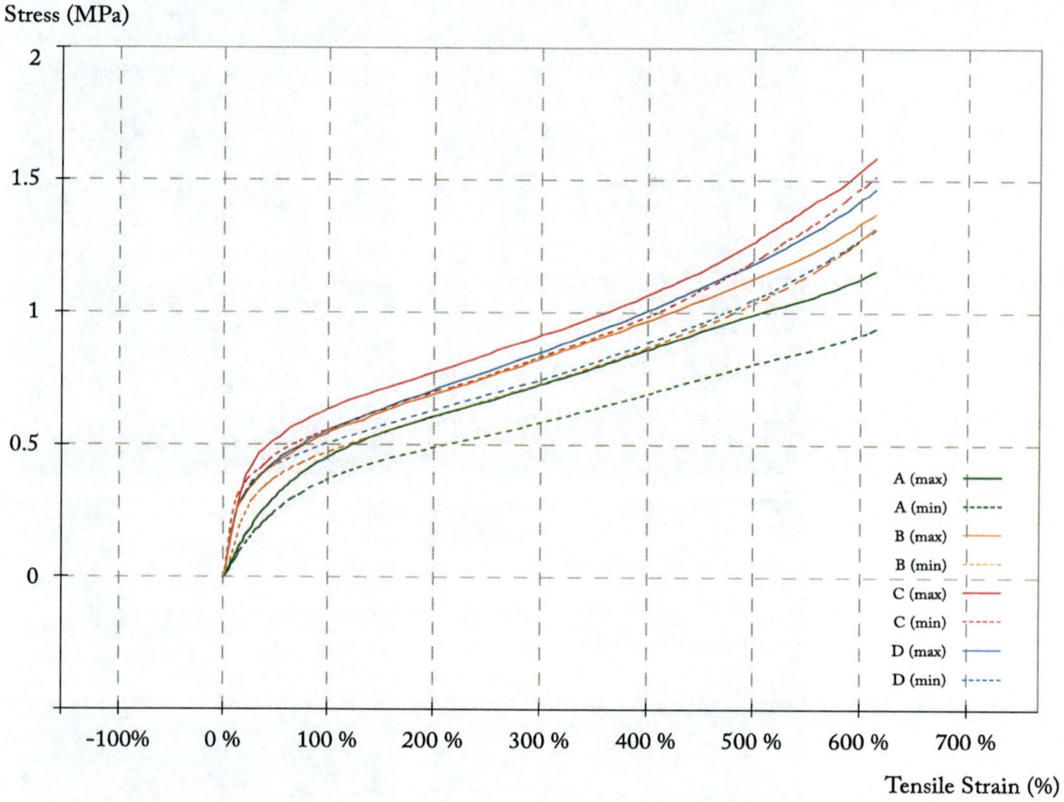

Fig. 33 Stress–strain curves for the max/min values per sample's specimens

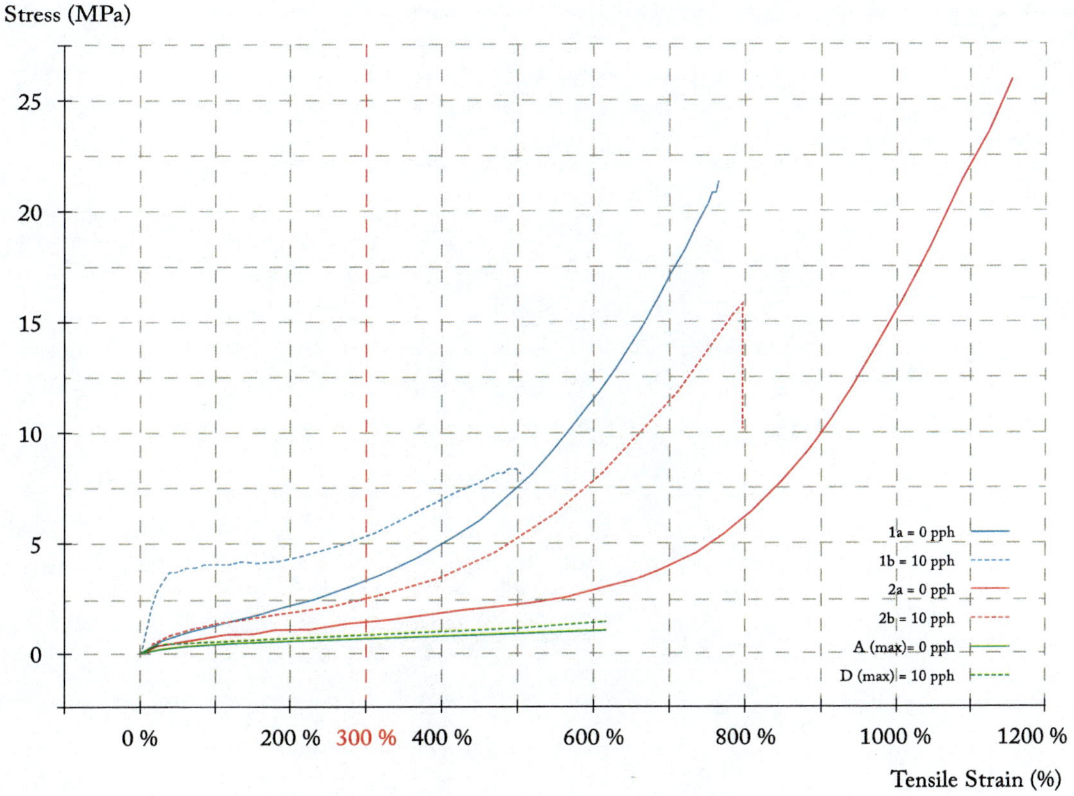

Fig. 34 Stress–strain curves for Pineapple leaf fibre–natural rubber composite (1a = 0 pph and 1b = 10 pph fibre samples) and coir macro-fibre–natural rubber composite (2a = 0pph & 2b = 10 pph fibre samples), adapted from Lopattananon et al.(2006), Parambath Kanoth et al. (2019), respectively, compared to the results obtained for the babassu fibre-reinforced natural rubber composite (A (max) = 0 pph and D (max) = 10 pph)

manufacturing low-cost materials locally/regionally and on demand, benefiting the extractivist communities directly and moving away from an industrialised and centralised model.

The outcome of the material testing does follow broadly the trends highlighted in previous studies on natural rubber biocomposites. Generally, the addition of fibre increased the stiffness of the material and tensile strength at lower elongations; the 7.5 pph sample shows a more significant increase of strength of about 40% at a 600% elongation compared to the non-reinforced sample. Nevertheless, the overall performance was well under the expected, and the observed in the previous studies.

It is possible to extract the fibres out of the shells with a relatively low-tech approach, though, machinery to crush and break the shells before or during the retting process will likely be required. The endocarp fibres seem to be stronger than the endocarp fibre and harder to extract. A more detailed analysis of the fibres is needed to understand the effect of chemical treatments on their surface and identify what processes enhance their adhesion to rubber.

The low-tech hand lay-up technique adopted for this study does not provide a uniform distribution of the fibres as anticipated. It was challenging to eliminate air bubbles and avoid fibre agglomerations beyond a fibre content of 7.5 pph. Microfibres may work better with this method for higher fibre loadings, as identified by Parambath Kanoth et al. (2019). Other methods, perhaps the already used during the processing of the liquid field latex into vulcanised rubber sheets, may be more appropriate to guarantee a

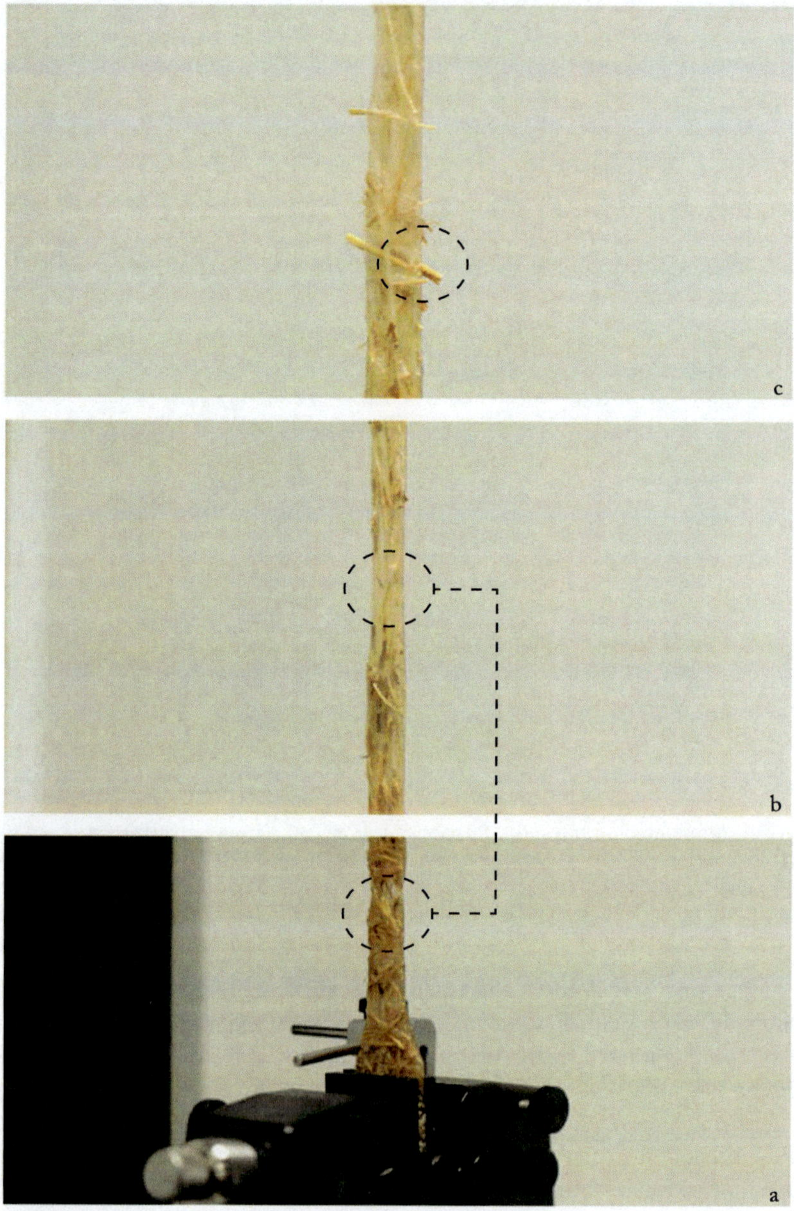

Fig. 35 **a** Initial state of the sample. **b** Halfway the testing of the sample and visible clusters with no reinforcement. **c** Fibre seems not to be fully adhered to the rubber

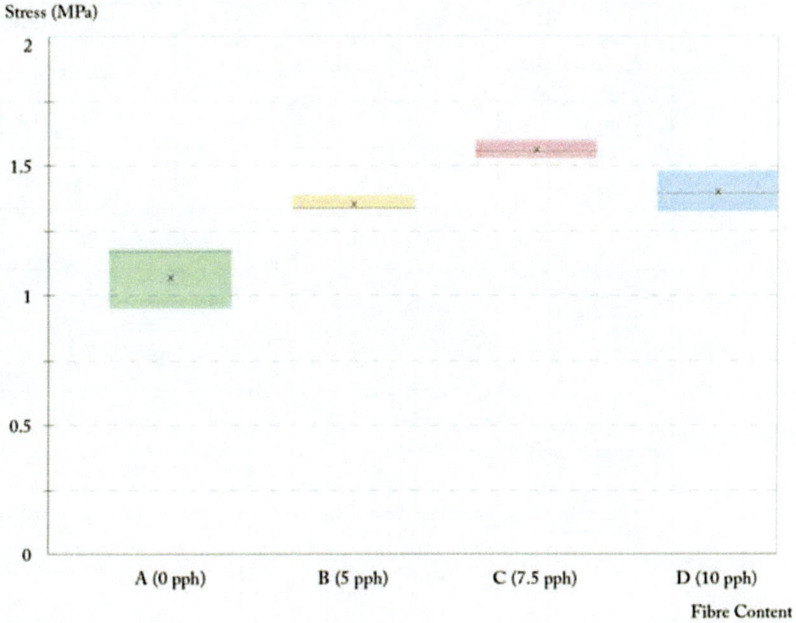

Fig. 36 Tensile stress—Fibre content. Tensile strength reaches a maximum at 7.5 pph. An increase of about 40% over the non-reinforced sample (X = Mean values)

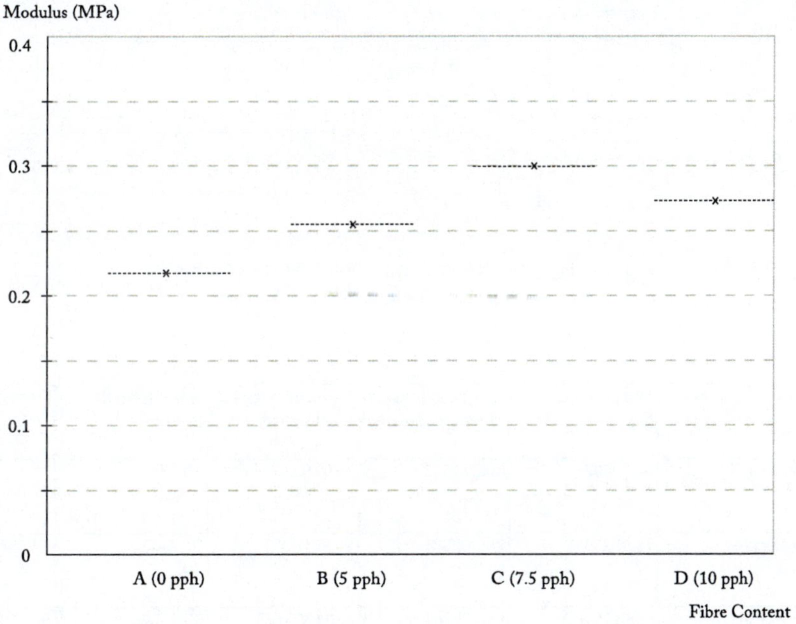

Fig. 37 Tensile modulus—fibre content

Year	Authors	Natural Fibre	Sample	Fibre Treatment (Type x Time)	Fibre Size	Load	Orientation	Modulus at 10% strain (MPa)	Modulus at 20% strain (MPa)	Modulus (300% elongation MPa)	Tensile strength (MPa)	Tear strength (MPa)	Elongation at break (%)	Hardness (Shore A)
1995	**Geethamma, Reethamma and Sabu, 1995** *Tensile Strength ASTM D412-68* *Tear Strength ASTM D624-54*	Coir	Gum			0%	L	0.31	0.42		17.95	35.05	1119	
							T	0.31	0.44		20.06	37.93	1141	
			Untreated Fibre		10 mm	35%	L	1.88	1.69		6.69	30.07	777	
							T	0.55	0.7		4.35	20.2	708	
			Treated Fibre	5% NaOH x 48 h	10 mm	35%	L		6.28		6.98	54.38		
							T		1.97		3.98	42.43		
2006	**Lopattananon et al., 2006** *Tensile Strength ASTM D638-96* *Thermal Aging ASTM D573-81*	Pineapple Leaf Fibre	Gum			0%	L				22.71		755	
							T				21.28		733	
			Untreated Fibre		75 ±20 um	20%	L				7.2		43.75	
							T				5.54		321.43	
			Treated Fibre	5% NaOH x 18 h	75 ±20 um	20%	L				10.02		105	
							T							
2007	**Lovely and Rani, 2007** *Tensile Strength ASTM D412-68* *Tear Strength ASTM D624-54* *Compression test ASTM D395-86 (B)* *Hardness ASTM 2240-81*	Isora	Gum			0%	L			2.3	25.9	35.1	1045	
							T			2.3	25	34.5	1045	
			Untreated Fibre		10mm	30%	L			3..2	9.8	46.8	437	69
							T			2.2	6.8	43.9	480	
			Treated Fibre	5% NaOH x 4 h	10mm	30%	L			3.8	10.7	47.9	400	73
							T			2.8	8.2	44.2	412	
2017	**Maslowski, Miedzianowska and Strzelec, 2017** *ISO 37*	Corn	Gum			0%					11.5 ±0.5		701 ±6	
			Fibre			20%					16.4 ±0.8		632 ±10	
2019	**Parambath Kanoth et al., 2019** *Tensile ASTM D412*	Coir	Gum			0%					25	45	1100	
			Treated Fibre	5% NaOH	6 mm	10%					14	50	800	
					400-550um	10%					30	80	950	
2021	Babassu Coconut Fibre reinfroced Natural Rubber Composite	Babassu Coconut Fibre	Gum							0.217	1.0957	28.09	515.56 (no break)	
			Treated Fibre	5% NaOH x 72 hs & 3% H2O2 x 8 hs	2-12 mm (approx)	5%	N/A			0.255	1.3477	37.98	515.74 (no break)	
						7.5%	N/A			0.3	1.5567	33.45	515.12 (no break)	
						10%	N/A			0.273	1.3941	43.06	515.52 (no break)	

Fig. 38 Results from the testing in context with previous research on natural rubber biocomposites

Specimen Label	Maximum Load (N)	Tensile Extension at Maximum Load (mm)	Nominal Strain of Fracture (%)	Tear Strength (N/mm)	Strain at Max Extension	Strain at Max Extension (%)
A (0 pph)	56.19492	402.73316	1610.932645	28.09746	7.94963	794.96258
B (5 pph)	75.96239	384.02508	1536.10033	37.98120	7.53389	753.38907
C (7.5 pph)	66.91211	350.83342	1403.333665	33.45605	6.79630	679.62981
D (10 pph)	86.09293	402.98329	1611.933175	43.04646	7.95518	795.51842

Fig. 39 Tear testing data

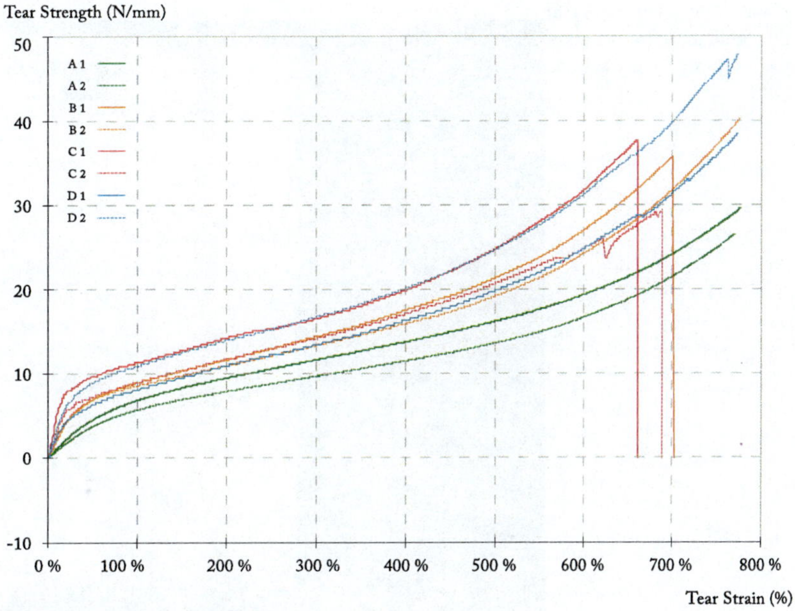

Fig. 40 Tear strength–strain curves for both specimens tested per sample

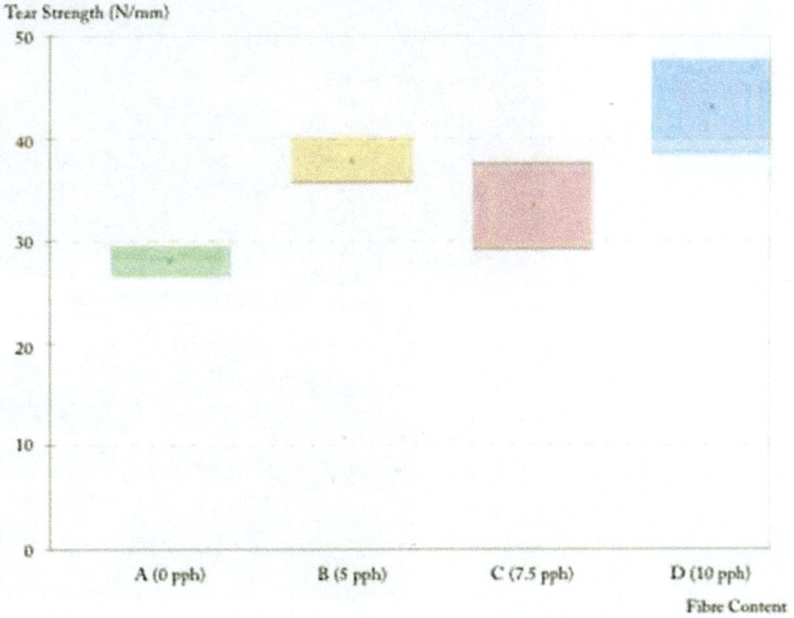

Fig. 41 Tear strength—fibre content. (X = Mean value/— = indicates fracture)

homogenous distribution of the fibres, plus it would rely upon existing craft skills. Different fabrication methods and post-treatment of the composite should be tested and compared.

It is envisaged this material, in the form of sheets or tiles, could replace the polycarbonate sheets currently used as part of the building envelope (Fig. 43). Though in early stages and with much refining required, the outcome of the material testing is promising and allows to start thinking about how the sheet material could be shaped or manipulated to maximise its mechanical performance to respond to more specific applications for upcoming design projects.

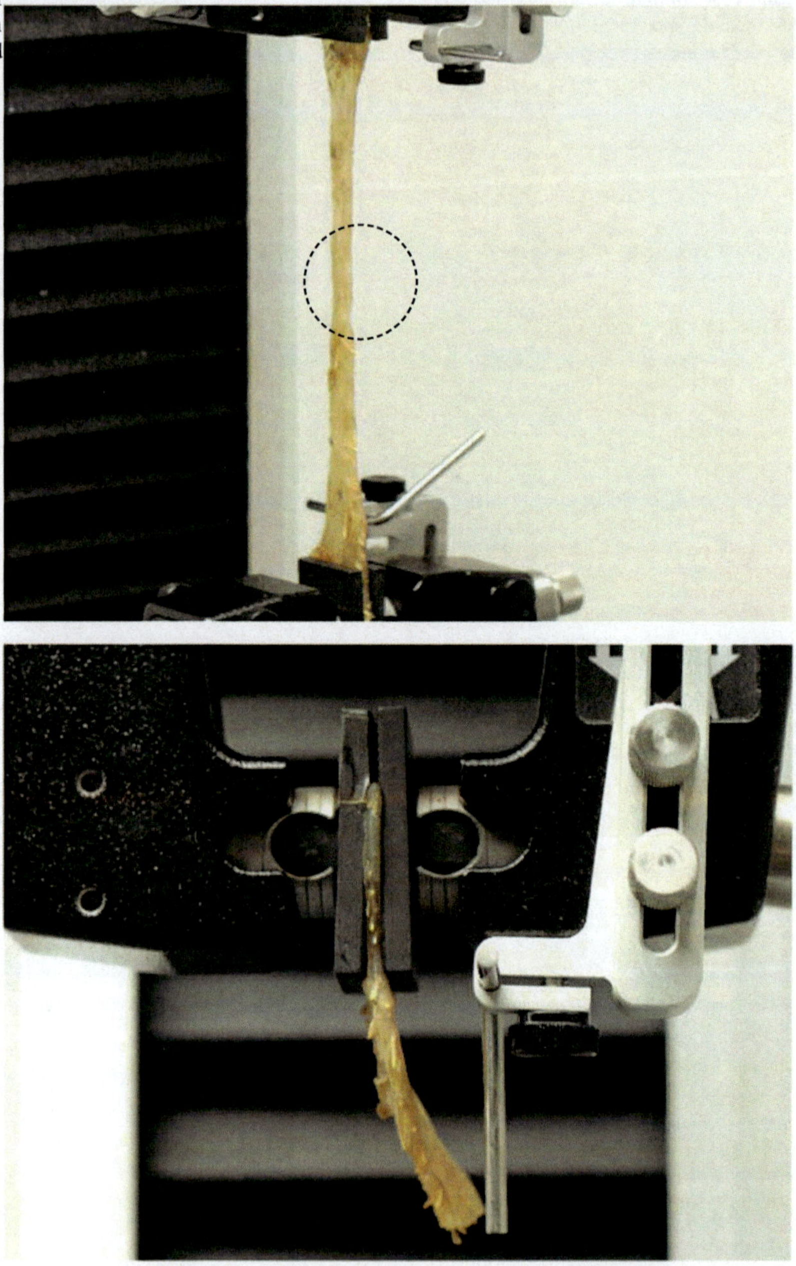

Fig. 42 Photos from the tear testing indicating the location of the nick (top) and fractured sample

Fig. 43 Potential application —Concept

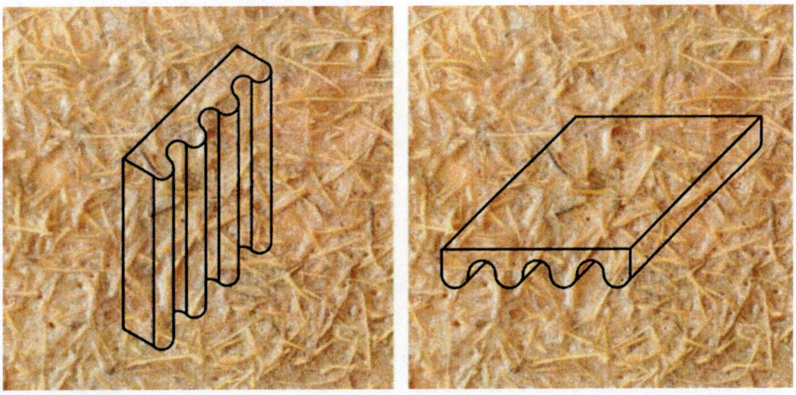

Acknowledgements The author gratefully acknowledges Dr Graham Ormondroyd and Dr Simon Curling from the Bangor Biocomposite Centre at Bangor University and Brenda Parker from Bio-ID at the University College London.

References

Abramovay R (2019) Amazonia: Por Uma Economia do Conhecimento da Natureza. Elefante Editora

Almeida Campos JL, da Silva TLL, Albuquerque UP et al (2015) Knowledge, use, and management of the babassu palm (Attalea speciosa Mart. ex Spreng) in the Araripe Region (Northeastern Brazil). Econ Bot. https://doi.org/10.1007/s12231-015-9315-x

Alves Lopes I, Coelho Paixão L, Souza da Silva LJ, et al (2020) Elaboration and characterization of biopolymer films with alginate and babassu coconut mesocarp. Carbohydr Polym. https://doi.org/10.1016/j.carbpol.2019.115747

Amaral HR, Cipriano DF, Santos MS et al (2019) Production of high-purity cellulose, cellulose acetate and cellulose-silica composite from babassu coconut shells. Carbohydr Polym. https://doi.org/10.1016/j.carbpol.2019.01.061

Beber VC, de Barros S, Banea MD et al (2018) Effect of Babassu natural filler on PBAT/PHB biodegradable blends: an investigation of thermal, mechanical, and morphological behavior. Materials (basel). https://doi.org/10.3390/ma11050820

Carrazza LR, Cruz e Avila JC, Lima da Silva M (2012) Manual Tecnologico de Aproveitamento Integral do Fruto e da Folha do Babaçu (Attalea spp.), 2nd edn

Clement CR (2016) 3. Demand for two classes of traditional agroecological knowledge in modern Amazonia. In: Human impacts on Amazonia

De D, De D, Adhikari B (2006) Curing characteristics and mechanical properties of alkali-treated grass-fiber-filled natural rubber composites and effects of bonding agent. J Appl Polym Sci. https://doi.org/10.1002/app.23305

de Lemos AL, Pires PGP, de Albuquerque ML et al (2017) Biocomposites reinforced with natural fibers: thermal, morphological and mechanical characterization. Rev Mater. https://doi.org/10.1590/s1517-707620170002.0173

de Paula PT, Trugilho PF, da Silva César AA et al (2014) Babassu nut residues: potential for bioenergy use in the north and Northeast of Brazil. Springerplus. https://doi.org/10.1186/2193-1801-3-124

dos Santos Silva ID, Schäfer H, Jaques NG et al (2020) An investigation of PLA/Babassu cold crystallization kinetics. J Therm Anal Calorim. https://doi.org/10.1007/s10973-019-09062-2

Ferrari RA, Soler MP (2015) Obtention and characterization of coconut babassu derivatives. Sci Agric. https://doi.org/10.1590/0103-9016-2014-0278

Fonteles CAL, Brito GF, Carvalho LH et al (2016) Composites based on thermoset resin and Orbignya phalerata (babassu coconut): Evaluation of mechanical properties, morphology and water sorption. In: Materials science forum

Geethamma VG, Pothen LA, Rhao B et al (2004) Tensile stress relaxation of short-coir-fiber-reinforced natural rubber composites. J Appl Polym Sci. https://doi.org/10.1002/app.20746

Gehring C, Zelarayan MLC, Almeida RB, Moraes FHR (2011) Allometry of the babassu palm growing on a slash- and-burn agroecosystem of the eastern periphery of Amazonia. Acta Amaz. https://doi.org/10.1590/s0044-59672011000100015

Gomes MSSO, Nascimento JR, E Silva M das GO (2016) MESOCARPO DE BABAÇU (ORBINYA SP) COMO ADSORVENTE DO DODECIL BENZENO SULFONATO DE SÓDIO (SDBS). HOLOS. https://doi.org/10.15628/holos.2016.3019

Hoffmann R, Morais DDS, Braz CJF et al (2019) Impact of the natural filler babassu on the processing and properties of PBAT/PHB films. Compos Part A Appl Sci Manuf. https://doi.org/10.1016/j.compositesa.2019.105472

Lewis SL, Maslin MA (2020) The human planet: how we created the anthropocene. Glob Environ. https://doi.org/10.3197/ge.2020.130308

Lima AM, Vidaurre GB, Lima RDM, Brito EO (2006) Use of babaçu staple fiber as alternative raw material for panel production. Rev Arvore. https://doi.org/10.1590/s0100-67622006000400018

Lopattananon N, Panawarangkul K, Sahakaro K, Ellis B (2006) Performance of pineapple leaf fiber-natural rubber composites: the effect of fiber surface treatments. J Appl Polym Sci. https://doi.org/10.1002/app.24584

Machado NAF, Furtado MB, Parra-Serrano LJ et al (2017) Painéis aglomerados fabricados com resíduos do coco babaçu. Rev Bras Ciencias Agrar. https://doi.org/10.5039/agraria.v12i2a5434

Masłowski M, Miedzianowska J, Strzelec K (2017) Natural rubber biocomposites containing corn, barley and wheat straw. Polym Test. https://doi.org/10.1016/j.polymertesting.2017.08.003

Mathew L, Joseph R (2007) Mechanical properties of short-isora-fiber-reinforced natural rubber composites: Effects of fiber length, orientation, and loading; alkali treatment; and bonding agent. J Appl Polym Sci. https://doi.org/10.1002/app.25065

Nair AB, Joseph R (2014) Eco-friendly bio-composites using natural rubber (NR) matrices and natural fiber reinforcements. In: Chemistry, manufacture and applications of natural rubber

Nunes MABS, Castro-Aguirre E, Auras RA et al (2019) Effect of babassu mesocarp incorporation on the biodegradation of a PBAT/TPS blend. Macromol Symp. https://doi.org/10.1002/masy.201800043

Parambath Kanoth B, Thomas T, Joseph JM, Narayanankutty SK (2019) Restructuring of coir to microfibers for enhanced reinforcement in natural rubber. Polym Compos. https://doi.org/10.1002/pc.24667

Peres JBR, Pastore F (2019) Amazon rubber, a potential yet to be rediscovered. J Polym Environ. https://doi.org/10.1007/s10924-019-01381-7

Reul LTA, Pereira CAB, Sousa FM et al (2019) Polycaprolactone/babassu compounds: rheological, thermal, and morphological characteristics. Polym Compos. https://doi.org/10.1002/pc.24861

Sreekumar PM, Gopalakrishnan P, Saiter JM (2013) Biofiber-reinforced natural rubber composites. In: Polymer composites, biocomposites

Stelescu MD, Manaila E, Craciun G, Chirila C (2017) Development and characterization of polymer eco-composites based on natural rubber reinforced with natural fibers. Materials (basel). https://doi.org/10.3390/ma10070787

Teixeira PRS, Teixeira AS do NM, Farias EA de O et al (2018) Chemically modified babassu coconut (Orbignya sp.) biopolymer: characterization and development of a thin film for its application in electrochemical sensors. J Polym Res. https://doi.org/10.1007/s10965-018-1520-8

United Nations (2019) Global Sustainable Development Report 2019: The Future is Now – Science for Achieving Sustainable Development

Vadjunec JM, Gomes CVA, Ludewigs T (2009) Land-use/land-cover change among rubber tappers in the Chico Mendes Extractive Reserve, Acre, Brazil. J Land Use Sci. https://doi.org/10.1080/17474230903222499

WWF (2020) Living planet report 2020—bending the curve of biodiversity loss

The Scope of Egg Waste Use in the Built-Up Environment: A Study on the Viability of Eggshell Waste as an Organic Building Material

Esther Kiruba Jebakumar Clifford

Abstract

USA, India, and China are the three largest producers of eggs and annually India produces an average of 83 billion eggs. The egg waste generated from this massive production is sent to either landfills or incinerators. The EPA declared eggs as the 15th largest waste causing pollution. Companies pay up to $100,000 to dispose of egg waste in the form of landfills and this is an unsustainable and expensive way of disposing of waste. Egg waste in the form of shells and whites can be rather used as a building material that is cost-efficient and energy-efficient. Eggs have been used as a building material in the past millennia throughout Asia, Africa and parts of Europe. In the local vernacular architecture of India, its properties have been known to help in cooling the built-up environment and also provide a smooth finish to walls. In modern usage, eggshells have been used as an additive in concrete and have proven to add strength to the concrete admixture. Rethinking the way egg waste is repurposed can help in creating resilience through building materials. This paper will help understand the scope and viability of using egg waste in architecture by studying and recording the amount of egg waste generated and the logistics involved to help procure and use eggshells and white waste. This study will take an example of an Indian district to help demonstrate the viability of the practical usage of egg waste in the built-up environment.

Keywords

Eggshell waste · Rethinking egg waste · Eggshell in construction

1 Introduction

The management of large amounts of waste in the form of eggshells is a tough challenge for both developing and developed countries. An efficient way of repurposing this waste can be converting it into valuable and useful resources such as converting it for the built environment as a building material or use in a building construction technique/method. This will help improve both the sustainable development of the particular area and contribute toward becoming an adequate waste management strategy.

In 2020, the production volume of eggs worldwide exceeded 86.67 million metric tons. This is up from 74.14 million metric tons in 2016. Since 1990, the global egg production

E. K. J. Clifford (✉)
Sathyabama Institute of Science and Technology, Architecture, Chennai, India
e-mail: archiesther@gmail.com

M. R. Thomsen et al. (eds.), *Design for Rethinking Resources*, Sustainable Development Goals Series,
https://doi.org/10.1007/978-3-031-36554-6_13

volume has increased by over 100% (OECD-FAO 2022). This increases the scope to work with eggshell waste as a building material or used in building techniques.

This study on repurposing waste focuses on the usage of the organic waste of eggshells in the built-up environment. Historically speaking egg has been used for the past centuries from Europe to Africa to Europe (Gubba and Sasikala 2022). There have been studies surrounding the use of eggshells in construction (Gubba 2019), mainly around the use of them as a fine aggregate for concrete (Raji and Tomi Aina 2015). This paper helps weigh the viability and sourcing of eggshell waste and then connect the same to various possible end products such as concrete, cement, wall plaster, and paint.

The inquiry into the scope of eggshell waste used as a building material is important because it caters to the following aspects:

(1) The fact that there is an acute shortage of resources in the construction industry.
(2) There is an urgent need to understand and rectify the way certain wastes are being disposed of and the reduction of landfills.
(3) The research would help in getting in line to help achieve Sustainable Development Goal 11 (make cities and human settlements inclusive, safe, resilient and sustainable), Sustainable Development Goal 12 (ensure sustainable consumption and production patterns), and Sustainable Development Goal 13 (take urgent action to combat climate change and its impacts).

This research will help understand the viability of using egg waste in architecture by studying and recording the amount of egg waste generated and the logistics involved to help in procuring and using eggshell waste. One other aspect will be the way that eggshell waste is used in different sectors and how it will help in the reduction of energy usage in terms of carbon footprint, waste disposal, and procurement of new building material resources.

2 Methodology

See Fig. 1.

2.1 Collection of Eggshell Data

Eggshell Waste Sources

Eggshell waste needs to be sourced from the most common producers of egg waste. During the study on the district of Kanyakumari, discussions brought forward the fact that one of the main sources of eggshell waste is from the academic sector. This is not as surprising since the district has the highest in the state and among all the other states as well. In 1956, then Chief Minister of Tamil Nadu, M. Kamaraj introduced the Mid-Day Meals Scheme in a move to encourage children from villages and rural areas to attend schools and get proper education instead of going into the fields and working with their families. This has been one of the most successful schemes to enhance education and literacy rate. The scheme has been renamed the Pradhan Mantri Poshan Shakti Nirman and has been successfully implemented in most of the states. This scheme is instrumental in the distribution of eggs as part of lunch at least 3 days a week. This particular region has up to 74,241 enrolments per day to take advantage of the scheme. This accounts to up to 44,500 eggs being consumed (Ministry of Human Resource Development 2022) (Figs. 2 and 3).

There are around 200 retail commercial bakeries and home bakers in the region producing on average around eggshell waste from 80,000 eggs per day and 7000 eggs waste from local food stalls (Google Maps 2022).

In India, the average egg eater eats around 45 eggs per year according to surveys and polls (Lokniti 2017). 48.5% of the population eats at least one egg per week (Sharma 2021). With these statistical data, it can be inferred that the biggest source of eggshell waste comes from the 260,000 egg eaters in the district. It is interesting

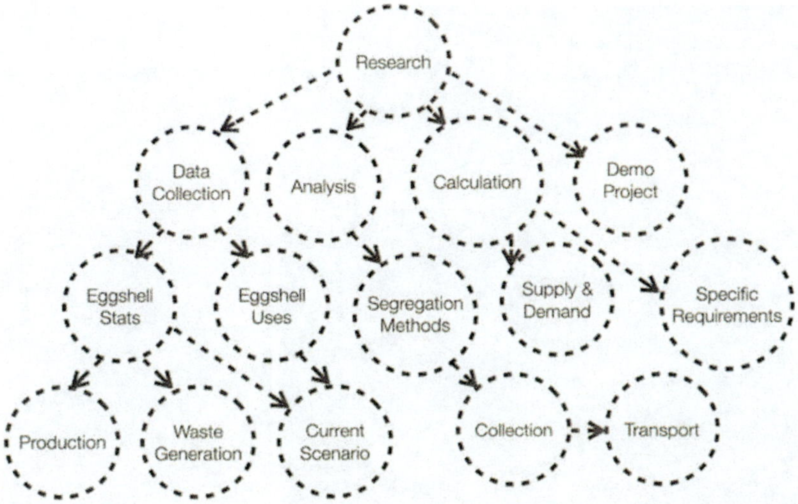

Fig. 1 Research methodology

Fig. 2 Primary school children getting ready for lunch with eggs in the Kuruntankode sector school under Pradhan Mantri Poshan Shakti Nirman

to note that every person in the district has access to at least six eggs per day.

India is known for its lavish weddings and grand feasts. On average 30,000 weddings take place every day. Taking the context of the Kanyakumari District and the population (Kanniyakumari District Administration 2022), along with the wedding seasons and weddings serving non-vegetarian means (survey taken with local wedding businesses and event planners), an average of 50,000 eggs generates egg waste at the venues and kitchens. 53,800 eggs are produced on poultry farms in the district. Hatcheries in the district produce around 53,800 eggs (Animal Husbandry Department 2022) which leads to the waste from eggs that are damaged and broken which ranges from 6 to 8% of the total (Table 1).

Fig. 3 Unsegregated and segregated eggshell waste in food stalls and commercial retail bakeries in Kanyakumari District

Table 1 Eggshell waste data which will help in connecting the eggshell source with prospective building sector as per the weight and the source

Source	Average no. of eggs/Day	Average weight (Kg)	Current disposal method
Schools	44,500	333.75	Natural fertilizer/incineration/landfills
Bakeries	80,000	600	Incineration/landfills
Road side stalls	7000	52.5	Incineration/landfills
Wedding caterers	50,000	375.0	Incineration/landfills
Human consumption	260,000	1950.0	Natural fertilizer/incineration/landfills
Hatchery waste	4000	30.0	Composting/rendering

2.2 Eggshell Morphology

Eggshells make up about 9.0–12.0% of the entire egg weight and considering the chemical constitution of an eggshell, 98% of eggshell comprises dry matter and 2% is composed of water. The dry matter is constituted of mostly ash (93.0%) and crude protein (5.0%). Eggshell weight on average is 10% of the entire weight of the egg which in turn is around 70.0 to 80.0 gms (Van Immerseel et al. 2016). This means the average weight of eggs shell waste from one egg would be 7.5 gms (Waheed et al. 2020).

3 Current Scenario for Waste Disposal

The management of the large amounts of eggshell waste daily produced in the district is problematic because this is only disposed of in landfills and incinerators. Landfills having egg waste cause the issue of odor production and microbial growth and incinerators are major air pollution contributors (Mignardi et al. 2020).

Eggshells are collected along with other organic wastes and taken to municipal solid

Fig. 4 Waste eggshell in a landfill (Shiferaw et al. 2019)

waste incinerators. The main advantage of this method is that it saves the eggshell from being sent to landfills which would cause accumulation and piling up of waste in mass and volume since eggshells cannot break down easily in the soil or allow themselves to compost. It is also one of the effective methods for destroying potentially infectious agents while eliminating the threat of disease. But the reduction is only 7.5% of the original volume of the eggshell waste that is disposed of. One of the biggest disadvantages is the air pollution caused by the burning of the material. The air emission and the disposal of solid residue need to be strictly controlled in proper conditions (Singh et al. 2018). It also needs expensive fuel and supervision and is thus an expensive method.

Landfills: Most of the eggshell waste produced in the district are sent to landfills. Even with the segregation of waste and cleaning at the source and after collection, the other wastes attached to the egg cause them to rot quickly, causing a smelly by-product and toxic bacterium. Studies indicate that landfills are the least sustainable and least efficient way of disposing of eggshell waste (Shearman 2016).

Natural fertilizer: This maybe the only sustainable and beneficial way of disposing of the eggshell waste by recycling it. Eggshells can be used to enhance soil fertility as it quickly decomposes. This adds nutrients like calcium and other minerals to the soil. A crushed and powdered eggshell is scattered around the potential landscaped site and this protects it from harmful slugs, snails, and cutworms that affect the plants, without using harmful pesticides. Another way is by using the biodegradable eggshell halves as an organic plant pot which is filled with potting soil to which seeds are added to each shell to start seedlings. This is an organic and simple way to propagate a seedling in shells. Once the seedlings are large enough to be transplanted to the outside, the shells are broken from the bottom of the eggshell and reared seedlings are planted (Fig. 4).

4 Connecting the Study to the Relevant Sustainable Development Goals

4.1 Sustainable Development Goal 11

One of the targets is by 2030 to, "Reduce the adverse per capita environmental impact of cities, including by paying special attention to air quality and municipal and other waste management" (11.6). Another way is by "Supporting least developed countries, including through financial and technical assistance, in building sustainable and resilient buildings utilizing local materials" (11.C).

Cement production accounts to 5% of the carbon emissions in the world (Barcelo et al.

2013) which is around 900 kg of carbon dioxide (CO_2) gas that is emitted into the environment for every ton of cement produced (Benhelal et al. 2013). The substitution of eggshell waste in place of cement has been experimented with and up to 20% replacement (Tan et al. 2018) in mixtures such as concrete is possible which will eventually help in reducing the otherwise produced high carbon emissions. This is also achieved through the reduction of clinker in cement production (Shiferaw et al. 2019) which is the main contributor to the high emission.

The environmental impact due to the eggshell waste is high risk. The US Environmental Protection Agency declared waste as the 15th largest food industry-produced pollutant (MA and RK 2003). The district selected for study is located in one of the largest egg-producing states (Tamil Nadu) of India. India itself despite being a developing country is the second largest egg producer in the world. The use of eggshell waste at every level of the community will help in creating resilient environments.

4.2 Sustainable Development Goal 12

This SDG works toward, "Achieving the sustainable management and efficient use of natural resources" (12.2), "Substantially reducing waste generation through prevention, reduction, recycling and reuse" (12.5), "Ensuring that people everywhere have the relevant information and awareness for sustainable development and lifestyles in harmony with nature" (12.8).

Waste management is an issue that countries like India have been facing in recent times, owing to the ever-growing population. 97.4% of the population in Tamil Nadu eats eggs and this itself leads to an exponentially increasing pile of eggshell waste not being utilized or properly disposed of. This method of using waste can help in reducing waste generated as well as creating awareness in places like public schools where eggs are distributed as a part of the Mid-Day Meal Scheme.

4.3 Sustainable Development Goal 13

One of the goals is to "Improve education, awareness-raising and human and institutional capacity on climate change mitigation, adaptation, impact reduction and early warning" (13.3).

Eggshell waste disposal can be a messy process and when sent to landfills can produce harmful toxins and odor in the environment when the residual dirt and matter are not cleaned or removed from the whole egg. This study helps in increasing awareness of the negative environmental impact of eggshell waste while helping the community understand and hopefully participate in alternative solutions.

4.4 Eggshell Waste Usage

See Fig. 5.

Examples of eggshell being used in building construction include but are not limited to:

Lime-Based Mortar

5–20% of the mixture with eggshell with 5% being the best mixture helps in creating good quality lime mortar mix (Pliya and Cree 2015).

Fly-Ash Blended Concrete

The strength and permeability properties of concrete significantly jump up to 30% of cement substitution with 5% eggshell powder (Mahendran 2014).

Fine Aggregate in Concrete

100% of fine aggregate substitution helps in bringing about a stronger, lightweight type of concrete (Raji and Tomi Aina 2015).

Filler in Concrete

Concrete with the eggshell powder addition of 10% as a filler in concrete showed the highest compressive strength. The inclusion of eggshell powder as filler into concrete also improved the flexural behavior of concrete by up to 22.9% compared to the controlled concrete (Doh 2014).

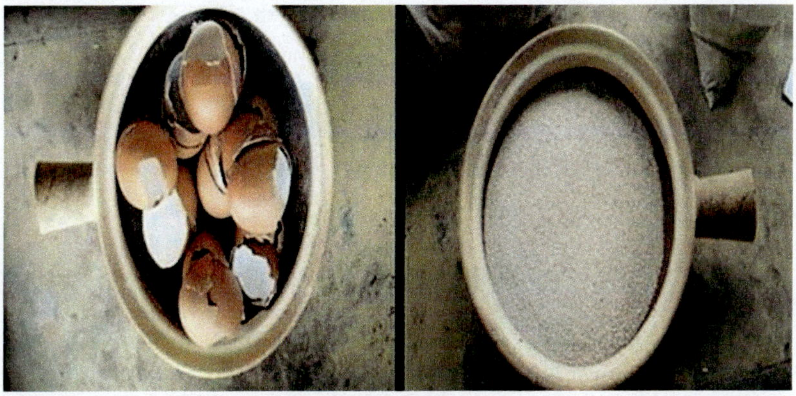

Fig. 5 Eggshell and powder (Tanyildizi et al. 2022)

The base for Road Construction

8% eggshell usage in the road base mixture helped in stabilizing the Laterite soil processes by helping in bringing it to the closest optimum moisture content and maximum dry density properties (Okonkwo et al. 2012) (Fig. 6).

Reducing the Setting Time of Concrete

Research shows that Cement-eggshells ash (CESA) paste when constituted using 2.5% of ESA by weight of cement can help in manipulating the setting time of concrete. A setting time test was conducted on the paste that has the eggshell waste. Results showed that the addition of eggshell waste powder to the ordinary Portland cement decreased the setting time of the cement. The conclusions drawn were that the eggshell waste in the paste acted as an accelerator and further experiments were projected to prove that the higher the eggshell content, the faster the rate of setting would be (Ujin et al. 2016).

Plaster for Walls

After having the inner film of eggshells peeled off in this experiment, the eggshells were cleaned, sterilized, crushed, and ground into powder. Then water was added along with gypsum and the mixture was then kneaded into a plaster-like texture. In this process, eggshell powder was the main ingredient with 70% of the volume of the plaster mixture (Thomas 2005).

Replacement of Portland Cement in Concrete

While using eggshell powder in quantities of 5–20% by volume to Portland cement (Tan et al. 2018), the results show that water-cured eggshell concrete greatly improves the compressive and flexural strength of concrete. This went up to 57.8% while also helping in the rate of water

Fig. 6 Eggshell concrete brick (Raji and Tomi Aina 2015)

Table 2 Eggshell usage ratios, properties, and improvement

Type of use	Ratio	Fire/water resistance	Improvement
Lime-based mortar	5–20%	F. R. properties	Improved strength and durability
Fly-ash blended concrete	6% by weight of cement	Permeability increase	Increase in strength
Fine aggregate in concrete	100%	NA	Reduction in weight/improvement in strength
Filler in concrete	10% by volume	Reduction in water absorption and water penetration	Flexural strength by 22.9%
Base for road construction	8% by volume	Improved W.R	Reduction in cost
Setting time of concrete	2.5% by weight	–	Reduction in setting time
Plaster for walls	70% by volume	Repels moisture	Decrease in temperature
Replacement of Portland cement in concrete	5–20% by volume	W.R. Increased by 50%	Compressive and flexural strength

absorption of eggshell concrete by reducing approximately 50%, as eggshell powder filled up the existing empty spots, making it an impermeable material (Table 2).

5 Results

Stakeholders: To understand how to practically bring about this research, the stakeholders must be discussed. Since the source is from where the process starts, the first stakeholders include the local business owners of the food stalls, bakeries, and eateries using eggs in their daily routines. Mid-Day Meals distributors and egg consumers at the home and local levels are also part of this stakeholder group.

The next level of stakeholder is the collector of eggshell waste whose job might include the jobs go segregation and cleaning of the eggshell waste as well. Collectors can be directly the source workers or workers at the collection points where the first stakeholders can come and give the eggshell waste they have accumulated.

Transporting Personnel, Cleaning, and Storage Parties have the important job of making sure the eggshell waste reached the end user without any problem (such as rot, bacteria, breakage, and waste through transit).

The penultimate stakeholder is the distributor who has to connect the eggshell waste collected to the end user.

The end user is the main stakeholder of the process since they will pay or barter for the eggshell waste which they use in the construction sector.

Connecting to the end user: The best way to explain the source to end user flow of the eggshell waste will be to understand Fig. 7 explaining the process along with the stakeholders and the activities involved.

6 Discussions

The final research when taken to the practical stage of connecting all the stakeholders will have an output that is beneficial for the construction industry in terms of sourcing readily available eggshell waste for usage as a building material or in building construction/finishing techniques. While used as a resource, the scope of eggshell waste would also help in the reduction of landfills and the use of incinerators. In addition, the conversion of eggshell waste into usable material will help in achieving the UN Agenda by directly and indirectly helping in attaining the Sustainable Development Goals (United Nations 2022).

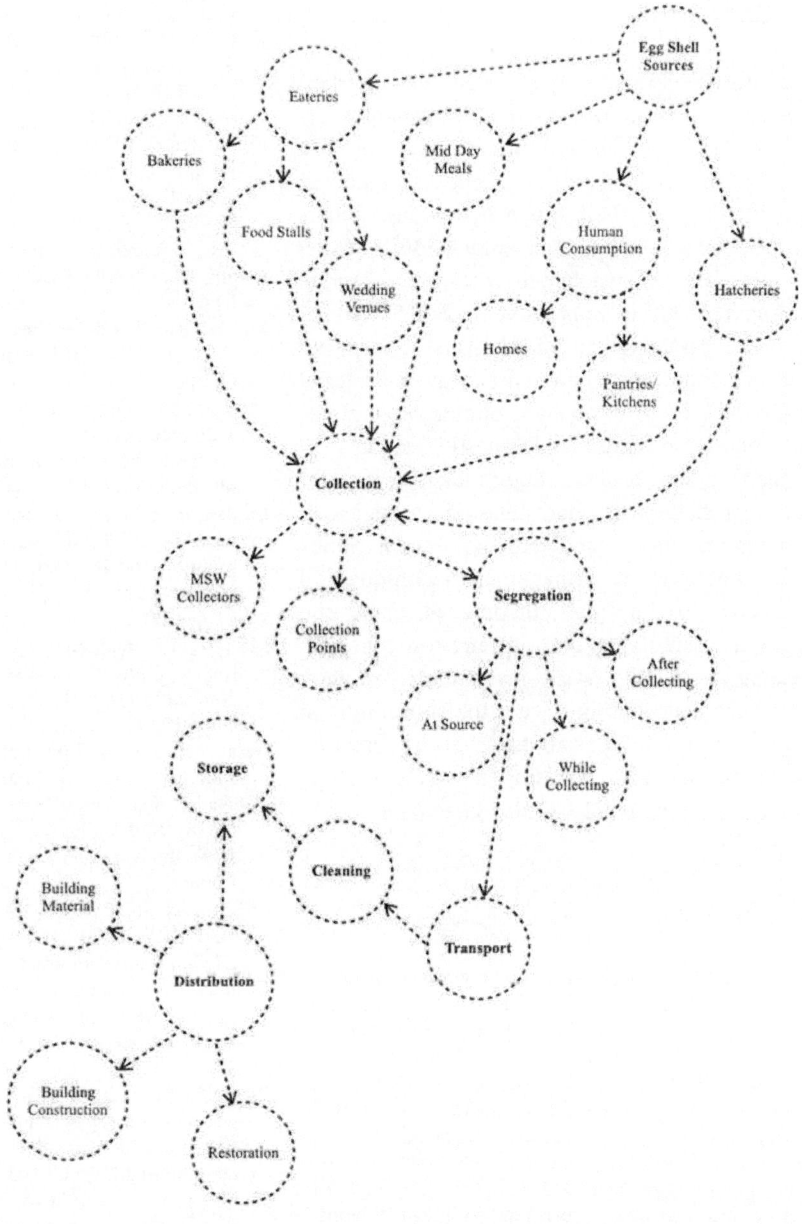

Fig. 7 Flowchart showing source-to-end use of eggshell waste

Cleaning and storage may prove a challenge since eggshells, while well preserved and cleaned, can only last 3–4 months. The storage space might prove a problem when space is tight in our cities.

With just the data collected for the research conducted on eggshell waste, and Kanyakumari as the main district of study has shown the sources to produce at least 3341.25 Kgs (around 3.5 tons) of eggshell waste through various methods of production. It's interesting to note that this can be substituted in terms of percentage weight for different uses in the building construction sector from 5% up to 100% of the entirety of the building construction technique/ building material.

7 Conclusion

The results of this study showed that eggshell biowaste is a suitable, sustainable way of sourcing material for the built-up environment as a building material as well an effective method of rerouting the eggshell waste that would otherwise be sent to incinerators and landfills which are not safe or sustainable methods of waste management. It is impressive that 3.5 tons of eggshell waste can be successfully repurposed and thus contribute to a circular way of material used which helps in creating circular economies thus increasing the Gross Domestic Product.

Starting this research, though taken in smaller steps has helped in understanding the eggshell production, supply, and demand, recent trends and different waste management techniques all connected to the final outcome of using the eggshell waste in the built-up environment thus creating an effective waste management strategy while also contributing toward the community at large and at the grassroots level by creating awareness and also choosing a welcome alternative to conventional building materials.

References

Animal Husbandry Department (2022) department | Tamil Nadu Government Portal. [online] www.tn.gov.in. Available at: https://www.tn.gov.in/department/3. Accessed 12 Oct 2022

Barcelo L, Kline J, Walenta G, Gartner E (2013) Cement and carbon emissions. Mater Struct 47:1055–1065. https://doi.org/10.1617/s11527-013-0114-5

Benhelal E, Zahedi G, Shamsaei E, Bahadori A (2013) Global strategies and potentials to curb CO_2 emissions in cement industry. J Clean Prod 51:142–161. https://doi.org/10.1016/j.jclepro.2012.10.049

Google Maps (2022) Kanyakumari. [online] Parotta stall near kanyakumari. Available at: https://www.google.com/maps/search/parotta+stall+near+Kanyakumari. Accessed 12 Oct 2022

Gubba N (2019) Egg as an organic building material a comparative study and understanding in Indian context. Key Eng Mater [online] 803:267–271. https://doi.org/10.4028/www.scientific.net/KEM.803.267

Gubba N, Sasikala R (2022) Comprehensive advanced specific summarised studies -for architecture studies (CASS Studies) an official publication of bureau for health & education status upliftment egg as an organic building material: usage in modern and historic context. Compr Adv Specif Summ Stud-For Arch Stud (CASS Studies), [online] 5(1). Available at: http://heb-nic.in/cassarc-issue/admin/paidPDF/md9mutlpebhleh2icrxj.pdf. Accessed 3 Oct 2022

Kanniyakumari District Administration (2022) Statistical Report|Kanniyakumari District, Government of Tamil-Nadu | India. [online] https://kanniyakumari.nic.in/. Available at: https://kanniyakumari.nic.in/document-category/statistical-report/. Accessed 12 Oct 2022

Lokniti (2017) Mood of The Nation 2017. CNN-IBN, New Delhi

Ma H, RK T (2003) Erratum. Bull Environ Contam Toxicol 70:188–188. https://doi.org/10.1007/s00128-002-0174-7

Mignardi S, Archilletti L, Medeghini L, De Vito C (2020) Valorization of eggshell biowaste for sustainable environmental remediation. Sci Rep [online] 10(1):2436. https://doi.org/10.1038/s41598-020-59324-5.

Ministry of Human Resource Development (2022) Pradhan Mantri Poshan Shakti Nirman—PM POSHAN. [online] 164.100.160.114. Available at: http://164.100.160.114/School_Reported_view.aspx. Accessed 12 Oct 2022

MS (2017) Production of eggs worldwide, 2017 | Statista. [online] Statista. Available at: https://www.statista.com/statistics/263972/egg-production-worldwide-since-1990/

Natalie Rodríguez Eugenio, Mclaughlin, M.J., Daniel John Pennock, Food And Agriculture Organization Of The United Nations and Global Soil Partnership (2018) Soil pollution : a hidden reality. Food and Agriculture Organization of The United Nations, Rome

OECD-FAO (2022) OECD-FAO Agricultural Outlook 2022–2031. OECD-FAO Agricultural Outlook. OECD. https://doi.org/10.1787/f1b0b29c-en

Okonkwo U, Odiong I, Akpabio E (2012)The effects of eggshell ash on strength properties of cement-stabilized lateritic. Int J Sustain Constr Eng Technol 3:2180–3242

Pliya P, Cree D (2015) Limestone derived eggshell powder as a replacement in Portland cement mortar. Constr Build Mater 95:1–9. https://doi.org/10.1016/j.conbuildmat.2015.07.103

Raji A, Tomi Aina S (2015) Egg shell as a fine aggregate in concrete for sustainable construction. Int J Sci Technol Res 04

Samrat Sharma (2021) India's meat map: 7 out of 10 people relish non-vegetarian items, East & South lead the way [online] India Today. Available at: https://www.indiatoday.in/diu/story/india-meat-map-people-relish-non-vegetarian-items-east-south-lead-way-1878313-2021-11-18

SC (2003) History of architecture and ancient building materials in India. Tech Books International, New Delhi

Shearman S (2016) Scotch egg company claims to have cracked problem of eggshell waste [online] the Guardian. Available at: https://www.theguardian.

com/sustainable-business/2016/jun/30/scotch-egg-company-cracked-eggshell-waste-problem-recycling-plastic. Accessed 12 Oct 2022

Shiferaw N, Habte L, Thenepalli T, Ahn JW (2019) Effect of eggshell powder on the hydration of cement paste. Mater 12. https://doi.org/10.3390/ma12152483

Shu Ing D, Chin SC (2014) Eggshell powder: potential filler in concrete

Singh DR (2019) Utilization of poultry by–product (Egg Shell) for sustainable poultry farming in India. [online] Pashudhan praharee. Available at: https://www.pashudhanpraharee.com/utilization-of-poultry-by-product-egg-shell-for-sustainable-poultry-farming-in-india/. Accessed 12 Oct 2022

Tan YY, Doh SI, Chin SC (2018) Eggshell as a partial cement replacement in concrete development. Mag Concr Res 70(13):662–670. https://doi.org/10.1680/jmacr.17.00003

Tanyildizi M, Karaca E, Bozkurt N (2022) Green concrete production with waste materials as cement substitution: a literature review. 1:33–45

Thomas J (2005) Using eggshells to build walls in Japan. [online] MetaEfficient. Available at: https://metaefficient.com/uncategorized/using-eggshells-to-build-walls-in-japan.html#:~:text=Making%20full%20use%20of%20the [Accessed 3 Oct. 2022].

Ujin F, Ali KS, Harith ZYH (2016) Viability of eggshells ash affecting the setting time of cement. World Acad Sci, Eng Technol Int J Civ Environ Eng 10(3):2016

United Nations (2022) United Nations Sustainable Development [online] United Nations Sustainable Development. Available at: https://www.un.org/sustainabledevelopment/

Van Immerseel F, Nys Y, Bain M (2016) Improving the safety and quality of eggs and egg products : egg safety and nutritional quality. Woodhead Publishing, Oxford

Waheed M, Yousaf M, Shehzad A, Inam-Ur-Raheem M, Khan MKI, Khan MR, Ahmad N, Abdullah, Aadil RM (2020) Channelling eggshell waste to valuable and utilizable products: a comprehensive review. Trends Food Sci Technol 106:78–90. https://doi.org/10.1016/j.tifs.2020.10.009

Heritage to High-Tech

Bricolage Sustainability: Addressing the Fundamental Misalignment Between Environmentalism and Patronage-Based Practice

Scott Shall

Abstract

Architects are pre-inclined, in identity, training, and legislation, toward patronage-based practices. This bias has had a significant impact upon the architect's pursuit of sustainability, causing the professional to emphasize environmental approaches rooted upon linear and hierarchical processes and specifiable material palettes. Although not without reason or benefit, this prejudice toward predictability causes the architect to struggle when attempting to realize environmental gain through the use of unconventional means, reclaimed materials, or localized processes. To realize more profoundly sustainable work demands a shift in practice. Specifically, architects must replace top-down patterns of engagement for those that incorporate local wisdom. They must question their allegiance to specifiable material palettes and embrace the potential held by reclaimed or idiosyncratic means. And they must challenge the notion that the needs associated with the construction of the project at hand are inherently more important than those associated with its inevitable de-construction. In short, they must begin to think less like engineers, and more like bricolers. To investigate this premise, this writing will study three community-based projects, each of which was recently realized using a distinct, bricoler-inspired approach. From globalized crowdsourcing to hyper-local, peer-to-peer production, ad hoc construction to digital simulation, and design-build to viral propagation, each of the offered studies will lend readers a unique perspective into new, more sustainable forms of practice. And, hopefully, incite a radical re-imagination of each.

Keywords

Bricolage · Environmentalism · Sustainability · Community-based projects

1 Bricolage: From Metaphor to Concept

Claude Lévi-Strauss' concept of the *bricoler*, originally described within his 1962 publication *The Savage Mind* (*La Pensée Sauvage*), operates as a metaphor, used to describe the generation and composition of mythical thought. To support his use of the term in this manner, Lévi-Strauss traces the concept of *bricolage* back to the fourteenth-century French verb "*bricoler,*" which was utilized to describe unanticipated reactions

S. Shall (✉)
College of Architecture and Design, Lawrence Technological University, Southfield, USA
e-mail: sshall@ltu.edu

M. R. Thomsen et al. (eds.), *Design for Rethinking Resources*, Sustainable Development Goals Series,
https://doi.org/10.1007/978-3-031-36554-6_14

to fixed, and seemingly known, settings, as would sometimes occur in activities like billiards (a ball rebounding oddly off the bumper) or riding (a horse diverging from the given path). He then posits that the "*bricoler,*" who he characterizes as a "devious craftsman," works in a similar manner, generating unanticipated outcomes using a pre-constrained set of resources (Lévi-Strauss 1968). Having thus defined the concept, Lévi-Strauss' applies his definition to mythical thought, arguing that this type of belief system always expresses itself through a heterogeneous repertoire, constructed from a pre-existing set of component beliefs.

Since that time, Lévi-Strauss' idea has transcended its etymological and metaphorical roots to become a much more universal concept, deployed to describe similarly inclined approaches in a range of fields including science; art; and, more recently, technology (Johnson 2012). In the sciences, the term *bricolage* was initially used within the field of evolutionary biology, to support an argument that evolution is not the product of intelligent design, but a series of makeshift adaptations. It is interesting to note that François Jacob, a colleague of Lévi-Strauss at the Collège de France, was the first to employ the term in this manner (Jacob 1977). Decades before this, the art world had been embracing the concept of *bricolage* to describe the work of artists associated with surrealism, dada, and even cubism. However, it was not until the rise of the Italian *arte povera* movement in the 1960s that *bricolage* was put forward with specific critical intent. Specifically, the *arte povera* movement, which was critical of contemporary arts' commercial bias, promoted work that was constructed with whatever rubbish was at hand, in order to devalue the art object (Dezeuze 2008). Many contemporary artists continue to embrace this mindset, most of whom explicitly acknowledge their approach as an act of *bricolage* (Fig. 1). In a similar vein, *bricolage* has been used to describe technology, as various authors have argued that software, biotech, and even the Internet are not produced through an orchestrated design effort, but are the result of unanticipated contributions from millions of people, each of whom was independently offering guidance using whatever finite resources were at hand. Thus, these innovations are, arguably, better identified as acts of *bricolage* than products of intelligent design or rational planning (Longo 2009).

From this, it is clear that *bricolage* is now commonly viewed as a universal concept, to be used to describe activities and products that share four core attributes. First, as all acts of *bricolage* are constructed from a finite palette of second-hand parts, the *bricoler* must be skilled at using idiosyncratic, heterogenous, and unforeseen components in order to produce new work.[1] This is the first, and foundational, attribute. Second, as the *bricoler* only produces work from previously collected componentry, all acts of *bricolage* will be combinatorial in nature—a reality places great importance upon not only the number of resources gathered, but the potential embedded in each one. Although understanding the former concern is a simple matter of accounting, addressing the latter concern requires the *bricoler* to engage in direct experimentation with their store of resources. The purpose of these experiments is to expand or even subvert the original purpose of the gathered means. This is why Lévi-Strauss (1968) refers to the bricoler as a "devious craftsman." The devious or subversive nature of the *bricoler's* experiments provides the third and fourth attributes of *bricolage*: the foundational nature of hands-on experimentation and the need for these experiments to create new intention within means designed for other purposes (Johnson 2012). Through this, the *bricoler* is able to not only realize new promise within each gathered resources, but also to establish useful dialogues between these component parts.

[1] "The set of the 'bricoler's' means cannot therefore be defined in terms of a project (which would presuppose besides, that, as in the case of the engineer, there were, at least in theory, as many sets of tools and materials or 'instrumental sets', as there are different kinds of projects). It is to be defined only by its potential use (...) because the elements are collected or retained on the principle that 'they may always come in handy' (ça peut toujours servir). Such elements are specialized up to a point (à demi particularisés) (...) but not enough for each of them to have only one definite and determinate use" (Lévi-Strauss 1968, 17–18).

Fig. 1 Artist Andy Goldsworthy uses the techniques of bricolage to generate his work. Shown here is a 2013 installation made for the Berrymore Foundation. Constructed of carefully arranged leaves, this piece was made on-site, in direct dialogue with the resource and context (Funderburg 2016)

When viewed in this manner, the work of the *bricoler* offers an epistemological challenge for all those who design and construct our built environments (Longo 2009). The fact that this approach has been used with some success by the residents of extra-legal settlements—communities that feature incredibly unforgiving and challenging built environments—only serves to underscore this challenge (Shall 2018). In each of these instances, the ability of the *bricoler* to quickly uncover and deploy the potential offered by reclaimed materials, occupied sites, and other idiosyncratic resources is a remarkable counterpoint to the time and cost associated with these operations when operating within the conventional design-then-build-then-occupy creative process.

2 Patterns of Practice

To fully understand the nature of this epistemological challenge, and any potential reformulation that might result, one must first establish the current patterns of practice, and how they came to be. For architecture, these routines arguably find their root in the fifteenth century, with the construction of the Brunelleschi-designed dome over Santa Maria del Fiore—a work which concretized a shift of architecture from a mechanical to a liberal art, and the architect from a master-craftsperson to an independent professional (Fig. 2) (Ettlinger 1977). As the profession of architecture evolved over the next few centuries, the distance between the architect and the trades would become more pronounced, as aesthetics, not technical competency or superior building, became the root of the field's identity—a transition that would compel the architect to trade the experimental, hands-on, and improvisational tactics previously used for the more theoretically motivated and detached concerns of the artist. (Crawford 1991). This attached the value of architecture to the field's ability realize larger, more complex work than the trades could accomplish, which helped to forge an alliance between the architect and those actors who could best afford this scale of work (Cuff 1998). Eventually, a symbiotic relationship between the field of architecture and those who wanted to leverage the prestige of the architectural product to establish their taste, reaffirm their rarified position, and extend their influence was established. The work of the architect thus became a luxury item, buttressed by ideological claims, and isolated from material, economic, or other technical concerns over which the field had very little control (Hollein 1962). In identity, training, and legislation, the field of architecture, like engineering and planning, became a patronage-based practice—a bias that predisposes the professional toward patterns of engagement that are hard-wired to serve their small, but powerful, client base (Stevens 1998).

This predisposition has had a significant impact upon the manner by which architects engage in environmental concerns, privileging approaches that can be realized using patronage-based palettes and practices. In terms of palettes, architecture, like engineering, is generally constructed from a predictable and specifiable

Fig. 2 The Brunelleschi-designed dome over Santa Maria del Fiore concretized a shift of design from a mechanical to a liberal art, and the architect from a master-craftsperson to a professional (Herb 2020)

inventory of materials. This allows the architect to guarantee a certain level of performance within the offered work—a critical asset when authoring the large, complex projects favored by the field's patrons. It also causes the architect to struggle when attempting to incorporate the environmental promise found within already inhabited sites, unconventional means, or reclaimed materials. The idiosyncratic nature of such elements prevents simple engagements when operating within the remote, linear, and highly choreographed practices favored by the field, as described below, resulting in the addition of time and cost to the effort. This reality encourages the architect to embrace the environmental benefit of heavily processed materials or open sites, both of which are easily incorporated into traditional patterns, instead of pursuing the more substantial environmental gains offered by reclaimed materials or occupied sites—a professional bias that subverts the ability of these professionals to create more sustainable consumption and production patterns, as advocated by the United Nation's 12th Sustainable Development Goal (SDG).

In terms of practice, architects and engineers are trained to operate within a carefully choreographed, linear, and hierarchical design process. The actors involved are identified well in advance and their insight offered at clearly defined moments within the effort. If the insight of others is included, it is either because the patron stipulated that the architect creates a more inclusive process or the architect determined it necessary and was willing to add uncompensated time to their effort. Yet, even then, the insight offered by the non-patron remains secondary to that of the patron. This arrangement, although useful in establishing the primacy of the patron in the process, prevents the professional from fully understanding or appreciating local concerns or vernaculars. It also makes it difficult to produce anchored and effective work when operating in contexts unlike their own, as is generally the case when working within extra-legal settlements. These limitations pose particular difficulties whenever the architect attempts to deploy participatory models or offer socially sustainable projects. In these instances, the distance between localized knowledge and the technocratic approach favored by the professional creates what researchers Theime and Kovaks term "*malevolent urbanism*," Thieme and Kovacs (2015, 1). Within such an arrangement, the architect cannot possibly create the type of resilient and sustainable settlements advocated for by the United Nation's 11th Sustainable Development Goal (SDG).

The costs of these professional limitations are significant to everyone involved. The members of under-served communities, most of whom could benefit greatly from the professional's insight, are forced to live within extremely unhealthy and unsustainable environments. As these settlements represent one of the fastest growing urban or peri-urban conditions on the planet, the impact of this shortcoming is not only quite substantial, but is becoming more so with each passing year (UN-HABITAT 2003, rev. 2010). The architect, who could enlarge their environmental and societal impact if they were able to better incorporate idiosyncratic conditions, is able to produce only marginally sustainable work. And the planet, which bears the cost of the massive environmental footprint of both legal and extra-legal settlements, continues to deteriorate. Perhaps by questioning the field's historic bias toward certain ways of working, the architect might move from exploitative, consumer-based patterns of resource management to approaches that more restorative, regenerative, and circular—a crucial evolution if the architect is to embrace resource guardianship and equity as fundamental to practice and to generate truly sustainable work. Perhaps by embracing the tenants of *bricolage*—a well-established approach to working with idiosyncratic means, materials, and contexts—the architect might expand practice, embrace the promise of new resources, and help to create a healthier planet for everyone (Shall 2021).

3 Bricolage Architecture: Three Case Studies

Selection To investigate this premise, this writing will analyze three community-based projects which the author helped to co-create with local citizens using, a distinct, bricoler-inspired approach. So that the offered study might permit a fair comparison, the author established three criteria for the selection of works. First, and most obviously, each project in the study had to be realized using the techniques of the *bricoler*, as earlier defined. Second, each project had to be designed and constructed using only modest resources. Although this is undoubtedly a relative parameter, for the purposes of this study a cap of $5,000 USD (2022 valuation) was established. And third, each of the selected works had to be designed and constructed in using primarily idiosyncratic means and localized practices.

Methodology Once the works were selected, a comparative analysis, using the collective case study methodology defined by R. E. Stake, could be performed (Stake 1995). So as to ensure a consistent and relevant assessment, each of the works was analyzed using four metrics: (1) the deployment of localized practices, (2) the use of found or reclaimed materials, (3) the ease with which the work can be deconstructed and amended, and (4) the ability of the work to propagate its core tenants or innovations. This final concern—the propagation of the offered insight—is arguably the most critical. This is because the cost of design practice renders professional services unobtainable for most people. Thus, if architects and engineers are interested in increasing their impact upon our shared environment, then they must find ways to better propagate their insight, at little cost. Viral propagation is thus included as a final metric within this analysis.

3.1 Control: Mansarovar Park Development (2018)

To establish a useful baseline for this study, this writing will first analyze a series of design proposals created in 2018, when a massive fire destroyed an extra-legal community in eastern New Delhi (Fig. 3). Immediately following this

Fig. 3 After a massive fire destroyed an extra-legal community in New Delhi, local designers volunteered their talents to help the residents reconstruct their community in a smarter, more sustainable fashion (Malhotra 2018)

event, a handful of local designers volunteered their time to help the residents reconstruct their community in a smarter, more sustainable fashion. Working within a traditional approach, these designers offered some very clever projects, each of which utilized locally available, cost-effective, and non-flammable materials to not only rebuild the community, but to allow the residents to easily move their homes whenever they were forced to relocate.

Analysis Surprisingly, very few of these ideas were ever implemented. From interviews, it appears that the offered designs were not rejected due to their lack of efficacy, but because their utility was measured using absolute terms, not those valued locally. For example, although the ideas were shared as quickly as normative design processes would permit, the pace of work was still too slow to keep up with that commonly used in the settlement. Thus, by the time the ideas were presented, the residents had already purchased materials or even begun rebuilding. Similarly, although the proposed material palette was inexpensive by the standards of the profession, they were measurably more expensive than the resources typically used within the settlement. This is likely because the designers, who were somewhat removed from the context of the work, did not fully appreciate the inventory of materials gathered by the community and thus based their work on an assumed palette. Finally, the offered designs were rejected because the residents feared that they would cause the authorities to view the community as permanent, which could prompt its removal. This was a particularly ironic shortcoming, as the offered works were intentionally designed to be deconstructed, and removed. However, because their transportability was not expressed clearly within the design, this intention was unclear. In the end, the only ideas implemented, such as moving cooking fires outside the home and widening the lanes between structures, were those that could be realized without upsetting the core values of the community. All other ideas, no matter how efficacious, were rejected (Malhotra 2018).

3.2 Case Study: Demountable Schools for Mumbai Mobile Creches (2008)

The first case study is centered on a series of temporary communities that are built within the construction sites of India. Designed to house those who are building the project, these settlements are typically built along the edges of the site with whatever construction debris is available—a situation that can be less than ideal for the children living there. To help address this, an Indian-based non-profit organization, Mumbai Mobile Creches (MMC), made a proposal to several local developers: if they would provide the physical infrastructure for a school, then MMC would provide the teachers and materials needed to run it. The idea took root, leading to the formation of schools on construction sites around Mumbai. It also led to an interesting dilemma regarding the design of these schools: how to establish a trustworthy and secure resource that was overtly temporary and built using only borrowed resources. To help answer this question, the author worked with a US-based non-profit and MMC to rethink the design of all educational structures that define a school, from the building itself to the furnishings and equipment (Figs. 4 and 5). From this rethinking, a number of projects emerged, including a $2 water filter, a portable earth wall, and a foldable play mat that allowed the children to easily, and independently, restructure their learning environment (IDC 2021).

Analysis In assessing this work, it is clear that localized practices were heavily used, as every project was developed in partnership with local officials and realized using only local resources. For example, the $2 water filter was fabricated using reclaimed tarps and a clear plastic storage bag, joined together using the sealer borrowed from a local spice merchant and gromets provided by a local tailor. Similarly, the foldable playmat was designed in coordination with a local autorickshaw upholsterer, who was excited to partner with MCC in this work, because he

Fig. 4 To generate the work in India, the design team first documented how the local population built their shelters—approaches that prioritized the practices of the bricoler. Shown here are some of the materials and details that were analyzed as a part of this investigation (IDC 2021)

viewed it as a promising business opportunity. In fact, almost all of the works were crowdsourced in this manner, inviting those impacted by the potential school to contribute their wisdom independently in a manner very much like author James Surowiecki describes in his book, *The Wisdom of Crowds* (Surowiecki 2005). This is likely why the designers working with MCC were able to offer work that was so locally rooted—their insight was supported by the unpolluted insight of hundreds of others with intimate knowledge of the situation.

Yet, despite these successes, none of the offered works were ever replicated by the host community. This is likely for two reasons. First, although the resources used were objectively modest, they were not free. The resulting price point presented a hurdle too high for MMC to either purchase additional copies of the work or to convince the developers to do so. Second, all of the offered projects were designed as a well-crafted solution to a given problem, instead of an open-ended approach to the issue which could be easily evolved. The resulting designs were unable to be amended or evolved by MMC, as is necessary when confronting unanticipated sites or concerns. These shortcomings likely prevented the work from propagating and compromising its impact upon MCC's situation.

3.3 Case Study: An Outdoor Classroom and Community Garden in Philadelphia (2011)

The second case study focuses on a small project at Bodine High School in North Philadelphia. Completed in 2011, the stated intent of this project was to develop an outdoor garden and classroom space for the school. Additionally, the project was intended to strengthen the relationship between the school and those living in the surrounding neighborhood—a relationship that had become quite fraught over the years, manifesting in the erection of a large, secure fence around the perimeter of the property. Inspired by a desire to address the message of distrust conveyed by this fence, the large number of abandoned lots in the surrounding neighborhood, and the modest budget for the project, the author, in coordination with local partners, endeavored to construct a response to this call from components common to an abandoned lot (Fig. 6). To do this,

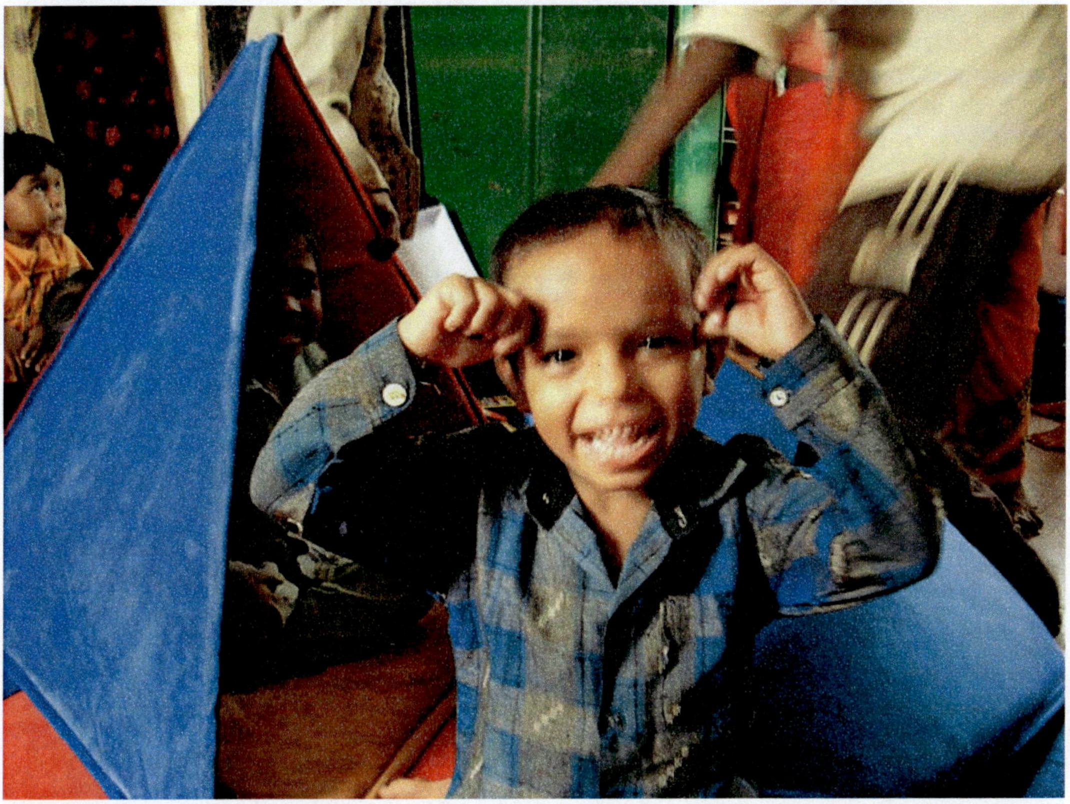

Fig. 5 Mumbai Mobile Creches (MMC) provides school services to kids living on construction sites around Mumbai. To help this effort, a US-based non-profit worked with MMC to rethink the design of all educational structures that define a school, generating projects like this foldable play mat through which the children might easily restructure their learning environment (IDC 2021)

this team had to first gather a massive inventory of these materials. They then had to experiment with these resources to find new uses in order to meet the demands of the given program and site. Through these efforts, the design team discovered that by challenging the orthogonal geometry of the fence typology, common fence posts and 90° elbows could be utilized to create the structure for all of the programmed elements, including benches, retaining walls, raised garden beds, and a large overhanging canopy. The complex geometry thereby generated caused a provoking distortion of the other elements used, including standard, chainlink fencing, and reclaimed wood (Fig. 7). The resulting work offered a playful, inviting space to both the school and the surrounding neighborhood, prompting those living around the school—many of whom volunteered to provide scraps of wood or other assets to the project—to ask if they might help to maintain the new gardens or use the space. In this way, the project shifted the prevailing message of the fence typology from one of protectionism to one of welcome. It also offered a compelling testimony to the potential of reclaimed materials common to Philadelphia's abandoned lots (IDC 2021).

Analysis From the various accounts of the process and work, it is clear that the project successfully adopted many strategies of the bricoler, including the gathering of resources "just in case," the open-ended experimentation with these reclaimed resources and the devious craftsmanship that these activities are intended to inspire (Fig. 8). In each stage of the work, the design team appears to have made a conscious

Fig. 6 The design team responsible for the project in North Philadelphia began their effort by creating an exhaustive inventory of materials that could be found or reclaimed and then experimenting with these assets to uncover new potential. Shown here is a local vendor donating construction debris to the effort—materials that would prove instrumental to the design response (IDC 2021)

decision to minimize their reliance upon highly predictable material palettes and embrace the potential held by local, reclaimed, and idiosyncratic means.[2] This stance permitted the designers to minimize the budget and environmental footprint of the work—achievements that are well-aligned with the intention of many acts of bricolage. The project's resulting emphasis upon local, reclaimed materials seems to have been complemented by a commensurate emphasis upon local practices and tools, rooting the project to its context materially, tectonically, and programmatically. Yet, like the project in India, this project also failed to provide obvious means or incentive to possess and evolve the work. As a result, the project remains a compelling one-off, despite the fact that it featured many innovative strategies to build with common, local, and inexpensive means. The reasons for this are twofold. First, although the project utilized the components of an abandoned lot, gathering the massive amount of materials required for the work was not a small task. To accomplish this, a small team of designers had to scour the local area for weeks, organizing, at key moments, several larger scale volunteer efforts to find, receive, and organize the rubble, earth, construction debris, and other reclaimed resources. The scale of this effort, although necessary for the project, created a hurdle for those who might

[2] This is a central tenant of bricolage. As stated by Levi-Strauss: "For the bricoler, the design and construction of any work comes from whatever is at hand; for the engineer, what is at hand is specified by the professional to meet the largely quantifiable demands of the project, and their patrons" (Levi-Strauss 1968).

Fig. 7 *ChainlinkGREEN* uses components common to an abandoned lot in order to provide outdoor classroom spaces for a North Philadelphia High School, and to strengthen the relationship between the school and the surrounding community (IDC 2021)

wish to develop a related work. Second, and perhaps more importantly, the skill needed to fabricate this project was not nearly as accessible as the tools or materials used. To point, although a common, accessible material (reclaimed wood) was used to fabricate the surfaces for all of the benches, the complex geometry of the project required that these resources be cut at precise lengths and angles, a skill not possessed by most amateur builders. Thus, although the community embraced the work as an artifact and expressed a desire to build similar versions around their neighborhood, they have, to date, been unable to do so.

3.4 Case Study: A Community Event and Makerspace in South Africa (2019)

The third, and final, case study analyzes a community event and makerspace recently completed in South Africa. Supported by a partnership spanning two continents and involving two universities, two non-profits and a host of volunteers, the now-completed Port Elizabeth Makerspace offers community members a physical and inspirational base of operations to support the use of reclaimed materials in construction. Physically, the makerspace offers the community a place to store tools, organize, and refine scavenged materials and construct work. As a precedent, the makerspace provides a compelling proof-of-concept, clearly demonstrating the value and potential of creatively reusing found objects in construction (Figs. 9 and 10). The fact that the makerspace was completed by a team of 11 in only 10 days—3 days scavenging materials and designing and 7 days constructing—and with a budget of only $1500 makes this potential resonate. For the citizens of Port Elizabeth, many of whom live and work within post-Apartheid communities of great need, this frugality is not a nicety, but an essential attribute. By embracing this restriction, the designers endeavored to shift the work from a tantalizing one-off to a blueprint

Fig. 8 To design and construct the Port Elizabeth Makerspace the design team had to leverage the potential within a variety of reclaimed materials, including the packaging for automobile engines, shown here. Donated by a local Isuzu assembly plant, these metal panels proved critical to the design response, providing the structure for both floors and walls as well as all of the wood used to fabricate the glue-laminated members that supported the roof (from IDC 2021)

through which the community might realize new project for years to come (IDC 2021).

Analysis Yet, did the project achieve this lofty goal? The answer is mixed. There is clear evidence to suggest that the tactics used to create the makerspace have been subsequently deployed in the community. For example, a local architect who was involved in the initial work has since used the strategies developed through this project to design and construct several new works, including a home for a local citizen living within a neighboring township. In this structure, many of the materials and approaches used in the initial work are clearly evident, including the prolific use of many of the same reclaimed materials and even the specific manner by which these assets were refined (Fig. 8). In so doing, the architect was able to create a work that was much more efficient, sustainable, and economical than could have been otherwise achieved. Yet, it seems that such propagation remains rare, as there is no evidence to point to their widespread use by others. Although the reasons for this deficiency are difficult to establish definitively, it seems logical to assert that the cause is not related to the means or materials used, as all of the resources deployed by the work remain quite accessible. After all, a designer involved in the initial work has had no trouble using these resources to serve other clients and develop other projects within communities of great need. Nor can it be attributed to the scale of the effort, as the projects that were developed from this precedent were constructed using smaller teams and no heavy equipment. Instead, it seems that the reason for this lack of propagation is related to how difficult it is to fully understand the offered insight. This would explain why someone who was directly involved in the initial work had no problem using its tactics to improve the financial

Fig. 9 The Port Elizabeth Makerspace offers community members a physical and inspirational base of operations to support the use of reclaimed materials in construction. Completed by a team of 11 in only 10 days and with a budget of only $1500, this project has resonated through the local community (from IDC 2021)

Fig. 10 Since its completion, the 2019 Makerspace has reverberated through the community, as local citizens leverage its unique material approach to realize new clinics, homes, and schools. This residence was designed and built by a South African architect, who leveraged many of the approaches developed through the 2019 project to reduce its environmental and financial footprint (photo courtesy of Kevin Kimwelle)

and environmental footprint of several other unrelated projects. And why there is no evidence of anyone else doing so.

Addressing this flaw is likely complex, as a designer can only directly involve so many individuals in the construction of a project of this scale without adding undue complexity to the effort. And having three or four individuals with such exposure is only a marginal improvement from one. Thus, if the goal is widespread propagation, then a different model of engagement is likely required. This could, theoretically, assume many forms, from a series of dedicated workshops wherein local citizens are invited to build under the guidance of someone who has more intimate knowledge of the work to a system encouraging more exponential propagation, whereby those involved in the first work each take on a second project, inviting two others to partner with them, who then take on a third project, and so on. If these processes are repeated enough, then the number of people with direct knowledge of the offered innovations will increase exponentially, theoretically creating enough knowledge and interest to promote large-scale propagation.

4 Conclusion

This hypothetical will be tested quite soon, as I have recently been invited to return to South Africa in order to design and construct a school in a low-income community outside Port Elizabeth. This work will provide me an opportunity to offer even more accessible versions of bricolage practice, where they will be tested once more by the rigors of their context.

And, judging from the projects highlighted above, it seems that such tests could be important. From the case studies, it seems evident that the tactics of the bricoler can be leveraged by the architect to more effectively incorporate idiosyncratic, reclaimed material into their efforts. From these works, it is also clear that the direct, experimental nature of the bricoler's approach can be successfully used by the architect to unlock unanticipated potential within unconventional resources, and to create more radically sustainable work. Regrettably, it is equally clear that these tactics will remain largely isolated to a single project, unless more robust patterns for vestment and education are established—a supposition that is currently under review in the township where the house based upon the Port Elizabeth Makerspace was developed. If these works are successful, the argument for a much more radical reimaging of practice will likely become warranted.

Yet, even if these experimental works demonstrate some measure of success, wide-scale implementation of their approach remains a difficult proposition, given the historic biases of the field, and the well-established academic and professional systems that have been created to support them. To challenge this professional inertia requires that those working within the field establish alternative support systems, including, but not limited to, the creation of new academic structures, alternative resource banks, and revised measures for professional success. The architect must complement academic pursuits that privilege theoretical mandates—exercises that will often assume unfettered access to an immense inventory of predictable resources—with those that are anchored upon limited means and idiosyncratic materials. They must commit to the prioritization of local, reclaimed, and scavenged resources within their work, so that the entities providing these assets and services might be better supported, and thrive. And award programs and publishing houses must shift resources from the promotion of theoretically rigorous, complex, and large-scale projects to those that herald work anchored upon the innovative use of localized assets, idiosyncratic resources, and alternative practices. They must also celebrate acts of de-and re-construction with as much energy as they do new construction. In short, architects, and the structures that support them, must embrace the tenants and works of bricolage with as much fervor as they do the products of current, patronage-based practices. Although undoubtedly a challenging endeavor, expanding practice in this manner seems warranted. After all, as evidenced by the works analyzed previously, there exists ample proof that by thinking less like engineers, and more like bricolers, the architect can offer much more radically sustainable work—to the benefit of both the field and all those served by it.

References

Crawford M (1991) Can architects be socially responsible? In: Ghirardo D (ed) Out of site: a social criticism of architecture. Bay Press, Seattle, WA, pp 27–45

Cuff D (1998) Architecture: the story of practice. MIT Press, Boston, Massachusetts

Dezeuze A (2008) Assemblage, bricolage, and the practice of everyday life. Art J 67(1):31–37. College Art Association, New York. https://doi.org/10.1080/00043249.2008.10791292

Ettlinger L (1977) The emergence of the italian architect in the fifteenth century. In: Kostoff S (ed) The architect. Oxford University Press, London, pp 96–123

Funderburg L (2016) Andy goldsworthy: natural man. Architectural Digest. [online]. Available at: http://www.architecturaldigest.com/gallery/andy-goldsworthy-book-ephemeral-works. Accessed 3 Oct 2020

Herb A (2020) Double helix of masonry—researchers uncover the secret of Italian Renaissance Domes. Princeton University News. https://www.princeton.edu/news/2020/05/21/double-helix-masonry-researchers-uncover-secret-italian-renaissance-domes

Hollein H (1962) Absolute architecture. In: Conrads U (ed) Programmes and manifestoes on 20th century architecture. MIT Press, Cambridge, Massachusetts, p 181

International Design Clinic (2021) What we've done. http://www.internationaldesignclinic.org/what-weve-done

Jacob F (1977) Evolution and tinkering. Science 196 (4295):1161–1166. https://doi.org/10.1126/science.860134. Accessed 1 Oct 2020

Johnson C (2012) bricoler and bricolage: from metaphor to universal concept. Paragraph 35(3):355–372. https://doi.org/10.3366/para.2012.0064

Levi-Strauss C (1968) The savage mind. The University of Chicago Press, Chicago

Longo G (2009) The epistemological turn: technology, bricolage and design. https://noemalab.eu/ideas/essay/theepistemological-turn-technology-bricolage-and-design/. Accessed 1 Sept 2022

Malhotra A (2018) Devasted and destroyed: delhi slums struggle to recover from frequent fires. The Guardian, September 5, 2018. https://www.theguardian.com/cities/2018/sep/05/devastated-destroyed-delhi-slums-recover-fires. Accessed 15 Sept 2022

Shall S (2018) Guerrilla architecture and humanitarian design. In: Arefi M, Kickert C (eds) Bottom-up urbanism. Palgrave-MacMillan, New York

Shall S (2021) Online bricollage: toward an architecture of scavenged means, improvisational methods, and decentralized processes. In: Jarrett C, Sharag-Eldine A (eds) Performative environments: ARCC 2021 international conference. Architectural Research Centers Consortium, Inc., Washington D.C., USA, pp 337–344

Stake R (1995) The art of case study research. Sage Publications Ltd., London

Stevens G (1998) The favored circle: the social foundations of architectural distinction. MIT, Cambridge

Surowiecki J (2005) The wisdom of crowds. Anchor Books, New York, NY

Thieme T, Eszter K (2015) Services and slums: rethinking infrastructure and provisioning across nexus. Econ Soc Res Council. https://www.researchgate.net/publication/279174398_Services_and_Slums

UN-Habitat (2003) Updated 2010. The Challenge of Slums–Global Report on Human Settlements. United Nations Human Settlements Programme. https://unhabitat.org/the-challenge-of-slums-global-report-on-human-settlements-2003

Leaving No Maker Behind: Cultures of Tile Vault Making for Situated Design

Wesam Al Asali

Abstract

Although recent calls in digital fabrication acknowledge local construction methods, the "local" in this framework is either a passive recipient of technological tools or a passive source of inspiration. The difference between high- and low-tech is much more than switching from robots to hand-making, as the two domains represent equal (but almost contradictory) modes of imagination, planning and decision-making This paper aims to reflect on the possible contribution of architectural practice and technologies to support vernacular building crafts. It reflects on four experimental case studies of tile vaults made in collaboration between the master vault makers and an architect to push the technique towards new applications and approaches to resourceful construction. The paper shows that the link between architectural practice and local building crafts is a threefold dialogue. First, designers should engage with the complexity of craft by illustrating its processes, not products. Second, designers should work with local building crafts communities beyond formal institutional channels. Third, designers should include vernacular knowledge in the current research on manufacturing and prefabricating replicable low-carbon building components. These threefolds can be translated to methods to engage socially and environmentally with local materials and knowledge. "Leaving no maker behind" acknowledges the overlooked human dimension of local construction: working directly with makers.

Keywords

Building crafts · Tile vaulting · Vernacular architecture · Low-carbon construction · Experimental pavilions

1 Introduction: On Withoutcisim in Architectural Praxis

Construction is one of the most environmentally extractive activities. It accounts for 40% of global energy consumption, 38% of greenhouse gas emissions, and 40% of waste generation (UNEP 2021). The increasing requestioning of 'how we build resourcefully under climate emergency' has fuelled two strands of critique. The first seeks to introduce digital tools, smart applications and data-driven planning into design and construction (Agustí-Juan and Habert 2017; Beorkrem 2017). The second emphasises the role of vernacular knowledge, sometimes called 'indigenous', 'situated' or 'traditional', that relies on local materials and building skills (Sayigh 2019;

W. Al Asali (✉)
IE University, Madrid, Spain
e-mail: wesam.alasali@ie.edu

M. R. Thomsen et al. (eds.), *Design for Rethinking Resources*, Sustainable Development Goals Series,
https://doi.org/10.1007/978-3-031-36554-6_15

Watson 2019). However, relatively little has been said about the relationship between the two.

This paper aims to reflect on the shared terrain between these two domains: the vernacular and the technological. We explore experimental and collaborative projects of tile vaults—a craft that incorporates design and engineering to make structural building elements (Collins 1968). The article starts by introducing tile vaulting as a vernacular technique that both outlived and co-lived with industrial construction for its efficiency and resourcefulness. These features are then examined in four case studies of collaboration between an architect and a vault master. After reflecting on these case studies, the article concludes with a methodological framework on 'how to work with local crafts?' through three sets of activities: (1) modelling existing processes of vernacular making, (2) collaborating with the crafts community within and outside formal and institutional structures and (3) working with artisans to develop replicable building components.

Since the establishment of architectural design as a modern profession, its relationship with other forms of making outside its recognised circles has been a space of tension and ambiguity. Between dismissing or patronising how non-architects build, studies of architectural theory and history understood architecture by the presence or absence of the centralised design agent—*with* or *without* architects (Rudofsky 1964). Under this self-centred framework, design studies coined multiple terms such as adhocism (Jencks and Silver 2013), design-in-use, unintentional or unselfconscious design (Alexander 1964) for describing the alteration or creation of the built environment beyond the formal architectural profession (Kuijer et al. 2017). This *withoutcisim* to define the vernacular and indigenous built environment represents a twofold problem in architecture praxis. The first is limiting design thinking to a 'single mode of processes', where steps and procedures of illustrating the building to its finest details precede its construction. The second is limiting design operations to a 'single mode of agency' where such illustrations are conditioned with verification procedures that are complexified and regulated through institutional channels. This twofold problem has led to limiting the creation of building components to a list of studied, tested and verified elements. It, therefore, pushed for an architecture that, due to its strict systematisation, excludes forms of building that are more open to adopting local conditions, materials and knowledge.

This is changing now. With the urgency to rethink how architecture can rethink its resources, emerging environmental approaches call for engaging with indigenous and vernacular architecture which are tied with the notion of local materials and techniques. However, with the excitement to link the high-tech and low-tech in construction cultures, there is a lot to examine on how such bridging culturally, socially and economically impacts both design and craft communities. Therefore, the question of 'how can we **learn from** indigenous or local knowledge of making?' is not enough because it treats such knowledge as static sources for a one-way understanding of learning. It limits the exchange to extrapolating modes from craft to architecture and deprives architects and local makers of a fruitful discussion on the future of both professions. The question, therefore, is how can we **work with** indigenous or local knowledge of making?

2 Methodology: Building in Conversation

Tile vaulting, also called Catalan, Timbrel and Guastavino vaulting, is a Mediterranean technique of laminated vaults (Fig. 1). This method uses lightweight terracotta tiles and fast-setting mortar such as plaster of Paris or rapid cement (Collins 1968; Huerta 2003). After finishing one layer of tiles with the fast-setting mortar, it serves as the formwork for subsequent layers, infill or lime concrete, or ribs that converts the thin shells into floor systems for multistorey building (Ochsendorf 2014).

Tile vaulting is associated with resourceful building techniques. First, by being a vault, it relies on geometry, not materials, to gain its

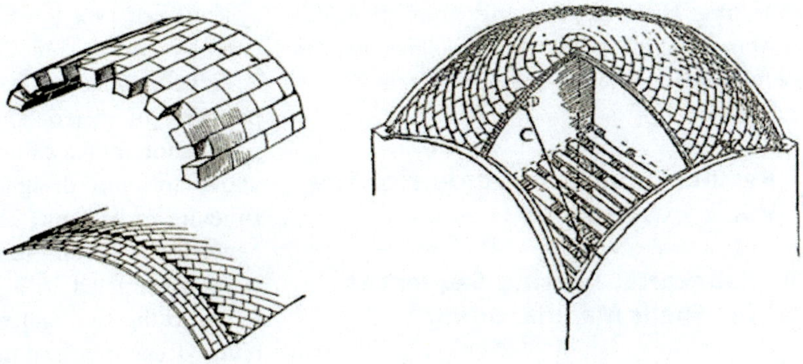

Fig. 1 Three types of vaults—clockwise, from top left: conventional stone, tiled dome and tiled vault. Luis Moya Blanco 1947

strength. This method leads to minimum use of steel and concrete, especially when constructing floor systems in multistorey buildings where a concrete slab has the highest carbon footprint among other building elements (Jayasinghe et al. 2022). Second, tile vaulting uses minimum formwork and relies on a builder's skill guided with light structures known as guidework. As an architectural element, it interlaces many aspects. It strongly connects with tacit knowledge, inherited from being a craft first and foremost, and structural design governed by rules of gravity. Approaching tile vaulting for contemporary application engages with both craft and structural design (Davis and Block 2012; Ramage et al. 2019; López et al. 2019).

The traditional knowledge of tile vaults could survive standardised construction. With the wide use of standardised reinforced concrete construction, formal architectural practice pronounced tile vaulting nearly dead in the 1950s (Truñoi Rusiñol et al. 2004, p. 4). However, this vernacular technique was kept alive thanks to the masters who specialised in building vaulted stairs (Al Asali 2016). Because they are faster to build than reinforced concrete stairs, vaulted stairs were, and are still, favoured in Spain in multistorey housing buildings.

Tile vaulted stairs have never been studied within building regulations. It is approved by empirical load testing after it is built. Further, while architects draw stairs in their layouts, their drawings are more figurative and approximative for building masters to rearrange the design on site before construction. It can be stated then that what was lost was not tile vaulting technique but its relationship with structural and architectural design. To restore this relationship with the recent attention to this resourceful technique, the role of these tile stairs master builders was crucial in training others to build tile vaults, contributing to new vaulted projects, and building several expressive structures in architectural exhibitions.

A snippet of these roles will be shown in the following four collaborative projects of tile vaulting. We will examine how to work with local craft by reflecting on design-build examples and architectural projects developed as part of a 5-year collaboration between an architect researcher and a master builder of tile vaulting. The experiments have been developed through small-scale commissions or self-funded initiatives. Each experiment had to respond to specific questions about the role of structural craft in contemporary construction, which includes vault prefabrication, developing tools for in situ form-finding, co-designing of shells and using accessible modelling technologies with vaulting craft. The analysis forms the experiment was based on notes, records of discussions and noting the development of the designed element in the experiment. Co-making as a methodological approach proved very useful in researching architecture and building crafts. Given the limited scale and scope of the experiment, they do not necessarily represent the complexity and intricacy usually found in the practice of architecture. However, learnings from them can be extrapolated into actionable strategies that can hopefully be explored in future research.

3 Results: Four Case Studies of Tile Vault Experiments

3.1 Fabricarte: Altering Sequences for Shells Manufacturing

In 2018, we were invited to design a pavilion for a project during the annual Expo of Spanish Ceramics (Cevisama), coordinated by the Instituto de Tecnologia Ceramica (ITC) and curated by the Asociación Española de Fabricantes de Azulejos y Pavimentos Cerámicos (ASCER). We proposed to make a walk-through pavilion that celebrates the craft of structural tile vaulting in Spain, inspired by late-gothic vaulting in Valencia's 13-fifteenth centuries. The pavilion FabricArte's is based on a groin vault composition resulting in a shell that changes from flat systems of shallow slabs to a full-height expressive vault (Fig. 2).

Expo regulations permit 2 weeks for transportation of materials, onsite construction, and preparation for the exhibition. This period did not allow for the designed 7.5 m walk-through structure to be built. The challenge prompted a new experiment in tile vaulting, in which vaults were constructed in a workshop, sliced, transferred to the site, and reassembled for the exhibition. Repetition and modularity were central to the design, where two moulds only of quarters of the vaults can generate the entire pavilion. The slices in each quarter were made to avoid linear horizontal cuts and maintain interlocking joints between the pieces (Fig. 3). The calculations of the cuts were studied based on the height and weight of parts to keep the manufacturing process manageable without machinery, each piece was calculated to be held by two workers on the ground (maximum of 120 kg) or one person on a scaffold (maximum of 50 kg). The structural analysis of the vault was made using graphic

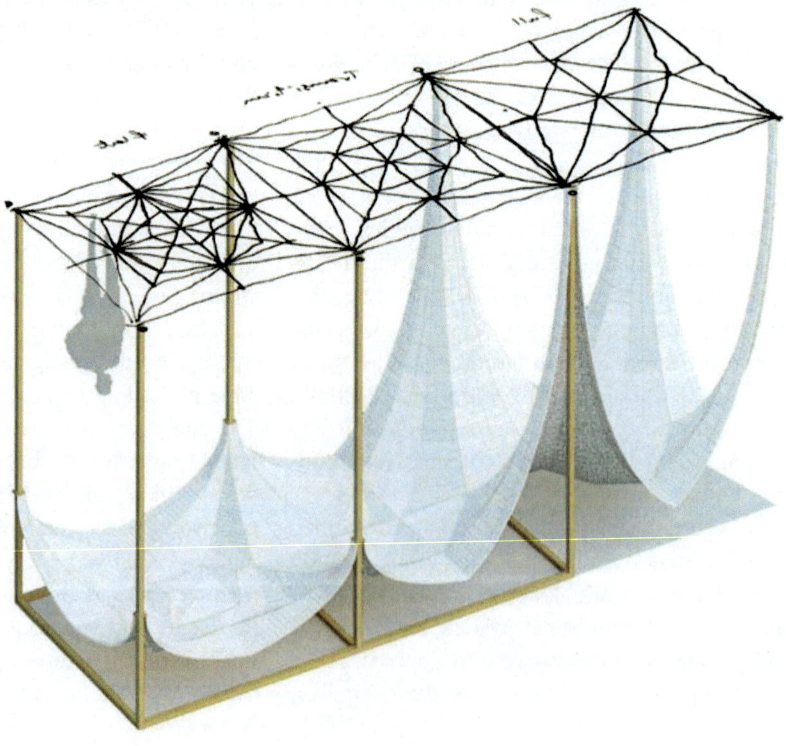

Fig. 2 Fabricarte design and construction concept (inverted)

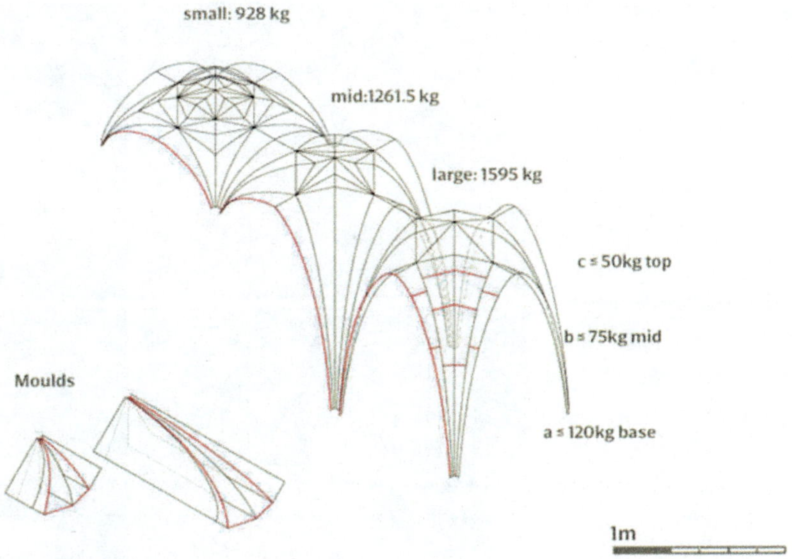

Fig. 3 Construction and manufacturing strategy, calculating the weights of pieces. Mould design and assembly

statics with a particle-spring system for form-finding, verification and maximum loads (Fig. 4). The vault was made of three layers of tiles with a maximum thickness of 100 mm. While the first and last symmetrical vaults were calculated by graphic statics for parallel sections, the forces in the middle transitional vault were studied in both directions to ensure that the modular construction of the vault would not result in discrepancies between thrust lines and the geometry of the vault. Moulds were designed for modular quarters and cut using computer numerical control (CNC) for screw-less easy assembly and disassembly without drawings.

In a workshop, the moulds were assembled and one layer of tile vaults with plaster was built; only side arches had a second layer to add stability to the pieces during onsite assembly. After the completion of a quarter, it was sliced by radial saws using the suggestive pattern as a guide. The pieces were transported on a mid-sized truck, and the moulds supported the shells' reassembly. The building team completed all loading and unloading; the mould/formwork

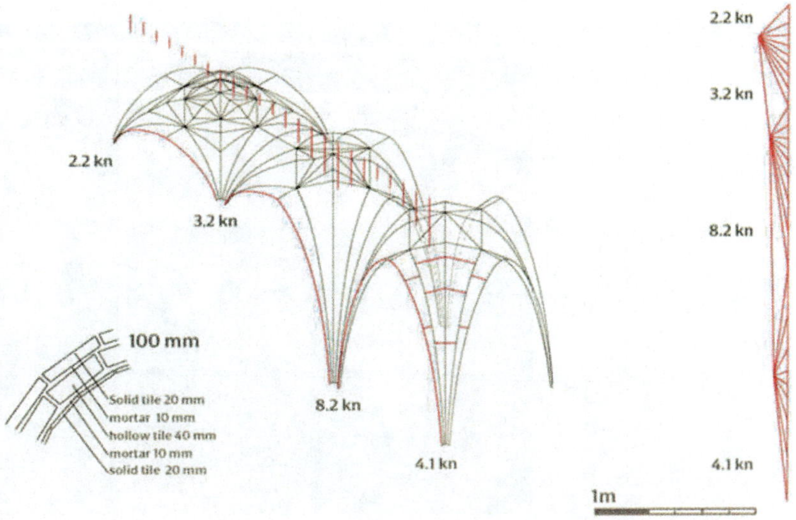

Fig. 4 Structural design and analysis using graphic statics

Fig. 5 Construction process of fabricarte

installation took 1 day and the vault assembly took 5 days. The assembly of the shell was made with plaster glueing its parts again. The tiles were hollow and the plaster at the cut edge did not need to be thick, making only minimal changes in the original geometry. Strips of fibreglass textile with plaster were added under the primary diagonal edge connections (Fig. 5). Once the pieces were joined in the three vaults, two layers of tiles were added at the top and bottom of the vault. The lower layer was made with plaster, and the joints between the tiles were filled with cement mortar (Figs. 6 and 7). The pavilion's flooring reflected the cutting lines of the vaults as a diagram of the process of manufacturing.

FabricArte examined the possibility of tile-vault manufacturing. It showed a system of pre-fabrication that does not rely on the high-precision fabrication of building components for rapid onsite construction. Instead, precision is inherent in the cutting of the shells. Tile vaulting remained the primary technique used in the building.

Fig. 6 FabricArte exterior views

Fig. 7 FabricArte interior view 1

3.2 Bending Parabolas

Inspired by the recent advances in the research on controlled bending, we wanted to explore bending-active structures for regenerative formwork or guidework for load-bearing shells (Tamke et al. 2013; Lienhard et al. 2013; Alexandrou and Phocas 2017). Traditionally, vernacular tile vaults' construction incorporates bent elements of reed or steel bars wedged between the vault's corners to mark the curvatures for sail or cross vaults. However, this method is only useable when the bending is minimal, where the parabolic, catenary and elastic curves are very similar. When bending with more acute angles, the elastic and parabolic curves begin to diverge significantly, and the bent steel bar is no longer valid as a reference for vaulting (Fig. 8). Our approach is to control the bending by changing the stiffness of the material along the strut by adding and subtracting material at a given location.

Two main equations were used to find the stiffness variation, the buckling equation and the parabola equation. The material variation can be solved and approximated when the two are equal at a given span-to-height ratio. The process resulted in a strip with a maximum thickness in its first and third quarters and a minimum thickness in the middle (Fig. 8).

$$y = a - bx^2 \tag{1}$$

$$R(x) = \frac{\left(1 + (2bx)^2\right)^{2/3}}{2b} \tag{2}$$

The tool in hand, now called the *bending parabola*, can be used to devise an elastic guidework for a parabolic arch as it closes and opens, resulting in a valid shape as guidework for vault builders (Fig. 9).

The structural and design property vaults become inherent in the tool. Both can be made without designing a geometry-specific structural analysis. Therefore, the strip offers the possibility for an autonomous in situ form-finding of structures that can be built with unskilled or novice labour (Fig. 10). Based on the bending parabola tool, we created three vaults using engineered bamboo strips with varied laminations. The method successfully described the geometry and served as a learning tool for the

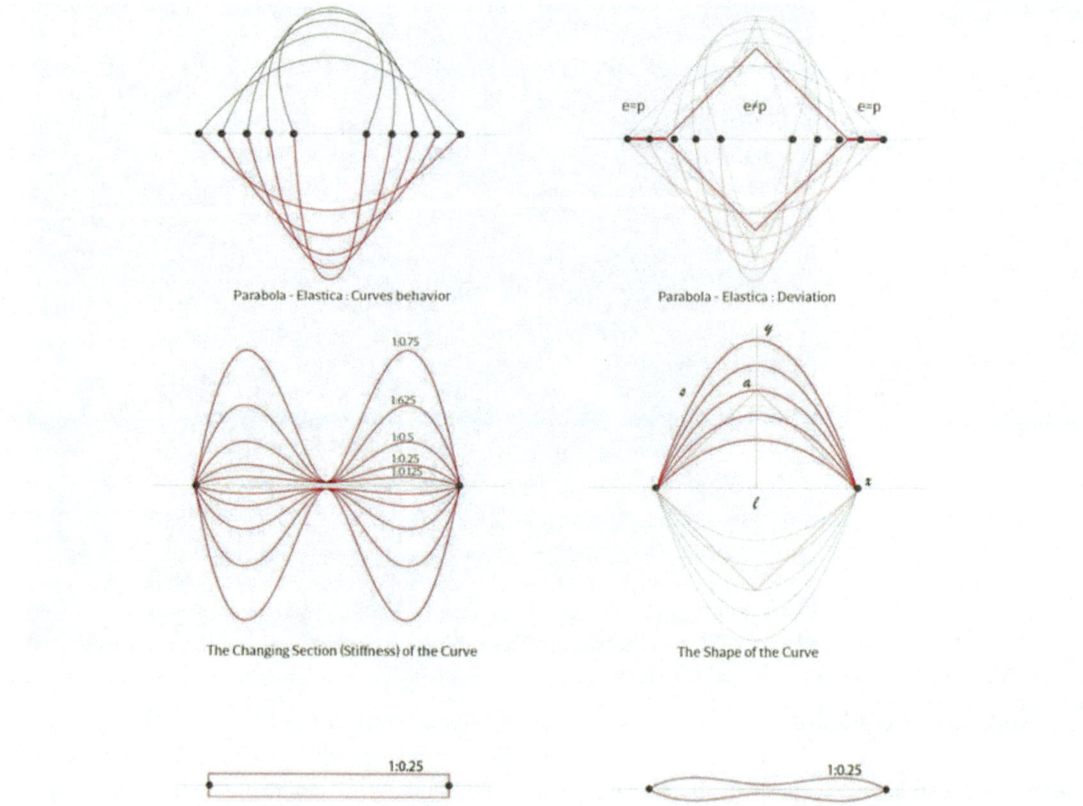

Fig. 8 Top: Elastica and Parabola behaviours. Middle: Results from approximation by changing stiffness concerning the height of the parabolic arch. Down: The resulting average bending parabola tool (right) compared to a typical strip

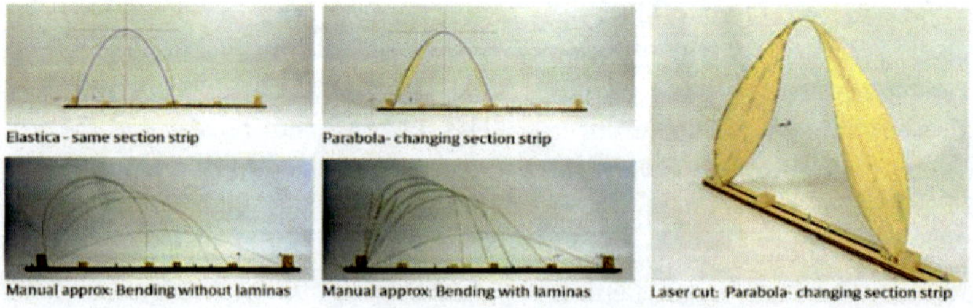

Fig. 9 Physical testing of bending parabola. Top left and right: Using laser cutting. Down left: Adding laminas of rectangular strips

builder to build an unconventional vault (Fig. 10).

However, this project was more intriguing because it moved into new design territories. The focus on the tools pushed for a departure from bending linear elements to combine networks of bending-active systems with flexible formwork for concrete slabs or lightweight flat-packed structures. In this system, the changing stiffness of each component in the structure not only drives the bending but also allows for a sagging effect of the flexible formwork to make the stiffening ribs. Following the pattern of the internal stresses inside a sail vault, the voids were reconfigured to align to the force's direction and enhanced by changing the number of layers of

Fig. 10 In situ, vaults prototype by bamboo strips as guidework 2016

textile to control the sagging depth (Fig. 11). Two 1/100 prototypes were made with wood plate and spandex and showed a possibility for developing this approach towards making concrete floor system from flatpack formwork. Because it is a pattern, the system can be translated into an object by various techniques, ranging from advanced computer-aided fabrication to a saw in the hands of a skilled carpenter. It can be reused or left as a stay-in-place formwork; be an offsite manufacturing industry or made on the move (Fig. 11).

While the core of this project was about tile vaulting, the technique itself was secondary, and the focus shifted to a study on describing the load-bearing shell by bending. Hence, what was examined was the tool to build and not the building. This project aimed to make a

Fig. 11 Floor system from flatpack sheets from drawing to casting

'generative' guidework whose inherent properties can find many shells. Although some limitations were encountered in the curved bending strips, such as the limited range of parabolic curves a strip can make, they expand the possibility of a formwork beyond solving one shell only.

3.3 La Hoja: Swapping Roles

In 2021, we got a commission to design an outdoor pavilion in a chef's house. However, unlike the traditional architectural design path, the initial ideation of the pavilion's geometry was proposed by the master builder with a shell representing an olive leaf, a cherished tree in the Spanish Levantine food culture. His design was developed through a sketch model (Fig. 12). Moving onwards, we wanted to develop this idea structurally and architecturally to host a space for a dining table for ten people. The builder developed the initial model but this time using small representative tiles, which inferred both a level of abstraction of the geometry and an approximation to a load-bearing shell structure.

After the conceptual design, we started a structural form-finding to make a pavilion from

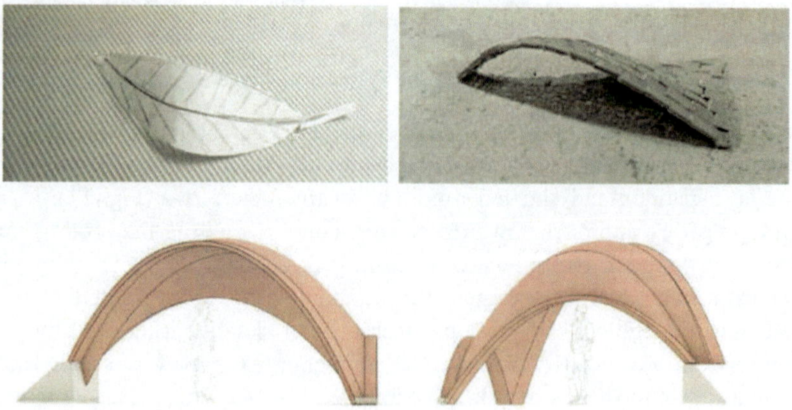

Fig. 12 Design development of the Leaf. Top: Early models by the vault maker. Down: Modelling developed by the architect

three edge arches where the variation of the height of each arch results in a leaf representation. This was achieved by extruding one pillar of the pavilion above ground level so that the catenary arches resulted in asymmetrical curves. The height of the arches also responded to the orientation of the pavilion to shelter from the afternoon sun but let the late evening sun enter. The play of the different heights gave an abstracted leaf-shaped shell seen from a specific angle facing the entrance of the chef's property. The resulting thrusts from the pavilion were calculated using graphic statics, with three sections in the edge arches and a section in the centre of the vault (Fig. 13). The horizontal thrusts were contained in reinforced limecrete foundations of 200 by 200 mm with rebars of 12 mm diameter. Three vertical pillars were extruded to receive and anchor the edge arches. The thickness of the pavilion was determined by the maximum thrust, resulting in a maximum thickness of 120 mm and a minimum thickness of 100 mm, which translated into three layers of tiles with different mortar thicknesses.

The construction of the pavilion was facilitated through a tile vaulting workshop. Participants of builders, architects and structural engineers worked together under the supervision of the master builder. The pavilion's construction started with the installation of formwork for the edge arches, which was reduced to the minimum through bending metal strips of 5 by 500 mm sections supported by extendable scaffolding props with changing height to govern the geometry of the edge arch (Fig. 14). An additional strip was installed inside the vault to provide a visual guide for the trainees.

During the workshop, the vault's construction started spirally from the edge arches towards the centre of the vault. After the workshop, the master builder will add two more layers of tiles, one exterior and one interior. For the interior coursing, we will use fast-setting plaster of Paris to add the tiles and then refill the grouts with the waterproofing lime-based mortar. The extra- and intrados layers will need carefully executed coursing which is hard to achieve in a workshop with novice builders and trainees (Fig. 14).

The design and construction of the leaf pavilion show that while the building site is one of the most probable spaces of interaction and decision-making in tile vaulting, other dynamics

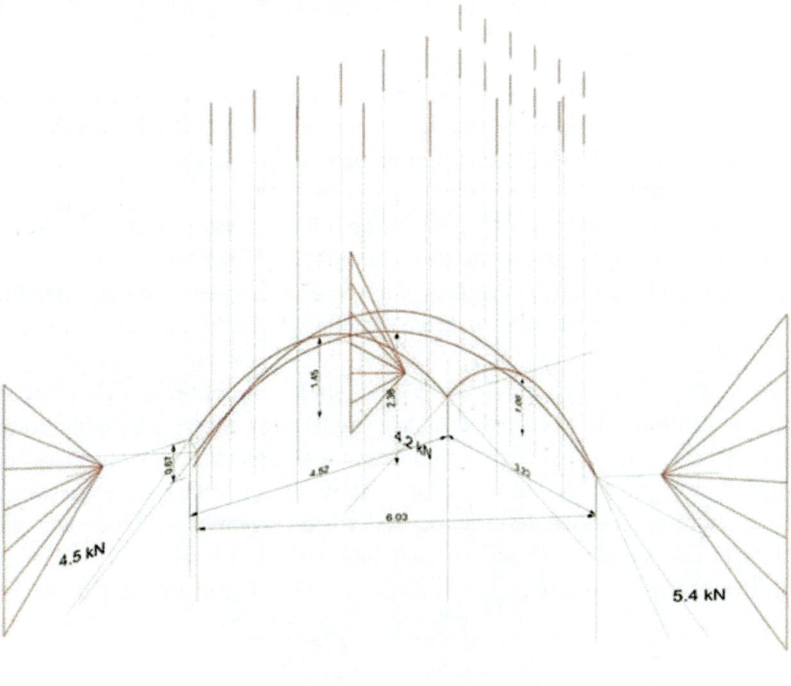

Fig. 13 Structural design and analysis of the Leaf using graphic statics

Fig. 14 The leaf after the workshop

happen outside the site too. Alteration in the hierarchy of design decision-making occurs at different stages and the design prompt for the project was initiated by the master builder, not the architect. However, through co-working with small-scale conceptual design models, the initial and literal configuration of the Leaf was developed into an architecturally useable pavilion, governed by its function and structural constraints as an unreinforced load-bearing shell generated through a series of catenary arches. This process suggests a shift from the usual design flow depicted by the institutional profession, to more open-ended design processes where architects can supplement and support builders with their design knowledge.

3.4 Las Cuevas: Digital Modelling for Site Uncertainty

Digital technology can offer a much-enhanced environment for designing and construction. However, this does not always have to result in offsite complex operations or onsite machinery. Technological tools such as digital modelling, VR and 3D Scanning can push traditional construction systems to new possibilities without radically changing their processes. In the following example, we used photogrammetry to understand the existing conditions of our intervention and base our design decisions on a more informed study. The project is located in the mountains of Xixona in Spain, where our vault design is intended to rehabilitate a cave dwelling. The cave-dwelling typology is ubiquitous in mountainous Spain from Malaga to Valencia; their sizes and complexities depend on the mountains' geology and the site's typography. The caves in our site were excavated in the 1940s out of a small fold in the mountain formation. The mountains belong to the Iberian Subbaetic System, most of which is a mix of soil, marl and marly limestone. Therefore, after the caves were abandoned for a long time without any maintenance, they suffered local collapses of lumps of soil and marl (Fig. 15). The two caves were originally purposed as barns for animals and space for food stock. Each had an average of 10 by 4 m in which two tile vaults were needed, firstly as a structural support of the inside the caves to prevent any further collapses and secondly to provide a usable space as a multipurpose hall.

The existing conditions of the caves and their natural formation resulted in many challenges. First, it was hard to get an accurate survey, but we had a schematic drawing only. Second, while the master builder can build the curved parabolic vaults without much guidework, a detailed study was needed to plan how this section varies to accommodate the changing height of the cave. Finally, because of the cave's delicate situation, it was essential to avoid heavy excavations during the construction. To respond to all these challenges, we started by making a three-dimensional model of the caves using photogrammetry. We used Meshroom software to compound the images into a model and simplify it to an overall surface and a series of sections (AliceVision 2018). The 3D scanned model became the envelope of the design inside which we studied, with the client and master builder, iterations of vaulting solutions (Fig. 16). The iterations were guided by the existing geological features of the site to respect the current section of the two caves, avoid heavy excavation and heavy infill operations between the cave and shells. The developed design comprised a domical vault geometry in the first cave with three arches at the entrance to elevate its height to the maximum. In the second cave, we proposed a parabolic undulating barrel vault. The two caves were connected using an existing door that was also vaulted (Fig. 17).

The structural design was developed using graphic statics of sections in the cave. The result from the study had details, plans, sections of the shells and zones of their different layers to vary their thickness according to the calculated weight of soil applied to them. The vaults had between three or four layers of tiles of 40 mm with around

Fig. 15 Caves interior before the rehabilitation, screenshots are taken from a photogrammetry model

Fig. 16 Design iterations in dialogue with the client and cave geology

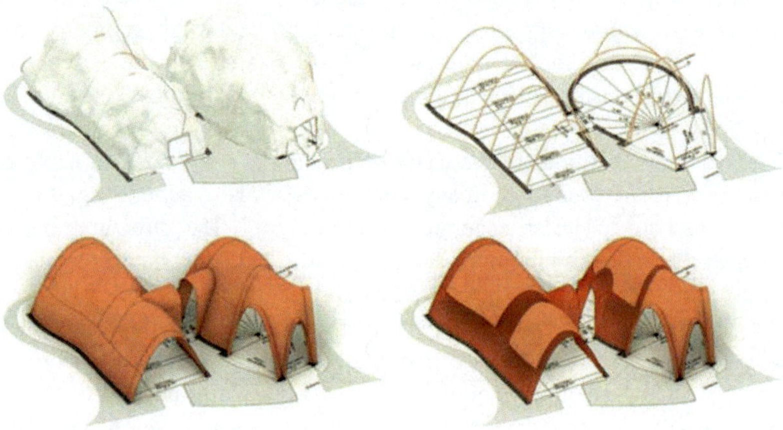

Fig. 17 Design of the two caves. Top: the guidework in caves, down: geometry and layers of the proposed vaults

10 mm of mortar. The infill between the cave and the tiles was made with construction waste of the tiles and plaster. Ribs above the tiles were also made in locations where the cavity between the shell and the caves was superior to 400 mm. The construction started with implementing a detailed plan of the foundations of the vaults, where the builders excavated a 250 mm deep canal and filled it with lime concrete. Because of the constant parabolic section of the vaults, a series of span and height ratios were sufficient to describe the geometry, which the skilled master builder could describe with light fibreglass guidework he occasionally used to mark the section of the vault (Fig. 18). For a formwork, the same walls of the caves were used to make the first arches. The use of local materials, construction waste and infill, and the minimal reliance on falsework have all resulted in a near-zero waste construction of structural shells as a rehabilitation project of the caves. The vaults were eventually plastered with earth rendering and finished with white lime wash, floors of the caves were paved using compacted mud flooring (Fig. 19).

The cave project of Alicante presents an example of the potential support to vernacular construction through utilising tools from digital technology. While photogrammetry is accessible with photos, its instant use paved for more informed decision-making that gave both architects and builders the power to accommodate and implement the design of the shells. The digital space created using this technology became the

Fig. 18 Vaults under construction

canvas for sketching and commenting on how to make a minimal but innovative intervention that lessens the use of material and reduces building operations. With demystifying the cave's complexity, the design development output was reduced to a floor plan and a diagram showing the zones of three or four layers in the vaults and supplemented by bent rods to describe the changing section of the vault.

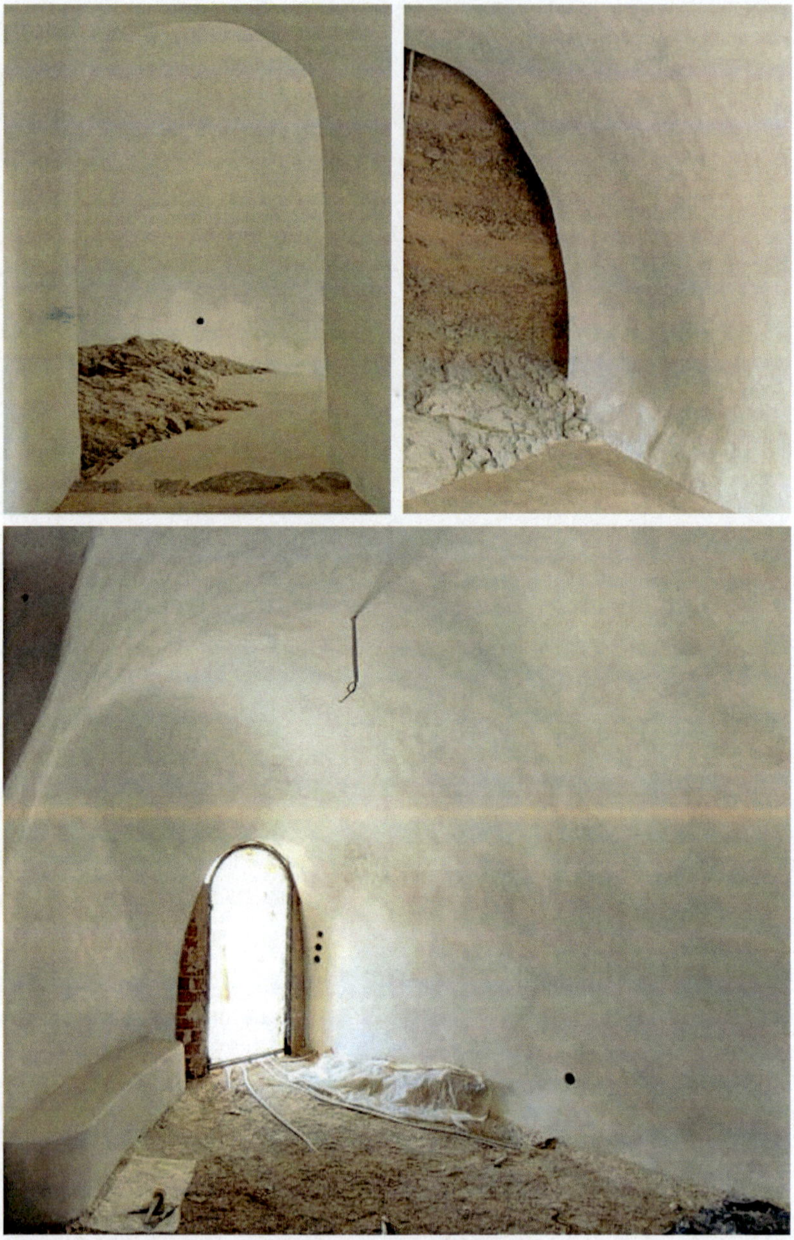

Fig. 19 Vaults after plaster rendering

4 Discussion: From *Withoutcisim* to the Architecture Relational Agency

In this essay, we examined four cases of tile-vault experiments to illustrate several possibilities of working with crafts communities and local making methods (Table 1). In each experiment, the focus was on one aspect of 'craft'. In FabricArte pavilion, the primary concern of the investigation was the 'time' of construction, which pushed for an unusual fabrication method in tile vaulting. In Bending Parabola, the scheme was to devise a 'tool' that helps builders visualise complex vault shapes. In la Hoja, the goal was to decentralise the architects' design process and adopt a 'design' initiated by the builder. In the

Table 1 Summary tile vaulting experiments

Project	Aim	Challenge	Method	Output
Fabricatre	Craft-based vaults prefabrication	Accelerate construction time	Designing cutting pattern	Manufactured floor system
In situ	Regenerative formwork	Tools for onsite form-finding	Controlled bending as a regenerative tool	Bending-active formwork for floor system
La Hoja	Co-design	Develop design ideas from Artisan's input	Design through models	Curriculum of tile vaults training
Cuevas	Design for uncertain scenarios	Drafting instructions without plans and sections	The use of 3D photogrammetry	Working in digital spheres

Caves of Xixona, the aim was to support the traditional system of tile vaults with the 'technology' of 3d photogrammetry in cases where conventional methods of designing are complex. Time, tools, design and technology are the main challenges that face vernacular building crafts in today's highly industrialised construction. Yet, they are also the very same features that render building crafts resourceful in their use of local materials and low-carbon processing methods.

Studies of architectural theory and history understood architecture by the presence or absence of the centralised design agent, *with* or *without* architects, resulting in dismissing or patronising how non-architects build. In contrast, the overall learning from these experiments shows that it is possible to move beyond this *withoutcisim* to an explorative domain between crafts and architectural design through a consistent, long-term and objective-oriented collaboration. This collaboration is conditioned with a relational, not hierarchical, agency. In psychology and education studies, the relational agency is 'the capacity to offer support and ask for support from others, which expands the resources available to actions on objects' (Edwards 2005). This methodological framework of collaboration between disciplines consists of establishing three main action points:

1. **Establish a common language**: Designers should engage with the complexity of crafts by illustrating their processes, not products. A new wave of such documentation is needed beyond documenting to 'save the past' but to open crafts to change and mutate in line with the transition to green building strategies. Architects can expand ethnographic studies on crafts to physical and digital modelling drawings of steps of making and computational coding of craft grammars and typologies. National and international institutions can contribute to establishing this common language by encouraging designers to work on encyclopaedic and digital databases. The outcome of this methodology can enhance the understanding of policymakers and designers about the potential future role of the existing network of makers as small enterprises and holders of cultural significance. Such understanding is crucial in the transition to a greener construction to secure decent work and growth of these enterprises.
2. **Aligning oneself in joint actions**: Designers should include crafts communities and vernacular knowledge in the current research on the manufacture and prefabrication of replicable low-carbon building components such as floor and façade systems. This includes shifting the focus from a site or project-specific collaboration to endeavour building systems through iterative prototyping. Craftspeople can contribute with more than their handwork making; they can expose nuances of how existing digital and industrial fabrication methods can be applied in different ranges of technologies. Therefore,

working with craft does not mean relinquishing the essential tools of architectural modelling and engineering testing but being preparade for a recalibration of these tools as the design grows. With such inclusion, research institutions in public and private domains can include low-tech and small-scale factories currently facing post-pandemic challenges. These small-scale factories can be an active part of developing sustainable cities and communities.

3. **Work beyond formal institutional channels**: To move to more amble and environmental matters, architects should engage with practices embedded in many forms of habitation ranging from cities' historical centres to informal settlements and from urban-to-rural settings. There is a need to expand the definition of building crafts beyond the heritage institutions' approach that only grants them value because of their cultural-historical significance. This collaboration can operate in various spaces, ranging from digital spaces of open-source building libraries to settings of informal construction. This role of designers will be needed in the highly urbanised globe where they can engage with builders and self-construction actors in informal areas to deal with improvisation, quick test-and-approve and modifications. Keeping an open-ended system with an error threshold is crucial for building craft inclusion in architecture. The skill of the builder and the preparation of empirical models by architects determine the elasticity of this system.

The proposed threefold dialogue explained several goals in the UN 2030 Sustainable Development Goals, from those related to the decent work and economic growth of local enterprise specified in goal 8 to the enhancement of innovation and supporting the small factories in SDG 9 (industry and innovation). Moreover, the dialogue main core is to bridge design for climate action, specified as SDG 13, with sustainable cities and communities. While a lot is being explored about participatory design and the inclusion of users in the planning process, there is a lot yet to be done on participatory construction, where local actors can also be part of shaping cities built environment. The work along these three points could link the thinking of the future architecture of resourcefulness with the societal and economic conditions of today's building practices. Future research is needed to understand and develop how to implement the three points at national and policy-making levels. Similarly, future research on different crafts and cultures of making is required in order to expand and build on the findings of this essay.

Acknowledgements The author would like to thank the master vault maker Salvador Gomis Aviño for his collaboration during the research. Fabricate was developed in collaboration and support of Light Earth Design. In Situ was supported by funds from the Department of Architecture, University of Cambridge and Clare Hall, the University of Cambridge.

References

Agustí-Juan I, Habert G (2017) Environmental design guidelines for digital fabrication. J Clean Prod 142:2780–2791. https://doi.org/10.1016/j.jclepro.2016.10.190

Al Asali MW (2016) Tools and technology in traditional architecture: a study of thin tile vaulting. Master of Philosophy, University of Cambridge

Alexander C (1964) Notes on the synthesis of form. Harvard University Press

Alexandrou K, Phocas MC (2017) Hybrid bending-active structures with multiple cables. In: Proceedings of IASS annual symposia. international association for shell and spatial Structures (IASS), pp 1–9

AliceVision (2018) Meshroom: A 3D reconstruction software

Beorkrem C (2017) Material Strategies in digital fabrication, 2nd edn. Routledge, New York

Collins GR (1968) The transfer of thin masonry vaulting from Spain to America. J Soc Archit Hist Soc Archit Hist 176–201

Davis L, Block P (2012) Earthen Masonry Vaulting: technologies and transfer. In: Cherenet Z, Sewnet H (eds) Building Ethiopia: Sustainability and innovation in architecture and design. Ethiopian institute of architecture, building construction and city development. Addis Ababa, pp 219–232

Edwards A (2005) Relational agency: learning to be a resourceful practitioner. Int J Educ Res 43:168–182. https://doi.org/10.1016/j.ijer.2006.06.010

Huerta S (2003) The mechanics of timbrel vaults: a historical outline. In: Becchi A, Corradi M, Foce F,

Pedemonte O (eds) Essays on the history of mechanics. Birkhäuser Basel, pp 89–134

Jayasinghe A, Orr J, Ibell T, Boshoff WP (2022) Minimising embodied carbon in reinforced concrete flat slabs through parametric design. J Build Eng 50:104136. https://doi.org/10.1016/j.jobe.2022.104136

Jencks C, Silver N (2013) Adhocism, expanded and, updated. MIT Press, The Case for Improvisation

Kuijer L, Nicenboim I, Giaccardi E (2017) Conceptualising resourcefulness as a dispersed practice. In: Proceedings of the 2017 conference on designing interactive systems. ACM, Edinburgh United Kingdom, pp 15–27

Lienhard J, Alpermann H, Gengnagel C, Knippers J (2013) Active bending, a review on structures where bending is used as a self-formation process. Int J Space Struct 28:187–196. https://doi.org/10.1260/0266-3511.28.3-4.187

López DL, Mele TV, Block P (2019) The combination of tile vaults with reinforcement and concrete. Int J Archit Herit 13:782–798. https://doi.org/10.1080/15583058.2018.1476606

Ochsendorf J (2014) Guastavino vaulting, Reprint. Princeton Architectural Press, New York

Ramage M, Hall TJ, Gatóo A, Al Asali MW (2019) Rwanda cricket stadium: Seismically stabilised tile vaults. Structures 18:2–9. https://doi.org/10.1016/j.istruc.2019.02.004

Rudofsky B (1964) Architecture without architects: a short introduction to non-pedigreed architecture. The Museum of Modern Art: Distributed by Doubleday, Garden City, N.Y, New York, NY, USA

Sayigh A (2019) Sustainable vernacular architecture: how the past can enrich the future. Springer

Tamke M, Stasiuk D, Thomsen M (2013) ALIVE: designing with aggregate behaviour in self-aware systems, pp 257–275

Truño i Rusiñol À, Huerta Fernández S, González Moreno-Navarro JL, et al (2004) *Construcción de bóvedas tabicadas*. Instituto Juan de Herrera, Escuela Técnica Superior de Arquitectura, Madrid

UNEP (2021) 2021 Global status report for buildings and construction

Watson J (2019) Lo-TEK. Design by Radical Indigenism. Taschen GmbH

Re-use in Danish Vernacular Architecture: Examples and Their Future Versatility

Birgitte T. Eybye

Abstract

With climate change, there is an urgent need to reduce CO_2 emissions, for instance, the ones related to the building sector. Contemporary building has developed into having a larger climate impact in the construction phase than in the operational phase which highlights the necessity to rethink resources. At the same time, research into connections between sustainability and Danish preindustrial vernacular architecture shows a widespread re-use of building materials and components. As a logical consequence, this paper investigates vernacular approaches to resource management in the form of re-use with a view to discuss the versatility of such in future building. The first part of the paper explores examples of re-use in preindustrial vernacular architecture. To achieve this objective, the paper initiates the methodic and analytical framework. Next, a literature review is presented to provide general information on re-use in architectural heritage and to focus on the two case studies. Through the case studies, examples of re-use and recycling are identified and outlined. The second part of the paper discusses the extent to which the vernacular approaches to re-use have the potential to contribute to future sustainable building and the possible challenges.

Keywords

Re-use · Vernacular architecture · Resource management · Architectural heritage · Sustainable building · Denmark

1 Introduction

For many years, Danish buildings focused on energy consumption in the operational phase for which reason demands on energy performance have been tightened several times in the building regulations. Hence, contemporary buildings developed into having a larger climate impact in the construction phase than in the operational phase (usually based on a study reference period of 50–60 years) because of the extraction, production and transportation of building materials (Andersen 2020). Concurrently, a larger number of buildings are demolished every year, thus generating enormous amounts of waste that usually downcycle to filling material as the base in road construction (Andersen 2014).

With the growing focus on climate change and subsequently, the urgent need to reduce CO_2 emissions, the Danish government has enacted legal requirements on CO_2 emissions targeting

B. T. Eybye (✉)
Aarhus School of Architecture, Research Lab 1/Transformation, Aarhus C, Denmark
e-mail: bte@aarch.dk

M. R. Thomsen et al. (eds.), *Design for Rethinking Resources*, Sustainable Development Goals Series,
https://doi.org/10.1007/978-3-031-36554-6_16

buildings. These requirements concern mandatory life cycle assessments (LCA) of new constructions and stepwise reductions of the emission limits from 2023 to 2029. Moreover, refurbishments and demolitions should become more sustainable, yet no precise demands have been defined. Finally, the agreement requires a survey with a view to clarify why buildings of technical quality are demolished (Valdimarsson and Kristensen 2021).

The above highlight the need for solutions that reduce the climate impact of building. One focus area concerns the construction phase, e.g. by application of materials with a small climate impact such as renewable or re-used materials as illustrated in the so-called 'pyramid of materials' (materialepyramiden) (Beim et al. 2019). Such an approach will be in line with researcher Michael Lauring that predicts building materials will become scarce and expensive in future (Lauring 2014, p. 185) and the Waste Framework Directive of the European Union (European Union n. d.), since the building sector produces 25–30% of all waste in the EU (Beim et al. 2019, p. 17). Relatedly, UN projections expect the global population to be around 8.5 billion in 2030 and around 9.7 billion in 2050 (United Nations 2022, p. i) which underlines the environmental as well as ethical needs for the rethinking of resources.

In parallel to this, a research project has investigated connections between sustainability and Danish preindustrial vernacular architecture. Apparently, sustainability in such dwellings manifests itself in the form of resource savings and prolonging the lifespan. The surveyed findings are of different kinds and, e.g. concern climate responsive design, passive energy strategies, knowledge of materials and their specific qualities in regard to durability, design for disassembly and building materials with a low carbon footprint such as renewable or re-used materials (Eybye 2016).

This knowledge of a widespread re-use of materials in preindustrial vernacular architecture contrasts the current throwaway culture that also includes buildings. With the need to rethink resources and reduce climate impacts, it seems logical to examine the management of resources in the heritage buildings, particularly preindustrial vernacular architecture, for the purpose of investigating whether such knowledge may contribute to sustainable solutions in future buildings. In doing so, this paper addresses the UN Sustainable Development Goals number 12 (Responsible Consumption) and number 13 (Climate Action) (UNDP 2022).

2 Materials and Methods

2.1 Objective and Methods

The objective of this paper is to investigate heritage, especially preindustrial vernacular, approaches to resource management in the form of re-used building materials in order to explore and discuss their potential in future building. Consequently, the primary research question of this paper is as follows:

How does re-use of building materials and components manifest itself in preindustrial vernacular buildings?

With a view to discuss the versatility of the possible findings of the primary research question, a secondary research question is posed:

How can such approaches to re-use inform future building and what are the challenges?

The paper is structured around answering these two questions. The first part of the paper takes its point of departure in research question one and examines how the re-use of building materials and components occur in heritage building, particularly preindustrial vernacular architecture, since this knowledge is essential to any discussion on the potential application of preindustrial resource management in future building. First, the paper presents a literature review with a view to explore accounts of re-use in heritage buildings. This review aims to provide general information on re-use as well as to focus on the case studies. For the purpose of investigating the primary sources of re-use, the paper performs two case studies of preindustrial vernacular dwellings, an agricultural worker's dwelling situated in Tåstrup, South Zealand and a farm situated south of Hvide Sande at the West

Coast. Finally, this section elaborates on the identified examples of re-use in preindustrial vernacular architecture. Consequently, the first part of the paper makes up the chapter on materials and methods and, the first section of the results. The second part of the paper aims to answer research question two. In doing so, the paper discusses the potential application of preindustrial approaches to re-use in regard to future sustainable building.

Regarding the delimitation, this study mainly focuses on buildings designed for habitation, since a large number of such buildings are preserved from the late medieval times and thus, provides an adequate study material for the first part of the paper. Furthermore, the study delimits 'preindustrial' building as constructed before 1850 and 1880 in the case of rural building. After 1880, socio-economic factors had influenced rural building, such as the co-operative movement that made farming more lucrative and the railroads that facilitated easy transport of building materials (Eybye and Vestergaard 2017, p. 591).

The main sources utilized comprise of author's PhD thesis for which the PhD degree was obtained. This paper is partly based on a cultivation and elaboration of particular findings described in chapters four and six which were not unfolded and developed in the thesis and, two case studies. Neither the thesis nor these particular findings have been subject to publication before. Moreover, it is important to note that there are no directly translated paragraphs of the thesis in this paper. Additionally, literature on architectural theory and history, architectural heritage, architectural conservation, sustainability and sustainable architecture, research methodologies as well as other relevant material have been used. Regarding the case studies, sources include literature, drawings, photos and empirical studies in the buildings.

Applied methods include literature review and case study research. The aim of the literature review is to look into re-use in architectural heritage, to focus and delimit the empiricism of the case studies. Methodologically, the search is not comprehensive for temporal reasons. Regarding appraisal, authors recognized for their research and/or experience from conservation practice is chosen. Finally, the synthesis is narrative (Grant and Booth 2009). Findings of the literature review may be applied as data triangulation in relation to the case studies. The choice of case study research is reflected in this methodology as being well qualified to provide in-depth information on a phenomenon (re-use of building materials and components) through the investigation of a small number of dense case studies (preindustrial vernacular dwellings) (Flyvbjerg 2006; Yin 2014). Furthermore, the case selection strategy is based on what researcher Bent Flyvbjerg calls 'information-oriented selection' since the aim of such strategy is '*To maximize the utility of information from small samples and single cases. Cases are selected on the basis of their information content*' (Flyvbjerg 2006, p. 230). Additionally, the two cases are selected with a large variation among them in order to survey diverse examples of re-use.

The choice of the two selected cases is based on the following criteria: (A) each case must be an exemplar of preindustrial vernacular architecture (see Sect. 2.2), (B) each of the two cases must be unique compared to the other in regard typology and geography in the aim of securing a thematic dispersion and subsequently, diverse findings, (C) each case must be well preserved, since examples of preindustrial re-use should be recognizable and (D) each case must be well documented for temporal reasons. The chosen cases are Trines Hus (prospectively Trine's House) and Abelines Gård (prospectively Abeline's Farm) that both meet these criteria, as they are (A) exemplars of preindustrial vernacular architecture, (B) differ in regard to typology and geography, (C) listed and hence, well preserved and finally, (D) well documented.

Lastly, re-use is addressed with a view to inform the discussion. Oxford Learner's Dictionaries defines re-use as '*to use something again*' in its current state and describes recycle as '*to treat things that have already been used so that they can be used again*' (Oxford University Press). For the purpose of promoting transition

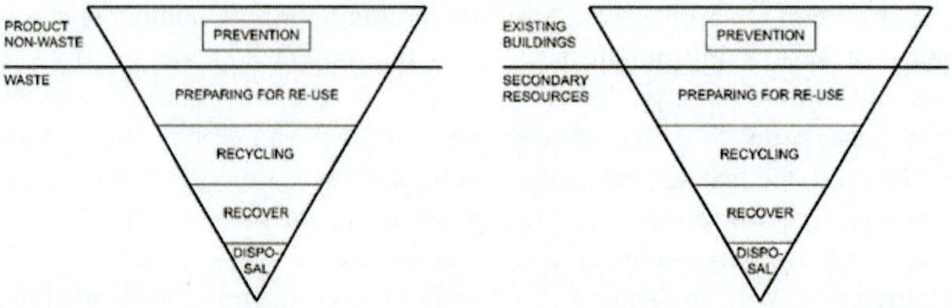

Fig. 1 The Waste Hierarchy. To the left, the Waste Hierarchy as introduced by the EU (European Union n. d.). To the right, the Waste Hierarchy elaborated onto buildings by Inge Vestergaard (Vestergaard 2022). Drawing by Birgitte Eybye

into a Circular Economy (CE) by preventing and reducing waste and, improving resource efficiency, the EU Waste Framework Directive has introduced the so-called 'Waste Hierarchy', see Fig. 1. This top-down triangle comprises of five categories or steps ranging from 'prevention' to 'preparing for re-use', 'recycling', 'recover' and 'disposal', in which the first is a non-waste product and the latter four categories are waste (European Union n.d.). Researcher Inge Vestergaard has elaborated the EU Waste Hierarchy into building resources and, prevention means '*that buildings should be utilized in its full consequence*' (Vestergaard 2022, p. 2). Below categories are referred to as 'secondary resources' comprising re-use, recycle, recover and disposal (Ibid). The discussion draws on the Waste Hierarchy as elaborated by Vestergaard, and in this connection, the paper applies 're-use' of the Waste Hierarchy as materials and components taken from one building and re-used in another with no changes nor substantial processing, whereas 'recycling' is a processing of a waste product. Prospectively, the term 're-use' is applied in the description of general dealings with materials and components that have been used in more buildings.

2.2 Literature Review: Focusing the Empiricism

The aim of the literature review is to focus the empiricism by looking into different types of heritage buildings designed for habitation and explore to what extent re-use took place.

According to the review, re-use occurred in all sorts of residences and dwellings. In posh buildings, structures were re-used, such as townhouses and manors built on medieval basements. Moreover, in such buildings, it seems that re-use of high-quality materials was preferred (Romanesque ashlars, baked bricks, sometimes timber) and re-use was likely 'hidden' (e.g. the basements) (Stilling 1999; Vadstrup 2004; Slots- og Kulturstyrelsen 2015).

Regarding vernacular architecture, several sources describe re-use as prevalent. Looking into the smallholding called Stine's House (Stines Hus), re-use includes timber, doors, windows and joinery (Sebro and Realdania By & Byg 2020). This account of extensive re-use in vernacular architecture is in keeping with other sources. For instance, researcher Bjarne Stoklund studied the vernacular architecture of 1700–1850 and expresses that '*... only in exceptional cases, a farm was constructed of exclusively new materials*' (Stoklund 1969, p. 22).[1] Similarly, researcher Erland Porsmose states that '*People economised in most parts of the county and re-use was absolutely necessary*' (Porsmose 2008, p. 77).[2] Finally, contemporary sources written around 1789–90 also mention re-use in farm buildings (Lerche 1987).

Based on the above review, it is presumed that the most and diverse examples of re-use will be found in vernacular architecture. For this reason, the following case studies are chosen among such buildings.

[1] Translated by the author.

[2] Translated by the author.

Fig. 2 Collage of Trine's House. 1. Trine's House seen from the south. 2. Re-used, leaded window. 3. Re-used door between the chamber and living room. 4. Timber number VI (six) though it is bay number four. 5. Re-used timber with different joints between post and tiebeam (left: halved joint, right: pierced beam). 6. Collar beam with traces of former use. All photos are taken by Birgitte Eybye

2.3 Case Study I: Trine's House (Stevns, Sydsjælland)

Trine House is an archetypical agricultural worker's dwelling situated in the village of Tåstrup, see Fig. 2. The building was most likely constructed between the years 1809–13 or shortly after, since it is situated on a farm plot which became vacant after the relocation of four Tåstrup farms in the context of the agrarian reforms.

Originally, this half-timbered building comprised of six bays made of oak timber that was re-used (probably from one of the relocated farms). The timber has traces of the former joints and the timber numbers[3] feature in a non-numerical order. Around 1850, the building was extended with another three bays made of re-used pinewood, as the posts and corresponding beams are joined using more principles (pierced beams and halved joints). Examinations of the roof construction also show re-used timber due to traces of holes for tenons and staves.

It is possible that the small leaded window by the front door and (parts of) the window in the west gable are relicts of the original windows that highly likely were re-used from one of the relocated village farms.[4] Moreover, the windows of the larders and the stable are probably also re-used.

In the interior, more doors such as the door between the living room and the chamber strongly indicate re-use due to its historic design and profiles. The other doors are in most cases batten doors. It is difficult to determine possible re-use for more reasons, e.g. the mix of old and newer fittings such as hinges and locks. Finally, usable parts of worn-down components have been used for repairs, e.g. repairs to the bottom of the doors or the wooden divider in the stables (Varmings Tegnestue 1988; Jensen 1990; Eybye 2016, pp. 180–189).

[3] In historic timbered constructions such as half-timbering and roof constructions, all timber pieces belonging to a particular bay would be marked with a number. Dimensions of the timber pieces were heterogenous, as wooden resources were used in a rational manner. Thus, a specific piece of timber would fit into a specific part of the construction and, the timber numbers prevented disorder. In the case of re-use, original timber numbers were usually ignored.

[4] In Denmark, so-called 'English windows' with wooden bars gained forward from the early eighteenth century, thus replacing leaded windows (Kulturstyrelsen 2012).

2.4 Case Study II: Abeline's Farm (Hvide Sande, Jylland)

Abeline's Farm is a four-winged farm situated in the dunes between the West Coast and Ringkøbing Fjord, see Fig. 3. The farm was built in stages from 1854 to 1871 and, it follows the traditional layout of such farms. This farm replaced the old buildings situated just south of the current farm site.

The buildings are built of baked bricks and adobes, as the latter are used in less exposed building parts. The timber of the roof construction comes in more types such as roundwood and blocks and with traces of re-use such as holes for tenons and staves. It is likely that some timber and wood are re-used from shipwrecks, as this was common practice in the area due to the lack of forests.

In the outbuildings, windows and doors are heterogenous, which indicates re-use. Additionally, the outer door between the dwelling and the courtyard is re-used. It has distinct traces of being one part of a former double door and most probably originates from one of the front doors of the south façade.

In the interior, the farmhouse can be divided into the distinguished rooms (where guests would be) and practical rooms (where the family and their servants would go). In the latter rooms, a large number of homogenous doors exist that point to re-use, many of them most likely from the former farmhouse, since their design, profiles and fittings such as hinges highly indicate manufacturing much earlier than 1871. Finally, wood was re-used for different purposes, e.g. window casings and fittings (Jensen 1975; Alsted 1985; Andersen 2000; Eybye 2016, pp. 144–157).

3 Results

This chapter comprises the findings of the case studies and of the discussion. Section 3.1 elaborates on re-use of materials and building components in preindustrial vernacular buildings. Section 3.2 concerns the versatility of the examples of preindustrial vernacular re-use in future building.

3.1 Findings of the Case Studies: Re-use in Preindustrial Vernacular Buildings

The two case studies showed examples of re-use in both agricultural workers' dwellings and farmers' buildings, respectively, in the east and

Fig. 3 Collage of Abeline's Farm. 1. Abeline's Farm seen from the southwest. 2. Exterior door re-used from a former front door. 3. Re-used door from the former farmhouse. 4. Re-used roundwood. 5. Re-used timber. 6. Window casing made of re-used wood, maybe from shipwrecks. All photos are taken by Birgitte Eybye

the west of Denmark and in half-timbered as well as brick buildings. Based on a cross-case generalization, re-use was common in preindustrial vernacular dwellings which is consistent with the sources of the literature review.

Examples of re-use manifest themselves in the following ways; (A) general re-use of materials, (B) separability of building materials and (C) design of building components for the purpose of possible re-use.

Re item A, a number of particular conditions influenced possible re-use. Generally, it goes for preindustrial building that it primarily applied 'raw' materials instead of 'compounded' materials. Moreover, the investigated buildings are characterized by being constructed of a small number of different building materials.[5] Both conditions made re-use easier and, as the procurement of building materials and manufacturing of components involved hard work by hand, re-use was rational. Thus, worn-down buildings equaled 'resource banks'. In relation to this, building materials and components appear to have 'degraded' slowly, e.g. doors and windows were re-used in buildings or rooms of lower status or, usable parts of worn-down materials and components were applied for repairs, thus corresponding to what we may call a 'hierarchy of resources' and prolonging the lifetime of such. Re item B, the joints between different materials were mostly 'weak', meaning that, e.g. timber and wattle and daub easily could be separated from each other. Similarly, doors and windows could be taken out without damage and then be re-used. Re item C, timber joints in half-timbering and roof constructions made it possible to disassemble and reassemble the timber multiple times (elaboration of Eybye 2016, pp. 102–103).

[5] Researcher Søren Vadstrup has explored building materials in a preindustrial farm situated in Viby, Fyn. He makes a list of 22 different materials including several types of different wood species used in the original farm of 1736 (Vadstrup 2004, p. 14) which confirms the small number of different building materials in preindustrial vernacular architecture.

3.2 Findings of the Discussion: Future Versatility of Preindustrial Vernacular Re-use

The evaluation of the findings based on the Waste Hierarchy indicated strong resource management in preindustrial vernacular architecture, since the examples are categorized as re-use and recycling. More building materials were biodegradable such as wood and thatching materials, thus reducing waste. Earth used for wattle and daub could recirculate into the soil.

Concerning the future versatility of the preindustrial vernacular resource management, the discussion calls for a new approach to resources, i.e. extensive re-use of buildings, materials and components (resource banks and long lifetimes). Additionally, demands on a reduced number of types of materials for the ease of re-use and separability should be made. The discussion also outlined challenges regarding re-use of legislative, technical, toxicological and socio-economical art.

4 Discussion and Conclusion

4.1 Discussion

The discussion opens with reflections on the applied methods. First, the literature review looked into re-use in heritage buildings and focused on the case selection accordingly. This type of review was chosen for temporal reasons and, though it was not exhaustive on the subject, the sources showed vernacular architecture as most relevant in regard to explore re-use in heritage. Concerning the sources on vernacular architecture, these converged and were both contemporary and modern. A more comprehensive review would probably have provided more information on re-use in different types of heritage, yet it is unlikely that the conclusion and thus, the choice of cases had changed.

In relation to the case studies, the selection was information-oriented with a view to maximize variation among the two cases. It seems, however, that the identified examples in the two cases tend to be somewhat homogenous, leading to the cross-case generalization that re-use was extensive in preindustrial vernacular architecture. Such conclusion is supported by the literature review.

The case studies were based on literature, drawings, photos and empirical studies in the buildings for the purpose of data triangulation. The empirical studies comprised of non-destructive observations, and none of the written sources gave accounts of the construction of the specific buildings at the time of their construction. Consequently, there may be an unreported number, as examples may be hidden in the construction or overlooked. Elements of doubt were left out. There is also a small possibility that materials and components have been misjudged as re-used due to an appearance of historic design and detailing. Dendrochronology or carbon dating could have provided exact dating and thus, verified or falsified possible cases of re-use. The economic aspects of such approach were, however, not possible. Conclusively, the surveyed examples of re-use are considered to be reliable since cases of doubt are left out and, the findings are consistent with the data of the reviewed literature.

The findings of the case studies comprise timber, doors, windows and usable materials taken from worn-down components such as wood and fittings. Compared to the 'Waste Hierarchy' (see Sect. 2.1), timber for the constructions, doors, windows and fittings are considered as re-use, since these materials and components were taken from one building and re-used in another. It appears that such re-use was performed with no processing of the element (there might have been minor adjustments to, e.g. the timber). Furthermore, usable parts of building materials and components are considered as recycled in the Waste Hierarchy, as they were sorted from waste and then processed to a new function. It is not possible to draw any certain conclusions on the waste, yet materials degraded very slowly (the hierarchy of resources) and, more were biodegradable such as wood and materials for thatching, i.e. no waste. Hence, the evaluation of the findings in regard to Waste Hierarchy indicates strong resource managements in preindustrial vernacular architecture. Moreover, some of the material properties and approaches resemble the biological circles of CE.

This discussion now proceeds to the secondary research question concerning the future versatility of preindustrial, vernacular approaches to re-use. First, the discussion looks into re-use in contemporary (Danish) architecture. Companies such as Genbyg trade re-used building materials and, Gamle Mursten is specialized in re-used bricks. Such bricks are, for instance, applied in Frederiksbjerg Skole (2016), Aarhus and the multi-storey car park (2016), Billund. Yet, the bricks are used as mere facing and thus, illustrate sustainability as a 'valued achievement' (Eybye 2016). Pilot projects like Upcycle House (2014), Resource Rows (2020) and The Swan (2022) by Studio Lendager concern extensive re-use. Furthermore, re-use is common in architectural conservation for aesthetic and authentic reasons. From the European context, the Delftse Poort is a remarkable example of an almost 100% circular office interior renovation (van den Heuvel 2022). Conclusively, the present-day re-use concerns prestige or idealism and, these examples constitute a minority, since the building sector produces 25–30% of all waste in the EU as already mentioned (Beim et al. 2019, p. 17). This topic was also highlighted by the exhibition 'Die Schweiz: Ein Abriss' in 2022, as the construction sector produces 84% of Swiss waste (Lundberg 2022). Consequently, there is a major difference between the resource management in vernacular versus current buildings. Moreover, many preindustrial building materials were biodegradable, which also reduced waste.

Based on the above, it is relevant to discuss what possibly prevents extensive re-use. First, the present-day building has changed substantially from preindustrial times. Building is now subject to a large number of legislations, e.g. regarding design, technology and, materials. Moreover, construction has become professionalized compared to preindustrial vernacular owner- or community building, the number of applied building materials and components has increased and become more complex with many compounds and, materials origin from all over the planet.

Initially, the main learnings for future building are the perception of discarded buildings as resource banks and subsequently, slow degradation and long lifetimes of materials and components. Re-use, recycling and repair will become easier if the number of different building materials and components, including compounds is reduced as was the case in preindustrial building. Solutions with increased use of biodegradable, untreated materials will decrease amounts of waste in accordance with CE. Moreover, the building sector must improve its handling of used building materials and components. The reduced separability of current constructions calls for changes to enable re-use. Today, design for disassembly and other mechanical joints are easily made with robot tools and, should be mandatory in construction. Challenges of legislative, technical, toxicological and socio-economical kind concern these proposals.

Following the EU Waste Hierarchy and Vestergaard, demolitions should be prevented to reduce waste in building (Vestergaard 2022). Concrete buildings in particular calls for retention (equaling principles of slow degradation and long lifetime), yet this approach contrasts the current politically motivated demolitions in areas appointed ghettos (Ibid.) and the enacted survey clarifying why buildings of technical quality are demolished, see Chap. 1. Thus, socio-economic factors may promote demolitions as in the much talked-about example of Robin Hood Gardens.

Regarding legislations, all building materials and components, re-used/recycled or not must meet the requirements of the Danish building regulations. Moreover, CE certification is very important with regard to the responsibility of the client consultants, the contractors and building owners. Despite the political intentions of reducing waste of demolished buildings, there is a gap between the certification and building regulations (Oberender 2017). For instance, the certification of Gamle Mursten was called a 'breakthrough in re-use' (Andersen 2018). New or tightened demands may also challenge the transformation of existing buildings such as the transformation of Hochbergerstrasse 158, Basel (a 1965 concrete structure), in which the staircase had to meet new criteria for earthquake protection (Lundberg 2022).

More challenges of technical and toxicological art concern the existing building stock and thus, re-use etcetera. In the Danish context, this goes especially for the buildings constructed after 1950–60. Around this time, many new materials were introduced in construction (Vadstrup 2019, p. 69). Some materials are difficult to re-use, recycle or recover for reasons of tectonics, e.g. concrete panels or masonry with cement mortar, others are toxic and, both conditions challenge possible re-use. Chemical compounds such as PCB evaporates into the surroundings and constitutes a global problem in building (Melymuk et al. 2022). Before any re-use is achievable, possible toxins must be identified.

Economically, it is not yet cost-effective to dismantle buildings for re-use and recycling in most Danish cases. For the purpose of illustration, re-used bricks of Gamle Mursten are much more expensive than new ones. Similarly, the Delftse Poort interior renovation was not cheaper than standard solutions (van den Heuvel 2022, p. 473).

In summary, there is both a large wish and need for extensive re-use in future building as in preindustrial resource management. Yet, such re-use calls for new structural approaches to building, including economy, legislation, technology and materials. The example of Delftse Poort shows that non-critical interior parts as well as exterior parts is a point of departure for extensive re-use.

4.2 Conclusion

The primary objective of this paper was to investigate re-use in architectural heritage, particularly preindustrial vernacular architecture. To achieve this objective, the paper has presented a literature review and two case studies. In doing so, the paper identified and discussed a number of examples of re-use and recycling of building materials and components that rated well in the Waste Hierarchy assessment. The secondary objective was to explore how these examples could inform future building as part of the discussion. Suggestions for increased re-use and recycling in future building were posed, most importantly were probably the long lifetime through slow degradation, a reduced number of building materials and the perception of discarded buildings as resource banks. The discussion also highlighted major challenges of legislative, technical, toxicological and socio-economic kinds which complicate or even prevent the increased application of re-used materials and components.

During the study, it became clear that investigating re-use as a solitary element in preindustrial vernacular architecture is difficult. The re-use is connected to, e.g. maintenance (and likely also other factors) among which the long lifetime of building materials and components are achieved. More research into this topic seems highly relevant, both for the purpose of understanding architectural heritage and informing future sustainable building.

Acknowledgements This study was conducted as part of author's assistant professorship at the Aarhus School of Architecture.

References

Alsted G (1985) Abelines gård: en strandfogedgård på Holmsland Klit. Den selvejende institution "Abelines Gård", Hvide Sande

Andersen PD (2000) Den vestjyske klitgård: fra Nymindegab til Nissum Fjord. Poul Kristensens Forlag, Herning

Andersen U (2014) 20 millioner ton forurenet byggeaffald under danske huse og veje. Ingeniøren 24.1.2014. https://ing.dk/artikel/20-millioner-ton-forurenet-byggeaffald-under-danske-huse-veje-165805. Accessed 4 Oct 2022

Andersen U (2018) Genbrugsgennembrud: Gamle mursten kan være lige så gode som nye. Ingeniøren 21.1.2018. https://ing.dk/artikel/genbrugsgennembrud-gamle-mursten-kan-vaere-lige-saa-gode-nye-209975. Accessed 11 Oct 2022

Andersen U (2020) Danmark sakker bagud: vi ignorerer nybyggeriets største klimabelastning. Ingeniøren 8.1.2020. https://ing.dk/artikel/vi-ignorerer-nybyggeriets-storste-klimabelastning-231059. Accessed 4 Oct 2022

Beim A, Ejstrup H, Frederiksen LK, Hildebrand L, Madsen US, Munch-Petersen P, Sköld SR (2019) Circular construction: materials, architecture, tectonics (Cirkulært byggeri: materiale, arkitektur, tektonik). Royal Danish Academy of Fine Arts, København

European Union (n.d.) Waste prevention and management. https://ec.europa.eu/environment/green-growth/waste-prevention-and-management/index_en.htm. Accessed 6 Oct 2022

Eybye BT (2016) Bæredygtighed i Danmarks førindustrielle bygningskultur og dens aktuelle relevans: belyst gennem studier af seks boliger. Dissertation, Aarhus School of Architecture

Eybye BT, Vestergaard I (2017) A survey of Danish earthen heritage for sustainable building. In: Mileto C, Vegas López-Manzanares F, García-Soriano L, Cristini V (eds) Vernacular and earthen architecture. Taylor & Francis Group, London, pp 587–592

Flyvbjerg B (2006) Five misunderstandings about case-study research. https://doi.org/10.1177/1077800405284363

Grant MJ, Booth A (2009) A typology of reviews: an analysis of 14 review types and associated methodologies. https://doi.org/10.1111/j.1471-1842.2009.00848.x

van den Heuvel M (2022) Delftse Poort. In: Baumeister R, Petermann S (eds) (with additional contributors) Back to the office: 50 once revolutionary office buildings and how they sustained. nai010 publishers, Rotterdam, pp 468–477

Jensen KV (1975) Bebyggelse og landskab på Holmsland Klit: egnspræg – bvaring. Arkitektskolen Aarhus, Aarhus

Jensen JH (1990) Trines Hus: en beretning om livet i en landsby på Stevns, som det formed sig for Trine i de "gode gamle dage" fra 1882 til 1962. Amtscentralen, Næstved

Kulturstyrelsen (2012) Datering af vinduer 1700–1850. https://slks.dk/fileadmin/user_upload/SLKS/Omraader/Kulturarv/Bygningsfredning/Gode_raad_om_vedligeholdelse/5.4_Datering_af_vinduer_1700-1950.pdf. Accessed 6 Oct 2022

Lauring M (2014) Fremtidens bæredygtige bolig og bebyggelse – et scenarie. In: Holm J, Søndergaard B, Stauning I, Jensen JO (eds) Bæredygtig omstilling af bolig og byggeri. Frydenlund Academic, Frederiksberg, pp 163–193

Lerche G (1987) Bøndergårde i Danmark 1789–90: byggeskik på Landboreformernes tid. Landbohistorisk Selskab, Odense

Lundberg KH (2022) Ned fra 84%. Arkitekten 124 (09):26–30

Melymuk L, Blumenthal J, Sáňka O, Shu-Yin A, Singla V, Šebková K, Fedinick KP, Diamond ML (2022) Persistent problem: global challenges to managing PCBs. https://pubs.acs.org/doi/10.1021/acs.est.2c01204

Oberender A (2017) Sådan er reglerne for genbrug og genanvendelse. https://bygherreforeningen.dk/saadan-er-reglerne-for-genbrug-og-genanvendelse/. Accessed 11 Oct 2022

Oxford University Press. Oxford Learner's Dictionaries. https://www.oxfordlearnersdictionaries.com. Accessed 6 Oct 2022

Porsmose E (2008) Danske landsbyer. Gyldendal, København

Sebro L, Realdania By & Byg (2020) Stines hus: lollandsk egnsbyggeskik. Realdania By & Byg, sine loco

Slots- og Kulturstyrelsen (2015) Fredningsværdier, Karnapgården, Viborg Kommune. https://www.kulturarv.dk/fbb/downloaddokument.htm?dokument=124437320. Accessed 4 Oct 2022

Stilling NP (1999) Danske herregårde: arkitektur, historie og landskab. Nyt Nordisk Forlag Arnold Busck, København

Stoklund B (1969) Bondegård og byggeskik før 1850. Dansk Historisk Fællesforening, København

UNDP (2022) What are the sustainable development goals? https://www.undp.org/sustainable-development-goals?utm_source=EN&utm_medium=GSR&utm_content=US_UNDP_PaidSearch_Brand_English&utm_campaign=CENTRAL&c_src=CENTRAL&c_src2=GSR&gclid=Cj0KCQiA5NSdBhDfARIsALzs2EDw29c7t21ukf-ttaTWxv9HGn7FOJ9ik5Vm-4-OEjpd87DC1vXb33gaAkWwEALw_wcB. Accessed 13 Jan 2023

United Nations Department of Economic and Social Affairs, Population Division (2022) World population prospects 2022: Summary of results. UN DESA/POP/2022/TR/NO. 3

Vadstrup S (2004) Huse med sjæl. Gyldendal, København

Vadstrup S (2019) Bygningskulturens vedvarende holdbarhed. In: Høi A (ed) Levende bygningskultur: en essaysamling. Realdania, København, pp 67–81

Valdimarsson E, Kristensen FB (2021) Bred aftale om bæredygtigt byggeri fremrykker CO2-loft. Byrummonitor 5.3.2021. https://byrummonitor.dk/art8125285/Bred-aftale-om-b%C3%A6redygtigt-byggeri-fremrykker-CO2-loft. Accessed 4 Oct 2022

Varmings Tegnestue (1988) Trines Hus; Matrikel 4b, Tåstrup by, Hellested (non-published material)

Vestergaard I (2022) Reusing concrete panels from the industrial mass housing of the 1960s. https://doi.org/10.1088/1755-1315/1122/1/012009

Yin RK (2014) Case study research. SAGE, Los Angeles

(Re)making the Haubarg—Towards Sustainable Dwelling on a Bounded Earth

Nicolai Bo Andersen
and Victor Boye Julebæk

Abstract

In contemporary architecture, there is a renewed focus on resource production, consumption and use. Wood may be considered a renewable resource, potentially abundant, carbon neutral and recyclable. However, the environmental benefits of using timber are not straightforward. Also, it seems as if the architectural implications of designing with wood have not been fully articulated. As such, it seems necessary to rethink the potential of wood in contemporary architectural design practice. The aim of this paper is to investigate the topic of Rethinking Resources with a specific focus on the qualitative potential of wood as informed by traditional building culture. It is asked how knowledge embodied in crafts tradition and local vernacular may inform a contemporary architectural design practice and inspire the development of (more) sustainable building culture(s). A traditional farmhouse has been investigated and analysed with regard to technical properties, cultural-historical qualities, and experiential effects. An experimental timber structure has been built by students at the Royal Danish Academy, and the results have been described and analysed seen through a phenomenological-hermeneutic lens. It is pointed out that technical properties, cultural-historical qualities and experiential effects as found in crafts tradition and local vernacular may inform a contemporary design practice. It is argued that questions concerning sustainability should be seen in a holistic perspective and the building as situated within a larger material, environmental and social (eco)system. In this perspective, traditional building culture may supplement and qualify contemporary design strategies and may contribute to inspiring future (more) sustainable building culture(s).

Keywords

Materials · Tectonics · Architecture · Aesthetics · Sustainability

N. B. Andersen (✉)
Royal Danish Academy—Architecture, Copenhagen, Denmark
e-mail: nande@kglakademi.dk

V. B. Julebæk
Centre for Sustainable Building Culture, Royal Danish Academy—Architecture, Copenhagen, Denmark
e-mail: victor.julebaek@kglakademi.dk

M. R. Thomsen et al. (eds.), *Design for Rethinking Resources*, Sustainable Development Goals Series,
https://doi.org/10.1007/978-3-031-36554-6_17

1 Introduction

1.1 Background

Architecture may be understood as a material practice where resources extracted from nature are deployed in a building system and cultivated to make a dwelling. However, on a bounded planet, material resources are limited (Daly 2007) and the safe operating space of numerous planetary boundaries are long exceeded (Rockström et al. 2009; Steffen et al. 2015; Raworth 2012, 2018). Accounting for 36% of European CO_2 emissions and 40% of the total European energy consumption (European Commission 2021), the construction industry constitutes a major part of the problem and in consequence—if any hope of meeting the Paris Agreement (UN FCCC 2015) should be kept alive—architectural design processes must be fundamentally revised. Wood has been used as building material in vernacular architecture around the world for thousands of years. The material may be considered a renewable resource and potentially abundant, carbon neutral and recyclable and as such the only widely used building material that is sustainable when coming from truly sustainable forestry. As such, enhancing carbon uptake and storage through bio-based building materials in construction may be one effective mitigation strategy (IPCC 2022). However, the environmental benefits of using timber are not straightforward (Ramage et al. 2017; Dooley et al. 2018) just as traditional knowledge of how to design with wood seems inadequate in contemporary architectural design. It seems as if contemporary sustainable design strategies are less concerned with the qualitative potential of wood, just as it seems as if the work of architecture is regarded a conceptual exercise, detached from tectonic, cultural-historic or contextual considerations. As such, it seems necessary to rethink the qualitative potential of wood in contemporary architectural design practice aiming at (more) sustainable building culture(s).

1.2 Research Question

This paper understands sustainable building culture as the meaningful synthesis of technical properties, cultural-historical qualities and experiential effects—in careful consideration of the planetary boundaries. Through the (re)making of the Haubarg at the Danish Open Air Museum, this paper aims at investigating the topic of Rethinking Resources, supplementing and qualifying the qualitative potential of wood as a sustainable building material as informed by traditional building culture—seen in a holistic perspective. It is asked how knowledge embodied in crafts tradition and local vernacular (with specific focus on timber construction) may inform a contemporary design practice and inspire the development of (more) sustainable building culture(s). The (re)making of the Haubarg—understood in and of itself as a production of authentic architectural knowledge—is considered an empirical finding that is described and analysed seen through a phenomenological-hermeneutic lens. The significance of the results is discussed in relation to the overall question of sustainable building culture(s). It is argued that technical properties, cultural-historical qualities as well as experiential effects must be taken into consideration when building with wood. Conveying architectural meaning as *dwelling*, the (re)making of the Haubarg may thus inspire renewed sustainable building culture (s) in careful consideration of the biophysically bounded Earth.

2 Materials and Methods

The research method in this paper is a combined strategy (Groat and Wang 2013), involving a qualitative, in-depth analysis of an existing building and the design and construction of an experimental timber structure. The analysis of the existing building is seen from the perspectives of technical properties, cultural-historical qualities and experiential effects, all following a phenomenological-hermeneutic approach aiming to identify architectural motifs that may point towards new architectural interventions.

The phenomenological method (Andersen 2018) comprises five stages: (1) experiencing an architectural phenomenon; (2) investigating the architectural phenomenon; (3) hermeneutical reflection; (4) describing the architectural phenomenon and (5) architectural phenomenological representation. The phenomenological descriptions developed in the article build on the framework detailed by the phenomenological method. The method is used as a way of thinking the world through experience, aiming at articulating, structuring, operationalizing and presenting experienced architectural phenomena in text and drawing. The method is based on the phenomenological-hermeneutic philosophy as developed by Edmund Husserl, Martin Heidegger and Maurice Merleau-Ponty and the phenomenology of practice as described by Max van Manen in combination with the concept of embodied communication as developed in the new phenomenology by Schmitz (2014, 2016, 2019).

It is important to underline that a purely qualitative approach does not in itself lead to a tangible and measurable sustainable building culture. As pointed out by ICOMOS (2019), however, climate science can tell us that adaptation and mitigation are necessary, but it cannot tell us what adaptation options are most workable within any given human system. Balancing economic, social and environmental concerns, the UN Rio Conference on Environment and Development highlights the need for qualitative perspectives in a future sustainable development as does the UN Sustainable Development Goals (SDGs). In this perspective, cultural heritage may, according to ICOMOS, be considered "[…] a source of creativity and inspiration for adaptation and mitigation actions that are responses to the findings of climate science" (ICOMOS 2019, p. 14). In continuation, this paper aims at inspiring future sustainable building culture(s) based on the findings of climate science, in this case through pointing at the need to rethink the use of wood in an architectural design practice—all seen in a holistic perspective. As such, the aim is not to exclude, but rather to supplement and qualify contemporary discussions on the climate crises, including the question of carbon footprint.

The design of the experimental timber structure is considered a "reflective practice" (Schön 1986, 2001) involving the continuous analysis and action performed in working with a complex and/or unique problem—in this case, the design of an experimental timber structure as informed by an existing building. It concerns the architect's experience, the understanding of the specific situation and a reflection on the presumed outcome. A "reflective practice" comprises "knowing-in-action", the general, practical knowledge we exhibit in our intelligent, physical performance; "reflection-in-action" in which experience, knowledge and intuition work in exchange with the action itself and "reflection on reflection-in-action", which is the retrospective analysis, which again indirectly may influence a future action (Schön 1986, 2001).

First, characteristic motives from the historic Ejdersted Farmhouse, originally called *Rothelau* and today located at the Danish Open Air Museum, have been identified, described and organised (Figs. 1 and 2). Aiming to get a better understanding of a given architectural phenomenon, the motives relate to technical properties, cultural-historical qualities and experiential effects. Second, selected motives have informed an architectural design (Figs. 3, 4, and 5), constituting a contemporary re-interpretation of the traditional marsh farmhouse. Through a "reflective dialogue with the situation" in a larger "network of choice", this "reflective practice" investigates the different so-called "Normative/Descriptive Design Domains" (Schön 2001), in this case, related to technical properties, cultural-historical qualities and experiential effects. The aim of the architectural design has been to make a new architectural entity, clearly relating to the motifs identified in the historic building, yet unmistakably autonomous.

Third, the experimental timber structure has been built by students at the Royal Danish

Fig. 1 Rothelau Farmhouse, 1651. *Photo* The Authors

Academy—Cultural Heritage, Transformation, and Conservation as part of the master programme curriculum (Fig. 6). The Haubarg has been described and documented photographically (Figs. 3 and 4) and the material—which in itself may be understood as authentic architectural knowledge—is considered empirical findings that have been described and analysed as a "reflection on reflection-in-action" (Schön 1986, 2001). Finally, the significance of the results of the (re)making of the Haubarg is discussed in relation to the overall research question regarding the development of (more) sustainable building culture(s) and a conclusion is made.

Fig. 2 Rothelau Farmhouse, 1651. *Photo* The Authors

3 Results and Analysis

3.1 Cultural-Historical Qualities

The historic Rothelau Farmhouse was originally located in the tidal marshland of the Ejdersted Province on a reclaimed area protected from the sea by dikes. The landscape was structured by a large patchwork of dams, divided by drainage canals, sluices and ponds. To further protect the buildings against floods and potential breaches of the dike, the Farmhouse itself was built on a *warf*, a large, human-made dwelling mound. Built in 1651, the Rothelau Farmhouse was one of the oldest in Ejdersted (Pedersen 2004, p. 44). The building is characterised by a single large roof supported by four tall wooden posts, called the *vierkant*, surrounded by the living quarters, stables and threshing floor. Being used for storing hay during the winter, the central square gave name to the building typology *haubarg* [German *Heu zu bargen*].

The typology presumably came to Ejdersted from Holland in the sixteenth century and the building typology gradually became considered the culturally significant way to build (Pedersen 2004, p. 43). The owners of the Rothelau Farmhouse belonged to the elite of the Ejdersted population that was divided into four social groups: the large landowners, the smaller milk farmers, the workers and the artisans (Pedersen 2004, p. 27). Being the largest contributors to establishing and maintaining dikes, the largest landowners had control of the administration of the landscape. As such, it is impossible to imagine the Rothelau Farmhouse without both its geographic and administrative landscape (Pedersen 2004). The Rothelau Farmhouse thus conveys the historically created material, political and economic values, just as dikes, canals and buildings may be understood as scenes of cultural meaning (Pedersen 2004, p. 82).

According to Tim Ingold, landscape may be understood as a temporal process that is continuously transformed by activities, i.e. “perpetually under construction”, always “work in progress” (Ingold 1993, p. 162). Landscape not only comprises related elements and features, but likewise related activities or “tasks” that are understood as constitutive acts of “dwelling” (Ingold 1993, p. 158). To Ingold, “landscape” is “*continuously going on*”, in the sense that hills, valleys, paths, tracks, trees, crops, buildings and people are understood as engaged in mutual “resonant” relations. As such, the materials, practices as well as the presence and character of landscape may be understood in a “dwelling perspective", suggesting agency of the elements that constitute "landscape" through rhythmic interrelations (Ingold 1993, pp. 160–164).

As with the relationship between the Rothelau Farmhouse and its geographic and administrative landscape, this perspective entails moving beyond a division of "inner and outer worlds", "mind and matter", "meaning and substance" (Ingold 1993, p. 154). Dwelling is, according to Ingold, "*with* us, not against us" as "landscape" is understood as the lived involvement in a temporal world (Ingold 1993, p. 154). In this perspective, forms of buildings, landscapes and relations do not arise from nowhere, but "grow from the mutual involvement of people and materials" in an interweaving that may soften a distinction between "artefacts and living things" (Ingold 2000, pp. 339, 347). With functional, cultural and historic significance and considered as a physical manifestation of lived involvement in a temporal world as dwelling, the (re)making of the Haubarg may thus be understood as a means of *communication* through which cultural-historic values and meanings are conveyed.

3.2 Experiential Effects

From a distance, the Rothelau Farmhouse is characterised by a large, hipped roof that sits on low, heavy set masonry walls elevated on a dwelling mound. The thatched roof expresses a softness in character, while also producing articulated edges with defined shadows at the footings. Entering through a low opening under the eaves, the interior space is dark, and one feels the uneven brick floor underneath one's feet. As one's eyes adjust to the dim raking light, an unexpected tall space lit only by a single opening at the ridge of the roof is revealed. Entering this central space, a large load-bearing structure of squared timber posts becomes visible. The structure is experienced as an upright, steady support to the tent-like drape of the roof and produces an enclosure around which the walls are both permeable and closed. Towards the living quarters, a double wall containing alcoves separates the residential spaces from the barn. On one side, the alcoves are sombre with a clear structural layering. On the other side, they are more elaborate, finished in planed timber with painted sliding doors that are articulated by delicate profiles that catch the light.

From a distance, the (re)made Haubarg is characterised by a steep hipped timber roof that extends to just above the ground, elevated on a dwelling mound. The roof is made of overlapping planed wood boards that produce folds and tucks with distinct shadows, adding depth to the sharp figure. Entering through a low opening, that protrudes outward above the terrain, the interior space is warm, and one feels how the structure lightly gives way under one's feet. As one's eyes adjust to the flickering light coming through the loose-fit cladding, an unexpected tall space, articulated by a single pointed aperture at the ridge of the roof, is revealed. The space is made up from a clearly layered load-bearing structure of rough sawn squared timbers. The upright, steady structure supports an inclined roof structure, in between which the entrance, the aperture and an alcove are located. Sitting down in the alcove, the delicate planed timber of the wooden lining feels soft to the touch, providing pause from the wealth of structural elements.

As the above phenomenological descriptions point out, the seemingly contradictory "ways of working" (Leatherbarrow 2009) at play in the Rothelau Farmhouse—the settled and closed character of the building in relation to the open march landscape, the stability and upright articulation of the timber structure against the enveloping drape of the roof, the opposition of ceiling heights, material finishes, sheen, mattness, softness and sharpness of light—all contribute to the production of distinct bodily felt experiences which are re-interpreted in the Haubarg. To Hermann Schmitz, the body is conceived as the basis for human experience and philosophy defined as one's contemplation of how one finds oneself in one's surroundings (Schmitz 2014, p. 9). To Schmitz, *dwelling* is understood as the *cultivation* of emotions in an enclosed space, that may take place through the articulation of *suggestions of movement* and *synesthetic characters,* which may be sensed on

Fig. 3 Haubarg, 2022. *Photo* Lars Rolfsted Mortensen

both one's own felt body and perceived in figures, whether static or in motion (Schmitz 2016).

Suggestions of movement are signs of imminent movement, without actual movement, or gestures that go beyond the limit of movement, such as "the gait of a person" (Schmitz 2016, p. 5); "the space spanned by the rhythmic and tonal movement suggestions of the sound, such as piercing noise, diminishing echo, rising and falling, pressing and circling" (Schmitz 2016, p. 3) or the broadening and narrowing of space. *Synesthetic characters* are qualities "which run through all specific senses and often, but not always, bear the names of specific sensory qualities" (Schmitz 2016, p. 5), such as "the sharp, bright, gentle, pointed, hard, soft, warm, cold, heavy, compact, delicate, dense, smooth, the roughness of colours, sounds, smells, sound and silence, bouncing and trailing gait, joy, zeal, melancholy, freshness, and fatigue" (Schmitz 2014, p. 31).

In an architectural perspective, *embodied communication* may lead to "the formation of atmospheres of emotion" and to the "tuning of the occupants and/or visitors into these atmospheres" (Schmitz 2016, p. 15). As such, things,

Fig. 4 Haubarg, 2022. *Photo* The Authors

materials and spaces may become bearers of atmospheres of emotion so "that the person can attune to them in harmony with his *corporeal mood*" (Schmitz 2016, p. 14). This includes the experienced material qualities of the walls, the ceiling and the floor as well as the furnishing and control of incoming light, temperature and sounds. In this perspective, the Haubarg may be understood as a new interpretation of a bodily experienced spatial sequence enacted between the closed and open, dark and light, matte and sharp, that may be considered a mean of embodied *communication* as atmospheres through which experienced architectural meaning as dwelling may be conveyed.

3.3 Technical Properties

In the traditional Haubarg post-and-beam typology, the timber structure comprises four, sometimes six or even eight posts, that are connected by longitudinal and transverse beams, stabilised by diagonal bracing, and joined with traditional wooden joints. Independent from the outer brick walls, the timber frame is resistant to the forces of nature, especially storm surges, just as it is protected from the weathering effects of the environment. The (re)making of the Haubarg is executed entirely in locally sourced Douglas fir. The timber was provided by *Bondeskovgaard*, a third-generation sawmill established in 1900 which is located about 50 km from the building site. The sawmill combines inherited knowledge of timber with the use of modern machinery, securing a recourse-efficient use of the entire trunk. The timber was grown in Danish forests and sawed to specified dimensions as either rough or planed (PAR and PSE) lumber. In the building, the main structure comprises three modules of eight 5 × 5" timber posts in total, stabilised by diagonal bracing and leaving a rectangular 4 × 8 metre large floorplan. The 4 × 5" roof rafters, supported by the timber frame and fitted with battens, are clad with overlapping planed wooden boards that serve as a contemporary reference to the historic building's distinctive thatched roof. The entrance, alcove and skylight, constituting re-interpretations of three spatial situations identified in the historic Farmhouse, are made using 1" planed wooden boards, supported by a slender exterior structure.

Wood has been used as building material around the world for thousands of years and the technical properties of wood in historic buildings are well described. Not only is the molecular and cellular structure of wood fundamental to its use as a material well suited for building construction (Ramage et al. 2017), also the selection, processing and treatment may be of critical

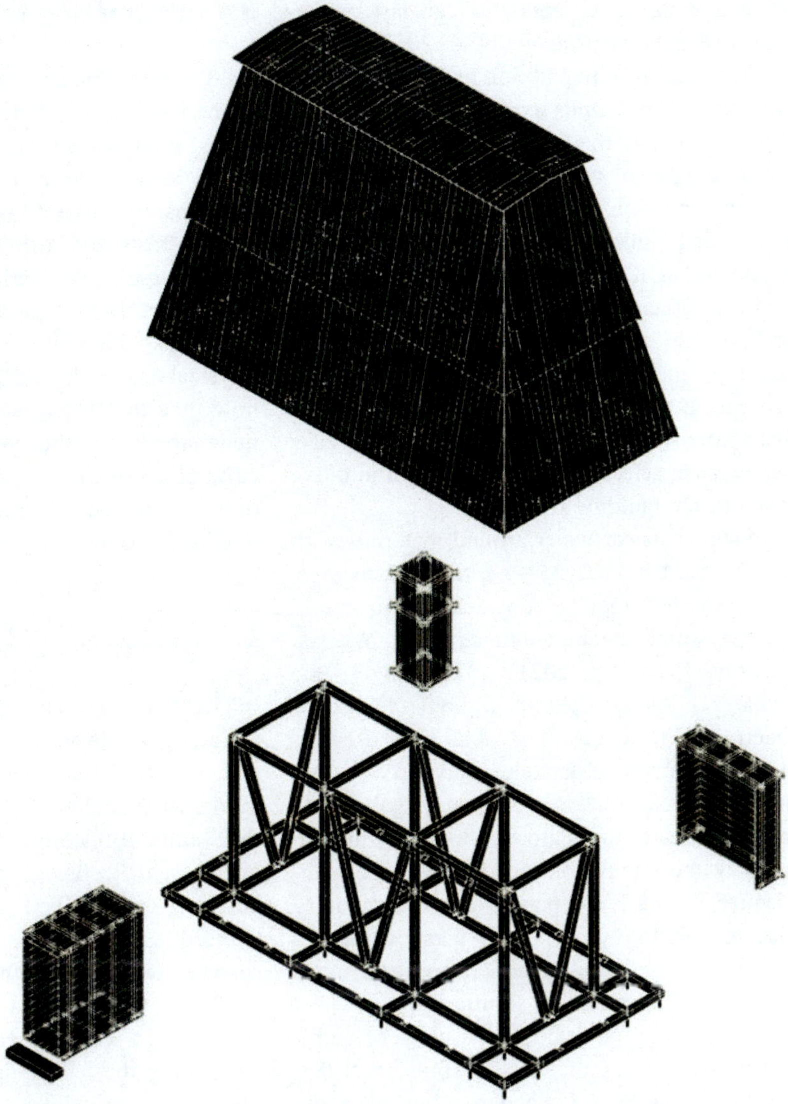

Fig. 5 Haubarg, 2022. *Illustration* The Authors

importance as a way of craftsmanship to improve the properties of the material (Glarbo 1959; Vadstrup 2021). As a building material, wood has some very specific properties that are completely different from, for example, concrete or bricks. Thus, the opposition between the tectonic culture of the filigree *light* to the stereotomic culture of the *heavy* (Semper 1989) is clearly articulated in the Rothelau Farmhouse as well as in the Haubarg. In addition to the structural effect, the timber structure—including diagonal bracing, battens and cladding—makes visible the "[…] tectonic statement: the noble gesture which makes visible a play of forces, of load and support in column and entablature, calling forth our own empathetic participation in the experience" (Sekler 1965, p. 93).

Because unprotected structural timber is likely to be exposed to elevated levels of moisture, making it susceptible to fungal degradation, *wood protection by design* details such as raising the timber structure above ground level and providing overhanging roofs that limit the exposure to wetting and direct sunlight may ensure that timber components can last, potentially for centuries (Ramage et al. 2017, p. 351; Vestergaard 2000). In addition to the geometrical configuration that limits exposure to wetting and shows water off, the clear tectonic articulation and layering—i.e. visually separating the structural timber frame, the secondary members and the cladding—may allow the visitor an intuitive understanding of how the building is built as well as of the structural hierarchy and varying temporality of its "shearing layers" (Brand 1995). As such, the structural configuration of the Haubarg may lead to an *engaging capacity* (Verbeek and Kockelkoren 1998) that may support easy maintenance, selective replacement and intuitive repairs to the building over time.

Condensation of water around materials with high thermal conductivity—i.e. metal fastenings, nails and bolts—may be considered "poisonous" to timber structures and counteractive to material longevity (Vadstrup 2021). Accordingly, the Haubarg is joined together without the use of modern steel fastenings, just as all nails used for cladding are intended made entirely out of wood. Only the ground screws are made of galvanised metal, reducing the carbon dioxide emission regarding the foundation (according to the manufacturer) by 89% compared to a contemporary concrete solution. In a detail perspective, the joints themselves may be understood as a minimal unit in the process of signification, as "[…] the 'construction' and the 'construing' of architecture are both in the detail" (Frascari 1983, p. 325). As such, the wooden joints have the double effect of connecting the individual structural members using durable wood-on-wood details as well as being the place where architectural meaning is created. The physical properties of the material itself, the processing and treatment, including the significant joint, the tectonic articulation and the static principle may thus be understood as a means of *communication* through which material and technological qualities and meanings are conveyed.

3.4 (Re)making the Haubarg

As described and analysed above, the (re)making of the Haubarg has been extensively informed by crafts tradition and local vernacular as manifested in the historic Rothelau Farmhouse, both in terms of cultural-historical qualities, experiential effects and technical properties. All three aspects are characterised by *communicating something*, both regarding the relation to the landscape, the spatial character as well as the physical material, inviting visitors to reflect on how they dwell. All in all, the Haubarg may be understood as the re-making of technical, cultural-historical and experiential characteristics of the Rothelau Farmhouse conveying values, qualities and meanings as dwelling.

4 Discussion and Conclusion

In light of the accelerating ecological crisis including sea level rise, extreme weather events and loss of biodiversity, all leading to higher mortality (Kemp et al. 2022) the question is how the cultural-historical qualities, experiential effects and technical properties of a historic building, as described above, may become reactualised as part of the development of (more) sustainable building culture(s).

The concept of sustainability was used for the first time in 1713 by Hans Carl von Carlowitz advocating the balancing of growth and harvest through the principles of rationalisation, substitution and limitation as a reaction to the acute scarcity of timber caused by the heavy exploitation of forests by the mining industry. Even though contested, the most widely used definition of the concept of sustainability today is the one offered by the Brundtland Commission Report, defining sustainable development as "development that meets the needs of the present without compromising the ability of future generations to meet their own needs" (UN 1987). In continuation, the sustainable development goals, SDGs "provides a shared blueprint for peace and prosperity for people and the planet, now and into the future" (UN SDGS 2022). The (re)-making of the Haubarg supports primarily SDG 11 (sustainable cities and communities), SDG 12 (responsible consumption and production) and SDG 15 (life on land). More specifically, the following targets may be identified: 11.4, strengthening efforts to protect and safeguard the world's cultural heritage; 11.c, building sustainable and resilient buildings utilising local materials; 12.2, sustainably managing and efficiently using of natural resources; 12.5, substantially reducing waste generation through prevention, reduction, recycling and reuse and 15.2 sustainably managing of all types of forests and the halt of deforestation. As the project aims at inspiring future sustainable building cultures, SDG 3 (good health and well-being) may also be considered relevant. Similarly, SDG 4 (quality education) and SDG 13 (climate action) may be considered pertinent, as the project is part of a master programme curriculum (UN SDGS 2022). It is, however, important to underline that even if individual goals have been identified, the SDGs should be considered in a holistic perspective since optimization at sector level will most likely fail as the individual sectors may compete with each other at the expense of the whole.

It has been argued that the SDGs prioritise economic growth over ecological integrity as they fail to monitor absolute trends in resource use (Eisenmenger et al. 2020). On a bounded planet, material resources are limited (Daly 2007) and the safe operating space of numerous planetary boundaries are long exceeded (Rockström et al. 2009; Steffen et al. 2015; Raworth 2012, 2018). In the construction industry, the concept of absolute environmental sustainability requires actions to respect the planetary boundaries and to stay within the safe operating space (Hauschild et al. 2020). In this perspective, sustainable building culture(s) must prioritise the balancing between the just demand for welfare among the living creatures and the bounded biophysical capital seen in a planetary perspective.

As a crucial part of the carbon cycle, wood accumulates and stores carbon dioxide while growing and acts as a carbon storage as long as it maintains its chemical form. When rotting or burned, carbon dioxide is released into the atmosphere again (Riebeek 2011). As such, wood may be considered a renewable resource and potentially abundant, carbon neutral and recyclable. With recommended rotations for forestry harvests ranging from 35 to 70 years depending on species and location, wood—compared to mineral resources like rocks, ores and soils—has a very short geological timescale and may as such be considered the only widely used building material that is truly sustainable (Ramage et al. 2017, p. 340). However—with an alternative response to climate mitigation and adaptation—the Climate Land Ambitions and Rights Alliance (Dooley et al. 2018) argues in favour of approaches that safeguard food security and food sovereignty, land rights and biodiversity. According to this, major shifts in today's land use and land management are required—including end of deforestation, forest ecosystem restoration, natural forest expansion, agroforestry, improved management of forests for timber and reduction in wood production (Dooley et al. 2018).

Understanding the building as a physical manifestation of lived involvement in a temporal world as dwelling, the historic Rothelau Farmhouse as well as the experimental timber structure have been informed by a large number of parameters, including material, political and economic values that may hold a number of potential sustainable potentials. The position of the building—protected from the sea by dikes and placed on top of a human-made dwelling mound—may in itself become reactualised as a necessary strategy in a near future with sea level rise and extreme weather events. The small size of the building may potentially inspire living on fewer square metres. In the building scale, traditional timber framing may be considered significantly more economical than the contemporary massive CLT construction, regarding the amount of wood used. With the recommendation that wood be employed in products with a design lifespan that (at least) matches timber rotation periods (Ramage et al. 2017, p. 351), wood utilisation should move to longer lived products (Dooley et al. 2018) and building longevity. In this perspective, the significant joint, the tectonic articulation and the static principle as means of empathic participation and conveyor of meaning have an engaging capacity, that may potentially invite maintenance, reuse, refurbishment and recycling, according to the principles of a circular economy (Ellen MacArthur Foundation 2022).

One study investigating the value of building heritage concludes that, in Denmark, listed buildings have a higher economic value than comparable not-listed buildings (Incentive 2015) suggesting that architectural and cultural-historical qualities may have a positive influence on building lifespan. According to the European Environment Agency EEA, the ecological crisis is closely linked to economic growth, including increase in production, consumption and resource use. It is pointed out that 100% circularity is impossible, just as full decoupling of economic growth from environmental pressures and resource consumption is not possible. As such, a sustainable future requires change of qualitative aspects, such as consumption and social practices, not only a change of technology. As pointed out by EEA, "[w]hile the planet is finite in its biophysical sense, on a biophysically finite planet, infinite growth in human existential values, such as beauty, love, and kindness, as well as in ethics, may be possible" (Strand et al. 2021). In a cultural-historic perspective, vernacular building culture manifests an embodiment of both material and landscape conditions, cultivated by using the ability of a given society. This involvement in the temporal world may be described as a meaningful material practice where resources extracted from nature are deployed in a building system and cultivated to make *dwelling*.

From a phenomenological perspective, the fundamental existential structure indicating how one feels is characterised by *attunement* [*Befindlichkeit*]. According to Heidegger, "[i]n attunement lies existentially a disclosive submission to world out of which things that matter to us can be encountered" (Heidegger 1996, pp. 129–130). As such, attunement makes it possible to direct oneself towards something, to be touched and have a sense for something. The making as disclosure of landscape characteristics, material qualities and static principles experienced through embodied communication through which architectural meaning as dwelling is conveyed may thus potentially invite "[…] staying with things for a longer while" (Andersen 2022, p. 335). In this perspective, it may be argued that longevity seen from both a technical, cultural-historical and experiential perspective is dependent on "[…] maintaining and reinforcing the meanings in an object" (Muñoz Viñas 2005) that may potentially contribute to a resource-saving strategy and sustainable development by ensuring maximum meaning for present and future generations.

In continuation of the above, it is recommended that sustainable design strategies include material parameters that may enhance the engaging capacity, such as the selection, processing and treatment, wood-on-wood joints, wood protection by design, separation of temporal layers, clear tectonic articulation and structural configuration; environmental parameters that may enhance the sense of interrelation, such as administration, geography, topography, ground, vegetation, weather and climate conditions; and spatial parameters that may enhance emotional attachment such as bodily experienced spatial sequence enacted between the closed and open, dark and light, matte and sharp. All parameters are conveying meaning through communication and may as such highlight the lived involvement and the capacity for maintenance and care that may support building longevity.

It may be concluded that technical properties, cultural-historical qualities and experiential

Fig. 6 Haubarg construction process, 2022. *Photo* The Authors

effects as found in crafts tradition and local vernacular, as in the case of *The Rothelau Farmhouse*, may inform a contemporary design practice, exemplified in the specific case of (re)making the Haubarg. In this, architecture should not be understood as a building in and of itself, but rather as situated in a larger material, environmental and social (eco)system. As reduction in wood production is required in order to safeguard food security and sovereignty, land rights and biodiversity, a holistic approach focusing on building longevity should be observed. Made with a potential abundant, carbon-neutral and recyclable bio-based material (if used correctly) and as conveyer of technical, cultural-historical and experiential values, qualities and meaning, the (re)making of the Haubarg may supplement and qualify contemporary sustainable design strategies. As *embodied communication* through which meanings as *dwelling* are conveyed, the (re)making of the Haubarg may thus inspire future (more) sustainable building culture(s) in careful consideration of the bio-physically bounded Earth.

Acknowledgements This paper is a part of the research project "Where We Live, Now, Then and in the Future" which is conducted in collaboration with the Danish Open Air Museum and Roskilde University (RUC) with financial support from the Velux Foundation. Ground screws were generously sponsored by Fremtidens Fundament ApS and wooden nails by BECK Fastening.

References

Andersen NB (2018) Phenomenological method—towards an approach to architectural investigation, description and design. In: Lorentsen E, Torp KA (eds) Formation—architectural education in a Nordic perspective. Architectural Publisher B, Copenhagen, pp 74–95

Andersen NB (2022) An ecopoetic architect: the resonate presencing of material properties and environmental conditions in the case of Jørn Utzon. In: Hvejsel MF, Cruz PJS (ed) Structures and architecture: a viable urban perspective? Proceedings of the Fifth International Conference on Structures and Architecture (ICSA 2022). CRC Press, Balkema, pp 329–336

Brand S (1995) How buildings learn: what happens after they're built. Penguin Books, London

Daly H (2007) Ecological economics and sustainable development, selected essays of Herman Daly. Edward Elgar Publishing Limited, Cheltenham

Dooley K et al (2018) Missing pathways to 1.5°C: the role of the land sector in ambitious climate action. Climate Land Ambition and Rights Alliance

Eisenmenger et al (2020) The sustainable development goals prioritize economic growth over sustainable resource use. Sustain Sci 15:1101–1110. https://doi.org/10.1007/s11625-020-00813-x

Ellen MacArthur Foundation (2022) Let's build a circular economy. https://ellenmacarthurfoundation.org/. Accessed 3 Oct 2022

European Commission (2021) Energy performance of buildings. https://ec.europa.eu/energy/en/topics/energy-efficiency/energy-performance-of-buildings. Accessed 3 Oct 2022

Frascari M (1983) The tell-the-tale detail. In: Deely JN, Lenhart MD (eds) Semiotics 1981. Springer, Boston, MA. https://doi.org/10.1007/978-1-4615-9328-7_32

Glarbo O (1959) Træ. Teknisk Forlag, København
Groat LN, Wang D (2013) Architectural research methods. Wiley, Hoboken
Hauschild MZ, Kara S, Røpke I (2020) Absolute sustainability: challenges to life cycle engineering. CIRP Ann 69(2):533–553. https://doi.org/10.1016/j.cirp.2020.05.004
Heidegger M (1996) Being and time. State University of New York Press, Albany
ICOMOS Climate Change and Cultural Heritage Working Group (2019) The future of our pasts:engaging cultural heritage in climate action. ICOMOS, Paris
Incentive (2015) Værdien af Bygningsarven. https://realdania.dk/publikationer/faglige-publikationer/v%C3%A6rdien-af-bygningsarv. Accessed 3 Oct 2022
Ingold T (1993) The temporality of the landscape. World Archaeol 25(2):152–174
Ingold T (2000) The perception of the environment. Routledge, London
IPCC (2022) Climate change 2022. Mitigation of climate change. https://www.ipcc.ch/report/ar6/wg3/downloads/report/IPCC_AR6_WGIII_Full_Report.pdf. Accessed 3 Oct 2022
Kemp L et al (2022) Climate endgame: exploring catastrophic climate change scenarios. PNAS 119 (34). https://doi.org/10.1073/pnas.2108146119
Leatherbarrow D (2009) Architecture oriented otherwise. Princeton Architectural Press, New York
Muñoz Viñas S (2005) Contemporary theory of conservation. Elsevier, Oxford
Pedersen MV (2004) Ejdersted. Frilandsmuseet, København
Ramage et al (2017) The wood from the trees: the use of timber in construction. Renew Sustain Energy Rev 68 (Part 1):333–359. https://doi.org/10.1016/j.rser.2016.09.107
Raworth K (2012) A safe and just space for humanity. https://www.oxfam.org/en/research/safe-and-just-space-humanity. Accessed 3 Oct 2022
Raworth K (2018) Doughnut economics: seven ways to think like a 21st-century economist. Cornerstone, New Orleans
Riebeek H (2011) The carbon cycle. https://earthobservatory.nasa.gov/features/CarbonCycle. Accessed 3 Oct 2022
Rockström J, Steffen W, Noone K et al (2009) A safe operating space for humanity. Nature 461:472–475. https://doi.org/10.1038/461472a
Schmitz H (2016) Atmospheric spaces. Ambiances. https://journals.openedition.org/ambiances/711
Schmitz H (2014) Kort indføring i den nye fænomenologi. Aalborg Universitetsforlag, Aalborg
Schmitz H (2019) New phenomenology. A brief introduction. Mimesis International, Milan
Schön D (1986) Educating the reflective practitioner. Jossey-Bass Publishers, San Francisco
Schön D (2001) The reflective practitioner—how professionals think in action. Klim, Århus
Semper G (1989) The four elements of architecture. Cambridge University Press, Cambridge
Sekler EF (1965) Structure, construction, tectonics. In: Kepes G (ed) Structure in art and science. G. Braziller, New York, pp 89–95
Steffen W, Richardson K, Rockström J et al (2015) Planetary boundaries: guiding human development on a changing planet. 347(6223). https://doi.org/10.1126/science.1259855
Strand R, Kovacic Z, Funtowicz S, Benini L, Jesus A (2021) Growth without economic growth. European Environment Agency EEA, Copenhagen
UN (1987) Our common future. United Nations. http://www.un-documents.net/our-common-future.pdf. Accessed 3 Oct 2022
UN FCCC (2015) The Paris agreement. United Nations. https://unfccc.int/sites/default/files/english_paris_agreement.pdf. Accessed 3 Oct 2022
UN SDGS (2022) Sustainable development goals. United Nations. https://sdgs.un.org. Accessed 18 Sept 2022
Vadstrup S (2021) Nye bæredygtige træhuse—helt af TRÆ, med vedvarende holdbarhed 2019–2021. Søren Vadstrup. https://www.bevardithus.dk/wp-content/uploads/Nye-traehuse-helt-af-trae-dec-2021-Kopi.pdf. Accessed 3 Oct 2022
Verbeek PP, Kockelkoren P (1998) The things that matter. Des Issues 14(3):28–42. https://doi.org/10.2307/1511892
Vestergaard M (2000) Dokumentation af konstruktiv træbeskyttelse. Miljøprojekt Nr. 515. Teknologisk Institut

High-Tech Meets Low-Tech

Andrea Veglia and Francesca Thiébat

Abstract

The quest toward a sustainable architecture from the early 70s onward highlights two seemingly non-compatible streams: the believers in the salvific potential of high technology and the partisans of low-tech, artisanal techniques. Those approaches were clear in the pages of Architectural Design, which published a melting pot of projects pushing for precision engineering and off-site fabrication along with experiments of constructions in garbage and salvaged materials. Designers in the first batch gave birth to the high-tech of the 80s, while those who dirtied their hands with mud looked naive dreamers. Are those two approaches irreconcilable, or through a fusion of low-tech and high-tech, can we find a way to solve some of our struggles? This is the question we faced when we set to design a farmhouse. Life cycle thinking informs the process and gives life to an approach where the building is seen as an active system of material, energy, and social transformation, an act that requires the involvement of the architects in understanding the production chain of the building, activating dynamics of circular economy. The walls of the ground floor exploit the combination of two local crafts: lime and hemp, a century-old cultivation typical of the area that is now being reintroduced contributing to biodiversity. By placing the sleeping quarters in a lightweight, kit of parts enclosure designed for deconstruction on top of massive living areas, the design looks for creative ways to reduce operational energy while at the same time minimizing embodied carbon.

Keywords

Sustainable design · Low-tech · High-tech · Natural resources · Circular design

1 High-Tech Meets Low-Tech

By the end of the 60s, it was clear that our spaceship earth has a finite amount of resources and cannot be resupplied (Fig. 1). The cry for help resonated through the architectural press (Figs. 2 and 3). An overview of the quest toward a more sustainable architecture from the early 70s onward can highlight two major streams, seemingly non-compatible in philosophy and aesthetically at odds: the believers in the salvific potential of high technology, innovative

A. Veglia (✉)
PAT. architetti associati, Torino, Italy
e-mail: andrea@patdesign.it

F. Thiébat
Department of Architecture and Design, Politecnico di Torino, Torino, Italy

M. R. Thomsen et al. (eds.), *Design for Rethinking Resources*, Sustainable Development Goals Series,
https://doi.org/10.1007/978-3-031-36554-6_18

Fig. 1 By the end of the 60s, it was clear that our spaceship earth has a finite amount of resources and cannot be resupplied. R. Buckminster Fuller, *Operating Manual For Spaceship Earth*, 1969

Fig. 2 The cry for help resonated through the architectural press. Cover of Domus 487, June 1970

materials, and new mechanical systems and the partisans of low-tech, almost forgotten, artisanal techniques on the verge of extinction.

Those approaches were clear in the pages of Architectural Design (Figs. 4, 5, 6, 7, and 8), which published a melting pot of projects pushing for precision engineering and off-site fabrication along with experiments of constructions in garbage and salvaged materials. Designers in the first batch, radical yet pragmatic, followed in the footsteps of visionary technocrats such as Buckminster Fuller and apologists of technology such as Reyner Banham and Cedric Price. They would give birth to the high-tech of the 80s (Banham 1982) that went on to become the eco-tech of the 90s (Slessor 1997), through the ever-evolving incorporation of elements of bioclimatic design and passive solar control who came to define the aesthetics of buildings much in the same way as exposed pipes or finely crafted joints had done in the previous two decades.

In the meanwhile, those who dirtied their hands with mud and straw bales would often result looking like just a bunch of outdated hippies, dropouts, or—at best—naïve dreamers, producing small buildings that looked more apt to the world of Peter Rabbit than to the twenty-first century. And yet, their poor, locally sourced materials are now the stuff of dreams in a world that has finally digested the concept of circular economy and has moved on to fight public enemy number one: embodied carbon. Are really

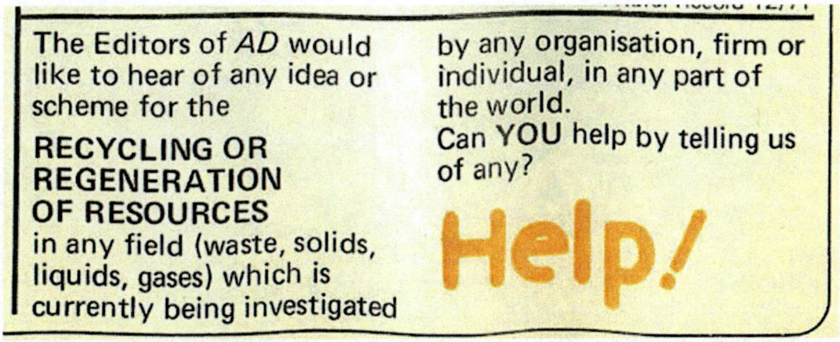
The Editors of *AD* would like to hear of any idea or scheme for the

RECYCLING OR REGENERATION OF RESOURCES

in any field (waste, solids, liquids, gases) which is currently being investigated by any organisation, firm or individual, in any part of the world.

Can YOU help by telling us of any?

Help!

Fig. 3 Call on Architectural Design, March 1972, p. 141

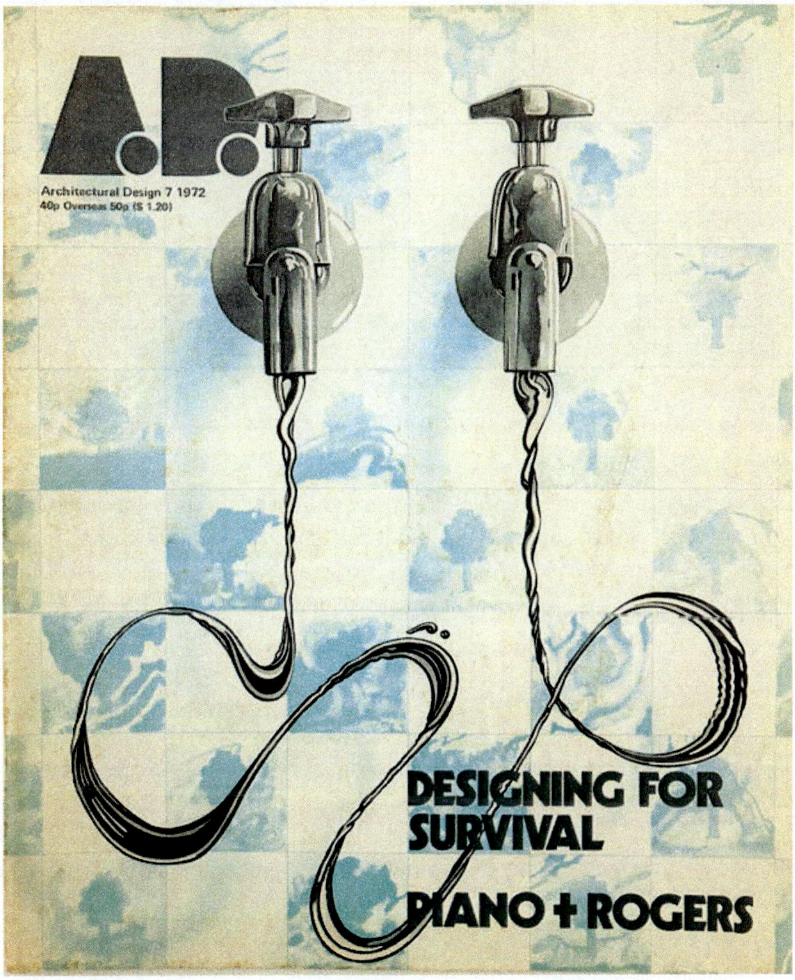

Fig. 4 High-tech versus low-tech: radical magazines such as Architectural Design published an exhilarating melting pot of projects from the likes of Foster, Hopkins, Piano, and Rogers—all pushing for precision engineering and off-site manufacturing of components in metal, plastics, and glass to be dry assembled on site and sealed through the help of kilometers of neoprene gaskets—along with experiments of constructions in mud, garbage, and salvaged materials. Cover, Architectural Design, July 1972

Fig. 5 Architectural Design, November 1970, pages 545–546

these two approaches irreconcilable, or through a fusion of low-tech and industrial techniques, can we find a way to solve some of our present struggles?

2 Bricolla: A Case Study

Bricolla is a country house project located on the hills of Northwestern Italy (Briaglia, Cuneo) (Fig. 9). Let's address here the first issue: does it make sense to dedicate time to the problem of detached single houses, as densification seems a mandatory prerequisite of sustainable dwelling models? The answer is: YES.

Let's look at the data (Eurostat 2022): detached single family houses in the US were 81.5 millions in 2015 and are expected to reach 85 millions in 2023.

In Europe, 35% of the population lives in detached houses (Fig. 10). If we look at Italy they represent around 23% of the housing stock, but one which has received significant attentions by the politics over the last 15 years, with laws such as "*Piano Casa*"—also dubbed "*Piano Villetta*" (*Small House Plan*) since it targeted relatively small detached or semi-detached dwellings—granting incentives to extend or to demolish and rebuild with an increased volumetry. On top of this, since the COVID outbreak in early 2020, demand for such kind of dwellings, isolated and with access to a garden, increased by 70%. If detached single family homes still are an in-demand typology in a comparatively dense country like Italy, counting 200 inhabitants per square kilometer, this is even more so in France, the biggest EU country by area and one of the less dense with 118 inhabitants per square kilometer. Here they represent 57% of total homes and their number increase at a rate of 130.000 units per year. The reality—as noted in the introduction to *Le Tour de France*

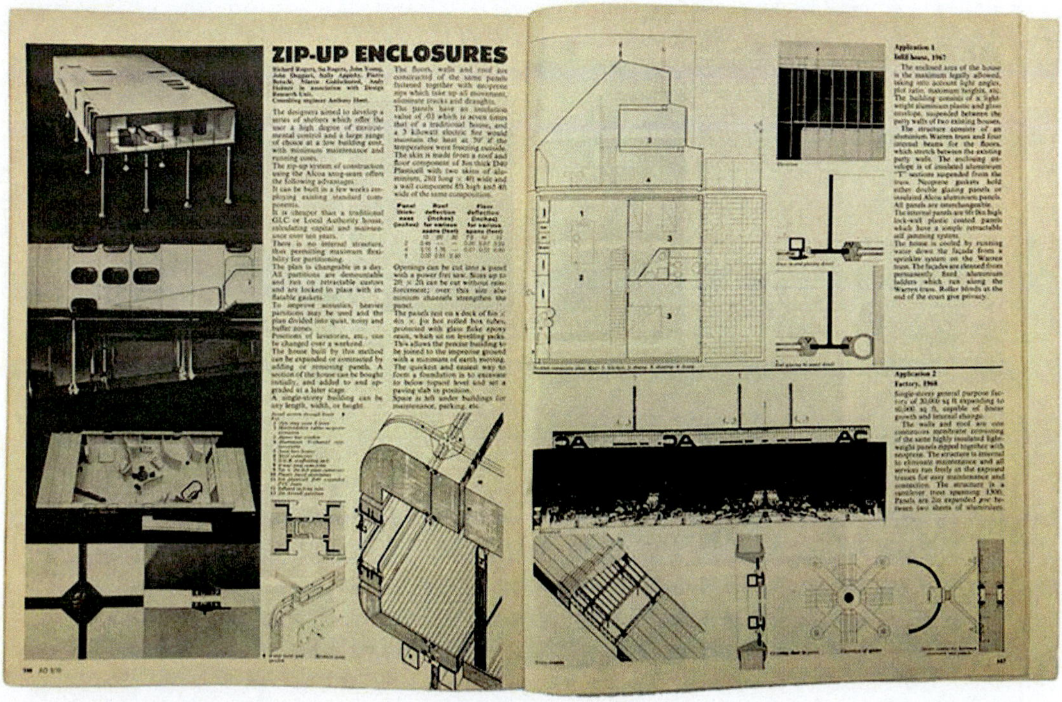

Fig. 6 Architectural Design, March 1970, pages 146–147

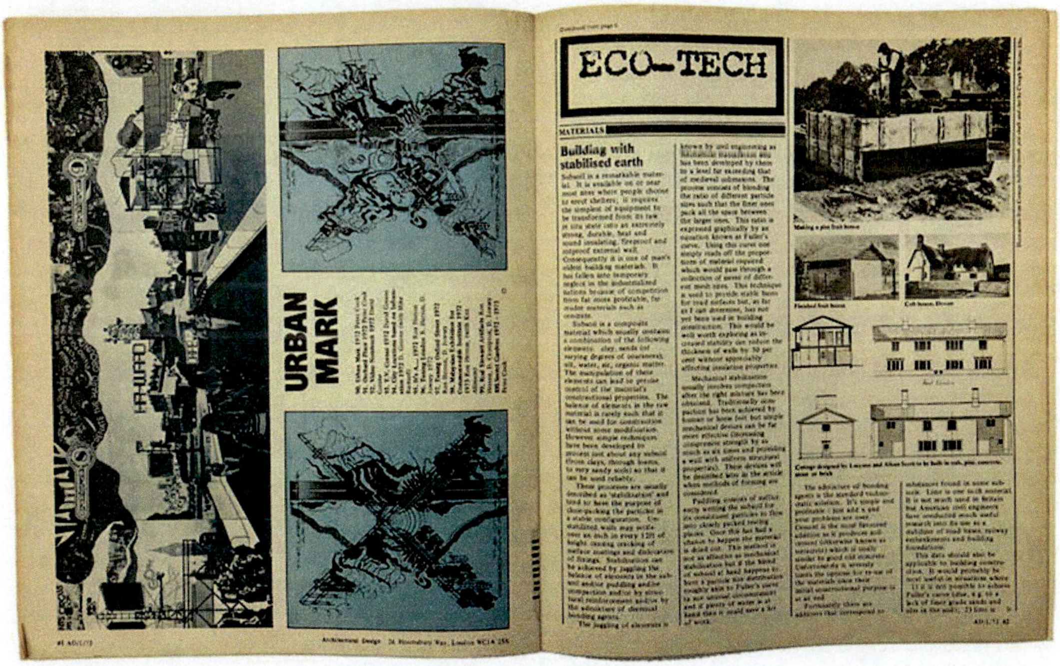

Fig. 7 Architectural Design, January 1972, pages 41–42

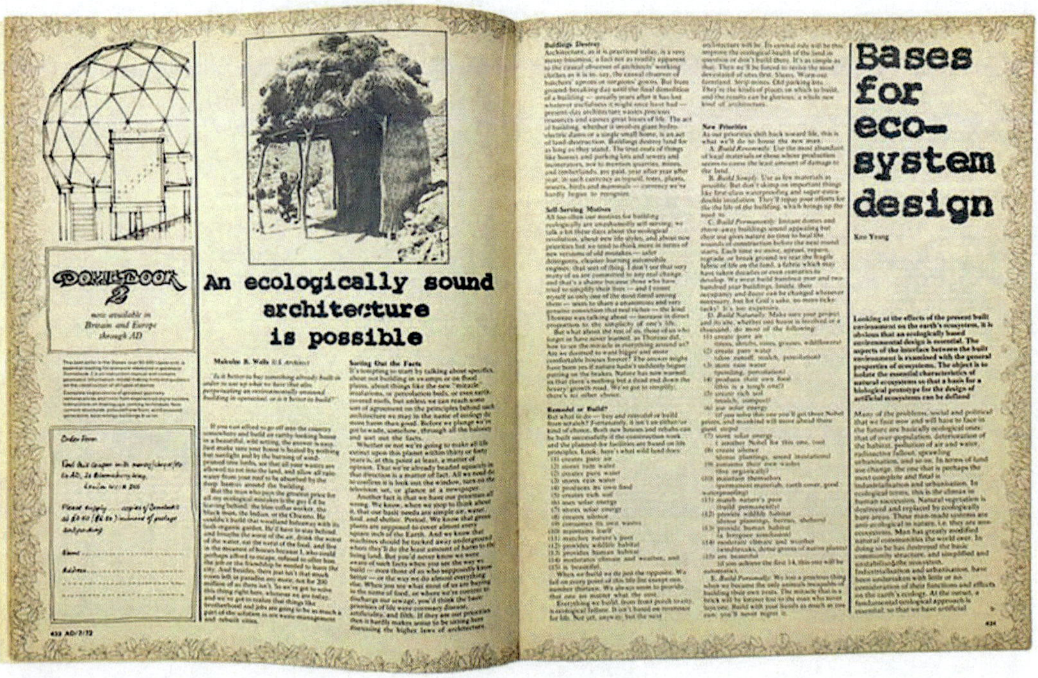

Fig. 8 Architectural Design, July 1972, pages 433–434

des Maisons Ecologiques (Rager et al. 2020)—is that "*the single family house still occupies a dominating place in the construction industry*".

For this reason, it is still important to reckon with this typology and its implication on the whole building sector. All the more so if we think that single family houses, or even villas, has been a testing ground for innovations in processes and technologies that have been subsequently appropriated by the market at large. Much like Formula 1 cars, some individual houses—given their limited size and scope—can benefit of a special attention providing they can count on the rare mix of a passionate, adventurous client and a design team willing to go the extra mile to tap into new instruments, be they technologies or processes, hardware or software, in order to push for an evolution of the species. Building on those premises, Bricolla turned from being a single family housing commission to being a complete research project, starting from the quest for non-conventional building solutions aimed to promote decarbonization by employing local/natural techniques and materials and combining them with off-site fabrication.

At Bricolla two ways of living and building merge, coming from two different cultures: the local *Monregalese* farm as a pedestal, half-sunken in the slope with massive heat-performing walls made of lime and hemp, and a light shelter atop of it, made of an exposed steel structure and enclosed by wooden panels (Figs. 11 and 12).

Life cycle thinking (Thiébat 2019) informs the process and gives life to an approach where the building is seen not as an object, but as an active system of material, energy, and social transformation, an act that requires the active involvement of the architects in understanding and governing the production chain of the building.

Because of its ecological virtues and the appeal of building with a material readily available on site, rammed earth was the first choice for the massive walls of the ground floor (Fig. 13).

Fig. 9 High-tech meets low-tech? Bricolla is a country house project located on the hills of Northwestern Italy (Briaglia, Cuneo). The architectural intent was that of deeply integrating the ground floor into the hillside, setting a compact, linear volume parallel to the level curves of the slope and the southwest facing vineyards below. On top, a long, thin roof above is the organizing element of the house. Beyond the vines, the city of Mondovì and then the Alps, on subsequent parallel planes. Model view and landscape of Bricolla © PAT.

Although *pisé* was a once popular technique in the area of Alessandria, 100 km to the northeast, it has been neglected since the construction boom of the second half of the twentieth century, leaving no artisans behind. Although now back into fashion in architectural circles, trying to use stabilized earth in a place which has no tradition for it can quickly prove impractical and expensive, thus turning a philosophically noble idea in an architectural whim that only wealthy clients can afford.

For the massive walls of the ground floor, an alternative emerged in the possibility to exploit an option made available by the combination of two local crafts: lime and hemp, a century-old cultivation typical of the area that had fallen out of use and is now being reintroduced contributing to biodiversity. The Carmagnola area, only 40 km away from Briaglia, has been known for centuries for the cultivation of hemp (Fig. 14), used to produce fabrics and manufacturing ropes and rigging. After 60 years of neglect, Assocanapa—established in 1998 in Carmagnola—has worked for the reconstruction of the once thriving production chain. Limestone from nearby quarries has been transformed into lime in

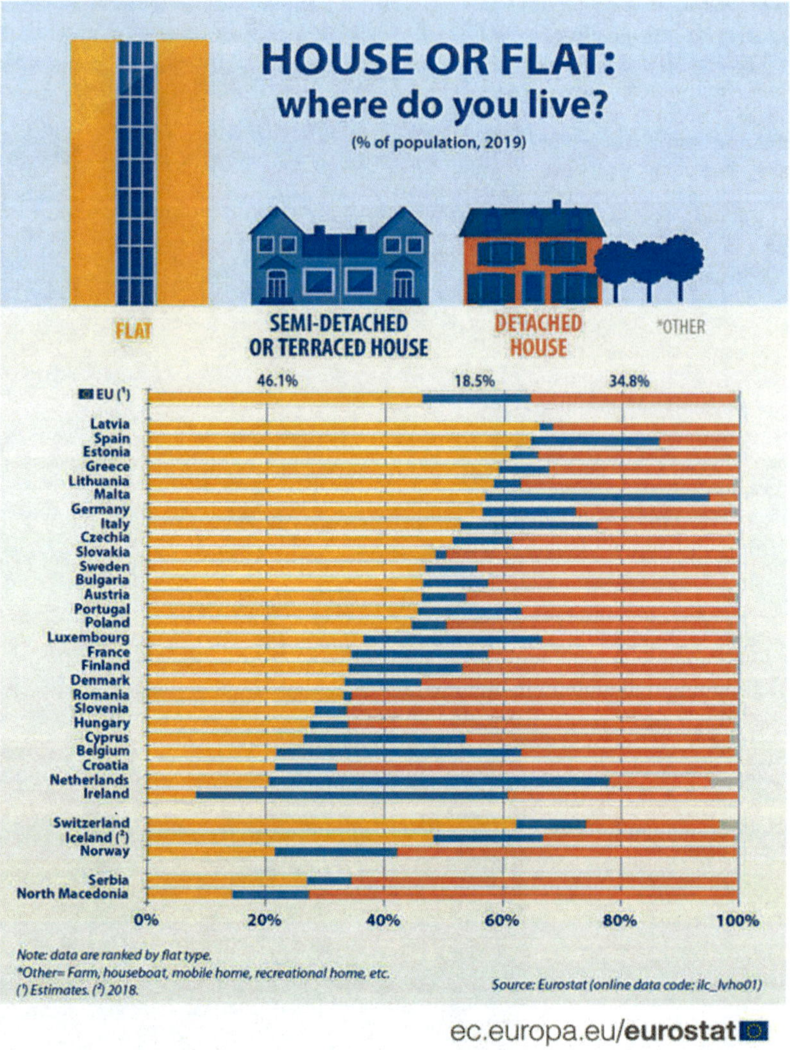

Fig. 10 Does it make sense to dedicate time to the problem of detached single houses? In Europe, 35% of the population lives in detached houses. The reality—as noted in the introduction to *Le Tour de France des Maisons Ecologiques* (Rager et al. 2020)—is that "*the single family house still occupies a dominating place in the construction industry*". Housing typology comparison by flat type, from Eurostat. *Distribution of population by degree of urbanisation, dwelling type and income group—EU-SILC survey ILC_LVHO01 (2022)*

Piasco (CN) for over 150 years. This made the right milieu for appropriating a relatively recent technology (Fig. 15). Hempcrete—or hemplime—has been used in France since the early 80s, originally to add insulation to historical buildings. Combining life cycle assessment with life cycle costs analysis of possible building components and bringing it down to site visits of local production sites has allowed to focus choices (Figs. 16 and 17).

Given the amount of effort put in selecting low carbon, locally sourced materials, the choice of steel for the framing of the lightweight upper floor might seem weird. Nevertheless, since nineteenth century the slabs of many local houses and farms were done combining iron I beams and bricks or flat tiles of local production. And although there are strong arguments for wood construction, in a life cycle perspective steel is still a viable choice in the long run. If construction wood is mostly incinerated at the end of its life, steel can be re-used over and over again and in the end results being as green as the energy used to transform it. Understanding and control of the supply chain on the part of architects is also in this case crucial, and still not common. Generally speaking, the architect delivers his concept to the structural engineers, which

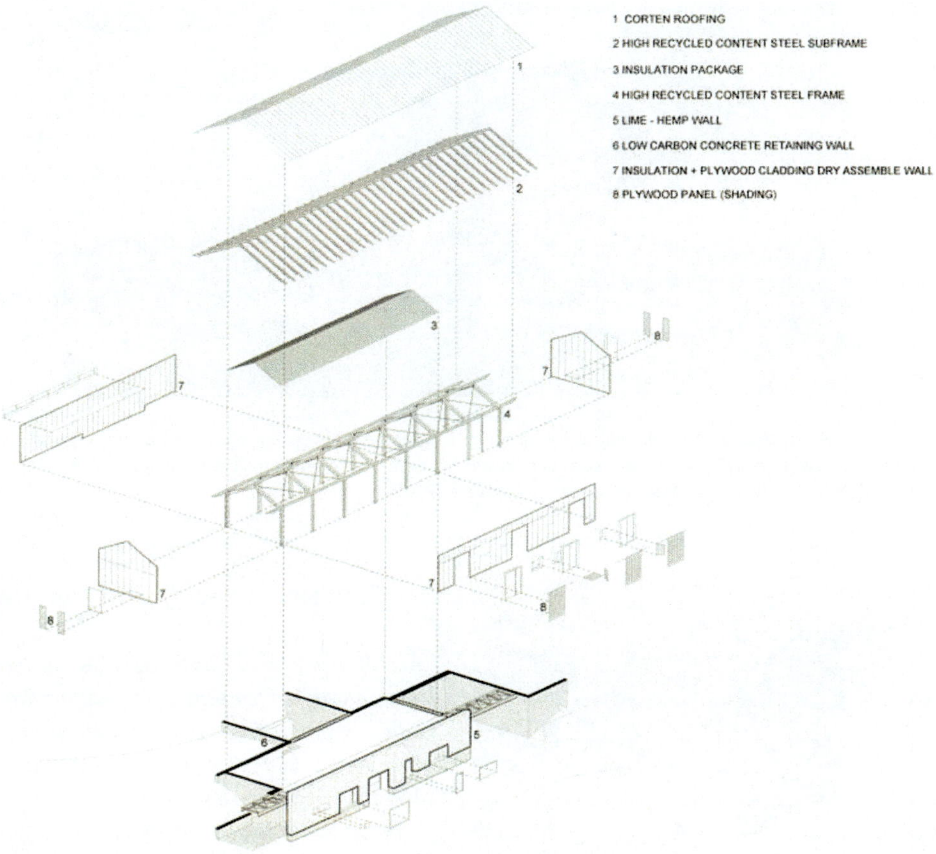

Fig. 11 By placing the sleeping quarters in a lightweight, kit of parts enclosure designed for deconstruction on top of massive living areas, the design looks for creative ways to reduce operational energy while at the same time minimizing the embodied carbon of the building. Exploded axonometric scheme of Bricolla © PAT.

calculates, optimizes, and sends the specification to a detailer, who then sends them to a fabricator that requires the product to a service center who buys the steel from the mill. In current practice, the architect has no idea of where the steel comes from. Tracking down the supply chain, making sure of where the steel comes from and what is the recycled content and energy mix that was used to transform the steel from scrap to beams is the way forward, especially if we think that in the long run still mills will be powered with green energy. In the case of Bricolla, steel beams will arrive from Pallanzeno (250 km), where a laminated steel plant is operated using a certified quantity of post-consumer recycled content of at least 92% (Fig. 18). The same company is now building a new construction plant totally powered by wind. If we think that most of the supply of structural wood in Italy comes from abroad, using a local product made by transforming scrap makes sense, and we can have the benefit of the reduced section of a steel frame.

Fig. 12 The massive ground floor allows the living areas to be warm in the winter and cool in the summer, while the well-insulated, carefully shaded, and less massive upper floor is easier to cool off at night during the summer season through cross ventilation. View of Bricolla © PAT.

Fig. 13 Rammed earth was the first choice as a material to build a ground floor seamlessly emerging from his setting, much as the calcareous soils left bare by the fractures in the land that abruptly cut the rolling hillside in this area of the *Langhe*. Elevation details of Bricolla © 2022 PAT.

3 Context, Impact, and Perspective

"*Architecture will only survive in something better than its present forms and practices by radically engaging in its most basic working procedures and disciplinary acts*", states Kiel Moe in *Unless* (2020).

Bricolla will be the heart of a family-run small farm producing hazelnuts, honey, and wine. It might not be a coincidence that Bra, the small town where the *Slow Food* movement originated, is a mere 40 km away. The ethos of going to the origin of products and understanding their supply chain permeates the culture of this corner of Piedmont. The jump from understanding the food supply chain to the construction supply chain is rather easy in this context (Fig. 19), and is further facilitated by a culture of work and doing business where circularity was a necessity even before it became a fashionable statement to boast. Bringing this ethos to the realm of architecture leads us back to the initial argument of *Unless: "Describing the non-abstract terrestrial activity of architecture permits architects to finally address the ecological and political basis of architecture's formation as a central project. (...) Arguably, deep descriptions of—and enchantment with—architecture's formation is far more ambitious as a theoretical and practical*

Fig. 14 "*Life cycle design not only satisfies the user's complex requirements, but also promotes sustainable development vis-à-vis three important fields: the economy, the environment and society*" (Thiébat 2019). Forty kilometers away from Briaglia, the Carmagnola area has been known for centuries for the cultivation of hemp, used to produce fabrics and manufacturing ropes and rigging. Hemp ropes have a greater durability than ropes in synthetic fibers and resist well to humidity, as proven by their traditional use on boats. Harvesting hemp near Carmagnola. © Alice Belcredi

Fig. 15 The core of an hemplime wall is made of hemp shives mineralized with lime, thus making them resistant to fungi. Hemp shives guarantees high insulation values. The hemp shives and lime mix is then finished by hemplime thermoplaster, made by adding small hemp fibers to the traditional, breathable lime wall protection that has been used for centuries before being reduced to a small segment of the market by cement-based plaster, less virtuous but quicker and easier to apply by unskilled labor. Mock-up for hemplime insulated walls, Piasco (CN). © Andrea Veglia/PAT.

WALL			MATERIAL FEATURES				EMBODIED CARBON (CO_2)
code wall_op_01_Bricks		Construction Layer (internal to external)	Thickness [cm]	Density [kg/m³]	Weight [kg/m²]	Weight/Tot [%]	Total EC [kg CO_2/m²]
Ground floor West Poroton/Insulation/Lime Plaster	1	Plaster - lime	2,00	1300,00	26,00	9%	10,24
	2	Poroton light P600	30,00	630,00	189,00	63%	45,36
	3	Rockwool	6,00	150,00	9,00	3%	10,08
	4	Poroton light P600	8,00	630,00	50,40	17%	12,10
	5	Plaster - lime	2,00	1300,00	26,00	9%	10,24
	Total		48,00	4010,00	300,40	1,00	88,02
code wall_op_02_Hemplime		Construction Layer (internal to external)	Thickness [cm]	Density [kg/m³]	Weight [kg/m²]	Weight/Tot [%]	Total EC [kg CO_2/m²]
[brief description]	1	Plaster - lime	2,00	1300,00	26,00	22%	10,24
	2	Wooden frame (1,12 kg/m - no Carbon Storage)		450,00	1,76	1%	0,46
	3	Calce canapa intercapedine (no Carbon Storage)	40,00	160,00	64,00	54%	5,12
	4	Plaster - lime	2,00	1300,00	26,00	22%	10,24
	Total		44,00	3210,00	117,76	1,00	26,07
code wall_op_03_rammed earth		Construction Layer (internal to external)	Thickness [cm]	Density [kg/m³]	Weight [kg/m²]	Weight/Tot [%]	Total EC [kg CO_2/m²]
[brief description]	1	Rammed Earth Stabilized 5% cement	15,00	240,00	36,00	10%	2,20
	2	Rockwool	12,00	150,00	18,00	5%	20,16
	3	Rammed Earth Stabilized 5% cement	15,00	2100,00	315,00	85%	19,22
	Total		42,00	2490,00	369,00	1,00	41,57

Fig. 16 After evaluating costs and Embodied Carbon (€CO tool, Thiébat 2019), business as usual practice such as bricks and alternative options such as rammed earth have been opted out in favor of hemplime (wall_op_02). The table shows the materials data and the Embodied Carbon (EC) for the three alternative stratigraphies

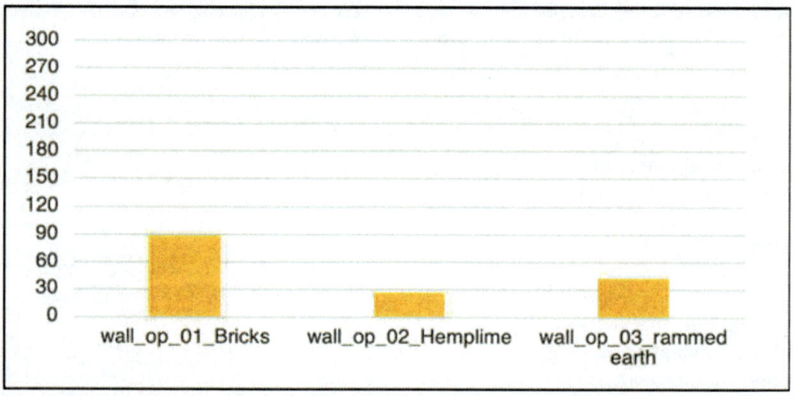

Fig. 17 Environmental impact assessment results related to the three alternative options. Embodied carbon values are expresses in kg CO_2 per square meter of the wall

Fig. 18 Given its endless recyclability steel is a perfect material for circular economy. In the case of Bricolla, steel beams will arrive from Pallanzeno (250 km), where a laminated steel plant is operated using a certified quantity of post-consumer recycled content of at least 92%. Scrap metal junkyard. © Donald Trung Wikimedia Commons

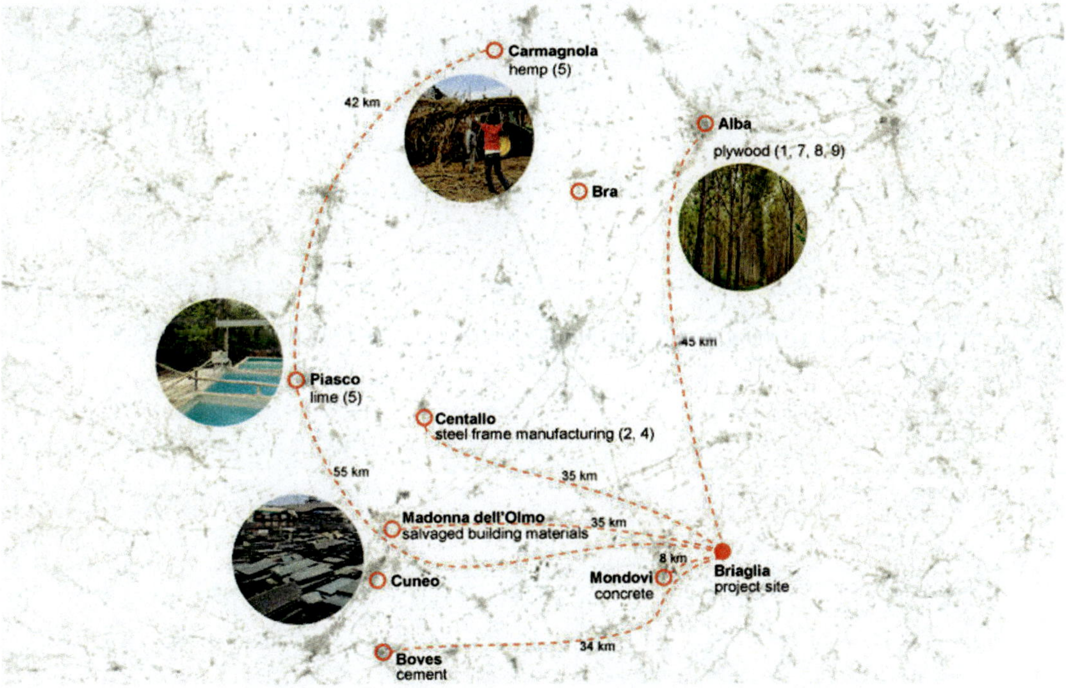

Fig. 19 Designing the construction process of Bricolla has been a research on the reduction of the embedded energy of the whole process, so that its ecological footprint could be lowered while at the same time a whole network of local knowledge and productions could be (re)activated, with hemp, lime, steel components being processed no farther than 60 km, and with the possibility to reuse materials such as the wood boards for floors from formerly dismissed buildings, thus activating multiple dynamics of circular economy around a single project

project than the purview of autonomy and autarky of recent architecture".

In the end, Bricolla is an attempt to go beyond all low-tech and high-tech narrations, discovering the possible synergies between different architectural cultures that used to co-exist since the very beginning of the 70 s radical era, to finally operate a synthesis in a perspective of operating sustainability.

We need to bring building decarbonization and building craftsmanship from the realm of eco-luxury market to a possibly replicable everyday of construction, where the building supply chain is not something mysterious and removed from the control of the architect.

And as much as the jargon has evolved over the years, it all sums up to a phrase in which Buckminster Fuller summarized his philosophy: "If we do more with less, our resources are adequate to take care of everybody" (Farrell 1972). Which sounds very close to the imperative of the piedmontese of old: "*Non sprecare*" (Do not waste).

References

Banham R (1982) Art and necessity. Architectural Review, Dec 1982

Eurostat (2022) Distribution of population by degree of urbanisation, dwelling type and income group—EU-SILC survey ILC_LVHO01. https://data.europa.eu. Accessed 11 Oct 2022

Farrell B (1972) R. Buckminster Fuller. A candid conversation with the visionary architect/inventor/philosopher. Playboy 2/72:59

Moe K (2020) Unless. Actar, Barcelona, pp 42, 44

Rager M, Stern E, Walther R (2020) Le Tour de France des maisons écologiques. Gallimard, Paris, p 13

Slessor C (1997) Eco-tech. Sustainable architecture and high technology. Thames & Hudson, London

Thiébat F (2019) Life cycle design. An experimental tool for designers. Springer, Cham, p XI

Learning from Vernacular Architecture: The Essence of Remakri Area in Bangladesh

Anindita Laz Banti

Abstract

For thousands of years, vernacular architecture has been experiencing limitations that are emotionally associated with its aesthetic variety, self-regulating construction, invention, and adaptation to its surroundings. Vernacular's authenticity describes the shared culture of people in a specific geographic region, including their language, heritage, religion, and customs to show the importance of its identity and existence in historical context. Even though vernacular architecture faces limitations, its adaptability and open-ended approach inspire modern designers to embrace traditional structures, which are highly sustainable and energy efficient. Approximately, how many vernacular buildings have been lost from a community? Furthermore, how does vernacular architecture respond to specific conditions in this contemporary era that affect people and places all over the world? This paper examines the current condition of vernacular buildings in *Remakri* area inside Thanchi Upazilla at Bandarban District in Bangladesh, their existing role in society, and studies the theories and practices of contemporary vernacular that are shaping and examining the community. As technology advances rapidly, traditional structures still provide valuable lessons in terms of their creation and ethnicity. By working at multiple scales, this paper explores and integrates contemporary strategies that will not only increase the quality of life in the community, but also greatly enhance the cultural values in the global platform.

Keywords

Vernacular architecture · Contemporary · Region · Tradition · Culture · Community

1 Introduction

"*Awareness of the past is essential to the maintenance of purpose in life. Without it we would lack all sense of continuity, all apprehension of causality, all knowledge of our own identity. What previous groups identifies and sanctify as their pasts become historical evidence about themselves*"—*D. W. Meinig* (Meinig and John 1979).

The importance of local influences on the lifestyle of people of a region and its past historical culture significantly express the existence of that region's identity. The adaptation of cultural and natural arrangement is expressively distinct and directly relates to a region's

A. Laz Banti (✉)
Miami University, Oxford, OH, USA
e-mail: lazbana@miamioh.edu

M. R. Thomsen et al. (eds.), *Design for Rethinking Resources*, Sustainable Development Goals Series,
https://doi.org/10.1007/978-3-031-36554-6_19

environment, people, local context, and local technology. Living in a culture that has its own deeply rooted heritage, there is always a sense of belonging to the community and society from the past, present, and future; it gives the society values, meanings, customs, and beliefs, which shapes the thinking process of an individual. Preserving cultural identity supports individuals and communities in a balanced, unchanging, persistent frame of reference and meaning.

To cherish something old or venerable, the most common way of protecting and preserving culture is by discontinuing deterioration and extending the culture's life span as long as possible. However, cultural preservation does not always support cultural longevity and often coincides and conflicts with esthetic, environmental or practical values. Fundamentally, cultural preservation relates to history and societal values of a specific region, which is influenced by the physical appearance, ancestral origin in a society, and demonstrates shared identities among communities. Diverse cultural experiences bring together multiple cultural identities among communities that integrate with a site context and natural landscape consistent with social changes. The distinct cultural experiences correspond with regionalism; that strives to sustain and refine successful design strategies; that are culturally embedded within a region; that emanate from the landscape; and that speak to the values, customs, and needs of its inhabitants (Heath 2009). Additionally, the preservation of a community's cultural experiences is reflected in the architectural fabric of a region. In the current era of globalization, can vernacular architecture inform and impact the specific shifting of social and environmental conditions if local regions are affected by specific cultural building fabric that is defined vernacular to a place? As the prime case study, this paper discusses vernacular architecture in Bangladesh as a way to consider the importance of cultural and architectural heritage. Furthermore, in an increasingly interconnected, yet independent world, the focus on a vernacular building's life cycle that responds to environmental and cultural contexts concerns the ability for the preservation of the architectural fabric of a place, of a region.

2 Methodology

Through analyzing the current models of vernacular buildings in Bangladesh as a standard form, reviewing scholarly works of architects, theorists, and critics, and observing current building conditions and the architect's role in local communities, suggests that is why a vigorous respect for vernacular architecture is needed. Moreover, in rural/local regions the local level of particular vernacular conditions is being threatened. A consideration of critical regionalism through the work of Alexander Tzonis and Liane Lefaivre ("Why Critical Regionalism Today?") (Tzonis and Lefaivre 2003), Kingston Wm. Heath ("Vernacular Architecture and Regional Design: Cultural Process and Environmental Response") (Heath 2009), and the architectural precedents such as the present crisis of identity in Indian architecture (Panicker and Ostwald 2002) is considered to best understand the experience of the current local building practices in Bangladesh. To consider contemporary ways of design and construction that learns from vernacular and honors the traditions of local communities, project such as Umubanu primary school in Rwanda is discussed. To understand the global phenomena of vernacular architecture and why the need for the contemporary architecture to learn from the vernacular is needed, this study surveys comprehensively Bangladeshi heritage studies and its regional vernacular practices.

3 Vernacular and Critical Regionalism

The term "Vernacular" means native or unique to a specific place, produced without the need for imported components and processes (Brown and

Maudlin 2012). According to Paul Oliver, traditional vernacular buildings are a rich cultural resource: highly complex objects that can express multiple meanings through form and decoration, enclose inhabitable space, and frame human ritual and the performances of daily life (Brown and Maudlin 2012). Vernacular results from the locality of a specific region with a distinct regional identity. This specific locality is the reflection of its history, culture, tradition, climate, geography, topography, and lifestyle. It is the simplest form of addressing human needs by embracing regionalism that engages its particular geographical and cultural circumstances in deliberate, subtle, and vaguely politicized ways (Eggener 2002). Vernacular architecture addresses the issues of design and materials, at the same time it follows the common form of building in a given place and time. Therefore, it must encompass easily available materials both handmade and industrially produced, depending on the particular circumstances in this contemporary world.

On the other hand, regionalism is concerned with differences between rural and urban settings, cosmopolitan progressivism, and sophistication versus rural tradition and conservatism. It also provides a method that attempts to understand buildings using the contextual forces which surround their production. According to Kingston Wm. Heath, critical regionalism designates a form of architectural practice that embraces modern/contemporary architecture critically for its universal unifying qualities while simultaneously responding to social, cultural, and climatic contexts of the region in which it is built (Heath 2009). In addition, critical regionalism is concerned with how a set of problems are dealt with, which tend to be more about design attitude rather than location; critical regionalism concerns fragmentation and alienation versus community; mobility versus stability; and technology and industry over and against traditional approach, where new/contemporary design is not necessarily a good idea to work with (Curtis 1986). Sometimes professional designers or architects disregard vernacular buildings and ignore the social structures which are constructed on traditional values based on local design attitudes, customs, and methods. Every country, every area has a unique culture and taste, and architecture cannot be separated from its context. Architects and professionals should be concerned about the vernacular, local culture, and their influence on contemporary architecture instead of other countries' language to design projects. The way critical regionalism formulates architecture within its universal integrated skills, at the same time responding to socio-cultural context of the region, there will be always a need to construct social settings differently without forgetting the effect of it. For example, participation of the local constellation of ideas could emphasize the construction processes that relate social activity to ecological conditions. Therefore, the approach could be to regenerate through integrated production and reproduction of independent and life-enhancing practices. Finally, it is clear that vernacular architecture and its language speak about a set of objects that describes the common buildings of a given place and time, a collection of buildings or vernacular landscapes, which are the products of a particular community, and last but not the least, an approach to studying buildings as cultural manifestations (Carter and Cromley 2008). These concepts significantly offer a useful way in the world of ordinary buildings and continuously reappear as analysis begins and interpretations evolve (Fig. 1).

4 Topography and Climatic Context in Bangladesh

Bangladesh has a wide variety of vernacular architecture based on the topography and climatic conditions of the region. Here, vernacular buildings follow traditional practices and patterns in indigenous styles that are constructed from locally available materials related to the native context. Although Bangladesh is a small country, it has significant topographic diversity, which includes three distinctive features: (a) a broad muddy plain land that concern to frequent flooding, (b) a slightly elevated terrace land, and (c) a small hilly and upland area drained by

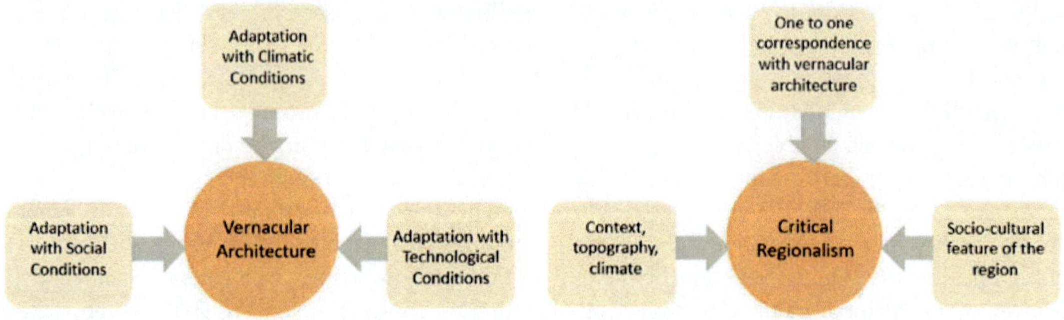

Fig. 1 Paul Oliver's conceptual illustrations of vernacular architecture (left); conceptual maps of critical regionalism viewed by scholars (right), produced by author

flashy rivers (Fatemi and Islam 2011). Bangladesh has about 80% flat land, 12% hilly areas, and 8% terrace land. The southern part has a highly irregular coastline of about 600 km that is ruptured by rivers and canals flowing into the Bay of Bengal. The hilly areas of the southeastern region of Chittagong, the northeastern hills of Sylhet, and highlands in north and northwest are of low elevations. The settlement pattern near the river side is linear, which is also found in the spring lines of the Chittagong Hill Tracts. Whereas the remainder of the country shows the clustered or scattered settlements. Climatic context in Bangladesh is different throughout the year. It has a subtropical monsoon climate featured by wide seasonal variations in rainfall, moderately warm temperatures, and high humidity. Due to the warm humid climate, it is important to have thermal comfort in the built environment and protection from rain in the design. This paper attempts to investigate the conditions of vernacular buildings in Bangladesh; analyzes the typologies of traditional building forms in different areas in terms of geography and topography; and observes their traditional vernacular housing technology, local perceptions of environmental behavior, and economic aspects of building construction in Bangladesh in terms of eco-adaptation and sustainability (Fatemi and Islam 2011) (Figs. 2 and 3).

In Bangladesh, the floodplains comprise over 76% of the country's rural settlement. These settlements are two types of Elongated Linear Settlement and Amorphous/Irregular Settlement. They are both constructed on elevated platforms or ground that is higher above the annual flood level. The elongated linear settlement is usually constructed on high land alongside river or water channels. Such settlement patterns are actually determined by the location of rivers or waterfalls. In contrast, the amorphous settlement is spread throughout the plot and has a clustered or scattered pattern on a raised platform to prevent annual floods. When high land is not accessible, soil from digging ponds or channels is raised into a mound that is 2–3 m or more above the water level nearby, on which a farmhouse or *BARI* is built (Fatemi and Islam 2011). This type of homestead, or *BARI*, starts out as a single household and later grows into a settlement of multiple dwellings and additional multigenerational homes, or *GHORS*. As the settlement grows, more and more earth is put to the mound to extend it. As a result, settlements are built on growing mounds, which act as an island during the rainy season.

5 Typologies in House Form: Bangladesh

Under a system of geographic controls, housing in a specific area develops in accordance with local resident needs, which evolves together with their requirements at various phases of socioeconomic and cultural development. A key consideration when choosing a house location is the

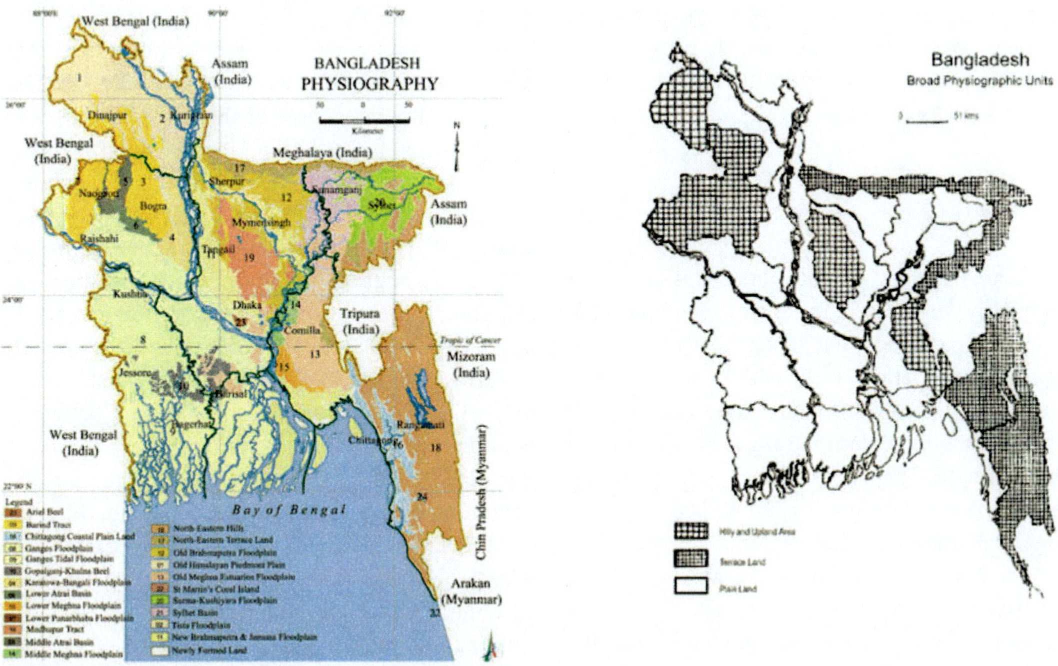

Fig. 2 Bangladesh Broad Physiography (pattern of topographical features or landforms). *Source* Banglapedia

Fig. 3 Rural settlement and built form in Bangladesh. *Source* Drawn by author

landform. Due to the nearly flat and deltaic terrain that makes up the majority of Bangladesh, dwellings must be built on higher ground or on land that has been purposely raised, as is the case in "*haor* regions," to avoid floods and inundation during the monsoon (Fatemi and Islam 2011). In terms of physiography, Bangladesh can be divided into three separate regions: (a) floodplains, (b) terraces, and (c) hills, each of which has distinctive qualities of its own. There are several house types that have been seen in those regions including houses on terrace land, houses in hilly and upland areas, and buildings in plain land classified as floodplain.

5.1 Plain Land

Houses that are located on plain land are more affected by flooding. Moreover, the floodplain means comparatively flat lands adjacent to that formed by alleviating rivers which are subject to overflow. In this type of land, houses are built in an open planning pattern in response to the warm humid climate. This type of houses can be found at some parts of Dhaka, Gazipur, Narayanganj, Tangail, Joydevpur, Sunamganj, Mymensingh, Kushtia, Jessore, and Cox's Bazar. The house has a separated rectangular form of living zones that extended to east–west direction and the layout is rectangular, with an open courtyard. This type of house has two types of roofs: *Chouchala* that means the roof has four roofs together and *Dochala* that means there are two roofs in total. The building is planned in a scattered way with large, free spaces between them to allow air circulation that also provides ventilation for cooling. Buildings are built on a raised earth plinth, which is common in Bangladesh and is a characteristic feature of indigenous architecture.

5.1.1 Analysis

The topography of this area shows that the landform is relatively at a lower elevation that is shallowly flooded. Most of the ridges and all the basins of this floodplain region are flooded more than 0.91 m deep for about 4 months (mid-June to mid-October) during the monsoon (Fatemi and Islam 2011). The houses in this region are made of bamboo matting and mud, which are attached to plinths as they have lower elevation and have proximity to flood level. The houses are arranged in scattered ways to allow free movement of air through buildings and through spaces between buildings. Moreover, the outdoor spaces are used for different activities that take place out-of-doors. The shape and form of the building is rectangular, which is elongated along east–west direction. The north–south orientation of the building creates an opportunity to get natural ventilation. Additionally, the shaded verandas and the shaded outdoor spaces are constructed for climatic reasons. Steep roofs are planned to protect against high rainfall, which is a regular phenomenon of this region. The warm, humid climate is the reason for the dispersed housing arrangements for various functions into diverse structures. Materials are environmentally friendly and can adapt to the surroundings; for instance, perforated bamboo matting walls can operate as a breathing wall and provide passive ventilation (Figs. 4 and 5).

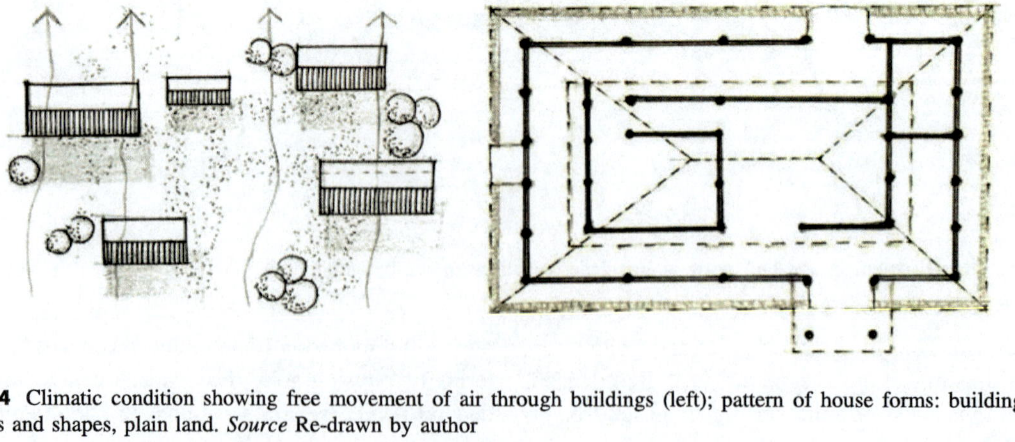

Fig. 4 Climatic condition showing free movement of air through buildings (left); pattern of house forms: building forms and shapes, plain land. *Source* Re-drawn by author

Fig. 5 Housing unit and higher plinth at plain land. *Source* Md. Nawrose Fatemi and Nabanita Islam

5.2 Terrace Land

Houses on terrace land are found in parts of greater Dinajpur, Bogra, Pabna, Joypurhat Rangpur, Rajshahi, and Naogaon Districts of Rajshahi Division. The planning pattern of buildings in this region is compact, which is appropriate for the hot dry climate of the region. The houses are rectangular in shape that is usually elongated along east–west direction and they orient in north–south direction. These houses also have courtyards around their perimeter, which partially shields them from the wind's full force and the effects of the outside air. Buildings are also packed and introverted in nature, with enclosed areas between them. Additionally, the placement of structures and their orientation in space influence the microclimate at each location (Fig. 6).

5.2.1 Analysis

Terrace land is comparably at a higher elevation than the nearby floodplains, according to the geography. Nearly 47% of this region is highland, while nearly 41% is medium highland. Two terrace levels exist: one is at a height of 4 m, while the other is between 1.98 and 2.29 m (Fatemi and Islam 2011). As a result, the surrounding land does not flood when the floodplains are submerged during the rainy season. Due to the lateritic component of the soil and the higher elevation, which is above the average flood level, mud is utilized in this area to build walls and dwellings. The houses are usually compact and joined together, which gives them a certain character. The basic structure is laid out so that a central courtyard is encircled by walls, partially shielding it from the effects of the outside air. Each site has a unique microclimate

Fig. 6 Housing unit at terrace land. *Source* Doza, S. B. and Razzaque, M. Z. I., 2008

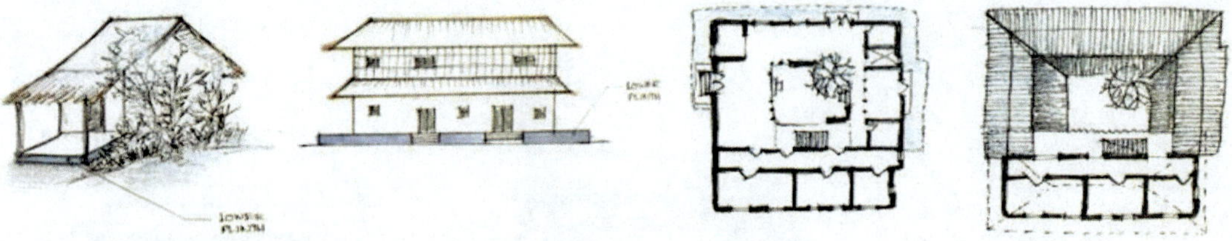

Fig. 7 Settlement pattern: topography showing lower plinth (left); pattern of house forms and shapes (right), terrace land. *Source* Re-drawn by author

created by the layout, orientation, and design of the structures. Terrace land follows a consistent geometry that is primarily squarish for building form and shape. All of the functions are arranged around a central courtyard, on which the house's design is specifically focused. Additionally, due to climatic factors, the houses are small and introverted in style. As a result, the created form is shielded from the effects of the hot air from the outside. The walls function as a thermal mass when it is heated outside (Fig. 7).

5.3 Hilly Area

Houses on hilly areas are found in Chittagong, Chittagong Hill Tracks, Mainamoti Hill Tracts, and Parts of Sylhet Division. To allow for unrestricted airflow around the building, buildings in this area are spread out far enough from one another. Adversely, a distinctive feature of rural housing in plain areas, the courtyard, is absent from this house. The houses are arranged in a tight planning pattern on elevated buildings, allowing cross ventilation that is a suitable response to the temperature and geography of the mountainous location. Buildings are extrovert in nature and essentially lifted off the ground, creating an elevated platform known as *MACHANG* to facilitate cross ventilation beneath the spaces. They built high platforms to prevent flooding and get rid of wild animals. The lower portion of those compartments is occasionally used for storage or housing domestic animals (Fatemi and Islam 2011).

5.3.1 Analysis

The hilly area's topography denotes sharply rising mountain ranges that resemble hogback ridges. In regions where the hill ranges have an average elevation of more than 300 m, the river valleys and the hill ranges are typically longitudinally aligned. In this region, houses are raised above the ground on stilts to provide security, protection from dangerous animals, and protection from flooding during the rainy season. Since there is a shortage of drinking water, the position of *JHIRI* or waterbody essentially controls the settlement pattern in this area (Figs. 8 and 9).

The dwellings are arranged beside the *JHIRI* direction in an irregular pattern (Fatemi and Islam 2011). The building forms in hilly areas are generally in square shape. Sometimes the extended portion of the house is also followed by the existing squares, whereas the other spatial divisions that they made by dividing the main house or any part are not square. In contrast to rural living in the plain lands, *PAHARI* or tribal communities don't have courtyards in their houses. With each built form, a raised platform known as *MACHANG* is produced at the front of the house and serves as a courtyard. These are present in houses on plain or terraced land. Furthermore, the living areas of these houses are

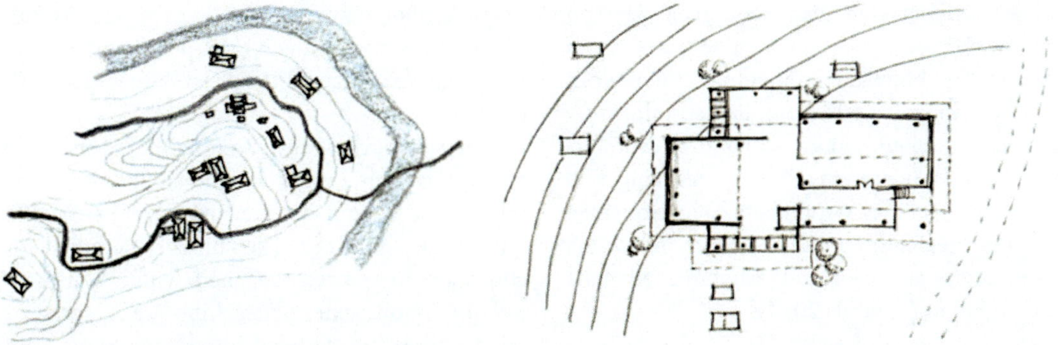

Fig. 8 Climatic condition showing location of *Jhiri* / waterbody determines the settlement pattern in hilly area (plan). *Source* Re-drawn by author

Fig. 9 Topography showing house with bamboo slit in hilly area (left). *Source* Re-drawn by author; housing unit at hilly area (right). *Source* Md. Nawrose Fatemi and Nabanita Islam

also too small because of the lack of available space, they are extrovert in nature which are constructed on stilts (Fatemi and Islam 2011).

6 Materials and Transportation

In vernacular architecture, structures are created using regionally specific and easily accessible materials as a necessary response to the local geology, climate, and customs. Due to their energy efficiency and consideration for the local eco-systems, these structures are also environmentally friendly. Local materials also have less energy-intensive production processes and don't require any transportation costs, which results in lower embodied energy and CO_2 emissions. Additionally, local materials have a low maintenance environmental effect because they are natural, organic, renewable, and biodegradable (Fernandes et al. 2014). Mud, rammed earth, bamboo, thatch/straw, and wood are the most prevalent native materials in Bangladesh since they are made from resources that are readily available there. In the plain area, people use mud as foundation and bamboo, bamboo mats for walls, which are then tied and connected to building frames. The roof is covered by bamboo and thatch and then processed by the locals. Locals employ their own local building methods and knowledge, which was passed down from their ancestors over time.

In order to encourage the development of sustainable communities, the ecological building sector needs to locate its production facilities close to where its products will be consumed, make use of locally available renewable resources, and prioritize low-energy and low-pollution processes. Supporting sustainable local development means also preserving a cultural heritage of construction knowledge inherent to regions. Therefore, in order to accomplish sustainability, architecture should find a way to combine ancient and modern methods by utilizing both techniques for environmental improvement employing technology and materials that would benefit regional social and economic advancements (Fernandes et al. 2014).

7 Case Study: Umubano Primary School, Rawanda

Umubanu primary school, Kigali, Rwanda is designed by MASS design group (Construction: March 2009–July 2011); the firm's main goal is to create an architecture that requires the form to improve the lives among the school and community. The architects focused on the site context with an aim of designing a school that fits in its context to inspire, merge, and integrate the neighborhood community called Kabeza (Mutuli 2016). They worked with local people, used local materials, and adapted their techniques (Figs. 10 and 11).

The school provides seven buildings and nine classrooms on a significantly sloped site; this school serves quality education for over 300 vulnerable children. The unique settings for education have been designed within a mix of interior rooms, exterior teaching areas, some of these spaces are covered by sloping roofs, and terraces are designed for the play area of the children. Local materials such as brick and bamboo are composed into walls with built-in shading screens. These local materials, shading, and natural ventilation are relied upon to reduce energy consumption. Inside the classrooms, there are clerestory windows that help to bring light into the classroom. The design of Umubano Primary School merges with the hilly landscape in which it is gracefully falling and rising with the contours. In the hilly landscape of Kabeza neighborhood, people travel across terraced

Fig. 10 View of the buildings on the sloped site. *Photo* Iwan Baan

Fig. 11 View of the library, a papyrus-reed dropped ceiling provides natural ventilation and diffused light (left) and classroom, handmade brick coursing also helps move air through the space (right). *Photo* Iwan Baan

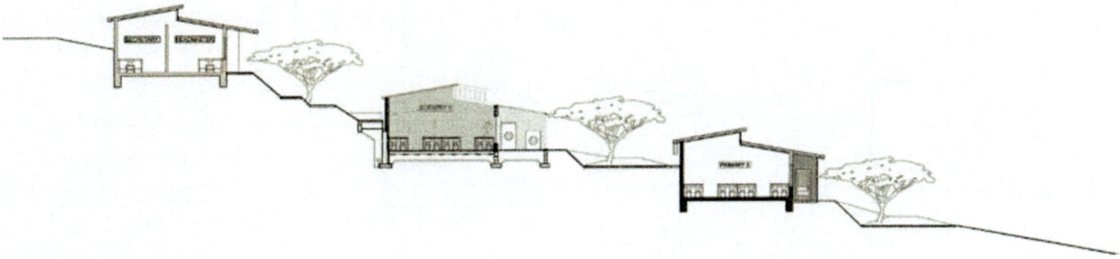

Fig. 12 Site section, the ramped network of the complex was informed by the stepped agricultural landscape. *Source* MASS Design Group

agricultural land to complete their journey. The school mirrors this kind of living, with outdoor classrooms providing the children with a school that is embedded in their daily lives. The main concern of this design was to create a school that would totally relate to the community (Menocal 2013) (Figs. 12 and 13).

The materials used in this project are brick and papyrus reeds that are found within the neighborhood and help to cut down the transportation cost. Moreover, the construction improved the regions' economy by encouraging construction professionals and residents to embrace the local market. Some local expertise is also included in this project that inspires natural ventilation strategies comprising thatched doors and brick jail, basically holes in the brick wall patterns across the structures. Furthermore, the limestone blocks are used to design the terracing and retaining walls that give the school a unique aesthetic as the school is a landmark in the Kabeza community.

The MASS Design group works with the community, understands them, and determines the needs and desires of their clients. They involve the local people in their design and construction phases, listen to the community's ideas and design accordingly, and build structures that are thoughtfully considered by addressing the site and use of material. Everyone in the community appreciates the level of commitment MASS Design group has to the community through their engagement. Moreover, the structures and materials implemented in this project have a unique integration of surroundings to reflect the cultural and vernacular context of the Kigali region.

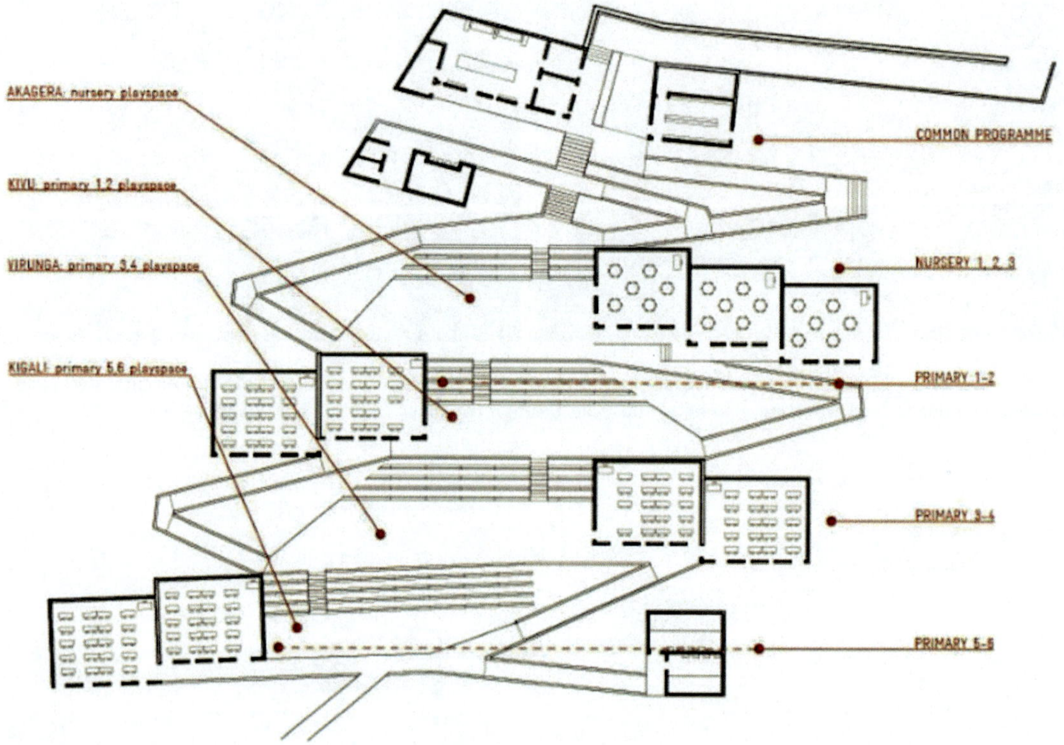

Fig. 13 Umubano Primary School building diagram. *Source* MASS Design Group

7.1 Site Location

Bandarban District is located in the southeastern part of Bangladesh and a part of the Chittagong division and Chittagong Hill Tracts, which has an area of 4479.03 km^2. Geographically, Bandarban is surrounded by the Rangamati District on the north, the Arakan (Myanmar) on the south, the Chin Province (Myanmar) and Rangamati District on the east, and the Chittagong and Cox's Bazaar District on the west. The population is around 298,120. This upazila is home to indigenous tribes such as the Marma, Chakma, Bawm, Murong, Tripura, Khyang, Khumee, and Lushei (Bandarban District 2021) (Fig. 14).

Bandarban District is home to Raikhiang Lake, the country's highest lake, as well as Bangladesh's three tallest peaks, Tahjindong (1280 m), Mowdok Mual (1052 m), and Keokradong (1230 m). The district's other two well-known landmarks are Boga Lake and Chimbuk Peak. River *Sangu*, also known as Sangpo or Shankha, is another prominent feature of the district and the only river to originate in Bangladesh (Bandarban District 2021). This district is also one of the most exotic tourist attractions and best hiking spots in Bangladesh.

7.2 Thanchi

Thanchi Upazila is inside Bandarban District and has an area of 1020.82 km^2. Thanchi is bordered on the north by Ruma Upazila, on the south by Arakan State of Myanmar, on the east by Belaichhari Upazila, and on the west by Alikadam and Lama Upazilas. It has a total of 16,992 population in which male population is 9438 and female population is 7554 (Fig. 15).

This upazila is home to indigenous populations such the Marma, Murong, Tripura, and Khyang (Thanchi Upazila 2021). Thanchi Bazar is a prominent bazaar and fair in this region.

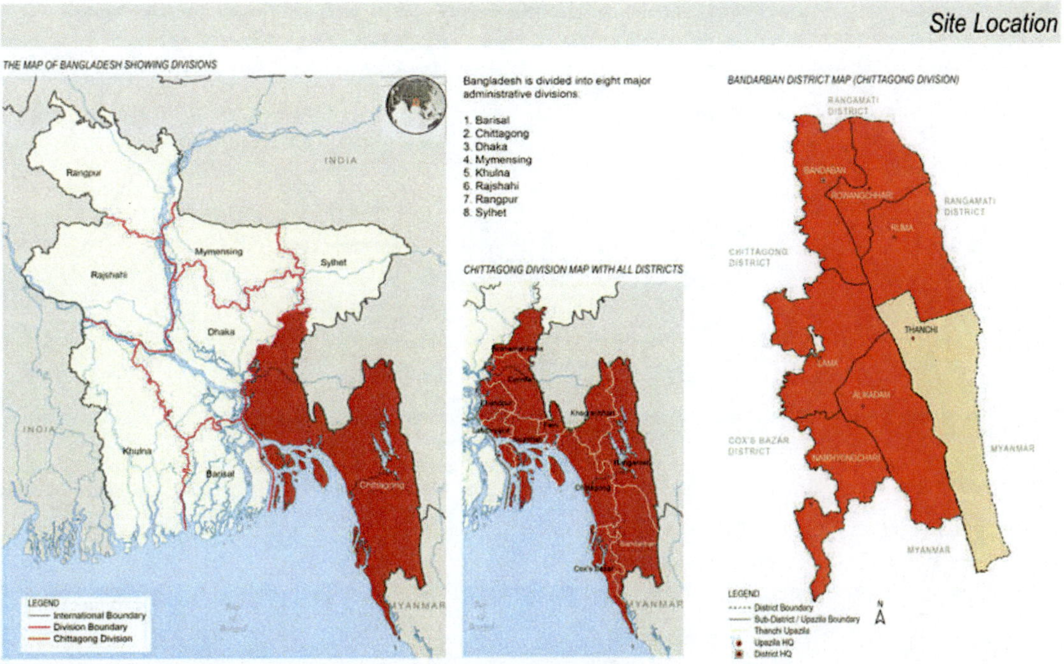

Fig. 14 Map showing Bandarban District. *Source* Created by author

Fig. 15 Thanchi area photos showing boat stoppage spot, different building structures, and bazar. *Source* Author

Additionally, it has some well-known hospitals and educational institutes where people from many villages and para travel for better services. (*Source* Bangladesh Population Census 2001, Bangladesh Bureau of Statistics; Cultural survey report of Thanchi Upazila 2007.)

7.3 Remakri

The site is situated in *Remakri* area, inside Thanchi Upazilla, which is at Bandarban District, Chittagong Division, in Bangladesh. It is also near to Myanmar boarder at east and west side of

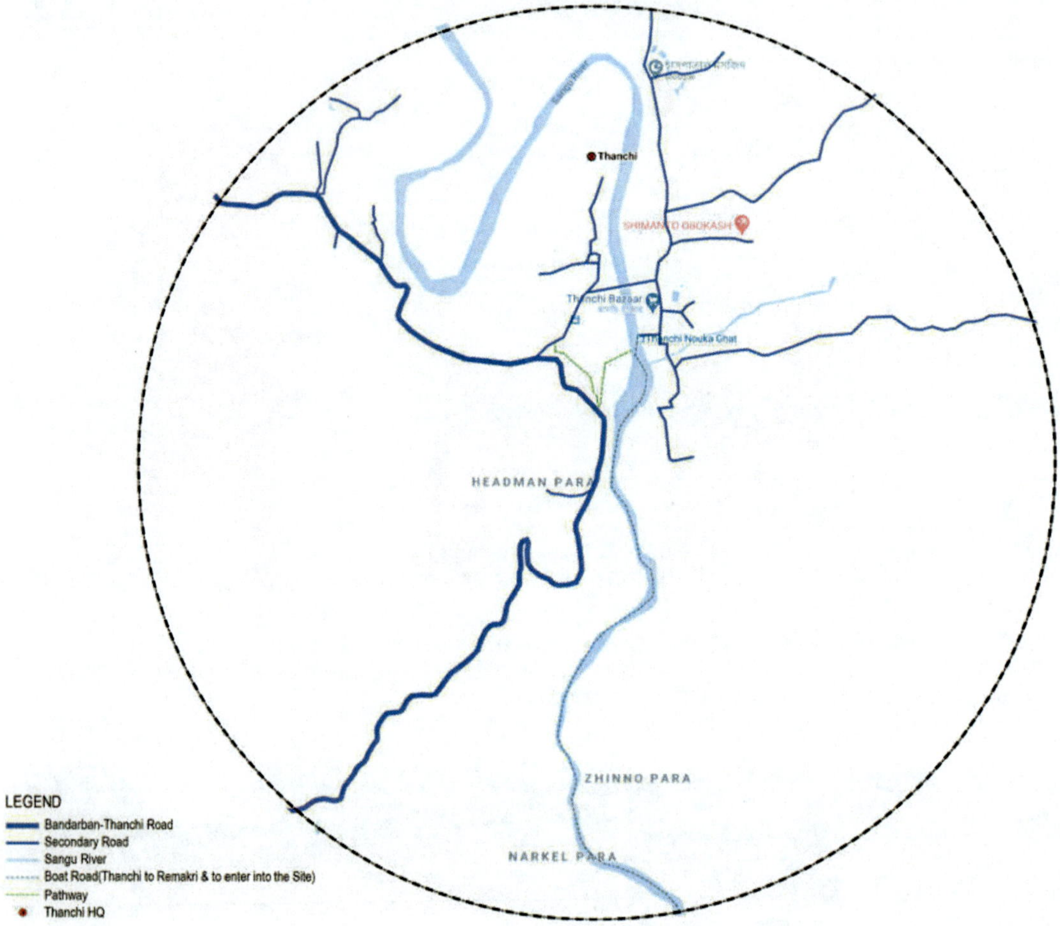

Fig. 16 Entry and circulation, showing Bandarban-Thanchi road. *Source* Created by author

the site. The only mode of transit from the Thanchi area to the site *Remakri* is by boat, which takes approximately 2 hours along the *Sangu* River. The *Sangu*, a cross-border river, rises on the Indo-Burma border (North Arakan Hills of Myanmar: at 21°13′N 92°37′) and flows through Chittagong before entering Burma to the north. It is one of the challenging watercourses with occasionally a river, a torrent, and mostly a stream (Hossen et al. 2019) (Figs. 16 and 17).

Despite not being a river of the Himalayas, this river depends on rainfall, and conveniently, its watershed is located in an area with substantial annual rainfall (Adnan 2019). The Sangu River has year-round flow due to the water sources and the trees that assist in retaining the water in the soil. The river has a total length of 294 km, of which 173 km are in Bangladesh. Furthermore, the *Sangu* River plays a significant role in the lives of the major ethnic communities of Marma, Murong, Tripura, Bawm, Tanchanga, Chakma, Chak, Khyang, Khumi, Lushei, and the Pankho (Bandarban District Statistics 2013) (Fig. 18).

Thanchi upazila and Bandarban Districts are linked by the Bandarban-Thanchi road, which also links to Chittangong Division. In Thanchi, there are many secondary roads that connect with different services and programs. In the Thanchi Region, there is also a "*ghat*," or boat stoppage place near the *Sangu* River, where traditional boats transport both people and things to the desired location. With so many linkages and ties between people and their rivers in river-based livelihoods, the *Sangu* River serves as the backbone of the entire region (Figs. 19 and 20).

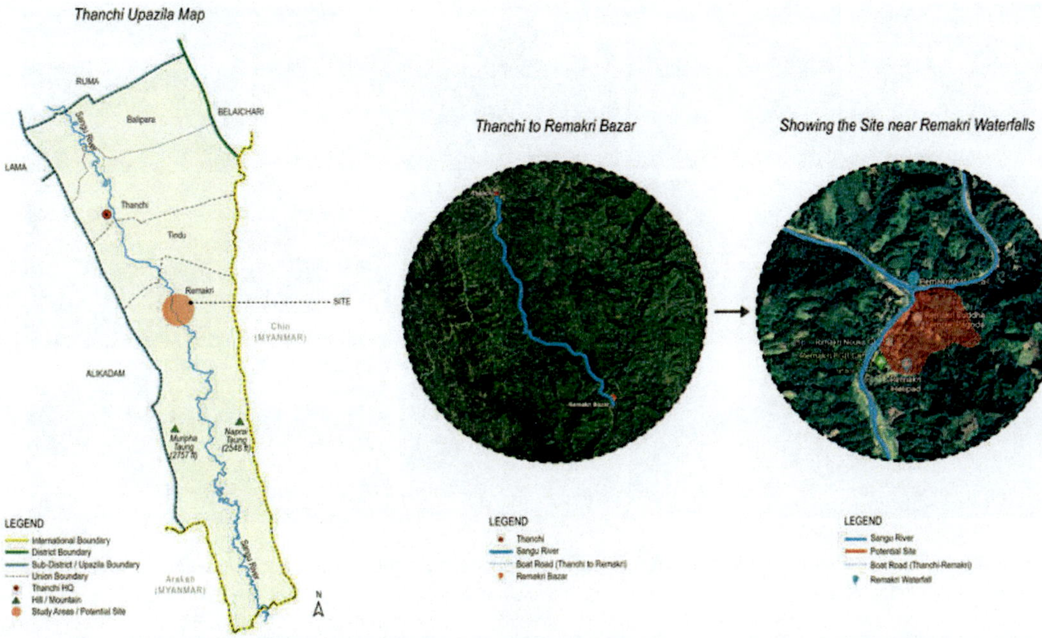

Fig. 17 Site access. *Source* Created by author

Fig. 18 *Remakri* area photos showing boat stoppage zone/*ghat*, buddhist temple, and bazar and gathering space. *Source* Author

7.4 Site Analysis

According to the building usage of the study area in *Remakri* most of the housing settlements are 1–2 storied. The area has some guest house for tourists, temporary structures for bazar both at the bottom and in the uphill, around 30-year-old Buddhist temple, BGB camp, and helipad. In terms of materials and construction methods, all building structures share a number of similarities.

Fig. 19 Site surrounding photos showing different significant spots and existing structures in *Remakri*. *Source* Author

Fig. 20 *Remakri* Waterfall, bazar, and boat stoppage area near the *Sangu* River. *Source* Author

The flood plains are basically near to the edge of the river, which rises up to 10–15 feet during rainy season. The average water level of the river is around 3–5 feet. This area is regularly affected by floods because of heavy rainfall, landslides, and riverbank erosion (Adnan 2019). Since the river controls the settlement patterns in hilly areas, circulation or movement mostly developed from rivers. The intersection point where the *Sangu* River divides into three directions serves as a junction for the local population of that region as well as for people from the surrounding Arakan State of Myanmar, which acts as a major factor (Figs. 21, 22, 23, 24, 25, and 26).

7.5 Existing Building Form Study

The settlement pattern in the *Remakri* area is determined by the existing building forms, particularly in hilly areas that follow waterways/ rivers or contour landscapes. Since there is a shortage of drinking water, the placement of the *JHIRI* or waterbody basically controls the settlement layout. The dwellings are arranged beside the *JHIRI* direction in an irregular manner. The hilly region's topography refers to mountain ranges that rise abruptly, where the river basins and hill ranges are longitudinally aligned, and the hill ranges have an average

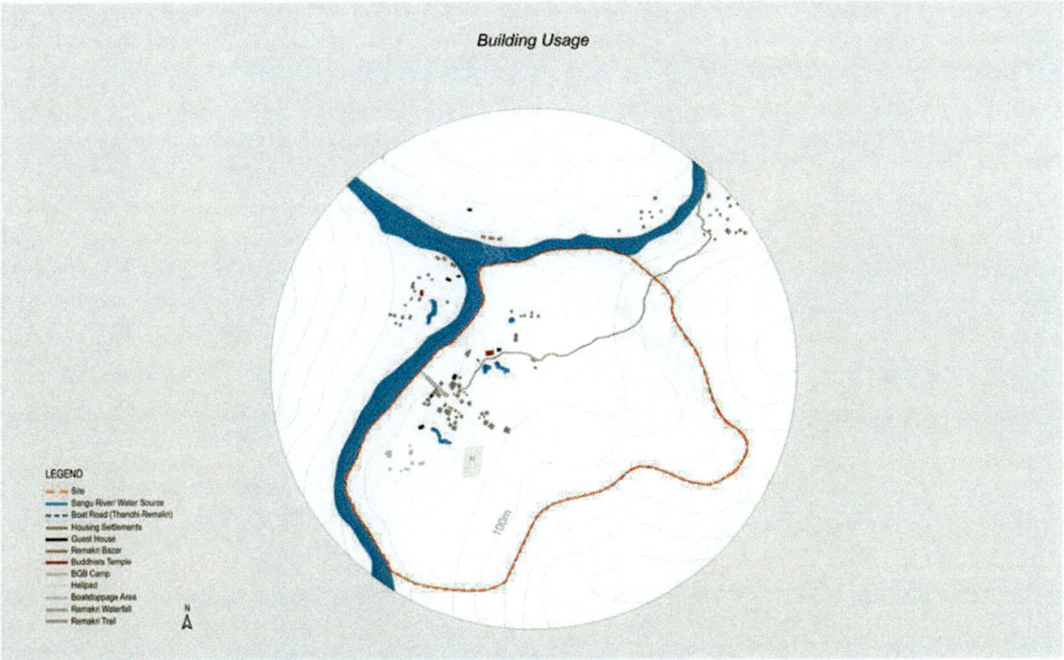

Fig. 21 Building usage at *Remakri*. *Source* Created by author

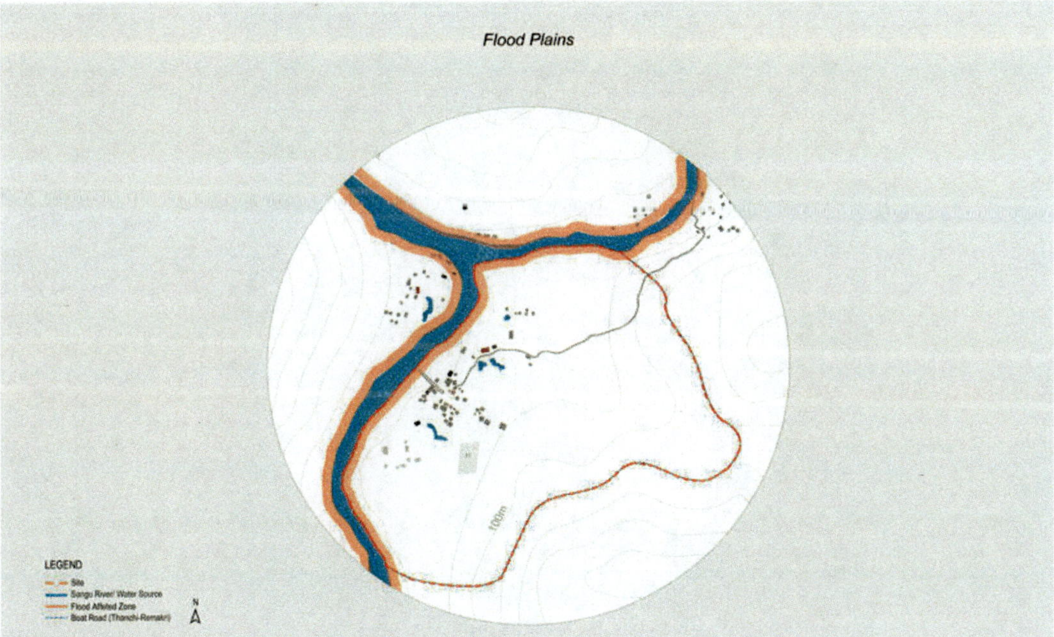

Fig. 22 Flood plains at *Remakri*. *Source* Created by author

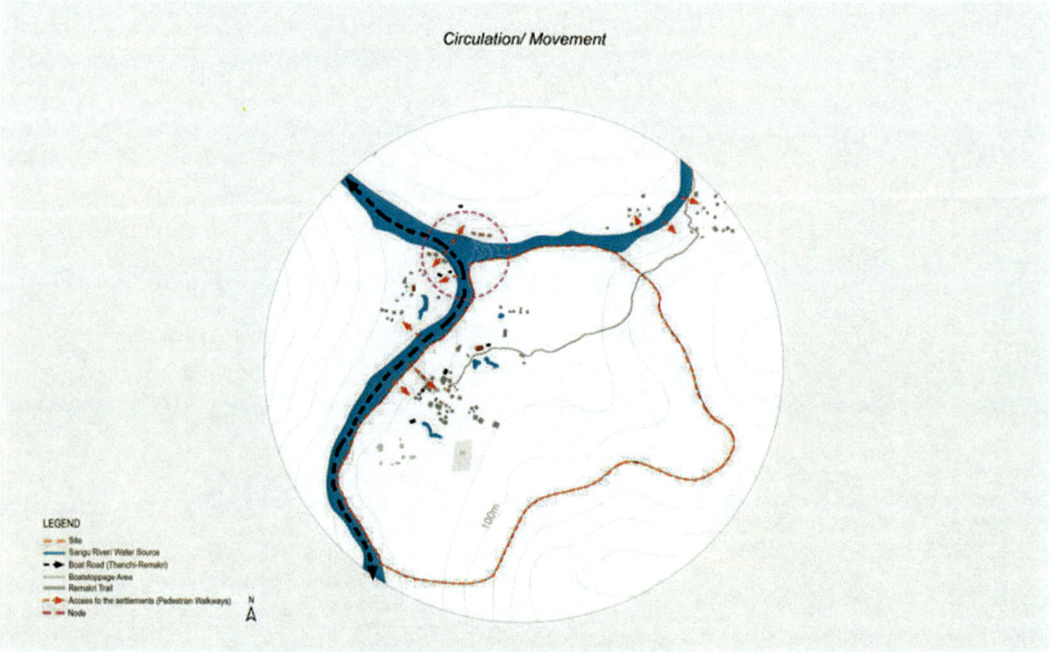

Fig. 23 Circulation/movement at *Remakri*. *Source* Created by author

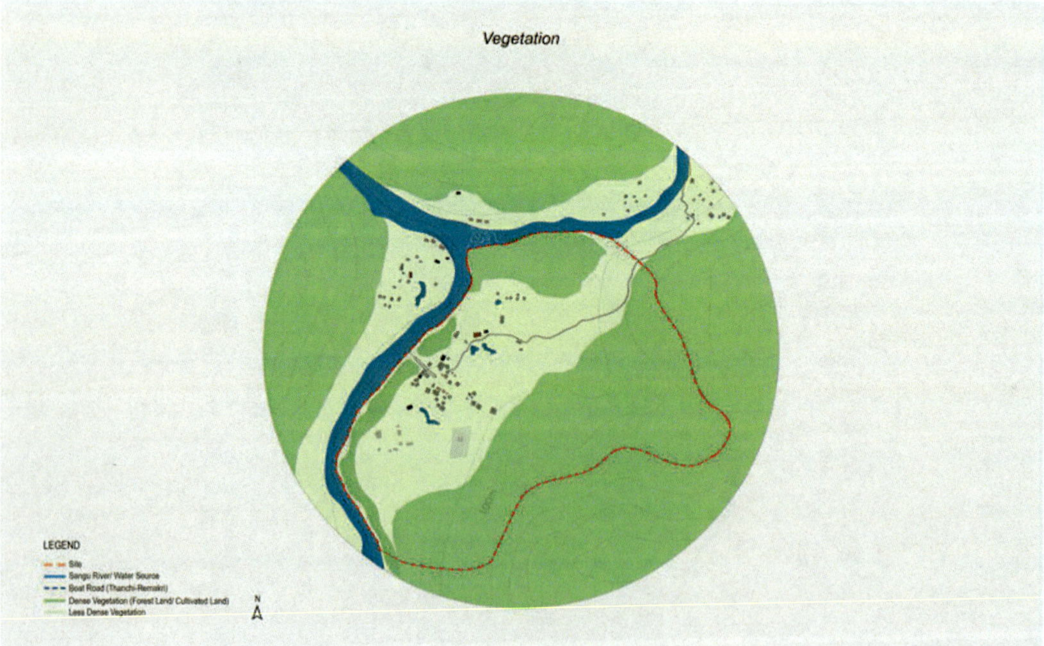

Fig. 24 Vegetation at *Remakri*. *Source* Created by author

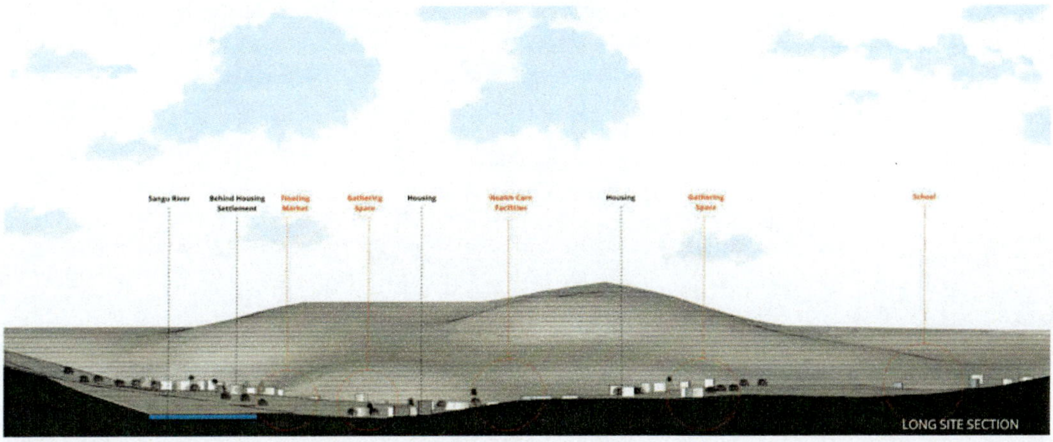

Fig. 25 Long section of the site. *Source* Created by author

elevation of more than 300 m. The building forms are generally in square shape, built on stilts, extrovert in nature, and living zones are too small to live in because of the limited space. In addition, houses are also constructed on a raised platform known as a *MACHANG*, which serves as the home's front courtyard. The elevated platform is also used to control flooding, get rid of wild animals, and as a storage area or gathering place. The materials that are used for the building are bamboo, wood, thatch, corrugated sheet, etc. (Figs. 27, 28, and 29).

Bamboo has been essential to the growth of businesses, building construction, and the diversification of the agricultural environment in the *Remakri* region. Since bamboo is frequently utilized for a variety of purposes, it will only remain 2–3 years underground if it isn't preserved. Houses on stilts are supported by bamboo posts. The strongest culms are those with tightly spaced joints and the biggest diameter. They are joined into columns if there are only smaller shafts available. In addition, the roof's frame is made of bamboo. The roof is covered with a variety of materials, including corrugated sheet, and culms cut in half to create bamboo tiles (Fig. 30).

Due to the typical warm and humid climate in *Remakri*, where bamboo is produced, it is necessary to utilize building materials with low thermal storage capacity and design structures in such a way as to effectively cross-ventilate. Ideal bamboo structures meet these criteria, which explains why they have long been adopted as housing in hot, humid settings (Humanity Development Library 2.0, no date). Some bamboo species' culms can be split apart and flattened to create floors. For houses on stilts, this particular kind of bamboo floor is elevated (1.5–2 m) above the ground to allow for the storage of equipment or animals beneath. Thinner culms are flattened for the floor, thicker culms are utilized as column supports, and woven mats are used as floor covering (Fig. 31).

This form of bamboo craftsmanship requires specialized skills, yet these have historically been available in most bamboo-growing locations. Certain patterns, styles, forms, and shapes used in this traditional technique are exclusively produced by locals, who depend on them for their livelihood. Moreover, these techniques are very hard to translate to the modern context. In order to improve their lives, many governmental and private organizations have promoted skilled craftsmen to manufacture goods for the market. As a result, locals are gaining benefits and are adopted new tools to increase their productivity.

Undoubtedly, bamboo is growing in popularity among contemporary designers as a

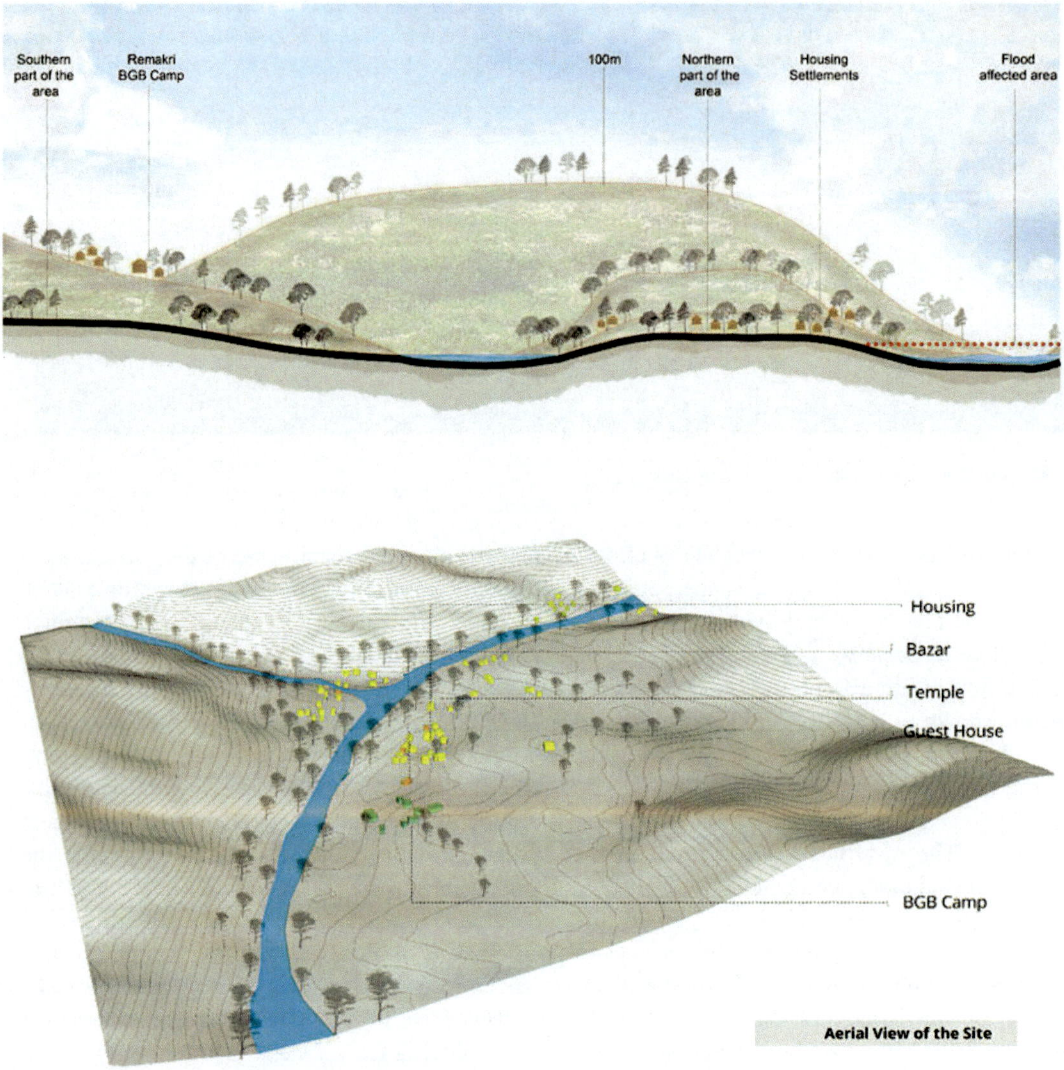

Fig. 26 Closeup section and aerial view at *Remakri*. *Source* Created by author

sustainable material and for its aesthetic appeal, cultural significance, and environmental advantages. Therefore, when working with local carpenters globally, it's important to learn about their skill set in order to design for them and produce construction documents that they can understand (Fig. 32).

There are three types of houses found in the *Remakri* area. Type 1 building form uses a combination of sustainably grown timber, bamboo, natural materials, and recycled materials to construct much of the framework. It contains a spacious living/communal space in the front with a *MACHANG* space which acts as a courtyard/ terrace, private rooms, and a kitchen and eating area with a backside *Machang* in the back. A split-roof and operable shutters allow natural ventilation to flow through the house, while the angled and overhanging roof litigate solar heat gain (Figs. 33, 34, and 35).

Fig. 27 Existing buildings at *Remakri* showing different shapes and forms, roads and pathways, and roofs. *Source* Author

Fig. 28 Existing building on stilts at *Remakri* (left); bamboo construction details (right). *Source* Created by author

Fig. 29 Beam and post details (left); roofing details (right). *Source* Created by author

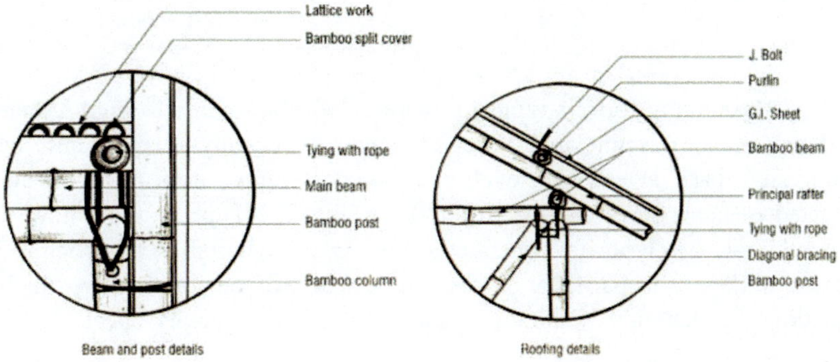

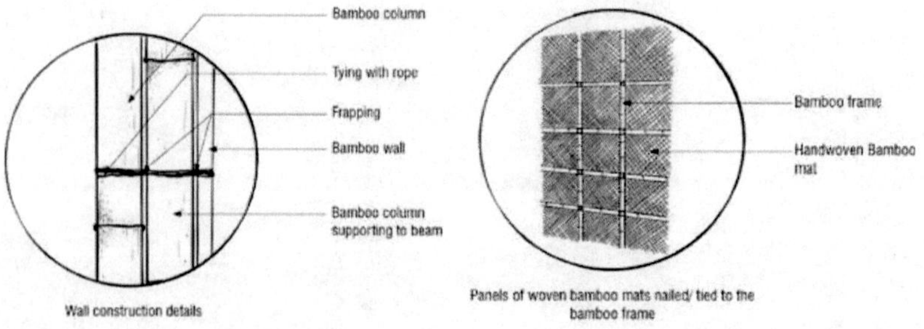

Fig. 30 Wall construction details (left); panels of woven bamboo mats/strips tied to the bamboo frame (right). *Source* Created by author

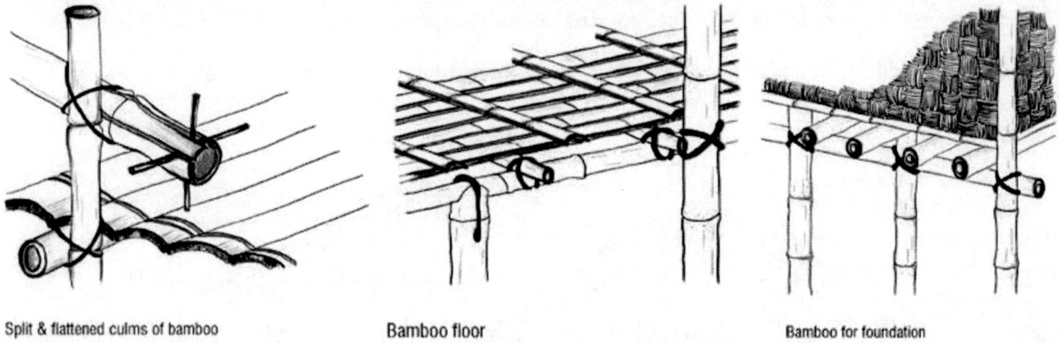

Fig. 31 Split and flattened culms of bamboo (left); bamboo floor (middle); bamboo for foundation (right). *Source* Created by author

Fig. 32 Handwoven bamboo wall (left); vertical halved bamboo culms (middle); whole bamboo culms (right). *Source* Created by author

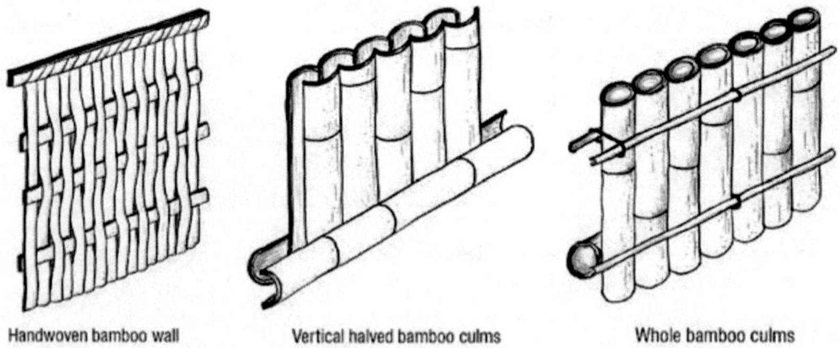

Type 2 building form is typically two-storied and square or rectangular in shape. It has a foundation base at ground level that is constructed of concrete, a durable material. On the ground floor, they typically construct a shop, and next to it, there is a space that serves a variety of functions, including storage and guest accommodations. A spacious living area that can occupy a large family or guests is located on the upper floor (Figs. 36 and 37).

Type 3 is a new structure that uses new materials. The structure is designed at the center of the community, which acts as a marketplace or community space. New materials such as brick,

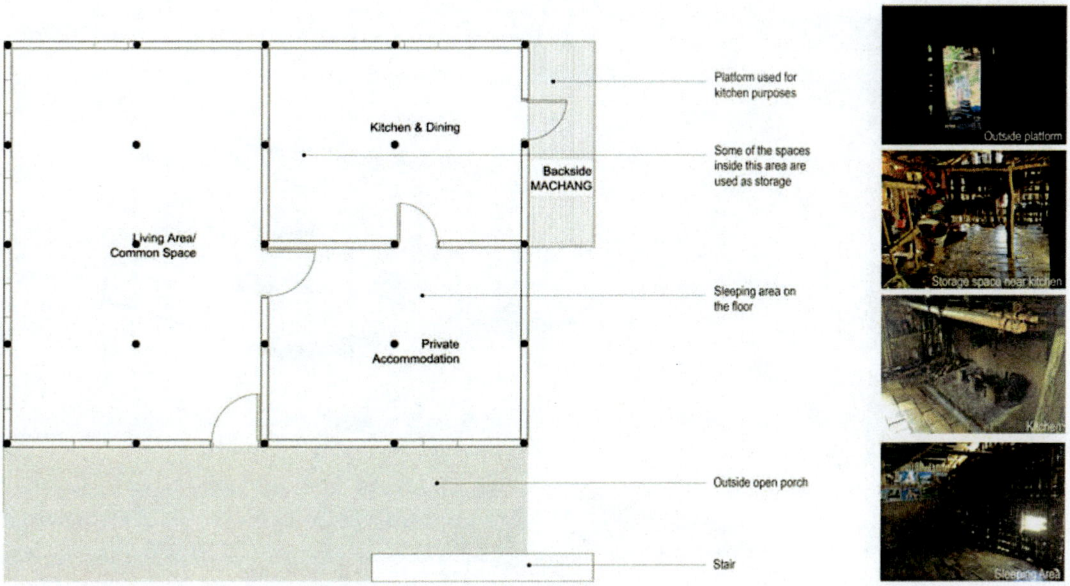

Fig. 33 Floor plan: building type 1; Drawn by author

concrete, steel, and corrugated sheets are used. In order to reduce the risk of flooding and promote natural ventilation, the construction is elevated above precast concrete pillars and designed for resilience (Fig. 38).

8 Inclusive of Politics

Working with local government is the key to supporting local communities. The creation of community infrastructure depends upon the local government, who is in a way responsible for society's growth and value. Moreover, working together with local people and local government will construct social settings in an integrated way. This also underscores the aspect of regenerative regionalism that helps to participate in local constellations of ideas, which describes the tectonic history of a place through the construction of integrated cultural and ecological processes. Basically, by definition regenerative system/regionalism means to provide continuous replacement through its own functional processes of the energy and materials used in its operation (Moore 1993). Moreover, it brings the force of the idea of fostering, converging human agreement through cultural identity that prefers a creation of critical and historically instructive places. In addition, regenerative regionalism gives importance to local labors and their participation in the building process while relying upon technologies of everyday life through democratic means that reveal the manner of their making to magnify local labor knowledge and local ecological conditions (Moore 1993). The idea of participation is emphasized in the case study—Umubano Primary School, Rawanda that is designed by Mass Design Group, where the architects not only worked with locals but also adapted their skills, knowledge, and techniques and incorporated it into the design. Therefore, the whole design reflects Kaigali region's vernacular, their own identity, and cultural pattern that significantly justify their belongings of that place.

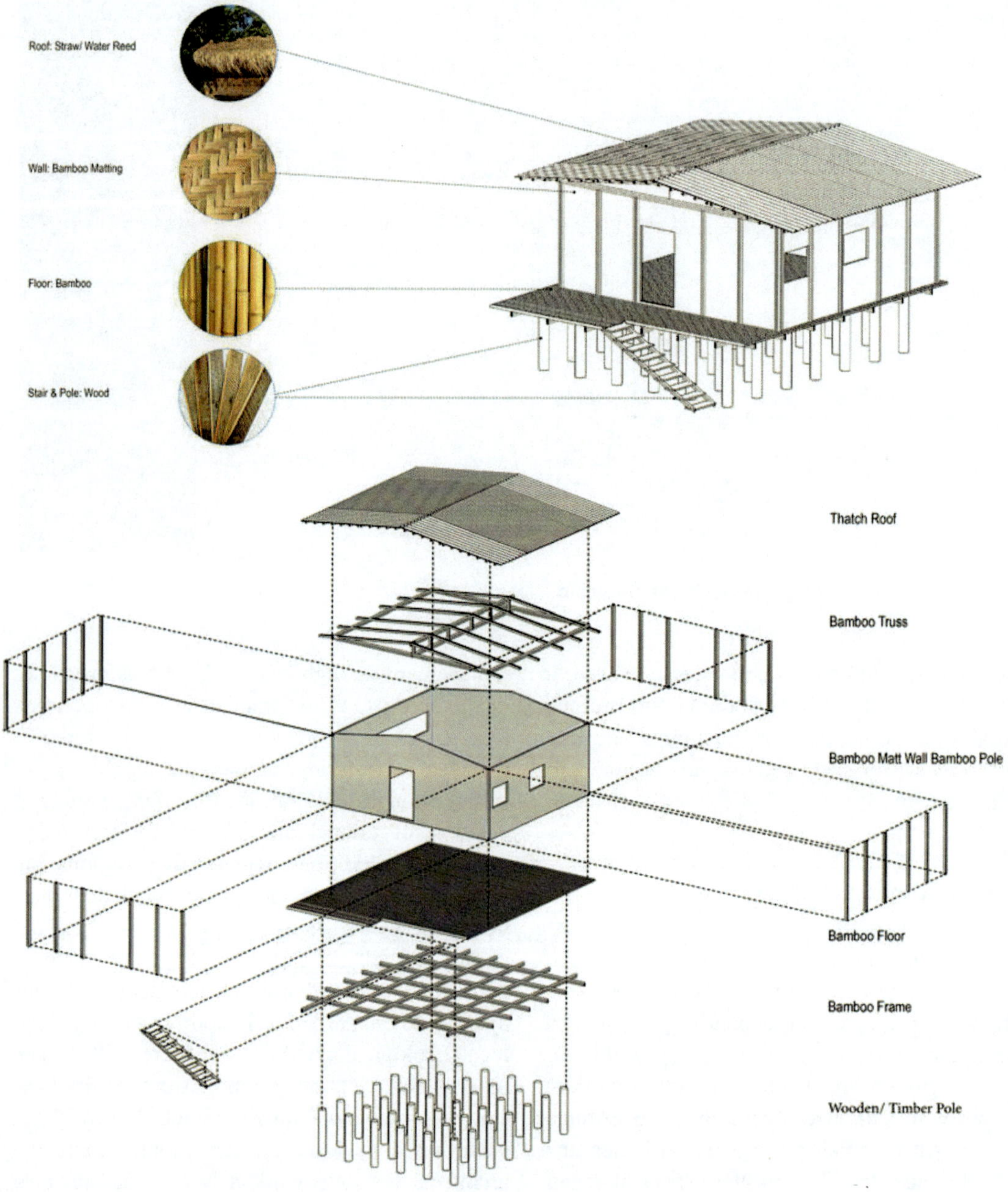

Fig. 34 Materials used in building type 1; Drawn by author.

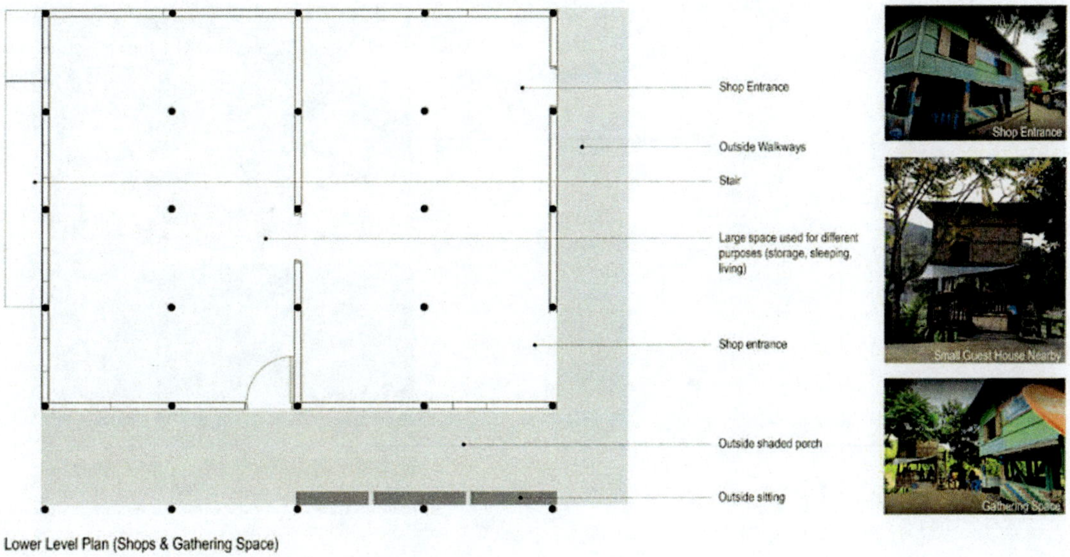

Fig. 35 Floor plan: building type 2; Drawn by author

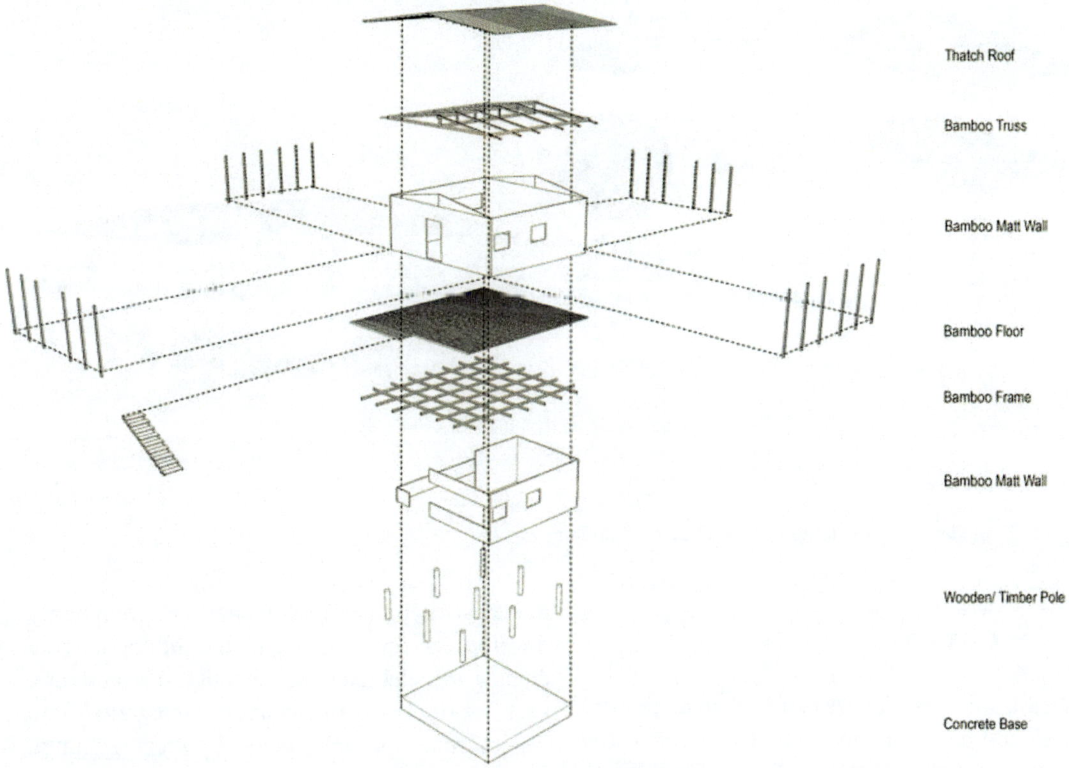

Fig. 36 Exploded axon: building type 2; Drawn by author

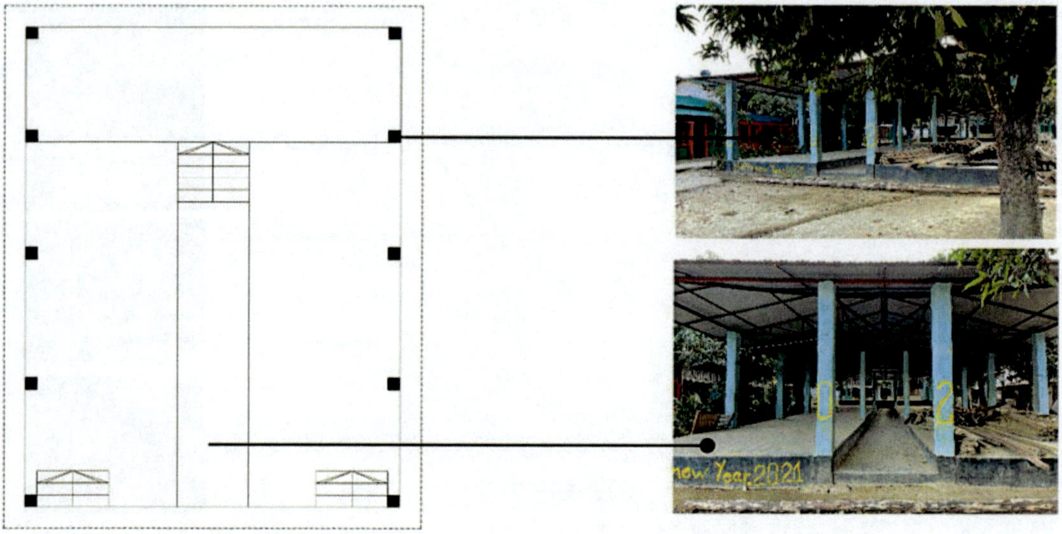

Fig. 37 Floor plan: building type 3; Drawn by author

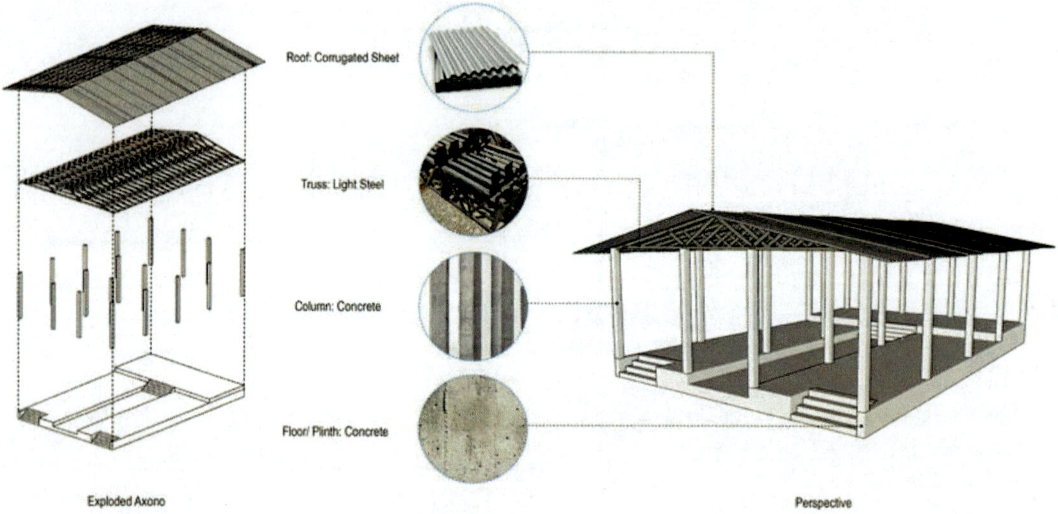

Fig. 38 Exploded axon: building type 3; Drawn by author

9 Conclusion

The case study emphasizes an important element that should be considered not only in vernacular architecture but also those elements needed to learn from the vernacular itself. By emphasizing the site context and including the community in the design process, it helps the community develop its own language and offers them a chance to celebrate their identity. Furthermore, each of these aspects can be integrated into design that not only helps to create cultural identity of a region but also learn from it. In Bangladesh, different regions have their own cultural identity, although they are all together

Bangladeshi culture. To understand this culture, tradition, their effect, and the result of globalization over the regional vernacular practices, a structure or design specific to Bangladeshi region's community should be introduced. Perhaps, the introduction should focus on overcoming generalizations about the region, its vernacular building, and most importantly the need for the contemporary to learn from vernacular itself and work with each other. Moreover, a language should be developed through the professionals' and architects' firsthand experience with the community by recognizing their wants and needs. Based on the community's need, a design solution could be proposed to help the struggling community to create an identity fitting to their existing predicament.

References

Adnan S (2019) LULC of Bandarban. Academia.edu. https://www.academia.edu/41183970/LULC_OF_BANDARBAN. Accessed 12 Jan 2023

Bandarban District (2021) 'Banglapedia: National Encyclopedia of Bangladesh, 18 June. https://en.banglapedia.org/index.php/Bandarban_District. Accessed 19 Aug 2022

Bandarban District Statistics 2011 (2013) Bangladesh Bureau of Statistic (BBS). Ministry of Planning, Government of the People's Republic of Bangladesh, Dhaka

Brown R, Maudlin D (2012) Concepts of vernacular architecture. In: The SAGE handbook of architectural theory, pp 340–355

Carter T, Cromley EC (2008) Invitation to vernacular architecture: a guide to the study of ordinary buildings and landscapes. University of Tennessee Press, Knoxville

Curtis WJ (1986) Towards an authentic regionalism. Mimar19:24–31

Eggener KL (2002) Placing resistance: a critique of critical regionalism. J Arch Educ 55(4):228–237. https://doi.org/10.1162/104648802753657932

Fatemi NM, Islam N (2011) Sustainability and eco-adaptability in vernacular housing in Bangladesh. In: International conference on society, technology & sustainable development. Kochi, India

Fernandes JEP, Mateus R, Bragança L (2014) The potential of vernacular materials to the sustainable building design, pp 623–629

Heath KW (2009) Vernacular architecture and regional design: cultural process and environmental response. Routledge, UK

Hossen MS, Hossain MS, Rahman MM (2019) Sangu River's contribution to the livelihood of local people. ResearchGate. In: River: a living being, 4th international water conference 2019. https://www.researchgate.net/profile/Mohammad-Hossain-144/publication/331715237_Sangu_River%27s_Contribution_to_the_Livelihood_of_local_People/links/5c890d22299bf14e7e799f01/Sangu-Rivers-Contribution-to-the-Livelihood-of-local-People.pdf. Accessed 12 Jan 2023

Humanity Development Library 2.0 (no date) Appropriate building materials: examples of wall materials: bamboo walls. Humanity Development Library 2.0 for sustainable development and basic human needs. https://www.nzdl.org/cgi-bin/library?e=d-00000-00—off-0hdl—00-0——0-10-0—0—0direct-10—4————0-1l–11-en-50—20-help—00-0-1-00-0-0-11-1-0utfZz-8-00&cl=CL1.1&d=HASH70c81f6386a2600bdfdd3f.8.7&x=1. Accessed 14 Jan 2023

Meinig DW, Jackson JB (1979) The interpretation of ordinary landscapes: geographical essays. Oxford University Press, New York

Menocal CG (2013) MASS Design Group: Umubano Primary School, Kigali, Rwanda, 08 May. https://www.designboom.com/architecture/mass-design-group-umubano-primary-school-kigali-rwanda/. Accessed 25 Mar 2020

Moore SA (1993) Technology, place, and nonmodern regionalism. In: Architectural regionalism: collected writings on place, identity, modernity, and tradition. Canizaro, pp 433–442

Mutuli I (2016) Umubano Primary School: a project after its own context in Kigali by Mass Design Group, 10 July. https://www.archute.com/umubano-primary-school-project-context-kigali-mass-design-group/. Accessed 25 Mar 2020

Panicker SK, Ostwald MJ (2002) Underlying ethos in Indian architecture: critical regionalism in the age of globalization

Thanchi Upazila (2021) Banglapedia: National Encyclopedia of Bangladesh, 18 June. https://en.banglapedia.org/index.php/Thanchi_Upazila. Accessed 20 Aug 2022

Tzonis A, Lefaivre L (2003) Critical regionalism: architecture and identity in a globalized world (architecture in focus). Prestel

Ancestral Earthen Construction Techniques Updated to the Needs of the People in the Central Andes of Peru, an Experience of Research and Training of Architecture Students Based on Community Service

Vladimir Simon Montoya Torres

Abstract

The objective of the research was to study and update ancestral construction techniques using raw earth as a fundamental material, these techniques were taken as a reference to different pre-Inca cultures that used this material as an alternative to construction needs, as is the case of the Chan Chan city 1200 to 1465 AD belonging to the Chimú culture and the architectural remains of the Huaca Puccllana 200 to 700 AD of the Lima culture. These cultures left us a legacy of construction techniques that serve to develop a learning and teaching methodology dedicated to architecture students from the city of Huancayo who, in the experimental earth construction laboratory, replicate these techniques applied in the towns of the Central Andes. From Peru. Peru, through participation based on community service, promoting SDG 11 based on sustainable cities and communities, as a result of the methodology being based on learning, practicing, and teaching these construction techniques to the communities of Palian and Cochas that they belong to the territory of the Mantaro Valley located at 3200 m above sea level. Finally, as a result of this academic and social process, local acceptance for the use of raw land as a constructive alternative will increase, it is within these communities that the most abundant resource is used and consequently construction costs are reduced and greater access is achieved. Houses by the neighbors were formed in this process of investigation and technological extension, finally these years of work achieved the development of several investigations related to the construction on land that helped the students to graduate as architects.

Keywords

Earth construction · Ancestral techniques · Sustainable communities · High Andean cities

1 Introduction

For centuries the Peruvian territory housed different archaeological remains that in their construction process used the earth as the predominant material, depending on the geographical context and climate, some constructions were located on the Peruvian coast and others in

V. S. M. Torres (✉)
Universidad Continental, Colegio de Arquitectos Junín, Junin, Huancayo, Peru
e-mail: vmontoya@continental.edu.pe

M. R. Thomsen et al. (eds.), *Design for Rethinking Resources*, Sustainable Development Goals Series,
https://doi.org/10.1007/978-3-031-36554-6_20

the Andes, however no record was found in the Peruvian Amazon, we consider that it is logical that in a rainy climate with high levels of humidity, it would be difficult to maintain earth-based constructions, this relationship between material and geographical context also has a direct relationship with the predominant construction system in the Inca culture in which many of its most representative constructions are built with carved stone, Machupichu located at 2430 m.a.s.l. and the city of Choquequirao at 3033 m.a.s.l. both located in the high jungle allow us to understand that a citadel built in stone is more suitable in tropical rainy contexts where the land would be vulnerable to climatic conditions.

On the other hand, the construction records built on land give us a totally different approach when located on the Peruvian coast, as is the case of the Pachacamac temple belonging to "the Lima culture that emerged in 100 B.C. and remains until 600 A.D." (Bouso 2012) this site is located south of the capital city of Lima and within a climate in which there are no rainy precipitations, this temple that belonged to the Lima culture and subsequent Wari expansion is tangible evidence that the criteria Ancestral constructions still endure and are sustained over time, the construction technique is that of stacking earthen adobes with clay plastering in the ceremonial temples, the rest of the architectural complex uses adobe without plastering, a similar development was generated in "the Moche culture in which constructions were made on the land of temples and pyramids that implemented polychromatic friezes and plasters, building a doctrine" (Morales 2007) this culture was located north of the city of Lima.

It is also important to compare that in the high Andean territories earth-based constructions were developed, as is the case of the Raqchi temple located in the department of Cuzco at 3550 m.a.s.l. this temple shows us a very different context from the coastal cities, this as a result of the fact that in the high Andean mountains there are rainy seasons and the particular case of the Raqchi temple is that it boasts of being the tallest construction on earth reaching 20 m in height, built with crossed adobes and on a stone wall as a foundation, this temple belonged to the Inca culture and was developed during the period of Inca expansion under the governments of the Incas Huiracocha, Pachacútec, and Túpac Yupanqui during the years of 1450 to 1532 A.D.

All this historical context represents the theoretical and motivational basis to involve young students in research and constructive experimentation based on land, visiting these archaeological remains centers and visualizing their magnitude and their perseverance over time are guidelines that drive and motivate students. Students to get involved in this research, the material resources for the construction of earth-based temples and sanctuaries are the same as we now have, they are mostly the use of raw earth, this means that the construction material is accessible, low cost and can be replicated, the conditions of use changed like the current times, however, the needs of roof and home by vulnerable groups still continues to be a problem in my country, that is why several generations of architecture students joined this experimental workshop on land "The recovery of ancestral knowledge guarantees its transcendence and can affect the development of sustainable housing" (Guerrero 2018) and managed to get involved with the proposal of construction alternatives inspired by ancestral techniques of handling raw earth for construction purposes, all based on an education methodology for service, in which students collaborate in the construction processes together with the residents and high Andean communities.

2 Materials and Methods

For the development of this research, the methodology based on education for service was selected, this was determined by the content of the subjects in which conventional construction processes were put into practice, this was achieved thanks to the fact that the academic curriculum in the architecture school proposed an experimental workshop under construction in raw earth that initiated the first academic impact on architecture students, the subject was elective, however, a complete section was enrolled and

Photograph 1 Community work with the students of the construction course in the community of Cochas, district of El Tamb

with this generation in 2016 the first intervention began within of geographical contexts in the Andean mountain range, two interventions were carried out within the Mantaro Valley, it is important to point out that this methodology "Allows to verify the effects produced in different dimensions related to personal-student and social competences: interest, autonomy, self-evaluation, connection theory–practice, reflection and satisfaction of students" (Mayor 2018) (Photograph 1).

The first was carried out in the district of San Agustín de Cajas in the Hatun Cajas neighborhood and the second intervention was carried out in the district of Rio Molinos, province of Jauja, in the first intervention called "Pachacútec 1.0 Workshop: Tradition, culture and landscape in Hatun Cajas (Montoya 2017). A socio-academic intervention was developed in collaboration with the population of the Hatun Cajas neighborhood, this peasant community is a fragment of architectural and cultural nuances, on the one hand it is located on the slopes of the hills of the sierra where pre-Columbian archaeological remains called "El Obraje" are found, additionally it has a constructive miscegenation, since most of the houses are built with earth-based materials but in a central patio design, a heritage of Spanish vernacular architecture adapted to the conditions of the central Andes, in this context the students developed a set of proposals based on the spatial and constructive needs of the inhabitants, the main difficulty was identified because many of the inhabitants end up demolishing their old houses built on raw earth to build new buildings built with brick, reinforced concrete, this seriously affected to the landscape context, as a result of these actions the inhabitants disfigure the urban image of their context with the insertion of industrialized materials, which break the constructive harmony that was identified in the place, additionally these new constructions have a deficient behavior with respect to thermal conditioning in relation to the climate and geographical context (Photograph 2).

This first contact with the population of the community determined a great learning for the students who became involved in a reality very different from that of the city, these peasant communities are vulnerable, because their economy is still based on agricultural production and some artisan activities Talking with the users and understanding their real habitability and shelter needs was surely one of the best lessons learned. The recognition of raw earth as a predominant, ecological and economic construction material was a first step to understand this material in relation to the training of young students of architecture, whose reasoning changed when interacting with the inhabitants of these communities, finally this first work was based on the fulfillment of SDG 11 of sustainable cities and communities, because in community work-family groups were prioritized in the situation

Photograph 2 Hatun Cajas community where the first community meeting and service-based learning developed by architecture students took place

of vulnerability due to the precariousness of their spaces has bitables.

The second intervention was developed within the framework of the academic work developed in the subject of environmental conditioning, this intervention was carried out in the communities of Río Molino, District of Pomacancha, the provinces of Jauja, this intervention based on service and learning had as its objective the to analyze the thermal comfort of high Andean dwellings in one of the districts where, due to its location and height above sea level, temperatures drop rapidly in the autumn months, taking into account that in the southern hemisphere autumn is a sunny month and without the presence of rain, but due to the geographical conditions of the place, low temperatures are reached at night due to the presence of frosty nights that decrease the temperature below −10 °C, the problem in this context is that the house built with raw earth based on adobes, it does not have insulation systems and they have too many empty spaces that they generate thermal leakage, for which the students led by their teacher generate a proposal for large panels built from empty PET plastic bottles which will be attached to the ceilings to avoid thermal leakage, this task represents an arduous work in collecting of recycled bottles for the construction of the panel, in a second moment the transfer and installation of these panels in the houses identified within the intervention zone, this task is in conjunction with the residents of the community.

This joint work proposal of students and residents of the community of Rio Molinos achieved the result that was defined in "two well-defined stages: The current stage, whose objective is to develop and implement a thermal insulation system that increases by 5 or 6 °C the temperature inside the high Andean dwellings and the second stage, which consists of massifying the system by bringing "THERMAL COMFORT" to a regional scale, where the aim is to replicate what has been learned in other communities, intervening as much as possible of households" (Moncloa 2017) (Graphic 1).

After the two interventions described, the school of architecture focuses its efforts on reinforcing this methodology based on education for service and promotes practical activity in revaluing ancestral techniques for construction based on raw earth, this decision is supported by the criterion of using the most abundant

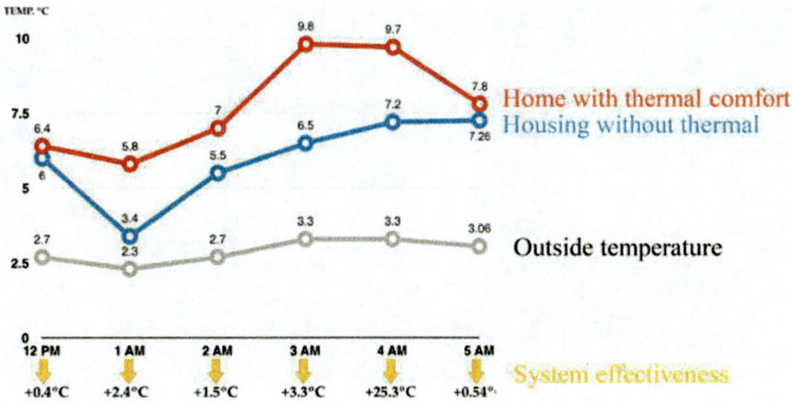

Graphic 1 Summary of the thermal gain at 3.3 °C in the time interval of 3 AM graphic from the publication Thermal Comfort: An insulation system for Alto Andina housing made with recycled materials (Moncloa 2017)

resources in the high Andean communities and avoid the consumption of sources of heat, electricity, and artisanal or industrial combustion for the manufacture of materials, this in relation to the location of the construction contexts and the costs that in most cases, the communities do not have enough money to spend on energy resources, for which construction techniques based on the use of raw land are put into practice. These techniques were classified as follows.

3 Construction of Molded Earth Bricks

The use of clay soil in the community of Chochas was an alternative to curb the growing activity of clay exploitation for the informal manufacture of baked bricks, which generates environmental pollution by burning harmful products such as the use of tires for cooking. the bricks for more than a week of firing, this merited that we visit this community and that we use the natural resource of earth, clay, and natural fibers to put into practice the construction of small adobe bricks such as those used in the construction of the Huaca Puccllana in the technique of booksellers (Mendoza 2015), for our context we refer to the current technical standards such as the case of the E-080 standard for construction on Earth of the National Construction Regulations of the Ministry of Construction, Housing, and Sanitation (RNC 2016), a technical standard that provides us with general guidelines, but that students need to experiment with, is thanks to the academic study of the school of architecture that an agreement is reached with brick manufacturers to access their workshops and develop there our practices of earth molding in small adobe formats, which are more suitable for the light construction process (Photograph 3).

4 Construction with Packed Earth

The construction of circular structures in which packed earth modules were used allowed us to experiment with a different technique in which we use easily accessible materials and supplies, such as the case of polymer bags that, after being packed and arranged in their alignment, are compacted. Conserving a slight level of humidity that allows its molding and compacting, we put this technique into practice in the rural community of Palian as it is a place in which clay soil abounds and the labor link between the students and the neighbors of the community was generated. peasant community who collaborated with the extraction of the land from the same construction site, this agreement was strengthened by finishing these structures and handing them over to the residents to be used as seed storage spaces, by guaranteeing dry spaces during the rainy months. The work of building the packaged earth modules required a transfer of more circular, small entrances and a coating that allows insulation in the rainy months (Photograph 4).

Photograph 3 Adobe manufacturing process after a process of treading and molding we share the experience of the community members with the students

Photograph 4 Construction process of circular modules based on packed earth developed in the peasant community of Palian—Huancayo

5 Earth Walls on Wooden Structures

The mixed earth walls on wooden structures are a technique that was used in the Spanish colony in other contexts it is known as Quincha, or Bahareque, in Peru there is a record of this technique only applied on the Peruvian coast, there are no historical records of this technique replicated in the mountains or in high Andean cities, this absence of constructive practice is attributed to the fact that in the Andes there are rainy seasons during the summer, we have to remember that in the southern hemisphere, the seasons have a different behavior and in the summer season the high Andean climates present their greatest rainy presence between the months of December to March, it is one of the reasons why there is no evidence of this technique in our context, this does not mean that it is not viable, one of the reasons goals of the experimental workshop on land was to test the feasibility and construction of these walls to subject them to impermeability and resistance tests. Resistance to rainy climates in real conditions. Additionally, the architecture students proposed that the walls could have texture and color as a reagent to measure the level of erosion by rain or wind, for which the students proposed to build full-scale modules and rehearse in the designs based on the iconography of pre-Inca cultures and that have historical representation, as shown in the photographs (Photograph 5).

6 Compacted Earth Walls

Another technique was the use of compacted raw earth in a mold and this construction process allowed experimenting with constructive and

Photograph 5 Earthen walls on wooden structures covered with earth plus polychromatic designs on earth in reference to pre-Columbian iconography

design possibilities, for which different colors of natural earth were applied, which when combined allowed the result of natural skins built with earth. raw, this formal possibility was the product of a process of trials and errors that were developed within the experimental laboratory on land, this was taken in reference to vernacular construction techniques on land that used compacted earth inside mobile and tamped wooden boxes with a mallet that compacted the earth by impact in a semi-humid state, it was a very innovative contribution that students can propose color designs by combining the different possibilities of natural color of the earth, in some cases it was combined with ferrous oxides to improve pigmentation, and the result was a block of compressed and pigmented earth, this technique It is proposed as an alternative to the use of coatings on earthen walls. In many communities, after completing the walls, they have to be plastered with plaster or cement and sand mortar. These final finishes are expensive due to the use of materials that are difficult to access in high Andean areas. From Peru, additionally these plastering materials affect the earthen walls because by sealing and covering the pores of the adobe walls they accumulate humidity and prevent the breathing and drying of these walls in months when the environmental humidity increases (Photographs 6 and 7).

Photograph 6 Compacted and pigmented earth tests inside modules of 15 cm on each side

Photograph 7 Student-developed full-size and conventionally proportioned tests for rammed-earth wall construction

7 Discussion and Conclusions

Finally, the last 6 years allowed us to achieve a repository of research and finished products based on experimentation in the construction process based on land. The results of these interventions were reflected in seven high Andean communities that now know the value of architecture based on community service, in ten thesis investigations presented during these 6 years of work and finally in more than one hundred and eighty students who participated in these workshops and continue to apply their knowledge in real contexts within Peru and outside the country.

The investigations carried out allowed us to have a better context on the needs of the communities, and being precarious populations, the research proposals focused on those needs, as is the case of the need for housing, in the case of the investigation entitled "BTC impermeability in rainy climates in the Cochas Grande annex" developed by a graduate of the architecture school who presents us with this research proposal in the places where the workshops were held before (Mallma Espinal 2017), her research brings us closer to the use of local land for the construction of compressed earth block as a masonry unit, reducing construction costs, but more importantly the possibility that these respond to rainy weather conditions without being eroded.

Complementarily, another graduate of the school developed an investigation related to the construction of houses in areas of sloping land, something very common in the Andes due to the hills of the mountain ranges, for which the author developed an investigation entitled "The blocks of compressed earth (BTC) and its influence on the cost of construction of social housing in development areas on the slopes of the city of Huancayo—2018" (Montes Galarza 2018), in this research the student proposes to provide the constructive technical possibilities for these houses in conditions of land on slopes and specifically analyzes how much the construction cost is when using this material, which benefits users by using a cheaper material, this is directly related to SDG 11 on Sustainable Cities and Communities.

Another investigation was also developed in which the issue of color in earth constructions is addressed, for which the author analyzed the constructive possibilities based on compacted earth and from her previous experiences in the experimental laboratory on earth, she developed the following investigation entitled "Use of ferrous oxides as a pigment in exposed mud walls and its influence on the cost of work for the construction of rural houses in the Cotay Annex, district of Huancavelica 2017" (Ventosilla Cruz 2018), its proposal is developed in a community. It is far from the Mantaro Valley because her city of origin is located in the department of Huancavelica, and it is the reason why she replicates the techniques learned and takes them to another context of high Andean cities, as in previous cases a total of a dozen investigations related to vulnerable contexts and the use of up-to-date techniques in earthen construction, that it was the most ambitious goal to apply this methodology based on community service. Students from then to now architects already have a human, social, and environmental approach to the architecture they propose, they have the capacity and knowledge to replicate these techniques in the geographical contexts of whoever they have to work, it is our greatest achievement to have been able to share these experiences with the residents who now have a different way of seeing an architect and his field of work.

References

Bouso JL (2012) La Construcción en el Antiguo Perú: Un país enigmático aún sin descubrir, Primera edn. Editorial Académica Española, Lima

Guerrero LF (2018) Arquitectura de tierra en América Latina, Una obra colectiva. Arquitectura en Tierra: Tecnologia sostenible y reutilizacion patrimonial, XIV CIATTI 2017. Congr Int Arquit Tierra, Tradic Innovación (1):35–44

Mallma Espinal PJ (2017) Impermeabilidad del BTC en climas lluviosos en el anexo de Cochas Grande. Universidad Continental, Huancayo

Mayor D (2018) Aprendizaje-Servicio: una práctica educativa innovadora que promueve el desarrollo de competencias del estudiantado universitario. Rev Electrónica "Actualidades Investig Educ" 18(1):1–22

Mendoza A (2015) La puesta en valor de la Huaca Pucllana y su. Facultad de Ciencias de la Comunicación, Turismo y Psicología Universidad de San Martin de Porres, Lima

Moncloa C (2017) Confort Térmico: Un sistema aislante para la vivienda alto andina fabricado con materiales reciclados. MODULO Arquit CUC 18(1):73–90

Montes Galarza JA (2018) Los bloques de tierra comprimida (BTC) y su influencia en el costo de construcción de viviendas sociales en zonas de habilitación en laderas de la ciudad de Huancayo – 2018. Universidad Continental, Huancayo

Montoya V (2017) Taller Pachacútec 1.0: Tradición, cultura y paisaje en Hatun Cajas. MODULO Arquit CUC 18(1):147–162

Morales R (2007) Arquitectura prehispánica de tierra: conservación y uso social en las Huacas de Moche, Perú. Apuntes 20(2)

RNC (2016) Norma E.080 diseño y construcción con tierra reforzada. Ministerio de Construccion y Saneamiento, Lima

Ventosilla Cruz NK (2018) Uso de óxidos ferrosos como pigmento en muros de tapial expuesto y su influencia en el costo de obra para la construcción de viviendas rurales del Anexo de Cotay, distrito de Huancavelica 2018. Universidad Continental, Huancayo

Plektonik—Active Yarns for Adaptive Loop-Based Material Systems

Daniel Suárez and Natalija Miodragović

Abstract

This research investigates contemporary textile techniques and plant-based "active" yarns to describe a new material system and fabricate lightweight structural prototypes. We define "active yarns" as bio-based structural yarns, filaments that store elastic energy, and flexible stems from short rotation cop-pice (SRC) plantation species. SRCs are woody fast-growing tree species often cultivated to produce high biomass yields in a short period. This research fo-cuses on Willow (Salix spp.), the most spread SRC in Europe. Due to their fil-ament structure, SRC plantations produce rod-like flexible stems with finite lengths. To overcome length limitations, we industrially produced a cord-like yarn with bundled filaments and materials with different properties. Wood-bending techniques can shape these active yarns into warp-knit loops and later be assembled into larger structures. This lightweight construction method lev-erages sustainability thanks to the combination of a fast renewable source of material like SRCs with efficient use of raw material resources granted by the ability of knitting to produce the form with almost no waste. Moreover, this material assembly method of rigid loops allows for re-assembly. As a result, the material would recover energy as part of the biomass production chain.

Keywords

Renewable resources · Short rotation coppice · Circular economy · Willow · Structural textiles · Spatial warp-knitting

1 Introduction

Textile techniques involve immense knowledge implemented through centuries. An alternative route to reach the architectural scale with textile materials is to upscale their structures (Beyer et al. 2019; Sauer 2019). The research aims to systematically look into contemporary textile techniques, yarns, and biological structures to develop active structural textile prototypes. This research emerged from interdisciplinary work from the architecture and textile design (Singer 2020) fields and through the dialogue with microbiologists and material scientists. It started in 2020 with the development of the Active Curtain Project exhibited at HU Lab. It evolved in 2021 for the exhibition Stretching Materialities at Tieranatomisches Theater, both in Berlin.

D. Suárez · N. Miodragović (✉)
"Matters of Activity" Image Space Material, Humboldt Universität Zu Berlin, Berlin, Germany
e-mail: natalija.miodragovic@hu-berlin.de

M. R. Thomsen et al. (eds.), *Design for Rethinking Resources*, Sustainable Development Goals Series,
https://doi.org/10.1007/978-3-031-36554-6_21

Using willow filaments is gaining momentum in architectural production. Previous research has also promoted its use as a sustainable alternative material. In particular, researchers at TU Kassel have developed a method to produce a solid endless wood filament made from willow twigs suitable for use within a robotic fabrication framework (Eversmann et al. 2021). Furthermore, in the Inter Twig project (2022), students and researchers from the Karlsruhe Institute of Technology have recently explored the potential of a weaving technique to inform a digital manufacturing process using willow sticks.

We see its application on the architectural scale as a stance in the circular economy. Regular uses of Short-Rotation Coppices (SRCs) are as biomass, its plantation as a regeneration driver to remediate soil (e.g. from former coal mining lands) (Castaño-Díaz et al. 2018), also an emergent use in a wide array of practical applications to restore damaged ecosystems (Kuzovkina and Quigley 2005) or willow's bark uses in medicine. Willow shows the capacity for guided growth, and we can envision the pre-formation on the plantation or construction site. In this sense, the standardisation of forest rest is an answer to growing concerns for the future of our planet and human-made rest in a post-waste society.

SRCs are fast-growing plantations in high density (willow and poplar). They show required mechanical and thermal properties for application in reinforcement, are biodegradable and low-cost (Oktaee et al. 2017). High-yield biomass production, high height growth through short rotation and high planting density, sustained vegetative growth, resilience to pests and diseases, and excellent wound healing after biomass harvesting are some of SRC's most notable characteristics. Furthermore, long-term usage of SRC also helps conserve groundwater, minimise soil erosion, and improve soil quality and biodiversity. Besides that, short rotation willow coppice (SRWC) can be established in soils unsuitable for agricultural exploitation (Castaño-Díaz et al. 2018).

Regarding sustainability, SRCs plantation removes CO_2 from the atmosphere in growth cycles of 4–6 years before being used as biomass. Due to this short and closed cycle, biomass from SRC is renewable and helps to protect our climate. However, this energy source is not entirely "CO_2-neutral" because of fossil energy sources used for harvest or transport (Dimitriou and Rutz 2015). Therefore, its use as an architectural material would improve the CO_2 footprint before being combusted as biomass.

Due to their filament structure, willow twigs and rattan liana are rod-like flexible stems, rare, bendable and conformable materials (Szczepanowska 2018). Moreover, willows are fast-growing and more malleable than timber wood. The material thickness determines the bending radius, which can be reduced with wetting. The obtained loop is stabilised by drying on pre-forming boards. The loop diameter (bending radius) is related to yarn thickness. The bending radius is reduced through wetting, and the filament "remembers" the form when dried. Hence, the wood is plasticised.

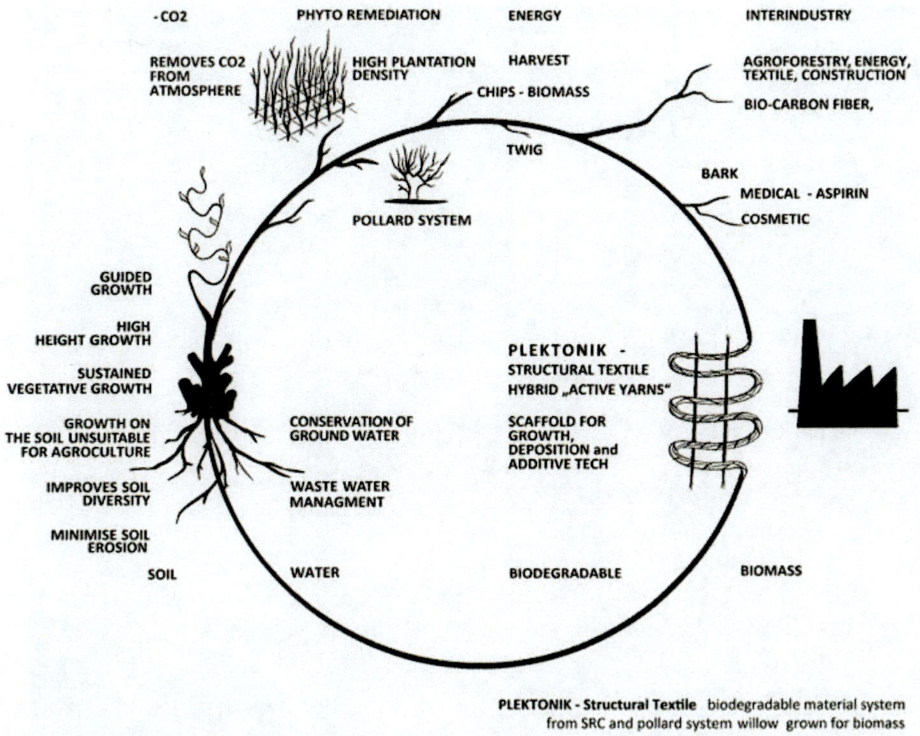

Fig. 1 Beyond the regular uses of SRCs as biomass, we see an application for this biodegradable material system on the architectural scale as a stance in the "circular economy". While the willow bark is used for some applications, the peeled twigs can be used as alternative "yarns". By starting research (and production) from the plantation, the project has the potential to obtain holistic knowledge about all phases of the material life cycle. Different growth information, such as mechanical properties, can be obtained by expanding the use of SRC from just biomass to design. Since willow is very suitable for guided growth, the pre-formed twigs could be grown using moulds, like in the *Full Grown* project (https://fullgrown.co.uk/). The research aims to contribute to the certification of wood filaments for construction and, thus, toward the post-waste society © Autors Plektonik Miodragovic, Singer, Suarez

Fig. 2 (*top*) SRC plantation, willow (Salix spp.) fields in Cañamera, Cuenca, Spain (© Mimbres Corredor), (*middle*) Harvesting a 4-year old willow plantation (© Lignovis_GmbH) and (bottom) bare white willow twigs harvested ready for use (authors)

Fig. 3 In the exhibition After Nature at Humboldt Forum in Berlin, the Active Curtain project (https://www.matters-of-activity.de/en/research/laboratories/49/humboldt-lab) showcases research in bacterial cellulose. It explores material activity between construction and emergence. We present a series of interdependent structures, similar to other case studies introduced in previous research practice in textile material systems (Ramsgaard Thomsen et al. 2012). In these interconnected systems, each part of the structure is intrinsically weak, and the structural whole is only stiffened and strengthened by the specified system of assembly and interlocking loops. The curtain model is digitally informed about overseeing dimensions and distributing voids where to display exhibits; however, its construction grew organically from the hands of experienced crafters. Oversized knitting loops form tubular paths. As in previous contemporary research (Harding et al. 2015), our method combines innovative materials, crafting knowledge, physical testing and digital design tools to generate a novel form of architectural structure. As in the case of the Active Curtain project, it renders a geometry that becomes a fibrous scaffold to display ongoing research and experimentation. Architectural/Design Team: Prof. Christiane Sauer, Natalija Miodragovic, Bastian Beyer PhD, Iva Resetar, Nelli Singer and Daniel Suarez. Science Team: Prof. Dr. Regine Hengge, Skander Hathroubi PhD © M. Mantel MoA

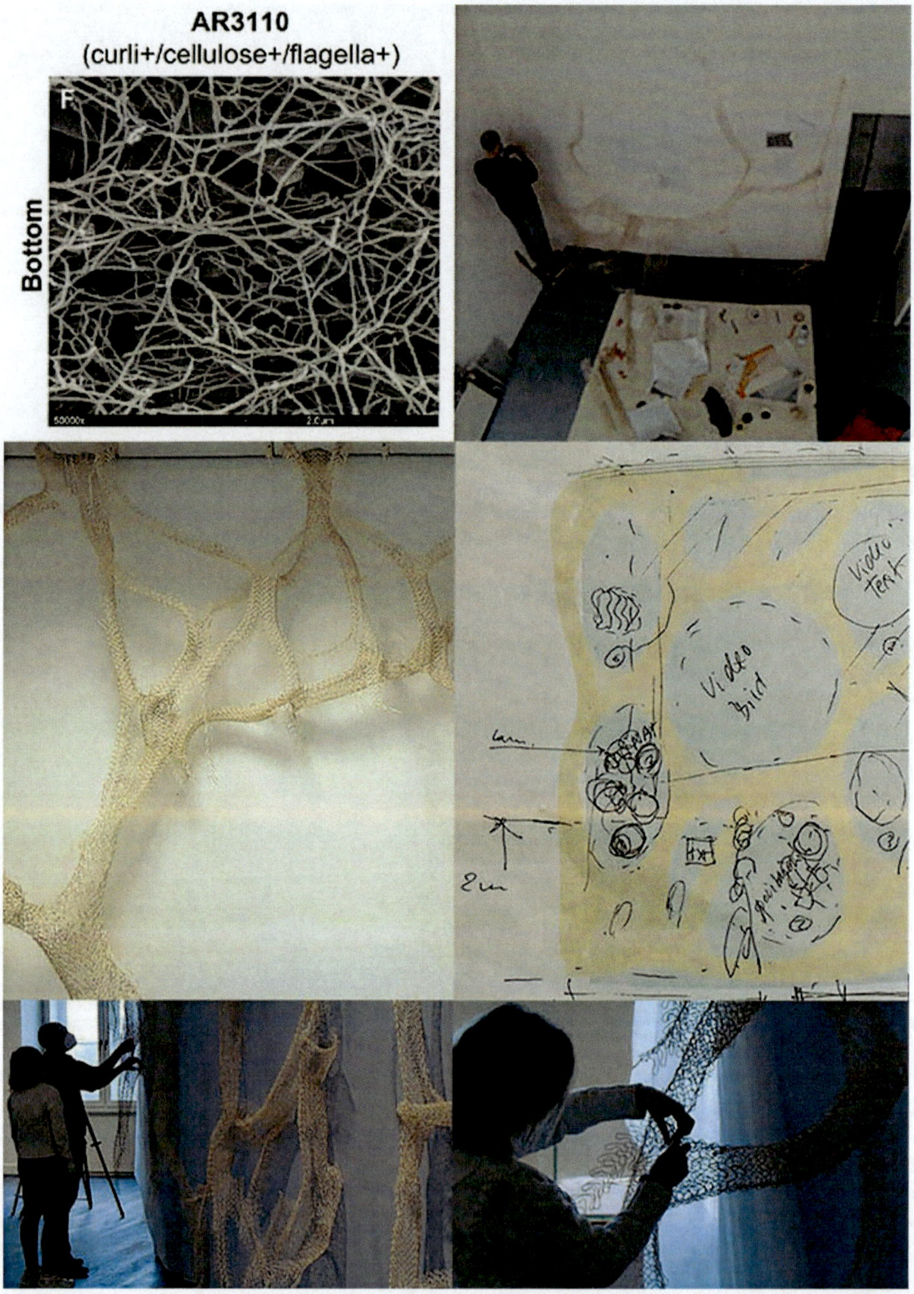

Fig. 4 Active Curtain project design and crafting process. This social knitting mimics biofilm formation and collective filamentation of bacterial population *(top left)* (Reproduced from Serra et al. 2013). From a drafted 1:1 model to the construction in place. *(top right, middle, bottom)* The final curtain design is a balance between aesthetic appeal and structural stability. Furthermore, the research is not only interdisciplinary but combines different knowledge cultures. It combines the tacit knowledge of material growth with wood-bending crafts such as basket or canoe making

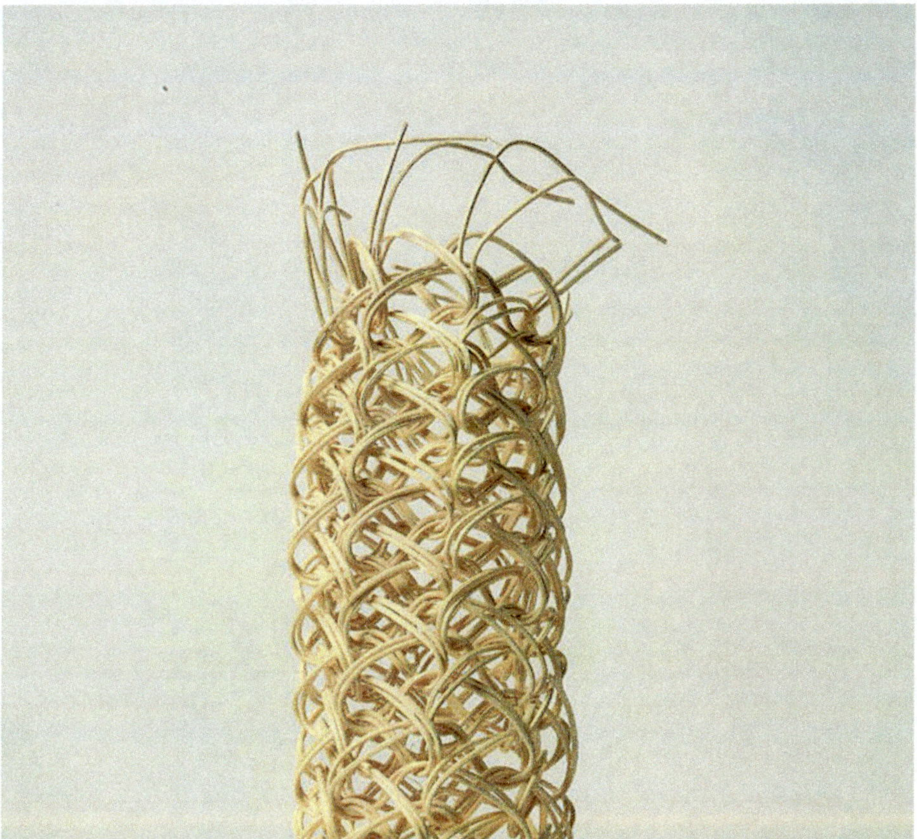

Fig. 5 Tubular warp assembly with two 2 mm rattan strands. Sample height approx. 400 mm

Fig. 6 The project aims to develop wooden-based textile-informed semi-finished products for architecture and construction. We introduced a new material system based on textile assembly methods with wooden threads. The material system can replicate complex textile structures such as spacer fabrics, hence providing higher strength and stability. Spacer fabrics are sandwich-like structures formed by two slabs of fabric and a third layer that connects them vertically (Bruer 2005). These interesting structures are suitable for innovative lightweight structural applications. Depending on the density and the type of filament chosen, the resulting structure offers a diverse range of compressive strength. *(top)* The inner layer made with flat willow stripes presents a better cushioning effect, whereas 4 mm willow withies *(bottom)* make the structure able to support small compressive loads

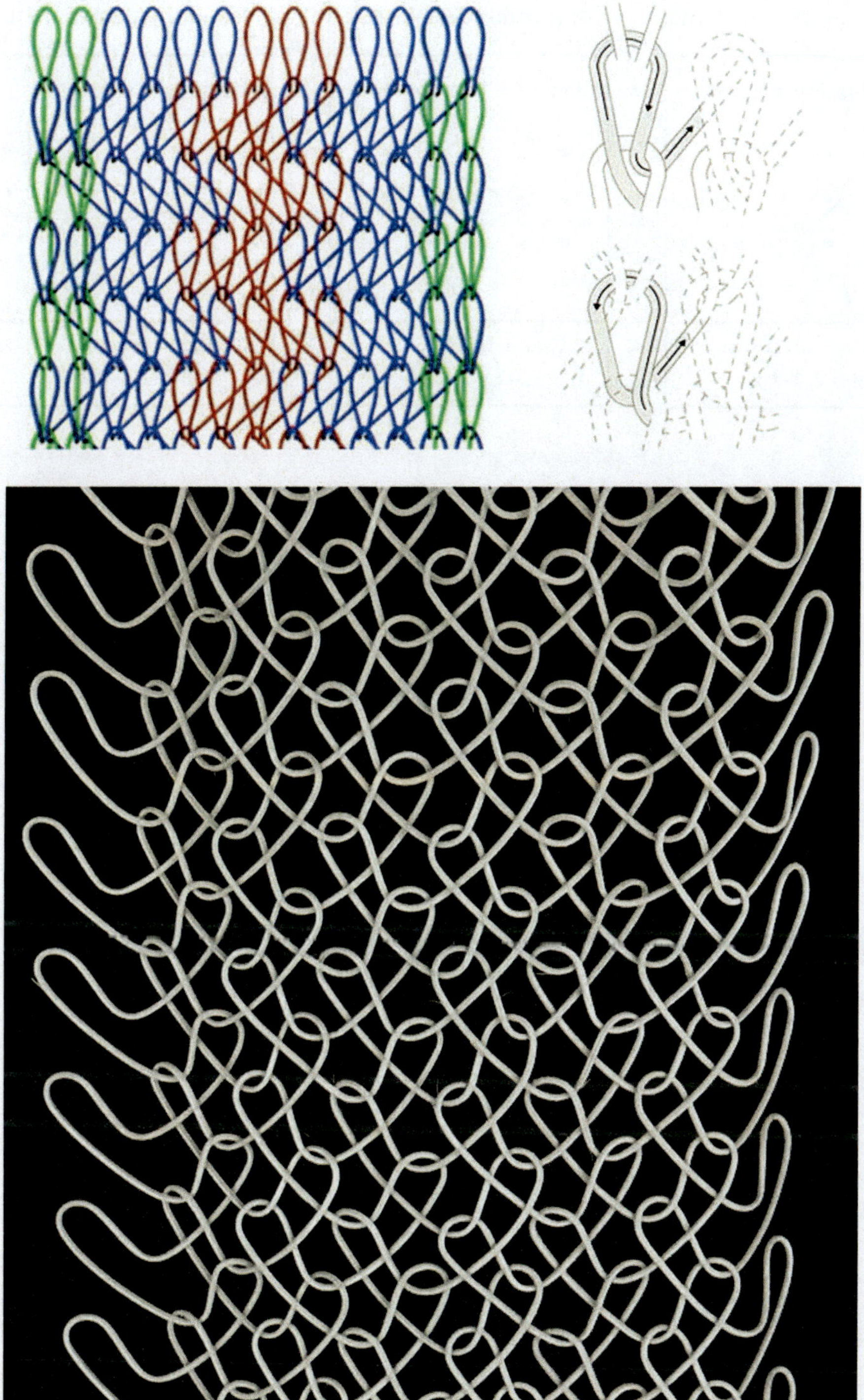

Fig. 7 *(top left)* In the warp knitting technique, loops are interlocked vertically. (Texmind.com, Kyosev 2019). Attending to the loop topology, we can differentiate between open *(top right)* and closed lap *(middle right)* (Reproduced from Spencer 2001). *(bottom)* We translated the open lap to work with resilient fibres, rattan and willow

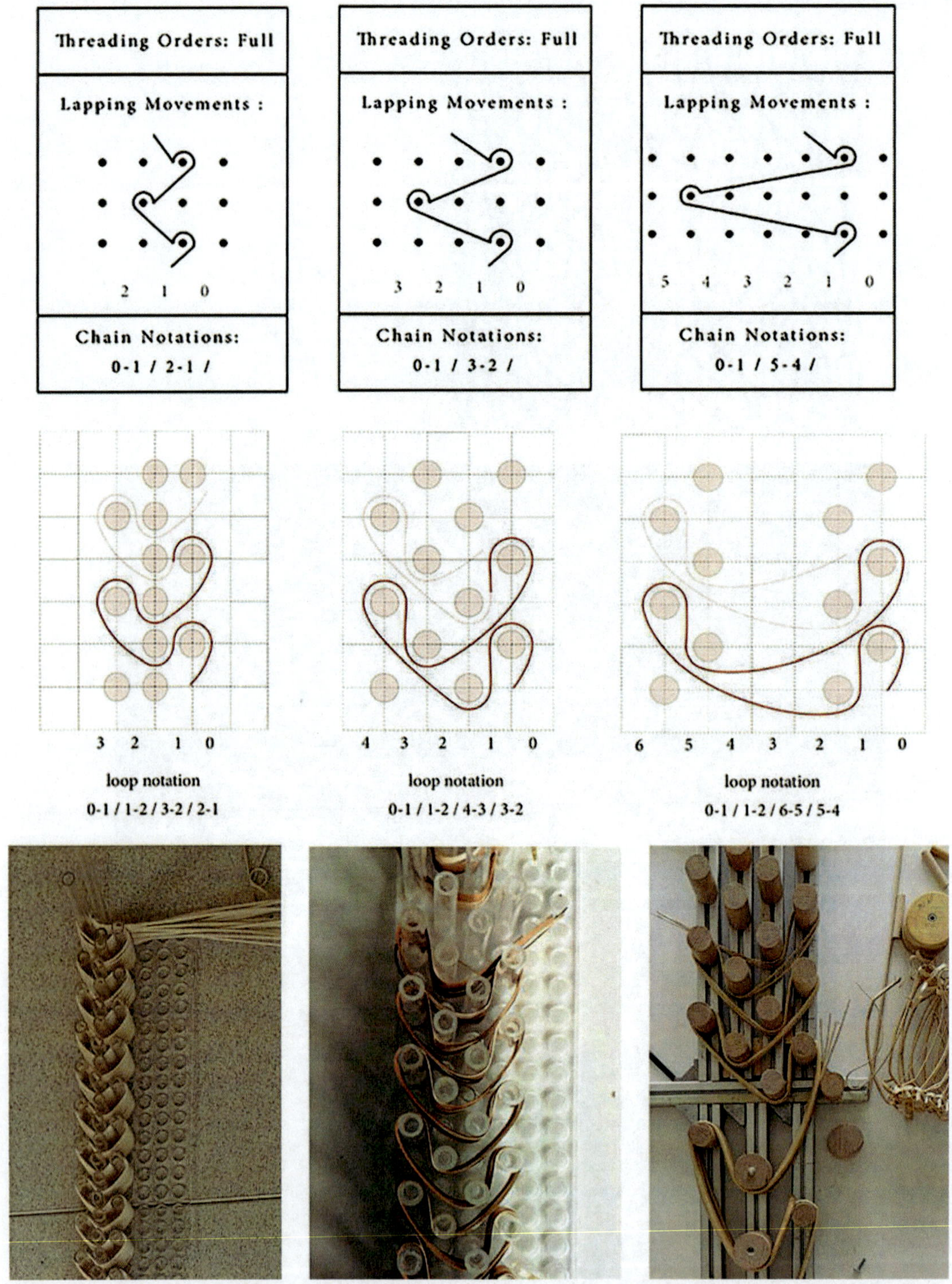

Fig. 8 Knitting instructions of open-lap single bar warp knitted fabrics: **b** 1 and 1 open-lap; **c** 2 and 1 open-lap; **e** 4 and 1 open-lap.(Reproduced from Kurbak 2019) *(top)*. The previous instructions were translated into our notation system for pre-forming boards developed as a graphic script in Grasshopper for Rhino3D *(middle)*. In this project, many different types and board sizes have been developed. The looping in textile techniques is a negotiation between material elasticity and the technique itself. This filament property provides specific stability to the textile itself *(bottom)*

Fig. 9 The project translates the industrial warp-knitting techniques with resilient wood-based filaments for application in construction and design. The size and spacing of the loops can be varied to determine the strength and stability of the structure © Lara Ladik MoA

Fig. 10 The rattan is pre-formed in a wetted state, in different "frequencies". Knit density defines the branching tree-like structure. A graded density of loops is distributed along the height of the curtain. Fabrics with the highest density, hence stiffer, no. 1, occupied the higher part, then density decreased gradually to medium, no. 3, and finally the no. 5 on the lower part, in a total height of five meters

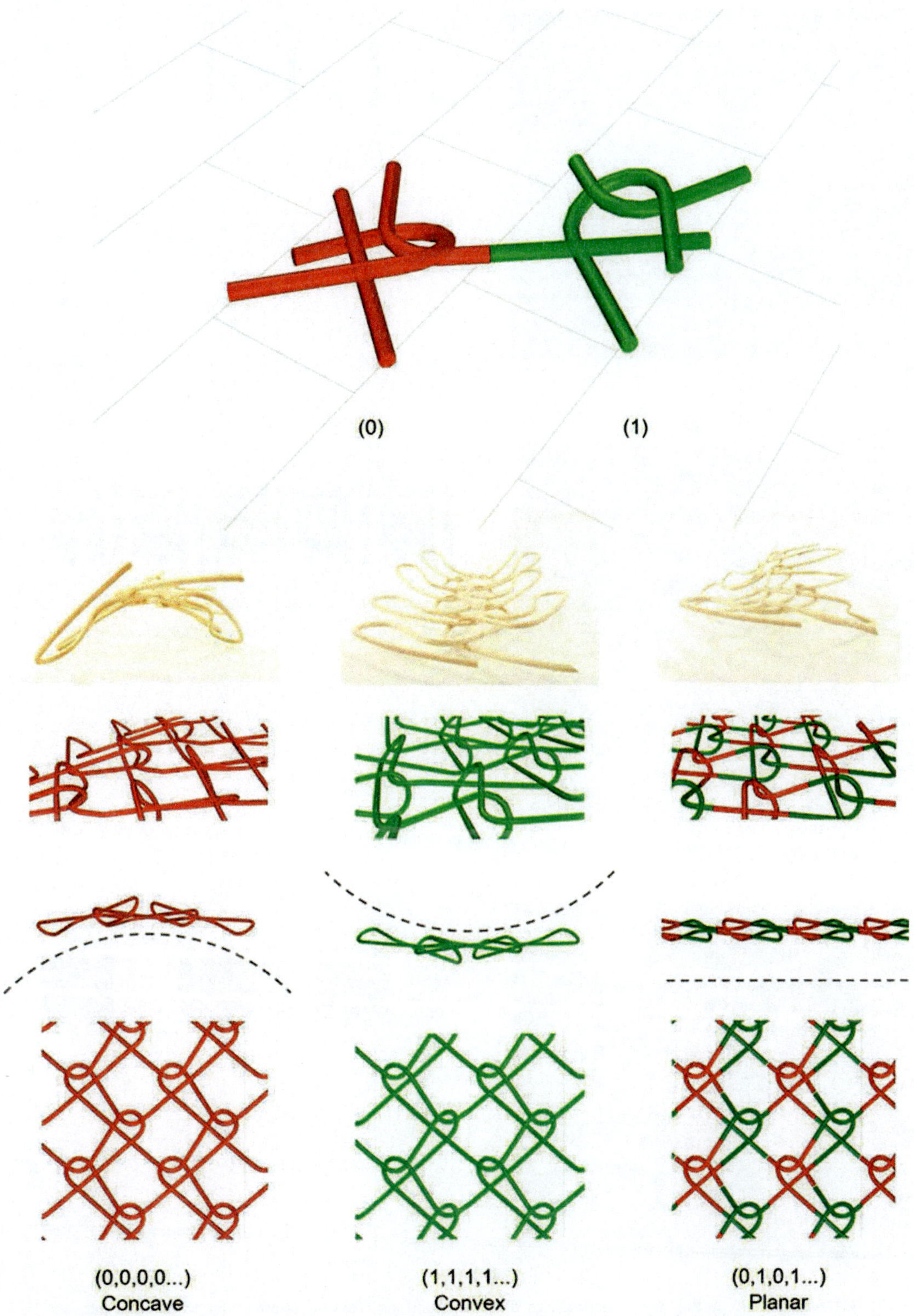

Fig. 11 Wood and non-wood warp-knit structures lend themselves well to doubly-curved forms. The surfaces positive or negative curvature is achieved through a simple hand gesture when making the textile structure. Depending on the direction in which one loop interlocks with its neighbour, i.e. from above or underneath, the loop will hinge in one order, leading to a convex or concave surface. That can be efficiently coded using a binary code, therefore programming the surface into a particular form. 3D modelled with Grasshopper for Rhino3D

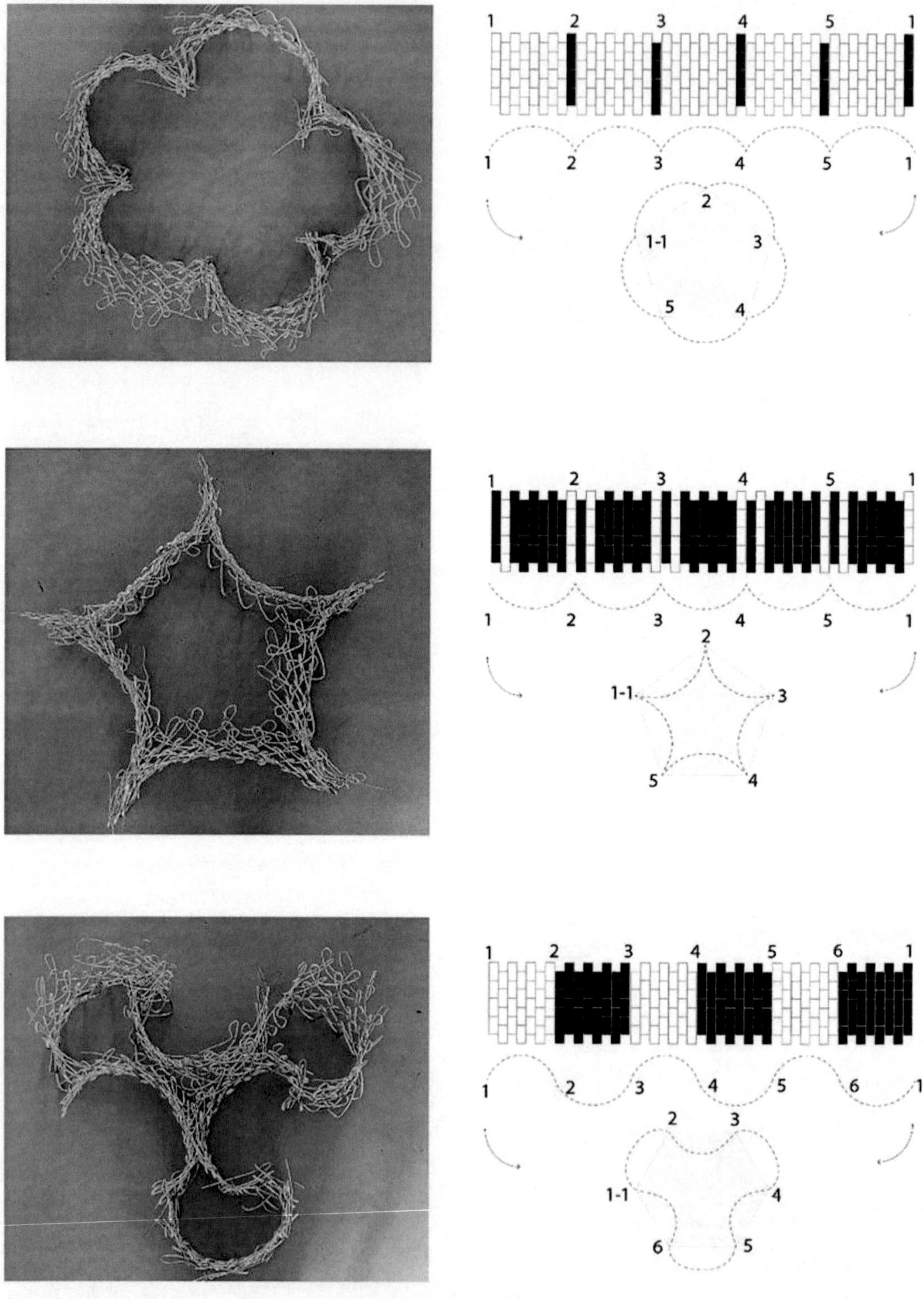

Fig. 12 Different self-supporting fabric geometries are achieved by alternating loops, like tubular or bi-directional surfaces *(left)*. The interlocking (i.e. knitting) of the discrete loops creates lightweight and stiff structures that are shaped into complex curved forms and can be assembled easily using only modular components without needing specialised equipment. *(right)* Assembly instructions. Textile patterns offer an exciting perspective on how architectural representation can function differently from traditional geometric measurement while illustrating the logic of assembly and the structural integrity of the resulting fabric (Ramsgaard Thomsen et al. 2012)

Fig. 13 Bending drier wood with the help of heat and water is centuries old (Wright et al. 2013). Steamed wood is less rigid since adding moisture and heat to wood results in plasticisation. During our explorations, we designed a collection of apparatus to aid in wetting, steaming and bending thin wooden rods-section. We achieved bends with the tightest thickness/radius ratio of almost 1:2 without a compression strap, however, the most frequently used ratio was 1:4

Fig. 14 The finite wood fibre filament is extended through the vertical interlock of multiple pre-formed filaments. This process defines a new crafting glue-less construction technique where the loop is observed as a building block

Fig. 15 Willow wood threads consist of individual sections cut tangentially from shrub willows, joined at the front. Their cross-sections in the millimetre range are adjusted differently through width and thickness processing. Willow twigs have finite lengths. We work with core-sheath yarn structures bundled with Kemafil® (KEM) technology to overcome this limitation. Six threads are bundled with the help of a horizontal KEM machine at Saxon Textile Research Institute e.V. STFI (Arnold et al. 1994)

Fig. 16 The hierarchical structures from nature can be translated with textile techniques. In particular, cross sections of both plant and wood fibre filaments informed a manufacturing method for continuous wooden-based yarns using Kemafil® (KEM) technology. KEM yarns are core-sheath cord-like structures, hybrid yarns engineered to fulfil different performance and functional requirements (Gong 2011). Hence by programming the yarn, the coding complexity expands from geometrical to more complex behaviours and applications. As Vincent noted, there is a considerable potential to obtain new or unusual combinations of material functions/properties by structuring a given material rather than changing its chemical composition' (Vincent 2008). In fact, textile fibre assemblies can readily provide an ideal test bed for this concept

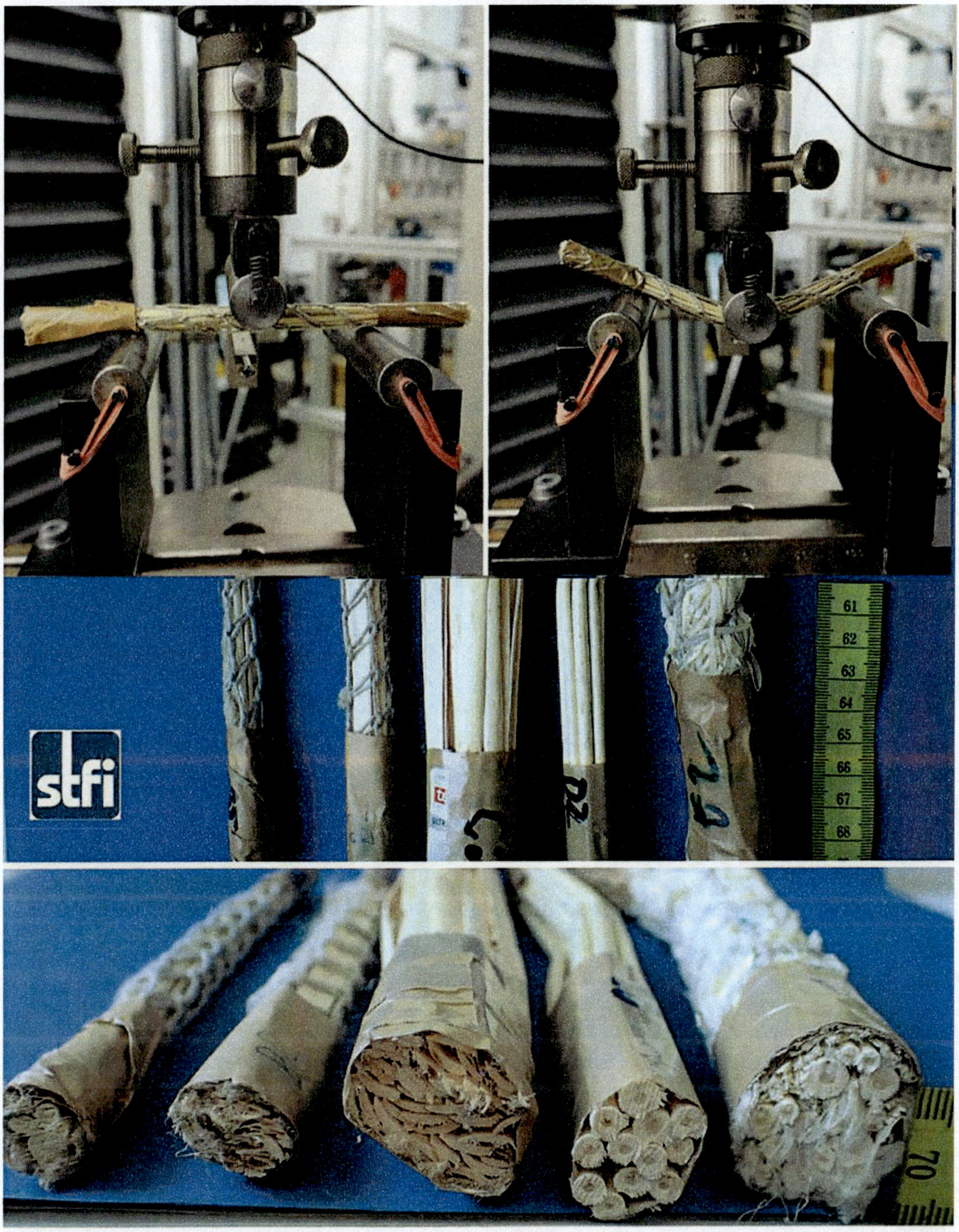

Tabelle 2: Biegerversuch an Stäben i.A. an ISO 3597-2, n=...; σEε

Bezeichnung	Feinheit (tex)	Höchstbruchkraft (N)	σ_F (MPa)	E_f (MPa)	σ_{x1} (MPa)	σ_{x3} (MPa)	σ_{x5} (MPa)	ε_M (%)	D_0 (mm)	A_0 (mm²)
01/A	28.355	56,7 ± 18,4	20,8 ± 2,15	553 ± 179	5,15 ± 1,82	14,8 ± 3,89	19,3 ± 2,46	6,6 ± 1,6	8,621 ± 1,091	59,19 ± 15,12
02/B	33.564	28,9 ± 3,74	7,69 ± 1,64	110 ± 26,8	1,79 ± 0,5525	4,26 ± 0,735	5,98 ±1,13	11,0 ± 2,5	9,896 ± 0,7759	77,31 ± 11,99
03/C	122.124	175,4 ± 11,98	4,53 ± 0,232	47,8 ± 14,6	0,526 ± 0,236	1,66 ± 0,486	2,5 ± 0,664	20,4 ± 4,7	21,43 ± 0,3348	360,90 ± 11,26
04/D	65.472	295,8 ± 130,29	21,4 ±1,87	417 ±189	3,92 ±1,82	11,3 ±4,5	15,5 ±5,88	12,1 ± 5,0	15,04 ± 2,644	181,20 ± 63,97
05/E	110.452	237,8 ± 77,96	6,85 ±1,99	60,0 ±19,5	0,235 ±0,452	2,41 ±0,906	5,3 ±25,64	20,8 ± 5,3	20,57 ± 0,9838	333,10 ± 31,13

Fig. 17 Work in progress for a first attempt to characterize fast growing wooden material as active yarn. © Jens Mählmann, Sächsisches Textilforschungsinstitut e.V., Chemnitz

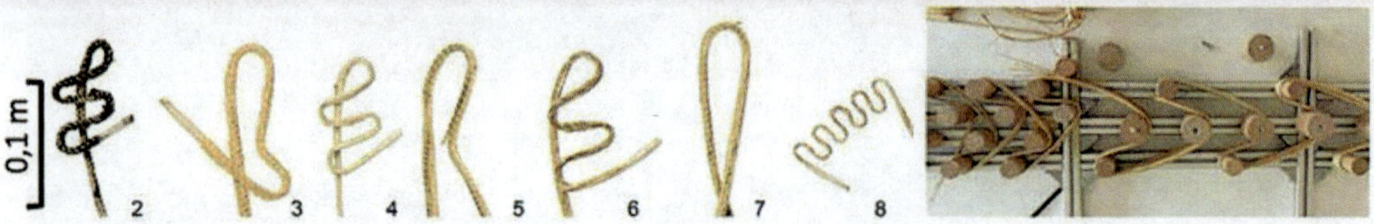

	Material	Number	Diameter	Core	Mantel/Viskose Density	Diameter KEMAFIL	BendingRadius/Dry	BendingRadius/Wet
Modell K1	Willow 100 cm	x 12	1 - 4 mm	flax	10 mm	20 mm	85 mm	30-40 mm
Modell K2	Willow 100 cm	x 6-7	1 - 4 mm	wool	8-12 mm	12 mm	75 mm	10 - 20 mm
Modell K3	Willow 100 cm	x 8-12	1 - 4 mm	wood wool	12 mm	20-22 mm	80 mm	20 - 30 mm
Modell K4	Willow 100 cm	x 7-5	1 - 3 mm	flax bleached	5-8 mm	8 -12 mm	60 mm	10 - 20 mm
Modell K5	Willow 100 cm	x 10-18	1 - 4 mm	flax chemnitz	15 mm	16-22 mm	100 mm	30 - 40 mm
Modell K6	Willow geschnitten	x 25-32			26 mm	18 mm	50 mm	10 mm
Modell K7	Willow 100 cm	x 16-22	1-4 mm		5-8 mm	20 - 22 mm	70 mm	30 - 40 mm
Modell K8	Willow geschnitten	x 10-14			26 mm	8-12 mm	30 mm	5 mm

Fig. 18 *(top)* Tubular structure of a 4 × 6 warp mesh. Loop width is approx.. 20 cm and total height 60 cm. Yarn is made out of 4–6 willow twigs. Sheath uses a 1 mm flax thread. (bottom) Experimental results for minimal bending radius of 8 hybrid yarns. Defining the minimal bending radius of produced dry and wetted hybrid active yarns and the adjustable pre-forming board

Fig. 19 Stretching Materialities, interactive installation at Tieranatomisches Theater, Berlin, 2021 © Landesman MoA

2 Results and Discussion

Although the focus is on willow, the research originated from the work with cellulose fibre yarn. Initially, using rattan calamus (plant fibre) threads and later willow twigs (wood fibre) (Madsen and Gamstedt 2013). Rattan material is a much softer filament and less resistant than willow. Also, rattan does not come from local production, being an imported material. Since one of the research aims was to adhere as much as possible to local resources, using willow material seemed to be the right decision.

Nevertheless, warp knitting with rattan also showed some benefits. On the one hand, rattan lianas are much longer than willow twigs. That eventually rendered longer chains of loops and consequently made it easier to achieve longer geometries (e.g. tubular constructions). Moreover, the thread's cross-section remains constant along the length of the rattan thread. In contrast, the willow twigs gradually decrease, creating some unbalance structures. However, this last issue led us to the decision to overcome the limitation of engineering our wood yarn. That, in the end, presents a promising potential as thread material for larger-scale architectural prototypes.

From a socioeconomic and cultural perspective, the interest of the material system we proposed is in using SRCs as an innovative regeneration driver for the land and the people involved in its cultivation, preparation and processing. Beyond the regular uses of SRCs as biomass, we see an application in the architectural scale as a stance in the circular economy.

3 Conclusion

We have introduced a novel material system to process wooden filaments into thicker yarns—borrowing thread bundling techniques from the textile industry. Through customising technical textiles fabrication techniques in combination with the wood pre-formation and plasticisation process, we are gaining insights into new lightweight construction applications in architecture, such as non-load bearing secondary components in the building shell (façade, roof, sun protection), also for inner scaffolds for cement-like matrices.

Future work will continue exploring the opportunities for novel fast growing wooden filament active yarns by tackling mechanical adjustments and conversion measures on the knitting machinery, developing the loop formation process and structural design and developing test methods for monitoring the process and quality. Besides, computational mechanics methods are also being developed. To use them to investigate the role of structure and material properties in determining the deformation behaviour of knitted materials to provide quantitative feedback to design and manufacturing.

Acknowledgements The research is conducted with Textile Designer Nelli Singer, awarded for her wood-based textile research Living Beings, Master Thesis, weißensee kunsthoschschule berlin (Berlin 2020). Collaboration and mentoring Prof. Christiane Sauer, weißensee school of art and design berlin. The authors acknowledge the support of the Cluster of Excellence »Matters of Activity. Image Space Material« funded by the Deutsche Forschungsgemeinschaft (DFG, German Research Foundation) under Germany's Excellence Strategy—EXC 2025—390648296. In addition, the authors want to thank the photo studio weißensee school of art and design berlin Dr. Michaela Eder, Max Planck Institute Potsdam—Golm and the support received at Sächsisches Textilforschungsinstitut e.V. (STFI) in Chemnitz.

References

Arnold R, Bartl A-M, Hufnagl E (1994) Production of Cord and Narrow Fabric Products with Kemafil Technology. 31:48–52

Beyer B, Suárez D, Palz N (2019) Microbiologically Activated Knitted Composites—Reimagining a column for the 21st century. In: Sousa, JP, Xavier, JP and Castro Henriques, G (eds.), Architecture in the age of the 4th industrial revolution—Proceedings of the 37th eCAADe and 23rd SIGraDi Conference—vol 2, pp 11–13. University of Porto, Porto, Portugal, pp. 541–552. CUMINCAD

Bruer S (2005) Three-Dimensionally Knit Spacer Fabrics: a Review of Production Techniques and Applications. 4(4):31

Castaño-Díaz M, Barrio-Anta M, Afif-Khouri E, Cámara-Obregón A (2018) Willow short rotation coppice trial in a former mining area in northern Spain: effects of clone, fertilization and planting density on yield after five years. Forests 9(3):154. https://doi.org/10.3390/f9030154

Dimitriou I, Rutz D (2015) Handbook on sustainable short rotation coppice

Eversmann P, Ochs J, Heise J, Akbar Z, Böhm S (2021) Additive timber manufacturing: a novel, wood-based filament and its additive robotic fabrication techniques for large-scale, Material-efficient construction. 3D Print Addit Manuf. https://doi.org/10.1089/3dp.2020.0356

Gong RH (2011) Specialist yarn and fabric structures. Elsevier, Developments and applications

Gries T, Stolyarov O, Quadflieg T, Raina MA (2016) Manufacturing of textiles for civil engineering applications. Elsevier [u.a.]

Harding J, Nielsen KP, Pearson W, Gmachl M (2015) Engineering Archilace. In: International association for shell and spatial structures (IASS) symposium 2015. Amsterdam.

Kurbak A (2019) Models for basic warp knitted fabrics Part II: single guide bar fabrics (closed-lap and open-lap). Text Res J 89(10):1886–1916. https://doi.org/10.1177/0040517518780001

Kuzovkina Y, Quigley M (2005) Willows Beyond Wetlands: Uses of Salix L. Species for Environmental Projects. Water Air Soil Pollut 162:183–204. https://doi.org/10.1007/s11270-005-6272-5

Kyosev Y (2019) Texmind.com, TexMind Warp Knitting Software

Kyosev Y (2019) Warp Knitted Fabrics Construction. CRC Press, Boca Raton

Madsen B, Gamstedt EK (2013) Wood versus Plant Fibers: Similarities and Differences in Composite Applications. Adv Mater Sci Eng 2013:e564346. https://doi.org/10.1155/2013/564346

Oktaee J, Lautenschläger T, Günther M, Neinhuis C, Wagenführ A, Lindner M, Winkler A (2017) Characterisation of Willow Bast Fibers (Salix spp.) from short-rotation plantation as potential reinforcement for polymer composites. BioResources 12(2):4270–4282. https://doi.org/10.15376/biores.12.2.4270-4282

Ramsgaard Thomsen M, Bech K, Sigurðardóttir K (2012) Textile logics in a digital architecture. In: New design concepts and strategies

Sauer C (2019) Entwerfen. Upscaling textiles. experimenteller materialentwurf im räumlichen kontext. In: Séverine Marguin/ Henrike Rabe/ Wolfgang Schäffner/ Friedrich Schmidgall (Eds.), *Experimentieren* (pp 51–66). Bielefeld: transcript Verlag. https://doi.org/10.14361/9783839446386-004

Serra DO, Richter AM, Hengge R (2013) Cellulose as an architectural element in spatially structured escherichia coli biofilms. J Bacteriol 195(24):5540–5554. https://doi.org/10.1128/JB.00946-13

Singer N (2020) Living Beings, Master Thesis, weißensee kunsthoschschule berlin (Berlin)

Spencer DJ (2001) Knitting technology: a comprehensive handbook and practical guide, 3 ed. Woodhead [u.a.], Cambridge [u.a]

Szczepanowska HM (2018) Deconstructing Rattan: morphology of biogenic silica in rattan and its impact on preservation of southeast asian art and artifacts made of rattan. Stud Conserv 63(6):356–374. https://doi.org/10.1080/00393630.2017.1404693

Vincent J (2008) Biomimetic materials. J Mater Res 23:3140–3147. https://doi.org/10.1557/JMR.2008.0380

Wright RS, Bond BH, Chen Z (2013) Steam bending of wood; embellishments to an ancient technique. BioResourcesd 8(4):4793–4796. https://doi.org/10.15376/biores.8.4.4793-4796

Fabrication Futures

Structural Performance-Based 3D Concrete Printing for an Efficient Concrete Beam

Hao Wu, Yu Li, Xingjie Xie, Xiaofan Gao, and Philip F. Yuan

Abstract

This paper presents the design and fabrication of an efficient steel–concrete composite beam prototype using structural optimization methods and innovative 3D concrete printing (3DCP). In traditional building industry, concrete objects like slabs or beams are cast by using low-cost standardized formwork, resulting in their high material usage and large carbon footprints. Although the development of large-scale 3DCP offers a formwork-free and rapid construction method, the elements created are like those created by standardized formwork. To achieve formwork-free construction of efficient concrete structures, this paper proposes the method of structural performance-based 3DCP. The work in this paper uses multi-material BESO technique and non-horizontal 3D printing to fabricate an efficient steel–concrete composite beam prototype with a span of 4 m. According to rough calculations, the optimized structure saves approximately 60% of concrete and 50% of steel compared to a conventional concrete beam. Combined with previous studies, this paper summarizes and proposes an important shift that is taking place in 3DCP. The general workflow of Structural Performance-Based 3D Concrete Printing is summarized, and future research topics are discussed. The study in this paper demonstrates the valuable outlook of the combination of structural performance-based design and precise material deposition methods, which could contribute to the UN Sustainable Development Goals.

Keywords

3D Concrete printing · Structural Performance-based · Steel–Concrete composite beam · Topology optimization · Non-horizontal 3D printing

1 Printing Better with Less

Portland cement, the main strength-giving component in concrete, has been widely used in the field of building construction, creating many shelters for mankind, since its invention in 1824 (Hall 1976). Most of the carbon dioxide emission from the concrete industry is from the production of Portland cement (Adesina 2020). It is estimated that approximately 7% of global carbon emissions originate from cement production (UN Environment Programme 2020). In addition, emissions from the manufacture of concrete formwork can't be negligible. Therefore, it is imperative to reduce the amount of cement and

H. Wu · Y. Li · X. Xie · X. Gao · P. F. Yuan (✉)
College of Architecture AND Urban Planning, Tongji University, Shanghai, China
e-mail: philipyuan007@tongji.edu.cn

M. R. Thomsen et al. (eds.), *Design for Rethinking Resources*, Sustainable Development Goals Series,
https://doi.org/10.1007/978-3-031-36554-6_22

formwork used in construction for developing a sustainable construction production pattern.

Some exciting progress has been made in reducing the consumption of formwork in concrete construction. 3D concrete printing (3DCP) technique, which is first proposed as Contour Crafting (CC), provides a formwork-free solution for concrete construction (Khoshnevis 2004). The past ten years have witnessed the development of this rule-changing technology from laboratory experiments to actual house fabrication (Cnet.com 2016; Mediaoffice.ae 2016; Totalkustom.com 2016), thanks to the development of digital design technique and large-scale automated truss equipment.

In traditional building construction, low-cost standardized formwork is used for casting non-optimized slab concrete structures with a high material usage and large carbon footprints, which is always widely criticized (Hawkins et al. 2016). However, the current construction application using 3DCP has also created non-optimized box-shaped structures that are similar to casting concrete. Many large-scale truss printing equipment designed for fabrication of large-scale house, with only three degrees of freedom, can only deposit concrete vertically to form non-optimized structures or deployable curved-surface structures, which is a 2.5D manufacturing process essentially. From a structural perspective, these non-optimized concrete structures consume a lot of excess material, and it should be questioned whether this way of building is worth promoting.

In another fabrication idea, researchers use additive manufacturing (AM) technology to produce innovative, material-efficient, lightweight formwork for concrete structures, thereby reducing the amount of cement in the structure. Inevitably, the use of AM to produce formwork generates additional formwork consumption. Fortunately, some designers and scholars begin to explore direct 3D printing of performance-optimized concrete structures. Bhooshan et al. built a concrete pedestrian bridge (2022) and Wu et al. fabricated a concrete spatial structure (2022), both applied 3D graphic statics in design process. A new type of prefabricated reinforced concrete beam using 3DCP, called minimass, has been invented and patented. Traditional 3D printing techniques (2.5D process) were used in this prototype. There is no built case study available (minimass.net). These cases rely on digital design and precise spatial deposition of concrete, providing a new paradigm for performance-based design and construction.

This paper presents the design and fabrication of a kind of innovative, efficient concrete beam structure using robotic 3D concrete printing, which is considered the next phase in the field of architectural 3DCP research. An innovative steel–concrete composite 3DCP beam with a span of 3.6 m was constructed, demonstrating the multi-material topology optimization process and toolpath generation. And a design-segmentation-printing-assembly workflow is proposed. The resulting prototype shows high structural efficiency, saving approximately 60% of concrete and 50% of steel compared to a conventional concrete beam with the same load-bearing capacity. Combined with previous research, this paper summarizes and presents a new paradigm—structural performance-based 3D concrete printing, discusses the advantages and challenges of this new fabrication approach, and future research outlook.

2 Non-Horizontal 3D Concrete Printing

Non-horizontal 3D concrete printing refers to the printing method in which the nozzles are not kept vertical during the printing process, and it is the key to fabricate efficient 3DCP structures. Previous studies prove that for concrete printed structures, they are more stress tolerant in the direction perpendicular to the printed layers than in the direction parallel to the printed layers, i.e., they maintain anisotropic in terms of forces (Rehman and Kim 2021). Non-horizontal 3DCP precisely controls the deposition of material during the manufacturing process, allowing the concrete to build up to the proper location for its stress condition in the structure.

Nowadays popular 3DCP technology was developed from the Contour Crafting (CC) technology first introduced in 2004 (Khoshnevis 2004), where objects are obtained by depositing material layer by layer. These layers obtained by intersecting with the object to be printed are horizontal, and the nozzles are always vertical during the printing process. Some scholars introduced the tangential continuity method in which the angle of the nozzle is continuously rotated according to the geometry (Gosselin et al. 2016). The deposition angle of the concrete material depends on the nozzle orientation, which requires 3D printing system with at least 6 degrees of freedom. Unlike the 3-axis truss system in which material can only be deposited in the horizontal direction, the 6-axis robotic arm-based 3DCP system can control the deposition of material in the non-parallel layers to match the force flow of the printed object in the structural system.

Accurate control of the material deposition rate is the key to achieving non-horizontal printing of an object, which requires consideration of system parameters such as material pumping rate, nozzle angle, and nozzle speed. Due to the uncertain pressure variations in the material delivery tube of 3DCP system, it is not feasible to control the nozzle discharge volume by changing the speed of the pumping motor. Keeping the pumping motor speed constant and varying the robot speed is another method of changing the material deposition rate, which has proven to be feasible (Yuan et al. 2022). In previous research, variable width printing was achieved by changing the robot speed (Zhan et al. 2021). While in this study, a similar approach is applied to achieve multi-angle printing with variable layer heights. The robotic trajectory and the nozzle speed for printing is defined with the toolpath generation method that is described in Sect. 4.

3 Material and Robotic 3DCP System

Non-planar 3DCP technique relies on precise management of nozzle speed and layer height to control the deposition of cementitious material, thus achieving the desired geometry. Buildability is an important factor in the 3DCP process. High buildability relies on low layer deformation rate. In other words, there should be a quick shear stress gain after the extrusion process to minimize the deformation effect (Reiter et al. 2018; Kruger et al. 2019).

A set-on-demand 3D printing mortar is designed by referring to the study of (Reiter et al. 2020). The primary printing material is a kind of fiber-reinforced high-strength cementitious mortar, and an alkali-free accelerator is added to control the set time. The major components of the mortar are listed in Table 1. By applying the proper dosage of the accelerator, the object can be printed with relatively low deformation and a short layer time gap. The fast build-up and low deformation of set-on-demand material are beneficial to maintain the geometry of non-planar 3DCP.

The robotic 3DCP system setup in this experiment is shown in Fig. 1. The material preparation for printing starts with mixing powder mixes encapsulated by material suppliers (including cement, sand, silica fume, and superplasticizer), water, and fiber in a planetary mixer (up to 120L of slurry). Then the slurry is poured into storage hopper which is combined with slurry pump and the planetary mixer as a mobile mix-to-pump machine. Then the slurry and the accelerator are pumped respectively via slurry pump and accelerator pump at a specific ratio

Table 1 Material and Mixture design

Material	Quantity (kg/m^3)
Ordinary portland cement (P. O. 42.5)	835
Silver sand (0-2 mm)	850
Silica fume	85
Water	250
Superplasticizer	0.65%
Alkali-free accelerator	2%
PVA fiber (12 mm)	2.5

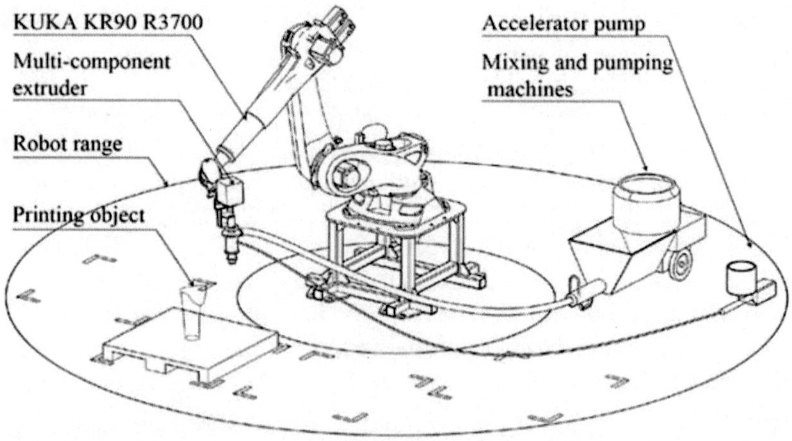

Fig. 1 The robotic 3DCP system setup

into the mixing chamber, where the material is mixed at high speed before deposition through the nozzle. The nozzle is mounted on a KUKA KR90 R3700 robotic arm established on a 750 mm-high base. The lifting of the base allows the robotic arm to achieve more complex spatial motion without joint limitation during 3DCP process, which is beneficial to the non-planar 3D printing of complex geometries.

During each printing process, this setup produces 120L cementitious slurry in a single mixing in 6 min. When the material in the storage hopper is about to run out, the next mixing is performed, and the slurry is replenished to the storage hopper in time to guarantee continuous printing. The set-on-demand concrete is deposited with 90–230 mm/s robotic printing speed with an extrusion rate of 2 L/min. After printing, the elements are covered with plastic film and cured in 80% relative humidity for the first 3 days, and then in 40% relative humidity for an additional 20 days without plastic film covering.

4 Digital Design of Beam Structure and Toolpath

4.1 Multi-Material Topology Optimization

Topology optimization is an effective means to achieve structural performance-based design, which includes solid isotropic material with penalization (SIMP) method (Rietz 2001), level set method (van Dijk et al. 2013), the bi-directional evolutionary structural optimization (BESO) method (Yang et al. 1999), etc. However, most of the traditional topology optimization techniques are based on a single material, and in the design of steel–concrete composite beam in this research, there are both tensile and compressive parts in the structure. In contrast, the multi-material BESO technique recently proposed by Li and Xie (2021a, b) is able to separate the tensile and compressive parts of the structure, enabling the topology optimization of materials with extinct different tensile and compressive properties, such as concrete. Based on this method, it is possible to design innovative and structurally efficient steel–concrete composite structures (Li et al. 2022).

In the multi-material BESO method, the first invariant of stress tensors I1 is used as the criterion for judging the FEA element in tension or compression and achieved excellent results. For a 3D element, the first invariant of the i-th element can be calculated as

$$I_{1,i} = \sigma_{11,i} + \sigma_{22,i} + \sigma_{33,i} \ (i = 1, 2, ..., N)$$

where $\sigma_{11,i}$, $\sigma_{22,i}$, $\sigma_{33,i}$ are the maximum principal stress, second principal stress, and minimum principal stress, respectively. N is the total number of elements in the FEA mesh. For a 2D

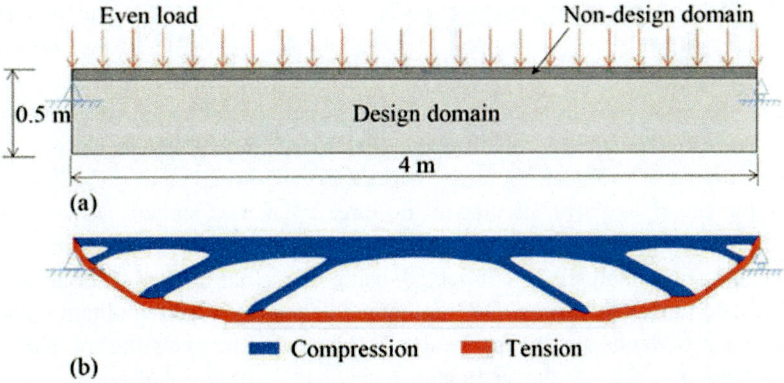

Fig. 2 Performance-based design of the beam via multi-material topology optimization. **a** design domain and load conditions, **b** optimized result

problem, only the maximum principal stress and the minimum principal stress are needed.

As shown in Fig. 2a, a steel–concrete composite beam with a span of 4 m is used as the prototype for this study, and the deck region is set as the non-design domain with a uniform load on it. There is a hinge and a roller at the left and right ends of the bridge. Based on this design model, multi-material topology optimization is performed, using material utilization as an indicator for material elimination and addition.

During the topology optimization process, concrete is distributed only in the compressive part of the structure, thus ensuring that the printed concrete parts in the final structure are only in compression. The mesh for topology optimization is set as 400 × 50 elements and the target volume of the final structure is 0.4, which means the total volume of the final structure is 40% of the original design domain. The final topology optimization results obtained for the dual material are shown in Fig. 2b, where red indicates the tensile region and blue indicates the compressive region.

4.2 Post-Design Process

Based on the 2D result of the multi-material BESO method, certain post-design processes are artificially performed to finally obtain a steel–concrete composite beam. The constructability of the experimental prototype, the design of the steel–concrete nodes, and the executability of the assembly are all considered in the post-design process. Fig. 3 shows the digital model after the post-design. The cross-sectional dimensions of the concrete structure were adjusted due to the 3DCP constraints, such as four legs of the beam being appropriately enlarged. The tension part is made of 12 mm diameter steel cables, and the steel–concrete connecting nodes are specially designed. Each end of the beam has a casting concrete member to anchor 3DCP objects and steel cables.

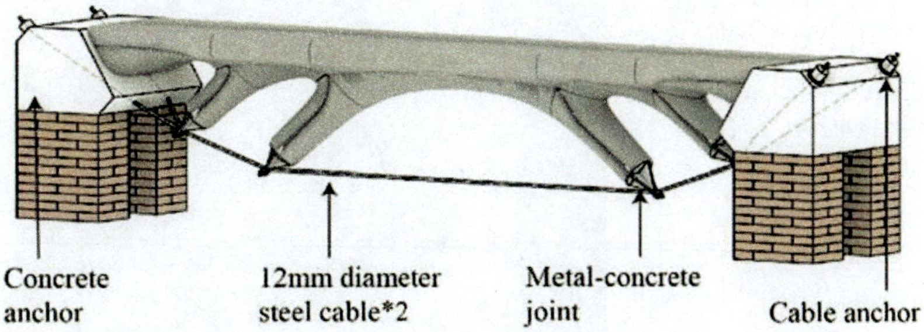

Fig. 3 Demonstrator model after post-design

4.3 Variable-Layer-Height 3DCP Toolpath

Since the compressive strength of 3DCP structures remains anisotropic in different directions, the principle of toolpath planning is to keep the printing layers as perpendicular to the direction of the structural stress line as possible. The Finite Element (FE) analysis is conducted using the Millipede plugin for Rhino3D/Grasshopper. The load case is defined with an evenly distributed area load of 10kN/m, the estimated dead load, and 6 fixed supports, one in 4 legs and 2 ends of the beam. According to the resulting stress line pattern, the segmentation of the beam prototype is established. Printing layers are also generated for each element, which are oriented as perpendicular to the stress lines as possible. The prototype is divided into 9 units, four of which are legs, the other five of which form the top surface of the beam. Figure 4a shows the stress line pattern and Fig. 4b shows the segmentation of the prototype.

To fabricate any element of this prototype using 3DCP, the conventional horizontal layering method could not satisfy the principle that printing layers should be perpendicular to stress line. What's worse, traditional 3DCP strategy could cause Step Effects on the contact surface of the element. However, by using 3DCP with non-horizontal and variable-height layers, it is possible to deposit concrete in the direction of stress flow and to obtain a flat contact surface. Figure 5 illustrates the toolpath diagram of one leg using two 3DCP layering methods. Undoubtedly, this innovative path planning approach poses challenges for accuracy layering of the element and precise control of robot speed.

In previous studies (Zhan et al. 2021), for filaments extruded from a circular nozzle, the instantaneous cross-sectional shape can be shown below in Figure X. For a solid extrusion without an inner cavity, the functional relationship between nozzle speed and filament height can be described as

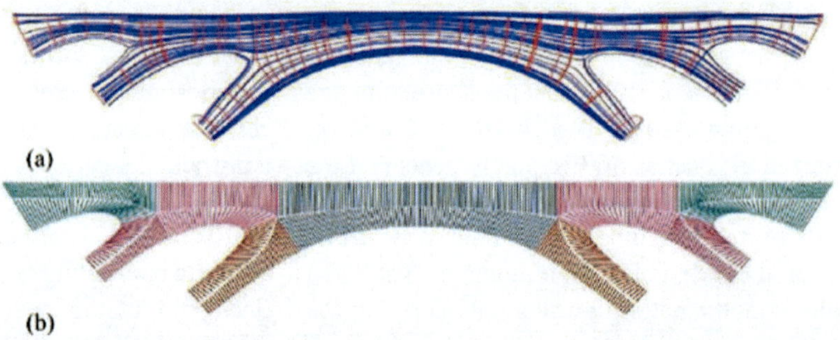

Fig. 4 Result of FEA and the segmentation of the prototype. **a** Stress line pattern, **b** Segmentation and toolpath of the prototype

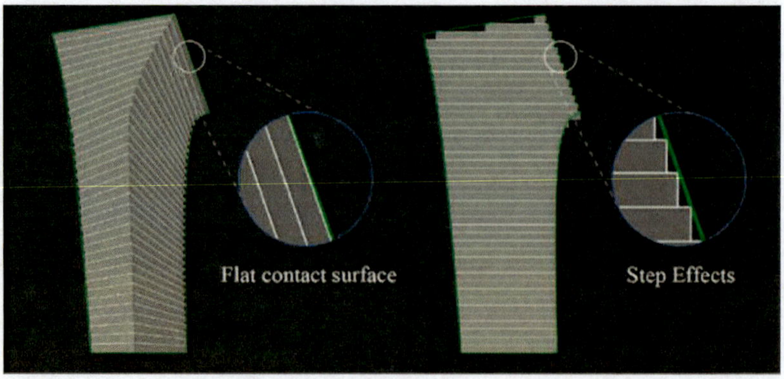

Fig. 5 The toolpath diagram of one leg using two 3DCP layering methods. Left: Variable-layer-height 3DCP toolpath, Right: traditional 3DCP toolpath

$$v = f(w) = \frac{Q}{\left(\frac{\pi}{4} - 1\right)h^2 + wh}$$

In this formula, v is the nozzle velocity, Q is the material flowrate, w is the filament width, h is the filament height. In this experiment, the material flowrate Q is controlled to be constant by controlling the speed of the pumping motor and the filament width w is set to be 25 mm. Thus, the relationship between the filament height h and the nozzle velocity v is established. Real-time relation between robot speed and layer height is achieved to accomplish precise concrete deposition.

5 Fabrication of the Prototype

The steel–concrete composite beam is fabricated in four major steps: printing the concrete elements, curing the concrete, preparation work for assembly, and assembly of the beam (Fig. 6).

First, 9 segmental elements were printed one by one on 4 print beds for a total of 5 h (excluding intervals). It took 15 min for the smallest elements printing and 80 min for the largest element. Although the printing process was fast for each element, only 4 components could be printed in a day because it took 8 to 10 h for element to get enough strength to be moved after printing. Eventually, the printing of the 9 components took 3 days to complete. Each print process contains two parts, printing of the base and printing of the element. During each element printing, a pause is set after base printing, at which time a plastic film is placed on the base to separate it from the element above, so that they can be separated later.

Second, 8 to 10 h after printing, each element reaches its early strength and can be carried at this time. They are moved to the curing area, covered with plastic film, and cured at 80% relative humidity for 3 days. The plastic film is then removed and cured at 40% relative humidity for another 20 days to finally reach full design strength. Its compressive strength can reach the level of C40, and it's suitable for general house construction, according to Chinese concrete industry standard.

Third, the preparation work before assembly includes casting the concrete anchors at the end of the beam, establishing scaffolding, and installation the steel–concrete nodes for each leg. Two concrete anchors were obtained by casting concrete in customized timber formwork and located in 500-cm-high base made of bricks. The bricks are not cement-bonded, which proves the beam structure has no horizontal force on the ground. Two linear holes of 20 mm diameter were left in each concrete anchor to facilitate subsequent crossing of the steel cables. Then a steel scaffold system that matches the button form of the beam was established, ensuring that

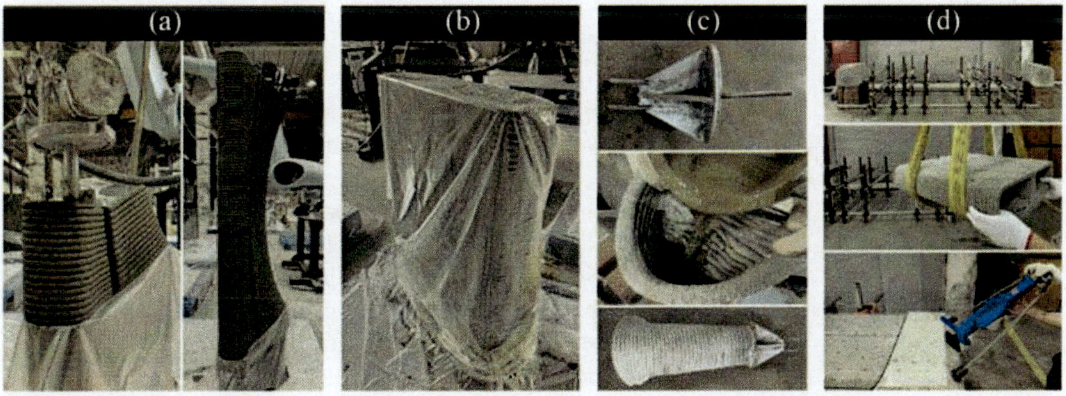

Fig. 6 Fabrication steps: **a** printing concrete elements, **b** curing concrete, **c** casting concrete to install metal-concrete joints, **d** assembly and tensioning

Table 2 Time and Labor consumption

Steps	Time	Labor
3D Printing of 9 elements	3 Days	2
High humidity curing	3 Days	1
Low humidity curing & Assembly preparation	20 Days	1
Assembly	5 h	3

each concrete element can be perfectly located before tensioning the steel cables. Rubber was applied in the middle of each steel tube to avoid direct collision between the 3DCP elements and the tubes. Concrete was poured into the four 3DCP legs to connect the previously customized metal-concrete joints, and the curing time of the casting concrete overlapped with that of the 3DCP elements.

Fourth, the gantry was used to position the elements sequentially from one side of the beam to another. Before locating, rubber strips are pasted on the elements to provide a cushioning effect. During the installation, a laser level was used for positioning, and each element is adjusted to the proper position before cable installation. Then two 12 mm diameter steel cables are passed through the two concrete anchors and four legs one by one. Metal latched were installed on the metal-concrete joints of each leg to keep the cable fixed. Finally, the two cables were joined into a loop in the end and tension was performed using a customized jack on each side.

Thanks to the digital design and additive manufacturing technology, a rapid and low labor cost construction of the experimental prototype was achieved. The time spent in each step and the labor invested are shown in Table 2 (each labor work 8 h per day).

6 Discussion and Future Work

6.1 Result

Several analytical methods were employed for this new steel–concrete composite structure, including 3D scanning to verify its geometric fidelity, and loading experiments of 1t. The result of 3D scanning showed that the non-horizontal 3DCP ensured the geometry of the prototype, and the loading experiment verified the load-bearing capacity of this structure.

This prototype was designed for a total load of 10t. And it was fabricated using $0.21m^3$ of 3D printed concrete material (9 hollow elements) and $2.9e{-}3m^3$ of steel (5.6 m * 12 mm diameter steel cable * 2 & 4 metal-concrete joints). According to the Chinese code for reinforced concrete beam, a conventional reinforced concrete beam structure was designed with a total design load of 10t. This beam has a span of 4 m, a section size of 200 * 600 mm, 4 m * 22 mm diameter cable * 4, and the concrete grade is C40. $0.48m^3$ of concrete and $6.1e\text{-}3m^3$ of steel are required to construct this conventional beam. It indicates that this innovative structure saves approximately 60% of concrete and 50% of steel compared to a conventional beam under the same load-bearing capacity. And about 175.35 kg cement is used in this prototype while 245.76 kg in traditional beam (512 kg cement is needed per cubic meter of C40 concrete). Therefore, this new structure is conducive to reducing carbon emissions. And it can be applied to the efficient construction of small bridges and beams in house (Figs. 7 and 8).

6.2 Challenges in Practical Applications

The successful construction of the experimental prototype has demonstrated the soundness of this steel–concrete composite beam and the feasibility of the fabrication strategy, but there are still lots of challenges to be faced if it's applied in building industry. For the detailed design of the structure, connection between concrete (in compression) and cables (in tension) is a key concern.

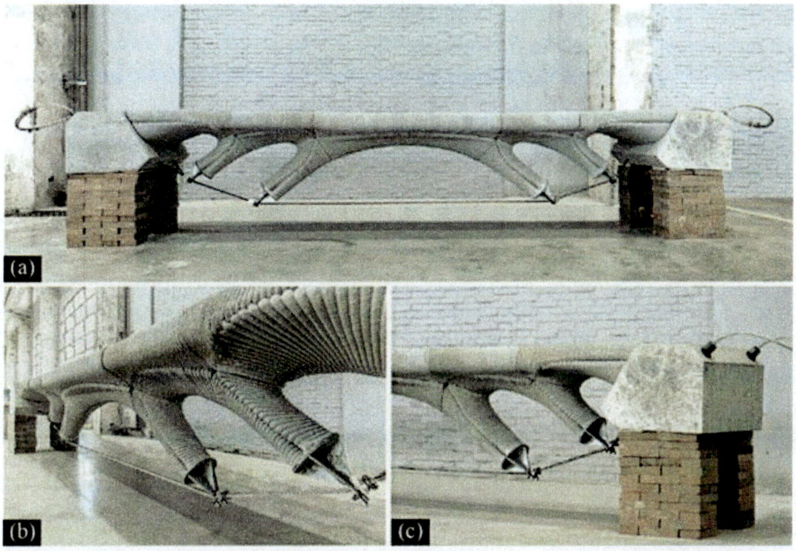

Fig. 7 The finished beam prototype. **a** Overall perspective, **b–c** Details closeups

Fig. 8 Interior scene of building with this new prototype

The clamps between metal-concrete joints and steel cables need to be improved, to avoid stress concentrations and their slipping on the steel cables. For the application of pre-stress, although the construction is completed by jacks, more reliable and regular tensioning methods need to be used in practice, such as tensioning measures applied in construction of traditional pre-stressed beam structure.

In addition, the impact of this structure on the architectural space and its safety are also worth being discussed. Due to the limitations of the research funding and the experimental site, the prototype beam fabricated in this study is only 4 m in length. The span is supposed to be extended to 6–8 m in practical applications. Even for the 4 m span, the height of the experimental beam reached 50 cm due to the segmenting angle of the geometry and the limitations of non-horizontal 3D concrete printing. In the current printing, objects with a tilt angle of up to 70 degrees can be fabricated with high quality, beyond which it would be difficult to print, meaning that as the span increases, so does the height of the beam, which may reach 70 cm for a beam with 6–8 m span. The increased height of the beam means that a larger floor height is required; however, the bottom form of this new kind of structure can be considered aesthetically as a unique interior feature. It is also important to ensure the strength and safety of the steel cables to avoid them being cut by someone or burnt by fire, which also poses challenges for building maintenance.

6.3 A Shift in 3DCP

In the field of concrete engineering, scholars have long criticized the drawbacks of standardized formwork as they result in structures with high material usage. Various kinds of flexible molds have been developed to manufacture efficient concrete structures with less material (Hawkins et al. 2016). Nowadays, in the context of large-scale 3D concrete printing, similar attempts to pursue efficient structures are taking place. This is an optimization process for both design and construction, in which digital modeling methods and digital fabrication techniques are involved. Recently, several projects with structural performance-based 3D concrete printing have been constructed, proving that this shift is taking place and valued by designers (Bhooshan et al. 2022; Wu et al. 2022). These emerging practices signal a shift in 3DCP from an emphasis on formwork-free, rapid construction to an emphasis on less material usage and sustainability. These innovative practices create a smart approach to concrete deposition, with goals consistent with flexible molds, showing a new paradigm for sustainable design.

6.4 Workflow

Combined with previous studies (Wu et al. 2022) (Fig. 9), a workflow applicable to structural performance-based 3DCP is summarized, and it is divided into the following steps.

Fig. 9 The previous spatial concrete pavilion using 3D Graphic Statics design methods. **a** Overall perspective, **b-c** Details closeups

- The structural design should be based on the mechanical properties of concrete to reduce the use of redundant materials. Common optimization methods include 3D graphic statics, which can be used to generate compress-only spatial structure, and topological optimization, which eliminates redundant materials.
- Post-design needs to be done to establish the final structural form.
- Segmentation is performed according to the range of the printing system and the direction of stress flow within the concrete structure. Connecting joints between segmentations are also taken into consideration.
- The toolpath for each element needs to be precisely planned, and then elements are built using a robotic 3D printing system. The processing of the connecting joints can be carried out simultaneously with concrete curing.
- With the support of scaffolding, the positioning of elements and the assembly of the overall structure is completed, relevant structural tests can be employed if required.

6.5 More In-Situ Material to Explore

Structural performance-based 3D concrete printing can be extended to other materials such as clay, gypsum, and other in situ materials. Thermoplastics, on the other hand, are not suitable for this method, because large errors will occur due to the excessive thermal stress after material extrusion, especially in non-horizontal 3D printing. Several scholars have attempted to use in situ or recycled materials for additive manufacturing (Romani et al. 2021), but there are few attempts to combine the unique properties of in situ materials with structural performance-based AM, which would be a worth-exploring new paradigm for in situ construction. As long as these materials can be printed by extrusion, then structural performance-based design methods can be incorporated into fabrication, and this reflects the generic value of this research.

7 Conclusion

This paper presents a novel design-to-fabrication method based on 3DCP for producing an efficient concrete beam with low material usage. The resulting prototype is an innovative combination of steel cable and 3D-printed concrete that shows great potential for saving materials and embodied carbon emissions. A workflow combining morphology design, robotic toolpath planning, and non-horizontal 3D printing could be achieved with more stability and applicability. In the era of additive manufacturing (AM), a new paradigm for the efficient use of concrete, the most widespread building material, is being established. This novel 3DCP strategy defines the optimal locations where concrete is deposited and creates new aesthetic forms as well as innovative fabrication modes. Structural performance-based 3D printing, combined with more consideration of the use of in situ materials, could provide a lighter and less impactful material practice for future sustainable construction.

Acknowledgements This research was supported by National Key R&D Program of China (Grant No.2018YFB1306903), National Natural Science Foundation of China (Grant No. U1913603), and Shanghai Science and Technology Committee (Grant No. 21DZ1204501).

References

Adesina A (2020) Recent advances in the concrete industry to reduce its carbon dioxide emissions. Environmental Challenges 1:100004. https://doi.org/10.1016/j.envc.2020.100004

Bhooshan S, Bhooshan V, Dell'Endice A, Chu J, Singer P, Megens J, Block P (2022) The Striatus bridge. Arch, Struct Constr:1–23.

Cnet.com (2016). Available from http://www.cnet.com/news/dubai-unveils-worlds-first-3d-printed-office-building/

Gosselin C, Duballet R, Roux P, Gaudillière N, Dirrenberger J, Morel P (2016) Large-scale 3D printing of ultra-high performance concrete–a new processing route for architects and builders. Mater Des 100:102–109

Hall C (1976) On the history of portland cement after 150 years. J Chem Educ 53(4):222. https://doi.org/10.1021/ed053p222

Hawkins WJ, Herrmann M, Ibell TJ, Kromoser B, Michaelski A, Orr JJ, West M (2016) Flexible formwork technologies–a state of the art review. Struct Concr 17(6):911–935

Khoshnevis B (2004) Automated construction by contour crafting—related robotics and information technologies. Autom Constr 13(1):5–19. https://doi.org/10.1016/j.autcon.2003.08.012

Kruger J, Zeranka S, van Zijl G (2019) An ab initio approach for thixotropy characterisation of (nanoparticle-infused) 3D printable concrete. Constr Build Mater 224:372–386

Li Y, Xie YM (2021b) Evolutionary topology optimization for structures made of multiple materials with different properties in tension and compression. Compos Struct 259:113497. https://doi.org/10.1016/j.compstruct.2020.113497

Li Y, Lai Y, Lu G, Yan F, Wei P, Xie YM (2022) Innovative design of long-span steel–concrete composite bridge using multi-material topology optimization. Eng Struct 269:114838. https://doi.org/10.1016/j.engstruct.2022.114838

Li Y, Xie YM (2021a) Evolutionary topology optimization of spatial steel-concrete structures. J Int Assoc Shell Spat Struct 62(2):102–112. https://doi.org/10.20898/j.iass.2021.015

Mediaoffice.ae (2016). Available from http://mediaoffice.ae/en/media-center/news/23/5/2016/3d-printed-office-build

Minimass (2022) From https://www.minimass.net/

Rehman AU, Kim JH (2021) 3D concrete printing: A systematic review of rheology, mix designs, mechanical, microstructural, and durability characteristics. Materials 14(14):3800

Reiter L, Wangler T, Roussel N, Flatt RJ (2018) The role of early age structural build-up in digital fabrication with concrete. Cem Concr Res 112:86–95

Reiter L, Wangler T, Anton A, Flatt RJ (2020) Setting on demand for digital concrete–principles, measurements, chemistry, validation. Cem Concr Res 132:106047

Rietz A (2001) Sufficiency of a finite exponent in SIMP (power law) methods. Struct Multidiscip Optim 21 (2):159–163. https://doi.org/10.1007/s001580050180

Romani A, Rognoli V, Levi M (2021) Design, materials, and extrusion-based additive manufacturing in circular economy contexts: from waste to new products. Sustainability 2021(13):7269

Totalkustom.com (2016). Available from http://www.totalkustom.com/3d-castle-completed.html

UN Environment Programme (2020) Global status report for buildings and construction. Retrieved from https://globalabc.org/sites/default/files/inline-files/2020%20Buildings%20GSR_FULL%20REPORT.pdf

Van Dijk NP, Maute K, Langelaar M, Van Keulen F (2013) Level-set methods for structural topology optimization: a review. Struct Multidiscip Optim 48 (3):437–472. https://doi.org/10.1007/s00158-013-0912-y

Wu H, Li Z, Zhou X, Wu X, Bao D, Yuan PF (2022) Digital design and fabrication of a 3d concrete printed funicular spatial structure. In: Proceedings of the 27th CAADRIA conference. Sydney, pp 71–80. https://doi.org/10.52842/conf.caadria.2022.2.071

Yang XY, Xie YM, Steven GP, Querin OM (1999) Bidirectional evolutionary method for stiffness optimization. AIAA J 37(11). https://doi.org/10.2514/2.626

Yuan PF, Zhan Q, Wu H, Beh HS, Zhang L (2022) Real-time toolpath planning and extrusion control (RTPEC) method for variable-width 3D concrete printing. J Build Eng 46:103716

Zhan Q, Wu H, Zhang L, Yuan PF, Gao T (2021). 3D Concrete printing with variable width filament. In: Proceedings of the 39th eCAADe Conference, vol 2. pp 153–160. https://doi.org/10.52842/conf.ecaade.2021.2.153

Integrated Design Models for Materially Differentiated Knitted Textile Membranes as the Means to Sustainable Material Culture Within Membrane Architecture

Yuliya Sinke, Mette Ramsgaard Thomsen, and Martin Tamke

Abstract

Material design and production must be rethought in light of limited resources, contributing to UN Sustainable Development Goals, promoting sustainable industrialization and fostering innovation. Currently, building practices rely on composite materials to meet a variety of performance demands, by bonding different materials together. This results in problems of reuse and recycling at the end of life, as they are hard to separate. As an alternative, functionally graded materials (FGMs), such as CNC-knitted membranes, allow the steered intensification of a single material in order to meet multiple performance requirements and ease their further reuse. Current architectural design practice conceptualises matter as homogeneous, while novel FGMs, and knitted architectural materials in particular, require radically different methods and tools for their design and digital representation. Given their extreme heterogeneity, complex material behaviour prediction and additive manufacturing nature, there is currently no ready-to-use software to address these challenges. In order to improve the quality of taken design decisions, better reflect knitted FGMs characteristics and pave the way for more sustainable production using FGMs, it is essential to create integrated design models, considering material heterogeneity, structural performance and fabrication constraints of such materials. In this research, we examine how CNC-knitted membranes, having versatile material structures, can be designed with heterogeneous composition. It is done through the development of custom digital tools that reflect their differentiated structure, analyse tensile performance and perform materially oriented simulation, while taking into account fabrication constraints (figure below). We describe a novel approach to inform the differentiation of knitted membranes by structural analysis and a method for calibrating heterogeneous membrane simulations. By utilising computational tools for structural performance evaluation, Graded Knitted Ceiling panels demonstrate how custom-made bespoke membranes can be materially differentiated and how such designs can then be simulated and calibrated based on the physical results.

Y. Sinke (✉) · M. Ramsgaard Thomsen · M. Tamke
Royal Danish Academy, Institute of Building Technologies, CITA - Center for IT and Architecture, Copenhagen, Denmark
e-mail: ybar@kglakademi.dk

M. Ramsgaard Thomsen
e-mail: mette.thomsen@kglakademi.dk

M. Tamke
e-mail: martin.tamke@kglakademi.dk

M. R. Thomsen et al. (eds.), *Design for Rethinking Resources*, Sustainable Development Goals Series,
https://doi.org/10.1007/978-3-031-36554-6_23

Graphical Abstract

A digital integrated model for graded knitted membranes incorporating aspects of material heterogeneity, digital simulation and fabrication constraints

Keywords

Integrated design models · Simulation of functionally graded materials · CNC-knitted architectural materials · Textile architecture

1 Introduction

1.1 The Scarcity of Resources Incites the Rethinking of Material-Making Culture

Resources scarcity forces us to rethink how materials are designed and made in contemporary design and building culture. This challenge is inciting research into new material practices that optimise resource allocation and reduced waste (Ramsgaard Thomsen et al. 2019a), contributing to the UN Sustainable Goals. Following of these goals ensures responsible production and good use of resources, promoting inclusive and sustainable industrialisation and foster innovation as described in detail in Sustainable Development Goals 12 and 9 (United Nations, Department of Economic and Social Affairs 2015). By moving from the paradigms of standardisation and mass production, these new practices position the making of hyper-specified and functionally graded materials for site- or use specific applications as central means of building smarter with less (Fig. 1). The advent of advanced fabrication techniques, supported by computational technologies, has enabled a differentiated distribution of resources for making novel high-performance materials. Despite these advances, design culture often remains disconnected from truly exploring the material understanding and opportunities for variation and mass-customisation digital fabrication enables.

1.2 The Contemporary Architectural Material Culture of Composites

Architectural material culture can be described through the implementation of technological innovation in the architectural design processes and applications. Today architectural materials and elements are mainly designed and made as *composites*, where several homogeneous matters are fused to achieve multi-functional properties. Despite their benefits for diverse applications, the composite nature of fused homogeneous materials reduces their circular economic value, as they cannot be separated for further recycling.

In lightweight membrane architecture composites are generally comprised of homogeneous woven substrates with a waterproof coating to provide strength and protection from sun and rain. The membrane property variation is achieved by sewing together multiple patches of

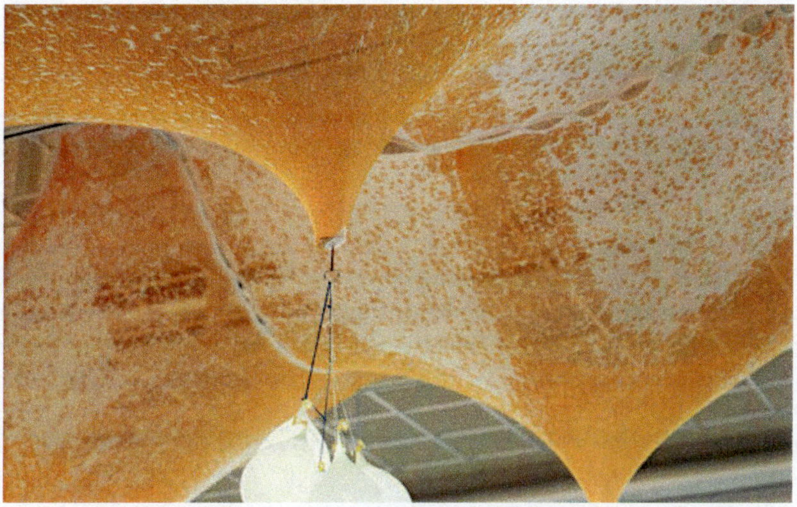

Fig. 1 Functionally graded CNC-knitted membranes use single yarn and are varied in their stretch properties

varying material properties, such as stiffness, fibre direction, transparency or permeability. The practice of pattern cutting and sewing leads to increased resource consumption as well as more waste during production. Furthermore, cut-out patches of conventional membranes can be hardly reused due to their composite structure and overspecified shape. Finally, seams produce undesired accumulations of internal material stresses, thereby reducing structural performance and creating a risk of material failure.

1.3 Future Architectural Material Culture of Functionally Graded Materials

The functionally graded materials (FGMs) are an alternative to composites. They have been the subject of research in material science, chemistry and computer graphics for many decades (Beukers and van Hinte 1998) and today are used in aerospace, nuclear energy, biology, electromagnetism, optics, energy and other fields (Li and Han 2018). The emergence of additive manufacturing, supported by computational interfaces, led to new fabrication processes of digitally controlled material deposition, enabling the production of FGMs of higher grading complexity (Richards and Amos 2014; Grigoriadis 2015). FGMs, therefore, have the advantage of integrating multiple properties and varying densities by grading the application of the same matter. In contrast to composites, FGMs allow the material properties to change continuously throughout the object, eliminating the problem of residual internal stresses and easing their further reuse. With the introduction of material heterogeneities in the design domain, various material advantages can be explored, while material incompatibilities can be alleviated with a gradual material variation (Kou and Tan 2007).

1.4 Knit as a Case of FGMs Within Structural Knitted Membranes

Within this research, CNC-knitted membranes are presented as examples of functionally graded materials in membrane architecture. By developing methods for utilising their heterogeneous compositional nature, they represent the practice of steered performance and reduced material use. The additive fabrication process of knit manufacturing offers opportunities to control membrane composition by strategic allocation of various stitches. Here, the membranes achieve their required three-dimensionality through material expansion, allowing geometrical deformation with reduced material use and less fabrication complexity in comparison to, e.g. 3D knitted textiles. Additionally, the homogeneous resource composition of continuous single yarn intertwined into various fibre configurations makes it suitable for recycling as no other additives are present and yarn can be simply unravelled back to the spool. This technology allows for scaling up and erect architectural structures, made with knit.

Tensile structures made with knit introduce a novel class of materially optimised structural membranes, expressing alternative to conventional woven membranes aesthetics. It has been demonstrated by experimental research projects on large-scale textile structures, where the membrane is employed structurally and materially graded (Deleuran et al. 2015; Ramsgaard Thomsen et al. 2019a; Sinke et al. 2022a, b) (Fig. 2).

FGMs can radically expand the material performances, as demonstrated in Isoropia, where more expandable stitch types are introduced in the areas where the membrane is required to stretch more. However, the heterogeneous nature of these membranes possess an unsolved challenge of predicting material behavior, which means that tools for a finer control of design and performance prediction of materially differentiated CNC-knitted membranes is necessary. Existing tools for design and manufacture of knitted materials are geared towards the garment industry, while architecture applications present another set of design, modelling and tolerance demands.

This paper contributes to the development of an integrated design model for materially differentiated knitted textile membranes. This is able to incorporate required demands of structural performance, the scale of application, and the fabrication limitations through the stages of conceptual design, structural and geometric analysis, materially informed simulation, and a direct link to digital manufacturing on industrial knitting machines. Using the methods and processes discovered in the presented research, we rethink the design of structural knitted membranes at the stitch level and propose new ways for working with heterogeneous custom-made materials in order to develop sustainable material cultures for membrane architecture and spatial design in general (Fig. 3).

Further a novel approach on how to inform the differentiation of knitted membranes by structural analysis and a method for heterogeneous membrane simulation calibration are presented. Mentioned findings are described, using a case study of *Graded Knitted Ceiling*—medium-scale installations, each made of 6 CNC-knitted differentiated panels. They were designed, fabricated and installed during the workshop for students of CiA Computation in Architecture Master Studio at Royal Danish Academy as a part of design workflow assessment and further development and evaluation of material grading and simulation prediction tools.

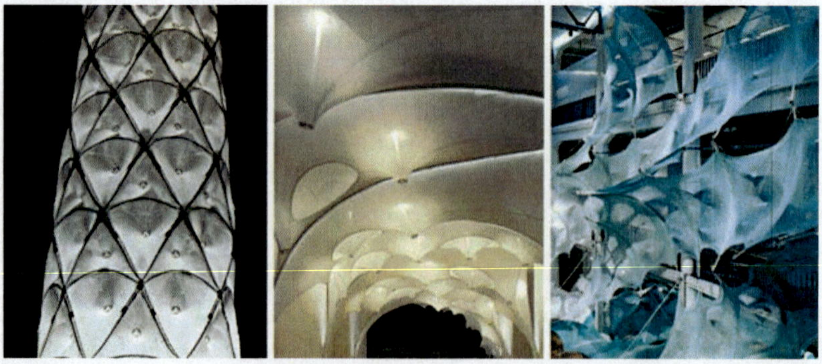

Fig. 2 Large-scale structures, employing CNC-knitted textile as a structural membrane. Left to right: Hybrid Tower 2016, Isoropia 2018, Zoirotia 2021. Authored by CITA—Center for IT and Architecture, Royal Danish Academy, Copenhagen, Denmark

Fig. 3 Close-up of the rich surface modulation of the CNC-knitted membranes through the variation of stitch composition

2 Textile Simulation Practices and Modelling of FGMs

2.1 Simulation in Membrane Design—FEM and Engineering Software

In conventional membrane architecture the canopy shape determines its structural performance. The simulations allow engineers to accurately assess how the membrane shape needs to be adjusted in order to achieve a stable three-dimensional membrane structure, that is able to accommodate different external forces, such as wind, snow and seismic loads. This enables engineers to design safer and more efficient membrane structures.

The membrane behaviour is described using partial differential equations in order to handle large-scale membrane geometries with differentiated loading setups. Due to the fact that there is no exact analytical solution to solve these equations (Chen and Hearle 2009), FEM (finite-element method) was developed. In FEM, the investigated surface is discretised into smaller elements (mesh segments), with a known finite size, where the node intersections are solved analytically (Kyosev 2012). As the nodes are shared between the elements, solving the equation is easier since the result should be the same for all linear elements meeting at the same node. A linear element is regarded as a mass spring, where the tension force of one segment is defined as a linear elastic force.

This method found its wide application within the structural engineering discipline being integrated into commercially available computer-aided engineering (*CAE*) software like Sofistik (www.sofistik.com), EASY.Form (Ströbel and Singer 2005), CADISI (TechNet GmbH), Formfinder and WinTess (Sastre 1992), where each is a stand-alone program with a focus on form-finding and simulation of woven-based structural membranes. In spite of their overall benefits over manual calculations, these software require engineering skills, thus reducing their accessibility to architects, who are trained to work in *CAD*-based (computer-aided design) programs like Rhino, VectorWorks, Sketchup or 3DMax.

2.2 Homogeneous Boundary Models Versus Material Heterogeneous Models

In current design practice multiple modeling environments are employed and several models are constructed. These include CAD (computer-aided design) models for design, CAE (computer-aided engineering) models for structurl evaluation and CAM (computer aided manufacturing) models for fabrication. Architectural design practice is bound to its means of representation. Today the dominant way to represent and model architecture is today digital, using homogeneous boundary models in conventional CAD-modelling enviroments. Here, the focus is on the geometrical representation of the designed object, described by the boundaries in available CAD software. When considering the variety of CAD software, architects and designers have at their disposal, there is a hierarchical distinction between boundaries and properties, shapes and composition (Tsamis 2013). Most of past and present CAD systems are based on the premise of material homogeneities. Designed objects are represented as uniform matter, defined by boundaries, modelled in 3D-Euclidean space (Requicha 1980). These lack the integration of material composition and structural performance. As a result, they fail to accurately represent material complexity of graded materials and hinder exploration of material possibilities for variation and mass-customization that digital fabrication can enable. Engineering softwares simulate object performance and propose its composition, however are rarely useful for design as aimed for structural evaluation rather than morphological exploration.

2.3 Challenges of Using Homogeneous Models When Designing Heterogeneous Materials

Knitted structural membranes occupy a new branch within membrane architecture as they represent radically different capacities if compared to conventional homogeneous membranes. The capacity of controlled stretch of graded CNC-knitted membranes defines their behaviour and shall be taken into account early on during the design process.

The separation of design process into several disconnected environments leads to the decoupling of design from material properties and fabrication constraints, thereby reducing the quality of design solutions, as each iteration and adjustment is cumbersome, slow to make and therefore avoided. Such approach demotivates design production improvements and hinders the opportunity to create sustainable material practices in architectural design. In order to improve the quality of taken design decisions, and better reflect knitted membranes heterogeneities. It is essential to create integrated design models. This would pave the way for more sustainable manufacturing and more efficient design processes.

2.4 Digital Design Chain Paradigm for Knits—Integrated Digital Design, Analysis, Simulation and Fabrication

The design of FGMs involves interdisciplinary knowledge and therefore integrated *CAX*-based (computer-aided technologies, *CAD* + *CAE* + *CAM*) design is envisaged as one of the promising directions for the future (Nayak and Armani 2022). A way to create a workflow between multiple codependent models of a designed object is presented in the paradigm of the digital design chain (Dohmen and Rüdenauer 2007). In this paradigm, several models coexist in the same or tightly interconnected environments, all cross impacting one another. This guarantees minimal or even no data loss when transferring from model to model. Here morphological design, structural evaluation through simulation, material grading and planning for digital fabrication happens simultaneously.

Analysis of structural performance and manufacturing specification integration in the design environment empowers collaboration between professionals and eliminates the possibility of mistakes (Bechthold 2008). It helps to bridge each separate field, enhance interoperability and facilitate design tool integration into the entire process from idea to making.

As a final evaluation of a mature design, the simulation is commonly performed at the end of the design chain to complete the necessary structural documentation. Backward design changes are avoided and made only in cases of severe problems. By incorporating material heterogeneity through materially oriented simulation, it becomes active in the design explorative journey. Here the simulation is a part of extensive design tuning for improved performances. For materially differentiated designs, simulation plays an essential role in evaluating material capacities and performances, which in turn is tightly connected to their manufacturing. When integrated into a design environment and used iteratively, simulation allows to introduce material heterogeneities, test of complex material behaviour exposed to different conditions and skim through the design options. Here, the simulation plays a crucial role in creating a property representational model, where it contributes to the exploration and creation of design possibilities (Ramsgaard Thomsen 2016). This can be especially relevant for CNC knits, where the material grading defines their behavior and shall be taken into account early on in the design process.

Given the novelty of CNC-knitted membranes as architectural materials, reflected in their extreme heterogeneity, dynamic shape, scaling challenges during fabrication, complex material behaviour prediction and additive manufacturing, there is currently no ready-to-use software to address these challenges. To solve this problem, researchers investigate the development of dedicated models that would cover the entire design chain for knitted structural membrane design, from conceptual design to structural analysis and material simulation, as well as the preparation of manufacturing files.

2.5 Environments for Differentiated Simulation of Graded Knits

The achievements of state-of-art digital simulations of materially graded knitted structures have been boosted by the concurrent development of design and modelling environments, supplemented with parametric tools, allowing architects and designers to construct heterogeneous material models. The software Rhinoceros (Robert McNeel & Associates 1980) with its embedded platform for parametric modelling Grasshopper (David Rutten 2007) has gained enormous popularity and created a wide community, actively contributing to custom open source plug-in development, oriented towards particular tasks. A physics-based Kangaroo plug-in (Piker 2013) for Grasshopper makes it possible to simulate real-world materials and objects, engaging an intuitive sense of the material world, and allowing designers and architects to construct materially oriented simulations (Deuss et al. 2015; Tamke et al. 2014; Zwierzycki et al. 2017).

Similar to FEM software, in Kangaroo, the membrane structures are modelled through cable net mesh-based geometries, informed by material properties of rest length and stiffness, assigned to each mesh segment. The discretised nature of mesh-based membrane representation, supported by the parametric capacities of Grasshopper, allows us to automate and steer the allocation of these properties across large membrane bodies. Once the approximate desired shape is form-found, a trusted FEM analysis that would output the stresses and deformations should be conducted. Until recently this was done in the mentioned above external engineering software for FEM analysis (Fig. 4).

Several plug-ins integrate structural analysis in the digital design environment. They can be classified into two distinct groups based on how the membrane is represented—(1) *discretised* and (2) *iso-geometric*, where the first uses *meshes*, while the second—*NURB*-continuum *surfaces*. Examples of the plug-ins that operate on mesh-based representations are Karamba (Preisinger and Heimrath 2014), SmartForm (www.smart.burohappold.com), FunnelWeb (Kimm 2018), Rhino Membrane (Philipp et al. 2015) and K2Engineering (Brandt 2016). The NURBs-based membrane design has found its implementation in the Kiwi3D plug-in (Philipp et al. 2016; Bauer et al. 2018). The choice of either representational method, and hence the tools, should be done based on the accuracy and capacity of each method to represent the material's varying properties. The *iso-geometric* method uses a smaller number of external control points in order to describe the geometry through B-spline formula, involving the degree, control points and knots (Piegl and Wayne 1997). However, considered more lightweight, it is not able to incorporate varied material properties into

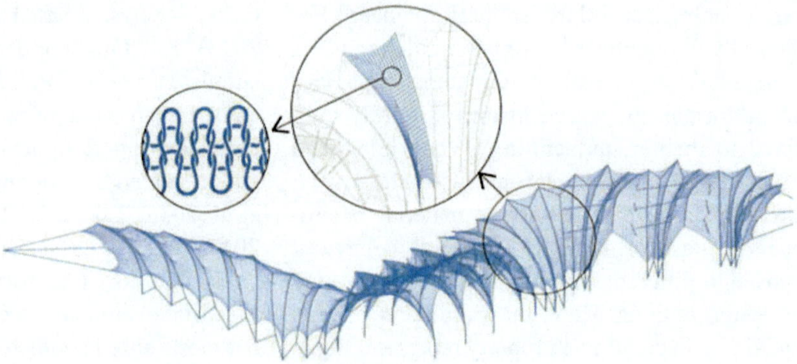

Fig. 4 Isoropia form-finding represented through quad-based mesh discretisations

the used NURBs surface. The *discrete* mesh-based representation uses a larger number of geometry subdivisions, allowing very fine changes in geometry properties to be constructed. This method is considered to be more applicable for modelling heterogeneous knitted membranes as it allows a detailed allocation of varied stretch properties, linked to mesh fine segments.

2.6 Properties Assignment Within Knit Simulations

Understanding property allocation mechanisms enables precise and accurate form-finding and simulation of differentiated membranes. A certain level of tolerance and imprecision is accepted within membrane simulations for representational and manufacturing purposes. Nevertheless, for some topological design solutions, a higher degree of geometric precision is required in order to avoid excessive membrane surface areas with reduced tension and the misalignment of compound elements due to surpassing of tolerances.

Before the emergence of plug-ins that allow for structural evaluation of membrane designs, mesh properties were assigned through an arbitrary cable stiffness value, which is comprised of *rest length* target values, calculated by Kangaroo Goal[1] as a percental reduction of the cable dimensions and its *strength* of achieving the target dimension. As a result, a differentiated mesh contraction is achieved during form-finding by varying these values using Kangaroo Solver (Mcknelly 2015; Baranovskaya et al. 2016; Karmon et al. 2018; Quinn et al. 2016; Hörteborn and Zboinska 2021). The results yield topologically similar geometries to physical outcomes as well as simulate the different degrees of mesh contraction for representing the different stiffnesses of the membrane. Form-finding simulations generated from these studies cannot be guaranteed to be geometrically accurate, since no quantitative comparison with physical models is made.

Alternatively, the use of K2Engineering cable goals allows to operate with the *E-modulus stiffness value* (Young Modulus[2]) in order to describe the material properties of the membrane, thereby departing from arbitrary stiffness value, used in the Kangaroo Cable goal. This allows for a more precise membrane material simulation, linked to the existing engineering standards of representing material properties through Young's modulus. However, the determination of the right Young's modulus value for the differentiated knitted membrane is a challenging task due to resolution misalignments and lack of studies. The

[1] *A component in the Kangaroo, K2Eng and Karamba plug-ins, that operates with the certain formula in order to describe the behaviour of the simulated object.*

[2] *The modulus of elasticity in tension or compression is a mechanical property that measures the tensile or compressive stiffness of a material when the force is applied lengthwise.*

differences occurring between sparsely discretised digital cablenet mesh geometry, which represents a much denser physical differentiated knitted membrane require the invention of *resolution translation concepts*. These are not present for heterogeneous knitted structural membranes in the available literature but are essential for building closer links between simulated and physically produced knitted membranes.

3 Extension of Design Models for Materially Differentiated Textile Membranes

The workflow presented here is based on the comprehensive design-to-production for knitted architectural membranes, developed over the past years by the authors (Deleuran et al. 2015; Ramsgaard Thomsen et al. 2019a, b; Sinke Baranovskaya et al 2020; Sinke et al. 2022a, b). This paper shows recent advancements especially the development of a *method for material differentiation informed by structural analysis* and a *method for heterogeneous membrane simulation calibration*, incorporating the data from structurally informed material differentiation and the values, derived from a physical experiment with homogeneous membranes of various densities. Both methods are closely interconnected as they inform each other, which is demonstrated in the example of design case study of Graded Knitted Ceiling panels.

3.1 Graded Knitted Ceiling—Design Case for Testing Digital Tools

Panels are three-dimensional, where the curvature is achieved by grading the programmed membrane pattern, based on the loading conditions of the membranes in the digital space (Figs. 5 and 6). The design intent drives the definition of the weight value in kg and their distribution across the membrane surface. Larger weight values produce deeper membrane geometry with a larger curvature, while the discontinuous edge clamping rearranges the tension distribution, where both directly reflect in the material grading patterns (Fig. 7). In order to achieve differentiatited stretch property two stitch types are combined—double jacquard stitch and unravelled version of it, where the stitch gets 3 times larger. We can control which stitches remain compact, and which get unravelled, achieving the controlled expansion of the fabric. Then panels are digitally produced with Serafil high tenacity polyester yarn on the industrial knitting machine STOLL 822 ki, using 12.8 N value for takedown and 9.8 N stitch length values. After that, panels are installed in space by clamping to the ceiling support structure (each frame 1.2 × 1.2 m), and loading with the weights in order to stretch the fabrics. In comparison to the digital environment, where the differentiated weights are used to trigger material differentiation, physical panels are loaded with the weights of same value. This is for activating

Fig. 5 Materially differentiated membranes, where the shape is informed by material differentiation, activated by the loading with weights and edge clamping conditions

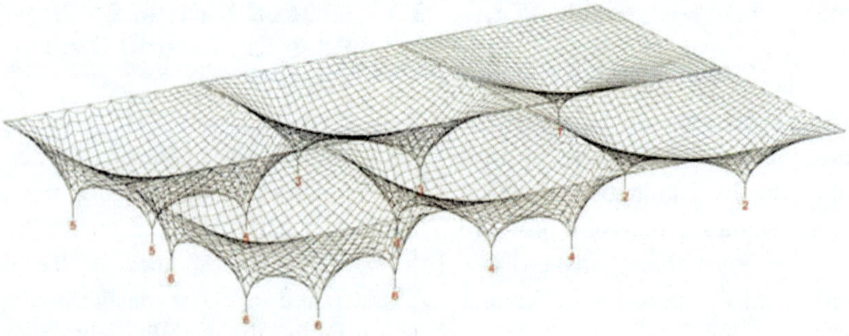

Fig. 6 Digital conceptual model of the textile ceiling panels with the increased surface depth transition. The arbitrary stiffness value used is 30 MPa in both U and V mesh directions

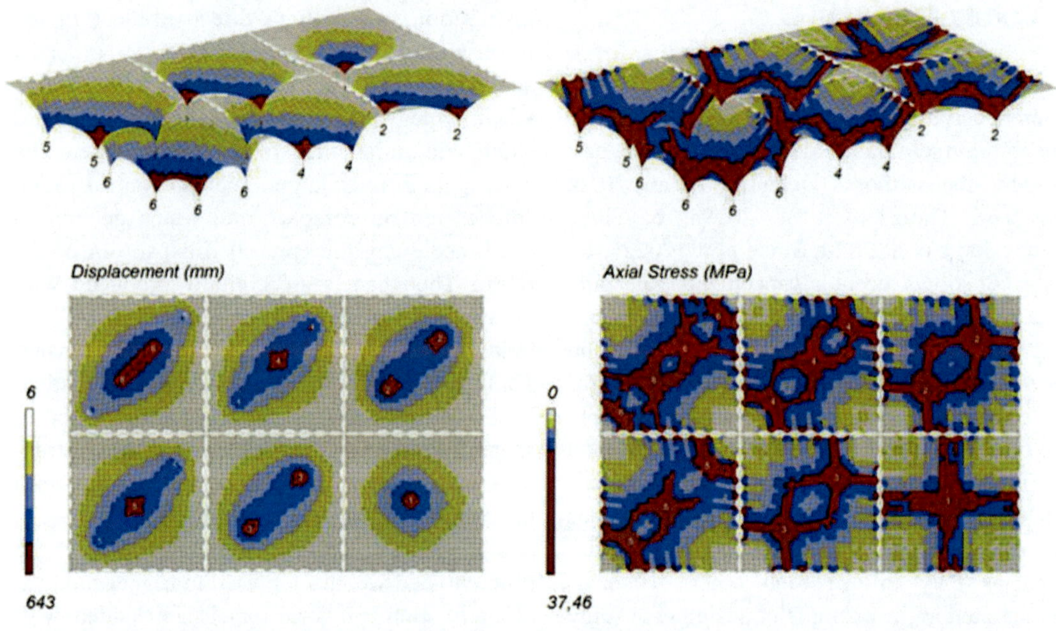

Fig. 7 Difference between displacement-based geometrical analysis (left) and tension-based structural analysis (right)

material differentiation, which is rather the result of membrane graded properties, rather than of differentiated loads.

3.2 Workflow Extension by Introducing Two-Stage Simulation

In order to determine the shape and grading within each membrane, we introduce a two-stage simulation, *A* and *B*. *Simulation A* is homogeneous in its set up and used to form-find a preliminary shape of a membrane in order to determine a plausible topology and conduct the structural analysis. The second—*Simulation B* – is materially informed differentiated simulation with distinct stiffness values per mesh area. It is built up on the results of structural analysis from Simulation A. The calibration of this model is done against the set of relevant knitted samples.

4 Finding 1. Structurally Informed Material Differentiation for Digital Manufacturing

4.1 Simulation A—a Homogeneous Form-Finding with K2Engineering Goal

As mentioned above, one of the findings described in this paper is an invention of a novel method for CNC-knitted membranes *material differentiation informed by structural analysis*. Its development is connected to the use K2Engineering goals already at the stage of preliminary form-finding during Simulation A. The membrane for Simulation A is constructed from a homogeneous mesh-based cable net with an *arbitrary* Young Modulus value, assigned evenly to each mesh cable segment. An arbitrary value is derived from standard handbook knowledge, as there is yet no information on membrane specification and its behaviour at this stage of the design process. Naturally, the designer operates with the parameters of membrane shape, clamping and prestress conditions, which in our case is the loading of membranes by weights, distributed across the surfaces and its attachment to the frame structure of the ceiling.

4.2 Structural Evaluation

After the preliminary form-finding (Simulation A) is converged, the *structural analysis* is conducted to extract the forces and the axial stress distribution across the resulting form-found mesh. It is a novel approach of membrane performance evaluation complementary to the earlier developed method of *geometrical analysis*, that calculates the surface displacement during the form-finding (Sinke et al. 2022a, b). The diagram in Fig. 7 compares these two analysis methods applied to the same geometry under the equal loading conditions.

In the current finding, the structural analysis is calculated with the aid of the BarOutput component of K2Engineering plug-in, where each mesh segment is associated with the axial stress value in MPa (Fig. 8A). Next, the axial stress values are interpolated into an average value per mesh face, allowing to cluster of the mesh *per mesh face*, rather than *per mesh cable* (Fig. 8B). This introduces the concept of treating the knitted membrane in simulation as a *shell* with distinct areal material properties, as opposed to a cable net made of single strands. This approach makes it possible to link the dense knitted physical surface to the sparsely subdivided mesh setup and introduces the concept of scalar translations between the low-resolution mesh and the high-resolution knitted fabric. As a result, meshes are clustered into zones, described by colour, where each colour is a result of a distribution of threshold tension data, present in the surface. Based on the resulting clustering arrangements, further material differentiation *for digital manufacturing* and for *heterogeneous simulation* can be performed on the mesh.

4.3 Knit Material Differentiation, Informed by Structural Evaluation

For further digital manufacturing of knitted membranes, the transition of differentiated material properties from the analysis is conducted through the dithering[3] graphic technique, due to the binary nature of knit. This permits the maintenance of the smooth variegation of material property for stretch while using only two stitch types for material differentiation—double jacquard (solid) and unravel double jacquard (loose). The dithering pattern is controlled through the percental density of double jacquard stitch type. This allows to correlate the tension-based clustering of the surface with the material density, where the areas of larger tension values receive the lowest material density, through the increased intensity of unravelling stitches,

[3] – *is an intentionally applied form of noise used to randomise quantization error, preventing large-scale patterns such as colour banding in images. A common use of dither is converting a grayscale image to black and white, such that the density of black dots in the new image approximates the average gray level in the original.*

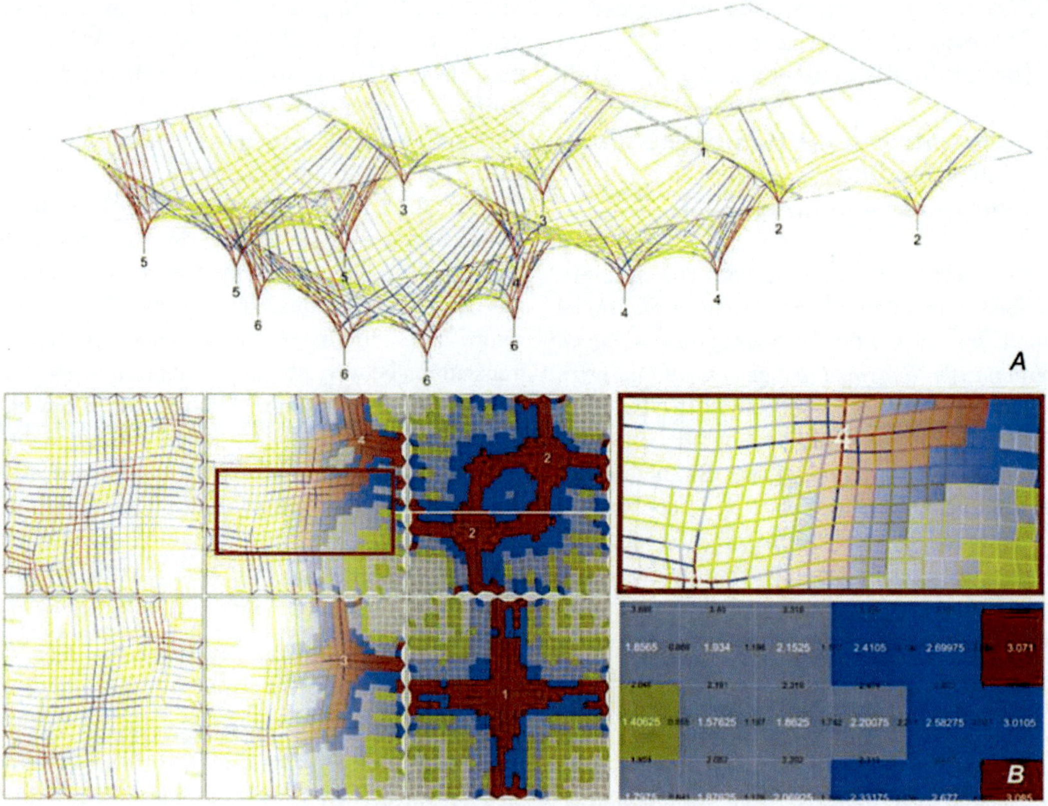

Fig. 8 Axial stress distribution in mesh cables and the translation of mesh cable tension value mapping per bar (mesh segment) to per mesh face

while the areas that experience less tension, remain dense, as they do not need to expand that much in order to achieve desired three-dimensionality.

This is counter-intuitive from the perspective of material performance for strength, as traditionally more dense textiles are perceived as more strong and expected to be in the areas of higher tensile forces. However, with the use of high-performance yarns, the integrity of the filament is guaranteed, which maintains the surface knitted strength despite stitch density. Here, the larger loops lock into their final locking position with the same strength properties as the smaller loops. Placement of larger loops in the areas of required larger deformation is very efficient in achieving material expansion and reaching required three-dimensionality with the fewer resources through the steering of material density. As the result, *cluster colour maps* are generated for each design, that are then translated to knitting machine-readable bitmaps (Fig. 9)

5 Finding 2. Structurally Informed Material Differentiation for Heterogeneous Simulation—Calibration Through Physical Prototyping with Homogeneous Membranes

With the membrane structural analysis, the cluster colour map is not only used for *material differentiation for digital manufacturing*, but also for *materially informed simulation*, incorporating these structurally derived surface gradings for more accurate digital representations of heterogeneous CNC-knitted membranes. Here each

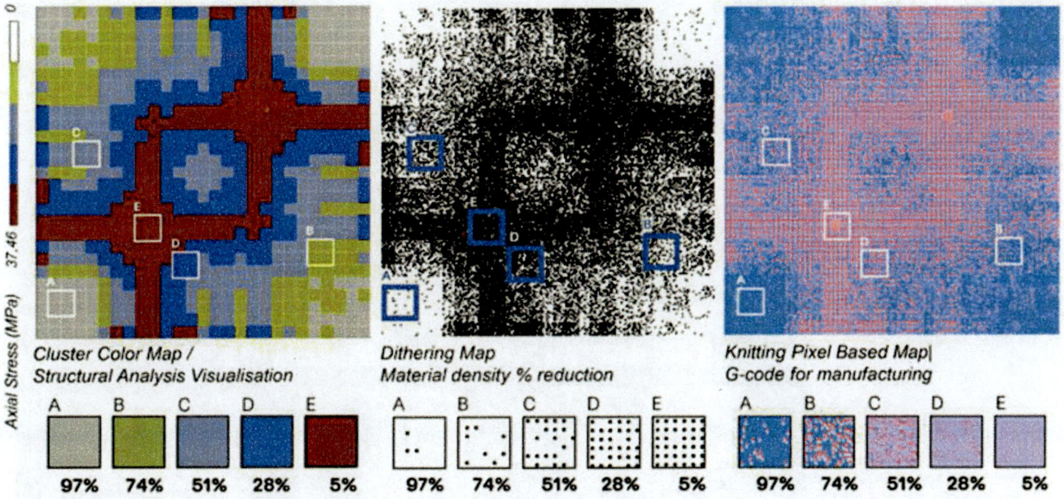

Fig. 9 Grading of the knitted heterogeneous membranes linked to the stress distribution in the surface and driven by the percental reduction of material density

colour zone of the mesh receives a specific stiffness value in relation to the corresponding material density of the knit.

The assignment of stiffness value for each area is not a trivial task, since no definitive understanding has been developed on how to transfer from an arbitrary stiffness value used in Simulation A, nor how to correlate material density in the form of stitches with a specific Young's modulus value.

Until now, the method of extensive physical prototyping has been used in order to tune the graded simulation. A 3D-scan data from the equivalent 1:1 physical prototype (Fig. 11), with corresponding graded areas as in the digital simulation, is compared to the digital simulation and in the event of deviation, the stiffness values are adjusted in order to fit the physical membrane. This can be done manually, by tuning the numbers incrementally, which is a tedious task, or with the use of evolutionary optimisation algorithms in order to find the right numeric values for stiffnesses (Sinke Baranovskaya et al. 2020; Oghazian et al. 2022). It is a valid method to gain more precise simulations. However, the need to produce the entire design first and only then be able to tune the simulation is not giving many benefits, as is expensive at first and does not inform the design, as structural deviations are only visible once fabricated. The results from such calibration also highly depend on many factors such as design topology and stitch types used, which are highly unlikely to be successfully transferrable so easily to other design topologies using alternative stitch types.

We find it important to develop a method that would aid the understanding of digital material properties of graded membranes early in the design process, without excessive 1:1 scale prototyping or the production of the final design. The aim is to find a way how to calibrate a computational system in a more generic way, that would improve the digital representation of materially graded CNC-knitted membranes. In this chapter, we present a proposal, on how to correlate the material density with the digitally assigned stiffness value with the aid of a supporting physical experiment.

5.1 Correlation of Material Density with the Stiffness Values

The differentiated knitted material has a certain limit of stretch and therefore a limit in stiffness. In order to understand the range of stiffness values possible to achieve with a certain material density, a few homogeneous panels representing

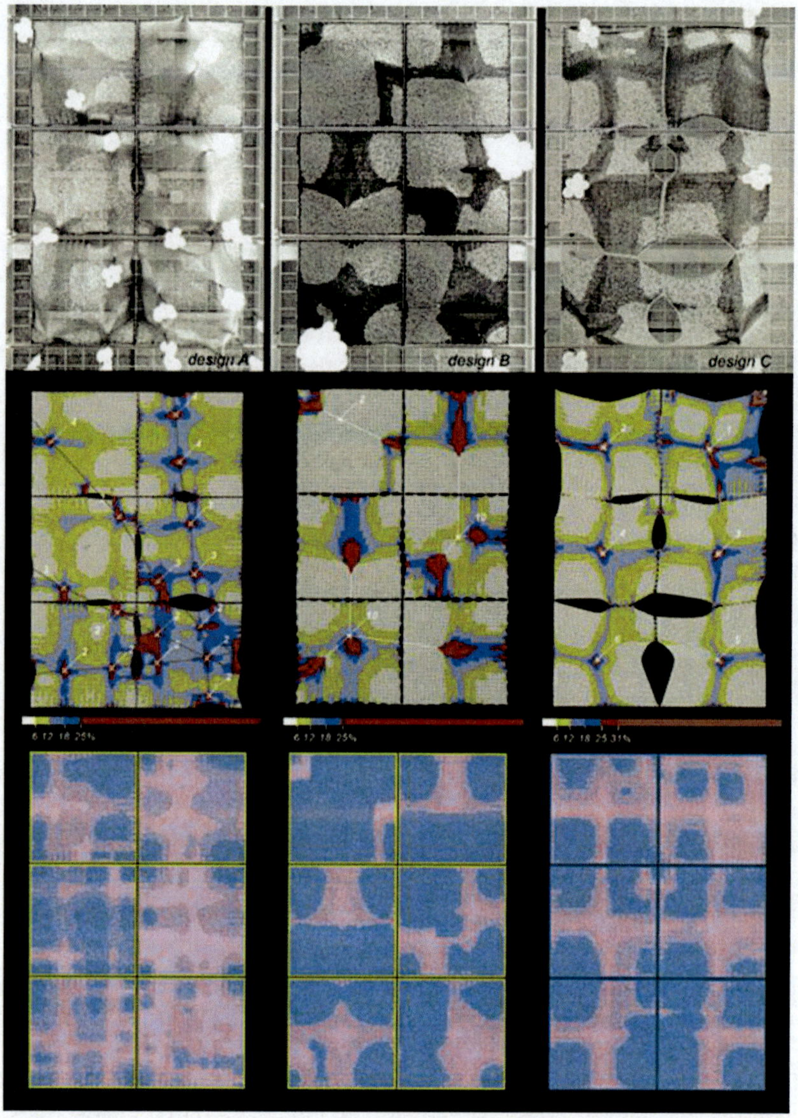

Fig. 10 Three installations of Graded Knitted Ceiling project, each with different loading conditions, reflected in variegated material differentiation of membranes (Fig. 10). Design C is used for further simulation calibration

the *maximum*, the *minimum* and the *medium* material densities are fabricated and loaded with the arbitrary weight of 2.9 kg. The exact loading value is not important here as it is important that it remains the same across all the membranes. All three membranes are produced with the same knitting machine settings, however with varied material density, reflected in the percental presence of unravel stitch—0%, 50% and 100% accordingly. The pattern density defines the textile's capacity for expansion under the load (Fig. 11). The dimensions and the material used for this experiment are the same as in the design that is intended for calibration.

5.2 Stiffness Values Discoveries

The results of the calibration of the homogeneous model through Simulation B are documented in the diagram in Fig. 12, where the digitally scanned physical samples are compared to the digital simulation B. With the aim to reduce the deviation between the scanned and the digital

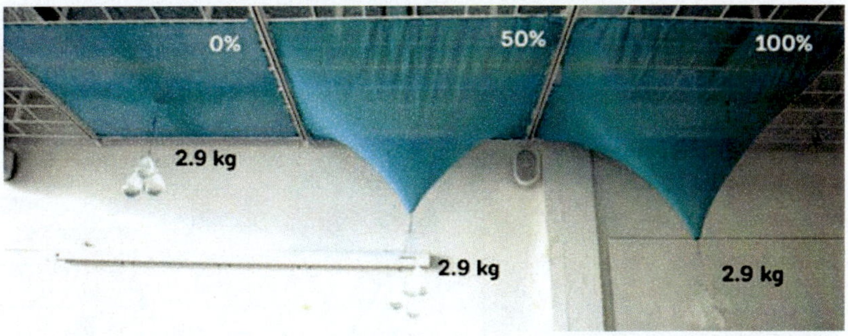

Fig. 11 Three homogeneous samples with various material densities loaded with the same weight

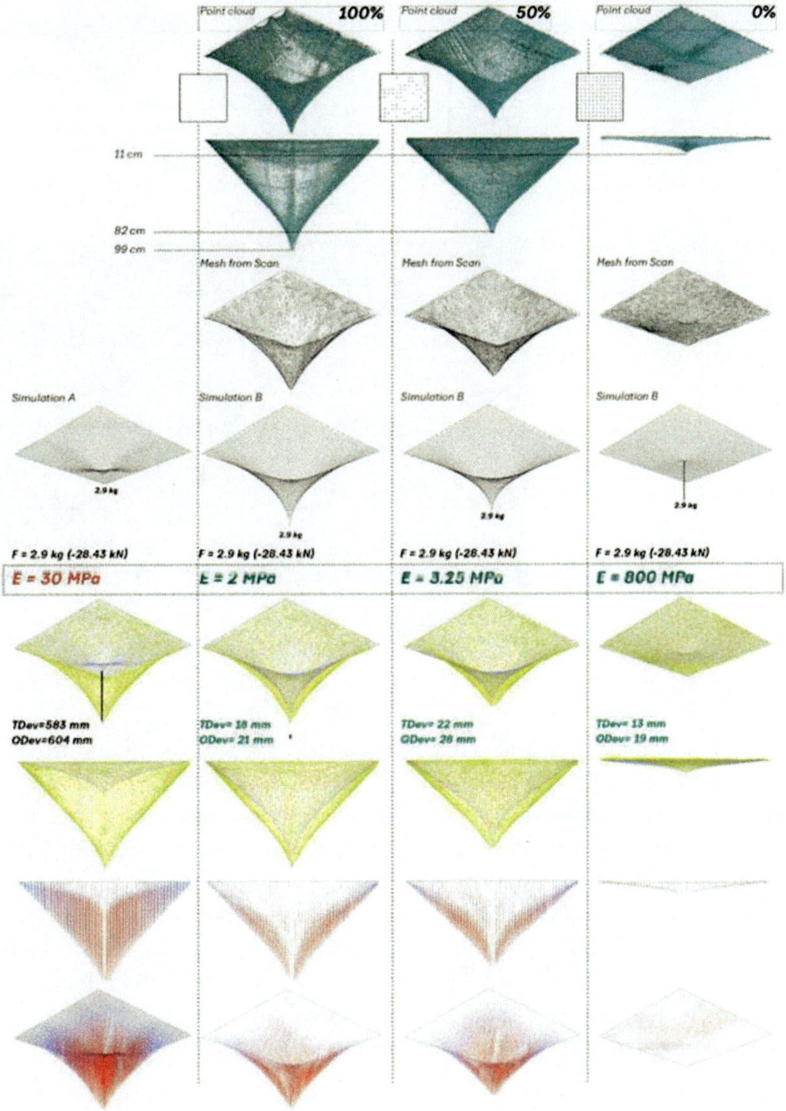

Fig. 12 Diagram showing the calibration of the model through the set of homogeneous samples, each of different material density—ranging from 0% unravel stitch reduction to 100% unravel stitch reduction. TDev—tip deviation, ODev—overall deviation

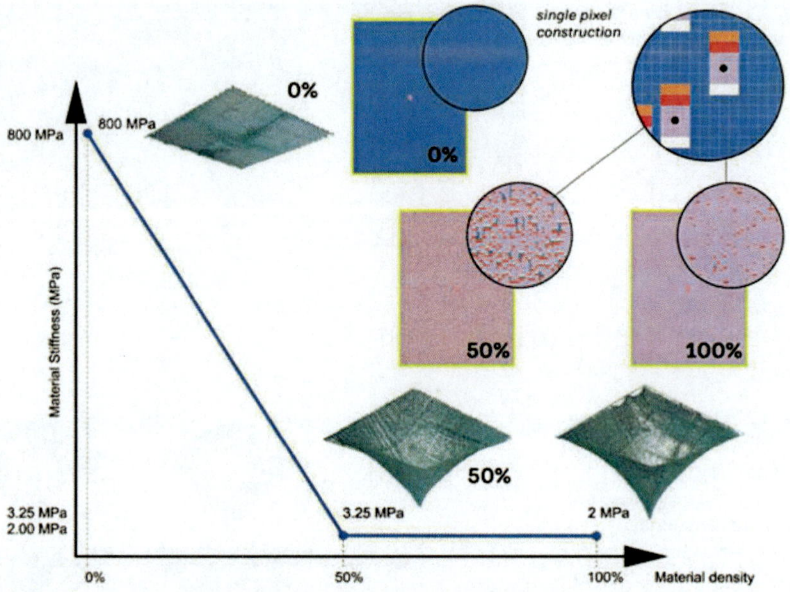

Fig. 13 The curve graph showing a non-linear relation between the material density vs material stiffness property

Fig. 14 Panels C design, used for the digital calibration of the model

mesh, the values of stiffness values are found for each corresponding membrane material density for 0%−2 MPa, 50%−3.25 MPa, 100% −800 MPa. The stiffness values discovered through this exercise demonstrate the non-linear relations between the material stiffness value and its material density (Fig. 13). Additionally, a more steep stiffness variation is expected between the 0% and 50% material density, while between the 50 and 100% the stiffness shows almost a flat curve. This can be explained through how the fabrication file is constructed, where each point from the dithering map, that is used to calculate the material density, is later assigned more pixels around it for technical reasons of executing the pattern on the machine. This results in an accumulated density of the unravelled stitch types. Despite that, the same principle is used to define the material densities in the ceiling panels intended for calibration, which guarantees the consistency of the experiment and the method (Fig. 14)

5.3 Calibration of the Heterogeneous Model

The graph of stiffness values, discovered during the homogeneous model calibration allows to estimate the stiffness values for other material percental densities (97%, 88%, 64%, 19%, 5%), present in the design intended for calibration (Fig. 15). The reapplication of corresponding intermediate stiffness values demonstrates a much more tuned digital simulation with visibly reduced deviation from the scanned data of the physical installation (Figs. 16 and 17), however, some deviation remains. Figure 16 demonstrates how resulted calibrated mesh of Simulation B correlates to the scanned data in comparison to same overlay of Simulation A with the scanned data. There it is visible that the calibrated mesh reaches much closer to the physical prototype in terms of depth, however, the curvature of the mesh is not entirely followed. This can be explained with the quad-based setup of the mesh, which lacks the springs in order to control the vertical displacement more precisely. Additionally, the method of Simulation B uses the discovered values on mesh bars, while there should be a backward method on how to translate values from mesh face to mesh bars as the latter ones are used for assigning the properties.

6 Conclusion

This paper describes a contribution on two aspects to design and fabrication of materially differentiated CNC-knitted membranes: (1) the material differentiation based on structural analysis and (2) the improved simulation of materially differentiated membranes by means of material probes.

The first contribution presented in this paper is the newly developed approach for material differentiation by using structural analysis. It allows for the introduction of free angles for membrane stretch, thereby reducing the unutilised areas of the membranes that would otherwise appear as sagging surfaces if another method for material differentiation is used (e.g. surface displacement strategy) (Fig. 18). Analysis-driven emerging patterns correspond to increased mesh geometrical complexity, while also providing pattern solutions that are not possible to predict manually or analytically. This leads to increased freedom in CNC-knitted membrane design, providing a more sustainable alternative to already employed woven textiles in membrane architecture.

Another contribution presented in this paper is a method for calibrating the simulation of heterogeneous membranes. This calibration

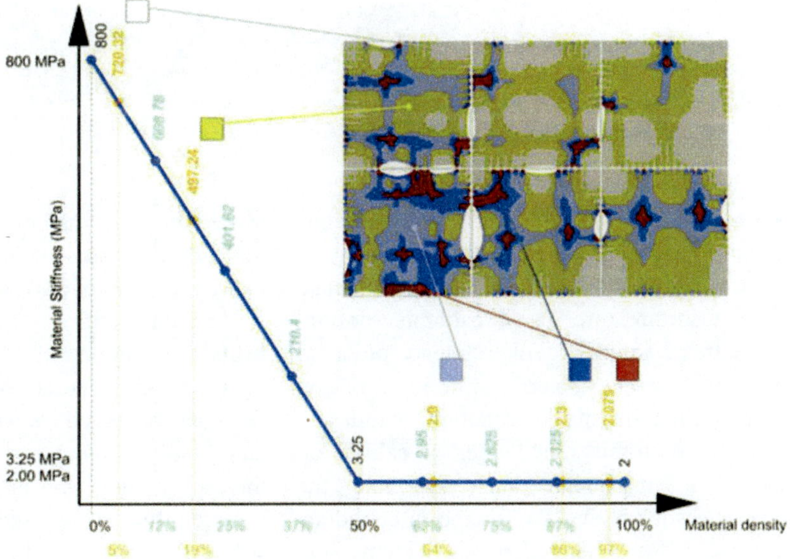

Fig. 15 Graph of correlated stiffness values related to the material density and linked with the material expansion capacity (new values for calibration are in orange)

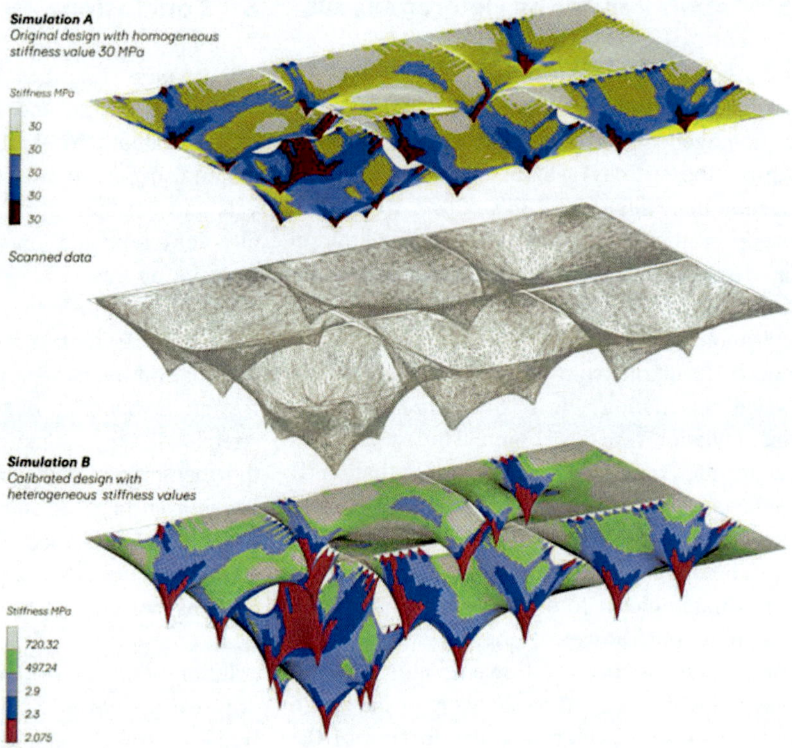

Fig. 16 Simulation A, Scanned data and calibrated Simulation B, applying discovered stiffness values, informed by the homogeneous membranes experiment

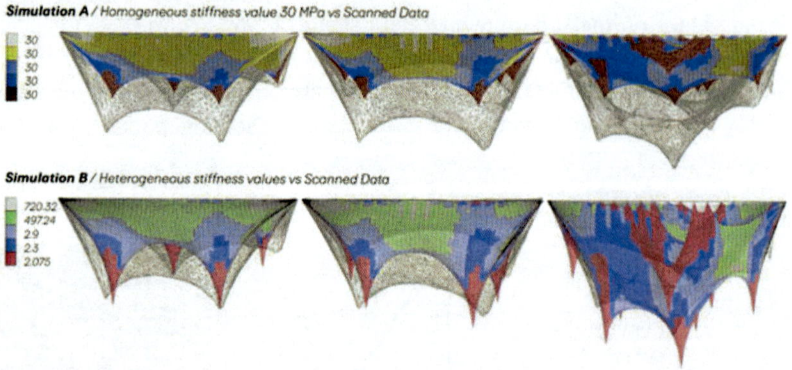

Fig. 17 Comparison of Simulation A and B to the physical prototype with the overlay of scanned data

process involves manufacturing homogeneous membranes with extreme densities, serving as reference points to determine the intermediate stiffness values required in the calibrated design. These extreme values provide reference points to understand how the material stiffness varies in relation to the density distribution within the membrane. With this knowledge, the calibration process can interpolate and determine the appropriate stiffness values for the intermediate zones of the heterogeneous membrane. Moreover, this method can be replicated with other design cases of alternative pattern combinations or employed yarns.

The found stiffness values are then used in the calibrated digital simulation to achieve more geometrically accurate outcomes. Certain deviations are observed between the calibrated mesh and the scanned data, which can be explained by the lack of prestress in the basic homogeneous samples, that were used to derive the extreme stiffness values. In other words, the

Fig. 18 The CNC-knitted textile panels design with angles loading directions demonstrates the benefits of using tensioned-based material differentiation strategy

homogeneous samples should have been loaded with the larger weight to provide sufficient pre-stress. As 2.9 kg used in the experiment might have been not enough to stretch the membranes to their maximum capacity, and therefore resulting three-dimensional geometry might be misleading from the start of the calibration process.

In general, this approach allows us to reduce excessive prototyping by avoiding creating the entire final design before we know the stiffness values that we can use to calibrate the simulation. Instead, it is sufficient to construct only three supporting samples to understand what stiffness range is possible with the given knitting setup and designed material density. To tune and confirm the values, additional 2-axis material tests of the smaller material samples of the corresponding densities can be conducted.

In the current situation of scarce resources, more materials, manufactured with additive tools will be invented. Additive technologies such as 3D printing or CNC knitting provide general access to methods of material grading. In order to be successful and efficient in designing with material gradings, we need to establish methods of how to speed up and ease the understanding and prediction of behaviour for these materials. Currently, physical sampling and prototyping can be considered the best way for understanding the material properties in order to tune simulations of FGMs. The presented method for simulation tuning uses only a few physical samples, which reduces the volume of physical prototyping and proves a better understanding of digital and physical material correlations of various material densities. The data from preliminary prototyping feedback into the construction of full representational models aid the implementation of these novel heterogeneous materials for real applications with the reduced risk of material failure. Our hybrid approach makes use of worldwide available digital fabrication technology and can in this way be used in a general way. An improvement and generalisation of the calibration of FGM simulation is part of future work.

References

Baranovskaya Y, Prado M, Dörstelmann M, Menges A (2016) Knitflatable Architecture—Pneumatically Activated Preprogrammed Knitted Textiles. In: Complexity & Simplicity—Proceedings of the 34th ECAADe Conference

Bauer AM, Längst P, La Magna R, Lienhard J, Piker D, Quinn G, Gengnagel C, Bletzinger K.-U (2018) Exploring software approaches for the design and simulation of bending active systems. In: International Association for Shell and Spatial Structures (IASS) Symposium

Bechthold M (2008) Innovative surface structures: Technology and applications. Taylor & Francis

Beukers A, Van Hinte E (1998) Lightness, The inevitable renaissance of minimum energy structures. Uitgeverij 010

Brandt C (2016) K2Engineering

Chen X, Hearle JWS (2009) Structural hierarchy in textile materials: An overview. In: Modelling and Predicting Textile Behaviour

Deleuran Holden A, Schmeck M, Quinn G, Gengnagel C, Tamke M, Ramsgaard Thomsen M (2015) The Tower: Modelling, Analysis and construction of bending active tensile membrane hybrid structures. In: Proceedings of the International Association for Shell and Spatial Structures (IASS)

Deuss M, Deleuran Holden A, Bouaziz S, Deng B, Piker D, Pauly M (2015) ShapeOp—A robust and extensible geometric modelling paradigm. In: Modelling behaviour: Design modelling symposium 2015. pp 505–15. https://doi.org/10.1007/978-3-319-24208-8_42

Dohmen P, Rüdenauer K, Dohmen P, Rüdenauer K (2007) Digital chains in modern architecture. In: Predicting the Future [25th ECAADe Conference Proceedings/ISBN 978–0–9541183–6–5] Frankfurt Am Main (Germany). CUMINCAD, pp 801–804. http://papers.cumincad.org/cgi-bin/works/paper/ecaade2007_136

Grigoriadis K (2015) Material fusion: A research into the simulated blending of materials using particle systems. IJAC 13 (3–4): 313–33. https://doi.org/10.13140/RG.2.1.3210.5682

Hörteborn E, Zboinska MA (2021) Exploring expressive and functional capacities of knitted textiles exposed to wind influence. Front Arch Res 10(3):669–691. https://doi.org/10.1016/j.foar.2021.02.003

Karmon A, Sterman Y, Shaked T, Sheffer E, Nir S (2018) KNITIT: A computational tool for design, simulation, and fabrication of multiple structured knits. In: Proceedings of the 2nd ACM Symposium on Computational Fabrication. ACM, Cambridge Massachusetts, pp 1–10. https://doi.org/10.1145/3213512.3213516

Kimm (2018) Funnel web

Kou XY, Tan ST (2007) Heterogeneous object modeling: A review. Comput Aided Des 39(4):284–301. https://doi.org/10.1016/j.cad.2006.12.007

Kyosev YK (2012) The finite element method (FEM) and its application to textile technology. In: Simulation in Textile Technology. Elsevier, pp 172–222e

Li W, Han B (2018) Research and Application of functionally gradient materials. IOP conference series: Materials science and engineering 394(2):022065. https://doi.org/10.1088/1757-899X/394/2/022065

Mcknelly CL (2015) Knitting behavior: a material-centric design process. http://hdl.handle.net/1721.1/99249

Nayak P, Armani A (2022) Optimal design of functionally graded parts. Metals 12(8):1335. https://doi.org/10.3390/met12081335

Oghazian F, Brown N, Davis F (2022) Calibrating a Form finding algorithm for simulation of tensioned knitted textile architectural models. Sydney, Australia, pp 111–120 https://doi.org/10.52842/conf.caadria.2022.1.111

Philipp B, Breitenberger M, D'Auria I, Wüchner R, Bletzinger K-U (2016) Integrated design and analysis of structural membranes using the Isogeometric B-Rep Analysis. Comput Methods Appl Mech Eng 303 (May):312–340. https://doi.org/10.1016/j.cma.2016.02.003

Philipp, Breitenberger, Wuchner, Kai-Uwe Bletzinger (2015) Rhino membrane

Piegl L, Wayne T (1997) The NURBs book. Springer, Berlin, New York

Piker D (2013) Kangaroo: Form finding with computational physics. Archit Des 83(2):136–137. https://doi.org/10.1002/ad.1569

Preisinger C, Heimrath M (2014) Karamba—A toolkit for parametric structural design. Struct Eng Int: J Int Assoc Bridg Struct Eng (IABSE) 24(2):217–221. https://doi.org/10.2749/101686614X13830790993483

Quinn G, Deleuran AH, Piker D, Brandt-Olsen C, Tamke M, Ramsgaard Thomsen M, Gengnagel C (2016) Calibrated and interactive modelling of form-active hybrid structures, vol 10

Ramsgaard Thomsen M (2016) Complex modelling: Questioning the infrastructures of information modelling. In: Herneoja A, Österlund T, Markkanen P (eds) Proceedings of the 34th International Conference on Education and Research in Computer Aided Architectural Design in Europe. eCAADe (Education and Research in Computer Aided Architectural Design in Europe) and ITU/YTU, Oulu, Finland, pp 33–42

Ramsgaard Thomsen M, Nicholas P, Tamke M, Gatz S, Sinke Y (2019a) Predicting and steering performance in architectural materials. In: Sousa JP, Xavier JP, Castro Henriques G (eds) Architecture in the age of the 4th industrial revolution—Proceedings of the 37th ECAADe and 23rd SIGraDi Conference—Vol 2. University of Porto, Porto, Portugal, pp 485–494. CUMINCAD. http://papers.cumincad.org/cgi-bin/works/paper/ecaadesigradi2019a_150

Ramsgaard Thomsen M, Sinke Baranovskaya Y, Monteiro F, Lienhard J, La Magna R, Tamke M (2019b) Systems for transformative textile structures in CNC knitted fabrics—Isoropia. In: Proceedings of the TensiNet Symposium 2019b Softening the Habitats (pp 95–110). https://doi.org/10.30448/ts2019.3245.08

Requicha AG (1980) Representations for rigid solids: Theory, methods, and systems. ACM Comput Surv 12 (4):437–464. https://doi.org/10.1145/356827.356833

Richards D, Amos M (2014) Designing with gradients: Bio-inspired computation for digital fabrication. pp 101–10. Los Angeles (Califormia), USA. https://doi.org/10.52842/conf.acadia.2014.101

Robert McNeel & Associates (1980) Rhinoceros 3D

Rutten D (2007) Grasshopper 3D

Sastre (1992) WinTess

Sinke Y, Tamke M, Ramsgaard Thomsen M (2020) Simulation and calibration of graded knitted membranes. In: ACADIA 2020: Distributed Proximities/Volume I: Technical Papers [Proceedings of the 40th Annual Conference of the Association of Computer Aided Design in Architecture (ACADIA) ISBN 978–0–578–95213–0]. Online and Global. 24–30 October 2020. Slocum B, Ago V, Doyle S, Marcus A, Yablonina M, Del Campo M (eds), pp 198–207. CUMINCAD. http://papers.cumincad.org/cgi-bin/works/paper/acadia20_198

Sinke Y, Ramsgaard Thomsen M, Albrechtsen DS, Tamke M (2022a) Design-to-Production workflows for CNC-Knitted membranes. In: Hybrids & Haecceities. Philadelphia, USA

Sinke Y, Ramsgaard Thomsen M, Tamke M, Seskas M (2022b) Strategies for encoding multi-dimensional grading of architectural knitted membranes. In: Towards radical regeneration. Springer, Berlin, Germany

Ströbel D, Singer P (2005) Computational modelling of lightweight structures. Textile Roofs

Tamke M, Quinn G, Evers HL, Deleuran AH, Gengnagel C (2014) The challenge of the bespoke: Fusion. In: Thompson EM (ed) Fusion, vol 2, pp 29–38

Tsamis A (2013) Software tectonics. MIT

United Nations, Department of Economic and Social Affairs (2015) Transforming our world: The 2030 Agenda for sustainable development. https://sdgs.un.org/2030agenda

Zwierzycki M, Vestartas P, Heinrich MK, Ayres P (2017) High resolution representation and simulation of braiding patterns: Acadia 2017. Acadia 2017: Discipl Disrupt MA 2–4 (November): 670–79

A Method for Designing with Deadwood for Architectural Acoustics

Isak Foged

Abstract

This paper presents the investigation and proposition of how to analyse and design with deadwood as a material resource, and how it can be applied as advanced acoustic design in architecture. The study is focused on Poplar wood in progressed decay, where visual, structural and sound characteristics are analysed and discovered through material studies, prototype studies, computational studies and measurement studies. The research findings have environmental, technical and aesthetical consequences for how we understand and rethink material resources, their structural state and how we can understand biogenic material transformations as part of the design process. Processes lead to a better understanding of using the regenerative materials we have available on the planet. The specific research contributions are increased knowledge of poplar density variances from natural decay and how density variance impacts sound absorption properties of the material and spaces. Furthermore, a new method for descriptive and prescriptive acoustic design processes is presented, based on image-analysis methods. This research and its findings are argued to open novel pathways for material practices as an approach that engage with biogenic material agency, which in turn empowers architecture to address urgent questions of material scarcity and material-climatic relations driven by the built environment.

Keywords

Rethinking biogenic resources · Architectural acoustics · Digital design method · Deadwood

1 Introduction

The current trajectory of using materials in the built environment does not align with the available materials. This is seen even in countries with a highly developed built environment, such as in Denmark, where the planetary limits of societal material use are reached by March (Lin 2021). Hence, even in well-informed countries with resources to transition to a sustainable society, we spend four times the resources for a balanced annual cycle. In a global context, each person on the planet spent on average 33 kg every day, with a projected increase to 45 kg every day by 2060 (OECD 2019). The material usage growth is coupled with another problem, which is that

I. Foged (✉)
Institute of Architecture and Design, The Royal Danish Academy, Copenhagen, Denmark
e-mail: iwf@kglakademi.dk

M. R. Thomsen et al. (eds.), *Design for Rethinking Resources*, Sustainable Development Goals Series,
https://doi.org/10.1007/978-3-031-36554-6_24

the increase will be non-metallic mineral materials constituting the vast majority of the development (ibid). These materials, such as concrete, have a high environmental footprint in its making, transport and recycling processes. Current discussion in Western countries encircle the argument for a de-materialisation and reduction as general processes for transforming to a sustainable situation (ibid). This will be a necessary strategy for countries that already spend too much, but in contrast, BRIC countries (Brazil, Russia, India, China), including the dominant part of the world population and population growth, the expansion of the built environment is driven by mechanisms which enable people to have the same living standards that others have enjoyed for decades (OECD 2019). Thus, while local reductions must take place, a global dematerialization appear less realistic until global equality in built environment quality is constructed.

Environmental responsible materials and viable quantities of materials through better use are therefore necessary. In this effort, research and industry efforts accelerate biogenic material investigations and applications, with construction timber being at the strategic foreground. In contrast, concurrently, forests around the world are decreasing in scale (Brack 2018; Eurostat 2016) removing the basis for wood-based material growth of raw materials. At the current level, global forests are producing 3912 million m^3 (2020) of raw timber (United Nations 2019), which means a large gap to the current and projected demand for construction and transformation of existing structures.

Forests consist, however, of more than raw material for certified construction timber. Deadwood, that is both lying and standing wood in decay, is a significant part of global forests, amounting to 20% of the forest material (Russell et al. 2015). Deadwood has important roles in maintaining the forest ecology, as it decays and rematerialises the forest bed, providing nutrients and homes for insects, plants and animals (Seibold et al. 2021). Hence, a significant deadwood capacity is a direct indicator of forest health and an important element for sustainable ecologies.

Yet, deadwood from forests emits 115% of the global fossil fuel emissions (Seibold et al. 2021). Rethinking deadwood as a material resource presents two technical and practical potentials, namely, a utilisation of materials that is readily available now and at the same time an immediate blocking of releasing the sequestered CO_2 into the atmosphere.

2 Material Characterisation

The study and application of deadwood in the built environment may offer additional potentials than a mere technical resource rethinking through quantity-based substitution of materials. As deadwood enters a process of material decay, it transforms its physical characteristics. This is seen in the change of colour, the material porosity, brittleness and density. Expressive qualities can be deciphered by visual observation and tactile registrations, and which are traceable by the observer back to the transformation process induced by local environmental forces. The expressed causality between the emerging material, its transformation and environmental forces drives the characteristics of both technical and aesthetical definition, such as a tree's positive and negative growth structures (Fig. 1). When considering an architectural form's articulation as a process of a materials response to a force, it is aligned with the theoretical notions of tectonics in architecture (Christiansen 2019; Foged & Hvejsel 2018; Frampton 1995; Hartoonian, 1994; Sekler 1965; Semper 1989). The advancement from common tectonic descriptions focusing on the articulation of gravitational forces is here based on environmental forces of wind, moisture, temperature and fungal growth processes. Hence, in material transformation processes, such as organic decay in the natural environment, a material and its form relation are at the same time structurally transformed and revealed. As an example, in the gradual decay of a tree, we can observe material distribution paths through

Fig. 1 Tree trunk in natural decay process revealing its inner structure and fibre composition

reversed growth processes combined with local environmental forces at play.

In this example, perceived material and form expression is the result of additive and subtractive formation forces. The material-environment processes undertake on-going articulations of structures beyond their initial state, where they become revealed through local, dynamic and complex force interactions. In such cases, a rethinking of materials as resources is based on a continuous understanding created by a temporal revealing of the relations between a material, a form and environmental forces (Foged 2015).

3 Material and Medium

Extending the position of environmental forces as drivers for new material practices in architecture, and as strategy for rethinking material resources, a force acts not only on the material, but also on the human. In the case of sound, an intimate and direct relation persists between the source of sound energy, a material form and the sound energy recipient. Sound is an environmental force manifested by pressure waves through air as the transfer medium (Cox & D'Antonio 2016; Long 2014). And while invisible, the perception of sound is formed by materials, when the pressure wave is either reflected, absorbed or transmitted by the material. Architectural theorist Steen-Eiler Rasmussen states:

> Can architecture be heard? Most people would probably say that as architecture does not produce sound, it cannot be heard. But neither does it radiate light and yet it can be seen. We see the light it reflects and thereby gain an impression of form and material. In the same way we hear the sounds it reflects and they, too, give us an impression of form and material. (Rasmussen, 1964).

Maybe less obvious than the visual, structurally expressed material-form convergence in a common tectonics discourse, material qualities and form definitions are captured and revealed by the sheer presence of a human, with sound being omnipresent, enveloping and partaking in defining the reading of a context.

4 Investigation

This study investigates how we can rethink resources by understanding, designing and applying naturally transformed wood by decay, both technically and aesthetically. This responds not only to a revealing of a material articulation, but also to a practical extension of wood's CO_2 cycle. Instead of organic material passing through a decomposing process, the uptake of CO_2 during its growth process, material and its embedded bonding of CO_2 is maintained as part of a carbon sink in the built environment. With the use of wood materials, which has entered a decay process, an explicit process of sustaining material for use, and CO_2 sequestering, resolves into a sustainable material discourse.

Specifically, the investigation explores and uncovers how a material in a decayed state, today considered non-usable, can be articulated with and for architectural acoustics, responding directly to the SDGs (11) sustainable cities and communities and (12) responsible consumption and production.

The paper presents experimental material-geometric studies using poplar wood in a progressed decay state. Through examination of the visually expressive material high density variance is discovered. This inhomogeneity is mapped and further investigated through acoustic impedance measurements to determine specific sound absorption coefficients. The knowledge derived from these studies are integrated into a computational design analysis model based on image-analysis techniques enabling the computing of highly specific acoustic properties when correlated with other findings in the literature and the impedance measurements, presenting a new method and model for material sound absorption estimation. Following the development of a full-scale cross-layered poplar prototype structure, room acoustic measurements are conducted in a testing space to understand the impact from experiential and aesthetic perspectives. The material-form and acoustic force findings are then presented, followed by a discussion and conclusion.

5 Methods and Materials

Studies are conducted through a mix-method approach with qualitative and quantitative investigations, including visual/tactile registrations of material samples, measurement of material properties, experimental prototype investigations, computational image-analysis with numerical filtering/mapping, absorption coefficient measurements, and room acoustic measurements and data analysis. As each sub-study includes specific methods and successive findings for progressive, experimental studies, some results are presented as a basis for following studies/methods description within this section.

6 Material Studies

The qualitative study of wood in a state of decay offers a direct decoding of the material condition. By opening the trunk by planar cuts along the grain, visual observation illustrates the process of fungal influence leading to spalting characterised by the noticeable black lines (Fig. 2). Spalting

Fig. 2 Piece of poplar with clear spalting patterns

occurs under specific environmental conditions, where fungi growth are provided by oxygen, moisture and temperatures between 20 and 40 degrees Celsius (Stange & Wagenführ, 2022). The pattern of fungi attack leaves the black colouring. The remaining wood is coloured in tones of yellow, brown, orange, green, white and light grey. Texture and patterns are particularly detectable close to the spalting, encircled by the dark lines. By touch, planned surface has varied smoothness, with brown and orange colour smoother and light yellow and dark-white-light grey with more rough surfaces. By pressing into the wood with a small blunt object, performing a sense-based Janka test of material hardness, brown-orange regions are significantly harder than the rest. A direct relation between hardness and smoothness appears to be correlated with visual decoding by colour of the material. A quantitative study determining material density characteristics is done by cutting smaller pieces from the plank, in dimensions 20 × 30 × 50 mm (Fig. 3).

The volume dimensions are found by analysing the wood for areas with consistent coloration/pattern, to create colour-based representative samples. 23 samples are made, which all are measured to 0.01 mm precision and weight with 0.001 g precision, Table 1. Densities, kg/m^3, for each sample is then calculated.

The study identifies a significant difference in material density within the samples, from the same plank of wood, ranging from 292 kg/m^3 to 521 kg/m^3. The common density description of Poplar is 350 kg/m^3 (Zhang et al., 2022). The deviation to the density norm and particularly between the samples studied indicates a material that is substantially transformed by the environment-driven decomposing and fungal decay process, when considering both its visual expressive character of colour, texture and pattern, and its physical properties of hardness and density.

7 Material Acoustic Studies

To further examine the material properties of the varied density of decay poplar wood, material acoustic measurements are conducted through Impedance Tube testing. The used instrument is a Brüel & Kjær Type 4206-T, with B&K LabShop signal generator, analysis and post-processing software, following the ISO 10534–2 testing standard. The impedance tube testing method is based on a specific instrument, using two different two-microphone procedures with small and large samples, Ø29 and Ø100 mm respectively (Fig. 4). Each measurement setup is based on calibrated microphones, noise-to-signal analysis and phase calibration through FFT analysis.

Fig. 3 Set of material samples that are examined for density variance, with lowest density in top left corner and highest density in lower right corner

Table 1 Material test sample measurements, weights and calculated densities

Sample	x	y	z	gr	mm^3	kg/m^3
1	29,49	19,82	49,97	9,917	29,207,05,525	339,5,412,484
2	29,2	21,78	49,95	10,872	31,767,0012	342,2,419,363
3	29,31	21,83	50,79	9,504	32,497,33,647	292,4,547,373
4	29,09	19,28	50,06	13,621	28,076,41,131	485,1,403,496
5	29,14	19,53	49,94	14,259	28,421,06,375	501,7,053,593
6	29,17	19,73	50,14	11,1	28,856,77,837	384,6,583,238
7	29,48	21,74	49,78	9,684	31,903,76,306	303,5,378,611
8	29,15	19,42	49,99	13,602	28,298,98,907	480,6,532,123
9	29,42	21,87	50,15	15,883	32,267,28,231	492,2,323,438
10	29,13	21,64	49,97	11,404	31,499,7488	362,0,346,331
11	29,22	21,76	49,98	16,044	31,778,64,346	504,8,673,655
12	28,97	21,43	50,24	15,616	31,190,3535	500,6,676,182
13	29,4	19,75	50,16	11,921	29,125,404	409,2,990,435
14	29,18	21,55	50,49	16,572	31,749,57,621	521,9,597,229
15	29,21	19,23	50,37	10,528	28,293,24,707	372,1,029,253
16	29,24	19,89	50,12	10,813	29,148,97,003	370,9,565,034
17	29,2	21,79	50,27	13,539	31,985,19,236	423,2,896,225
18	29,46	19,52	50,18	14,77	28,856,47,066	511,8,436,061
19	29,48	19,82	49,86	12,208	29,132,8789	419,0,454,381
20	28,97	21,84	49,95	13,55	31,603,60,476	428,748,559
21	29,19	21,71	50,13	12,97	31,768,12,794	408,2,708,312
22	29,27	21,77	49,86	12,572	31,771,18,589	395,7,044,613
23	28,8	21,74	49,9	14,195	31,242,9888	454,3,419,354

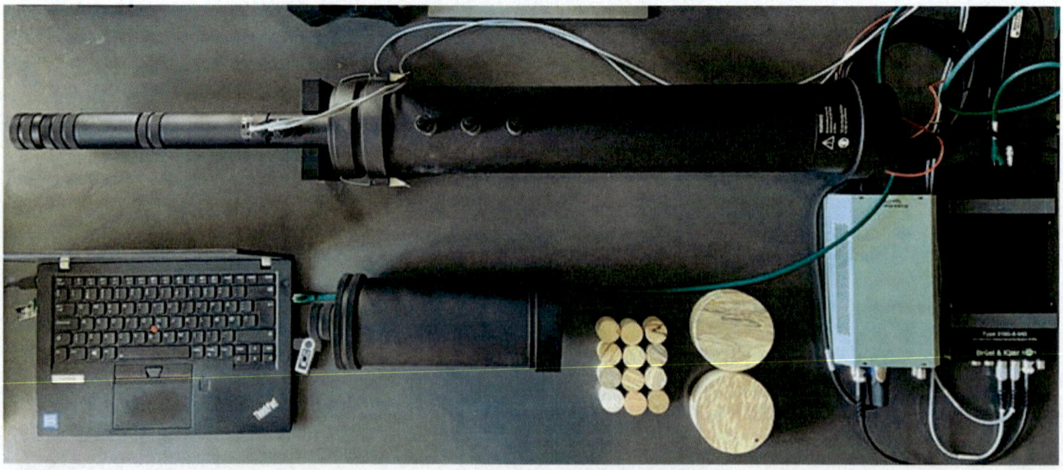

Fig. 4 Impedance Tubes testing instrument used for material acoustic analysis, including both large wide tube (low frequency analysis), small tube (middle and high frequency analysis), signal generator and specific computer/software for analysis, post-processing and export of results

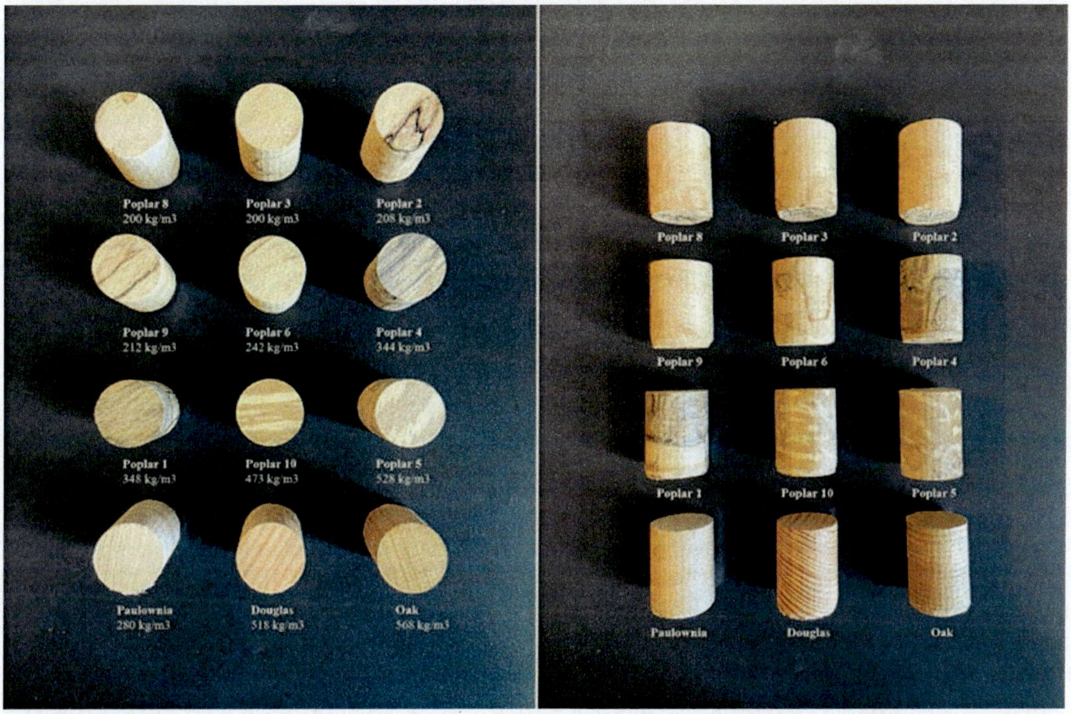

Fig. 5 Material samples (29 mm) prepared for Impedance Tube testing, including Poplar, Paulownia, Oak and Douglas fir wood species

Material samples are positioned in a measurement chamber, which is coupled with the signal generator tube housing the loudspeaker for the analysis signal produced. Measurements from 50 to 1600 Hz uses Ø100 mm samples/large tube setup, and analysis between 200 and 6400 Hz uses Ø29 mm samples/small tube setup. Given the large inhomogeneity identified in the density material study above, 9 different poplar samples are tested from 50 to 6400 Hz. For comparative analysis additional wood species are analysed, including Oak, Paulownia and Douglas samples. All samples have a depth of 40 mm (Fig. 5).

8 Computational Studies

With the discovered correlation between material colour and material density of the poplar wood samples, image-based material analysis can be used to capture, analyse and map material characteristics of the highly inhomogeneous material. Adding absorption coefficient data from the above study provides a dataset correlating colour analysis, material density and material acoustic absorption properties.

Image analysis has a wide variety and use in scientific investigations and applications, from engineering (Liu et al. 2021), psychology (Fleming, 2014; Frantz, 2003), computer science (Tanaka & Horiuchi, 2015) and architecture (Fragkia et al. 2021) among others. Image analysis provides the possibility for qualitative investigations by the observer, such as through direct pattern registrations and by indirect analysis of digital analysed images using false colour and brightness analysis methods, and as training data for artificial intelligence and machine learning methods. From the same image, quantitative investigations by computational analysis are possible, such as analysis of individual pixel properties. With each digital image pixel hosting four numerical values such as Red, Green, Blue and Alpha (transparency), a single image captures and holds a large number of material data. This study uses RhinoGrasshopper software for image analysis, image mapping, geometric modelling and sound absorption estimation based on the developed material dataset. The integrated image-analysis module (Image Sampler) is restricted to analysis of 40.000 pixel at a time. This results in analysis of 160.000 data points (RGB, alpha values) for each image-analysis processing pass.

With an image paired with the material dataset developed, correlating material colour and material density, a direct calculation and mapping of estimated density and absorption characteristics is made. The colour to absorption calculation uses data from the literature, establishing a relation between density and absorption coefficient from other studies (Nandanwar et al. 2017; Smardzewski et al., 2013, 2014) and the absorption coefficient measurements conducted above. A method for fast material analysis, material characterisation and material sound absorption properties estimation are then established to evaluate the highly complex material and compute material sound absorption relations by use of images (Figs. 6 and 7).

9 Prototype Experiments

To expand the studies from material probe investigation to full-scale prototype with room acoustic impact studies, a large acoustic surface of wood requires assembly. Experiments with cross-layer joining of poplar planks and boards are done by a three-directional fibre (wood grain) orientation approach. By this method, large, deep poplar planks can be positioned vertically, while front and back board elements, relative to the centre placed planks, can be angled freely to increase cross-weaving of the wood grain (Figs. 8 and 9). In the first full-scale prototype (Fig. 8) crossing is done with 45-degrees, with the front cross based on a 70 mm board height, extruding 60 mm from the plank surface. This technique increases stiffness and stability of the assembly, and more importantly introduces a varied deep structure that scatters sound energy in relation to frequency (Cox & D'Antonio, 2016; Cox et al. 2006). The back board element cross is 20 mm high, extruding 10 mm from the plank surfaces. Both crosses are recessed 10 mm into a milled track of the vertical planks that are 40 mm thick, 2000 mm high and 300 mm wide.

The poplar planks are almost planar in the first prototype experiment, yet much of the poplar

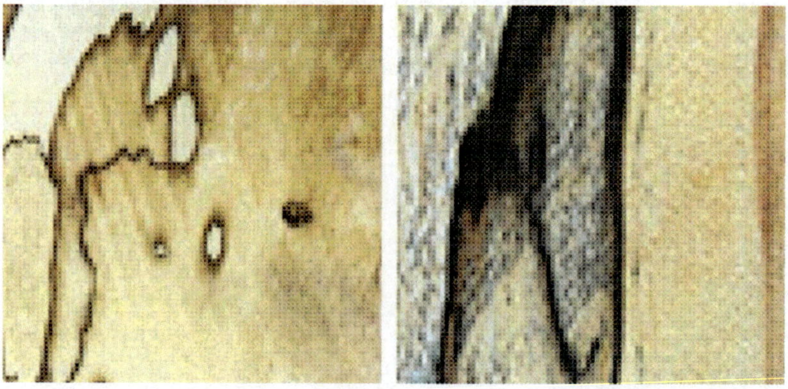

Fig. 6 Method for translating material visual character to material acoustic properties based on empirical studies of decay poplar wood

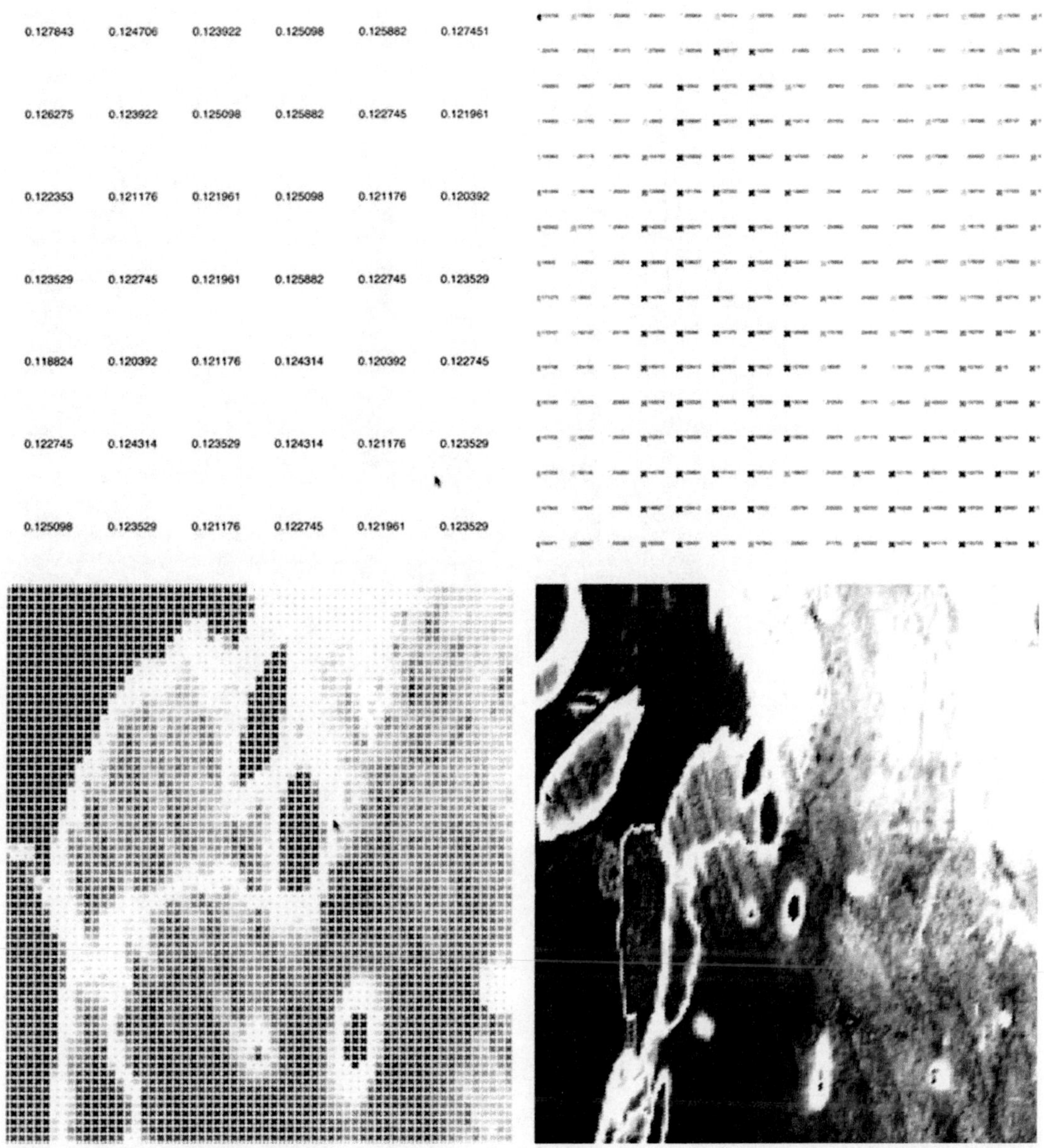

Fig. 7 Image analysis process showing the data resolution with a data point for each mm^2. Top left, zoomed in analysis (6 × 7 mm sample size), top right, zoomed in, pattern starts to be visible (15 × 16 mm sample size), bottom left, zoomed in, resolution and pattern is visible, but data points not visible, bottom right, pattern is clearly visible, data points not visible

Fig. 8 First full-scale cross-layer prototype with 45-degree crossing over two planks

wood has significant warping due to the moisture release process during drying of the anisotropic and highly inhomogeneous decayed material. The second full-scale prototype experiments with the assembly of three large planks (2000 × 400 × 400 mm) which warps more than 20 mm across the length of the plank (Fig. 9). The same three-directional fibre crossing method is applied, but with 60-degree angles to adapt to the large difference in local height differences between the planks. Using a deep front board element, as in the first prototype, the joining depth and structural overlay of the assembly elements are driven by the warping geometry's intersection with the crossing board. The same applies for the back crossing, albeit with the board being much lower in its protrusion height, relative to the plank surface.

Fig. 9 Second full-scale cross-layer prototype with 60-degree crossing over three planks

10 Room Acoustic Measurement Studies

The raw poplar planks and full-scale experimental prototypes of cross-structured poplar assemblies, with identified inhomogeneous material densities, are tested for material and room acoustic properties. The room used for measurements is a shoebox geometry, with dimensions 7.6 ×5.6x2.8 m. Ceiling and walls are in plaster and painted. Floor is hard linoleum. One wall is constructed from large glass panes. Three planar hardwood doors are placed in the room. The temperature is measured at 26 degrees Celsius and relative humidity is 55% during measurements. Prior to installing the prototypes for testing are 'clean' space test measurements conducted for comparison. Measurements are conducted with 6 measurement positions (averaged over 3 measurements at each position) and 1 source position, making a total of 18 measurements for each study (Figs. 10 and 11). Equipment used for measurements includes a calibrated Behringer 8000 measurement microphone and a Yamaha MSP5 active loudspeaker

Fig. 10 Room acoustic measurements with free standing poplar planks

Fig. 11 One of two positions of the prototypes placed in the measurement space

with a sound dispersion profile similar to human sound emitting patterns. Sound sweeping for impulse response is used with a 2 s period, and analysed from 125 Hz to 8 kHz, using Røde Fuzzmeasure software. The measurement setup complies to ISO3382-2 standard (ISO 2008). For analysis and results evaluation of the impulse response measurements, RT60 (T20 domain) and FTT Waterfall plots are produced and analysed.

11 Results

The findings from the investigations of rethinking resources using decay poplar wood for acoustic design propositions are outlined in categories of investigation.

Specifically

- Material and density

  Based on qualitative and quantitative investigations, the study finds that poplar in the decay state has significant density variance, ranging from 292–521 kg/m^3 in the specific study (Table 1).

- Material and sound absorption

  By quantitative investigations, the study finds that there is a direct, albeit complex material density relation with sound absorption qualities of the poplar wood studied (Fig. 12). In the graph including all material samples examined, two groups are identified. One group, based on higher material inhomogeneity (Poplar 1, Poplar 2, Poplar 3, Poplar 4, Poplar 6, Poplar 7, Poplar 9) as seen in the graph including all samples. Another group, based on the higher material homogeneity (Oak, Douglas, Poplar 5, Poplar 8) as seen in the graph with the four selected materials. As is documented, the Poplar 8 sample (200 kg/m^3) has significant higher absorption coefficient than Poplar 5 sample (528 kg/m^3).

- Material and Room Acoustics

  Based on quantitative investigations, the study finds that the poplar material and prototype assemblies have a significant impact on room acoustic properties and therefore the human perception of the material impact on experiential phenomena, with sound energy absorbed predominantly in the mid frequencies 250–1000Hz, and secondarily in the high frequencies 2000–8000Hz, dropping 0.5 and 0.3s in RT60, respectively (Fig. 13). This aligns with the material acoustic sample findings above.

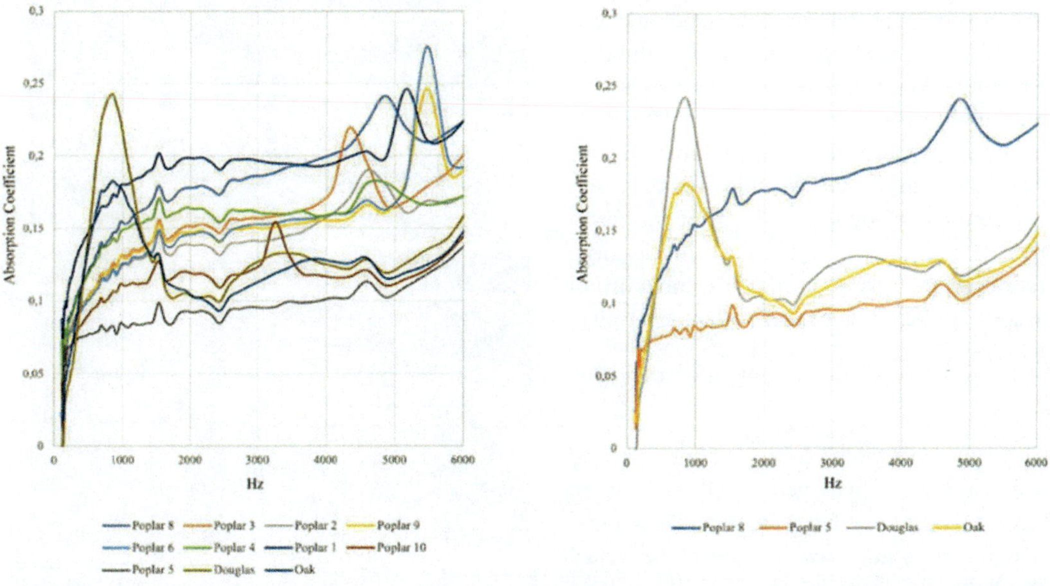

Fig. 12 Absorption coefficient results from impedance measurements. All samples to the left showing the absorption variance within the same material, and highlighted samples to the right showing the relation between density and absorption properties

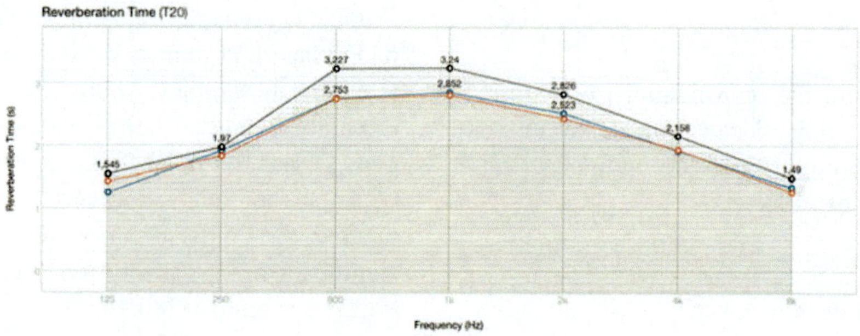

Fig. 13 Reverberation Time measurements with a 'clean space' (black line), prototypes against the wall (red line) and prototypes placed at an angle to one of the walls (blue line)

- Material and Design Analysis Method

  As the density of a fibrous material has a direct correlation with sound absorption properties documented, and in alignment with the literature, the study presents a computational design and analysis method for acoustic properties based on image-analysis techniques. The presented method is scalable in resolution in direct relation to the resolution of the sample photo of the material and the specific analyser used. The method can be used to analyse and inform the assembly logic of material complex bespoke elements, guiding design decision for a holistic intended sound absorption profile of a combined structure. While the detection of density and related absorption characteristics is identified, method estimation inaccuracy can be found when the material surface colour/texture does not continue through the depth of the material, leading to a 'uneven depth density', resulting in a less accurate estimation of absorption behaviour from surface analysis.

- Material and the Rethinking of Articulated Resources

  Through material, form, assembly and acoustic studies, synergetic relations between these aspects have been studied. The formative forces acting towards perceived visual and sound phenomena are driven by natural occurring and design-driven articulated material-environmental agencies. While structural integrity, assembly logic and

Fig. 14 Top: Poplar prototypes showing scale, back and front crossing. Centre: Detail of fibre crossing and gap between vertical planks. Bottom: Three-directions cross-structuring allowing highly warping elements to be joint through varied track/insertion depth

joining precision remain important, it is the revealing of material and form relations through visual-acoustic interactions that play the dominant part in the rethinking of resources, addressing environmental, technical and aesthetical aspects together. Rather than rejecting materials from structural inhomogeneity and high visual-acoustic variance, this study finds and demonstrates that material uncertainty can be considered and developed as a resource for unprecedented architectural articulations. As a result, this opens to a novel approach for studying and applying material resources that are in decay, uncertainty condition, or otherwise considered unusable in the built environment.

12 Discussion

This research expands the contemporary notion of material resources in architecture based on the idea and studies with poplar wood in a state of decay, specified for acoustic potentials and design. However, considering the importance of deadwood in the forest ecology, an expansive and unbalanced use of deadwood as a new material resource should be avoided, despite the presented potentials and findings. Instead, conventional timber and deadwood may go into a larger and systemic understanding and use of forests as providers of future biogenic material resources, following temporal material use structures that facilitate a more nuanced and potentially better relation between the building of forests and the building of architecture simultaneously.

13 Conclusion

Deadwood can become a new resource in architecture, with significant environmental, technical and aesthetical potentials. By rethinking material variance and material uncertainty as a resource for articulation, biogenic materials in decay offer novel potentials in general, and specifically in relation to acoustic design as findings in this study demonstrate.

Acknowledgements Will be added after final evaluation/acceptance.

References

Brack D (2018) Sustainable consumption and production of forest products. Backgr Anal Study 74

Christiansen K (2019) Tektonik - Den Tektoniske Fordring. Valville.

Cox TJ, Dalenback BIL, D'Antonio P, Embrechts JJ, Jeon JY, Mommertz E, Vorländer M (2006) A tutorial on scattering and diffusion coefficients for room acoustic surfaces. Acta Acust Acust 92(1):1–15

Cox T, D'Antonio P (2016) Acoustic Absorbers and Diffusers: Theory, Design and Application (3rd ed.). CRC Press Tayloy & Francis Group.

Eurostat (2016) Forestry, Statistics in. In *Forestry statistics in detail*. https://doi.org/10.1002/9781118445112.stat02649

Fleming RW (2014) Visual perception of materials and their properties. Vision Res 94:62–75. https://doi.org/10.1016/j.visres.2013.11.004

Foged IW, Hvejsel MF (2018) Reader: Tectonics in architecture (I. W. Foged (ed.)). Aalborg Universitetsforlag

Foged IW (2015) Environmental Tectonics: Matter Based Architectural Computation [Aalborg Universitetsforlag]. https://doi.org/10.5278/vbn.phd.engsci.00010

Fragkia V, Worre Foged I, Pasold A (2021) Predictive information modelling: machine learning strategies for material uncertainty. TAD - Technology | Architecture | Design 5(2):163–176 https://doi.org/10.1080/24751448.2021.1967057

Frampton K (1995) Studies in tectonic culture: The poetics of construction in nineteenth and twentieth century Architecture. MIT Press

Frantz R (2003) Herbert Simon. Artificial intelligence as a framework for understanding intuition. J Econ Psychol 24:265–277. https://doi.org/10.1016/S0167-4870(02)00207-6

Hartoonian G (1994) Ontology of Construction—On nihilism of technology in theories of modern architecture. Cambridge University Press

ISO (2008) ISO: 3382–2:2008 Acoustics—Measurement of room acoustic parameters —Part 2: Reverberation time in ordinary rooms

Lin D (2021) Estimating the date of earth overshoot day 2021

Liu J, Worre Foged I, Moeslund T (2021) Automatic estimation of clothing insulation rate and metabolic rate for dynamic thermal comfort assessment. Pattern Anal Applications. https://doi.org/10.1007/s10044-021-00961-5

Long M (2014) Architectural Acoustics (2nd ed.). Routledge

Nandanwar A, Kiran MC, Varadarajulu KC (2017) Influence of density on sound absorption coefficient of fibre board. Open J Acoustics 07(01):1–9. https://doi.org/10.4236/oja.2017.71001

OECD (2019) Global material resources outlook to 2060. OECD. https://doi.org/10.1787/9789264307452-en

Rasmussen SE (1964) Experiencing architecture. The MIT Press

Russell M, Fraver S, Aakala T, Gove J, Woodall C, D'Amato AW, Ducey M (2015) Quantifying carbon stores and decomposition in dead wood: A review. For Ecol Manag 350:107-$18. https://doi.org/10.1016/j.foreco.2015.04.033

Seibold S, Rammer W, Hothorn T, Seidl R, Ulyshen MD, Lorz J, Cadotte MW, Lindenmayer DB, Adhikari YP, Aragón R, Bae S, Baldrian P, Barimani Varandi H, Barlow J, Bässler C, Beauchêne J, Berenguer E, Bergamin RS, Birkemoe T, … Müller J (2021) The contribution of insects to global forest deadwood decomposition. Nature, 597(7874):77–81. https://doi.org/10.1038/s41586-021-03740-8

Sekler E (1965) Structure, construction, tectonics. Structure in Art and Science. http://en.cnki.com.cn/Article_en/CJFDTOTAL-SDJZ200902024.htm

Semper G (1989) The four elements of architecture and other writings. Cambridge University Press

Smardzewski J, Batko W, Flach A (2013) Experimental study of wood acoustic absorption characteristics. Holzforschung 68(November):467–476. https://doi.org/10.1515/hf-2013-0160

Smardzewski J, Batko W, Kamisiński T, Flach A, Pilch A, Dziurka D, Mirski R, Roszyk E, Majewski A (2014) Experimental study of wood acoustic absorption characteristics. Holzforschung 68(4):467–476. https://doi.org/10.1515/hf-2013-0160

Stange S, Wagenführ A (2022) 70 Years of wood modification with fungi. Fungal Biol Biotechnol 9 (1):1–6. https://doi.org/10.1186/s40694-022-00136-9

Tanaka M, Horiuchi T (2015) Investigating perceptual qualities of static surface appearance using real materials and displayed images. Vision Res 115:246–258. https://doi.org/10.1016/j.visres.2014.11.016

United Nations (2019) Food and Agriculture Organization—Global forest products. In: Forest Products and Statistics Team Forestry Policy and Resources Division FAO Forestry Department

Zhang Y, Fang S, Tian Y, Wang L, Lv Y (2022) Responses of radial growth, wood density and fiber traits to planting space in poplar plantations at a lowland site. J for Res 33(3):963–976. https://doi.org/10.1007/s11676-021-01382-0

Cap Ceilings Revisited: A Fabrication Future for a Material-Efficient Historic Ceiling System

Saqib Aziz, Emil Brechenmacher, Brad Alexander, Jamila Loutfi, and Christoph Gengnagel

Abstract

Ceilings are key to more sustainable and climate-friendly construction. Slab systems comprise the most embodied carbon in proportion to all component groups. The shortage of materials after World War II brought a brief renaissance for vaulted masonry ceiling systems. The simplicity and effectiveness of the purely compression-loaded caps enabled rapid reconstruction with the available material and rubble. These characteristics require the system to be re-examined in light of today's debates on resource scarcity and circularity. The research presents a LCA-Analysis, comparing six different ceiling systems under a uniform usage scenario. While a conventional concrete flat slab has a GWP of 136 $kgCO_2e/m^2$, the vaulted slab achieves a value of 64 $kgCO_2e/m^2$, representing a savings potential of 53%. Under the same conditions, masonry caps offer an operational solution that embodies less than half as much carbon as a conventional concrete ceiling. In addition, clear circularity properties can be demonstrated for the masonry cap ceilings. Circular economy principles are applied at both material and construction levels. The geometric precondition of cap ceilings with repetitive construction sequences lends itself to digital fabrication methods. This process enables further development of the historic form through multifunctional optimization. Hence, a 3D-printed acoustic brick was developed which enables the raw ceiling to meet a broad range of requirements. Our digital fabrication experiments show a unique combination in the joining of newly generated performative bricks and recycled material.

Keywords

Robotic fabrication · Additive manufacturing · Structural design · Acoustic optimization · LCA · Sustainable design

S. Aziz (✉) · E. Brechenmacher
Department of Architecture, University of Arts Berlin, Structural Design (Ket), Berlin, Germany
e-mail: s.aziz@udk-berlin.de

B. Alexander
TU Berlin, Berlin, Germany

J. Loutfi
Bollinger+Grohmann Ingenieure, Frankfurt Am Main, Germany

C. Gengnagel
Udk Berlin, Berlin, Germany

M. R. Thomsen et al. (eds.), *Design for Rethinking Resources*, Sustainable Development Goals Series,
https://doi.org/10.1007/978-3-031-36554-6_25

1 Introduction

1.1 Rethinking Geometry and Function of Structural Elements

As resources become scarcer, our construction methods must not only answer the pressing question of less grey energy, but they must also be more efficient in their use of available materials. The historic building type of masonry cap ceilings (German "Kappendecken") presents itself as a solution for a material-conscious construction approach. Due to its arched geometry, it combines two essential strategies: (i) material efficiency is increased due to the intelligence of the arched geometry. (ii) building materials with a poor life cycle assessment—such as concrete or steel—are substituted or reduced to the greatest possible extent.

This research project investigates whether the historical system of masonry cap ceilings can be of significance again today. Ceiling components are a big-ticket item, which is currently responsible for a large part of the embodied carbon in buildings due to the lack of good alternatives to reinforced concrete.

1.2 A Look Back: Constructive Diversity

The global scarcity of resources requires more constructive diversity. A look back into the past provides insightful approaches for this. With the advancing industrialization of construction in the nineteenth century, storey ceilings began to be constructed as flat-vaulted caps made of masonry between steel girders. The system is characterized by its simplicity, resistance to environmental influences, and high load-bearing capacity. However, the development of reinforced concrete soon began its triumphant march among the ceiling systems and quickly superseded a large variety of different construction methods. To this day, solid reinforced-concrete ceilings are predominant in contemporary building construction.

A return to the great constructive diversity predating the widespread use of reinforced concrete offers an opportunity to deal with the global material and resource problem. Many of the historical building forms can be easily repaired, converted, and deconstructed. We demonstrate this principle using the cap ceiling as an example.

1.3 A Look Ahead: Digital Transformations of a Traditional Form

For a long time, masonry cap ceilings were the only issue in the repair and renovation of existing buildings. However, our investigations and practical experiments show that, with novel technical transformations, the construction method could once again gain relevance in storey construction catering toward a more sustainable building practice. As part of a university research study, the historical cap ceiling was investigated and experimentally transformed.

First, a new type of wood-masonry hybrid system was developed. Historically, cap ceilings were built between parallel steel girders. However, medieval models also show vaults on wooden beams. To translate this technique into today, a construction system deploying glulam or Beech-LVL was developed. We see masonry cap ceilings between modern timber construction products as a further building block for constructive diversification.

The second transformation relates to the actual manufacturing process. One reason why vaulted constructions are rarely utilized nowadays is the high cost of labor and the need for skilled craftsmen. However, the elemental structure of cap ceilings opens a high prefabrication potential. The linear extruded form with its repetitive building blocks also lends itself to automated production using digital tools. We have therefore tested the prototypical and semi-automated production of cap ceilings exploring robot fabrication.

The third transformation refers to a multi-functional enrichment of the ceiling system. The aim was to integrate acoustic absorption properties into the raw ceiling. This requires answering the question of how an initially sound-reflecting surface can be transformed into a sound-absorbing surface. With the help of an interactive computer-aided process, we have developed special acoustic bricks for different frequency ranges. These can be produced using mineral 3D-printing processes and integrated into the digital fabrication process of the cap elements.

The notion to digitally transform this historic construction type was tested on 1:1 demonstrators and was presented to the public at a German festival on the future of construction in summer 2022.

2 Background

2.1 Embodied Carbon in Context

According to current figures, the building sector is responsible for 40% of global energy- and process-related CO_2 emissions. 28% of this is accounted for by the operation of buildings, 11% by embodied carbon, i.e., all emissions that are embodied in the materials and construction of buildings (United Nations 2020). Cement production alone accounts for 8% of all global emissions (Arup 2019). Thus, although decarbonization of the buildings sector is critical to success in meeting the 1.5 °C Paris target, emissions associated with the sector reached an all-time high of 10 $GtCO_2$ in 2019 (United Nations 2020). 60% of the structures and 25% of the infrastructure that will exist on Earth in 2050 have yet to be built. For a long time, decarbonization efforts have focused on reducing emissions in building operations. Especially with the increasing availability of emission-free energies, the operational efficiency of buildings can be improved over their life cycle and reduced through various measures. In the case of very efficient new buildings, embodied carbon emissions can already account for more than 50% of all emissions occurring in the life cycle. However, embodied emissions are for the most part already emitted at the beginning of a building's life cycle. Reducing embodied carbon in the built environment therefore represents a major lever for achieving immediate decarbonization success. Accumulated embodied carbon emissions will account for more than half of all emissions from the building sector from now until 2050 (World Green Building Council 2019). Thus, by 2030, embodied carbon emissions from all building activities will need to be reduced by 40%, according to recent calculations (World Business Council for Sustainable Development 2021). This will only be achieved with drastic changes in the way buildings and structures are designed and built.

2.2 Current Research Environmental Impact of Masonry Ceilings

There is one publication, which mainly deals with embodied energy in masonry vaults. De Wolf et al. (2016) examined three case studies in different locations around the world, find up to ten times lower global warming potential in this construction method compared to data from 260 conventional ceiling and roof structures. They highlight the importance of ceiling and roof structures in reducing embodied carbon emissions and discuss various strategies for doing so. The results from the analysis suggest investigating whether the large potential found can be replicated in the accounting of the floor ceiling system of the masonry cap ceiling (De Wolf et al. 2016).

A closer examination of the historical building type was the results of an architectural design thesis, in which the use of the construction form was analyzed and assessed for its ecological effects. The use of cap ceilings as floor slabs for a high-rise building with a height of more than 60 m was validated. In a simplified life cycle assessment, the investigated construction showed comparable good life cycle values as a timber hybrid ceiling (Apellániz et al. 2022) (Fig. 1).

Fig. 1 High-rise using cap ceilings, design proposition

3 Methodology

3.1 Brief History and Constructive Features of the Cap Ceiling

Barrel vaults have been a common form of roofing since antiquity. The flatter the vault, the greater the horizontal load component at the support. The particularly flat forms of barrel vaults are also called cap vaults and have a common stab height between 1/8 and 1/12 of their span. The designation "Prussian caps," which is common in Germany, should not obscure the fact that such structures were also used in other parts of the world.

Cap ceilings are considered the first solid ceiling system made of inorganic building materials (Marwede 2016). With the increasing industrialization of the iron and brick industry, they came to be used for entire buildings, especially in non-residential construction, as they already brought many of the advantages of today's solid ceilings in terms of fire protection and resistance to biological influences (Fischer 2008).

Cap ceilings were initially supported on brick belt arches. With the increasing availability of iron, the rowed bearing on T-beams developed, which made it possible to cover large spans with few support points. The greatest design challenge of the system to this day is the absorption of the horizontal thrust in the edge spans (Breymann et al. 1902). Similarly loaded adjacent bays cancel out the respective horizontal thrust of the other bay. In the edge bays, however, the horizontal load component must be absorbed differently.

3.2 Macro-Level: A Circular Baseline Design Concept for a Modern Cap Ceiling

Based on the historical construction rules, a novel cap ceiling system was conceptualized incorporating digital transformations (Fig. 2). On the construction level, we aimed to make the best possible use of the construction's potential for circularity.

Beyond the masonry, the cap ceiling is in principle free of composite materials in its construction. The load-bearing effect is essentially uniaxial. The load is initially taken up by the masonry vault caps and transferred to the beams purely via compressive forces. The steel (or timber) girders absorb the vertical load component from the vaults and in turn transfer it to the main load-bearing structure of the building. Where necessary, the horizontal load component is short-circuited into the ceiling system by tie rods. Each component has only one task and can be optimally materialized accordingly. This clear distribution of tasks improves above all the recyclability of the ceiling system. Easy accessibility to the different components enables better maintenance and interchangeability, as well as separation of the building materials by type at the end of their life cycle (Block et al. 2020).

3.3 Comparative LCA-Study for Ceiling Systems

To substantiate initial results from a design project (see 2.2), more detailed LCA analyses were conducted. Six ceiling systems for non-residential use were pre-dimensioned in a uniformly defined comparison benchmark. The systems compared are a reinforced-concrete flat slab, a prestressed concrete hollow core, a timber-concrete composite system, a timber hybrid system, and two forms of cap ceiling. An overarching life cycle analysis was validated, all built on the same axial grid and uniform span length of 8.10 m. The analysis results are shown in Fig. 3. While the conventional concrete flat ceiling has a global warming potential of 136 $kgCO_2e/m^2$, the cap ceiling achieves a value of only 64 $kgCO_2e/m^2$. It thus has a savings potential of approx. 53%. The examined cap ceilings are on a par with the timber systems or are even slightly better. Under the same general conditions, masonry cap ceilings are thus a

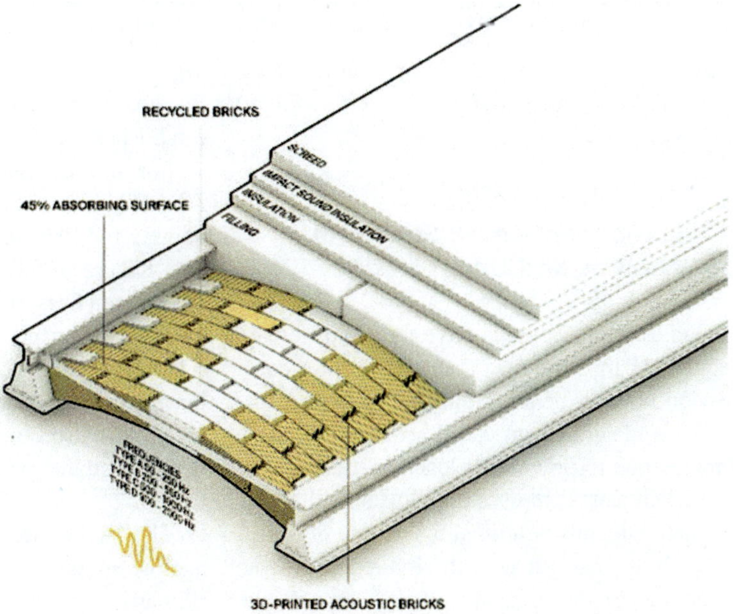

Fig. 2 Conceptual cap ceiling design with a span of 1.35 m

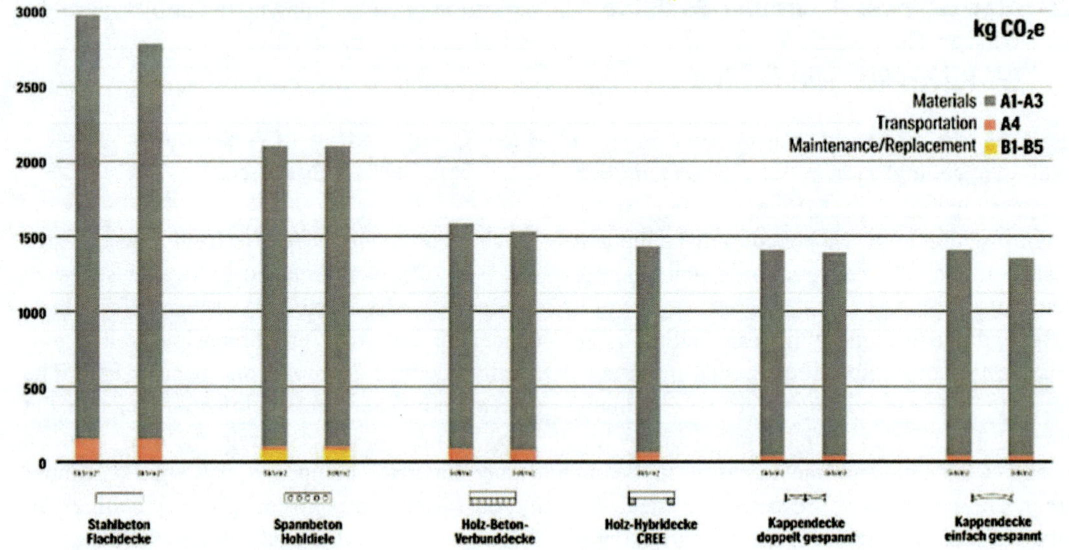

Fig. 3 LCA Comparison of six ceiling types

system that embodies less than half as much grey emissions as a conventional reinforced counterpart. Where there are still open questions regarding the end-of-life scenario of timber construction, clear circularity characteristics can be demonstrated for cap ceilings. Circular economy principles are applied at both the material and construction levels.

With these results, the question arises as to why this type of construction is hardly ever deployed in developed economies. The answer clearly lies in the great need for manual labor, i.e., high labor costs and the need for skilled craftsmen.

3.4 Component-Level: Structural Analysis and Grider Design

Vaulted masonry ceilings have the characteristic feature to carry downward loads by pure compression without bending moments or shear forces. To achieve this structural behavior, the flow of forces due to downward loads is studied. The loading creates compression thrusts within the arch span which is described by the line of thrust. If the thrust has no eccentricity regarding the center line of the arch, a uniform compressive stress distribution for this cross section occurs. As soon as the line of thrust does not pass along the center line, an additional moment occurs due to the thrust's eccentricity. This results in a linear stress distribution in the cross section. To achieve structural stability and no tensile stresses in the cross section, Rankine proposed in 1858 (Rankine 1858) that the shear line should not run outside the boundary lines of the middle third. These boundary lines are defined with a distance $e = d/6$ to each side of the arch center line as shown in Fig. 4 (Curtin et al. 2008). 1.35 m was chosen as the common span (L) for the vault, following established axial dimension in design practice.

For the design of demonstrators, a height $f = L/10 = 0.135$ m regarding the common heights for *Prussian caps* was assumed (Ahnert and Krause 1988). Figure 4 shows one of the determined lines of thrust visualizing how the line of thrust is derived using graphical methods. Here the line of thrust shown in dark green stays between the borderlines illustrated with light green. It is also shown that the resultant value of P depends on the magnitude of the horizontal reaction H.

The vaulted masonry ceiling thrust is pushed against single span girder that carries the loads primarily by bending moments. The horizontal

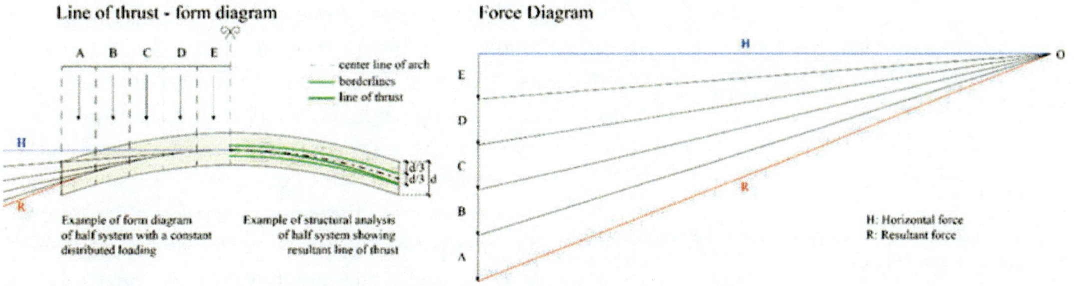

Fig. 4 Calculation of line of thrust—form diagram (left) and force diagram (right)

component evens out for the twin beams, whereas the edge beams need to absorb it. Horizontal steel bracings are designed in the edge vaults to counteract. For the sake of stability during transportation, every prefabricated vault possesses bracings. After the construction of the vaulted masonry ceiling at the building site, bracings of the inner vaults can be dismantled. This is achieved by the purely screwed connection of the inner tie rods to the beams. In the case of the steel variant, this is done by means of steel angles, in the case of the timber geometry, this is done by means of a milled attachment for the tie rods as shown in Fig. 5.

Due to the positioning of the tie rods in the height of the vault, torsion can occur in the girder so that the girders rotate slightly outwards due to the thrust load of the vault. However, this only occurs in the non-finally assembled phase and in the end fields. As soon as the girders in the inner panel are loaded from two sides, this torsion is canceled out. In the edge spans, this phenomenon must be considered by designing suitable supports.

Fig. 5 Illustration of the reversible connection of tie rod and beam on the prototype

During the design process, three beam designs were analyzed and compared.

Option 1: Steel profile.

Option 2: Glued Laminated Timber GL32h.

Option 3: Beech Laminated Veneer Lumber GL75h.

The structural analysis was followed by Eurocode 3, Eurocode 5, and the assessment manual for GL75h from the manufacturer Pollmeier (Blaß and Streib 2016). For fire protection, the steel profile is encased in concrete while the timber cross section needs to have a combustion height. The difference between the two options in timber (2 & 3) lies mainly in their respective strength. The Glulam GL32h is characterized by its wider availability and easier workability, while the beech LVL is a high-performance product with increased strength, but lower availability and more complex processing.

The calculation results showed that spans of 8.10, 6.75, and 5.40 m for design options 1, 2,

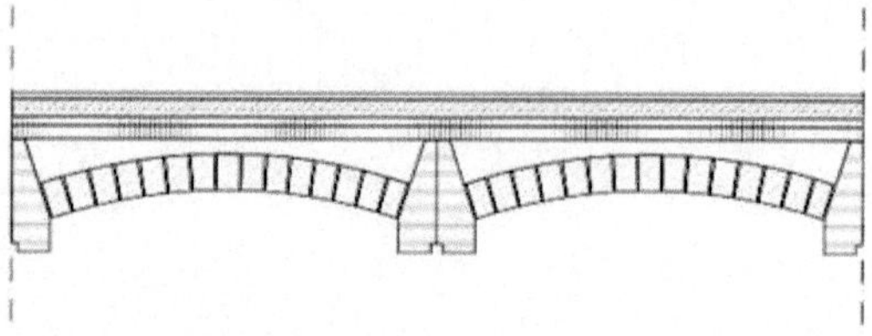

Fig. 6 Constructive section cap ceiling between specially milled timber cross sections

and 3, respectively, can be realized. The steel option stands out with its advantage to bridge wide distances due to its high resistance capacity. Furthermore, smaller dimensions of its cross section can be chosen. However, the LVL can be considered as a competitor over steel for smaller distances than 8.10 m. Timber provides diverse possibilities of three-dimensional shaping. This made it possible to design the shape of the timber cross sections as shown in Fig. 6. In addition, timber as a building material with its lighter density compared to steel makes it easier to fabricate.

3.5 Component-Level: Acoustic Bricks

To enrich the ceiling with additional multi-performative acoustic properties, the aim was to design bricks with integrated absorbers.

Programming the brick with acoustic properties was done according to three sometimes opposing objectives. The brick was to be designed with as small a mass as possible, while providing sufficient sound insulation, and act as a sound-absorbing surface. Conventionally a brick is considered to be sound hard. 3D-printing techniques, however, allow this limitation to be overcome by manufacturing the brick with a more articulated geometry enabling it to absorb sound via resonant cavities within the outer corpus. The resonant cavities act as sound absorbers, specifically Helmholtz resonators, which convert sound energy into heat as sound waves within various frequency bands accelerate a small air mass in the neck of the resonator against the upper cavity, which acts as a spring. The oscillations of the air mass within the necks result in an energy transfer from sound into heat, as the rapid air movements cause friction along the material boundary. This has two positive outcomes: (i) The reflected sound energy has a lower sound pressure due to the absorption, and (ii) The internal damping of the structure due to the dissipation results in a lower sound transmission across the structure into the next room.

To achieve the above-described effects, a numerical simulation of four different bricks with four resonators per brick was set up, and generatively optimized to be effective in four frequency bands from 100 to 2000 Hz. Using the finite element method within COMSOL Multiphysics including a general optimization study step, the neck lengths and radii, as well as the cavity depths and widths were adjusted until maximum absorption was achieved for the prescribed frequency range, by simulating the structures within a virtual impedance tube test setup (Aziz et al. 2022). The results of the simulations and optimizations produced four bricks which are able to absorb between 40 and 60% of incident sound energy. In a second step, the bricks were simulated for their sound insulation properties using FEM as well. Compared to a brick of equal mass, the acoustically programmed brick is able to provide 5 dB more airborne sound insulation, thanks to the internal damping of the resonators (Fig. 7).

3.6 Process-Level: Additive Manufacturing

The additive manufacturing process for the acoustically optimized bricks was controlled with a custom process utilizing grasshopper3D (Davidson 2022). To produce the acoustical optimized bricks, a print width of 10 mm and a readymade clay mixture were deployed. Due to the print width, small-scale structure of the bricks, and the selected print-path geometry, the cylindrical Helmholtz cavities were transformed into cubic volumes, hereby keeping their volumetric attributes for a more streamlined and controllable fabrication process (see Fig. 8a).

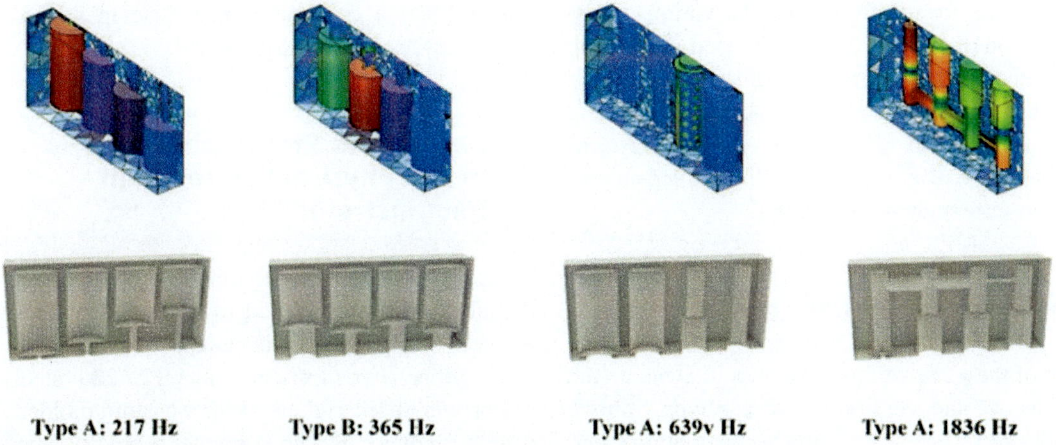

Fig. 7 Multifunctional bricks with integrated Helmholtz Resonators

A meandering print path was conceptualized creating adjacent print lanes, layer upon layer. The optimized brick geometry with integrated acoustical absorber cavities where first sliced vertically to accommodate the used print height of 3 mm. The custom g-code algorithm can assign, besides the relevant coordinate, also the speed and material inflow rate for each individual segment. With this strategy brick geometries comprising the unique cavities could be produced.

On the production site, a few trial-and-error approaches had to be cycled through to create an optimal output. With the used technical system each brick would have needed approximately 2 h to be finalized (see Fig. 8c). Due to the time constraint, it was not possible for the research team to produce all the bricks in time and an alternative solution for the construction of Demonstrator #1 had to be conceptualized.

The usefulness of the additive manufacturing process to produce the bricks is questionable against the background of the high quantities. In parallel, more traditional production methods for manufacturing them are being explored (i.e., casting). However, it is important to emphasize that the development of this complex geometry would not have been possible without the prototyping possibilities of mineral 3D printing. The option to create complex internal recesses that this process opens up got the research going in the first place (Fig. 8).

4 Case-Study: Demonstrators for Pop-Up Campus

As part of the Zukunft Bau Pop-up-campus on the future of construction in Aachen, the project was promoted to carry out tactile experiments on

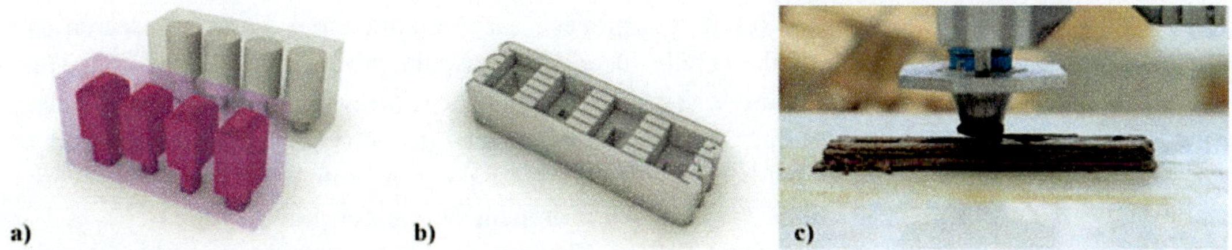

Fig. 8 Additive manufacturing process

a 1:1 scale for the first time. The various transformation strategies could be tried out using different demonstrators. In addition to designing and testing a digital manufacturing environment for masonry cap ceilings, traditional craft techniques were also explored with a skilled master mason in a student workshop.

4.1 Robotic Fabrication

One of the goals of this work was to evaluate the efficiency and feasibility of applying robotic fabrication (AF) for the construction and assembly of cap ceiling structures. As mentioned, the construction requires the experience of skilled labor and with the scarcity of specialized craftsmanship also the rich knowledge of construction methods for specific building component are diminishing. Prior to the digital initialization the research group therefore investigated various brickwork patterns of the cap ceiling (e.g., Stack-, Runner-, Herringbone-Bond). Further studies were carried out on how they used to be constructed by trained workers on-site or produced as prefabricated systems applying automation processes nowadays (Bruun et al. 2021).

First the brickwork patterns were digitized as algorithmic 3D models using the visual scripting editor Grasshopper3D inside the Rhinoceros CAD framework (McNeel 2022). This made it possible to design, set the base-case design scenario, and establish a benchmark to probe the robotic assembly process. Using Grasshopper3D and the Robots plug-in (Visose 2022) an interactive, dynamic, and algorithmic environment could be used to control and program the robotic fabrication at an academic facility in Berlin. The deployed robot arm was a Universal Robot 5. The maximum recommended reach of the robot is around 170 cm ø and the maximum load-bearing capacity is set at 5 kg making these preconditions (reachability and load-bearing capability) the main design restrictions for the further development for the programming. To pick and place the bricks a pneumatic gripper tool, Schunk OA-UR3510-PGN-plus-P 80–1 was utilized, and a customized metal clamp attachment was designed and produced.

5 Design and Prototyping of the Physical Environment—Demonstrator #1

To program and probe the robotic process first a digital production unit had to be designed, prototyped, and physically installed. For the single span girder, two 135 cm long IPE 220 steel beams were used and cast with a custom molded concrete covering. The concrete geometry was designed with chamfering side surfaces allowing to optimally dock or place the edge bricks on them.

Due to the constraints of the max. reachability, a solution had to be implemented enabling to sequence the assembly by building up to 6 rows in one process (one completion cycle) and then to move the sub-construction and continue the assembly. To this end, the edge beams were mounted on a set of uniaxial metal railway fixtures with roller bearings (a-b) allowing to move the sub-construction manually or automatically after each sub-set completion cycle. A wooden formwork cladding was designed as a fixture with the ability to mechanically raise or lower itself. The mechanism was powered through pneumatic cylinders and a simple valve switching system. This was important to ensure that after each completion cycle the formwork could be lowered to be moved alongside the sub-construction and the currently built cap structure to set/sink into its final form.

In this stationary proposal, the formwork always remains in the same place and the slab element continues to move on rails. This is a process for prefabrication. Further research is needed to make the process available on-site. The movement of the formwork could be an obstacle, but the beam geometry as a rail for the robot could hold great potential for controlled work environment on the construction site.

The robotic arm was mounted on to a custom build wooden table. The height of the table enabled the sub-construction and the formwork

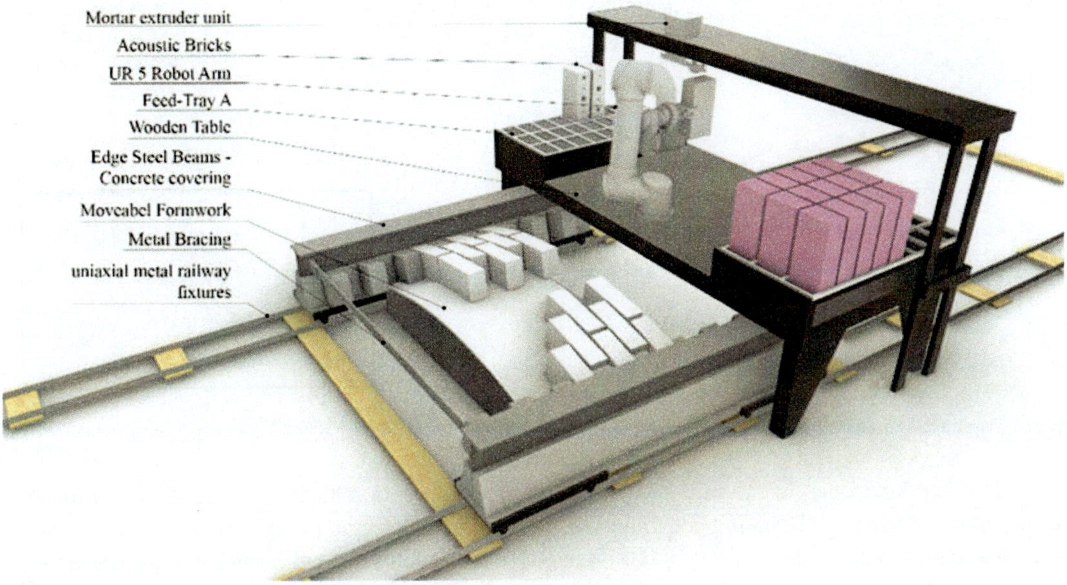

Fig. 9 Components of the digital manufacturing environment

to easily move beneath it. Alongside the robotic arm two wooden tiled bracket trays were installed. These functioned as the picking stations or inlay trays for the active feed bricks. On the one side reused and on the other side the additive manufactured (AM) bricks would be positioned. A secondary table was installed to host the extruder unit that would extrude the fresh mortar on to the bricks, powered manually or automatically with an electric drill tool (Fig. 9).

6 Robotic Programming and Experimental Implementation

The base-case geometry for the generation of the various brickwork patterns for the cap ceiling was generated on a single spanning vault extrusion. After initial testing of the various brickwork patterns, regarding easiness of assembly and aesthetic considerations the team decided to select the Runner-Bond for the further evaluation. The base geometry of the demonstrator was set with the following properties (Fig. 10).

Cap-span length and width: 135 cm.

Rise height: $f = L/10 = 13.5$ cm.

This generated an extruded surface for the placing and distribution of the bricks. The first digital process enabled to arrange the various acoustically optimized and reused bricks on to the structure. Each of the two main brick types would have its own feeding tray station. A standard brick size (NF) for both types was set with the dimensions: length: 24 cm, width 11.5 cm, height: 7.1, and standard physical weight of 3 kg.

The second sequence created a network set of three-dimensional planes for the brick layout in the longitudinal and transverse direction. For each direction also the joint/gap width could be assigned separately. The standardized brick was then algorithmically oriented onto the network of planes, hereby creating and visualizing the selected cap ceiling brickwork pattern. The acoustic bricks would be distributed along the wings of the vault and the reused bricks should be placed in the mid-section of the cap ceiling. This strategy was conceptualized together with the acousticians as more mass would be available above the wingspans, allowing further sound absorption qualities. A crucial addition to the program was the order of sequence each brick would be placed per row. First the both side wings would be subsequently assembled,

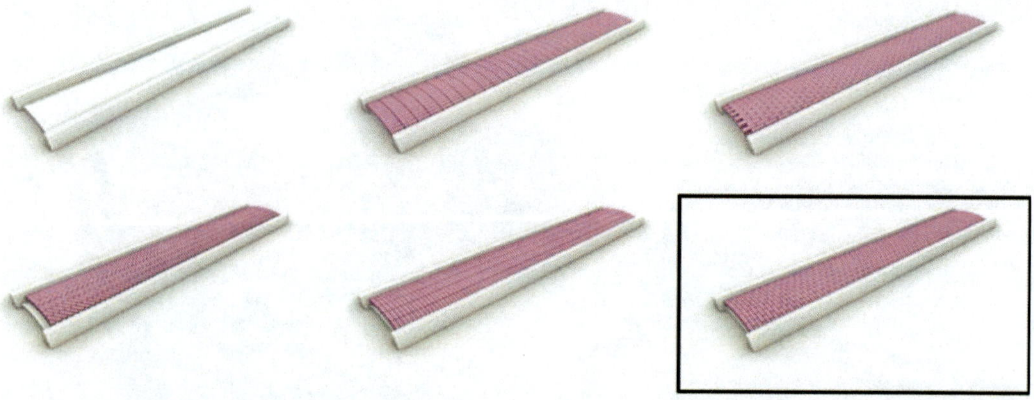

Fig. 10 Parametric modeling of the cap ceiling

followed by the keystones in the mid-section to optimally activate the load-bearing attributes of this construction type.

The sequence of the robotic pathway planning can be comprised of three main steps:

- Feed-station (FS): From an initial home position of the robot, the kinetic motion was set to identify the current needed brick type and then moved toward the associated feeding tray to grab and pick up the relevant input brick. The gripper was attached to a compressor in order to manage the pneumatic motion and was controlled (open, close) using a custom line of g-code, e.g., close ((set_tool_digital_out(0, False), (sleep(0.015)) and open ((set_tool_digital_out(1, False), (sleep (0.015)). To ensure a robust picking process the robot would first align itself with the selected brick, slide in with the Tool-Center-Point (TCP) at the exact center of the brick, grab the brick, and then slide the brick out toward the top out of the feeder station.
- Mortar-station (MS): The brick would then be moved toward the mortar station, consisting of a metal receptacle unit with a spindle screw mechanism that would be manually or automatically set to motion applying a simple drilling machine once the brick would be in a certain position. Mortar had to be mixed and filled into the receptacle prior to each deployment in a manual process. The outlet of the receptacle was stationary, meaning that the robotic motion had to revolve around it for the extruded mortar to be placed on two surface sides of the individual brick. The movement of the brick was programmed in a meandering motion to ensure an optimal placement of the mortar.
- Assembly-station (AS): After the mortar was applied/bonded to two sides of the brick the robot would then move toward the sub-construction and place the brick accordingly on to the formwork. Depending on the location of the placed brick certain conventions had to be established for an ideal placement. For instance, the side bricks that would be attached to the reinforced-concrete side beams would have to be moved in diagonally to avoid collision with the beam, whereas all other bricks would be placed from above parallel to the designated plane position. Due to the small gap sizes, a lot of trial-and-error efforts had to be tested to ensure a seamless bounding of the placed ricks. Once placed the gripper would open and move back to the home position for the next placement sequence (Fig. 11).

After these prerequisites were adequately set the production of the demonstrator could be finalized within 2–3 h by two of the research members. Each individual brick placing sequence (Feed-, Mortar-, Assembly-Station) took about 35 s to complete, so the construction sequence could be much faster. The issues in this first test were allocated to the manual labor

Fig. 11 Robotic sequence **a** Feed-station **b** Mortar-station **c** Assembly-station

needed to for instance supply the fresh mortar for the extrusion unit and to manually move the sub-construction and potentially recalibrate it for the next completion cycle to start. Apart from that the digital fabrication process seemed to work surprisingly efficient and robust, proving that further improvement can lead to a reliable and fast assembly process.

The development of a robotic workflow may seem inefficient for this small-scale task. After all, the creation of finished arch elements in the appropriate span is also possible. However, since one of the main goals of the project was the use of reclaimed material such as recycled bricks and not the development of a completely new arch element, the small-scale nature of the setup is necessary (Fig. 12).

6.1 Demonstrator #2—Classic Fabrication

In addition to exploring robotic manufacturing methods, learning about historical craft techniques was the focus of a student workshop. Guided by an experienced master mason, a second demonstrator for an exhibition was realized with a small group of students. Only recycled historical field bricks from the region were used.

The workshop made it clear how much embodied knowledge there is in the craft technique. The seemingly simple process of setting the stones together unfolds a great complexity in its details. The use of different tools and the possibility of responding individually to each stone are difficult to replicate through digital processes.

In addition to learning the craft techniques, a special focus was placed on the material science of mortar. Unslaked lime mortar was used, which was pumped in and processed on-site. Direct bricklaying as so-called hot lime mortar (Heisskalkmörtel) was also tried out. In this way, parts of the vault could even be built without formwork. The demonstrator in Aachen was built between specially made timber cross sections, which illustrate the newly developed timber masonry hybrid system (3.2) (Fig. 13).

7 Conclusion & Discussion

The construction of the two demonstrators has confirmed the underlying hypotheses. Thus, the specification of a masonry cap ceiling system between modern, specially milled timber beams is a valid and operational option. Compared to the classic version, it even has advantages in terms of structural fire protection and the general installation procedure. Due to the pre-processing of the beams, numerous finishing and connection details can already be integrated into the beam geometry.

The comparison of the two production methods, robotic-digital and traditional craftsmanship, has provided a great deal of knowledge. Only when the great precision and flexibility of manual production is understood can the digital

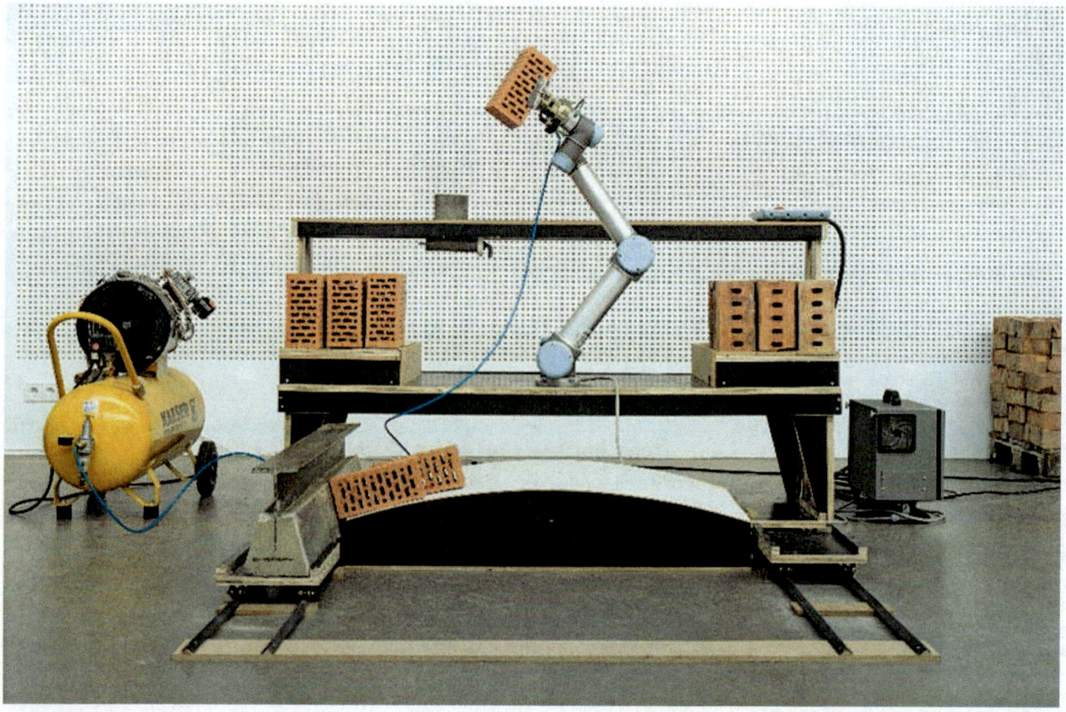

Fig. 12 Robotic setup

process be modeled accordingly. This always raises the question of the extent to which an imitation of manual activities by robots is a target-oriented approach at all. The possibility of individually carving and manually adjusting each stone is not possible with the current robotic setup. However, even with simple technical means, the digital production setup comes very close to the artisanal model. With minimal post-processing effort, a demonstrator could be produced that clearly shows the enormous potential of digital means of production in the construction industry.

The research project and the practical experiments were intended to investigate, whether the historical system of masonry cap ceilings can be of significance again today. This question arose on the one hand from a design interest in the historical technique, and on the other hand from the pressing realization that we lack practical answers to the question of how we can build in a more low-emission and circular way. While the debate on sustainable building often focuses on an ephemeral whole, the present approach deliberately tries to limit itself to a single building component. A focus on the real major emission drivers in the building sector is necessary to achieve the rapid reduction of grey energy and associated emissions that is the crucial task of the coming years. Ceiling systems offer great potential to drastically reduce the common practice of large-volume use of reinforced concrete. With the revival of masonry cap ceilings, we want to contribute to this (Fig. 14).

Fig. 13 Manual fabrication workshop

Fig. 14 Two demonstrators **a** Robotic fabrication on steel griders **b** Manual fabrication on timber griders

References

Ahnert R, Krause K (1988) Typische Baukonstruktionen-von 1860 bis 1960—zur Beurteilung der vorhandenen Bausubstanz - Gründungen, Wände, Decken. VEB Verlag für Bauwesen, Dachtragwerke

Apellániz D, Aziz S, Weber M et al (2022) Integration of life-cycle assessment in a multimodal building design approach. In: Sustainable built environment conference D-A-CH Berlin 2022

Arup (2019) Rethinking timber buildings: Seven perspectives on the use of timber in building design and construction. Arup Foresight, London

Aziz S, Alexander B, Gengnagel C, Weinzierl S (2022) Generative design of acoustical diffuser and absorber elements using large-scale additive manufacturing. In: Proceedings of the 16. International Conference on Architectural Acoustics and Sound

Blaß HJ, Streib J (2016) BauBuche Buchen-Furnierschichtholz—Bemessungshilfe für Entwurf und Berechnung nach Eurocode 5

Block P, van Mele T, Rippmann M et al (2020) Redefining Structural Art: Strategies, necessities and opportunities. The Struct Eng 98:66–72

Breymann GA, Köninger O, Lang H (1902) Die Konstruktionen in Eisen. Schäfer, Hannover

Bruun EPG, Pastrana R, Paris V et al (2021) Three cooperative robotic fabrication methods for the scaffold-free construction of a masonry arch. Autom Constr 129:103803. https://doi.org/10.1016/j.autcon.2021.103803

Curtin WG, Shaw G, Beck JK (2008) Structural masonry designers' manual. John Wiley & Sons, Chichester

Davidson (2022) Grasshopper. www.grasshopper3d.com. https://www.grasshopper3d.com. Accessed 12 Oct 2022

Fischer M (2008) Steineisendecken im Deutschen Reich 1892–1925. Bd. 1. Entwicklungsgeschichte, Typologie und Bewertung. Dissertation, Fakultät Architektur, Bauingenieurwesen und Stadtplanung der Brandenburgischen Technischen Universität Cottbus

Marwede H (2016) Preußische Kappendecken Analyse der Tragstruktur auf Grundlage von Belastungstests im Rahmen des Projekts Stadthöfe Hamburg. Master Thesis

McNeel (2022) Rhinoceros 3D. www.rhino3d.com. https://www.rhino3d.com. Accessed 12 Oct 2022

Rankine WJM (1858) A manual of applied mechanics. Charles Griffin, London

United Nations, Marrakech Partnership (2020) Climate action pathway human settlements. UNFCCC, Bonn

Visose (2022) Robots Plug-in. In: GitHub. https://github.com/visose/Robots. Accessed 12 Oct 2022

De Wolf C, Ramage M, Ochsendorf J (2016) Low carbon vaulted masonry structures. J Int Assoc Shell Spat Struct 57:275–284. https://doi.org/10.20898/j.iass.2016.190.854

World Business Council for Sustainable Development, Arup (2021) Net-zero buildings: Where do we stand? WBCSD, Geneva

World Green Building Council (2019) Bringing embodied carbon upfront

RE:Thinking Timber Architecture. Enhancing Design and Construction Circularity Through Material Digital Twin

Anja Kunic, Roberto Cognoli, and Roberto Naboni

Abstract

Wood is a fundamental resource for building the future environment and meeting CO_2 reduction goals. However, wood is truly sustainable only if it can be preserved for a long time and reused throughout several construction cycles. Previous research by the authors shows that algorithmic design and automation technologies are introducing novel possibilities for reconfigurable architecture, which is conceived to allow easy assembly and reassembly of wood components into different structures. In this context, dynamic and accessible material data is vital to enhance circular construction effectively. This paper introduces the concept of Material Digital Twin, an online material database directly connected to computational models and is used for real-time exchange of design, simulation, assembly and construction information. The outlined approach is applied to the realization of a laboratory-scale prototype structure, which is used as a testbed platform to verify and document the relevance of the accessible database in supporting the automation of reusable timber construction.

Keywords

Rethinking resources · Rethinking timber architecture · Material digital twin · Material circularity · Data circularity · Reconfigurable wood architecture

1 Introduction

1.1 Motivation

Circularity and renewable construction materials are necessary rather than a choice to address the climate emergency. Considering its carbon storage capacity, wood is seen as a natural asset for the future of sustainable construction. However, as the demand for a shift toward timber construction rises, the pressure on the forestal sources has increased as well. A new design paradigm that enhances circularity and reuse is urgently needed to balance the demand and supply of construction timber. In this context, computational tools and data-intensive Construction 4.0 challenge current building practices and 'end-of-life' concepts by enabling materials-and-data circularity through information-driven design and construction processes.

A. Kunic · R. Naboni (✉)
CREATE - University of Southern Denmark, Campusvej 55 5230 Odense, Denmark
e-mail: ron@iti.sdu.dk

R. Cognoli
University of Camerino - School of Architecture and Design, Viale della Rimembranza, 63100 Ascoli Piceno, Italy

M. R. Thomsen et al. (eds.), *Design for Rethinking Resources*, Sustainable Development Goals Series,
https://doi.org/10.1007/978-3-031-36554-6_26

Fig. 1 Interaction with the *ReconWood Proto 02* in Mixed Reality—an experimental prototype for reconfigurable wood construction from the *ReconWood* series

The introduction of design-for-renewable, -reusable, -reconfigurable architecture unlocks novel properties and aesthetics at different building levels. The CREATE Group at the University of Southern Denmark has been experimenting with combinatorial and reversible design (Naboni and Kunic 2019) and robotic reassembly (Kunic et al. 2021a, b; Naboni et al. 2021) approaches to extend the life cycle of wood-based materials, components and systems rendering them natively conceived for automated reconfiguration. Data-driven reconfigurability extends beyond a pure concept of circularity, suggesting higher design flexibility and shape transformability driven by the material's inherited performance throughout various life cycles. As part of this investigation, the *ReconWood* open project was initiated to study a series of methods and related architectural prototypes exploring the potential of circular construction through automation, particularly focusing on cyber-physical design, assembly and disassembly of reconfigurable structures. *ReconWood* is a system based on a construction kit of discrete wood elements used to build multi-resolution framing structures that are 100% reversible, structurally optimized for reusability, and usable to fulfil the needs of a wide range of architectural and structural typologies that can evolve over time (Fig. 1). Here, *ReconBlocks* are envisioned as materials that can 'travel' from one structure to another while carrying their associated datasets which allow them to be tracked, traced and reused in a data-informed fashion. Data-driven reconfigurability and reuse of building components can be effectively achieved only when material information is linked and accessible along with the material stock and shared flawlessly between different building phases.

1.2 Background

In recent years, research and practice-based studies aimed at reframing the construction value chain with the introduction of circularity principles. Existing buildings are now considered *material banks*, possessing a valuable material stock that can be mined for future construction up/recycles (Heisel and Rau-Oberhuber 2020; Honic et al. 2021). As part of the circular paradigm, a Material Passport (MP) was introduced to provide digital data sets containing information about the origin and construction history of the materials, products, and buildings and make them available for future construction uses (Heinrich and Lang 2019). Such material datasets are typically implemented at the end-of-life phase of a building, during the pre-demolition and demolition phases. Recent experimental work is instead looking into implementing MPs in the planning phase of a building, using BIM tools (Atta et al 2021; Copeland and Bilec 2020; Jayasinghe and Waldmann 2020). On the other hand, a significant stream of research looks into reclaiming and reusing existing building materials by acquiring data from existing buildings (Wibranek and Tessmann 2023), inspecting their maintenance conditions (Spotr 2022), and detecting healthy and reusable materials (Raghu et al. 2022; Àkànbí et al. 2020), and building online databases (Material Marktplats 2022; RotorDc 2015). All these approaches have, however, a common drawback: the information is statically loaded to the MP through an archive file format, permanently detaching the material information from its physical twin and design models. To enable higher levels of material circularity, more advanced data management and smoother interoperability between phases is auspicable, supported by MPs that are 'alive' platforms, constantly accessed, tracked and updated with emerging material data (Li et al. 2020).

1.3 Research Aim

Working within the premises of Industry 4.0 (I4.0), such as interconnectivity, online data exchange, and smart automation, this research engages with a cyber-physical environment for design and construction to extend the life cycle of wood building components through data-driven reconfigurability and reuse. Data circularity is as important as material circularity to enable reconfiguration as a design and construction process. In this sense, linking material data to a material stock is fundamental. To fulfil such requirements, this research introduces the concept of Material Digital Twin (MDT), a data-driven holistic approach to support the automation of reconfigurable wood structures. In comparison to the existing concept of MP, the MDT is characterized by the following features:

(1) It is established as a dynamic material database rather than a static one-shot descriptive material sheet. The database is constantly updated and incrementally loaded with new data throughout different building phases.
(2) It permanently links material data to a specific building material/component using QR (quick response code) codes.
(3) It is bi-directionally linked to a computational environment, from where the use-cycles are planned (design, simulation) and actuated (fabrication, assembly, disassembly).
(4) It is easily accessible by both human and machine users through an internet connection.

Based on the MDT, a construction workflow is applied and tested in a laboratory setup on the design and construction of *ReconWood Proto 02*, an experimental prototype of reconfigurable wood construction where the embedded information from individual material blocks is used to assemble structurally performative aggregations. Considering the vision for these structures to be reused, re-configured, and evolve in time, it is critical to create a watertight information loop with a shared database between different construction phases and actors.

2 Methodology

2.1 Material Digital Twin

The MTD is an online dynamic material database with a permanent link that enables cyber-physical

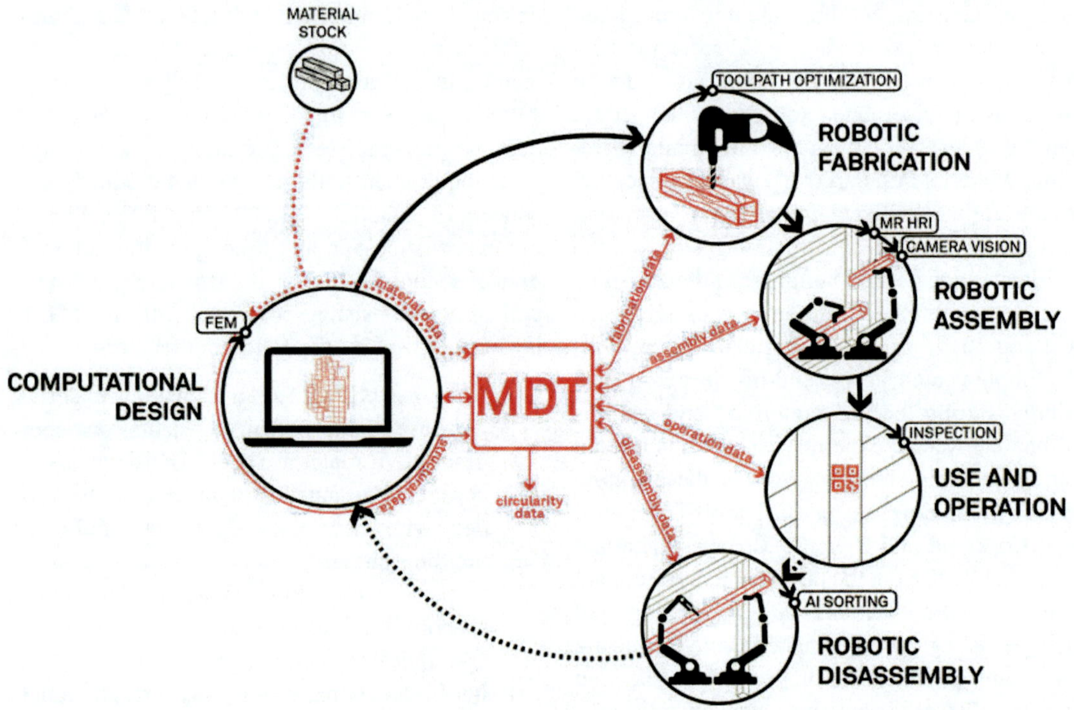

Fig. 2 Schematic representation of the data exchange and communication between different construction phases and MDT

interoperability and collaboration. It is used as a middleware to bridge the virtual and physical environments and thus foster the automation of material reuse. The MDT is linked to a parametric computational model, which exchanges data from/to the following life-cycle phases: computational design, robotic fabrication, robotic assembly, use/operation and robotic disassembly (Fig. 2).

In this research, the MDT is implemented using Speckle, an interoperability application used here to extract, exchange and visualize data in real-time (2020). A stream URL link is created and accessed through a dedicated Grasshopper 3D connector. The computational design model is developed in Grasshopper using dataset structures that refer to different construction phases, i.e. design and structural analysis, robotic fabrication, robotic assembly, use and operation, and robotic disassembly. These data are sent to the MDT and received back to Grasshopper as individual Speckle Branches, characterized by unique URL links associated with each block. QR code identifiers imprinted on the physical blocks enable the connection between physical material and its digital twin in the Speckle server. Throughout different design-construction phases, relevant datasets for circularity purposes—discussed later—are stored in the MDT to align material and data circularity within a cyber-physical loop accessible to different users, including designers, builders using Mixed Reality devices, fabrication and assembly robots which can seamlessly communicate and collaborate (Table 1).

From the computational design phase, six datasets are extracted and loaded into the MDT. *Identity/design* shows data for each block derived from the algorithmic design choices and material stock information; *Structural Performance* data obtained from the structural simulation is used to distribute the blocks, assuring to maximize their reusability; *Circularity* data contains predictions on the material expected performance for following cycles, combining simulation and

Table 1 Six relevant datasets for circularity purposes sent to MDT

MDT

IDENTITY/ DESIGN	STRUCTURAL PERFORMANCE	CIRCULARITY	ROBOTIC FABRICATION	ROBOTIC ASSEMBLY	USE AND OPERATION …
block ID	structural utilization	CLS (Circularity Limit State)	fabrication method	block assembly sequence No	block function in the structure
dimensions	joint mechanical performance	stored carbon	fabrication* tools data	block assembly coordinates (x,y,z)	replacement record
cross-section	material* characteristics	No of life cycles / uses record	fabrication* toolpath data	block assembly orientation	material health
material data*		RR (Reusability Rate)	fabrication* settings	holes coordinates	external treatments
material distribution		life expectancy		connections /* bolts data	environmental conditions exposure
location		future structural layer prediction		connections /* embedded nuts data	
connectivity					
reccurence in the structure					
structural layer					

observations on material imprecisions/decay. Datasets regarding the *Robotic Fabrication*, *Robotic Assembly* and *Use/operation* phases are fed to MDT during the corresponding project stages, providing a rich database to inform the subsequent reconfiguration process. The Robotic Disassembly phase is not discussed here, as it is out of the scope of this paper. However, the data required for its integration and closing of the *ReconWood* circular loop is considered and can be derived from the precedent phases.

2.2 Computational Design and Circularity

The *ReconWood* construction system is based on discrete wood blocks designed to be robotically assembled and (re)assembled. These blocks are computationally aggregated into multi-resolution frame structures of variable size, shape, density and strength. Computationally, multi-resolution layouts are obtained from a voxel-based discretization process, providing a targeted form. Resolution gradients are used globally to closely approximate the volumetric characteristics of input forms and locally as a means of structural strengthening through densification.

The *ReconWood Proto 02* is designed as a prototype structure integrating sheltering and seating areas. Here, Mixed Reality (MR) is used to start the design process with human inputs to the form-finding process (Kunic and Naboni 2022). Subsequently, a multi-resolution frame structure is generated from a three-level voxelization layout, resulting in a 12-block construction kit. In parallel, a structural optimization algorithm is run to evaluate the structural utilization of individual timber blocks and suggest a reconfiguration to maximize their reusability. The optimization process is developed by establishing a bi-directional link between the Grasshopper computational design environment and SOFiSTiK-based numerical analysis. Specific material and joinery characteristic values, such as stiffness and strength derived from mechanical testing and loading boundary conditions, are fed to the linear FEA calculation. As a result, structural utilization values for all the blocks and activated joints are extracted. Since the goal of the design process is that each block can be reused in the future, a *Circularity Limit State* (CLS) is defined and used as a critical design optimization parameter. The available material stock consisting of white pine, pressure-impregnated spruce and LVL wood was assigned to the blocks considering the stress distribution. In the first design iteration of *ReconWood Proto 02*, 11.2% of the structure exceeded the CLS (1), with a 204% maximum utilization value (2.04

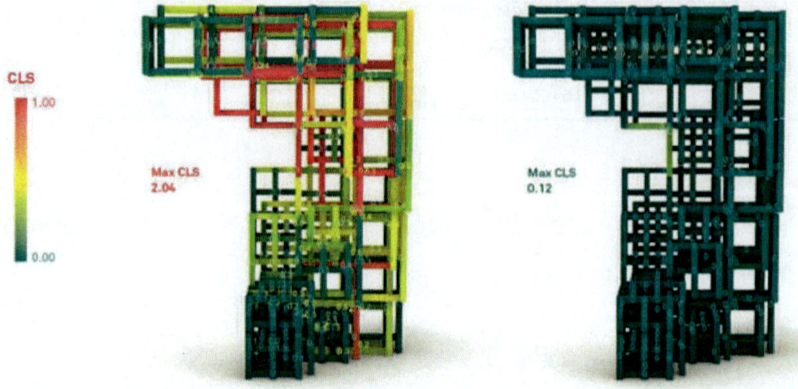

Fig. 3 Linear FEA outcomes of the *ReconWood Proto 02* before (left) and after (right) the optimization

Table 2 Detailed overview of the three MDT datasets derived from the Design and Structural Simulation

MDT

IDENTITY/ DESIGN	STRUCTURAL PERFORMANCE	CIRCULARITY
block ID	structural utilization	CLS (Circularity Limit State)
dimensions	joint mechanical performance	stored carbon
cross-section	material[+] characteristics (tensile strength, compressive strength, bending strength)	No of life cycles / uses record
material data[+] (species, grade, density, weight, volume, color)		RR (Reusability Rate)
material distribution		life expectancy
location		future structural layer prediction
connectivity		
reccurence in the structure		
structural layer		

CLS). After the optimization process, the structural density of the over-stressed area was increased to fit within the CLS, achieving a 12% utilization (0.12 CLS) (Fig. 3).

Through a unique identification tag, each block inherits estimated stress values for future evaluations of fatigue and the structural strength in following construction cycles (e.g. V1B4.7, where V1 identifies the voxel level 1, B4 defines the block typology 4 and 7 is a specific count identifier). To each identified block, related data are attached and fed to the MDT through organized datasets (Table 2).

Material information such as species, grade, density, weight and volume is received from the material stock sheets, while specific structural characteristics within the structure, such as compressive, tensile and flexural stress, are gained from simulation. On the other hand, circularity data are interpreted and predicted based on existing datasets. For instance, the *Reusability Rate* (RR) is calculated as an inverse ratio between the number of life cycles (NLC) and the CLS (1/(NLC*CL)). An individual block's life expectancy can be estimated based on simulated stresses across the various use cycles as a combination of fatigue and decay, which are stored within the MDT and made available for further elaboration.

2.3 Robotic Fabrication and Material Identification

The *ReconWood* construction kit is robotically fabricated with a mass-customization approach (Fig. 4) from three different wood types, namely, LVL spruce, white pine and pressure-impregnated spruce wood to simulate real-life conditions where the *ReconWood* blocks could be sourced from the disassembly of different constructions. The fabrication process involved using three different end tools with their corresponding toolpaths. The machining settings were adjusted depending on the wood type to ensure qualitative cuts and finishing. These parameters have been optimized through iterative milling tests and finally loaded to the MDT as *ReconWood* blocks *fabrication datasets* and made accessible for future re-fabrication purposes (Table 3).

During the fabrication process, the produced blocks were given an identity through a unique QR imprint, which links to their datasets in the MDT. The QR code was selected over the other tracking systems (barcode, NFC, RFID) due to the easy readability from different devices (smartphones, tablets, MR headsets, robots equipped with vision systems, etc.) and thus its predisposition for automation workflows. After comparing different techniques with hands-on tests, 35 × 35 mm QR codes were applied with laser-cut stencil and black spray paint (Fig. 5). Numerous scanning tests confirmed that applied QR codes are easily readable by both smartphones and MR HoloLens 2 devices. Even though this method is relatively time-consuming,

Fig. 4 Robotic fabrication of a *ReconWood* block

Table 3 Detailed overview of the MDT datasets derived from the robotic fabrication

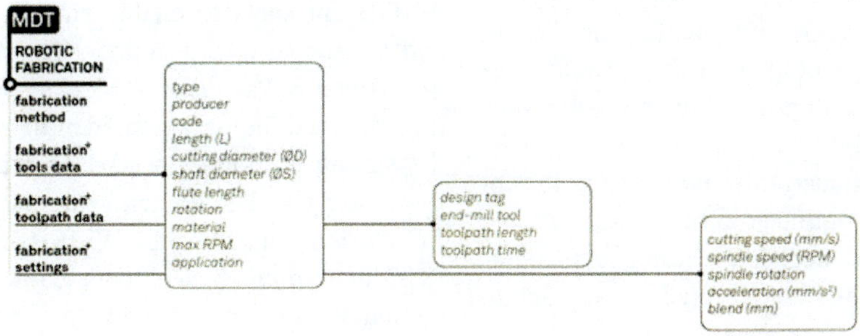

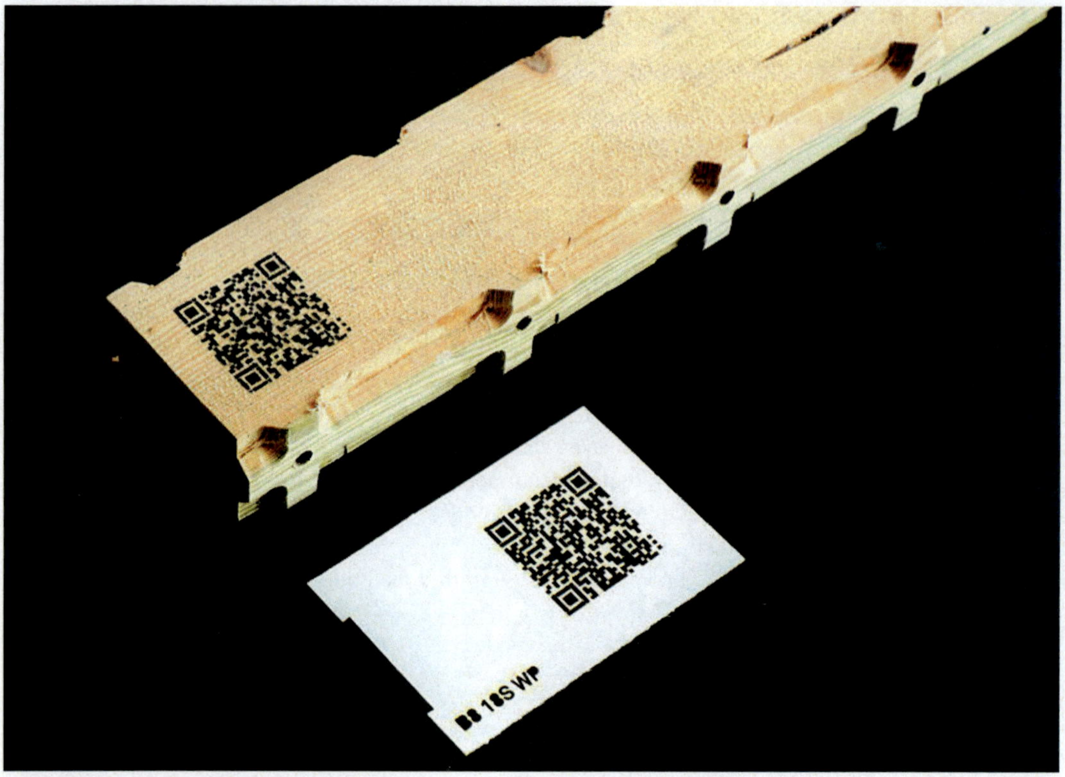

Fig. 5 *ReconWood* block with the applied QR code and the laser-cut stencil

it provides a cost-effective and permanent transfer of QR codes to wooden blocks, which can be easily automated in an industrial context.

2.4 MR-Aided Human–Robot Assembly

The readability of the QR code imprints by different digital devices allows various machines to access the MDT datasets. This supports the automation of assembly and disassembly processes and contributes to a more data-fluid cyber-physical construction environment. A previously established robotic setup for cyber-physical re-assembly of reconfigurable wood structures (Kunic et al. 2021a, b) is here enriched by MR-aided human assistance to provide (i) an instant QR reading and identification of the *ReconWood* blocks, (ii) critical decision-making and cognitive skills during unexpected construction circumstances and (iii) assist the robots by facilitating construction tasks that would require high-complexity and time-consuming programming and facilities.

Cyber-physical Setup—The *ReconWood* assembly setup consists of three main areas (Fig. 6). The *Assembly area* is shared between an MR-aided builder and two robots. It contains an assembly table with auxiliary materials and tools such as a bolts holder, picking/placing reference holders and a hand screwdriver. Holographic construction data and instructions are projected in different spots to enable smooth interactivity and assistance to the builder. The *Material station* contains *ReconWood* blocks sorted by type, material and holographic identifier tags. Since the station is mobile, an ArUco marker enables its tracking and referencing in space. Finally, a *PC station* connected to Wi-Fi consists of two PCs, one running Windows with Grasshopper computational model and the other running

Linux operating system with ROS. The Grasshopper computational model is the primary node that communicates bi-directionally with (1) Hololens 2 headset, (2) ROS on the PC2 and (3) MDT in Speckle.

Assembly process—The assembly instructions, such as block sequence and coordinates, and bolts and nuts locations, are communicated to the robots from the computational design model through ROS bridge connection (Kunic et al. 2021a, b) and to the builder through a custom-made MR User Interface (UI) (Kunic and Naboni 2022). The same information is loaded to the MDT as the assembly data tree (Table 4).

Upon receiving the assembly sequence information, the builder picks the required block from the *Material station*. In the first-time assembly, the sequence of the blocks in the computational

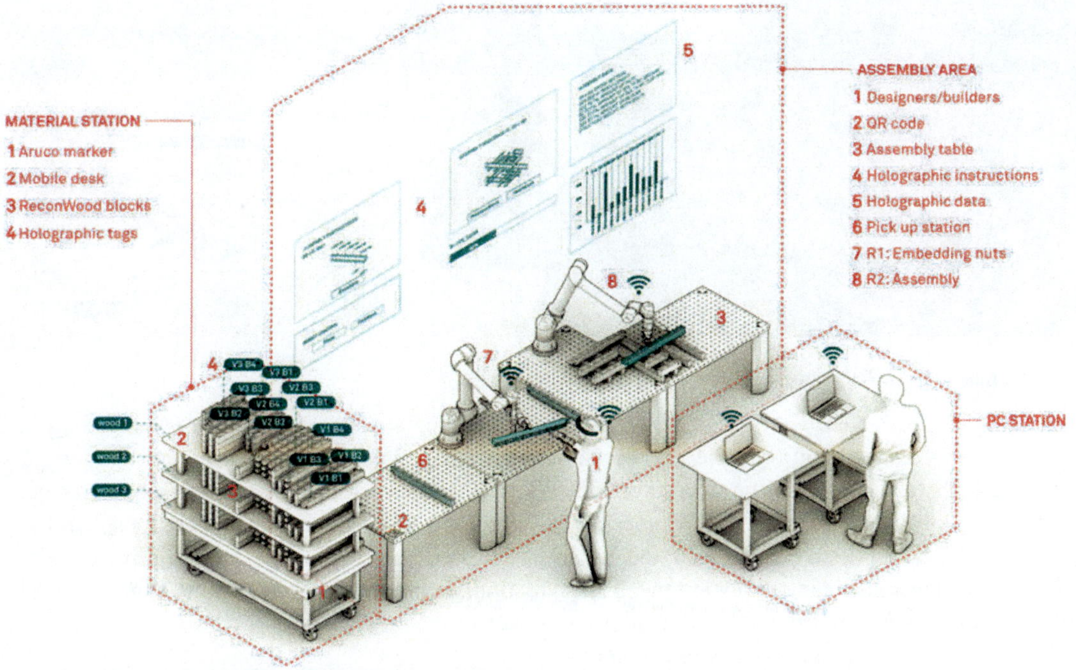

Fig. 6 Cyber-physical human–robot collaborative setup for the robotic assembly/disassembly of *ReconWood* structures

Table 4 Detailed overview of the MDT datasets derived from the human–robot collaborative assembly

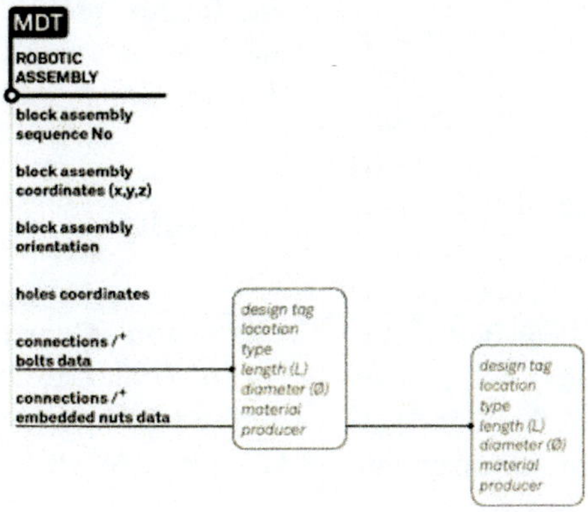

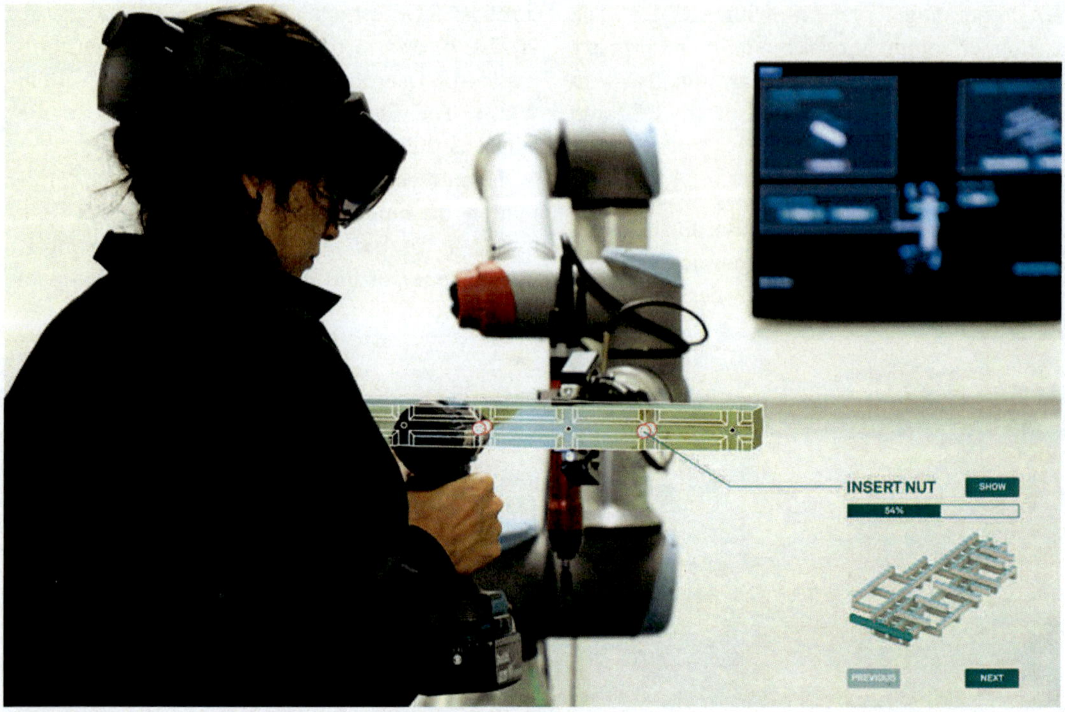

Fig. 7 Human–robot collaborative process of embedding nuts

model is simply incremental (e.g. the first block type V2B4 that is met has the identification number 1). Once the builder selects a specific block (e.g. V2B4 in LVL), they scan the QR code and read the specific identification tag. This information is then sent to the computational model and MDT to update the related design and assembly data. This operation is significant for three reasons: (1) each physical block inherits its performance data from the simulation considering its position in the structure (e.g. CLS structural data), (2) during the disassembly phase, the robots will be 'aware' of the structural performance of each block and will be able to sort them accordingly and (3) in the second-time assembly, the sequence will be defined more precisely, based on the inherited performance data from the first use of the blocks. In the following assembly step, the builder and robot 1 collaborate on the block preparation, embedding steel insert nuts into specific holes (Fig. 7). While robot 1 holds the block, the builder visualizes the nuts locations holographically and performs the operation. After this, robot 1 passes the block to robot 2 (Fig. 8), which completes the assembly process following a sequence of operations (block placement, holes detection and coordinates correction, screwing) described in Kunic et al. (2021a, b).

Use and Operation—Once assembled, the structure was installed upon a wooden base and exposed to the outdoor environment (Fig. 9). Additionally, the structure was spray-coated with 1–5% citric acid fire retardant and oil-based water/fungus protective paint. The used substances are recorded through an MR device to the MDT *operation data*.

3 Results

Design and Construction—The *ReconWood Proto 02* comprises 250 *ReconWood* blocks with a 125 m wood length of three different types and 612 reversible steel connectors. 23.2% of the

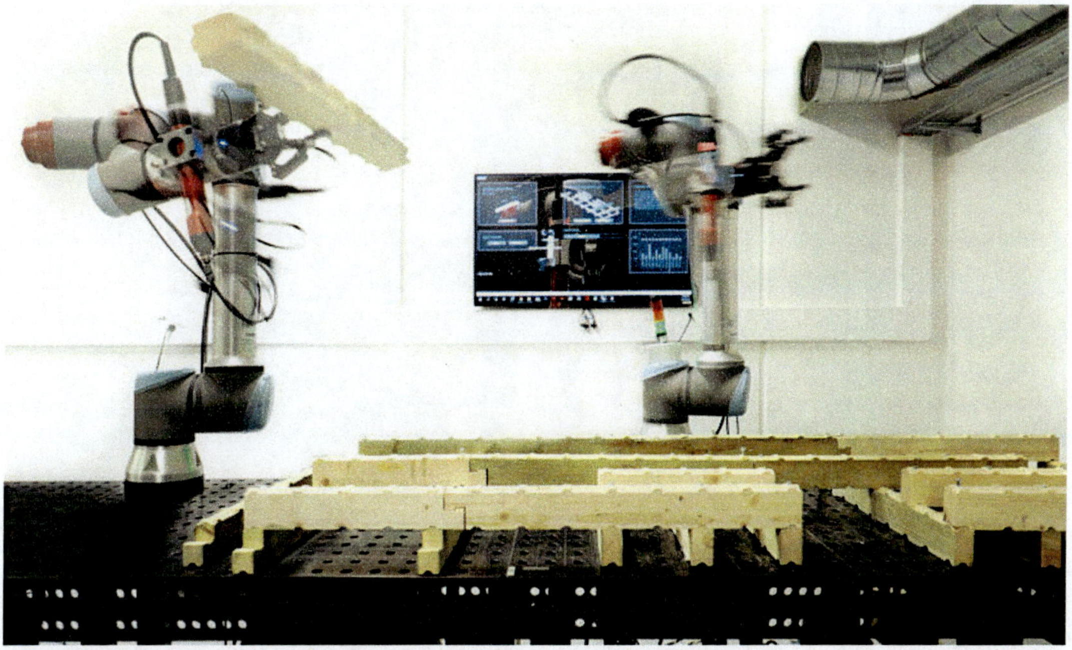

Fig. 8 Multi-robot assembly of the *ReconWood Proto 02*

Fig. 9 *ReconWood Proto 02* installed outside of the CREATE Laboratory at the University of Southern Denmark

structure is covered in white pine, 52.8% in impregnated spruce and 24% in LVL. Besides meeting the structural requirements for reuse, the variety of species provides a colour-rich gradient throughout the structure and introduces a new level of aesthetic detail. The prototype was successfully designed and constructed in a cyber-physical environment with human–robot collaboration, aided by MR. Each physical block carries data regarding its origin, structural performance, construction and circularity aspects, embedded through QR code imprints and stored in the MDT (Fig. 10). All the blocks have been used for the first time in this prototype, and they are ready for full reusability in future use cycles.

MDT—The *ReconWood Proto 02 MDT* database holds the structured data trees regarding each building element in the form of a Speckle online link (SPeckle Stream). These are nested into 12 Speckle Branches (block type) and 250 sub-

Fig. 10 Close-up of the *ReconWood Proto 02* showing applied QR codes on the blocks

branches (unique identifier), each with six additional data layers (data categories). To facilitate the interaction with the MDT of the blocks, the six data layers are translated into visual boards in Grasshopper and sent to corresponding Speckle Branches (Fig. 11). These data can be visualized and navigated by scanning QR codes on the physical blocks. Furthermore, iFrames of the selected Speckle branches (one per block typology with six categories) are integrated into the official projects' website page to provide easy access and interactivity with the MDT datasets to a larger audience. Even though Speckle was initially conceived as a tool that enables interoperability and collaboration between different users and software, it provided a solid base for this prototype exploration of MDT and its functionalities.

Accessibility—The MDT was accessed and updated several times throughout the design-to-construction workflow. Firstly, it received the basic material stock data coming from the producer. Secondly, the designer and structural engineers collaborated on the computational design, mechanical testing and structural optimization algorithms, where rich datasets were generated and sent to MDT. From the digital fabrication point on, the MDT could be accessed by the 'machine users' since a scannable QR code has been applied to the physical *ReconWood* blocks. During the robotic assembly phase, MDT received several assembly data updates on the actual sequence of the blocks from the MR-aided human operator and the updated blocks and holes coordinates through the parametric design model, which controlled the real-time adaptive assembly operations of robot 2. Subsequently, during the final installation of the prototype, material data was updated again with specifics about the fire retardant and water/fungus protective coating. Finally, lab users and visitors had easy access to the MDT data by scanning individual QR codes on their smartphones, tablets or MR headset devices.

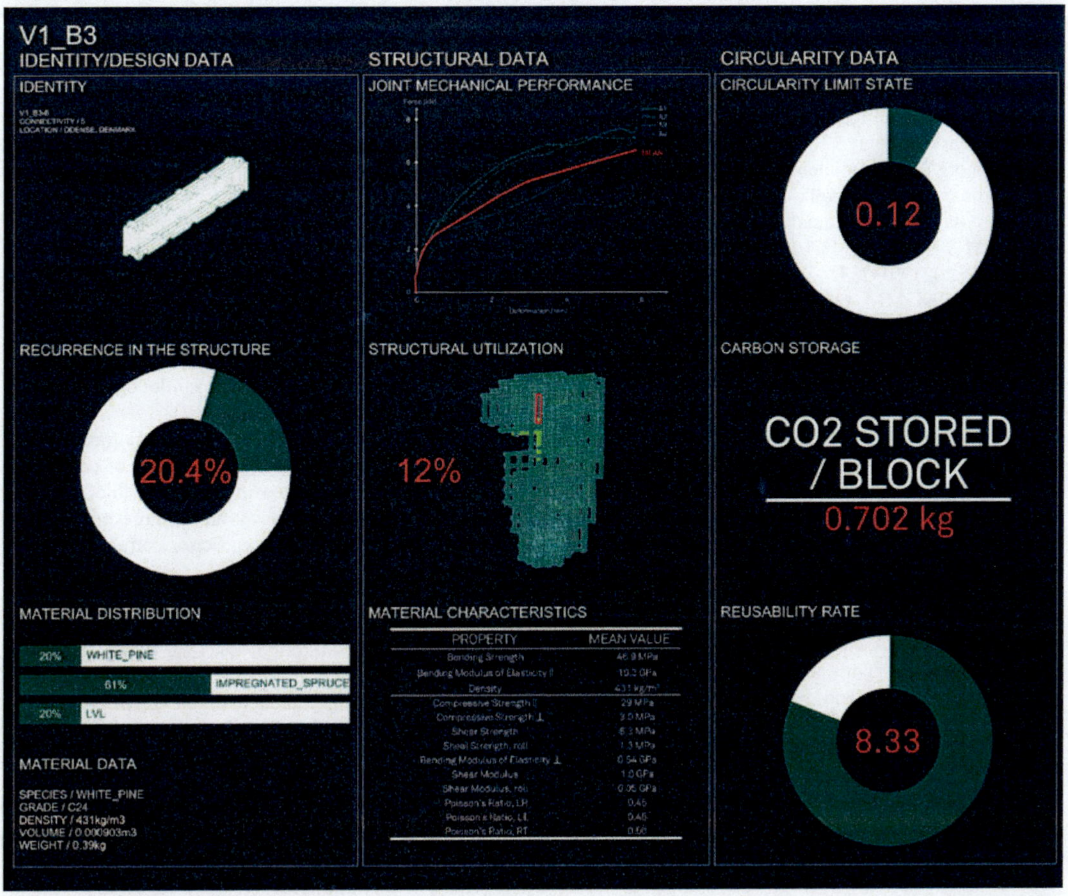

Fig. 11 Example of the MDT datasets and data visualization in Speckle regarding the Identity/Design, Structural Performance and Circularity information of the block V1_B3

4 Conclusions

Accessible, real-time information updates throughout the material's life cycle(s) are crucial to inform data-driven processes of circular construction. The presented prototype of MDT shows initial working functionalities that effectively support a circularity practice through the automation of design, fabrication and reconstruction phases. From a longer-term perspective, the MDT will be developed as an online platform that combines features—currently developed in Speckle—for collaboration and data sharing with a more accessible user interface for data visualization and interaction.

5 Project Links

Speckle link: https://speckle.xyz/streams/e25cf79b4b.

Website link: https://www.create-sdu.com/projects/reconwood02.

Acknowledgements The research was conducted at the CREATE Lab at the University of Southern Denmark. The authors wish to thank Angelina Garipova for her help in the fabrication and construction of the *ReconWood Proto 02*.

References

Àkànbí LA, Oyedele AO, Oyedele LO, Salami RO (2020) Deep learning model for Demolition Waste Prediction in a circular economy. J Clean Prod 274:122843. https://doi.org/10.1016/j.jclepro.2020.122843

Atta I, Bakhoum ES, Marzouk MM (2021) Digitizing material passport for sustainable construction projects using BIM. J Build Eng 43:103233. https://doi.org/10.1016/j.jobe.2021.103233

Copeland S, Bilec MM (2020) Buildings as material banks using RFID and building information modeling in a circular economy. Procedia CIRP 90(143):147. https://doi.org/10.1016/j.procir.2020.02.122

Heisel F, Rau-Oberhuber S (2020) Calculation and evaluation of circularity indicators for the built environment using the case studies of UMAR and Madaster. J Clean Prod 243:118482.https://doi.org/10.1016/j.jclepro.2019.118482

Heinrich M, Lang W (2019) Materials passports-best practice-innovative solutions for a transition to a circular economy in the built environment. Technische Universität München, Munich

Honic M, Kovacic I, Aschenbrenner P, Ragossnig AM (2021) Material Passports for the end-of-life stage of buildings: challenges and potentials. J Clean Prod 319:128702. https://doi.org/10.3390/su132212466

Jayasinghe LB, Waldmann D (2020) Development of a BIM-based web tool as a material and component bank for a sustainable construction industry. Sustainability 12:1766. https://doi.org/10.3390/su12051766

Kunic A, Naboni R (2022) Reconfigurable timber construction in an extended reality environment. In: Proceedings of the 25thSIGraDi conference: critical appropriations, Lima, Perú, November 2022

Kunic A, Kramberger A, Naboni R (2021a) Cyber-physical robotic process for re-configurable wood architecture. Closing the circular loop in wood architecture. In: Proceedings of the 39th eCAADe conference-towards a new, configurable architecture, Novi Sad, Serbia, September 2021a

Kunic A, Naboni R, Kramberger A, Schlette C (2021b) Design and assembly automation of the Robotic Reversible Timber Beam. Automation in construction. Elsevier. https://doi.org/10.1016/j.autcon.2020.103531

Li CZ, Chen Z, Xue F, Kong XT, Xiao B, Lai X, Zhao Y (2020) A blockchain- and IoT-based smart product-service system for the sustainability of prefabricated housing construction. J Clean Prod 286:125391. https://doi.org/10.1016/j.jclepro.2020.125391

Materialen Marketplaats (2022) Netherlands. Retrieved from: https://materialenmarktplaats.nl/

Naboni R, Kunic A, Kramberger A, Schlette C (2021) Design, simulation and robotic assembly of reversible timber structures. Construction robotics. Springer Nature Switzerland AG. https://doi.org/10.1007/s41693-020-00052-7

Naboni R, Kunic A (2019) A computational framework for the design and robotic manufacturing of complex wood structures. In: Proceedings of the 37th eCAADe and 23rd SIGraDi conference: architecture in the age of the 4th industrial revolution, Porto, Portugal, 11–13 September 2019, pp 189–196

Raghu D, Markopoulou A, Marengo M, Neri I, Chronis A, De Wolf C (2022) Enabling component reuse from existing buildings through machine learning-using google street view to enhance building databases. In: 27th international conference of the association for computer- aided architectural design research in Asia (CAADRIA) 2022, Sydney, Australia. https://doi.org/10.52842/conf.caadria.2022.2.577

RotorDc (2015) www.rotordc.com/. Accessed 7 Jul 2022

Speckle (2020) www.speckle.systems/. Accessed 7 Jul 2022

Spotr (2022) www.spotr.ai. Accessed 7 Jul 2022

Wibranek B, Tessmann O (2023) Augmented reuse. In: Gengnagel C, Baverel O, Betti G, Popescu M, Thomsen MR, Wurm J (eds) Towards radical regeneration. DMS 2022. Springer, Cham. https://doi.org/10.1007/978-3-031-13249-0_33

Printsugi: Matter as Met, Matter as Printed. Leveraging Computational Design Tools for a More Virtuous Material Extraction and End-of-Life

Nadja Gaudillière-Jami, Max Benjamin Eschenbach, and Oliver Tessmann

Abstract

You're outside. You want to build. There's earth. There's twigs and logs. There's stones. There's all you need. Our research proposes to leverage 3D-printing and computational design strategies to use matter as met, i.e., matter as found on site in building processes. Combined into a design method enabling disassembly and reassembly, this approach enables less transformation of the materials used and thus less energy consumption. It also renders possible the use of resources without a brutal extraction process, as well as a focus on materials that are locally available. We propose a design and fabrication workflow in answer to these requirements. This workflow starts with testing raw earth available on the site in order to devise a recipe for the raw earth to be 3D-printed. The second step is the collection, sorting, and 3D-scanning of other types of materials available on site, such as logs and stones. The third step is the resort to a computational design process placing these materials in the most relevant place within the shape that is to be constructed. The fourth step is to model connectors linking these pieces together. The fifth step is to print the connectors by using the recipe devised in the first step. To print the pieces, a low-cost 3D-printer is to be used on site, in order to remain as local as possible. As the connectors are to be only components of the structure, they are of small scale and the 3D-printer can thus have a relatively small printing area as well. The final step is to assemble the structure by bringing together connectors and materials. A prototype serves as a case study to demonstrate each of the steps within the paper, using raw stones and clay connectors.

Keywords

Scan-print-assemble · Life-cycle visualization · Clay 3D-printing · Raw materials

N. Gaudillière-Jami (✉) · M. B. Eschenbach · O. Tessmann
Digital Design Unit (DDU), Technische Universität Darmstadt, Darmstadt, Germany
e-mail: gaudilliere@dg.tu-darmstadt.de

M. R. Thomsen et al. (eds.), *Design for Rethinking Resources*, Sustainable Development Goals Series,
https://doi.org/10.1007/978-3-031-36554-6_27

1 Introduction Wood, Sand, Clay, Stone

Our traditional building materials are what can be called matter as met—matter as found on site. The term, coming from the French "matériau de rencontre" and sometimes also translated as "materials encountered by chance" (Bréchet 2015), has been coined by material scientist Yves Bréchet. Bréchet describes matter as met in opposition to what he calls architectured materials: tailored materials designed by humans to meet their needs. In his courses and publications, he gives many examples showing how various existing materials can be combined into new ones and how geometry at the nanoscale can be refined to obtain materials that perform extremely well in one very particular use. This however requires a large amount of energy as many processing and shaping steps are required in order to obtain the desired material structure. In contrast to these architectured materials, he explains, matter as met are materials that we initially simply found on the ground, before gradually developing methods to extract initially out of reach materials further and further (Bréchet 2015).

By now, matter as met is harvested in specific locations and in large quantities for our industrial use of it. In the construction industry, we seek to harvest vast quantities of wood, of bauxite, of lime, of sand, or of stones to provide for the rapid increase in size of the built environment. These materials are then transformed in building components and shipped to building sites all over the world. But they are also present raw and in smaller quantities at many of the building sites we choose before construction starts. What if we could use these resources directly in the building used, right there, right then? What if everything was already there on site? We could harvest rocks and logs directly there, devise construction processes that enable using them in buildings structures and envelopes raw, without any major transformations, and therefore significantly lower energy consumption. We could harvest earth, extract the clay, and resort to traditional recipes for turning that clay into building materials directly on site, as is still done in many places around the world (Fabbri et al. 2022).

In fact, scaling down and staying local by resorting to matter as met in such fashion holds major potentialities for lowering the impact of building activities—it would enable acting on manufacturing and transport phases and reduce their footprint significantly. This is what we aim at exploring within the Printsugi project. Our research proposes to leverage 3D-printing and computational design strategies to use matter as met in building processes.

2 Background. Strategies for Lowering the Impact of Construction

One does not need to stress again the urgency of finding ways of rendering construction more sustainable. While full assessments of the impact of buildings fortunately are on the rise, with more and more architecture offices integrating an LCA division to watch over the planning phases, tools to leverage individually by unspecialized architects are also on the rise. These are key aspects for rendering our practices more sustainable. They range from applying basic principles on some processes and materials (glass is recyclable, aluminum very costly environmentally, wood captures carbon, and so on) to integrating data relative to the impact directly in digital models from early design phases on, with tools such as One-Click-LCA (Apellániz et al. 2021) or RhinoCircular (Heisel et al. 2021).

In order to evaluate the environmental impact of construction, the life-cycle of a building is structured in six phases, each with its specific steps and their consequences: extraction, production, transportation, construction, exploitation, and end of life (ISO 14040-14044) (Abd Rashid and Yusoff 2015). As architects, we can act to reduce the overall impact by relying on three strategies, corresponding each to part of that life-cycle, as is summed up in Fig. 1.

The first strategy is to act over the production process of the building, which covers the extraction, production, transportation, and construction phases. During these, we can aim at

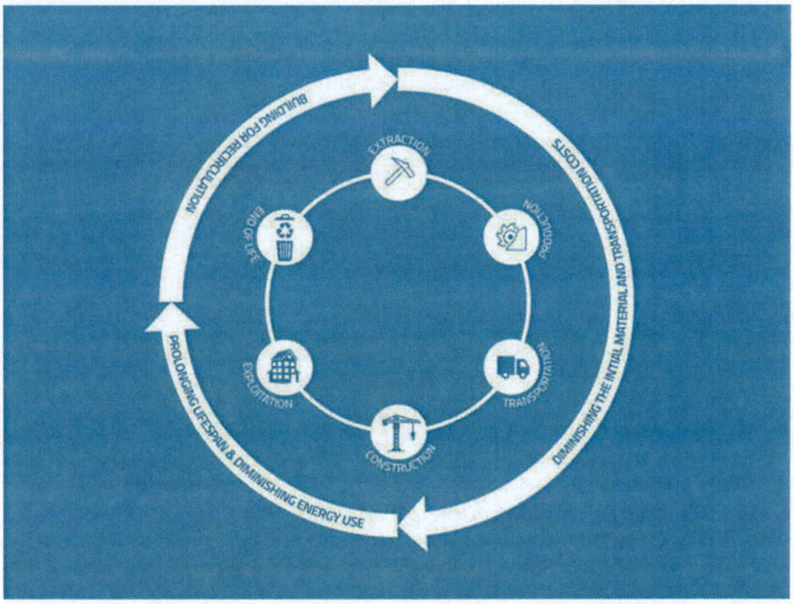

Fig. 1 Building life-cycle and acting strategies

diminishing the amount of material used. Multi-objective optimization methods have for example been developed by the MIT-based Digital Structures Group in order to balance structural requirements and others with minimizing the amount of material (Brown and Mueller 2015; Brown et al. 2015). During these phases, we can also aim at minimizing the distances over which the materials travel. As an example, the large-scale research project Earth of the Greater Paris (Terres du Grand Paris) brings together a large number of players of the French construction industry and from associated academic fields (Loiret 2021). The aim is to make use of the vast amount of raw earth excavated as part of the many new construction projects launched for the Olympics 2024 hosted by the city of Paris. Instead of throwing away the excavated materials, they are stored, cleaned, sorted, and used in local raw earth construction projects.

The second strategy is to better the performance of the building during its exploitation, prolonging its lifespan and reducing its consumption. For example, the project Cinder by Umbrellium aims at raising awareness among students by associating the energy consumption of their community college building with a virtual cat appearing on screens throughout the building (Umbrellium 2015). As long as the building has a positive or neutral energy consumption level, the cat can be fed and played with. However, if the building consumes too much energy, the cat withers and refuses to interact. The project Natural Fuse, by the same team, also aims at raising awareness, but this time in adults (Haque and Fuller 2010). Inhabitants of a neighborhood are provided each with a lamp and a plant. The amount of energy available to light the lamps depends on the carbon captured by the plants. However the energy reserves are shared by all inhabitants, meaning that each of them has to be careful in their use to not deprive others of light. It thus focuses well on the crucial role of communication between inhabitants regarding how to handle this step of the life-cycle.

The third strategy is to develop more virtuous ends of life for the buildings, focusing on recirculation by designing for deconstruction or for reuse. This is what the Circular Construction Lab of Cornell University has been working toward, documenting available resources within construction of Ithaca, the city the lab is located in. The team has then developed time-saving deconstruction methods in order for it to be economically viable (Heisel 2022). A number of computational design processes are also being developed to enable reuse with the least reshaping, as research on trusses by Corentin Fivet and his team shows (Brütting et al. 2019; Fivet and Brütting 2020). Daniel Marshall's research Computational Arrangement of Demolition Debris (Marshall et al. 2020) shows an alternative approach to these, while leveraging as well reused materials—although in a poorer state. This has also already been implemented in existing architecture projects, such as those created by PimpYourWaste (2022), or by the well-known Rotor architecture office (Chudoba et al. 2020). Finally, the origin of the reused materials here again matters. The Precast Concrete Components 2.0 Project developed by the Digital Design Unit aims at reusing parts of discarded concrete building structures, enabling a renewed building materials supply within cities (Eschenbach et al. 2023; Tessmann 2022). Another example is the 2022 Basel pavilion Slicing Time by Sterling Presser, only built with reused materials sourced from within the city itself (Sterling Presser 2022). This is in line with the recently passed bill for the city of Basel, which demands that new buildings are constructed first and foremost from reused materials. Regulation are in fact crucial in order to enable such strategy to be implemented. A great amount of research is therefore aimed at framing the type of information needed to enable this, the most adequate format to transmit those information and the most adequate methods to obtain them (Heisel et al. 2022; De Laforce 2020). Finally, a more general framework of design based on envisioning a more virtuous end of life has also been addressed (Crowther 2018; Gorgolewski 2008; Addis 2006).

3 Method Part 1. Leveraging Life-Cycle Visualization for a More Virtuous Design Process

Our team from the Digital Design Unit has been developing a workflow we believe can help act by leveraging these strategies together. Scan-Print-Assemble starts with collecting and 3D-scanning materials to be used in construction. The resort to a computational design process then enables placing these materials in the most relevant place within the shape that is to be constructed. Connectors are then modeled to link these pieces together and printed, and the structure is assembled by bringing together connectors and materials relying on Augmented Reality tools. The project Digital Rubble has resulted in multiple prototypes with polymer square-shaped connectors binding rocks into arches (Wibranek and Tessmann 2019). The project Net-Connectors resorted to tree branches binded with polymer connectors, exploring different geometrical requirements as the shape of branches requests an adaptation of the Digital Rubble process (Gehron et al. 2019). Both therefore leverage matter as met as their base material and explore the resort to 3D-printed bespoke geometries to avoid transforming the rocks and branches.

Our Scan-Print-Assemble workflow however raises many questions regarding its environmental impact. As is summed up in Fig. 2—first of the life-cycle visualizations in this research, each life-cycle phase has its associated issues, relating among others to the novel resort to 3D-printing in construction that has unfolded in the past decade. Some are questions relative to transport distances. Given the type of machines used for 3D-printing, one can choose to implement printing on site or prefabrication and transportation, a choice that will depend on the type of printer resorted to. The French company XtreeE has for example chosen to print near to the construction site instead of printing in their own facility in the Viliaprint project (Marais 2022). They thus leave materials and printers to travel from another part of France to the printing site but diminishing the distance traveled by the printed components, thus reducing the overall volume of matter moved. Some are questions relative to the production phases of the life-cycle. Depending on whether the building combines 3D-printing with other construction processes and on how much materials have to be transformed before being printed or becoming building components, the footprint may vary greatly.

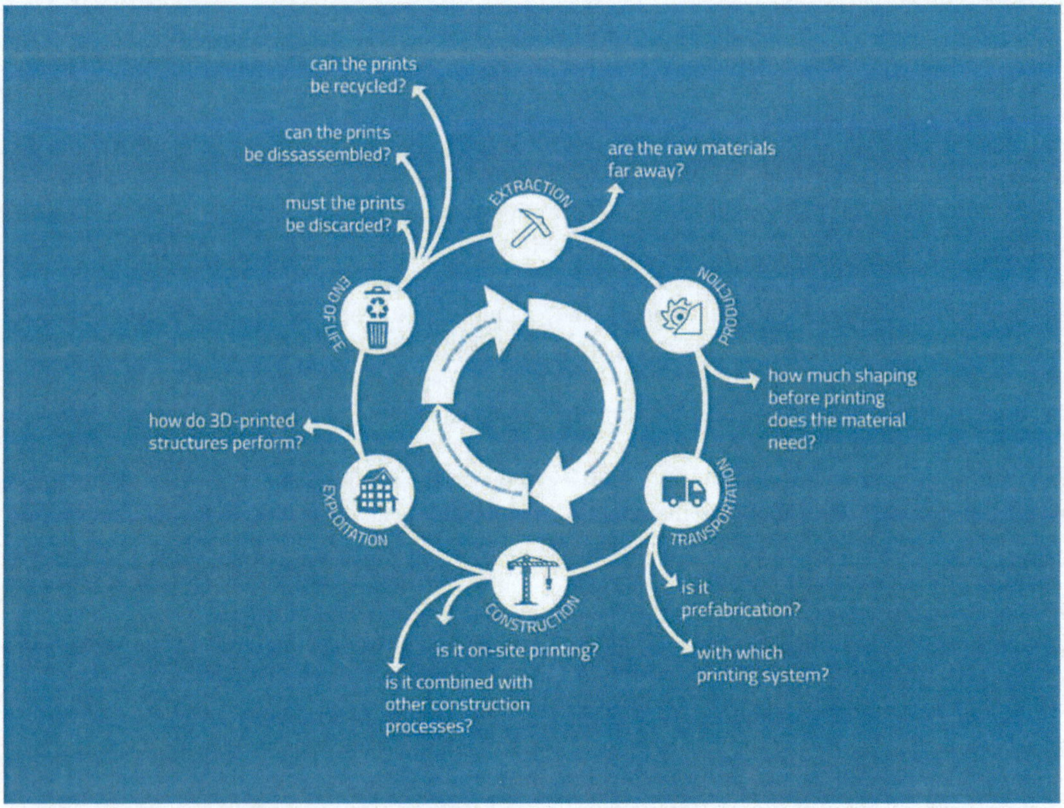

Fig. 2 Building life-cycle and issues relating to 3D-printing (based on Gaudillière-Jami 2022)

The choice of 3D-printing system may play a key role in the impact of the project: as research as shown, the weight of the printing system in the overall impact varies greatly, from insignificant (Agustí-Juan et al. 2017) to significant, in particular regarding metal depletion (Kuzmenko et al. 2020). Some are questions relative to the exploitation phase, as the assessment of structural and thermal performances of 3D-printed structures is still ongoing research, but has strong influence on its energetic performance during exploitation as well as on the longevity of the building. Finally, some are questions relative to the end of life phase, as 3D-printed structures have as open an end of life as other materials, with some design strategies being based on disassembly and reuse, some printing materials producing objects that can be recycled or that must be discarded.

But in the Printsugi project, Scan-Print-Assemble meets Matter as Met, enabling us to rethink parts of the workflow to improve its environmental performance. The analysis of the Scan-Print-Assemble life-cycle through its graphic visualization, and the study of existing intervention strategies highlighted above, has allowed us to determine leads to improve three indicators of environmental performance: energy use, metal depletion, and climate change. In order to keep the matter processing as low as possible, and therefore to keep a low energy consumption, we have chosen to keep working with matter as met. But we choose to modify the connector material, choosing clay, a material with low carbon emissions (ref), which also allows us to work with small-scale printers, therefore ensuring low metal depletion. This enables us to explore a material-system configuration that allows for using the geometric versatility of 3D-printing while attempting to diminish its environmental impact. Finally, aiming for a mix solely composed of clay and water allows us for an easy recycling through reprinting of the material. While not being a full life-cycle assessment, the visualization of the different life-cycle parts of the Scan-Print-Assemble workflow has therefore enabled us a critical overlook and has guided our decisions in updating this workflow. Issues to tackle in the Printsugi project have been identified through this visualization, which enables us to reflect upon the different stages and how they can be improved without having to resort to a full assessment.

4 Method Part 2. The Printsugi Matter as Met Fabrication Process

The fabrication process at large implemented in the Printsugi project relies on the same six steps as the Scan-Print-Assemble process. However some of these steps are carried out in ways specific to Printsugi. The first step is still to collect matter as met, but this time directly on the building site. The second step is still scanning, through a photogrammetry process. The third step, designing the prototypes and the connectors within, is specific to the Printsugi project. It draws its name from the combination of the words *3D-print* and *kintsugi*. While *3D-print* obviously refers to the additive manufacturing fabrication method developed since the 1980s, *kintsugi* is a Japanese word and concept of mending using pure gold to fix a snag or to put together again broken pieces. The use of pure gold is intended to confer renewed value to the objects and praise the extension of life of objects that such reparation results in. Strategies of design for the prototypes and connectors rely on similar aesthetic approaches in order to promote the same idea. This can either be done by mending the materials in order to recreate a previously existing volume, literally following the kintsugi logic, or by associating the materials within a new volume, but with their initial, raw aspect still visible. The connecting logic can vary and largely contributes to underlying the peculiar aesthetic of the Printsugi project design strategies, as is shown in Fig. 3.

Several computational design processes associated with placing raw materials already exist, such as the one implemented for the Woodchip Barn project at Hooke Park by the AA team, placing raw logs in the most adequate place of a roof structure to be able to minimize the manufacturing (Self et al. 2017). Since this project, work with raw wood has continuously been developed (Larsen and Aagaard 2019; Gehron et al. 2019; Vestartas and Weinand 2020). Similar processes have been implemented for stones as well, as is shown in The Cannibal's Cookbook (Clifford 2021) and in the ETHZ project Autonomous Dry Stone (Johns et al. 2020). The automation of raw materials placing strategies in Printsugi draws inspiration from these works and is based on previous research from the authors (Eschenbach et al. 2023).

The fourth step is the preparation of printing material. The connectors are printed with clay extracted from ground harvested directly on site. Successive finer and finer sieving with a set of sieves up to 0.002 meshes extracts clay, separating it from sand, dirt, and soil and giving access to printable material. Once the clay is extracted, however, the issue of defining the exact material recipe that will render that particular clay printable remains. As is already well covered in the literature, many different types of clay indeed exist and require different amounts of water and different kinds and amounts of additives in order to achieve a rheology suitable for printing (Fabbri et al. 2022; Perrot et al. 2018). A key particularity of the Printsugi Scan-Print-Assemble workflow is the set of material tests performed in order to determine this recipe. This is done building on tests traditionally performed to assess clay, in particular on the Atterberg limits test (Fabbri et al. 2022) as well as on preliminary research performed by the IFSTTAR institute on how to adapt the plunger test to printable rheologies (Tourtelot et al. 2021). The Printsugi project indeed aims at developing material tests and material recipes that enable

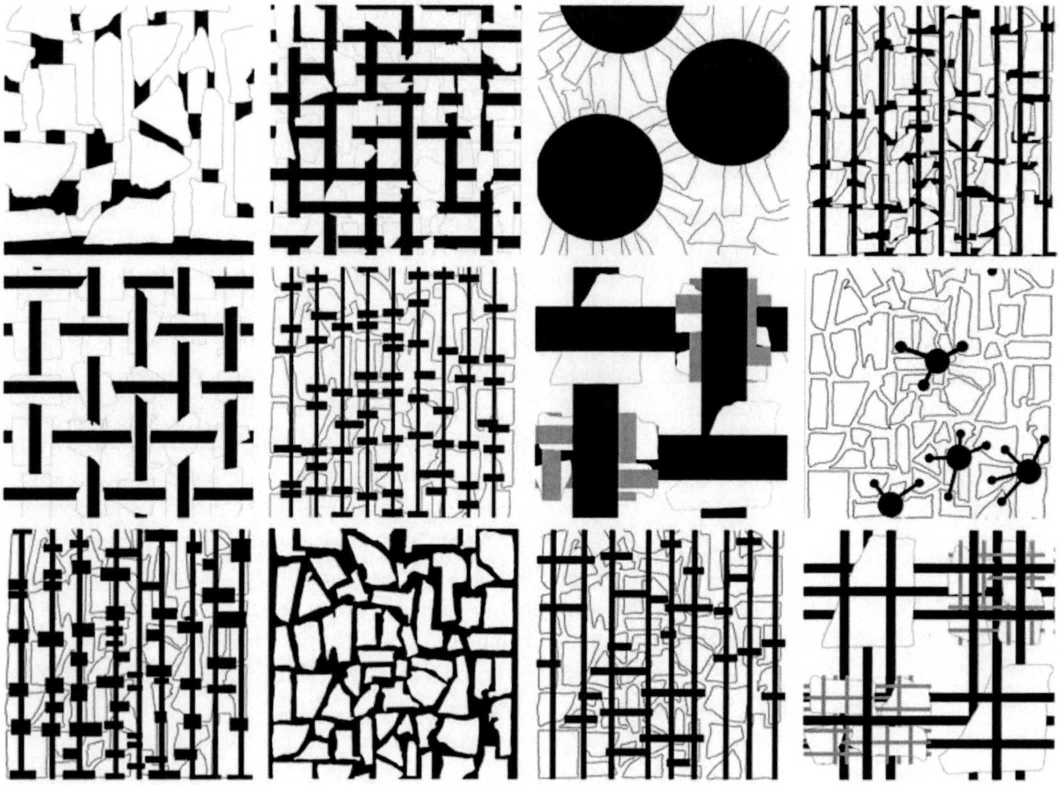

Fig. 3 Connector Design Variations. Black parts are 3d-printed while the white perimeters describe the matter as met. (Images by Zahra Babadi, Dominik Binzel, Mirko Dutschke)

printing any clay extracted that way, tailoring the ingredients of the recipe to the particular composition of the ground, adjusting water amount and the potential additives to it. However, a particular condition in the Printsugi material recipe development is the aim to radically restrain the resort to additives, to keep the recipes both as easy to implement and as clean as possible.

Step five, printing, is performed with an off-the-shelf clay printer, robust enough to handle the variety of recipes we have been experimenting with, but also small enough to be easy to transport on site. Instead of transporting materials, we thus can transport the printer only, bringing it to the site to produce the necessary pieces. The final step, assembly, had previously been made with the help of augmented reality (Wibranek and Tessmann 2019). Always in the spirit of remaining easy to appropriate, for Printsugi the assembly was done without this help, simply by hand.

Printsugi not only aims at making the best out of available resources, but also at being easy to share and implement. Each part of the workflow has been implemented with guidelines and tools that make them simple to grasp. Scan, Print, Assemble: our construction process relies on three core steps. Thanks to existing tools, here a basic camera and the Autodesk ReCap Photo software, photogrammetry is almost completely automated. The off-the-shelf printer makes 3D-printing straightforward as well if the material is well calibrated. The assembly is performed without resorting to Augmented Reality, relying on the interlocking provided by the peculiar shapes in our rocks and connectors, guiding the user in putting them together. Material tests are inspired from easily performed ones, necessitating only a cup, a ruler, and a wood block. To acquire the gesture, a few hours of training are needed but nothing more. Atterberg limits tested as part of the Rock Tower fabrication successfully provided data on the gathered clay. Here, working with clay and water only keep the mix simple to make once the Atterberg shrinked limits enable you to define the right water amount for a printable mix. The workflow is thus currently indeed easy to use, enabling us to follow up at every step on our logic of an easy to share and easy to implement process.

5 Results and Discussion. The Rock Tower Prototype: Less Transport, Less Manufacturing, Less Carbon

Once the Scan-Print-Assemble workflow adapted to Printsugi has been set, the research method in order to work with it is that of research by design, producing prototypes to test the workflow and assess its performance. The prototype, the Rock Tower, presented here is the first. Figure 4 highlights its different steps of fabrication according to the Printsugi Scan-Print-Assemble workflow.

The Rock Tower prototype shows promising results for the Printsugi project. As a reminder,

Fig. 4 Rock tower prototype fabrication steps

the evaluation criteria retained to assess the results are the following:

- material testing and used recipe;
- mechanical and geometric characteristics of the prototype;
- updates in the life-cycle and resulting environmental impact;
- ease of implementation of the process;
- aesthetics promoting the use of matter as met.

Regarding the physical characteristics of the prototype, considering the choice of material for the connectors and its generally low performance, the subject of structural capacity is subject to future work. However, the tower is self-supporting. In fact, the stones and connectors enabled this particularly well given the peculiarities of their geometries and the interlocking of their shapes, which, while not being planned as an interlocking system per se, did provide more stability and resistance. The precision of the scanning is entirely adequate, and the precision of the prints is satisfying as well. However, further experimentation on the nozzle, layer size, and layer orientation might enhance the precision of the latter.

Regarding the environmental impact, when looking closer at the life-cycle of our updated workflow, we observe reductions of the impact in several phases, as well as promising venues for further research in others. By using matter as met, we drastically reduce the manufacturing phase that building products require. By harvesting materials and 3D-printing on site, we shrink transportation to the lowest possible. By using only clay and water in the printing recipe, it becomes less pollutant and completely recyclable. By combining connectors and matter as met, we obtain an enhanced disassembling and recycling potential. These aspects are summed up in Fig. 6.

Our proposal also challenges aesthetic habits and performance expectations, opening up changes in thinking habits, toward more sustainable practices. Previously cited projects also have contributed to this, such as the Computational Arrangement of Demolition Debris, the Cannibals Cookbook, or the Autonomous Dry Stone project (Marshall et al. 2020; Clifford 2021; Johns et al. 2020). In the same line, Printsugi shows raw materials and their variability, brought together in a pleasing way and playing with the materials and texture contrasts between print and stone. In provoking by showing unusual things, we contribute to making them become usual visions little by little and thus deconstruct prejudiced views on the use of waste or raw materials (Rigobello and Gaudillière-Jami 2021).

6 Conclusion. Sustainable, Easy, Beautiful… Radical

In this paper, we show how to leverage life-cycle visualization as a starting point to steer research methods, workflow, and design and production of prototypes. This method is applied to the existing Scan-Print-Assemble workflow, adapted to serve the Printsugi project and its recourse to the untransformed matter as met. The collecting, designing, and preparing printing material steps of the workflow are updated into a more virtuous approach. A prototype demonstrates the viability of the renewed workflow, building further upon existing research in the domain of raw material use in architecture. Two of the UN sustainable development goals have been addressed in particular: n○11—sustainable cities and communities—through the ease of appropriation of the method and through working toward a more virtuous production of the built environment; and n○12—responsible consumption and production—through the material choice and the raise of awareness that our venture in new aesthetics can bring.

Further work envisioned includes scaling up and enhancing the structural performance of the connectors, or focusing on the geometries to compensate for the low performance of clay, as well as gathering more results on the material testing in order to define a shrinked interval of water amount, based on the Atterberg limits but specific to 3D-printing. A new prototype with waste rather than with matter as met would

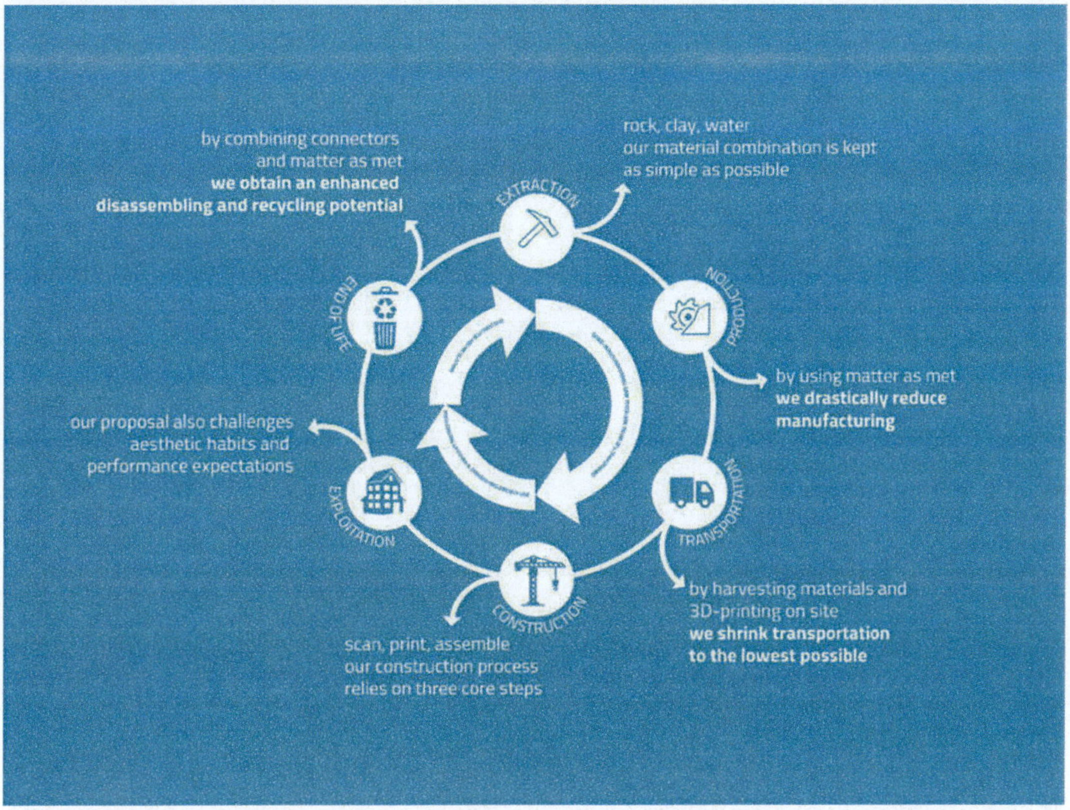

Fig. 6 Building life-cycle and enhancements brought in the Printsugi project

enable to explore further other design possibilities discovered within the design phase of the Printsugi project. A detailed LCA for both versions would then confirm findings regarding the footprint of the workflow. While the visualization method provides leads, proper quantification through LCA needs to validate the different choices made. In particular, whether resorting to a low-impact and harvested on site material such as clay would truly compensate for the impact of a high-tech system such as 3D-printing, even with a small printer, still needs to be evaluated. However the visualization provides directions for preliminary designs and for the definition of frameworks for LCA studies.

Finally, Printsugi represents a venture into a new, different thinking process for architects and designers. As put by German architect and professor Johanna Meyer Grohbrügge, it is now necessary to start projects based on a "radical acceptance of the existing" (Meyer-Grohbrügge 2022). This not only concerns existing built structures, as she demonstrates in her own practice, but more generally our entire environment. It has now become necessary to use what is already there as much as possible, to make do with it, and to get used to such an approach. Within Printsugi, the Rock Tower is just a first step toward a more radical positioning on building and on practicing architecture.

Acknowledgements Contribution to this research has been made thanks to funding from the Architecture of Order Research Cluster and the Hessian State Ministry of Higher Education, Research and the Arts: State Offensive for the Development of Scientific and Economic Excellence (LOEWE).Contribution to this research has been made thanks to funding of the research project Precast Concrete Components 2.0 (Fertigteil 2.0) from the Federal Ministry of Education and Research Germany (BMBF) through the funding measure Resource-efficient circular economy—Building and mineral cycles (ReMin).We thank the students that have participated in courses related to the research topic introduced in this paper: Zahra Babadi, Dominik Binzel, Mirko Dutschke, Jiangpeng Chen, and Yijun Wang.

References

Abd Rashid AF, Yusoff S (2015) A review of life cycle assessment method for building industry. Renew Sustain Energy Rev 45:244–248. ISSN 1364-0321. https://doi.org/10.1016/j.rser.2015.01.043

Addis B (2006) Building with reclaimed components and materials: a design handbook for reuse and recycling, 1st ed. Routledge. https://doi.org/10.4324/9781849770637

Agustí-Juan I, Müller F, Hack N, Wangler T, Habert G (2017) Potential benefits of digital fabrication for complex structures: environmental assessment of a robotically fabricated concrete wall. J Clean Prod 154:330–340

Apellániz D, Pasanen P, Gengnagel C (2021) A holistic and parametric approach for life cycle assessment in the early design stages. In: Proceedings of the symposium on simulation for architecture and urban design (SimAUD). https://www.oneclicklca.com/wp-content/uploads/2021/05/SimAUD-2021-Camera_Ready-LCA_early_design_stages.pdf. Accessed 31 Mar 2022

Bréchet Y (2015) The science of materials: from materials discovered by chance to customized materials: inaugural lecture delivered on Thursday 17 January 2013. Collège de France. https://doi.org/10.4000/books.cdf.3641

Brown N, Mueller C (2015) Optimization for structural performance and energy usage in conceptual building design. In: Presentation at the 2015 engineering mechanics institute conference. http://digitalstructures.mit.edu/page/research#brown-mueller-2015. Accessed 12 Oct 2022

Brown N, Tserandis S, Mueller C (2015) Multi-objective optimization for diversity and performance in conceptual structural design. In: Proceedings of the international association for shell and spatial structures (IASS) symposium 2015. http://digitalstructures.mit.edu/files/2015-09/ncb-iass-paper-final.pdf. Accessed 12 Oct 2022

Brütting J, Desruelle J, Senatore G, Fivet C (2019) Design of truss structures through reuse. Structures 18:128–137. ISSN 2352-0124. https://doi.org/10.1016/j.istruc.2018.11.006

Chudoba M, Hynynen A, Rönn M, Toft AE (2020) Salvage and integrity built environment and architecture as a resource. In: Proceedings series 2020–1. Nordic Academic Press of Architectural Research, pp 39–54. ISBN 978-91-983797-4-7

Clifford B (2021) The Cannibal's cookbook: mining myths of cyclopean constructions. Goff Books

Crowther P (2018) A taxonomy of construction material reuse and recycling: designing for future disassembly. Eur J Sustain Dev 7(3):355–363

Eschenbach MB, Wagner A-K, Ledderose L, Böhret T, Wohlfeld D, Gille-Sepehri M, Kuhn C, Kloft H, Tessmann O (2023) Matter as met: towards a computational workflow for architectural design with reused concrete components. In: Gengnagel C, Baverel O, Betti G, Popescu M, Thomsen MR, Wurm J (eds) Towards radical regeneration. Springer International Publishing, Cham, pp 442–455. https://doi.org/10.1007/978-3-031-13249-0_35

Fabbri A, Morel JC, Aubert JE, Bui QB, Gallipoli D, Reddy BVV (2022) Testing and characterisation of earth-based building materials and elements, state-of-the-art report of the RILEM TC 274-TCE. Springer

Fivet C, Brütting J (2020) Nothing is lost, nothing is created, everything is reused: structural design for a circular economy. Struct Eng 98(1):74–81

Gaudillière-Jami N (2022) Life-cycle analysis as a design tool_the case of 3D-printing. In: Tessmann O, Knaack U, Borg Costanzi C, Rosendahl PL, Wibranek B (eds) PRINT! architecture, AADR

Gehron M, Jiangpeng C, Wang Y (2019) NET-CONNECTORS 3D-scanned wood-branches with additive 3D-prints, 3D-scan Seminar at the DDU-digital design unit, TU Darmstadt

Gorgolewski M (2008) Designing with reused building components: some challenges. Build Res Inf 36 (2):175–188. https://doi.org/10.1080/09613210701559499

Haque U, Fuller M (2010) Networking overload, with potplants: an interview about the Natural fuse project. Thresholds, No 38, future. Massachusetts Institute of Technology, pp 40–45

Heisel F, Nelson C, Petrova V, Wu H, Ma H (2021) RhinoCircular [WWW Document]. https://labs.aap.cornell.edu/ccl/rhinocircular. Accessed 12 Oct 22

Heisel F, McGranahan J, Boghossian A (2022) ScanR: a composite building scanning and survey method for the evaluation of materials and reuse potentials prior to demolition and deconstruction. In: Sustainable built environment conference SBE 2022. IOP Publishing, Berlin

Heisel F (2022) After construction, inevitably comes deconstruction. In: Warke V, Petrie T, Black H (eds) Cornell journal of architecture: AFTER, vol 12. Ithaca NY, USA, pp 58–75

International organization for standardization: ISO 14040-14044

Johns RL, Wermelinger M, Mascaro R, Jud D, Gramazio F, Kohler M, Chli M, Hutter M (2020) Autonomous dry stone: on-site planning and assembly of stone walls with a robotic excavator. Constr Robot 4:127–140. https://doi.org/10.1007/s41693-020-00037-6

Kuzmenko K, Gaudillière N, Feraille A, Dirrenberger J, Baverel O (2020) Assessing the environmental viability of 3D concrete printing technology. Gengnagel DC, Baverel O, Burry J, Ramsgaard Thomsen M, Weinzierl S (eds) Impact: design with all senses. Proceedings of the design modelling symposium, Berlin 2019. Springer, pp 517–528

de Laforce G (2020) Diagnostiquer le bois centenaire de récupération de charpente en vue de son réemploi. Master Thesis. Séminaire de recherche sur l'éco-conception. ENSA Paris-Malaquais, Département Transitions

Larsen NM, Aagaard AK (2019) Exploring natural wood. In: ACADIA 19: ubiquity and autonomy [Proceedings of the 39th annual conference of the association for computer aided design in architecture (ACADIA) ISBN 978-0-578-59179-7] (The University of Texas at Austin School of Architecture, Austin, Texas, 21–26 Oct 2019), pp 500–509

Loiret PE (2021) Penser & construire avec les déblais de terre, ressource principale de nos villes. Le cas de Cycle Terre, première fabrique européenne de matériaux issue du recyclage des terres du Grand Paris, Les Cahiers de la recherche architecturale urbaine et paysagère, 11 | 2021. https://doi.org/10.4000/craup.7218

Marais F (2022) Reims: avec Viliaprint, le hors site gagne du terrain. https://www.lemoniteur.fr/article/reims-avec-viliaprint-le-hors-site-gagne-du-terrain.2209582. Accessed 12 Oct 2022

Marshall D, Meuller C, Clifford B, Kennedy S (2020) Computational arrangement of demolition debris. Detritus, 3–18. https://doi.org/10.31025/2611-4135/2020.13967

Meyer-Grohbrügge J (2022) Vom Wohnen und vom Teilen. Johanna Meyer-Grohbrügge im Gespräch mit Thorsten Jantschek · 27. August 2022, 17:30 Uhr. https://www.deutschlandfunkkultur.de/vom-wohnen-und-vom-teilen-die-architektin-johanna-meyer-grohbruegge-im-gespraech-dlf-kultur-205ab7a5-100.html. Accessed 12 Oct 2022

Perrot A, Rangeard D, Courteille E (2018) 3D printing of earth-based materials: processing aspects. Constr Build Mater 172:670–676

PimpYourWaste (2022) https://www.pimpyourwaste.com/. Accessed 12 Oct 2022

Sterling Presser (2022) Basel Pavilion–Circular building. https://www.sterlingpresser.com/circular-building/. Accessed 12 Oct 2022

Rigobello A, Gaudillière-Jami N (2021) Designing the Gross. In search for social inclusion. In: DesignCulture(s). Cumulus conference proceedings Roma 2021, vol 2, pp 811–827. ISBN: 978-952-64-9004-5

Self M, Vercruysse E, Menges A, Sheil B, Glynn R, Skavara M (2017) Infinite variations, radical strategies. In: Fabricate 2017. UCL Press, pp 30–35. https://doi.org/10.2307/j.ctt1n7qkg7.8

Tessmann O (2022) Fertigteil 2.0: Kreislaufwirtschaft im Bauwesen. In: DETAIL, (3). DETAIL Business Information GmbH, pp 16–20. ISSN 26272598

Tourtelot J, Ghattassi I, Le Roy R, Bourgès A, Keita E (2021) Yield stress measurement for earth-based building materials: the weighted plunger test. Mater Struct 54:6

Umbrellium (2015) Case study: engage students in energy & sustainability. https://umbrellium.co.uk/case-studies/trumpington-community-college-augmented-bms/. Accessed 12 Oct 2022

Vestartas P, Weinand Y (2020) Laser scanning with industrial robot arm for raw-wood fabrication. In: Proceedings of the 37th ISARC, Kitakyushu, Japan, pp 773–780. ISBN 978-952-94-3634-7, ISSN 2413-5844

Wibranek B, Tessmann O (2019) Digital rubble-compression-only structures with irregular rock and 3D printed connectors. In: Proceedings of the IASS annual symposium 2019–Structural Membranes 2019. Barcelona

Computationally Enabled Material Management—Learning from Three Robotically Fabricated Demonstrators

Jens Pedersen and Dagmar Reinhardt

Abstract

Addressing the 12th Sustainable Development Goal (SDG) is important and has become financially necessary over the last few years. The covid-19 pandemic and the consequences of war have led to material scarcity, making the global market respond with increased prices and longer delivery times. Consequently, this has inverted the cost relationship between construction material and labor cost, making construction material a more significant expense than labor. To better understand this challenge, this research developed a computational strategy for linear timber elements focused on minimizing material use through preplanned cutting sequences that indicate waste and off-cut data. Thus, the paper analyses the material use of three robotically fabricated timber case studies to demonstrate the process. Each case study is morphologically different, representing the following construction concepts (1) Different parts, simple joinery; (2) Similar parts, complex joinery; and lastly, (3) similar parts, simple joinery. The analysis indicates the material consequences of using different timber joinery strategies. Furthermore, the research suggests a process to handle off-cuts through labeling and databases, which can lead to new sources of revenue for an industry-leading fabricator.

Keywords

Robotic fabrication · Material planning · Timber construction · Timber joints

1 Introduction

The 12th SGD—*responsible consumption and production* (UNDP 2023), is an essential challenge for the construction industry due to high CO_2 emissions related to using specific construction materials. It is widely known that concrete emits much CO_2, where recent innovations in engineered timber and engineering calculations for timber construction have made it an alternative to concrete specifically in taller buildings (Dangel 2016). However, other research indicates that replacing merely 25 pct of current concrete consumption with wood would require a new forest one and a half times the size

J. Pedersen (✉)
Odico Construction Robotics, Odense, Denmark
e-mail: jenspedersen88@gmail.com

The Aarhus School of Architecture, Aarhus, Denmark

D. Reinhardt
The University of Sydney, Sydney, NSW 2006, Australia

M. R. Thomsen et al. (eds.), *Design for Rethinking Resources*, Sustainable Development Goals Series,
https://doi.org/10.1007/978-3-031-36554-6_28

of India to be planted every year (Zeitung 2023). In addition, history documents that construction material can suddenly be scarce for multiple reasons; today, we see it because of a global pandemic or as a consequence of war (Government of Canada, Statistics Canada 2022; NYT: Byggeomkostninger … 2022); previously it was decrees from kings that could ration material use (Benzon 1984; Jensen 1933; Vejlby 1991). This further cements that changes to how we consume material need to change while developing a sensibility about how we use resources. Coincidentally we are in a time where meeting SGD 12th comes with a financial incentive since the cost relationship between labor and material has been inverted, making construction materials more expensive than associated labor costs (NYT: Byggeomkostninger … 2022). In an initial attempt to address this challenge, this research focuses on timber construction, where the timber value chain offers potential solutions to material handling. Subsequently, this research learns from sawmills, where the implementation of digital and automation technologies has allowed fewer sawmills to achieve higher yields than more sawmills did in years passed (Fredriksson 2015). Here high-speed CT scanners are used to convert timber logs into construction lumber by using the CT data to predict grading and compute-optimized cutting sequences to maximize the yield from a log (Svilans et al. 2022a). Complex scanning technologies are required due to the many material unknowns, which greatly affect the final product. However, developing a method to maximize material yield through computational techniques and automation technologies presents great potential.

Achieving this requires shifting the current 3d-modeling paradigm from representation to fabrication, moving from modeling volumes with holes for windows or doors to modeling the discrete geometric elements that make up our designs. In doing so, information can be processed through algorithmic strategies that precisely control the fabrication sequence, labeling, and amount of waste generated from the system.

This paper limits its scope to focus on the discrete timber structure for three robotically fabricated timber demonstrators. Each demonstrator's geometry is analyzed and used to evaluate a computational method that describes a cut sequence and labeling and can handle waste produced by the system. Additionally, each demonstrator encompasses different fabrication and assembly concepts; (1) Different parts, simple joinery; (2) Similar parts, complex joinery; and lastly, (3) similar parts, simple joinery. The analysis of each concept gives an indicator of material use related to the complexity of geometry, which will be the subject of the discussion in Sect. 4.

The research presented in this paper is part of an ongoing industrial Ph.D. project at Odico Construction Robotics with the Aarhus School of Architecture. The research aims to offer a software solution for a transportable robot unit (Odico Construction Robotics 2023) developed with the industry partner to aid on-site timber construction. This robotic framework will be detailed in a future publication. This paper aims to present an initial answer to the naïve question; "what is the benefit of having robotic fabrication facilities on the construction site?". The initial response is efficiency and less strenuous work for craftspeople, but this research identifies that it can aid in managing the resources used on the construction site. Thus, the research unfolds through the following sections: Sect. 2 describes robotic timber fabrication's state of the art, emphasizing material management. Section 3 describes the different demonstrators and presents their fabrication data. Lastly, Sects. 4 and 5 discuss the proposed research and conclude and define the future work related to the submitted research.

2 State of the Art

Robotic timber fabrication using linear timber elements has been a growing field since the late 2000s (The Stacked Pavilion, Wettswil am Albis, Switzerland 2009; Helm et al. 2016), resulting in a ongoing research trajectory at ETH, Zürich. Parts of this trajectory were concluded with the completion of the sequential roof (Apolinarska et al. 2016a; b) in 2016. The sequential roof consists of multiple timber slats with varying cross sections resulting from a structural analysis (Apolinarska et al. 2016a). Although this could indicate a minimized material use, the structure is described as redundant, where elements can be removed without loss of structural capacity (Apolinarska et al. 2016b). However, the notion of structural optimization as a means of minimizing material use is interesting. Still, this research will focus on optimizing material use for already designed structures. This position stems from the research aiming to be part of an on-site robotic framework, where carpenters will use it to fabricate a given project. At the Center for Information Technology and Architecture in Copenhagen, researchers have taken a different approach centered around a material stock that is digitized through 3d-scanning, cataloged, and used when needed (Svilans et al. 2022a; b). Complex scanning technologies such as CT or MRI scans are performed during this process to better utilize the wooden material properties at specific positions within a designed structure. Their research aims to optimize material performance instead of material use, but the notion of cataloging material is of great inspiration. Other researchers have developed a technique titled "Projectables" to maximize the reuse of discarded sheet material. Their system scans discarded sheet material and identifies areas that can be used for future fabrication (Rasmussen 2017). Unfortunately, if only discarded sheets are used, a challenge emerges; pieces will progressively become smaller and smaller. Additionally, scanning elements before fabrication will be time-consuming and complex if used in on-site fabrication processes. But it does bring up an interesting question, how do we handle off-cuts, and what do we consider waste?

Following these references, this research proposes a strategy to manage the use of construction materials through the evaluation criteria stated below:

- Off-cuts should be placed in a database and physically marked.
- Off-cuts could become a resource for other fabrication scenarios.
- Material estimates should be accurate and part of the design work.

2.1 Method and Computational Framework

The paper presents a comparative study of three different computational methods; First Fit Bin Packing (FFBP) (Johnson 1973), Best Fit Bin Packing (BFBP) (Johnson 1973), and OpenNest (ON) (Food4Rhino 2018). The ON method has been developed by Petras Vestartas and is commonly used to optimize the layout of elements on sheet material—a task which performs well. This research also tests how it organizes linear data and tests it against two different bin-packing algorithms developed in c#.net (BillWagner 2023). The logic of these algorithms is described in Fig. 1, where "bin" refers to stock material, and "element" is the piece fabricated from the stock material. The FFBP organizes elements sequentially into "bins", and when an "element" doesn't fit into the current "bin", a new one is opened (Fig. 1, top). The BFBP method is similar, but previous "bins" are rechecked before new "bins" are opened (Fig. 1, bottom). This produces nonsequential cutting sequences that can be hard to predict and are only possible if a discrete 3d model of a designed structure exists. When construction material is cut on-site, it is believed to begin sequentially, like the FFBP method, and when new pieces are needed, previous off-cuts are remeasured to see if new pieces can be made from them. If true, this would be a time-consuming process, but this requires further research, as described in Sect. 5.1.

The research will use the FFBP method as a control measurement, serving as a benchmark for the two other methods. All elements will be presorted from long to short before being presented as data input for each method.

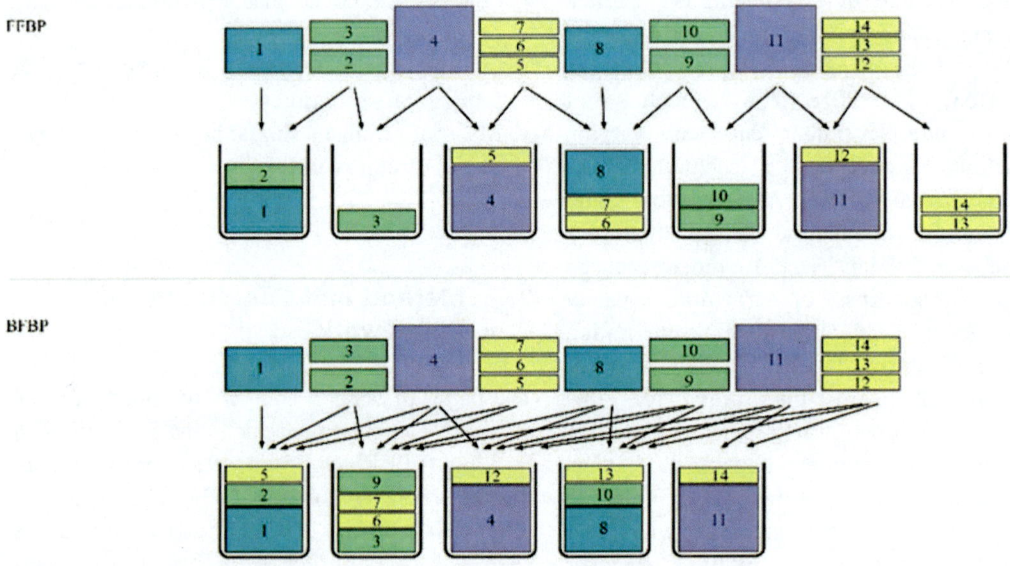

Fig. 1 A visual representation of the logic described above, where the BFBP method can produce better fabrication sequences than the FFBP method

3 Three Timber Demonstrators

The focal point of this paper is the analysis of material use in three robotically fabricated demonstrators shown in Fig. 2. The basis for the analysis is the CAD model representing the physical structure in Fig. 2 (right), which resides in the CAD software Rhinocerous3d (BillWagner 2022) (rhino). The parametric modeling extension in rhino McNeel (2022), Grasshopper3d (grasshopper), becomes a prototypical

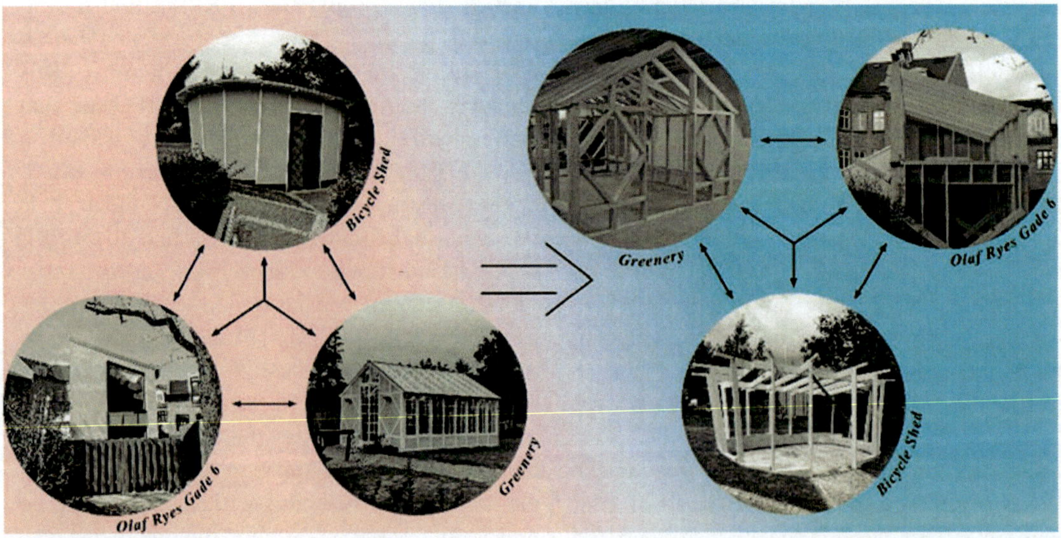

Fig. 2 Images to the left show the analyses projects once finished. The images on the right show the structures analyzed in this paper

canvas for the material management experiment. Preceding the material analysis are three short sections describing the fabrication system, the used timber material, and outlines how this research views material use, i.e. what is “waste” and “off-cuts”. Following this are sections describing each project and which material management strategy yielded better results.

3.1 The Robotic Fabrication Setup

The three demonstrators were made with different fabrication setups, where lessons learned led to the development a new timber processing unit with the industry partner, which address the 9th SGD—*Industry Innovation* (UNDP 2023). This paper will not describe the unit and its design process in detail since it is the focal point of a future publication. Still, it is an essential part of the thinking presented within this paper and is the framework for future demonstrators. It consists of an industrial robot ABB IRB 6700 equipped with a spindle using either a circular saw or a routing bit. Next to the robot is a linear axis with an automatic labeling setup. The fabrication setups for projects 3.4 and 3.6 were fabricated with a circular saw mounted on a spindle, and project 3.5 was made with a twenty- and ten-mm routing tool in a spindle. The robotic actions for the presented and future projects are developed within grasshopper, in a bespoke plugin developed with the industry partner. The presented research has become a toolkit for prefabrication and cost estimates, since it gives accurate material estimates and produces an optimized cutting sequence with label information for the fabrication setup.

3.2 Timber Material

The base material used for the three demonstrators is 1-dimensional wooden material, where projects 3.4 and 3.6 have used “off-the-shelf” construction C24 lumber (Træ 2022). This type of wood can be bought in uniform lengths from the interval 3.6–7.2 m, has a thickness of 45 mm, and the project were made using either 195- and 145-mm widths. Previous research has identified that conventional construction lumber will have imperfections along the length (Pedersen et al. 2023). Project 3.5 used an engineered wood, where smaller un-knotted pieces are finger jointed and glued together, resulting in form-stable timber elements of 120 by 67 by 6000 mm.

3.3 What is Waste?

Within the field of digital fabrication for architectural applications, there is a prevailing opinion that fabricating unique pieces comes at a small added processing cost compared to manufacturing multiple of the same piece (Iwamoto 2009; Kolarevic 2004; Mitchell 2001). There can be some truth to this statement. Still, it is relative to the associated programming and machining time for a specific fabrication task—the cost of machining can be related to how much material is removed with a particular type of tool, which can easily be calculated. However, returning to the initial statement, an essential factor to remember is the cost of material, where similar pieces often give a higher yield than different pieces would. Thus, presenting two aims for this research, (1) to see if the analysis indicates if unique pieces use more material, and (2) to define when something is considered ***waste***, an ***off-cut*** that can be used in future projects, or a ***cut-out*** (waste) due to a design choice (Fig. 3). What categorizes a “useful” off-cut (Fig. 3, dark green) will be correctly determined through future research, but at this moment, off-cuts larger than 300 mm are seen as potentially useful in other processes with the industry partner. Off-cuts less than 300 mm are considered waste since these can be dangerous to pass through other manual machining processes. Cut-outs, such as the pink areas, can be made with a circular saw but are considered waste due to their odd sizes, which are hard to implement in other processes.

In contrast, removing the light green (Fig. 3, bottom) requires a routing tool that converts the material to dust. Reintroducing dust into the value chain requires collaborations across

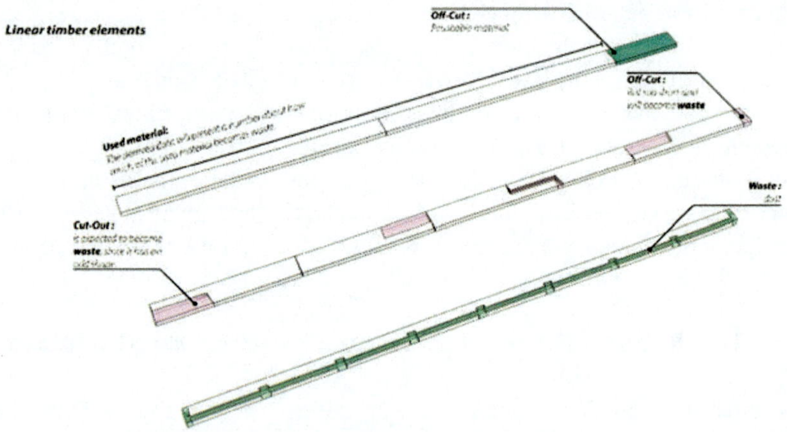

Fig. 3 A simple diagram indicating the multiple complexities that emerge when analyzing 2D material stock (left). More importantly, it describes a ledger that explains what is considered waste or an off-cut

industries, which is why it is considered waste for now. Within the produced data, two terms are used, "material use" and "material yield", and therefore essential to clarify:

- "*Material use*"; refers to the percentage of material used from a given stock material and is best visualized in Fig. 3, top.
- "*Material yield*"; refers to the percentage of material used that becomes the finished piece after robotic processing—see the middle and bottom part in Fig. 3 (pink + light green color). This number indicates the machining time associated with making a given element.

These terms categorize the use of 1-dimensional material for each demonstrator. Future research will dive into the complexities of 2d or 3d stock material (Fig. 3), where lessons learned from this research will be used to extrapolate new methods to meet other material layouts in the future—sheets or blocks.

3.4 Bicycle Shed

The bicycle shed was a job for a private client that wished to have a shed in the shape of a super ellipse (Weisstein 2022) with a flat roof rotated around two axes (Fig. 4). It was fabricated using a circular saw mounted on a robot since the geometry of joinery allowed for it. Using a circular saw comes with certain constraints, but it can remove much material in fever movements compared to milling, thus requiring less fabrication time.

The roof structure was challenging to solve since the building was circular. But we managed to develop a structure comprised of standard components (central box, Fig. 4, right) and unique elements to meet the edge and posts. These bespoke roof elements have a notch corresponding to a given post geometry, ensuring the correct positioning and orientation of individual components. The project was cut from conventional 5.4-m C24 timber (stock).

The shed's structure is simple and consists of three different components, made in either 195—or 145 mm construction lumber. The material analysis results are highlighted in Fig. 5. For this project, the ON method performed well due to the geometry of the custom roof elements could be reoriented and nested very tight, whereas the BFBP method positions elements one after another—leading to more waste (Fig. 4, bottom). However, the absolute difference in needed material is only 5.9%, where the BFBP and the ON method use a total of 17 vs. 16 pieces of timber—half is 195 mm wide, and the rest is 145 mm wide. The FFBP performs poorly and uses a total of 20 members, which is 20 and 15 pct more material than the ON or BFBP methods. Interestingly, the difference in produced off-cuts by the ON and BFBP methods are comparable,

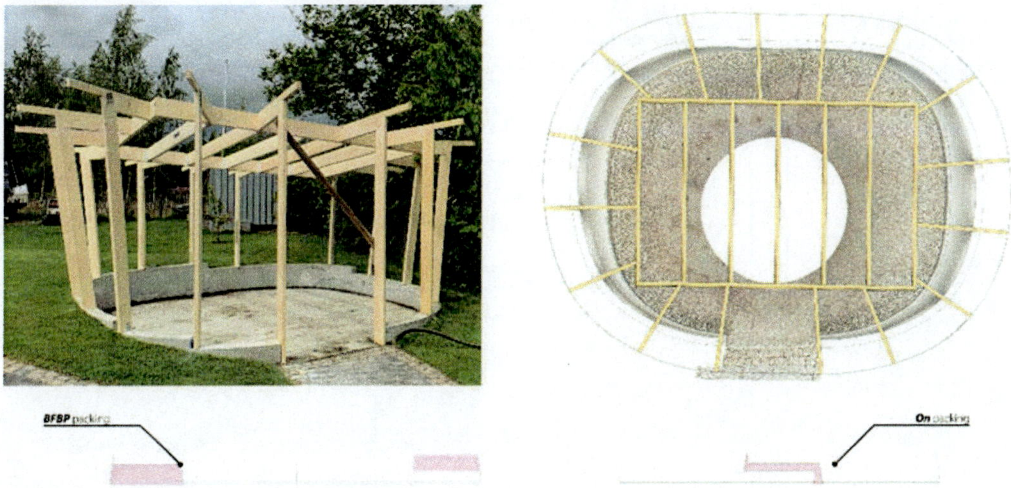

Fig. 4 Left image is a picture of the finished structure before being filled with the remaining structure. The image to the right shows a central box comprising standard elements with custom roof elements attached

especially if the 145 mm members are viewed; here, the BFBP method outperforms the ON method by a small margin. But the ability to nest geometry tighter with the ON method produces better results for the 195 members, albeit requiring some parameter tuning to achieve these results. Regarding the 195 members, it is interesting to see how the yield varies a lot, resulting from the large number of cut-outs produced, which becomes waste due to their odd shape and size. Ultimately, the three methods document that if an optimized cutting sequence is followed, you would average material use of >85% if outliers such as the last member are culled (Fig. 5).

3.5 Greenery

The greenery was a project for the same client as the bicycle shed. The project was required to be designed and fabricated with inspiration from the half-timbered construction system found in Jutland/Funen in Denmark (Benzon 1984; Jensen 1933; Vejlby 1991), which, unlike the shed, could not be cut with a circular saw. Instead, it had to be fabricated using two different diameter routing tools due to the complexity of the joinery. Besides dictating the fabrication tool, this construction required the timber to be straight and planned to certain dimensions, a quality found in engineered lumber. The acquired lumber was long and straight, making it possible to use continuous instead of discrete elements, as one would generally see in the half-timbered construction system. Unfortunately, this made assembly more difficult.

Each 99-timber element has a low material yield since each element has many joinery details, which, when milled out, becomes dust and waste (see Fig. 6).

For this project, the BFBP method outperforms the ON method, but only by eight pct or 0.4 m less generated waste or off-cut. The FFBP only uses 18 pct additional material compared to the other methods. However, the main interest is the variance in the amount of removed material compared to the shed; within this project, we see upwards of 20% of the material being removed before resulting in finished pieces. This, indicates a high machining time and higher machining cost compared to the type of joinery shown in the shed project. Each piece within this project had a milling time of between 30 and 90 min per piece relative to 15 min for the most complex pieces in the shed.

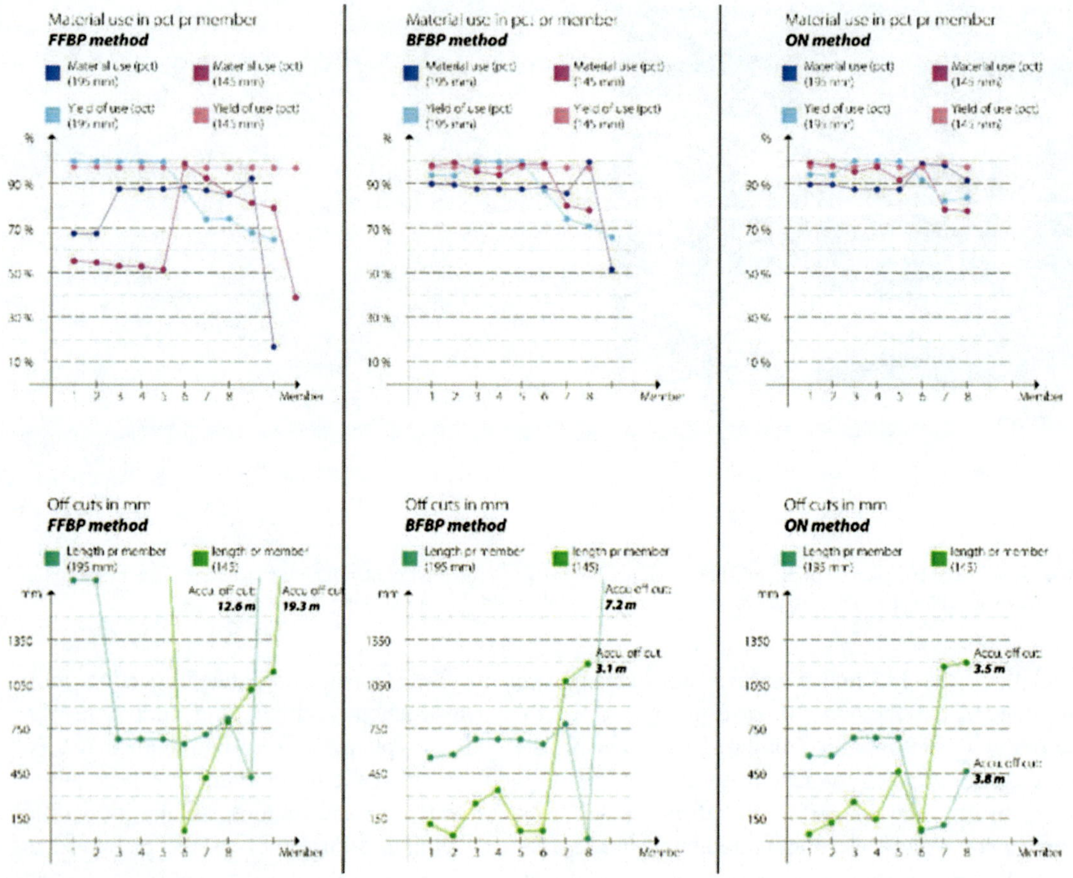

Fig. 5 Comparative data derived from the Shed structure. The horizontal axis in the graphs indicates which "member" the vertical axis data represents

3.6 Olaf Ryes Gade 6

This project was for a different client than the previous two projects and was fabricated using a mixture of milling and a circular saw, which is described in previously published research alongside fabrication performance numbers (Kolarevic 2004). The project was a 50 m^2 two-story building made from conventional 195—and 145 mm 5.4-m C24 lumber (stock) and consisted of multiple individual pieces. This project aimed to develop a construction system where a minimal percentage of material was removed from each component to have minimal fabrication times and still conveyed positional information (Fig. 8, right).

The analysis of this project revealed a similar trend as the previous projects where the ON and BFBP methods had near identical performances. Unlike the greenery project, these joint details were minimal, so less than five pct of material (Fig. 9) needed to be removed compared to 20 pct within the greenery project (Fig. 7). This is important to note since both projects use a version of timber joinery but with greatly varying production time and, thus, cost are vastly different. Off course, there are other qualities (aesthetics or tectonic) in joinery used in the greenery, namely that it could quickly be taken apart and reassembled elsewhere.

The system made it possible to utilize an average higher than >91% after culling outlies of

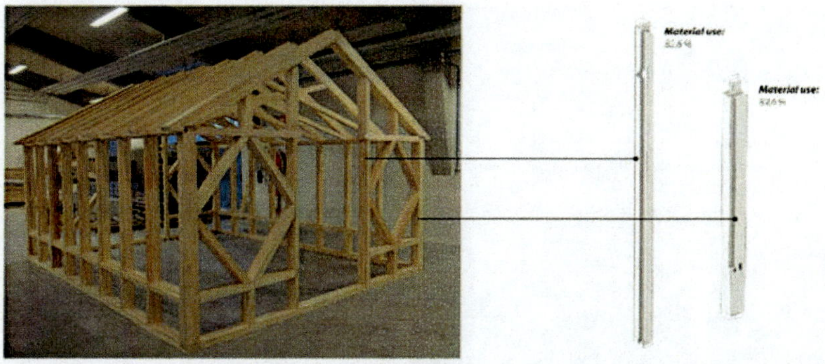

Fig. 6 The finished timber structure for the Greenery is in the left image and the diagram to the right shows the bounded volume of two finished elements—to document that between 10 to 20% of material is removed

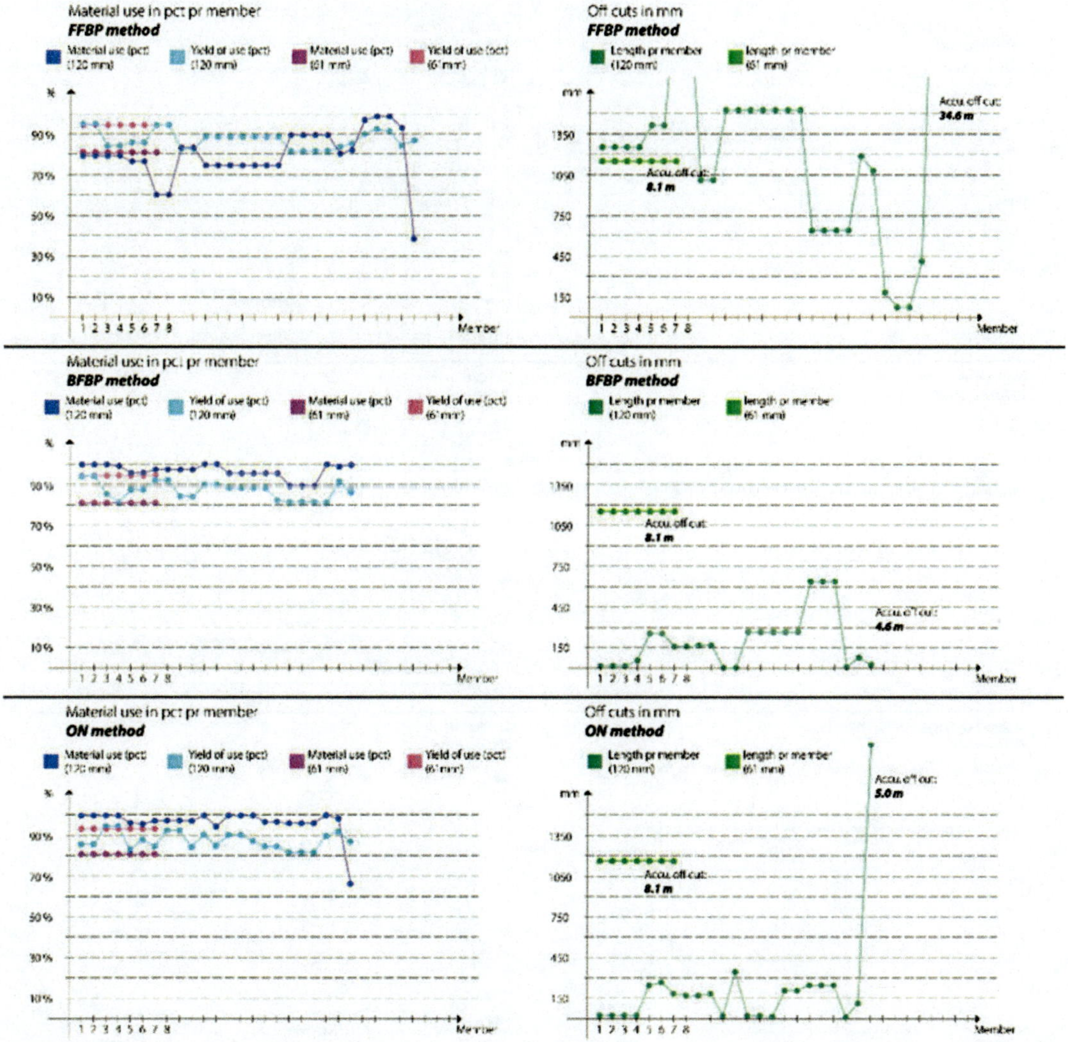

Fig. 7 Graphs describing the gathered data for the greenery project. The horizontal axis in the graphs indicates which "member" the vertical axis data represents

Fig. 8 Images to the left are from the finished structure. At the same time, the image to the right shows the system's simplicity, where simple notches are used to specify positions

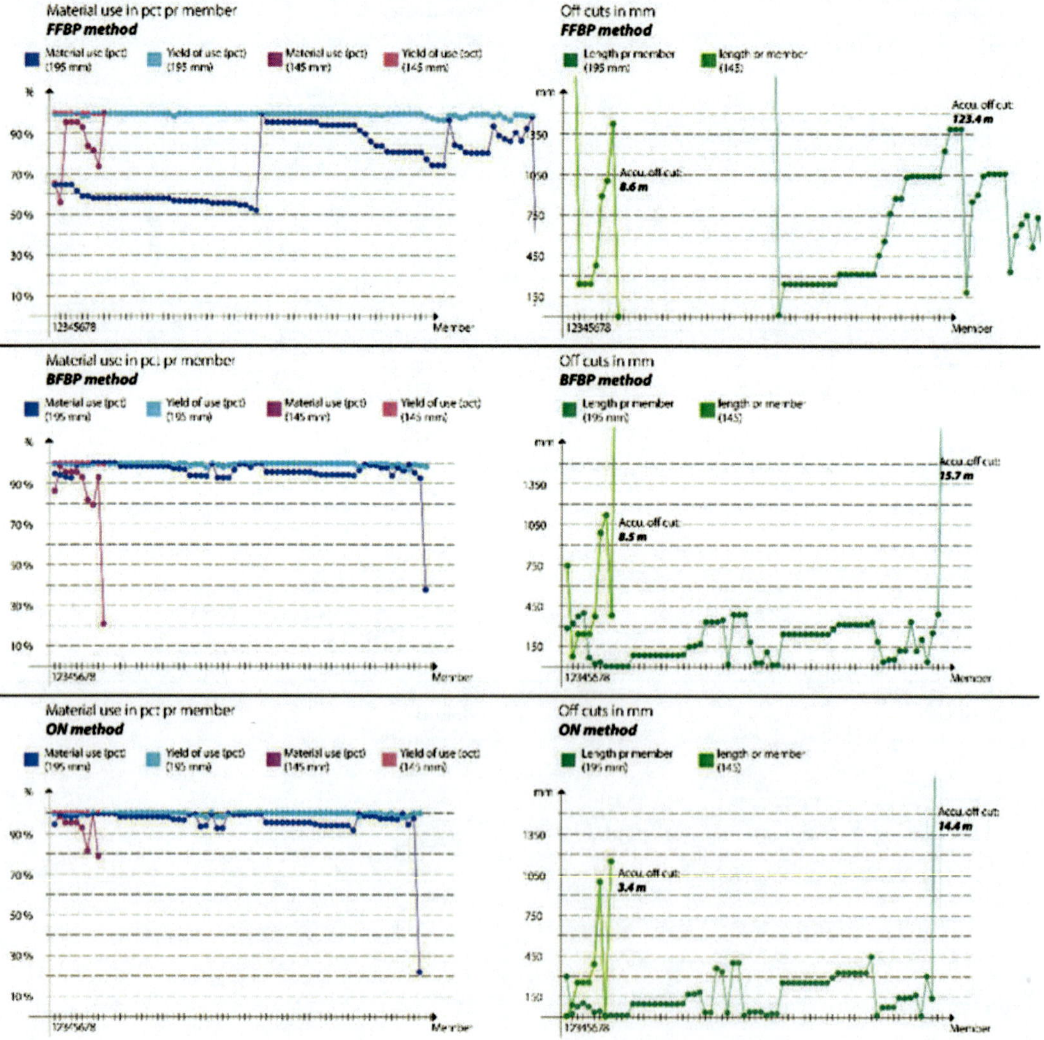

Fig. 9 Graphs depicting material use information about the Olaf Ryes Gade project. The horizontal axis in the graphs indicates which "member" the vertical axis data represents

less than 40% material used. Off-cuts generated from the system had a length domain between 1 and 2338 mm, which brings up two separate questions to address in Sect. 4, should off-cuts be cut to specific lengths? And what should it be used for?

4 Discussion

The research presented three methods to analyze timber structures to develop an optimized cut sequence. As described in Sect. 2.1, FFBP was a control method to see the effects of a sequential cutting sequence, similar to what is believed to be common construction site practice (a claim that needs documentation—sees Sect. 5.1). It was outperformed by the other methods, where both produced cutting sequences with similar performance data, with variations at approximately five pct. Therefore, which method is better boils down to preference, the ON method required parameter tuning, whereas the BFBP method produced its result in one shot. Despite outperforming the FFBP method, both methods come with a drawback; namely the sequence may not give the piece you need at a given instance. A feature that will need addressing to allow for just-in-time fabrication. Despite slight underperformance, the BFBP method will be the preferred method for linear material analysis in the future since it will be simpler to implement in other software situations.

In contrast, the ON method relies on rhino and grasshopper in its current configuration. Thus, the research indicates that preplanned cutting sequences can minimize material use within digital fabrication pipelines. Future research could test if it minimized material and time spent on a construction site by construction workers compared to their conventional practice. From the data presented within this paper, different or similar parts can be fabricated at a similar pace and with minimal material use. However, this is only true if pieces are identical to elements from the Olaf Ryes Gade project, it consisted of 195 elements, of which 57 are unique. But since fabrication tasks are simple and quick (primarily used a circular saw), it doesn't increase the additional cost. The same can be said about the bicycle shed since it used a similar fabrication method to the Olaf Ryes Gade project. Still, it does have a pourer material use percentage in comparison—85% versus 93%. The same cannot be said about the greenery project, which consists of 108 pieces with only 12 unique elements. But each element needs to be removed between 5 and 30 pct material, meaning little variance comes at a much higher cost, not due to increased material use but the more significant difference in machining time and programming. By comparison, the Olaf Ryesgade only needed to have a maximum of 5.5% removed from some elements. Therefore, the cost of variation depends on the fabrication task at hand. Still, this research indicates that variance, as we saw in project 3.4, comes with a higher material cost since it is hard to optimize cutting patterns, as shown in Sects. 3.5 and 3.6.

Common for all three projects is a known off-cut, which could be cut to a given modulus, marked by the fabrication system, or saved to a database, but the question remains—what for? Similarly to the argument presented in the state-of-the-art, regarding the "Projectables" project (BillWagner 2023), off-cuts can only produce smaller elements than the initial stock material. However, if contractors, builders, or fabricators could reuse the material in other processes, the material could re-enter the construction value chain. One could imagine that the off-cuts could be sold to a CLT fabricator or be glued together to form "blanks" for making molds for concrete. If the off-cuts were stored in a database, they could become a resource for people doing small-scale interior renovations. Essentially it has the potential to open new business revenues for a fabricator. However, for this to be established, some common elements need to be agreed upon:

- Modulus for off-cuts
- What is the minimum length of off-cuts
- Are off-cuts of a given length reintroduced to the fabrication process.

If these three elements are understood and introduced, they can be implemented immediately, granted there is storage for the off-cuts.

5 Conclusion

The paper has successfully demonstrated a method whereby it is possible to evaluate discrete timber structures on the fly, enabling to accurately predict the amount of material needed for a given project. Additionally, since the method produces data about how much material is removed, it can be used to develop an initial cost estimate for a structure if paired with data about volume removed per unit of time for a given machining process. Lastly, besides enabling accurate material coordination, it can open new sources of income by linking to another value chain—as presented in the following section.

5.1 Future Work

The research presented in this paper is yet in its infancy. Still, it gives a positive indicator for how we can implement material savings through a computational—and, in the future, on or near-site fabrication workflow. Thus, ensuring work for the foreseeable future, where some of it has already been discussed.

- Develop the methods to a stage where early-stage material evaluations can be made for both 2d(sheet) and 3d (box) material stock. Solving for 2d materials will use the current ON method, making the primary focus on techniques for box materials.
- As mentioned within the discussion, more research into how carpenters work on-site would be needed to understand better if this system improves the current status quo. This will initially be addressed through interviews with educators and carpentry schools.
- Propose and develop a specific use case for the reuse or repurposing off-cuts within the framework of Odico Construction Robotics and their transportable robot unit. This could revolve around forming industrial partnerships with engineered wood fabricators, where off-cuts of a given size could be introduced into their fabrication process.
- Develop a machining matrix to better utilize this data for early-stage cost predictions.
- Identify a new strategy that can allow for optimized and just in time fabrication.

References

NYT: Byggeomkostninger stiger for sjette kvartal i træk. https://www.dst.dk/da/Statistik/nyheder-analyser-publ/nyt/NytHtml?cid=35862. Accessed 4 Oct 2022

Apolinarska AA, Bärtschi R, Furrer R, Gramazio F, Kohler M (2016a) Mastering the "Sequential Roof" computational methods for integrating design, structural analysis, and robotic fabrication. In: Advances in architectural geometry 2016a***. VDF Hochschulverlag AG, pp 240–58. https://doi.org/10.3218/3778-4_17

Apolinarska A, Knauss M, Gramazio F, Kohler M (2016b) The sequential roof. In: Advancing wood in architecture-a computational approach, pp 45–59. https://doi.org/10.4324/9781315678825-4

Benzon G (1984) Gammelt dansk bindingsværk. Det Benzon'ske Forlag, Bogense

BillWagner. Rhinoceros 3D. https://www.rhino3d.com/. Accessed 29 Mar 2022

BillWagner. C# docs-get started, tutorials, reference. https://learn.microsoft.com/en-us/dotnet/csharp/. Accessed 14 Jan 2023

Dangel U (2016) Turning point in timber construction: a new economy. Birkhäuser, Basel

Svilans T, Gatz S, Tyse G, Thomsen M, Ayres P, Tamke M (2022) Deep sight-a toolkit for design-focused analysis of volumetric datasets, pp 543–55. https://doi.org/10.1007/978-3-031-13249-0_43

Food4Rhino (2018) OpenNest. Text. https://www.food4rhino.com/en/app/opennest. Accessed 15 Nov 2018

Fredriksson M (2015) Optimizing sawing of boards for furniture production using CT log scanning. J Wood Sci 61(5):474–480. https://doi.org/10.1007/s10086-015-1500-0

Government of Canada, Statistics Canada (2022) The daily—sawmill industry in Canada: 15 years in review. https://www150.statcan.gc.ca/n1/daily-quotidien/220711b/dq220711b-eng.htm. Accessed 11 Jul 2022

Helm V, Knauss M, Kohlhammer T, Gramazio F, Kohler M (2016) Additive robotic fabrication of complex timber structures. In: Advancing wood architecture. Routledge

Iwamoto L (2009) Digital fabrications: architectural and material techniques, 144th edn. Princeton Architectural Press, New York

Jensen CA (1933) Dansk Bindingsværk fra Renæssancetiden, dets Forhistorie, Teknik og Dekoration. Gad, Kunst i Danmark. Kbh

Johnson DS (1973) Near-optimal bin packing algorithms. Thesis, Massachusetts Institute of Technology. https://dspace.mit.edu/handle/1721.1/57819

Kolarevic B (2004) Architecture in the digital age: design and manufacturing. Taylor & Francis

Associates, McNeel R (2022) Grasshopper-New in rhino 6. https://www.rhino3d.com/6/new/grasshopper/. Accessed 29 Mar 2022

Mitchell WJ (2001) Roll over euclid: how frank gehry designs and builds, 1st ed. Guggenheim Museum Pubns, New York

The Stacked Pavilion, Wettswil am Albis, Switzerland (2009) Gramazio Kohler research. https://gramaziokohler.arch.ethz.ch/web/e/projekte/165.html. Accessed 16 Sep 2022

Pedersen J, Olesen L, Reinhardt D (2023) Timber framing 2.0. In: Gengnagel C, Baverel O, Betti G, Popescu M, Thomsen MR, Wurm J (eds) Towards radical regeneration. Springer International Publishing, Cham, pp 320–331. https://doi.org/10.1007/978-3-031-13249-0_27

Rasmussen T, Merritt T (2017) ProjecTables: augmented CNC tools for sustainable creative practices. https://doi.org/10.52842/conf.caadria.2017.757

Odico Construction Robotics (2023) Factory on the fly. Odico (blog). https://odico.dk/en/factoryonthefly/. Accessed 14 Jan 2023

Svilans T, Tamke M, Thomsen MR (2022) Integrative strategies across the digital timber value chain. In: Structures and architecture a viable urban perspective? CRC Press

Træ L (2022) Hvad er styrkesortering? c18–c24'. http://lavpristrae.dk/Blog/information/hvad-er-styrkesortering-c18-c24. Accessed 29 Mar 2022

UNDP (2023) Sustainable Development Goals | United Nations Development Programme. https://www.undp.org/sustainable-development-goals. Accessed 14 Jan 2023

Vejlby U (1991) Bindingsværkshuset: renovering, fugtskader, isolering. 1. udgave, 2. oplag. Skovlænge, Søllested

Weisstein EW (2022) Superellipse. Text. Wolfram Research, Inc. https://mathworld.wolfram.com/. Accessed 12 Oct 2022

Zeitung S (2023) Klimafreundlicher Beton gesucht. Süddeutsche.de. https://www.sueddeutsche.de/wissen/beton-bauen-klima-1.5299825. Accessed 14 Jan 2023

Comparative Experiment on Adaptive Reuse of Wood Stud Partition Walls: Integrating the DfD Concept into Building Component Design

Harrison Huang, Lu Li, Nan Xia, and Mengdi Zhao

Abstract

Resource consumption and waste generation during the construction, renovation and demolition of buildings are global problems. Therefore, architects are required to consider the feasibility of the future deconstruction of buildings in the design phase, that is, design for disassembly (DfD). This paper presents a comparative experiment on the assembly and adaptation of a wood stud partition wall, for which the wall component was designed via conventional construction and the DfD concept. Upon comparing the results, it was found that the experimental group with DfD integration consumed less raw material but required more time during the assembly process. However, the novel design of the wall demonstrated obvious advantages with regard to labour savings, waste generation and quality control as well as modification flexibility during adaptation. These results have shown the feasibility of integrating the DfD concept into building component design and validated its related potential under experimental conditions. This work also provides useful information for architects, engineers and other architecture, engineering and construction practitioners in designing resource-efficient buildings using innovative techniques of (dis)assembly in the future.

Keywords

Design for disassembly (DfD) · Wood stud partition wall · Comparative experiment · Assembly · Adaptation

1 Introduction

Due to function, aesthetic or natural depreciation factors, as well as the impact of constantly changing market demands on the effectiveness of a building, owners or operators must decide whether the building should be completely reconstructed or adaptively reused (Gaspar and Santos 2015; Bullen and Love 2010). The demolition of buildings and the construction of new ones generate considerable waste but also consume large amounts of building materials and energy (Huang

H. Huang (✉) · L. Li
Zhejiang University, Hangzhou, China
e-mail: harrison@zju.edu.cn

H. Huang
Center for Balance Architecture, Zhejiang University, Hangzhou, China

L. Li · N. Xia
Architectural Design and Research Institute of Zhejiang University, Hangzhou, China

M. Zhao
CCTN Architectural Design Co., Ltd., Hangzhou, China

M. R. Thomsen et al. (eds.), *Design for Rethinking Resources*, Sustainable Development Goals Series,
https://doi.org/10.1007/978-3-031-36554-6_29

et al. 2013). In comparison, retrofitting buildings for adaptive reuse produces relatively less waste and uses fewer resources (Itard and Klunder 2007). Therefore, adaptive reuse has increasingly been accepted by academics and practitioners as a responsive strategy to suit changing user needs (Wilkinson et al. 2009).

However, there are significant technical implementation barriers in the adaptive reuse process, including damage to components during deconstruction, unavailable building information, and architectural design challenges (Rakhshan et al. 2020; Huang and Li 2022). There are a large number of buildings that do not have detailed information to effectively support deconstruction, and there are no guidelines to help practitioners design operational strategies before deconstruction (Rakhshan et al. 2020). The introduction of the design for disassembly (DfD) method is regarded as an effective solution to these problems to optimise deconstruction operations for the reuse of building components (Kissi et al. 2019).

2 Literature Review

DfD refers to the design principles that facilitate and ensure deconstruction to allow for the reuse and recycling of building components at the end of the building's life (Jaillon and Poon 2014). Within the research field on DfD, many scholars have focussed on the design of structural connections in buildings. For example, Lin (2013) designed a prefabricated concrete beam-column connection based on the DfD concept. After structural analysis and experimentation, it was concluded that this type of connection is compatible with the application requirements of a typical high-rise residential building in Singapore. Studies have also developed a new system of dismantled bolt-assembled prefabricated concrete panels for the design of low-rise buildings in earthquake-prone areas (Cai et al. 2019). Subsequent studies have also been carried out on their seismic performance and damage modes (Xiong et al. 2022). In addition to DfD studies of joints in prefabricated concrete components, research has been conducted on steel connections used in timber components and the disassembled design of timber junctions. Examples include prefabricated bolts and swallowtail connections (Grmela 2020). Much of the research into these disassembled joints has focussed on whether they meet mechanical requirements or have the potential for improvement during construction and deconstruction.

Although several studies by structural experts have proven that the nodes designed based on the DfD principle can meet the structural requirements, architects recognised that buildings need to change during their lifecycle (Cruz Rios et al. 2021). In this context, DfD makes it easier for buildings to accommodate the adaptive needs arising from changes in building function, offers greater possibilities for the recycling of building materials and contributes significantly to sustainability. This has also been confirmed by other studies. Arrigoni et al. (2018) assessed the environmental sustainability of a temporary exhibition building through LCA and found that the ecological impact on the building's life cycle could be reduced by 30% through DfD. In addition, another study assessed the environmental benefits of modular buildings designed for disassembly. After comparison with buildings using the contemporary construction approach, it was found that designing and constructing reusable components could offset 88% of greenhouse gas emissions while also benefiting several other tested environmental indicators (Minunno et al. 2020).

However, few buildings in the world are designed for disassembly at present (Cruz-Rios and Grau 2020) because there are many barriers to the application. Particularly for architects, limitations in the use of DfD are inevitable: the design process itself is considered difficult because of the need to think about the end-of-life phase of the building and the reuse of materials and components (Salama 2017); neither internationally recognised guidelines for DfD design nor empirically based research can help architects understand the state of the art of DfD in current design practice (Kanters 2018; Cruz-Rios and Grau 2020); the additional design costs, the uncertainty of the economic and environmental benefits of applying DfD and the lack of consideration of reused building components in the current building codes are also unignorable barriers (Jaillon and Poon 2014; Salama 2017).

Therefore, to obtain a deeper understanding of the potential barriers to the practical application of DfD, an experiment is designed to simulate a hypothetical scenario that needs the adaptive reuse of building components. Two adjacent rooms are separated by a wood stud partition wall. For a certain reason, i.e. a functional change, it is required that a window opening be made in the wall connecting two rooms. In accordance with this scenario, the wood stud partition wall is adaptively reassembled.

The objective of this experiment is to explore the changes that DfD guidelines bring to the conventional techniques of wood stud walls. Comparisons are made for the construction and adaptation processes, material and labour consumption, waste generation and project quality. Through this redesign of the wood stud wall and our experimental transformation of the design into practice, we hope that more academics and architects will consider and explore the integration of the DfD process into building practices in the future.

3 Methods

3.1 Experimental Setup

This experiment was implemented in the carpentry working shop of Zhejiang University to imitate an indoor environment under stable conditions. To support the wood stud wall, we customised, in advance, an iron frame that was welded together with steel bars in the same profile (Fig. 1). Thus, a rectangular wall up to a size of 2456 × 2100 mm can be installed into this iron frame.

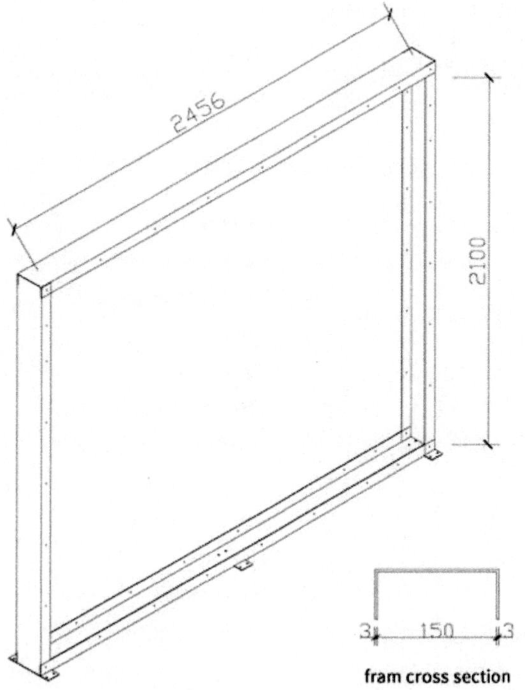

Fig. 1 Iron frame

3.2 Materials

The materials used for the experiments were mainly oriented strand boards (OSB), construction pine lumbers, gun nails and self-tapping screws. The tools used were a table saw, handheld circular saw, nail gun, air pump, hammer, pocket ruler and chalk line marker. Details of the materials and tools used are shown in Table 1.

3.3 Experiments

The experiments were divided into two groups to present comparative results. Both groups shared the objective of inserting a window opening into a whole wood stud partition wall. The opening was designed to a size of 1080 (width) × 1120 (height) mm and was 640 mm above the ground to duplicate a real building interior environment.

Table 1 Materials and tools used in the experiment

Material properties	
Material	Dimension (mm)
OSB (oriented strand board)	1220 × 2440 × 15 (Width × Height × Thickness)
Construction pine lumber	40 × 90 × 3000 (Width × Height × Length)
Steel nail	38 mm
Self-tapping screw	50 mm
Tool specification	
Tool	Model
Table saw	SawStop PCS31230-CH
Hand-held circular saw	LEISHI Z1E-LS-110
Cordless screwdriver	Deli DL600012
Nail gun	Dong Cheng FF-T50
Air compressor	OUTSTANDING 980W-30L
Hammer	–
Pocket ruler	DOSH 3 m/10FT
Chalk line marker	–

3.3.1 Group A—Conventional Construction

According to our field investigations on timber frame construction sites, working efficiency plays an important role in constructing wood stud partition walls. Using given standard materials, including oriented strand boards (OSBs) and lumber, the construction process can be facilitated, with or without fewer additional processing steps. Moreover, less time is required to use nails than to use screws when fixing studs and panels. Furthermore, within the constraints of the iron frame described above, the dimensions of the wall for the experiment were set at 2440 × 2100 (length x height) mm. Taking into account the standard size of OSB panels: 1220 × 2440 mm, a module of wall panel of 1220 × 2100 mm was defined. Studs with a length of 2020 mm and an axial distance of 600 mm and top and bottom plates with a length of 2440 mm were designed to be as close to the original size as possible. In addition, nails were used for attaching the OSB panels to studs.

Before the experiment began, the OSBs and construction lumber pieces were pre-processed with a table saw in the carpentry working shop. According to the design, each standard-sized OSB could be cut down to a 1220 × 2100 mm panel. For each stud (2020 mm long) and for each top and bottom plate (2440 mm long), lumber with a standard length of 3000 mm was consumed. In addition, 2 standard lumber pieces were sufficient for the preparation of a window header and a window sill (1160 mm long) and 2 trimmers (1120 mm long). The cutting scheme for the panels and lumber pieces is shown in Fig. 2.

After the materials were prepared, the wood stud partition wall in Group A was installed and then adapted in the following order (Fig. 3):

Step 1: The 2440 mm long top and bottom plates were fixed to the iron frame with self-tapping screws, and the 2020 mm long studs on the two sides were attached to the top and bottom plates;

Step 2: The three middle studs were secured in sequence with a nail gun so that they were positioned at a distance of 600 mm along the axis from each other. Three steel nails are used for each joint;

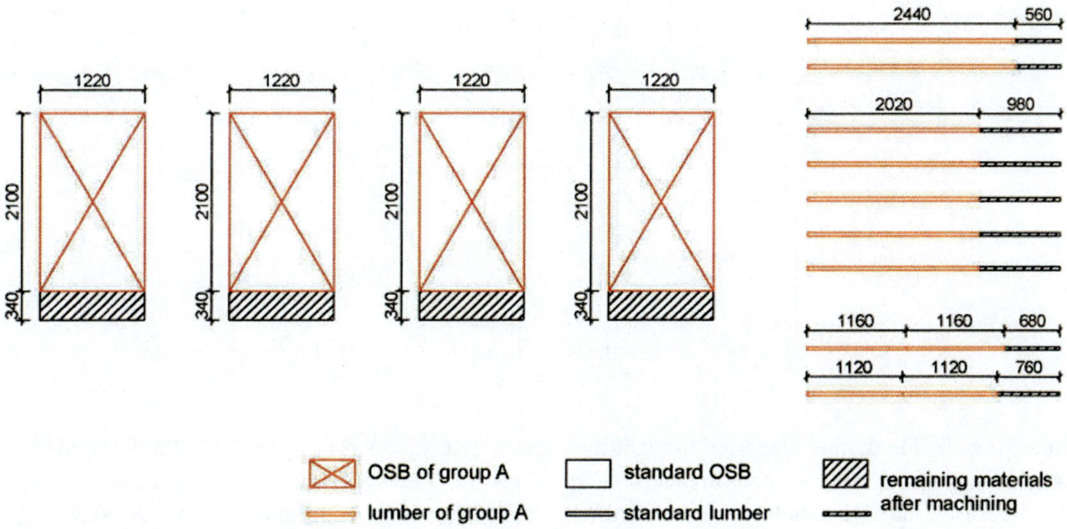

Fig. 2 Cutting scheme for Group A

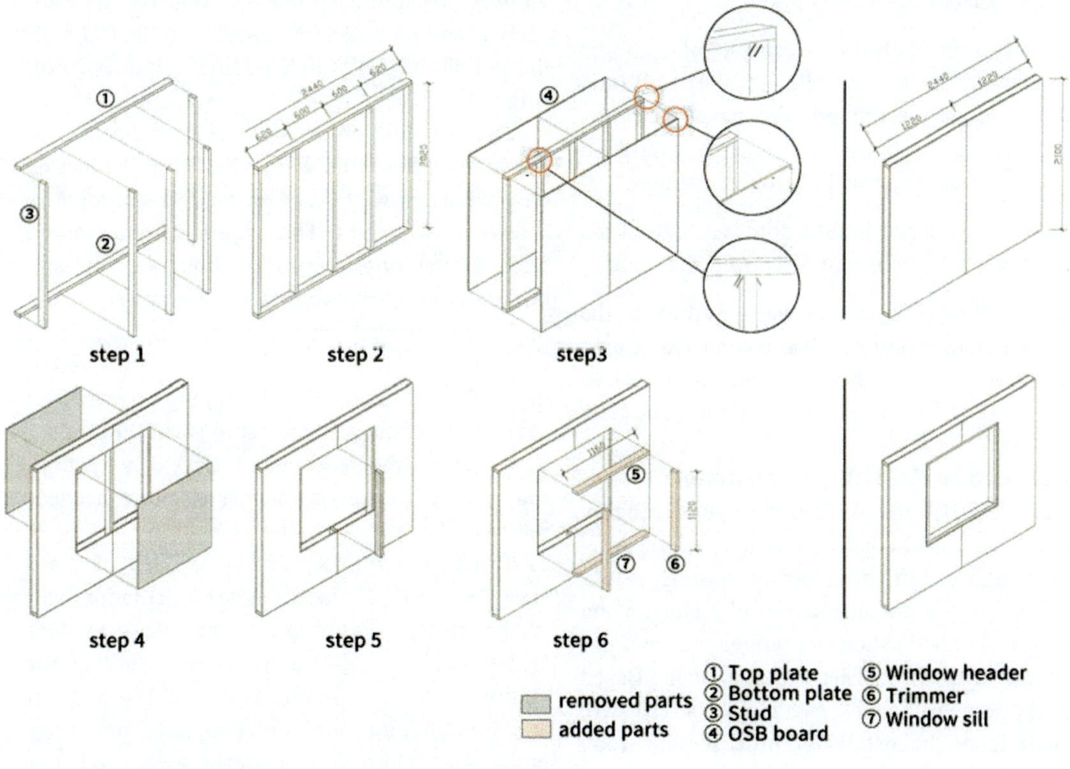

Fig. 3 Steps of wall assembly and adaptation in Group A

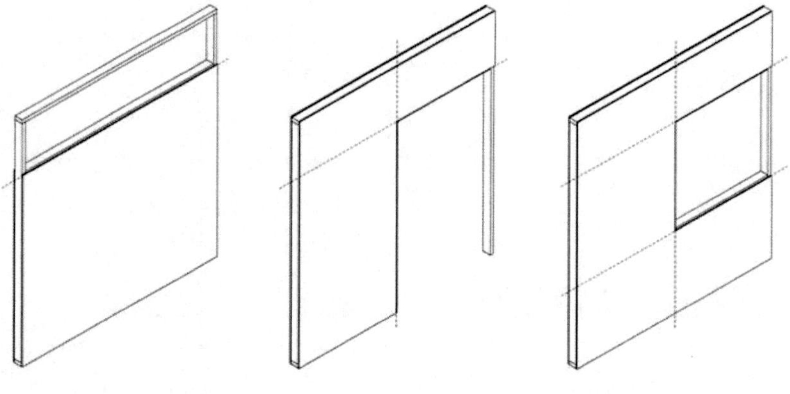

Fig. 4 Adaptation possibilities

Step 3: A chalk line marker was used to mark the locations on the OSB panels where the nails were to be driven in, and the panels were attached to the frame using nails according to the markings. Nails should be evenly distributed at the edges and in the middle of the panels. At this point, the wall was successfully assembled;

Step 4: A 1200 × 1200 mm rectangular outline was marked in the middle of the wall and 600 mm above the ground. The panels are then cut out along the outline by using a hand-held circular saw and eventually broken through;

Step 5: The exposed middle stud was cut with a circular saw and then removed from the wall;

Step 6: Self-tapping screws were used to fix the window header, window sill and trimmers, which were 40 × 90 mm in cross section and measured 1160, 1160, and 1120 mm in length, respectively.

3.3.2 Group B—DfD Construction

Kissi et al. (2019) analysed research studies highly relevant to the DfD concept over the period 2000–2018, and three DfD principles were most recognised as follows: documentation of materials and methods, standardisation of components, and use of mechanical joints instead of chemicals. Based on these three principles, especially the last two, we redesigned the installation process for a wood stud wall. The most significant changes were in the substructure made of studs and noggings, the size of the OSBs and the connections between elements. A full-length stud was divided into three parts by horizontal noggings. OSBs with dimensions of 1220 × 2100 mm were divided into smaller modules according to the new skeleton design. All nail connections were replaced with self-tapping screws for easier decomposition. All of these redesigns are for the consideration of the various possibilities for adaptation of the wall, such as adding a window opening at the top or in the middle of the wall or adding a door opening (Fig. 4).

The top and bottom plates were connected to the studs in the same way as in Group A; that is, they were screwed together using 2 self-tapping screws in each joint. The noggings were secured with self-tapping screws on both the top and bottom studs. Two screws were attached to the upper stud on each side, and only one screw was attached to the lower stud. Each OSB was fixed to the timber frame using self-tapping screws. The axis spacing between each stud was 600 mm.

Based on the redesign of DfD construction, four different sizes of wall panels were needed: 300 × 1220 mm, 620 × 1200 mm, 600 × 1220 mm and 600 × 1200 mm. Each of these sizes required 4 pieces. After calculation and arrangement, it was found that at least 4 standard OSBs (1220 × 2440 mm) were required for cutting all the panels out. For the 2440 mm top and bottom plate and the two 2360 mm noggings, 4 standard lumber pieces were used. For the remaining studs with lengths of 1160, 540 and 240 mm, it was possible to cut the parts from two lumber pieces with a sensible arrangement.

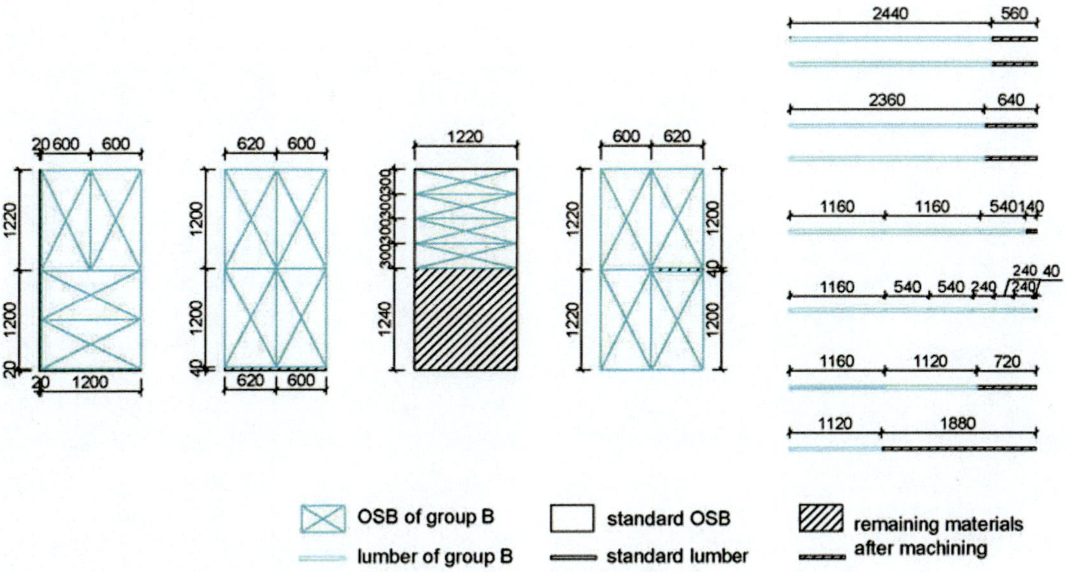

Fig. 5 Cutting scheme for Group B

In addition, the window header, sill and trimmers needed a total of two lumber pieces. The window header and sill had a cross-sectional dimension of 20 × 90 mm and a length of 1160 mm, and the trimmer had a cross-sectional dimension of 40 × 90 mm and a length of 1120 mm. The cutting scheme for the panels and lumber pieces is shown in Fig. 5.

After the materials were prepared, the wood stud partition wall in Group B was installed in the following order (Fig. 6):

Step 1: The 2440 mm long top and bottom plates were fixed to the iron frame with self-tapping screws, and 2020 mm long studs were attached on the two sides to the top and bottom plates. This step is the same as that in Group A;

Step 2: Self-tapping screws were used to fix the upper and lower noggings as well as the bottom, top and middle studs. The studs were 240, 1160 and 540 mm long from top to bottom and had a horizontal axis distance of 600 mm between each other;

Step 3: The OSB panels were installed in sequence from the bottom to the top using self-tapping screws evenly distributed on the four edges of each panel. The panels have three sizes of 600 × 1200, 620 × 1200 and 300 × 1200 mm to fit into different positions.

Step 4: The two 600 × 1200 mm middle panels were removed by pulling out the self-tapping screws on the four edges;

Step 5: The exposed middle stud was removed by using a cordless screwdriver for the self-tapping screws. At that point, all four edges of the opening had a 20 mm exposure made of studs and noggings;

Step 6: The window header, sill and trimmers were fastened with self-tapping screws. The window header and sill had a cross section of 20 × 90 mm and a length of 1160 mm; the trimmer had a cross section of 40 × 90 mm and a length of 1120 mm. At this point, the window frame had the same width as that in Group A.

4 Results and Analysis

4.1 Comparison of the Process Complexity

The procedure for both experimental groups can be divided into three phases: timber frame

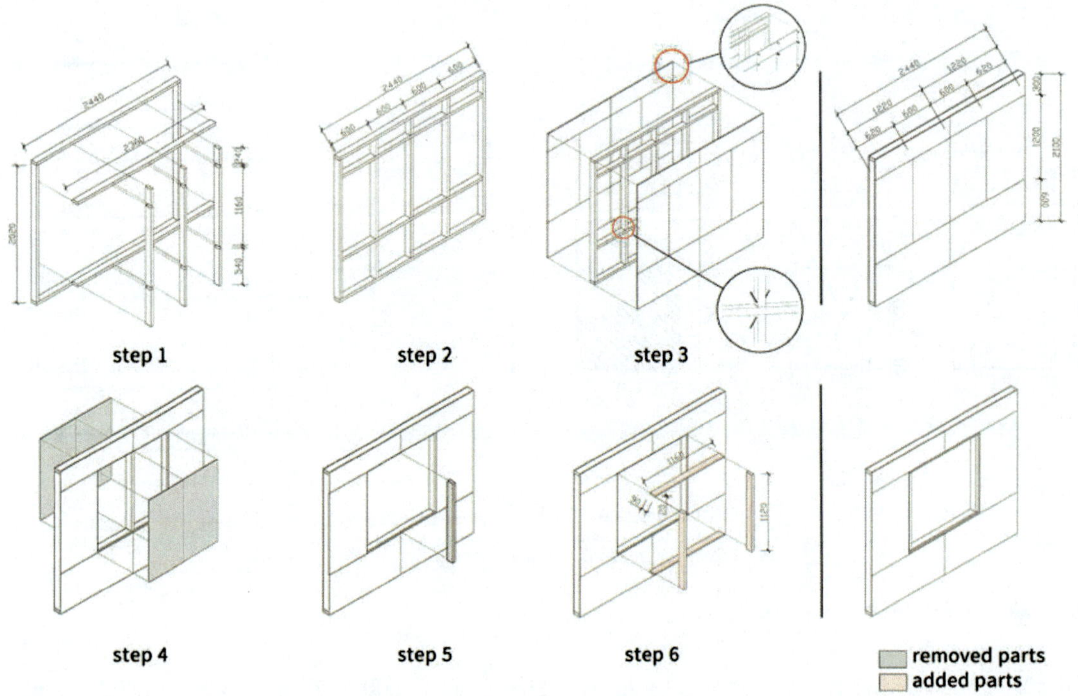

Fig. 6 Steps of wall assembly and adaptation in Group B

installation, OSB installation and adaptive reassembly. The processes are listed in Table 2.

By comparing the process of installing the timber frame with OSBs, notably, the assembly process in Group B was more complicated than that in Group A because the wall studs in Group B were all cut into three short parts. There were also two extra noggings and 6 more pieces of OSB. In contrast, the adaptation process of Group B was simpler than that of Group A. In Group A, additional work had to be conducted to mark the window opening outline on the surface of the panels and to cut them down. However, the panel could be removed by simply loosening the screws in Group B.

Furthermore, the experimental process in Group A required more complex tools to complete than that in Group B. The chalk line marker was used to assist in marking where to punch the nails. Additionally, an electric hand-held saw was applied to remove the panels. To avoid damage to the studs behind the panel, the saw blade also had to be adjusted to the same thickness as the panels. In contrast, the smaller size of the OSB panels in Group B allows one person to simply complete all steps with a cordless screwdriver.

Moreover, the circular saw in Group A generates much noise, and the flying wood chips can harm the eyes and respiratory tract of workers. Thus, there is a need to introduce additional vacuuming equipment and protection methods to protect workers from health injuries. Comparatively, these issues did not occur in Group B, in which the fixing and removal of the elements by using self-tapping and a cordless screwdriver created a noiseless working environment.

4.2 Comparison of Material Consumption and Waste Generation

Based on the statistics (Table 3, Figs. 2 and 5), during the preparation of materials for the wall installation, both Group A and B consumed 4

Table 2 Comparison of the installation and adaptation process

Group A	Group B
Timber frame installation	
(1) Top and bottom plates	(1) Top and bottom plates
(2) Right and left studs	(2) Right and left studs
(3) Three middle studs	(3) Lower nogging
	(4) Three bottom studs
	(5) Upper nogging
	(6) Three top studs
	(7) Three middle studs
 (1) (2) (3)	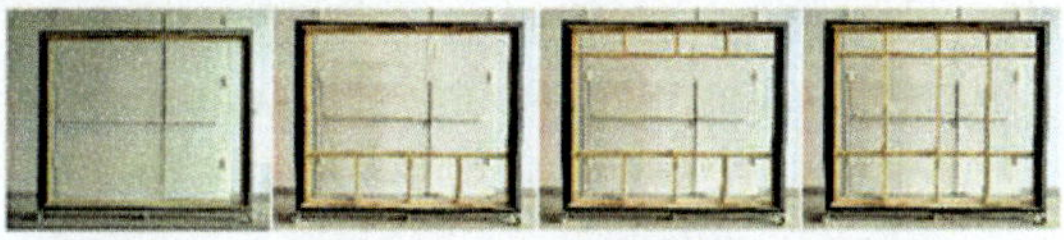 (1)(2) (3)(4) (5)(6) (7)
OSB installation (both sides)	
(1) Nail position marks	(1) Bottom panels
(2) OSB panels	(2) Middle panels
	(3) Top panels
 (1) (2)	 (1) (2) (3)
Adaptation with a window opening insertion	
(1) Window opening outline mark	(1) Middle panels removal
(2) Panel cut and removal	(2) Middle stud removal
(3) Middle stud cut and removal	(3) Window header, sill and trimmers
(4) Window header, sill and trimmers	
 (1) (2) (3) (4)	 (1)(2) (3)

standard OSBs (1220 × 2440 × 15 mm) for a total volume of 178.608 dm^3, of which 153.72 dm^3 was used for the wall installation (step 1–3) and 24.888 dm^3 was left after cutting the required panels. During these steps, 7 lumber pieces totalling 75.6 dm^3 were consumed in Group A, which was one lumber piece more than that in Group B. However, the net consumption of lumber pieces in Group A was even fewer than that in Group B, that is, 53.928 dm^3 relative to 55.512 dm^3. Consequently, more than 21 dm^3 remained in Group A but only approximately 9 dm^3 in Group B. This was due to the smaller dimensions of individual parts required for the stud-nogging-framework in Group B (Fig. 5), which led to efficient utilisation of each lumber piece.

During the adaptation process (steps 4–6), both groups consumed 2 lumber pieces of 21.2

Table 3 Material consumption and waste generation (dm^3)

	Step	Group A		Group B	
		OSB	Lumber	OSB	Lumber
Gross consumption	1–3 (Installation)	178.608	75.6	178.608	64.8
Nett consumption		153.72	53.928	153.72	55.512
Remains		24.888	21.672	24.888	9.288
Gross consumption	4–6 (Adaptation)	0	21.2	0	21.2
Nett consumption		0	16.416	0	12.24
Remains		0	4.784	0	8.96
Waste generation		43.2	4.32	0	0
Total consumption	1–6	178.608	96.8	178.608	86

dm^3 for the window frame, but Group A had a slightly higher consumption than Group B, that is, 16.416 dm^3 relative to 12.24 dm^3. This was because the cross-sectional dimensions of the window header and sill in Group B were 20 × 90 mm, in comparison with those of 40 × 90 mm in Group A. As a result, Group B provided slightly more leftover materials for processing the window frames than Group A. Because the remaining segments of OSBs and lumber pieces, including those after installation and adaptation, were mechanically processed with a table saw in the carpentry workshop, they were unpolluted and therefore retained the potential for reuse and recycling.

Furthermore, the removed panels with a volume of 43.2 dm^3 and the studs with a volume of 4.32 dm^3 in Group A could not be directly reused due to the damage caused by the hand-held circular saw and were classified as waste. In contrast, the removed parts in Group B were practically complete, and their condition was, thanks to the use of screws, similar to when they were originally installed (Fig. 7). According to their great potential for reuse, these materials were not categorised as waste. In other words, there was no waste generation in Group B.

4.3 Comparison of Project Quality

The edges of the window openings in Group A are rough, but those in Group B are very uniform (Fig. 8). This is because the window openings in Group A were cut using a hand-held circular saw, and the corners could not be cut consistently due to the round shape of the saw's blade. However, all the panels in Group B were machined by a table saw to the design dimensions before installation so that the edges of the window openings were unbroken when the middle panels were removed.

Fig. 7 States of the removed panels

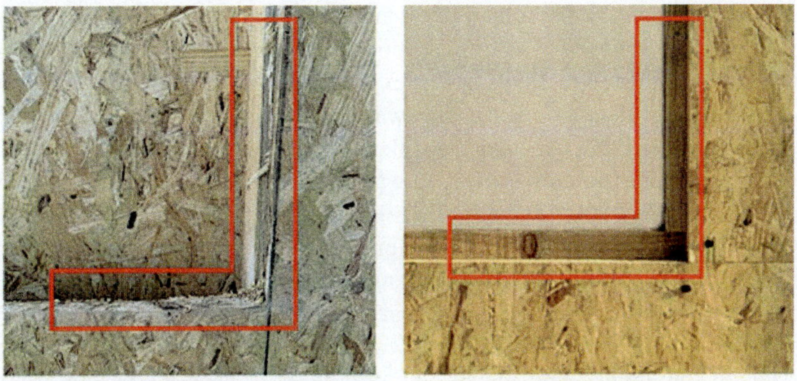

Group A Group B

Fig. 8 Comparison of the edge cuttings of openings

Group A Group B

Fig. 9 Comparison of stud exposures

After removing the OSBs, 20 mm of the stud was exposed on each side of the opening. In Group A, the left stud was exposed by 31 mm, and the right stud was exposed by 19 mm. In Group B, the left side stud was exposed by 24 mm, and the right stud was exposed by 18 mm (Fig. 9). The impreciseness of the exposures resulted mainly from the inaccuracy of the stud installation. Although there were more studs and installation steps in Group B, the stud exposures were even closer to the theoretical values. This is likely because all studs were installed carefully to ensure that each OSB was in the predesigned position. This also resulted in the adaptation process, where the distance was kept within 4 mm.

5 Discussion

The final purpose for wall reconstruction was known in advance, which guided us somewhat towards drawing support from the DfD concept in the redesign of the wall. However, the potential of this redesigned construction is not limited to adding window openings in the centre of the wall. This wall is also subject to other modifications, such as new windows at the edge or top, new door openings and an entire deconstruction or relocation. The flexibility of wall modifications increases the possibility of changing the spatial relationship between two rooms, which leads to an improvement in the building's resilience against functional changes.

In this experiment, a wood stud partition wall was installed, in which the wall component was designed via conventional construction and the DfD concept. Compared with the conventional method, installation utilising DfD guidelines involves more steps. This means that much more time can be expended when applying DfD construction in a much larger project. These additional costs for DfD application should be a large obstacle in designing buildings with deconstruction capacities. There is uncertainty as to whether these initial construction costs are counterbalanced by the potential environmental benefits and possible future savings.

Our findings showed that the DfD wall construction in Group B was more complex for the installation than the conventionally constructed wall in Group A, which coincide with the research conclusions by Eckelman et al. (2018), who used life cycle assessment (LCA) to evaluate the environmental benefits of a novel DfD flooring system. Their results showed that buildings with DfD flooring systems have a better environmental impact than conventional buildings, assuming that reuse occurs at least once in the future. However, when reuse is precluded, either due to component damage or lack of economic incentive, the DfD system causes greater environmental impact than conventional designs for the same building configuration. Therefore, controlling the construction costs of DfD components and increasing their durability and adaptability should be the focus of our subsequent research.

6 Conclusion

In conclusion, this study carried out DfD for a conventional wood stud partition wall and experimented on its adaptive reuse. The experiment was divided into two groups for testing the conventional construction and the DfD integrated construction. Both groups shared the objective of adapting the wall construction to insert an additional window into a whole wood stud partition wall. As shown in the results of the experiment, the DfD integrated wall construction was less labour- and material-consuming, more flexible for changes, higher quality and with less waste generated after adaptation. It is hoped that this experiment will provide useful information for architects, engineers and other architecture, engineering and construction practitioners who are willing to integrate the DfD concept into their work. If it is difficult to apply this theory to an entire building, then the design of simple building components can be performed first, as was carried out in this study. We firmly believe that such experimental attempts that translate theory into practice can make a solid contribution to building a sustainable future.

Acknowledgements This research was funded by the Start-up Foundation for Hundred-Talent Program, Zhejiang University and the Centre for Balance Architecture of Zhejiang University.

References

Arrigoni A, Zucchinelli M, Collatina D, Dotelli G (2018) Life cycle environmental benefits of a forward-thinking design phase for buildings: the case study of a temporary pavilion built for an international exhibition. J Clean Prod. https://www.sciencedirect.com/science/article/pii/S0959652618309077. Accessed 26 Dec 2022

Bullen PA, Love PE (2010) The rhetoric of adaptive reuse or reality of demolition: views from the field. Cities. https://www.sciencedirect.com/science/article/pii/S0264275109001450. Accessed 05 Sep 2022

Cai G, Xiong F, Xu Y, Si Larbi A, Lu Y, Yoshizawa M (2019) A demountable connection for low-rise precast concrete structures with DfD for construction sustainability-a preliminary test under cyclic loads. Sustainability. https://www.mdpi.com/2071-1050/11/13/3696. Accessed 28 Sep 2022

Cruz-Rios F, Grau D (2020) Design for disassembly: an analysis of the practice (or Lack Thereof) in the United States. In: Construction research congress 2020. https://doi.org/10.1061/9780784482889.105. Accessed 22 Dec 2022

Cruz Rios F, Grau D, Bilec M (2021) Barriers and enablers to circular building design in the US: an empirical study. J Constr Eng Manag. https://doi.org/10.1061/%28ASCE%29CO.1943-7862.0002109. Accessed 28 Sep 2022

Eckelman MJ, Brown C, Troup LN, Wang L, Webster MD, Hajjar JF (2018) Life cycle energy and environmental benefits of novel design-for-deconstruction structural systems in steel buildings. Build Environ. https://www.sciencedirect.com/science/article/pii/S0360132318304256. Accessed 29 Sep 2022

Gaspar PL, Santos AL (2015) Embodied energy on refurbishment vs. demolition: a southern Europe case study. Energy Build. https://www.sciencedirect.com/science/article/pii/S0378778814009712. Accessed 05 Sep 2022

Grmela V (2020) Towards zero-waste buildings-Building design for reuse and disassembly. Master, Chalmers University of Technology. https://hdl.handle.net/20.500.12380/301384. Accessed 28 Sep 2022

Huang H, Li L (2022) Study reviews and rethinking the key processes for managing building materials to enhance the circular economy in the AEC industry. Sustainability. https://www.mdpi.com/2071-1050/14/19/11941. Accessed 26 Sep 2022

Huang T, Shi F, Tanikawa H, Fei J, Han J (2013) Materials demand and environmental impact of buildings construction and demolition in China based on dynamic material flow analysis. Resour Conserv Recycl. https://www.sciencedirect.com/science/article/pii/S0921344912002273. Accessed 06 Sep 2022

Itard L, Klunder G (2007) Comparing environmental impacts of renovated housing stock with new construction. Build Res Inf. https://doi.org/10.1080/09613210601068161. Accessed 06 Sep 2022

Jaillon L, Poon CS (2014) Life cycle design and prefabrication in buildings: a review and case studies in Hong Kong. Autom Constr. https://www.sciencedirect.com/science/article/pii/S0926580513001556. Accessed 28 Sep 2022

Kanters J (2018) Design for deconstruction in the design process: state of the art. Buildings. https://www.mdpi.com/361594. Accessed 05 Sep 2022

Kissi E, Ansah M, Ampofo J, Boakye E (2019) Critical review of the principles of design for disassembly. In: Modular and offsite construction (MOC) summit proceedings. https://journalofindustrializedconstruction.com/index.php/mocs/article/view/101. Accessed 05 Sep 2022

Lin Z (2013) Designing Moment-resisting connection for disassembly for implementation in precast reinforced concrete buildings. Doctor of engineering, National University of Singapore. https://core.ac.uk/download/pdf/48680481.pdf. Accessed 28 Sep 2022

Minunno R, O'Grady T, Morrison GM, Gruner RL (2020) Exploring environmental benefits of reuse and recycle practices: a circular economy case study of a modular building. Resour Conserv Recycl. https://www.sciencedirect.com/science/article/pii/S0921344920301750. Accessed 26 Dec 2022

Rakhshan K, Morel J-C, Alaka H, Charef R (2020) Components reuse in the building sector–a systematic review. Waste Manag Res. https://doi.org/10.1177/0734242X20910463. Accessed 06 Sep 2022

Salama W (2017) Design of concrete buildings for disassembly: an explorative review. Int J Sustain Built Environ. https://www.sciencedirect.com/science/article/pii/S2212609016301741. Accessed 23 Dec 2022

Wilkinson SJ, James K, Reed R (2009) Using building adaptation to deliver sustainability in Australia. Struct Surv. https://doi.org/10.1108/02630800910941683/full/html. Accessed 06 Sep 2022

Xiong F, Zhao F, Cai G, Chen J, Si Larbi A (2022) Seismic performance of a bolt-assembled precast panel building with DfD: a quasi-static test and discussion on existing design codes. J Earthq Eng. https://doi.org/10.1080/13632469.2020.1820408. Accessed 28 Sep 2022

Research on 3D Printing Craft for Flexible Mass Customization: The Case of Chengdu Agricultural Expo Center

Tianyi Gao, Sijia Gu, Liming Zhang, and Philip F. Yuan

Abstract

Mass customization of prefabricated architecture is becoming increasingly crucial for developing the architectural industry. With the continuous iterative development of digital design tools and construction methods, paradigm innovation liberates social productivity significantly. This thinking of architecture allows architects to accomplish profound design concepts and intentions with sustainable production strategies. At the same time, digital design tools and architectural robotics also powerfully awaken the relevance of tradition and local culture, addressing the issue of the declining productivity of traditional construction methods. In the engineering practice of the Chengdu Agricultural Expo Center, robotic timber fabrication and plastic 3D printing craft efficiently completed the construction of complex structural elements. The 3D-printed roof covers an area of about 3,000 square meters. All the roof panels are prefabricated in the factory and transported to the site for assembly. It proves that 3D printing can meet the demand for the efficient construction of large-scale and high-dimensional geometric roofs. This paper introduces the integrated design-to-fabrication process and robotic construction method of the large-scale prefabricated roof, taking the Chengdu Agricultural Expo Center as an example.

Keywords

3D printing · Mass customization · Digital fabrication · Sustainable production · Labor cost

1 Introduction

In recent years, the rapid development of tools, including computers and robotics, has driven the digital transformation of the construction industry. Additive manufacturing, represented by 3D printing, is a notably developing advanced manufacturing technology in the last 30 years, with the advantage of free manufacturing of three-dimensional structures and material saving, which has been widely used in innovative product exploitation and small-scale manufacturing. 3D printing, as an emerging digital technology, uses materials with different properties as the basis and printing processes as the material reconfiguration

T. Gao
Shanghai Research Institute for Intelligent Autonomous Systems, Tongji University, Shanghai, China

S. Gu · L. Zhang · P. F. Yuan (✉)
College of Architecture and Urban Planning, Tongji University, Shanghai, China
e-mail: philipyuan007@tongji.edu.cn

M. R. Thomsen et al. (eds.), *Design for Rethinking Resources*, Sustainable Development Goals Series,
https://doi.org/10.1007/978-3-031-36554-6_30

method. The technical advantages are based on precise positioning and efficient construction. It is increasingly applied in agriculture, healthcare, automotive, locomotive, and aerospace industries since the amplitude of robot arm motion and material nozzle diameter can be controlled in conditions for different manufacturing dimensions. With the widespread use of 3D printing technology, the manufacturing industry may no longer depend on traditional factories, which centralize production factors such as labor, capital, and equipment on a large scale, but rather shift to a more flexible production method based on industrial robotics that requires less investment. Especially from the architectural perspective, 3D printing has the advantages of fast construction, fabricating beautiful forms, and material repurposing. Combined with the digital construction method of human–machine collaboration, 3D printing has significant potential for large-scale manufacturing scenarios.

After World War II, European countries, Japan, and other countries with severe housing problems demanded a practical solution urgently. The debate on rapidly building practical, reliable, and cheap housing boosted the development of prefabricated buildings. Prefabricated buildings were promoted in large numbers, resulting in many studies and practices in the 1860s. Generally, most attempts in the post-war scenario focused on cheap and rapidly replicable assembly strategies. It increased the production efficiency of the architecture industry remarkably, mitigating the pressure of post-war housing problems. However, current housing problems are not only limited to the scenarios above but transformed to the demand for better quality, low carbon emissions, and complex geometry. Thus, the contradiction between the need for customized design and industrial standardization in developing prefabricated buildings has been revealed. Specifically, the conventional production process of prefabricated elements can provide minimal combinations and often has low applicability. Sometimes to achieve universal promotion, additional costs could be added to projects. The relationship between customization and standardization is still critically antagonistic and mutually constrained. 3D printing technology can achieve flexible mass customization. It can use the same production logic and process for different components, which is consistent with the demand for mass customization in the architecture industry. This paper discusses architecture robotics 3D printing modified plastics technology application for mass customization based on large-scale roof construction of Chengdu Agricultural Expo Center (Fig. 1).

2 Background and Overview

As an additive manufacturing (AM) technology, 3D printing is used to create various structures and complex geometries from three-dimensional (3D) model data. The process consists of printing successive layers of materials that are formed on top of each other (Ngo et al. 2018, p. 1). With the rise of digital fabrication in the construction industry in recent years, 3D printing as an emerging construction technology is gradually entering the construction industry, creating new possibilities for architectural design and construction. Since Joseph Pegna proposed the system of quick hardening concrete layers (García-Alvarado et al. 2021), 3D-printed construction (3DPC) technology has shown rapid development through extensive research and practice. In 2003, Contouring Crafting (CC) technology was proposed as a new Layered Manufacturing (LM) process by Hwang and Khoshnevis (2005). In 2008, a free-form construction method was created by Loughborough University by using the layering technique and powdered sintering technique. One of the practices of the method is a 3D-printed bench, which is a highly customizable building component (Lim et al. 2011). Many other related types of research also have been conducted. At the same time, the concept of sustainability has become of interest to many disciplines, including architecture. The goal of sustainability in architecture is to minimize the number of resources consumed during the construction and operation process and to reduce the environmental harm caused by emissions, pollution, and waste (Beyhan et al. 2018). In this

Fig. 1 A photograph of the completed project of Chengdu Agricultural Expo Center

Fig. 2 3D printed components for the canopy (Griffiths 2013, News)

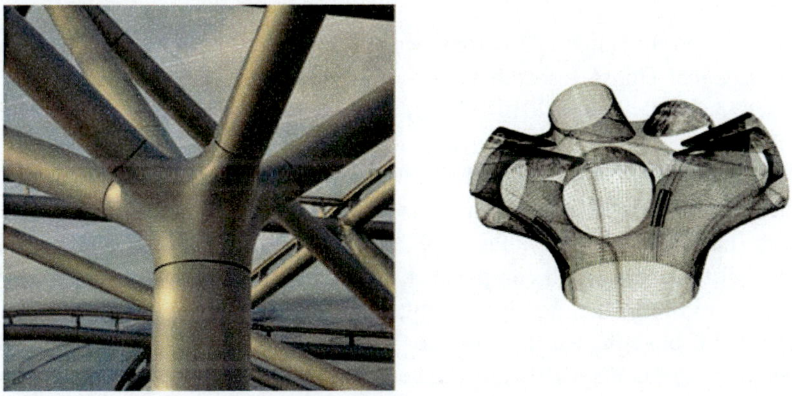

regard, as two concepts of interest to architecture in this century, how digital technology in architecture can be applied to sustainable development is a question worth thinking about. In the field of 3D-printed construction (3DPC), many pioneers have made efforts on this issue for different materials. In 2013, British architect Adrian Priestman printed components for a canopy on the roof of an office building in London using 3D metal printing (Fig. 2). As the primary material of the component, steel and aluminum can be recycled at a rate of 100% (Munn and Soebarto 2004, p. 162). This recycling resulted in a 72% reduction in energy consumption for steel material and 95% for aluminum material (Beyhan et al. 2018, p. 261), thus contributing to the development of sustainable buildings.

Fig. 3 3D printed canal house (Van der Veen 2014)

Fig. 4 3D clay printed acoustic column

Conceived by Dutch DUS Architects, the 3D Printed Canal House research project was performed in Amsterdam in 2014 (Fig. 3). The goal of the project is to print a three-story high, full-sized house on the canal side using a 3D printer called 'KamerMaker'. The printer uses a bio-plastic mixture of plastic fibers and 80% vegetable oil to produce wall components. The wall components were then interlocked together and filled with bio-concrete to provide structural strength (Van der Veen 2014). All materials used in the project are recyclable (Beyhan et al. 2018, p. 263), which means less environmental harm. Also, this approach will reduce construction waste and minimize transportation costs (Hager et al. 2016).

In 2015, Phillip F. YUAN's team at Tongji University carried out a 3D clay printed acoustic column project to aid spatial acoustic design through sound visualization studies (Fig. 4). The device was fabricated with a digital technique using robotic clay laminated contouring crafting technology. As a malleable and reusable material, the combination of clay and 3D printing techniques shows excellent potential for sustainable development. Robotic 3D clay printing is also suitable for the prefabrication of architectural components and skin panels due to its high accuracy, which has a relatively broad development prospect in manufacturing architectural components.

The Digital Design Research Center (DDRC) team at Tongji University has extensively researched the use of modified plastic 3D printing in architectural scenarios. Moreover, they completed a project at the east gate of the Nanjing Happy Valley theme park in 2020 (Philip et al. 2021). The project used modified plastic materials with a highly weather-resistant formulation, allowing the prefabricated 3D-printed panels to be used in outdoor environments for extended periods. Prefabricated assembly and remote positioning were also used to ensure construction efficiency during COVID-19 (see Fig. 5).

Fig. 5 The modified plastic 3D printed installation in Nanjing Happy Valley theme park

In addition to materials such as metal, plastic, and terracotta, there are cases such as the 3D printed house completed by WinSun Decoration Design Engineering Co in 2014 using a concrete-like material and the castle printed in a garden by Andy Rudenko using Contour Crafting (CC) technology.

The above studies show that the 3D printing technique can significantly contribute to sustainable development in the construction industry with the innovative development and application of materials and techniques. The contribution of the 3D printing technique to sustainable architectural construction can be summarized as follows:

(1) It enables prefabricated assemblies in the factory, reducing construction time.
(2) It gives architects different perspectives on building materials and chances for reducing material losses and construction waste.
(3) 3D printers can manufacture a variety of different material types, available in different states (powder, filament, pellet, granule, resin.), including a variety of recyclable materials such as glass, plastics, thermoplastic polymers (ABS), metals and ceramics (Beyhan et al. 2018, p. 258).
(4) It reduces labor costs and safety risks for on-site construction.

In summary, 3D printing has great potential to contribute to the sustainability of construction. Features of the 3D printing process, such as higher construction efficiency, better fabrication accuracy, lower labor costs, less construction resource requirements, and lower CO_2 emissions throughout the architectural lifecycle (Gebler et al. 2014), can provide revolutionary progress in the development of sustainability (Beyhan et al. 2018, p. 253).

3 Large-Scale Mass Customization 3D Printing Methodology

3.1 Design-to-Fabrication Workflow Based on FUROBOT Platform

The software platform is one of the core parts of mass customization production based on 3D printing craft. For architects, the design-to-fabrication workflow and collaboration between humans and new industrial robotics establish a new partnership. Architects are not limited to the characteristic of the designer just but also the fabrication logic of the parameterization in this partnership. Thus, the software platform contributing to the architect's design-to-fabrication interface is one of the most significant sessions of the entire design workflow. It plays a decisive role in whether the architect's concepts and schemes can be built correctly and remarkably.

FUROBOT is a promising robot programming software platform based on Grasshopper for designers in different fields to program industrial robots intuitively. Different from the conventional dialog interface of the robotic control system, it is a node-based programming software inheriting Grasshopper's parametric control mode (Fig. 6). Most of its parameter information is conveyed in the form of node input and output instead of the traditional dialog box input mode, enhancing the efficiency of parametric programming for architects (Fig. 7). Due to this friendly interface, the robot path planning for fabrication becomes simple for users without a holistic knowledge base. In addition, it accommodates multiple types of robots and combines various techniques (Fig. 8). Different robotics allows accuracy and stability in many degrees of freedom of the robotic arm. The path simulation and optimization in FUROBOT's platform allow designers to anticipate possible problems in manufacturing environments before formal production. The rule of 6-Axis robotic arms operation is the tool head coordinates move at spatial points, derived from the spatial translation of the 3D model in Rhino. The following three main steps are path simulation and start-stop point optimization after generating the 3D model.

Step 1: Select Breps in Rhino for Grasshopper to set 3D models.

Step 2: Use FUROBOT core to simulate the printing range for the sliced contour lines prepared in the parametric program.

Step 3: Collison and singularity check.

3.2 Robotic 3D Printing Hardware System and Material

The hardware system for 3D printing can be separated into three main parts: the material preparation system, the robotic extrusion system, and the post-processing system. The material preparation system consists mainly of the dosing, drying, and feeding of the printing material. The prepared material is crushed in a screw extruder and extruded through a toolhead with customized geometry and dimensions. Before actual production, a five-layer base is pre-printed to complete the automatic leveling operation based on previous project experience. The printing process is transferred to the robot according through an offline program with a programmed number. After printing, the printed components need to be cut and polished to meet the applicable standards for the engineering project. The detailed robotic 3D printing hardware system and workflow can be shown in Fig. 9.

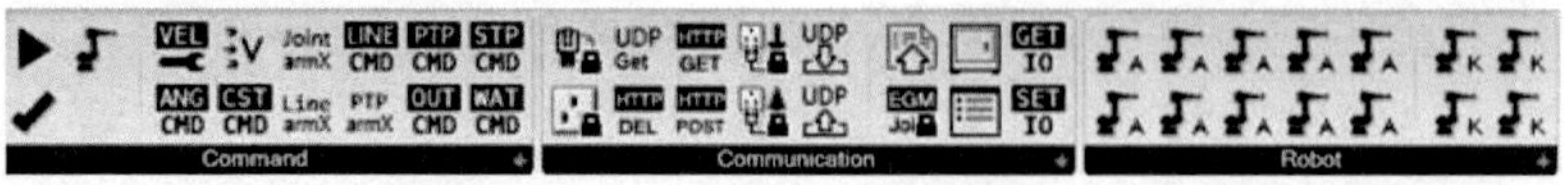

Fig. 6 Node-based programming software

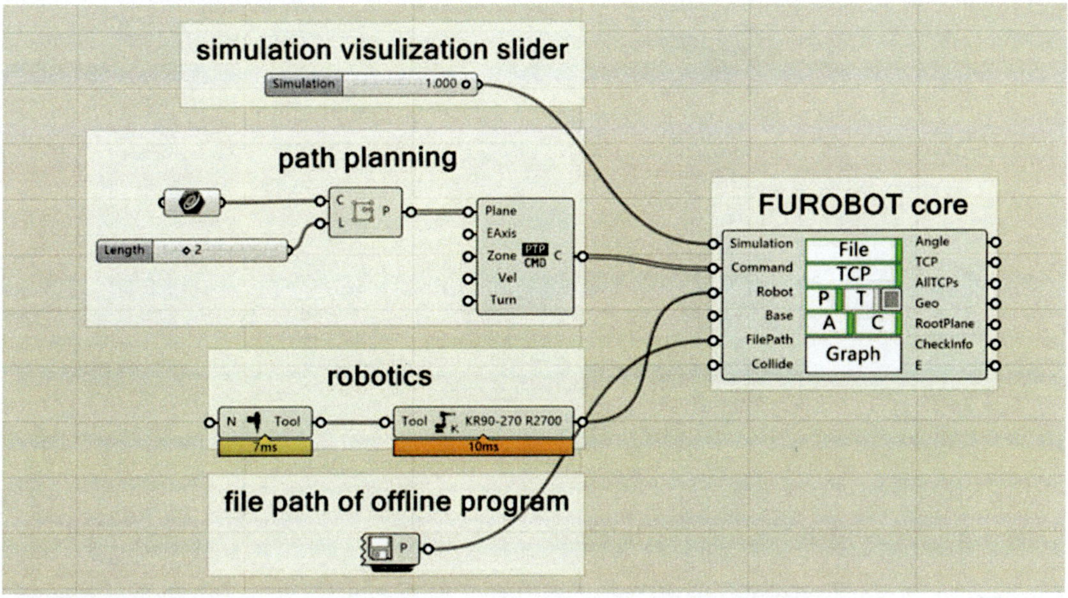

Fig. 7 FUROBOT parametric operation interface

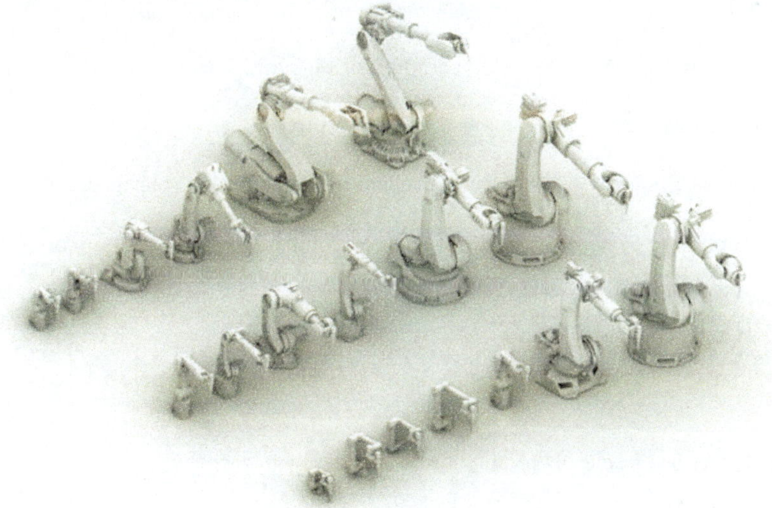

Fig. 8 Multiple types of robots integrated in FUROBOT

As a relatively low-cost and highly malleable raw material, plastics have significant advantages in constructing complex shaped components. Due to the varying levels of performance of plastics at this stage, their application in engineering projects is still limited to small-scale architecture. In this project, fiber-reinforced polymer-based composite material is introduced to explore the long-term use of modified plastics as the essential part of the building. This hybrid material has two main advantages. On the one hand, it effectively promotes stress relief during the melt processing of polymer materials. On the other hand, it significantly improves material rigidity and thus inhibits deformation. The composite modification technology with ABS

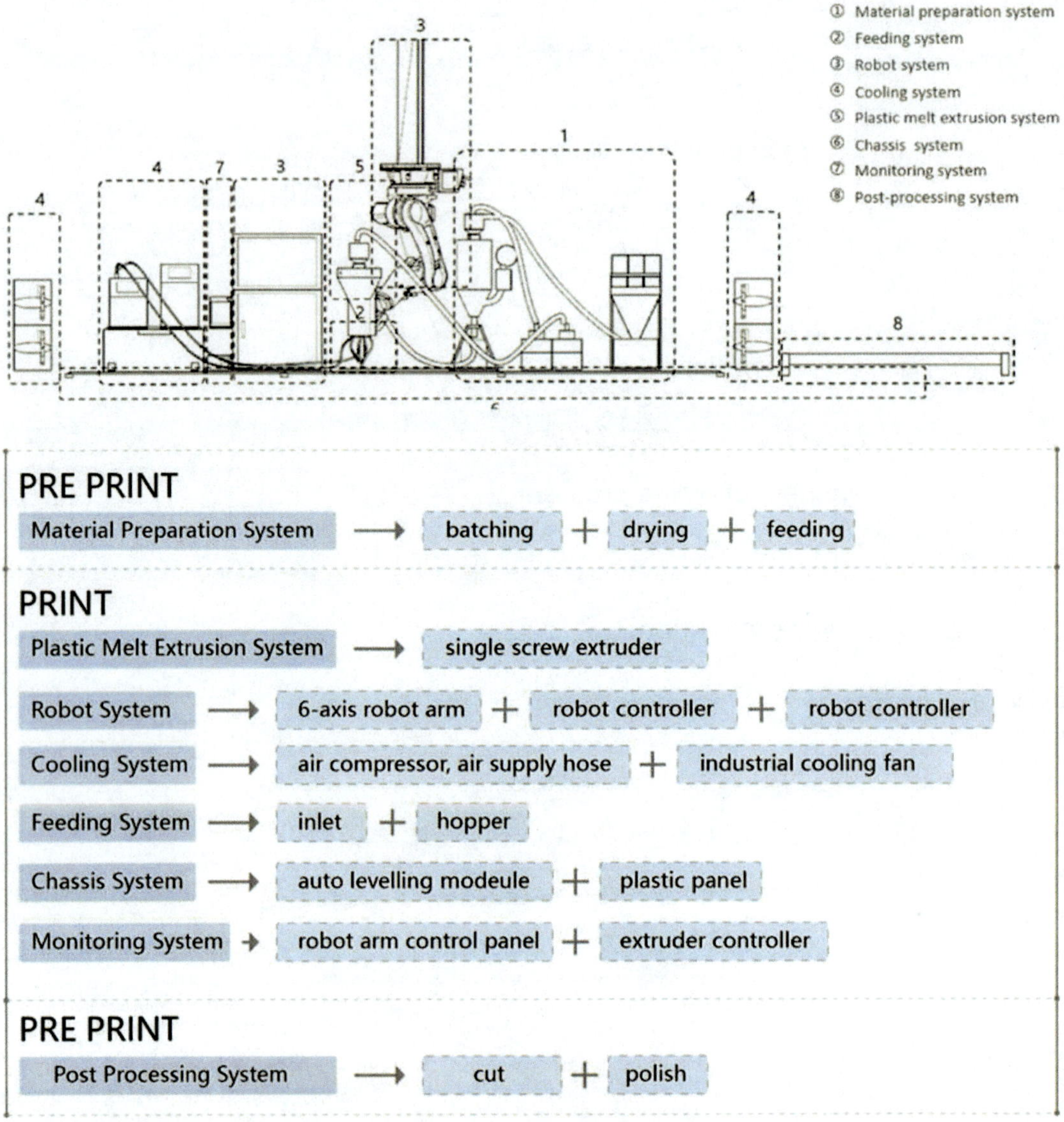

Fig. 9 Detailed robotic 3D printing hardware system and workflow

material can effectively enhance the rigidity of the composite material, inhibit the occurrence of warpage during the printing process, and improve the creep resistance.

For a more systematic study of fiber-reinforced plastic, the material performance is tested to get stiffness and linear expansion coefficient. Specific experimental results are shown in the following Table (see Fig. 10).

Based on the data in the table, this composite material's tensile and bending strength increased substantially after the glass fiber reinforcement (10 wt.%). The modification in material stiffness facilitates the resistance to the warpage caused by internal material stress during the printing process. At the same time, the linear expansion coefficient decreases significantly after glass fiber reinforcement (10 wt.%). It decreases the linear

Fig. 10 Material performance of ABS and fiber-reinforced ABS

Material	linear expansion coefficient(α)	tensile strength (MPa)	bending strength (Mpa)
ABS	6.6×10-5	40.7±1.0	55.3±1.0
ABS/GF(15%wt)	3.2×10-5	57.0±1.0	65.2±1.0

expansion coefficient, which can help to reduce deformation and thermal shrinkage. From the results, the glass fiber-reinforced ABS material is better than the non-reinforced ABS in terms of dimensional stability and is suitable for building roof cladding.

On the other hand, the material was modified with UV and oxidation resistance for the exterior scenarios by adding 2% UV absorber UV-P and 2% UV770. In the weathering experiments, the specimens in the experimental group were treated with 1828 h of UV irradiation, simulating 10-year UV irradiation. The experimental results showed that the test specimens' tensile strength and flexural strength decreased by 7.8% and 9.5%, which were within 10% and met the structural safety requirements. Therefore, this material modification is practical and applicable to the exterior roof of the Chengdu Agricultural Expo Center.

3.3 Fabrication Optimization of More Than 4000 Single-Layer Panels

One of the most prominent challenges for large-scale engineering projects is the mass generation of offline programs for robotics compared to the current small-scale 3D printing projects. Although robotics can work 24/7, offline programming often takes a lot of time and labor costs. In this project, there are more than 4,000 roof components with different geometries (Fig. 11). The previous manual programming workflow can hardly meet the project deadline. To overcome this challenge and ensure a logical production workflow, robot motion simulation, automated print component pickup and programming are two necessary steps. In FUROBOT, an automatic program is built which can generate all the printing files one by one every 30 s and check collision and singularity simultaneously (Fig. 12). Based on this effective digital-to-fabrication toolkit, all 4000 printing files are successfully prepared in 5 days before production.

Since the entire roof panels are prefabricated in the factory, the cost and carbon footprint in transportation need immediate concern. The project improves the closed profile path of the traditional 3D model of double-layer printed panels with a lightly shaped single-layer printed version for rapid construction. Intuitively, the layer change point of a two-layer printed element is to split the closed curve of each layer and then connect the adjacent curves, while the path of a single-layer component is an open, one-way round-trip curve (Fig. 13). In contrast, the single-layer printing process not only encourages designers to innovate with more complimentary geometry design but also significantly reduces the carbon footprint of the entire building life cycle. At the same time, detailed improvements at the joints allow for fabrication tolerances on the top after the printing is complete. Thus, the single-layer printing process is a more straightforward and efficient production method for mass customization scenarios.

However, additional optimization for print feasibility is still necessary in producing single-layer panels. Typically, each printed panel needs to couple the spatial plane of the printed bottom edge to the fabrication base plane in the factory.

Fig. 11 Over 4,000 roof panels with different geometries

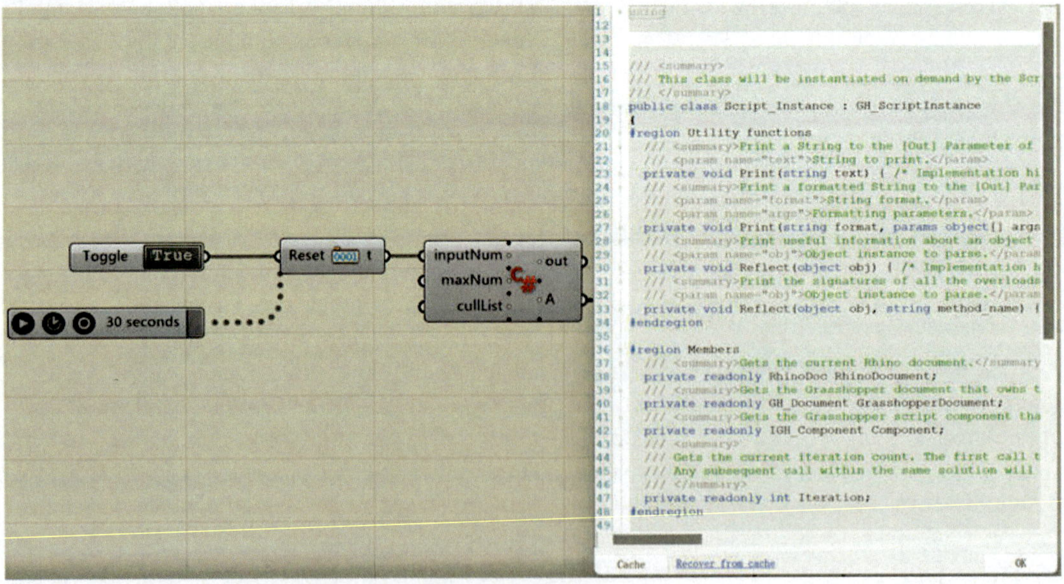

Fig. 12 Automatic printing file generation module

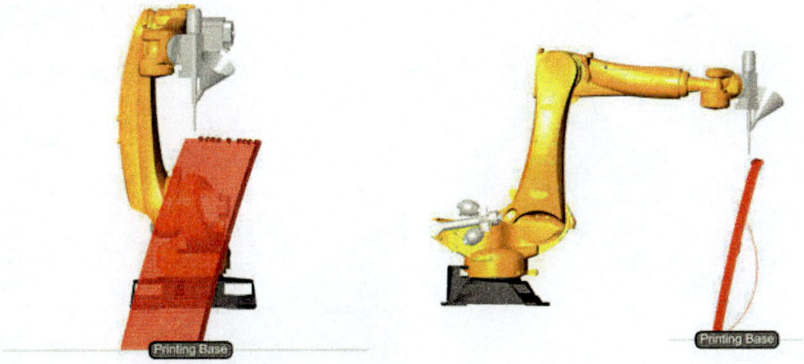

Fig. 13 Printing path of a single-layer panel

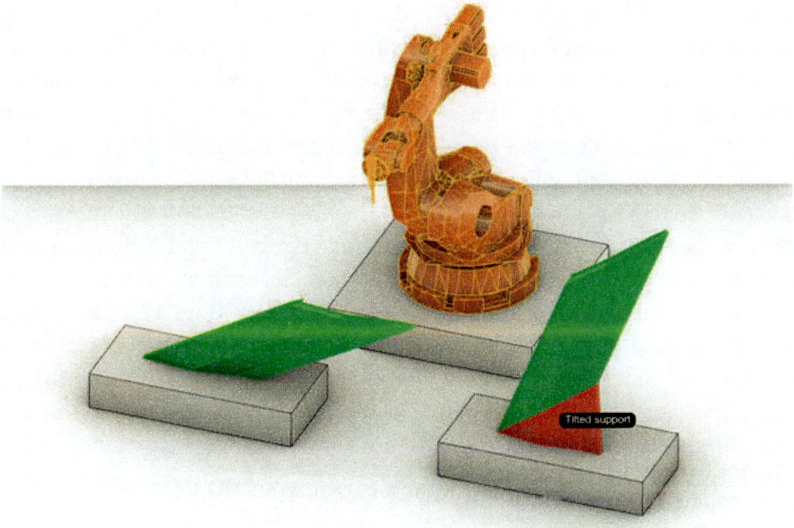

Fig. 14 Additional tilted support

In this project, the curvature of some roof panels is too large, causing the model to be placed on the fabrication base plane over-tilted. Experience has shown that when its angle to the base plane is below 45°, the quality degrades significantly or even fails to print. Therefore, to ensure the passing rate and reduce losses during transportation, a parametric program in Grasshopper is constructed to automatically add tilted supports (Fig. 14), which is used to ensure all the tilt angle of each printed panel is kept below 45°.

4 Construction Practices

4.1 Flexible Construction System of Prefabricated 3D Printed Roof

Although 3D printing can liberate some of the limitations of the design, its varying forms also cause an essential challenge for installation and construction, especially the accurate spatial

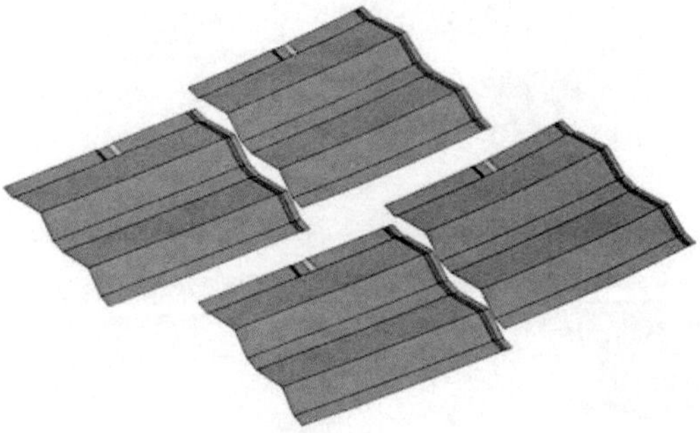

Fig. 15 Slotted lap joints

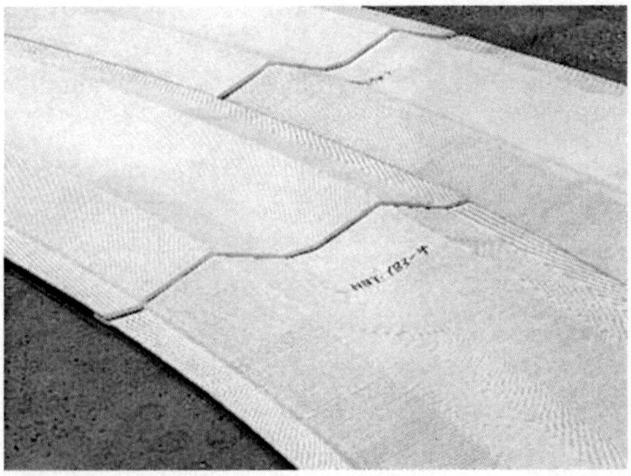

Fig. 16 Tile type installation concept

positioning of each component. In this project, we started to consider the construction process's tolerances from the early design phase. Predictably, suppose the traditional roof construction positioning method is used. In that case, cumulative tolerance will occur during successive installations, which may result in the eventual failure to close the head and tail panels. Thus, a single layer of slotted lap joints is used (Fig. 15). Specifically, one upper and side surface of each panel are partially overlapped with its adjacent panels so that each printed component can be adjusted within 1 cm. Considering the construction logic of single-layer panels, different slotted lap joints were designed and will be installed based on the traditional tile type installation concept (Fig. 16), which can simplify the positioning steps and improve construction efficiency. At the same time, the control parameters of each detail can be adjusted based on different conditions. This parametric design method also facilitates the architect to have more direct control over the project.

Regarding the roof structure, Chengdu Agricultural Expo Center adopts a layer-by-layer tolerance elimination strategy. As shown in the structural details (Fig. 17), the entire roof system consists of the central timber reciprocal structure, oriented strand board (OSB), SPF wood keel, adjustable metal joints, and 3D printed panels to eliminate the tolerance generated at different stages of construction. The installation and

Fig. 17 Roof structural details

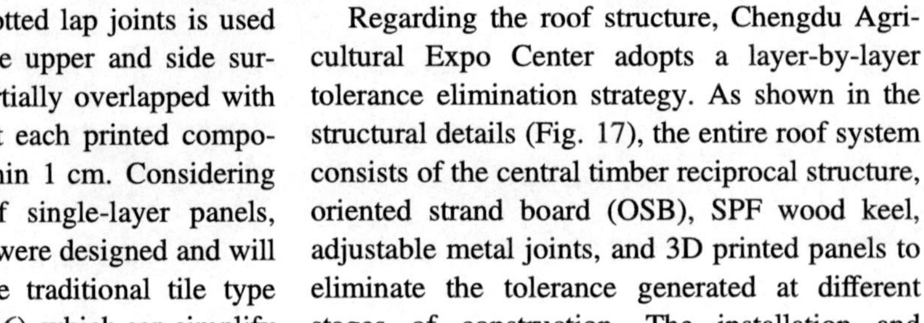

positioning of this flexible system can be divided into three steps. First, after the main timber structure is constructed, a through-length SPF keel is installed based on Real-time kinematic positioning (RTK) of the head and tail endpoints. Next, a set of L-shaped metal connections are fixed to the keel according to the distance from the digital model to accommodate the non-standardized distance between the panel and SPF keel. Each metal connection can eliminate about 8 cm of construction tolerance combined with the width of the SPF keel. Finally, the printed panels are installed in a zoned manner sequentially according to the position of the connections, with each panel secured by six screws at the edges. And it can also be used for mutual positioning reference based on the joint detail. Thanks to this method, the installation of 3D printed panels can be finished more smoothly, ensuring space for adjustment according to the specific site conditions. It effectively avoids the problem of cumulative tolerance to a great extent.

4.2 Panel Division System and Construction Process

Despite many advantages of 3D printing technology in producing complex building components, it still has certain limitations. One of the most important is that the robotic arm span determines the maximum range of parts that can be printed. The 3D printing machine used in this project is KUKA R2700-210, which means the length and height of each panel are limited to 2 m. However, the significant variation of the entire roof form at the widest and narrowest points made the problem unavoidable (Fig. 18). To address this issue, a panel division system is developed to handle panels of all different sizes and scales.

First, to maximize the success and efficiency of printing, the height of 80 percent of the panels was limited to 1.8 m. This step was accomplished by adjusting the range of UV subdivision lines on the surface. Secondly, for the unavoidable oversize panels, we subdivided each printed panel into two pieces and added folded edges at the joints to ensure the stability of the screw fixing. According to the model statistics, a total of 800 boards needed similar treatment. After determining the size of each panel, we divided them into five types according to the installation sequence (Fig. 19): regular panels, ridge panels, slit panels, cornice panels, and collision panels (Fig. 20). These panels differ in their joint configuration due to their different properties. For example, compared to the joints of the regular slab, the ridge slab does not have a groove on its side because it is at the top of the roof. To outline the smooth geometry of the whole roof, the side edges of the cornice panel have an extra folded edge. Slit panels are defined as the last to be installed after remeasurement to ensure that the construction will distribute the error evenly over each area. Collision panels are distributed around skylights and patios, all of which have completely different dimensions and shapes due to their collision with the structure. Each of the four panels above has a joint deepening program and a production component information sheet. In the

Fig. 18 Widest and narrowest part of roof

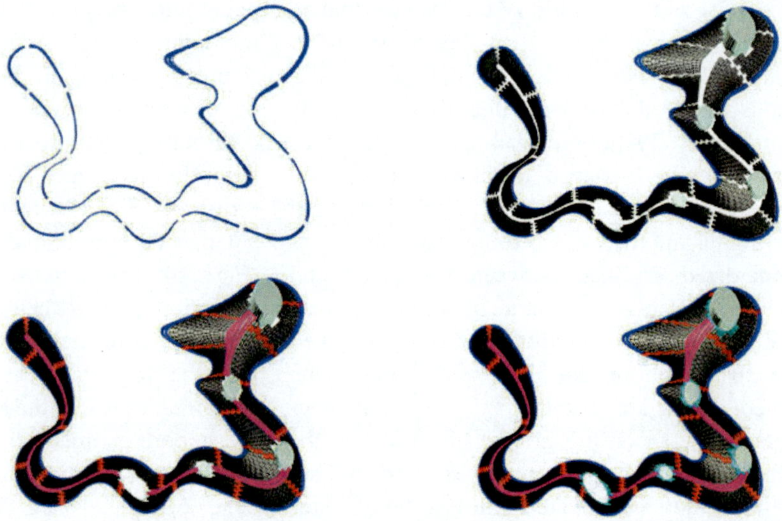

Fig. 19 Construction process

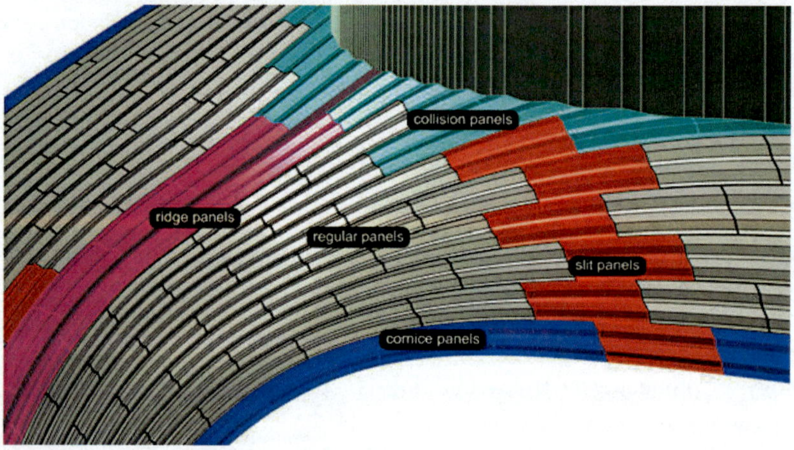

Fig. 20 Different types of panels

end, according to the parametric model, more than 4,000 different panels with unique label form the whole continuous and complex roof system (Fig. 21).

5 Discussion

5.1 The Feasibility of Large-Scale Mass Customization 3D Printing Craft

The complete 3D printing workflow and digital construction technologies presented in the paper have been fully validated in the Chengdu Agricultural Expo Center project. In particular, it has considerable potential to contribute to the sustainable development of prefabricated architecture. The series of digital tools and fabrication platforms utilized in the project have enabled an efficient design-to-fabrication integrated system at the levels of form generation, single-layer panel division, robotic 3D printing, and on-site construction. Thanks to the single-layer panels, the reduced printing time, lower transportation costs, and efficient and fast installation methods also significantly reduce the carbon footprint of the entire building cycle. The project was prefabricated by ten industrial 3D printing robotics in a digital factory, producing a total of over

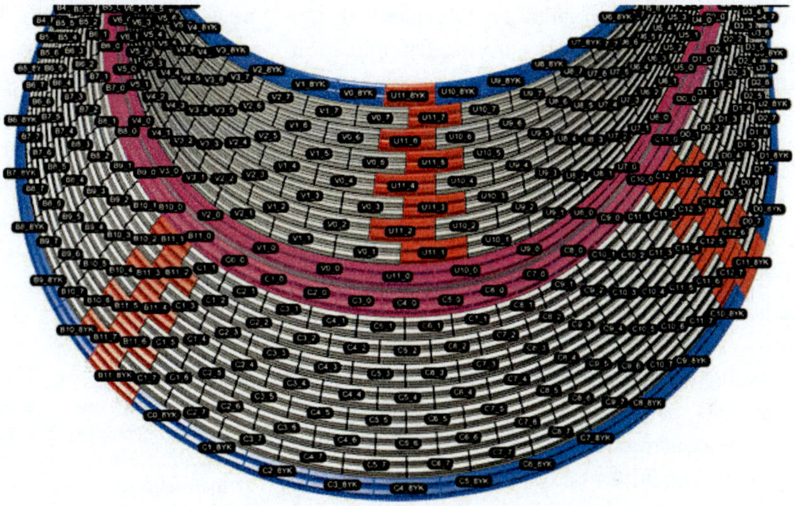

Fig. 21 Roof panels with labels

4,000 printed panels of various shapes and sizes in 3 months.

From a technical point of view, after the basic shell form has been generated in the architectural scheme phase, the roof panels are also generated through parameters based on the actual construction conditions. The model of each roof panel is then imported into FUROBOT to optimize the bottom support structure and special construction requirements. With the FUROBOT core module, the architect can program the 3D printing and independently complete the offline program for the robot control. The offline program is transferred from the person responsible for production to the robot in the digital factory. In the factory, each worker can be responsible for the control and operation of 2 or 3 robotics. Compared to the traditional manual construction method, the robotic arm can continuously work 24 h a day. The result is a process in which the interface between design and construction is intuitive and seamless. Designers can receive feedback on production quality at any time to optimize subsequent designs. This efficient prefabricated approach supports mass customization scenarios and ensures that projects are completed on schedule. Therefore, there is great potential and scope for exploring the feasibility of 3D printing in practical engineering.

5.2 The Limitation

However, there are some limitations to the workflow above. For instance, it asks architects to fully understand the craft and fabrication, which raises the level of the profession of architects. In addition, since FUROBOT has already been integrated into Grasshopper, many manual programming operations are still needed for a specific project, which also requires other skills for traditional architects. This also gives a new direction to the future development of FUROBOT: completing more intuitive and graphical-based toolkits based on different 3D printing crafts. By writing each manual step as a module inside, the design and fabrication efficiency will be significantly improved.

On top of that, the inconsistent human–machine collaboration approach leads to additional material waste during the digital fabrication process. And ways to reuse these defective products are in the process of being developed. According to the specific statistics on the project, about 15% of the panels need to be printed by adding a tilting base to ensure the quality of the print. Also, some errors that could be more intuitive in Rhino viewport, such as the discontinuous tool path, cause additional material waste and reduce the finished product rate. This problem will be remarkably alleviated if the maximum intelligent and automated application can be achieved in the future.

6 Conclusion

In conclusion, this paper demonstrated a holistic large-scale mass customization 3D printing methodology in FUROBOT platform, industrial 6-axis robotics, and fiber-reinforced plastic. According to these software and hardware platforms, this paper continued to introduce a 3D printing construction practice based on the Chengdu Agricultural Expo Center project. In the whole workflow, the basic form was generated and optimized in Rhino and Grasshopper. The initial geometry is calculated based on specific environmental conditions. Then, the roof is separated into different main areas according to the construction situation. UV direction decides the size of each panel. Next, FUROBOT transfers the model information into the robot path for 3D printing. Finally, all the panels with different details are assembled following a specific sequence through a flexible structural system. In this project, the proposed workflow ensures a coherent design-to-fabrication methodology based on 3D printing craft and can contribute to sustainable production practices.

Future research could focus on developing a complete plastic printing toolkit based on all the parameters and data we have already accumulated to prospect more diversity of projects in better quality. Some predictable research directions can be summarized as exploring more reusable and extrudable material, online robot programming modules for 3D printing, and craft toolkits. The development of architectural ontologies based on the technology above has created new possibilities for individuality and diversity. This new production system forms an integrated design and construction process, creating opportunities for architects to become deeply involved in the architectural production process. Nevertheless, it is also a challenge to rethink the materials, environment, and production process of architecture creation, not only to form more personalized expression but also to allow architects to accomplish better design concepts and intentions, such as performance, efficiency, collaboration mode, or green low-carbon goals. This is one of the most significant technological innovations for social development.

References

Beyhan F, Arslan Selçuk S (2018) 3D printing in architecture: one step closer to a sustainable built environment. In: Proceedings of 3rd international sustainable buildings symposium, vol 1(3). Springer International Publishing, pp 253–268

García-Alvarado R, Moroni-Orellana G, Banda-Pérez P (2021) Architectural evaluation of 3D-printed buildings. Buildings 11(6):254

Gebler M, Uiterkamp AJS, Visser C (2014) A global sustainability perspective on 3D printing technologies. Energy Policy 74:158–167

Griffiths A (2013) 3D-printed components for 6 Bevis Marks roof [Photograph]. Dezeen. https://www.dezeen.com/2022/01/10/adrian-priestman-3d-printed-building-components/

Hager I, Golonka A, Putanowicz R (2016) 3D printing of buildings and building components as the future of sustainable construction. Procedia Eng 151:292–299

Hwang D, Khoshnevis B (2005) An innovative construction process-contour crafting (CC). In: Proceedings of the 22nd international symposium on automation and robotics in construction. https://doi.org/10.22260/ISARC2005/0004

Lim S, Buswell R, Le T, Wackrow R, Austin SA, Gibb A, Thorpe T (2011) Development of a viable concrete printing process

Munn S, Soebarto V (2004) The issues of using recycled materials in architecture. In: The 38th international conference of architectural science association ANZAScA "Contexts of architecture", Launceston, Tasmania

Ngo TD, Kashani A, Imbalzano G, Nguyen KT, Hui D (2018) Additive manufacturing (3D printing): a review of materials, methods, applications and challenges. Compos Part B Eng 143:172–196

Philip FY, Liming Z, Tianyi G (2021) The future of robotic fabrication for flexible mass customisation. World Arch (07):36–42+128. https://doi.org/10.16414/j.wa.2021.07.007

Van der Veen AC (2014) The structural feasibility of 3D-printing houses using printable polymers. https://www.dezeen.com/2013/12/02/first-architectural-application-of-3d-printing-adrian-priestman-6-bevis-marks/

Restarting from Renewables

Prototyping Thatched Facades—Global Scaling of Local Knowledge

Henriette Ejstrup and Anne Beim

Abstract

The paper presents the results from a cross-disciplinary research project investigating clay as a fire retardant in vertically thatched facades, and it elaborates on its consequences on aesthetics and buildability by prototyping. The prototypes take point of departure in three different typologies: thatching on construction sites, thatched prefabricated façade elements, and thatched masonry. Initially, a mock-up of a reed-thatched façade sprayed with clay and with clay boards as fire stop every 100 cm was subjected to a Single Burn Item test (SBI test). The result showed a FIGRA at 25 W/s, which is positive. This raised the question on how these findings could form radical tectonic solution as a respond to the need for CO_2 neutral constructions. The prototyping of prefabricated clay-impregnated thatched façade elements seems to have multiple possibilities regarding time optimization, transportation, and (dis) assembly. Whereas thatching and impregnating on the construction site is possible, but slower than standard thatch construction therefore processes must be developed and tested further. An unanswered question is how the reed will respond to humidity, when thatched to a solid base of plywood without ventilation. Furthermore, questions about the ceramic curing of clay when burned with downfall as a result are unanswered.

Keywords

Thatched facades · Clay · Low emission constructions · Radical tectonics · Fire retardant · Cross-disciplinary

1 Introduction

Through centuries reed has been globally known as a building material. In a Danish context, the vernacular architecture was traditionally thatched with reed. This technique was mostly used for roofing, but also facades have been thatched. Reed on facades was used both as an inhibitor of weathering of the load-bearing construction, and as a provisory cladding when other materials were not at hand or extra ventilation was needed (Zangenberg and Ingemann 1982; Kirk 1979). During the twentieth century the use of reed has diminished. This was due to the fear of fire hazards, novel industrial materials, and decrease

H. Ejstrup (✉)
Royal Danish Academy, Institute of Architecture and Technology, Copenhagen, Denmark
e-mail: hejs@kglakademi.dk

A. Beim
Cinark/Institute of Architecture and Technology, The Royal Danish Academy, Copenhagen, Denmark

M. R. Thomsen et al. (eds.), *Design for Rethinking Resources*, Sustainable Development Goals Series,
https://doi.org/10.1007/978-3-031-36554-6_31

of the wetlands where reed was grown (Bak-Andersen 2020). New research points to that reed have multiple benefits. Locally grown reed has a positive impact on biodiversity, as well as it contributes to absorb CO_2 and NO_x (Andersen et al. 2021; Schipull 2013). Globally reed has been used as building material. This makes the material a clear candidate as a mediator between local building traditions, detailing and scaling, and generic technologies, and global infrastructure and building systems. This article takes point of departure in a cross-disciplinary research project that focuses on thatched facades and clay as a fire retardant. The project is funded by the Danish Environmental, Technology, Development and Demonstration Program (MUDP), and is a collaboration between DBI—The Danish Institute of Fire and Security Technology, Strågets Kontor (Joint Secretariate of the Danish Thatchers Guild), and a couple of thatchers, a clay mason, and a group of architectural researchers.

Historic investigations of thatched facades point to a previous tradition of façade thatching in Denmark, although it most likely was a provisional measure. Most known is the "Dutch mills", which originally had thatched facades as the material was economically affordable and within reach (Andersen 2011). In the Danish vernacular architecture, bundles of reed or straw have been fastened with battens to the facade.[1] This was done to protect the limewashed waddle and daub constructions from wind and rain prolonging the material lifespan and reducing the frequency of maintenance. This tradition still exists as a stylized architectural feature in the small costal village of Hesnæs on the Danish island Falster (Christensen 1984). Another vernacular tradition from Falster is thatched gables, where the thatched roof was prolonged to the gables ending in a small eave in the middle of the gable (Realdania 2021).

Only few traces were found in the historic building culture and literature regarding knowledge about reed and clay as a fire retardant. A building permit from 1889 issued in Haderslev, Southern Denmark states: "*The thatched roof ... (must be) sewn with galvanized iron wire, and above all 4... exit doors of the house, the roof must be ... underlined with clay*".[2] Although it is not specifically mentioned to be due to fire hazards, the context makes it plausible that the clay treatment is for fire safety measures. Furthermore, the construction manual, *Byggebogen ("The Building Book")*, from 1949 describes a historic fire retarding technique in roofs from Southern Schleswig, Germany. The technique consists of a mixture of clay and ammonia which was applied to the reed bundles before thatching (Kjærgaard 1949).

On the basis of the historic research, the procedural and anecdotal knowledge of the thatchers, and the fire and building technical knowledge of the engineers and architects, 14 small test mock-ups were constructed. The mock-ups were designed as 90-degree angles in calcium silicate plates measuring 235 × 15 mm at one side, 220 × 20 mm at the other side, and 600 mm in height. The mock-ups were held together by metal brackets. The reed was dipped, sprayed, and painted with different mixtures of limewash, moraine clay, clay adhesive and ammonia, and mounted with metal straps on the inward side of the angles (thickness of max. 50 mm). After drying, the weight of the mock-ups was noted, before they were subjected to a Mini SBI test (Single Burn Item test) (Figs. 1 and 2).

The best performing adhesive (sprayed moraine clay) was applied to two full-scale SBI tests. The mock-ups were designed as a façade section of 1200 mm width × 2400 mm height. The thatched layer of 220–250 mm reed was fastened

[1] As an example, in the 1920s the Danish architect and museum director, Halvor Zangenberg, documented vernacular farmhouses in photos and descriptions. His works contain multiple pictures of buildings with reed cladded facades. Danish National Museum, https://samlinger.natmus.dk/.

[2] From a building permit from the address, Ørby 20, retrieved from the municipal archive of Haderslev with thanks to architects Nik and Sigmundur Hyllestad.

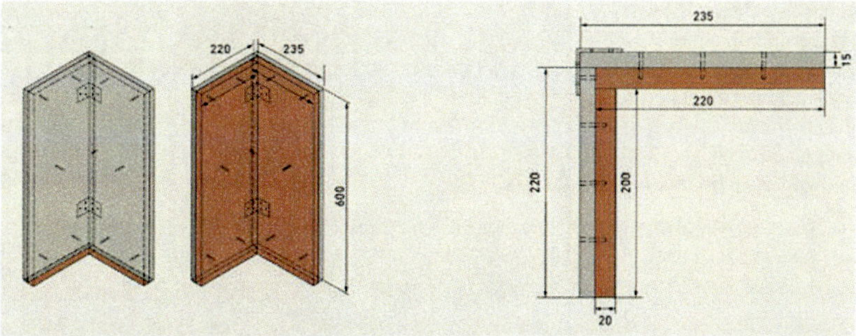

Fig. 1 Sketch of the angel model for the initial test burns (mini SBI)

Fig. 2 Thatching the mock-up

with metal screws and wires on a fireproof MDF board mounted on a wooden frame raised 400 mm above ground level. The mock-ups were closed on the sides and at the bottom with MDF boards. Test #1 (sprayed moraine clay) was incomplete as the prototype was not mounted properly onto the rig in the test chamber. The failure occurred, as the backside of the rig was not covered properly according to regular standards hence allowing a chimney effect on the backside of the mock-up. After 22 min the test was cancelled for safety reasons. Although the test was incomplete the charring and pockets of unburned clay inside the reed construction suggested that the clay had a positive fire preventing effect. The FIGRA of the incomplete test was measured at 140 W/s. Mock-up #2 (sprayed moraine clay and clay boards as fire stop) was mounted correctly onto the rig in the test chamber and was successfully burned showing a FIGRA-value of 25 W/s (Fig. 3).

Fig. 3 The three SBI test mock-ups. Left.: unimpregnated reed. Middle, mock-up 1: impregnated reed. Right, mock-up 2: impregnated reed and fire stops with clay boards in recesses

2 Prototyping at the Construction Site

The design of the mock-ups showed the potentials of different architectural expressions, building techniques, and building processes. To elaborate on the consequences of the reductionistic results, it was decided to do a full-scale construction, where problems and possibilities of the buildability and aesthetics could be investigated further. A prototype of a building corner with the dimensions of an average one-storey building was constructed. The corner represented multiple critical joints regarding the transition between the thatched layer and the load-bearing construction, as well as the scale made it relatable to the standard housing recognizable to the general public. The prototype only implied the roof construction and did not include the foundation of the construction.

The first building corner prototype was constructed for the 70% less CO_2 exhibition at The Royal Danish Academy in the fall/winter of 2021/2022. For the load-bearing structure, modular elements of wooden frames filled with compressed straws (Ecococon) were used. The walls were plastered with clay sprayed upon a hemp mesh. On the external side, the elements were clad with wood fiber boards (Agapan), which was intended to function both as wind breaker and the underlying construction layer for the thatch. The thatching was mounted horizontally in sections of 100 cm divided by a fire stopper (clay board). Between each layer of the reed it was sprayed with moraine clay.

The prototype pointed to several problems of the clay impregnation technique. The most grave was the interruption of the usual "workflow" of the thatcher as the clay mason had to intervein to apply the clay. An alternative workflow occurred, as the compressor that gave air to the mortar sprayer used for applying the clay was underdimensioned (ca. 24 l./1,500 W). This gave natural breaks when the compressor had to recharge. These breaks were utilized by the thatchers to mount the next section. The assessment of the process was that it came with a high labor cost, as the waiting time for either the thatcher or the mason made the process slow compared to standardized building methods. On the other hand, time was saved since the standard glass fiber fabric did not have to be mounted.

When the exhibition closed in March 2022, a reduced part of the prototype was placed outside to be exposed to weathering and human tear. In June 2022, it was transported by truck to Aalborg to an open-air exhibition facilitated by the 5th International Association of Structures and Architecture (ICSA). The prototype was deconstructed in mid-July. Visual inspections during the period from March to July showed that the clay did not wash out and tear as expected. Initially, it was perceived that the air-dried clay would wash off or crumble within weeks, when exposed to rain and human tear. When the prototype was deconstructed, the clay was still sticking on to both facades. The recesses and eaves had acted as a constructional protection and underneath eaves the clay visually seemed unaffected by weathering. This indicates that constructional protection like deeper eaves or recesses can hinder the weathering of the clay impregnation significantly. In addition, a hypothesis is that the mortar sprayer is central to the fastening of the clay, as it is driving the clay into the reed buts thereby securing a deeper binding.

The disassembly of the prototype was difficult, as there was no structured system of re-harvesting of the reed in place. Furthermore, it had to be done in layers, from the outside in, which made sectionized deconstruction difficult. This generated a great amount of material waste that could have been reused directly as thatch material. Instead, it was down-cycled to animal bedding and waste. The Ecococon elements were held together by wooden frames and screws, and they were easily disassembled and harvested for reuse. Metal scrap like screws, straps, and wires were collected and recycled as metal waste (Fig. 4).

Fig. 4 The first thatched construction at the exhibition 70% less CO_2 exhibition

3 Prefabricated Façade Elements

A second version of the project was accepted for the Architecture Triennale in Lisbon in September 2022, where new ideas for sustainable construction methods were called for. Due to transportation and costs it was important that the elements could be flat-packed and assembled on-site with only handheld tools. Furthermore, the economy and tools provided by the institution of the exhibition pointed to a manual approach for assembling the prototype. Therefore lifting, mounting, etc. of the elements should be manageable by two persons with handheld tools. Again, it was decided to use Ecococon elements for the load-bearing structure, as they could be flat-packed together with Agapan boards. The thatched façade was transformed into a modular cladding system. This consisted of cassettes made by the wooden boards, in which the thatching was mounted. Altogether 14 cassettes were constructed assembled by a thatcher and an assistant in 4 weeks in a workshop. The cassettes were mounted on the load-bearing construction by angle cut battens. The thatched layer was mounted with the same workflow as in the first prototype with clay boards as fire stops every 100 cm, and sprayed moraine clay on each thatched section. On each side of the cassette, the reed was angled outward to cover up the joints between the cassettes. This was both to suggest a thesis on how water could be handled technically in the construction in relation to the joints between the thatched prefabricated cassettes, and to gain an aesthetically unison expression. After

Fig. 5 The second thatched construction with prefabricated cassettes mounted on the Ecococon element

testing the system on the load-bearing structure in the workshop, the prototype was disassembled and flat-packed for transportation to Lisbon.

Although it was technically difficult it was possible to thatch within the confined space of the cassettes. The assembling on-site was also manageable for two persons, and the clay impregnation did not seem to suffer any particular damage during the transportation. One of the unsolved questions is, how the joining details of the prefabricated element will perform in open air, as the prototype only is exhibited in an indoor area. Furthermore, questions concerning the reed and how it responds to humidity both within the confined space of the cassettes and with the treatment of the clay remains unanswered. The construction has yet to be disassembled, but as it is mainly fastened by screws, we believe it is more successful and easier to harvest the construction materials than its predecessor (Fig. 5).

4 Sketching Thatched Masonry

With reference to the building technologies related to thatch from The Netherlands, an architectural proposal was sent to the Virserum Biennale in Sweden in April 2022. The paper project suggested a thatched and clay-impregnated isolating masonry construction (Porotherm clay blocks) for a one-storey building (single-family house). The thatching is fixed directly onto the masonry wall with no ventilated cavity.

In collaboration with RØNNOW, Leth and Gori Architects, the construction concept has been accepted for further development as part of the Danish competition: "From Four to One Planet" launched by the funding bodies of Realdania and the Villum Foundation in the autumn of 2022. The overall idea is to propose a four-storey apartment building in homogenous insulating block masonry with thatched facades. The project is still under development, so no real-time constructional experience has yet been harvested.

The expectation is that the experiment with the prefabricated cassettes can be further developed in the "From Four to One Planet" project. Again, the question about humidity is raised in regard of the overall function of the thatched layer on a construction type that by proxy maybe more relevant to cover with plaster or brick tiles. Nevertheless, the pavilion will be exhibited in open air during the summer of 2023 enabling the team to make further observations on the deterioration of the clay and the moisture in the reed.

5 Conclusions

The practice-oriented research collaboration between craftsmen, fire engineers, and architectural researchers has been central to the development of the three construction prototypes. Although the results of the SBI tests were confined to the fire-retardant effect of clay on reed, the elaboration of the results into full-scale building constructions has informed discussions regarding future scenarios. If this method can be translated into plausible (accepted) building techniques and how it can be handled or processed at today's construction sites. Furthermore, the prototyping pointed to how elements could be prefabricated in workshops manually or even further developed to a semi-automated or fully automated industrial level.

Questions about how mechanical tear and weathering of the clay and reed have been raised as a discussion in regard of all the prototypes. Secondly, questions about how the reed would react in regard of the hygroscopic properties of the clay over time. Another sub-question was how the overall biogenic construction would perform if exposed to fire? In this context it is mentionable, that in Denmark anecdotal stories exist suggesting that reed thatched directly on wooden boards without a ventilated cavity does not burn easily and has no increased tendency for rot. The Danish stories are backed by experiences from the Netherlands. Here fire prevention in new constructions is based on the method of thatching directly on a solid layer of plywood without any ventilated cavity (Janse 2010). Generally, Dutch literature does not mention any increased damage of the reed due to humidity, although this is contested in a Danish context by concerns about moisture in reed, if not ventilated on the backside. This points to that further documentation based on experimental testing is needed. Furthermore, a question on how the overall construction of the load-bearing structure with the thatched clay-impregnated façade would perform in a fire test is unanswered as well as it suggests another practice of assessing, as todays tests are done by the layer and not by the overall construction.

On basis of the protype constructions it is assessed to be possible to do the thatching and clay impregnation on a regular construction site although the processes are slowed down due to the layering of the materials in the construction. At its present state, it is only relevant for demonstration projects or financially exclusive construction project. The overall assessment is that the building techniques and methods can be further developed and trained thereby improving the buildability. However, thatching and impregnation of prefabricated cassettes is assessed as being more approachable for a broader market, as it can be produced off-site, easily transported and mounted at the construction site thereby fitting into existing efficient processes in the construction industry.

6 Perspectives

The SBI fire test prototypes have generated a series of full-scale construction prototypes that takes craftmanship and modern processes of fabrication into consideration. The cross-disciplinary approach had the outcome that the scientific results promptly could be discussed and assessed both from a technological perspective as well as a cultural point of view. This has resulted in various prototypes and full-scale constructions, which tests the theses about buildability and aesthetical expressions. It would be beneficial as part of further development, if the different construction prototypes could be submitted to an SBI fire test thereby instigating a second iteration of the prototypes. Furthermore, an investigation into different placements of the clay impregnation and its effect might again have an influence on the buildability and expression.

If clay as impregnation for reed can be a secure alternative to traditional fire retardants (glass fiber fabric) a new realm of biogenic construction could emerge. This path will require more research and probably result in a more radical tectonic approach than offered by modern constructions, also new on-site processes have to be developed. Furthermore, cross-disciplinary work in the early stages of research is essential to act both fast, with flexibility, and with as much relevant knowledge up front to secure the most informed and sustainable development.

Upscaling and variations of the prototypes have pointed to other questions present in the processes of the construction industry. As an example, the production of prefabricated elements could be inspired by established industries such as assembly-line fabrication. Although the production probably could be automated, a semi-automated version where cassettes are mechanically made but thatched by hand have other potentials. Fx. the production could also be decentralized where thatchers could bid into central tendered tasks (e.g., elements conveyed by drawings from central database). This would have the benefits when the weather does not allow to work on site, the thatcher could produce prefabricated elements in their shop thereby having the full benefit of their working hours.

In relation to this, a more widespread production and use of reed could benefit one or more perspectives of the 17 SDGs. Promoting a biogenic regenerative material as reed and making it an applicable building material by solving the problems of fire hazard would have several benefits in relation to SDG 6, 14, and 15. The production of reed could have a positive effect on local biodiversity and support reestablishment of wetlands. It could also be an agent for conversion of farmlands with low-lying fields and a high risk of flooding unfit for traditional crops for consumption. Instead cultivating reed would fit the circumstances of these types of fields. As reed also have a positive impact on the aquatic environment production of reed could also be beneficial to local water production. Furthermore, as reed and clay are known building materials across the globe, the technological innovations done in this project have the possibility to be globally distributed in a generic manner, as well as the local building culture can adapt the aesthetics to their specific context. Generally, reed has a low degree of processing before it can be used as a building material, hence a highly industrialized fabrication system is not needed to implement the technology. Also, the production could be subjected to an industrialized fabrication. This makes the methods of fabrication scalable to a great diversity of contexts and scenarios enabling local manufacturer to secure social, economic, environmental, and cultural aspects in the local production (SDG 11 and 12).

Acknowledgements This project is funded by the Environmental Technology Development and Demonstration Program (MUDP) under the Danish Eco-innovation Program administered by the Ministry of Environment of Denmark. It is a cross-disciplinary project involving The Danish Institute of Fire and Security Technology (DBI), Straatagets Kontor by Jørgen Kaarup og Sven Jon Jonsen, Tækkemanden Horneby by Thomas Gerner, Hemmed Tækkefirma by Ruud Conijn, Egen Vinding & Datter by Lasse Koefoed Nielsen and Martins Ozols, and Center for Industrialised Architecture (CINARK).

References

Andersen LH, Nummi P, Rafn J, Frederiksen CMS, Kristjansen MP, Lauridsen TL, Trøjelsgaard K, Pertoldi C, Bruhn D, Bahrndorff S (2021) Can reed harvest be used as a management strategy for improving invertebrate biomass and diversity? J Environ Manage 300(December):113637. https://doi.org/10.1016/j.jenvman.2021.113637

Andersen L (2011) Træk af dansk møllebyggeris historie. Nordjyllands Historiske Museum, Aalborg

Bak-Andersen S (2020) Gammel viden til nye bygninger: Traditionelle byggematerialer og håndværksteknik i nutidigt byggeri. KADK, København

Christensen TP (1984) Husene i Hesnæs. In: Hartmann S, Pedersen EK, Kronborg S (eds). Stubbekøbing

Janse EW (2010) Brandveilige Rieten Daken: Detaillering Gelijkwaardige Oplossing. SBR, Rotterdam

Kirk F (1979) Tre primitive sjællandske gavle. In: Arkitekturstudier tilegnede Hans Henrik Engqvist, 160–63. Arkitektens Forlag, Kbh

Kjærgaard P (ed) (1949) Byggebogen 348.91 Stråtag—Marts 1949. København: Nyt Nordisk Forlag. http://www.danskbyggeskik.dk/pdf/get.action?pdf.id=106

Realdania (2021) Restaureringen Af Stines Hus. Issuu. https://issuu.com/realdaniaby/docs/stines_hus_web/s/11110689

Schipull J (2013) LCA Tanghus Læsø

thatchers.eu (2022) Dutch Federation of Thatchers. International Thatching Society, ITS'. https://thatchers.eu/content/holland/

Zangenberg H, Ingemann H (1982) Danske Bøndergaarde: Grundplaner Og Konstruktioner: Foredrag Holdt Ved Det 3. Nordiske Folkelivs- Og Folkemindeforskermøde i København 1924. 2. udg. Foreningen Danmarks folkeminder, København

InterTwig—Willow and Earth Composites for Digital Circular Construction

Erik Zanetti, Eszter Olah, Tamara Haußer, Gianluca Casalnuovo, Riccardo La Magna, and Moritz Dörstelmann

Abstract

The construction sector has high resource demands and generates a significant amount of waste, a consequence of its linear approach. A shift towards renewable and local material sources and the implementation of closed material cycles represent a significant opportunity for the construction industry to curtail the depletion of raw materials. To address these challenges, this paper presents a strategy for a novel circular construction method that combines willow, a rapidly renewable material, with earth and is enabled by digital fabrication, which can sustain their industrialisation through tailored processes. Emerging from a materiality perspective, the research revisited vernacular building techniques that used plant- and earth-based composites, exemplified by the vernacular wattle and daub, to understand how these can be enhanced through digital design and digital fabrication. Willow (Salix) is a woody plant native to Europe whose stems can be harvested yearly, thanks to specific forestry practices, namely short rotation coppice, that allow the plant to regenerate in rapid cycles. To use willow for construction, geometry and textile techniques were implemented to create stable structures. In combination with earth, a finite but abundant and infinitely recyclable material, it creates a sustainable and circular composite that exploits the structural characteristics of each constituent material. Digital design methods enabled the exploration of different geometrical variations and ensured an increased degree of control over their complexity at different scales. The research results were tested in a full-scale prototype, demonstrating the principles of the envisioned construction systems.

1 Introduction

The construction sector has high resource demands and generates a significant amount of waste, a consequence of its linear model, in which resources are extracted, used and lastly disposed of as waste (Çetin et al. 2021). It is responsible for more than a third of global resource consumption (Klep 2015) and accounts for about 50% of all extracted material (European Commission 2020). In the EU, it accounts for over 35% of all waste generated (Eurostat 2022). A shift towards alternative, local means of sourcing and use of materials, as well as implementing concepts for assembly and disassembly,

E. Zanetti (✉) · E. Olah · M. Dörstelmann
Digital Design and Fabrication (DDF), Karlsruhe Institute of Technology, Karlsruhe, Germany
e-mail: erik.zanetti@kit.edu

T. Haußer · G. Casalnuovo · R. La Magna
Design of Structures (dos), Karlsruhe Institute of Technology, Karlsruhe, Germany

M. R. Thomsen et al. (eds.), *Design for Rethinking Resources*, Sustainable Development Goals Series,
https://doi.org/10.1007/978-3-031-36554-6_32

opens new possibilities for the construction industry to reduce the depletion of raw materials and implement circular material cycles.

Material flows in a circular economy can be divided into biological and technical cycles (Ellen MacArthur Foundation 2022). Technical cycles focus on extending the life of secondary raw materials through reuse or recycling. Biological cycles rely on renewable resources or biodegradable materials that are used, regenerated and finally returned to the biosphere. Considering the European context, wood is regaining importance and remains the only significant renewable building material in the industry for structural applications. However, within the repertoire of historic building techniques, other construction methods that use renewable materials can be found. One such construction method is exemplified by the European vernacular wattle and daub, in which renewable materials are combined with earth-based materials into a sustainable composite. Such vernacular techniques and materials, based on old craftsmanship and know-how, are of limited applicability for mass production and have been sidelined by the industrial revolution over the years.

To create novel circular material cycles, the material shift should therefore be supported by devising new fabrication concepts, tools and technologies. Digital fabrication, in particular, can enable custom solutions for novel material processes, providing in this way a plausible industrialised serial construction. With the help of digital fabrication, renewable materials other than wood (such as bamboo, willow, reed and flax) are being revised through new processing techniques. The approaches are disparate, ranging from post-processing the raw materials for use in additive manufacturing (Dawod et al. 2019) or filament winding (Gil Pérez et al. 2022), to using the material with minimal processing thanks to demountable joints (Bouza and Asut 2020; ETH Zurich 2022).

Combinations of renewable materials with other materials allow for the creation of composites that exploit the specific properties of each material. Such investigations combine, for example, willow and mycelium (Özdemir et al. 2022), wood and concrete (Dias et al. 2018), bamboo and earth (MAS Dfab ETH 2022) or wood and earth (Trummer et al. 2022).

This paper aims to investigate a novel circular construction method that combines willow, a rapidly renewable material, and earth. In particular, it aims to understand how this building composite, already present in the vernacular language, can be enabled and enhanced through digital design and digital fabrication. Starting from a materiality perspective, the research revisited plant- and earth-based composites by exploring novel building techniques through hand-made prototypes at increasing scales accompanied by the development of their respective digital fabrication concepts and custom machinery. This ultimately led to the fabrication of a full-scale demonstrator made of the combination of willow and earth.

2 Materials and Methods

To explore novel circular material systems, the research started by examining the properties of the materials. The objective was to assess their suitability for construction applications by reviewing relevant literature. This exploration aimed to gain insights into the materials' potential for circularity, their current applications, and their availability. Concurrently, the research iteratively developed initial concepts for utilising these materials, primarily focusing on architectural applications, structural characteristics, circularity, and digital fabrication prospects.

One selected concept was further developed as a construction system. Its fabrication system and details were prototyped and tested in different configurations, using both hand-made and digital techniques. Eventually, a full-scale demonstrator served to test the correlations between the different parts of the design, fabrication and assembly process.

This progression involved incrementally scaling up the research prototypes and employing a design-through-making methodology. Additionally, specialised subtopics such as material characterisation and the advancement of the digital fabrication system were explored, incorporating feedback loops with the primary line of investigation.

2.1 Material Selection

2.1.1 Willow

Plant-based materials have a high potential for introducing sustainable and circular principles into construction and for rethinking building materials as non-dependent on finite resources. In particular renewable materials, defined as "materials that are continually replenished at a rate equal to or greater than the rate of depletion" (Ellen MacArthur Foundation 2019), and rapidly renewable materials, defined by USGBS as "made from agricultural products that are typically harvested within a 10-year or shorter cycle" (USGB 2022), offer the potential to introduce circular systems that are based on biological cycles, returning the materials to the biosphere at the end of their lifecycle.

Specific forestry practices for rapidly renewable materials, namely short rotation forestry (SRF) and short rotation coppice (SRC), are currently used for energy production and can enable large-scale harvesting. Notably, SRC involves shorter cycles in which the plant is cut back to the stump and regenerates year after year (Verwijst et al. 2013). Although SRC is still marginal in Germany and most European countries compared to land used for cropland and forests (Faasch and Patenaude 2012), the area has been increasing in recent years and could replace marginal agricultural land with minimal effects on overall food and feed production (Aust et al. 2013).

Willow (Salix), in particular, can be coppiced every year and the plant remains productive for several decades (Fig. 1). Although willow species are present worldwide, their main natural distribution is in the northern hemisphere (Verwijst et al. 2013). When they are not used as energy crops, willow is sold in bundles after being left to dry for a year (Fig. 2).

Thanks to the combination of fast growth cycles, regional availability and the existence of forestry practices for its commercial availability, willow could become a local alternative to bamboo in Europe, albeit showing vastly different structural properties. For willow to be considered a building material, structural and consistent stiffness was introduced by exploiting geometrical and textile principles, such as weaving and braiding (Fig. 3).

These techniques utilise the bendability of the willow stems to form three-dimensional woven or

Fig. 1 Willow (*Salix*) plants during harvesting

Fig. 2 Bundles of willow stems after a 1-year growth cycle and a 1-year drying period

a b

Fig. 3 Exploratory prototyping implementing braiding (**a**) and weaving (**b**) techniques

braided structures that have a higher structural stiffness than the single stems, while still retaining tensile capacity. This study tested different types of commercially available willow to determine the best balance between flexibility, thickness, area of cultivation and growth cycle (Table 1).

Among the initial concepts (Fig. 4), additive willow weaving (Fig. 4b) was chosen as the

Table 1 Different commercially available willow types were tested and evaluated according to their flexibility, area of cultivation and length after a 1-year cycle

Plant	Length of stems (1-year growth cycle) (cm)	Average diameter of stems (mm)	Closest area of cultivation (commercially available)	Availability in large quantities	Flexibility after 14 days of soaking
Salix alba	250	9	Spain	+++	+
Salix purpurea uralensis	240	8	Germany (Rheinland-Pfalz, RP)	+++	++
Salix fragilis	220	9	Germany (RP)	+	+++
Salix purpurea	260	8	Germany (RP)	++	+++

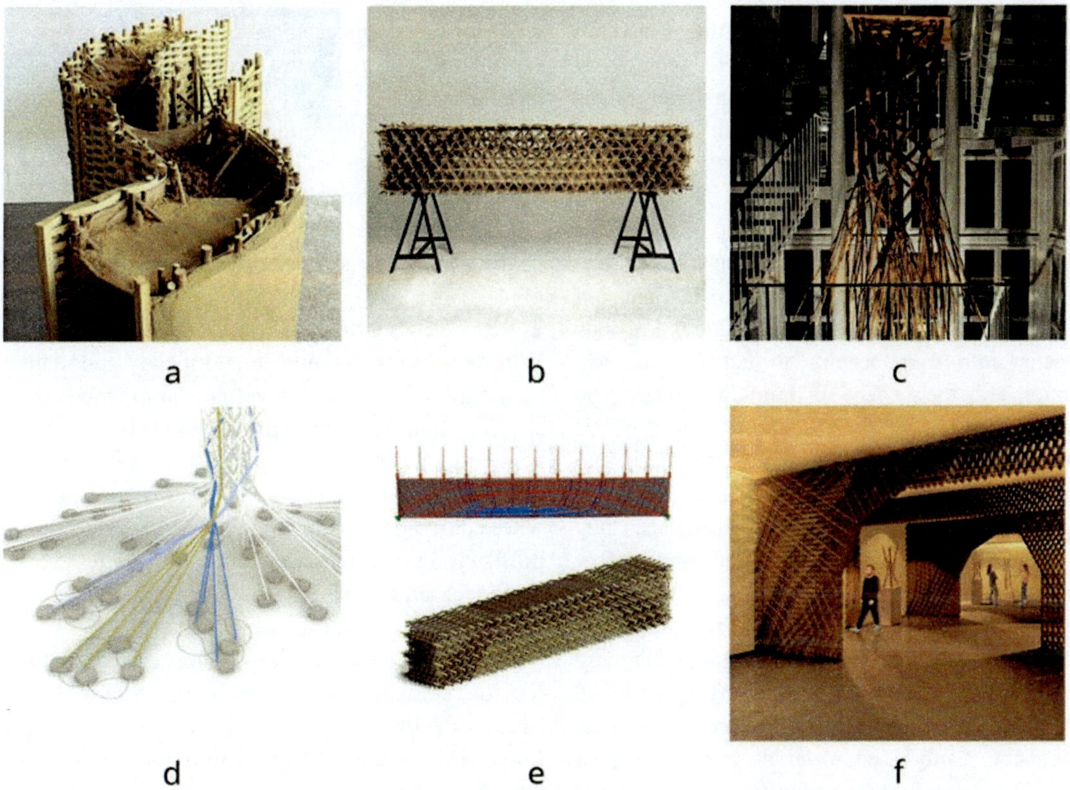

Fig. 4 **a–c** Initial 1:1 prototypes used to assess the initial concepts for **d** digital fabrication, **e** structural optimisation potential and **f** envisioned architectural application

Table 2 Earth techniques for construction

Technique	Processing state[1]	Processing intensity	Load-bearing
Mortar and bricks	Plastic	Low	Yes
Cob	Plastic	Low	Yes
Rammed earth	Optimum	High	Yes
Earth bag	Plastic	Low	Yes
Compressed earth blocks (CEB)	Optimum	Medium	Yes
Adobe	Plastic	Low	Yes
Wattle and daub	Plastic	Low	No
Light earth	Liquid	Low	No
Plaster	Plastic	Low	No
Deep-soil mixing	Plastic	Medium	Yes
Casting in willow formwork (InterTwig)	Plastic	Low	Yes

method with the most potential for further research thanks to its simple fabrication process, geometrical freedom for different construction applications and potential for localised earth infill.

2.1.2 Earth

Earth-based materials have been used in vernacular architecture both as individual construction elements (e.g. monolithic walls) and in combination with other materials (e.g. wattle and daub). Previous research into their benefits in terms of climate control, and their effect on humidity and energy efficiency (Fabbri et al. 2022) has renewed efforts for earth-based materials to be re-introduced into contemporary architecture. Offering an alternative to currently predominant aggregate materials, they require far less embodied energy during manufacturing than traditional building materials (Bruno et al. 2020). Earth-based materials have also been found to have a significantly lower global warming potential than traditional concrete (Hegger et al. 2007). Depending on the application and the pre-processing, earth-based materials can be crushed and mixed again for new applications or returned as soil without loss of value or generating waste (Hegger et al. 2007). In particular, if earth-based materials are used without stabilisers (such as cement)—which improve their properties but make the final material system irreversible (Zami and Lee 2010)—the clay binding allows a complete and low-energy recycling of earth (Linden et al. 2019) with no loss of mechanical strength. In this way, earth-based materials can be considered infinitely recyclable.

A selection of different earth construction methods is shown in Table 2, highlighting their differences in terms of the amount of processing required, whether the final structure is load-bearing and their state during construction.

2.1.3 Earth and Willow Composite

Willow as a fast-renewable and locally abundant material has been used in vernacular practices for several centuries. Examples range from basket weaving, where willow is used in a three-dimensional arrangement, to wattle and daub, where a two-dimensional woven lattice is used as infill within a timber frame structure. However, these techniques are generally small-scale and do not create load-bearing components.

Using woven willow as tension reinforcement for the compression-bearing earth components can create three-dimensional building elements for a self-contained construction system. Additionally, implementing the woven willow structures shown in Sect. 2.1.1 as an integrated permanent formwork for earth means that the

[1] Optimum refers to the optimum water content state where the soil reaches maximum dry density. Plastic state indicates an excess of water above this level. Liquid state is from the point when the earth begins to crumble when rolled into a specific cylindrical shape.

Table 3 Additives for earth (based on Vyncke et al. 2018; Cruz 2013 and own research)

Additive	Function	Relative cost	Circularity
Straw	Tension reinforcement, lighter weight	€	Yes, biodegradable
Sawdust	Shrinking improvement, lighter weight	€	Yes, biodegradable
Seagrass	Tension reinforcement, lighter weight	€€	Yes, biodegradable
Sand	Increased compression strength	€€	Yes, reusable after separation
Stones	Increased compression strength	€€	Yes, reusable after separation
Pozzolans	Increased compression strength	€€	No, irreversible chemical stabilising process
Biopolymer	Increased durability	€€€	Yes, recyclable
Cement	Increased durability	€€	No, irreversible chemical stabilising process
Lime	Increased durability	€	No, irreversible chemical stabilising process
Soda	Increased durability	€	No, irreversible chemical stabilising process
Sodium hydroxide	Increased durability	€€€	No, irreversible chemical stabilising process

placement of earth could be optimised thus saving material and energy, contributing to the overall sustainability of the composite.

Different techniques and material mixtures were explored to find those that could bond and be poured into the woven willow to provide structural stability, through a series of experiments informed by prior research.

Initial findings showed that clay was an essential part of the mixture as it introduced cohesion by acting as a binding agent between the willow formwork and the earth mixture. However, too much clay causes cracks to appear when drying as it increases the amount of shrinkage. Since the mechanical behaviour of earth depends not just on cohesion but also on friction due to the granular additives (Morel et al. 2021), sand or other aggregates were essential. While the addition of sand improved the shrinking behaviour, it decreased the plasticity of the mixture (Fabbri et al. 2022) thus making it more difficult to fill in the cavities. When choosing the additives, it was important to evaluate their circularity—avoiding using cement and similar additives—with a preference for locally sourced and minimally processed materials. Table 3 lists additives, in terms of their function, relative cost and circularity.

The final mixture was created by using commercial earth products, with known standards. These were tested in small prototypes in combination with different additives (Fig. 5). The two products, known as "plaster underlay" (Conluto 2022) and "building loam" (Claytec 2022), were mixed in a 10:1 ratio. The former contains a large part of sand and aggregates—for strength and to avoid shrinking—while the latter contains significant amounts of clay, which contributes to plasticity and cohesion. This mixture was then combined with straw fibres of 1–2 cm to further avoid shrinking (Fabbri et al. 2022).

The result was a mixture used in a cob-like technique in which the additive content and water content could be varied to achieve the

Fig. 5 Small prototypes combining commercially available earth products with different additives

optimum mixture properties. This technique is compared to other earth techniques used in construction in Table 2.

2.2 Fabrication and Geometrical Strategy

2.2.1 Additive Willow Fabrication

Material Preparation Before the willow can be used for weaving, it must be temporarily softened to enhance its flexibility by soaking in water or by steaming. The latter method is faster, taking six hours to reach the required flexibility for the foreseen application, instead of two weeks. Soaking offers, however, the potential for a less energy-intensive process, whose timing could be matched with the schedule of a fabrication line.

Setup for Fabrication The setup for fabrication comprised a bed of wood poles in which willow is arranged in an additive process to create a three-dimensional structure (Fig. 6). In this first exploration, a base grid of equilateral triangles was chosen in order to make use of their resistance to deformation. The distance between the vertical elements followed the need to minimise free-spanning willow stems, which act as unreinforced elements, while maximising the spacing for reachability during the weaving process.

Weaving Patterns The weaving process followed a sequence of patterns, which transformed the otherwise thin and bendable material into load-bearing configurations. Different approaches were explored, aided by creating various weaving sequences with digital techniques, which were then tested in prototypes (Fig. 7). Two main approaches emerged and were categorised into linear and closed patterns (Fig. 6). The final configuration consisted of a combination of the two, exploiting their individual strengths: linear patterns distribute the stress throughout the component, while closed patterns, which follow the triangular base grid, stabilise the component locally. Different patterns were alternated to create sequences that created cell-like configurations.

Several parameters were observed to influence the resulting strength of the prototypes (Fig. 8). The tightness of the weaving around the vertical elements was important as this increased the friction and the pattern would be closer to the shortest path, thus creating components more resistant to deformation. The prototypes also tended to delaminate under shear stress as no chemical binders were used between the layers, in order to maintain the circular potential of the final product. This was counteracted by adding strings and self-locking washers (Fig. 9).

2.2.2 Earth Casting

Earth Distribution The willow formwork consisted of distinct cells, bounded by layers of

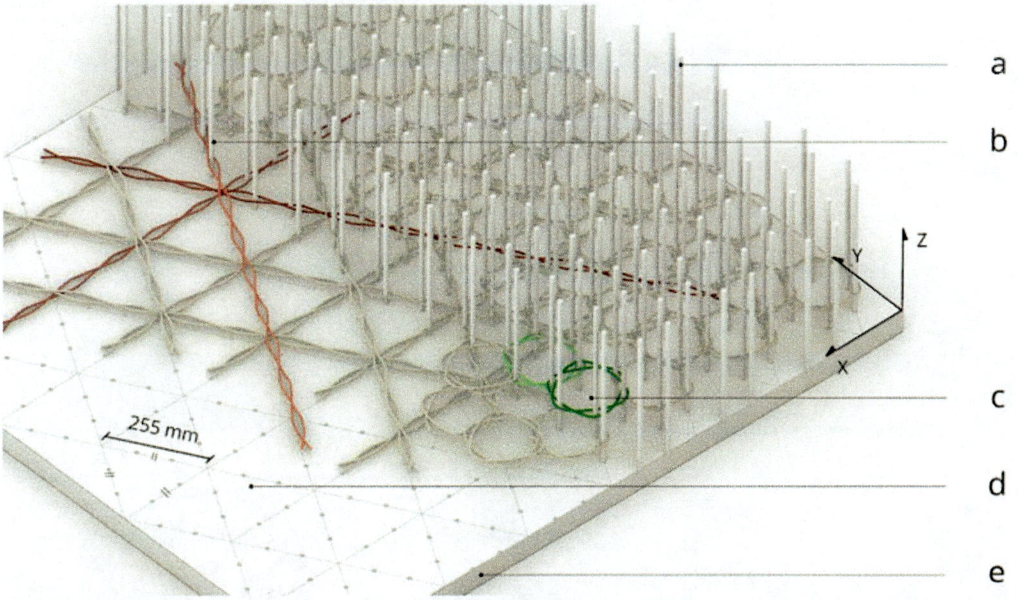

Fig. 6 Fabrication setup showing weaving patterns on poles (**a**), with linear (**b**) and closed (**c**) types, which create the cell-like structures (**d**)

Fig. 7 Different types of weaving pattern configurations, developed digitally and tested with hand-made prototypes

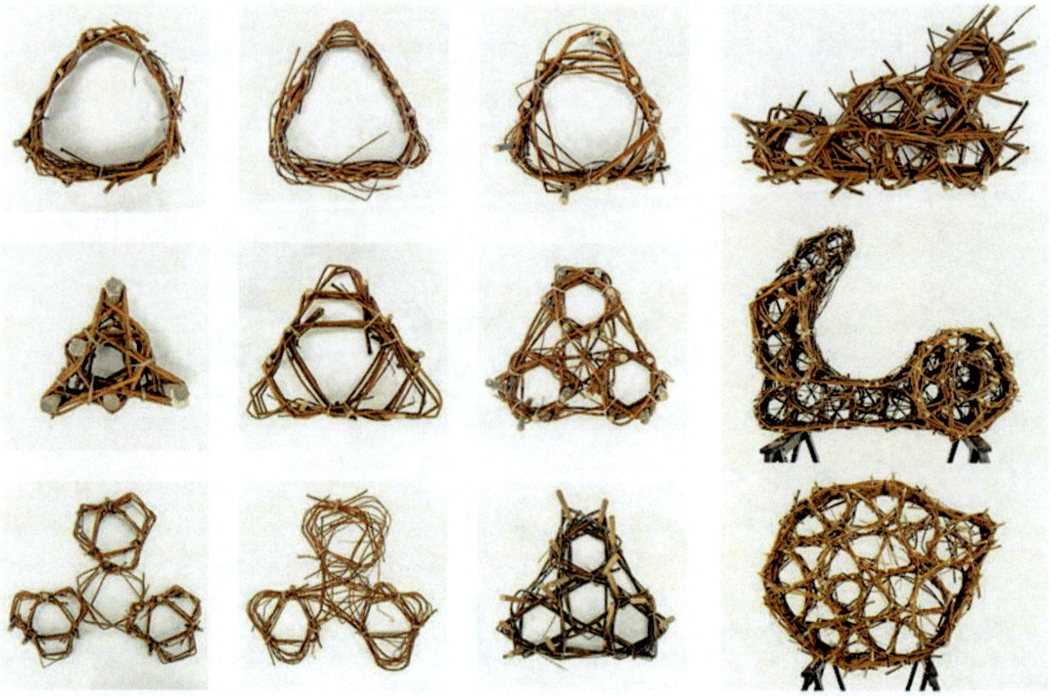

Fig. 8 Prototypes to test the influence of different parameters

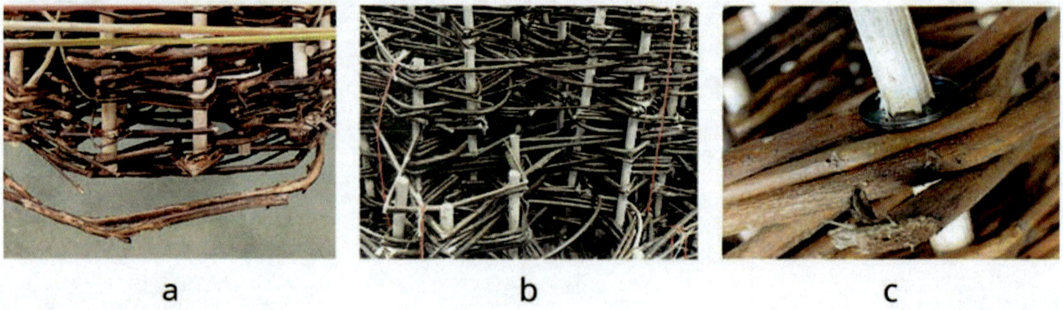

Fig. 9 Delamination (**a**), strings (**b**) and self-locking washers (**c**)

willow on each side, which enabled the use of plastic-state earth as a filling material.

Two parameters were considered when choosing which cells to fill. As earth was the main contributor to self-weight, minimising the earth infill meant easier handling and transportation. Secondly, the surface area exposed to air had to be maximised, in order to aid evaporation and reduce drying time (Fig. 10).

Setup for Fabrication There are a number of readily available tools and machines to aid the filling process; however, using them adds constraints on the composition of the earth mixture as the additive sizes and viscosity need to conform to the machine dimensions and the extrusion mechanism.

Long straw fibres reduce the amount of shrinking but will block the pipes used with

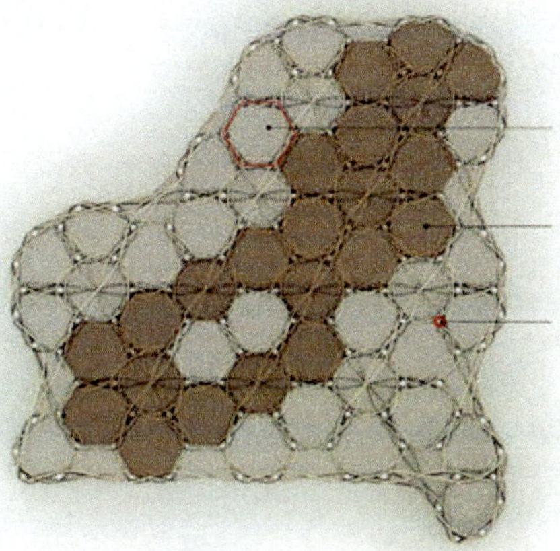

Fig. 10 Top view of a component showing willow cells (**a**) and earth-filled areas (**b**) defined by poles (**c**)

conventional mortar pumps and are also more likely to separate from the earth mixture during compaction. However, choosing a mixture using such tools can streamline the process and make the construction system more viable.

Process of Fabrication Compaction was done locally in each cell by a concrete vibrator. This was to ensure that the mixture reaches all the crevices in the components and that air bubbles are removed (Fig. 11). Thus, the entire process emerged, starting with pre-mixing the dry ingredients, adding the water in a concrete mixer to create a uniform material; filling was then done by hand in batches and followed by the compaction with a concrete vibrator.

2.2.3 Digital Fabrication Development

In order to prove the feasibility of the automated digital weaving process, a customised digital fabrication system was developed. This consisted of a two-axis, stepper motor-driven system, controlled by an FPGA-based controller running G-codes, which pulls willow stems within the bed of poles (Fig. 12).

The unpredictable behaviour of unprocessed materials gave rise to several challenges during the development of the digital fabrication system.

Firstly, analysing the process of hand-weaving versus machine-weaving highlighted some key differences. During hand-weaving, the willow is supported in two points and the bending location is controlled by the distance of these two support positions, whereas the machine pulls the willow stems through the bed of poles. This results in less control over the point where the willow bends and higher friction with the poles, with a consequent higher chance of stems breaking in case of patterns with small bending radii. While this aspect can be minimised, a specific repertoire of machine paths that avoided tight radii had to be developed.

Due to the changing diameter of the willow stems, the orientation for machine-weaving was crucial. The thinner sections were able to take on more bending but were also less resistant to the large stresses caused by pulling at the end effector

Fig. 11 Willow and earth combination in a component, top (**a**) and side (**b**) views

Fig. 12 View of the two-axis machine showing the bed of poles (**a**), the end effector (**b**) and the stepper motors (**c**)

which resulted in the willow breaking. Thicker parts of the willow were more resistant to tearing but also stiffer to bend which in turn meant that the forces were transferred to the wooden poles effectively bending them about their vertical axes. Finally, the varying lengths of the willow stems were also considered during the machine fabrication by designing the machine paths with overlaps between consecutive branches.

2.3 Component Design

The design language emerged from the possibilities of geometrical freedom that the fabrication process affords, enabling non-standardised elements that can be tailored to specific conditions. In the XY fabrication plane (see Fig. 6), a large range of shapes is possible, while the freedom is restricted to a variation of heights in the Z direction, up to a maximum of 50 cm to avoid bending the vertical elements excessively.

Seven final weaving patterns were repeated in different sequences until the desired height was reached to create a component (Fig. 13). In this way, redundancy was introduced in the system to counterbalance the inconsistencies created by the non-processed willow stems. A digital design process was developed that automated the workflow from overall volumetric geometry to the creation of the fabrication setup and patterns, which were finally exported either as instructions for a person weaving or as a coded path for the machine.

Two different component types were explored in this study.

Components Made of Willow and Earth These were planned to be used in a horizontal orientation with respect to the fabrication setup. Their maximum size was restricted by the weight of the final components, to ensure transportability. The geometrical freedom in the XY plane was exploited to create interlocking components (Fig. 14) and the sequence of weaving patterns was tuned to allow for bigger gaps so that the earth could flow between the cells.

Components Made of Willow These could be arranged more freely, horizontally or vertically with respect to the fabrication setup. In this case, a pattern sequence that created denser components was implemented to achieve more stable configurations.

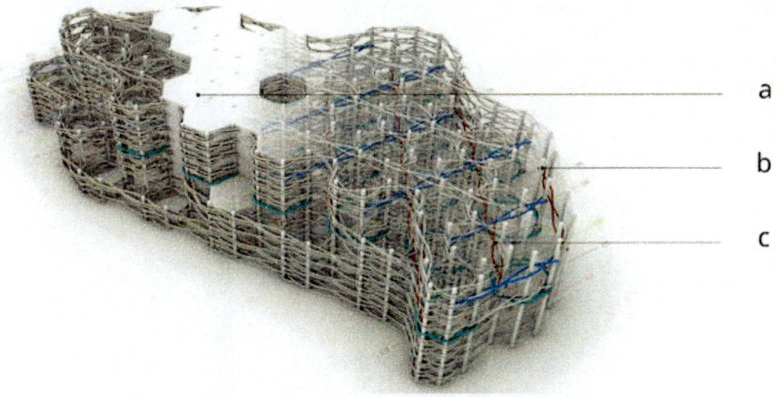

Fig. 13 Example of a willow and earth component, which shows earth (**a**), poles (**b**) and different weaving patterns (**c**)

Fig. 14 Interlocking component system

The component design included the integration of additional elements for joinery, transportation and assembly.

2.4 Material Characterisation

As this is a new material combination for which no standard exists, a wide variety of structural test scenarios were necessary to understand both the respective properties of the individual materials and the behaviour of their combination.

For the classification of willow, the focus was placed on different tensile test scenarios. Willow is a natural material that has a large variation in tensile strength due to material imperfections, so five identical samples were tested. The tensile test of individual willow stems (Fig. 15a) aimed to obtain an average value of the maximum tensile stress. Stems of different thicknesses were tested and the tensile force was converted to stress with respect to the available cross-sectional area. Three willow stems laid parallel to each other (Fig. 15c) provided a reference value to braided willow (Fig. 15b) to be able to record the influence of the braiding on tensile strength.

For the earth tests, cube samples were made with a side length of 20 cm (Volhard and Röhlen 2009) to test the compressive strength of the self-developed earth mixture (see Sect. 2.2.2) once the equilibrium moisture was reached (about six weeks) (Volhard and Röhlen 2009).

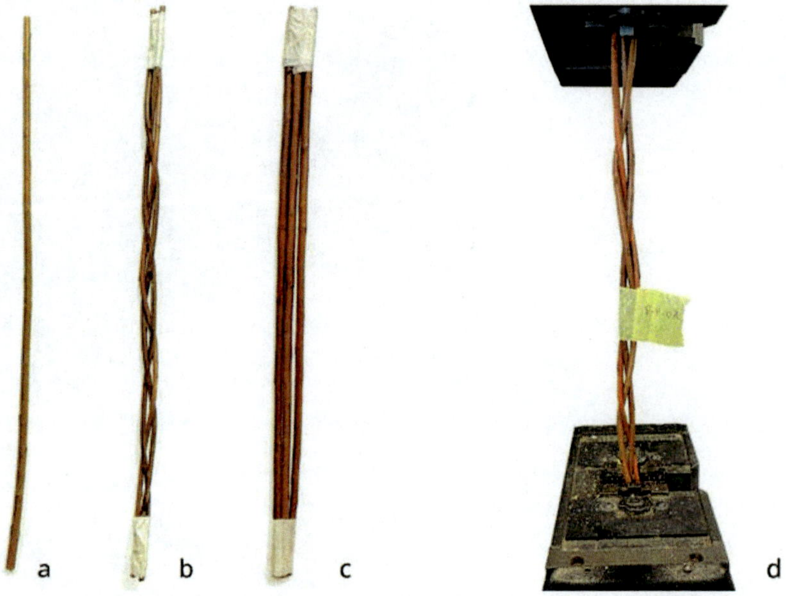

Fig. 15 Single willow stem (**a**), three willow stems braided (**b**), three parallel willow stems (**c**), braided willow in the testing machine (**d**)

To test the interaction between willow and earth as a composite material, rectangular test specimens with dimensions 10 cm × 10 cm 75 cm were made, which underwent bending tensile strength tests using a four-point bending setup.

Finally, two willow reinforcement types were examined. First, the willow was placed in the formwork in three layers (Fig. 16), where the reinforcement acted as a tie rod. The second type was a three-dimensional willow structure (Fig. 17). Due to the spatial reinforcement structure, a three-dimensional force transmission was expected (cf. truss).

The test scenarios described above were carried out qualitatively and still require further development steps to obtain quantitative results from them.

The preliminary tensile tests of the single willow stems showed that they can absorb an average tensile stress of about 70–90 N/mm^2, depending on the minimum and maximum cross-sectional area of the tested willow stems. The earth cubes absorbed an average compressive stress of 1.8 N/mm^2.

In the case of the braided willow stems, lower absorbed tensile force was observed in comparison to the willow stems that were laid parallel to each other. Although the braided willow had a reduced tensile force, the better bonding effect between willow and earth was a great advantage.

Furthermore, a tensile stress capacity was observed in the earth-willow test specimens, which withstood a load of up to 2.2 kN in the four-point bending setup before the beams failed or the deformation became too large, compared to specimens made of only earth with the same dimensions, which broke directly under their self-weight.

3 Results

A full-scale demonstrator named "InterTwig" was developed and built by an interdisciplinary team of students and researchers at the Karlsruhe Institute of Technology (KIT). It was presented in Karlsruhe in July 2022 (Figs. 18 and 19).

The demonstrator served as a proof of concept to test the whole process from design to fabrication to assembly, widening the scope of the research from components to a whole structure representative of a construction system. Specifically, it aimed to test component logics (interlocking components vs. joints) and different application potentials (load-bearing willow and earth components vs. lightweight willow

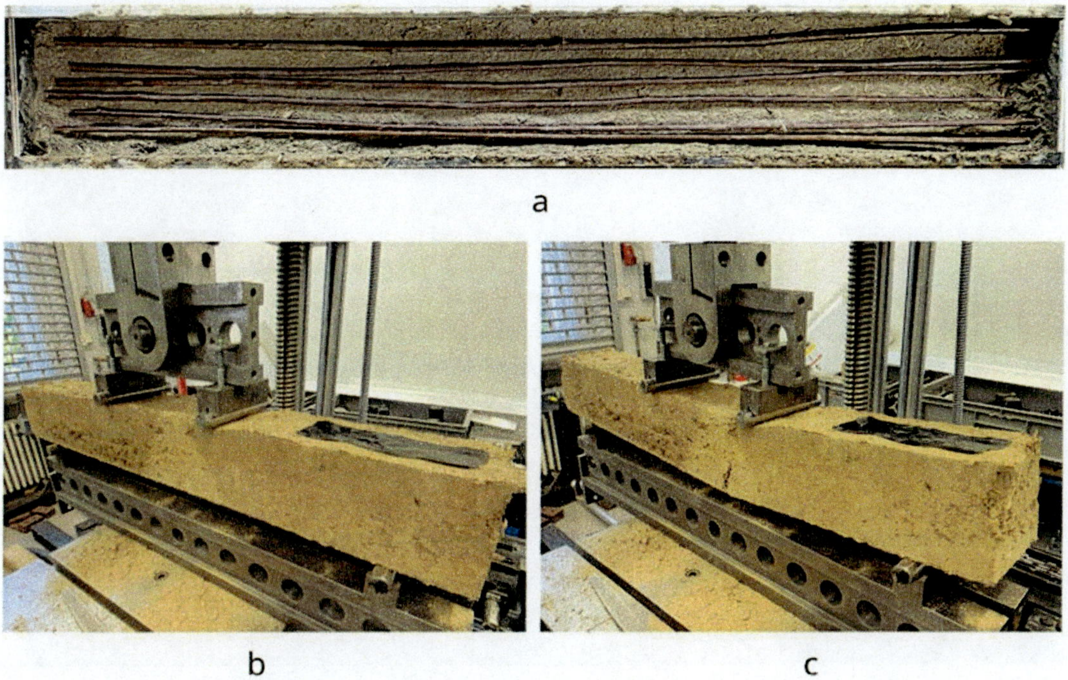

Fig. 16 Test specimen with willow laid in layers parallel to each other (**a**), specimen before loading (**b**), specimen at failure load (**c**)

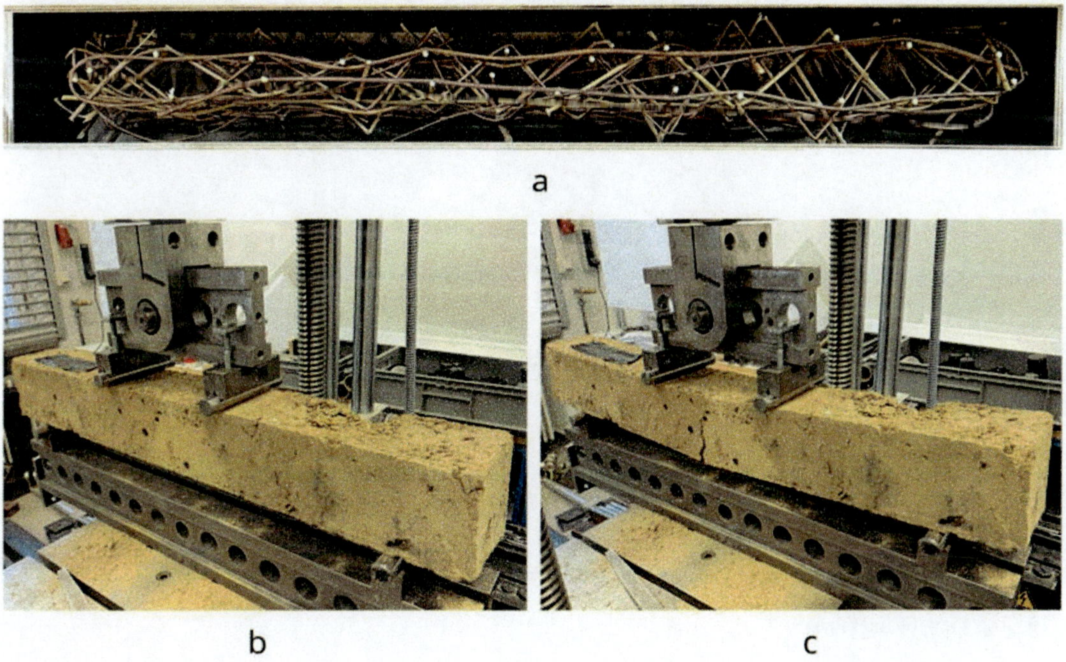

Fig. 17 Test specimen with spatially braided willow (**a**), specimen before loading (**b**), specimen at failure load (**c**)

Fig. 18 View of InterTwig

Fig. 19 Top view of InterTwig

Fig. 20 Exploded view of InterTwig

components), integrate a structural scenario for earth distribution and test the assembly process.

Designed as a freestanding structure, the demonstrator forms a tapered shape that rises to a height of about 4 m. Its core is articulated into four legs at the base with a ground plane area of 7.5 m^2.

It consists of 21 components that have been prefabricated, thus enabling a production that is independent of weather conditions and a construction system that optimises ease of assembly and disassembly (Fig. 20). The 16 components at the base of the structure, composed of willow and earth, are geometrically interlocked and stacked on site. Here, the willow acts as a permanent formwork and as a tension reinforcement in the composite material. The top five components, which are made primarily of willow and mechanically connected to the bottom, are arranged in different layer orientations to create a vertical segment that relies on component-integrated joints and showcases the intricate filigree of the material system (Fig. 21). The structure's materials are therefore functionally graded, with earth being predominant at the bottom to carry the loads from the top and distribute them over a larger area. To determine the proportion of earth to willow, each component was first analysed globally with reference to its local position, thus determining the approximate direction of the component's internal forces. This informed the overall earth distribution, which, thanks to the versatility of the willow formwork, can be placed only where needed, facilitating the implementation of material efficiency principles to create structural earth components that are lighter than usual applications like rammed earth. This resulted in earth and willow components with a total weight ranging between 150 and 430 kg and an average of 315 kg (Fig. 22).

By designing a construction and material system that does not use chemical binders, the components can be separated into renewable (willow, wood and straw) and non-renewable (earth) materials and circulated back through different cycles. By maintaining the earth in its raw form with no stabilising agents, it can be removed from the willow framework by saturating it with water and recycled without any significant loss of value. Once the earth is removed from the components, willow and wood can be returned to the biosphere through composting or anaerobic digestion.

The temporary purpose of the structure represented an opportunity to implement and test a construction system that was designed to work with reversible, easily accessible joinery systems and lifting anchors (Fig. 23). These connections are integrated into the components with wooden plates that are kept in place by the woven willow layers, while the lifting anchors for the components made of willow and earth additionally employ a threaded rod. A truck with a mounted crane was used to transport and instal the structure, which was assembled in 5 h and disassembled in 3 h (Fig. 24).

Digital design techniques were fundamental to handle and control the intricacy of different weaving patterns and their modification and optimisation according to the principles

Fig. 21 Detail of a willow component, showcasing two pattern types (orange and yellow) and an integrated joinery

Fig. 22 Earth and willow components before assembly

Fig. 23 Exploded view of a portion of InterTwig showing the joinery system

discovered in the iterations of exploratory prototypes. While the application of digital techniques in this study focused on automation aspects, this is the first step for a process that could inform the geometries, both global and detailed, according to structural principles, thus adapting them to specific design conditions.

A process that mutually informed handcrafting, digital design and digital fabrication was developed, exploring the innovation potential provided by their combination. Digital design and digital fabrication allowed the development of customised principles that deal with the imperfections and variations that are intrinsic in plant-based materials, by-passing the standardisation necessary to conform to existing fabrication processes. While the current prototypes were mostly hand-crafted, a digital fabrication system was developed using the principles discovered in the study that could enable automated and affordable applications.

4 Discussion and Conclusions

The presented work showcases a proof of concept for the development of willow and earth composites for circular construction and their potential for digital fabrication, as well as

Fig. 24 Assembly of InterTwig

providing relevant research questions for future investigations.

Further research needs to be conducted on collecting data for quantifying and validating the structural performance and mechanical behaviour of the composite material. Although the results of this work are to be considered ahead of application scenarios in industry, increased automation, especially regarding earth casting, can contribute to enhanced efficiency and facilitate their implementation in construction both in terms of technology and economics. A higher degree of geometrical freedom will enable refinements and optimisations according to structural and material efficiency principles, as well as novel architectural parameters to further expand a language that is driven by the materiality and the possibilities of resolution and articulation afforded by the material system.

This study demonstrates a model for broadening the current construction material boundaries by revisiting vernacular materials and techniques and redefining their contemporary potential. In particular, it contributes to extending the expressive, morphological and construction potentialities of renewable materials and earth composites towards the field of architectural exploration. By implementing a design-through-making methodology, in which strategies are developed with hands-on physical prototypes across different scales and result in a 1:1 experimental proof of concept, it showcases the potential for a unified understanding of conceptual, material and production aspects.

Acknowledgements This research was possible thanks to the invaluable contribution of various researchers and students at the Karlsruhe Institute of Technology (KIT). The initial concepts and prototypes were developed together with the students of the course "Digital Wicker". The research was deepened and expanded with the full-scale demonstrator together with the students of the course "Digital Wicker 2.0" Teodora Bondar, Elisabeth Genest, Shunze Hou, Alicia Pizzignacco, Cesar Requejo Peña, Lara Sodomann and Kalin Yanev. Finally, the authors would like to express their gratitude towards their fellow investigators Daniel Fischer, Fanny Kranz, Javier Fuentes and Michael Kalkbrenner for their support throughout the development of the research.

References

Aust C, Schweier J, Brodbeck F, Sauter U, Becker G, Schnitzler J (2013) Land availability and potential biomass production with poplar and willow short rotation coppices in Germany. GCB Bioenergy 6 (5):521–533

Bouza H, Asut S (2020) Advancing reed-based architecture through circular digital fabrication. In: Proceedings of the 38th eCAADe: anthropologic—architecture and fabrication in the cognitive age, vol 1

Bruno A, Scott B, D'Offay-Mancienne Y, Perlot C (2020) Recyclability, durability and water vapour adsorption of unstabilised and stabilised compressed earth bricks. Mater Struct 53(6)

Çetin S, De Wolf C, Bocken N (2021) Circular digital built environment: an emerging framework. Sustainability 13(11)

Claytec. Baulehm. https://www.claytec.de/de/produkte/ergaenzungsprodukte/baulehm_pid436. Accessed 01 Oct 2022

Conluto. Lehm-Unterputz erdfeucht. https://www.conluto.de/produkt/lehm-unterputz-erdfeucht/. Accessed 01 Oct 2022

Cruz P (2013) Structures and architecture. CRC Press, Boca Raton, Florida

Dawod M, Deetman A, Akbar Z, Heise J, Böhm S, Klussmann H, Eversmann P (2019) Continuous timber fibre placement. Impact, Design With All Senses, pp 460–473

Dias A, Schänzlin J, Dietsch P (2018) Design of timber-concrete composite structures. COST Action FP1402/WG 4

Ellen MacArthur Foundation (2019) Circular Economy Glossary. https://emf.thirdlight.com/link/vj6i9k5yax0n-1fkyvu/. Accessed 29 Sept 2022

Ellen MacArthur Foundation (2022) Circulate products and materials. https://ellenmacarthurfoundation.org/circulate-products-and-materials. Accessed 29 Sept 2022

ETH Zurich—dbt (2022) Digital bamboo. https://dbt.arch.ethz.ch/project/digital-bamboo/. Accessed 01 Oct 2022

European Commission, Directorate-General for Communication (2020) Circular economy action plan: for a cleaner and more competitive Europe. Publications Office of the European Union

Eurostat (2022) Waste statistics—statistics explained. https://ec.europa.eu/eurostat/statistics-explained/index.php?title=Waste_statistics#Total_waste_generation. Accessed 28 Sept 2022

Faasch R, Patenaude G (2012) The economics of short rotation coppice in Germany. Biomass Bioenerg 45:27–40

Fabbri A, Morel J-C, Aubert J-E, Bui Q-B, Gallipoli D, Reddy V (2022) Testing and characterisation of earth-based building materials and elements: state-of-the-art report of the RILEM TC 274-TCE. Springer International Publishing, Cham

Gil Pérez M, Guo Y, Knippers J (2022) Integrative material and structural design methods for natural fibres filament-wound composite structures: the LivMatS pavilion. Mater Des 217:110624

Hegger M, Fuchs M, Stark T, Zeumer M (2007) Energie atlas. Institut für Internationale Architektur-Dokumentation, München

Klep M (2015) A policy study on the sustainable use of construction materials. Organisation for Economic Cooperation and Development. https://www.oecd.org/officialdocuments/publicdisplaydocumentpdf/?cote=ENV/EPOC/%20WPRPW(2014)4/FINAL&docLanguage=En. Accessed 1 Oct 2022

MAS DFAB ETH (2022) MAS DFAB: mesh mould earth construction—D-ARCH. https://works.arch.ethz.ch/thesis/mesh-mould-earth-construction. Accessed 29 Sept 2022

Morel J-C, Charef R, Hamard E, Fabbri A, Beckett C, Bui Q-B (2021) Earth as construction material in the circular economy context: practitioner perspectives on barriers to overcome. Philos Trans R Soc B Biol Sci 376(1834):20200182

Özdemir E, Saeidi N, Javadian A, Rossi A, Nolte N, Ren S, Dwan A, Acosta I, Hebel D, Wurm J, Eversmann P (2022) Wood-veneer-reinforced mycelium composites for sustainable building components. Biomimetics 7(2):39

Trummer J, Schneider M, Lechner M, Jarmer T, Demoulin T, Landrou G, Nagler F, Winter S, Dörfler K (2022) Digital design and fabrication strategy of a hybrid timber-earth floor slab. IOP Conf Ser Earth Environ Sci 1078(1):012062

USGB. Rapidly renewable materials | U.S. Green Building Council. https://www.usgbc.org/credits/new-construction-schools/v2009/mrc6?return=/credits/new-construction/v2009. Accessed 29 Sept 2022

Van der Linden J, Janssens B, Knapen E (2019) Potential of contemporary earth architecture for low impact building in Belgium. IOP Conf Ser Earth Environ Sci 323:012018

Verwijst T, Lundkvist A, Edelfeldt S, Albertsso J (2013) Development of sustainable willow short rotation forestry in Northern Europe. Biomass now—sustainable growth and use

Volhard F, Röhlen U (2009) Lehmbau Regeln, 3rd edn. Dachverbands Lehm e.V, Weimar

Vyncke J, Kupers L, Denies N (2018) Earth as building material—an overview of RILEM activities and recent innovations in geotechnics. MATEC Web Conf 149:02001

Zami M, Lee A (2010) Stabilised or unstabilised earth construction for contemporary urban housing? In: 5th International conference on responsive manufacturing—green manufacturing (ICRM 2010)

A Study on Carbon-Neutral Biochar-Cementitious Composites

Nikol Kirova, Areti Markopoulou, Jane Burry, and Mehrnoush Latifi

Abstract

The global environmental challenges are calling for novel solutions and sustainable practices in the fields of architecture and construction. To respond to the SDGs agenda, the research investigates strategies for integrating biochar as an aggregate for carbon-neutral cementitious composites and to gain knowledge upon the effects that large quantities of biochar within the material system has on mechanical properties as well as carbon sequestration. The paper presents the development and assessment of a series of grades of Biochar Cementitious Mortars (BCMs). Assessment is structured upon (i) material consistency, (ii) structural performance and (iii) composite's embodied carbon. The results propose that biochar, derived from agricultural waste, and as a carbon-sequestering material, can be used to lower the net embodied carbon of cementitious composites when used as an amendment and cement replacement. As BCMs meet compressive strength standards and can provide building materials with a net-negative embodied carbon, the material research provides valuable input for applications in the construction sector. It is proposed that in the later stages of the research the developed BCMs can be applied within a materially driven design process trough functionally grading the material and consequently designs with optimised environmental impact can be achieved.

Keywords

Biochar · Cementitious composites · Carbon-neutral · Carbon sinks · Material-driven design

1 Introduction

The buildings and construction sector accounted for 39% of energy and process-related CO_2 emissions in 2020, 11% of which resulted from manufacturing construction materials (United Nations Environment Programme 2020). Compared to other construction materials, concrete stands out with its high use rate and environmental impact (Estokova et al. 2017). Due to the high CO_2 emissions associated with the

N. Kirova (✉)
Advanced Architecture Group, Institute for Advanced Architecture of Catalonia, Barcelona, Spain
e-mail: nikol.kirova@iaac.net

N. Kirova · J. Burry · M. Latifi
School of Design, Swinburne University of Technology, Melbourne, Australia

A. Markopoulou
Institute for Advanced Architecture of Catalonia, Barcelona, Spain

M. R. Thomsen et al. (eds.), *Design for Rethinking Resources*, Sustainable Development Goals Series,
https://doi.org/10.1007/978-3-031-36554-6_33

production of cement, the main ingredient of concrete, the cement industry alone accounts for approximately 5% of global anthropogenic CO_2 emissions (Birol 2019).

Various frameworks, policies and international agreements starting with the Kyoto Protocol (1997) and more recently the Paris Agreement (2015) and the Sustainable Development Goals (SDGs) have been put in place as acting measures to prevent resource depletion and a further increase in the CO_2 emissions that has been directly linked to temperature rise (Delbeke et al. 2019; European Commission 2019). The "Advancing Net Zero" global campaign conducted by the World Green Building Councils is working towards total sector decarbonisation by 2050. A net-zero carbon vision acknowledges the time value of carbon emissions from materials and construction. One of the proposed ways to achieve the Net-Zero vision is through radical cross-sector collaboration (WorldGBC 2019). In this context, the European Commission suggests that agricultural waste can be used to mitigate the large environmental impact by potentially linking the construction and agricultural industries. It is proposed that waste material from agricultural produce can be used within construction materials.

Building on the idea of cross-industry collaboration, the "From Farm to Facade" EU initiative proposes that by combining end-of-life products with new materials, the agricultural and construction sectors can mutually benefit (Polytechnic University of Bari 2021). For example, if waste produced by the agricultural sector is used as a product in the construction sector not only does waste obtain a newfound value and is therefore upcycled but raw materials are not extracted and therefore depleted in the same manner. This scenario aids both sectors to become more sustainable, producing less waste and decreasing their energy consumption and loss.

In 2016 a report by the European Environmental Citizen's Organisation for Standardisation (ECOS), included biochar among other waste materials that can be used as a fertiliser in the agricultural industry. Up until recently, there were no links between biochar and the construction sector, but this has changed. Research publications on the non-agricultural uses of biochar have increased exponentially since 2017. Several studies on cement mortars demonstrate the carbon sequestering potential of biochar when investigated as an amendment to the composite (Gupta et al. 2021a, b, 2018a, b, c, 2017; Gupta and Kua 2020; Rockwood et al. 2020; Li et al.; Nair et al. 2020; Suarez-Riera et al. 2020). The main value of research on biochar within the construction industry is that it is an abundant material with large carbon sequestering potential. Furthermore, it is predicted that the supply chain will increase soon as the interest in the properties and specifically its carbon sequestering potential creates incentives for its use outside of agriculture.

The properties of biochar have been explored for the first time in an architectural context by the Ithaka Institute where it was demonstrated, on a building scale, how the use of biochar plaster can regulate humidity, provide carbon negative finishing, and enhance thermal insulation (Schmidt 2014). Due to its non-structural application, the biochar-based composite was to be mixed with clay and lime instead of cement allowing for the material to be directly used as compost at the end of its life cycle. The Ithaka Institute of research laid the ground for further investigation of the applications and potential of biochar within the built environment.

2 Literature Review

2.1 Sustainable Concrete Amendments

There are several types of industrial waste or by-products that have pozzolanic properties, which can substitute cement or be added in concrete to reduce the amount of cement needed. Some of the most used by-products are blast furnace slag, which derives from the production of iron and steel, silica fume, a by-product of the manufacturing of silicone, and fly ash, a by-product of the coal industry. Fly ash is one of the most adopted in the industry amendments for concrete, however it has significant environmental limitations. Cement replacement and sustainable concrete amendments are slowly penetrating the concrete market but are limited by the associated economical costs and dependability on external, often highly polluting, industries. An alternative solution for a sustainable concrete amendment that derives from the agricultural industry is proposed, where biochar can be used within cementitious materials to decrease the carbon footprint of concrete and cementitious mortars. Biochars are produced throughout the globe by the local agricultural industries, making them an abundant product in comparison to some of the previously discussed industrial by-products.

2.2 Biochar-Cementitious Composites

The compressive strength of biochar cementitious materials has been extensively studied in recent years. Research has shown that the mechanical properties of biochar-cement composites can vary depending on the type, content, and pyrolysis temperature of the biochar used. Studies have revealed a range of outcomes when it comes to the impact of biochar addition on the compressive strength of concrete and mortars. In general, most studies concur that high concentrations of biochar can weaken the composite, while low concentrations (below 2%) can enhance strength through better hydration. Additionally, it is recognised that the source and type of biochar can greatly influence the mechanical properties of biochar-cement composites, and further research is needed to fully understand these differences. For instance, a study by Akhtar and Sarmah (2018) found that adding 1% biochar to concrete blocks made from rice husk, poultry litter, and paper mill sludge resulted in lower compressive strength than conventional concrete. However, after 28 days of curing, the study found that the 1% addition of biochar improved the strength of the concrete due to improved hydration during the curing process.

The study of Cuthbertson et al. (2019) supported the observations of Akhtar and Sarmah (2018). The compressive strength of concrete with the addition of biochar derived from dry distiller grains was investigated. In particular, the study looked at the effects of replacing sand and aggregate within the composite with 1.2 and 3% of biochar. The addition of 3% biochar yielded maximum strengths of 21 MPa (when replacing sand) and 22 MPa (when replacing aggregates).

In a study by Gupta et al. (2018a, b, c), the optimum concentrations of biochar for increasing the compressive strength of cement mortar were found to be 1 and 2%. When compared to standard mortar, a significant increase of 22 and 27% was observed. Following that, biochar addition beyond 2% resulted in a reduction in the strength.

Furthermore, Asadi Zeidabadi et al. (2018) reported maximum strength on the concrete samples with added 5% bagasse biochar. In contrast, Mrad and Chehab (2019) observed a decrease in compressive strength when high percentages of biochar as 5% were added. Based on these findings, we can conclude that the compressive strength of the concrete is highly influenced by the feedstock used in the production of the biochar. The Sirico et al. (2020), also agreed that the properties of the feedstock material have a great impact on the compressive strength of the composite.

A recent study by Navaratnam et al. (2021) examined the effects of higher concentrations of biochar on cementitious mortars. The study tested three different proportions of biochar (5, 10, and 20% of cement weight) and found that the compressive strength decreased as the biochar concentration increased. Specifically, the compressive strength of the biochar mortars at room temperature was recorded at 35, 39, 28, and 16 MPa for 0%, 5%, 10%, and 20% biochar addition, respectively. This confirms the findings of other studies that the addition of biochar in higher quantities than 5% leads to a decrease in the compressive strength. Additionally, it can be inferred that the feedstock type, pyrolysis temperature, and particle size play a significant role in the mechanical properties of biochar-cementitious composites, but high concentrations of any form of biochar will always lead to a decrease in the compressive strength. The composition of concrete directly impacts its compressive strength and introducing biochar in the mixture ultimately results in a decrease in that strength, however there is still room for investigation of cementitious composites with high concentration of biochar (above 20%) as well as how does that relate to the embodied carbon value of the material system.

2.3 Biochar in Architectural Composites

In addition to cementitious composites, biochar has also been utilised in other architectural composites, with one of the most notable being developed by the company Made of Air. Founded in 2016 by architects Allison Dring and Daniel Schwaag, Made of Air created a thermoplastic that can be easily shaped and contains a significant amount of biochar mixed with sugar cane. In April 2021, the thermoplastic was employed for the first time on a building, specifically as the facade cladding on an Audi dealership in Munich, comprising of seven tonnes of hexagonal panels. It was estimated that the entire facade cladding sequestered 14 tonnes of carbon until the recycling of the cladding tiles (Hahn 2021).

Another example of biochar-based architecture can be found in the U.S. based company "Interface', which produced its first carbon-negative carpet tile in 2021. The company specialises in carpet backings made of biochar and other composites. However, information on the material composition and carbon footprint of Interface's product is limited as it is currently on the market.

Currently, there are no biochar-based building or structural elements for the architectural sector. All the products mentioned above are related to either interior or exterior cladding systems. Nevertheless, it is noteworthy that all these products have reached the market after 2020, indicating a growing interest in biochar and its ability to sequester carbon as more companies, industries, and researchers are exploring the properties of this material.

3 Materials and Methods

The research proposes that pyrolyzed agricultural waste can be utilised within the construction sector as a carbon-sequestering amendment for material composites. Creating synergies between the construction and agricultural industries can help develop sustainable solutions to support the growing population and the unavoidable construction of new housing and infrastructure. Globally the agricultural industry generates about 1.8% of the total global waste. Most of this waste is either being left to either decompose or used to create compost allowing for most of the captured CO_2 to be released back into the atmosphere. However, through the process of pyrolysis, agricultural waste can be transformed into biochar, a stable carbon-rich material that sequesters large amounts of carbon and holds them for long periods of time. Pyrolysis is a process during which organic matter is being heated to high temperature in the absence of oxygen so that combustion does not take place and instead the matter thermally decomposes into gas and charcoal (Boslaugh 2022).

Biochar is commonly used for soil remediation (Sizmur et al. 2016; Matovic 2011), carbon sequestration (Tan, 2020), energy storage and conversion purposes (Li et al. 2017), etc. The most common use of biochar, currently, is as a fertiliser. The architectural and construction applications of biochar are underdeveloped and have been explored more rigorously only in the past several years.

The phase of the research disclosed in this paper looks specifically into the material development and studies how the introduction of biochar affects the mechanical properties of cementitious composites when added in large quantities. The research is inspired by the idea that the carbon cycle can be altered as excess CO_2 is removed from the atmosphere and sequestered within building structures and envelopes. The main premise is that as carbon is stored in buildings it converts them from "carbon sources" into "carbon sinks'. The research focuses on the mechanical properties of the developed composites when the quantity of biochar is progressively increased and used as a cement replacement.

Biochar as an amendment for cementitious composites has been explored from a material engineering perspective. The addition of small quantities of biochar within concrete has many beneficial properties such as carbon sequestration (Nair et al. 2020; Gupta et al. 2018a, b, c; Gupta and Kua 2020; Das et al. 2020), improved vegetation compatibility, fire resistance, thermal insulation, humidity regulation and acoustic dampening. However, the literature review indicates that by introducing biochar to the composite matrix the mechanical properties decline. It is demonstrated that the quantity of biochar is proportional to the decrease in both compressive and tensile strength. This decrease can also be associated with the decrease of cement within the mixture.

3.1 Research Methodology

The study relies on empirical research in developing a materials system of Biochar Cementitious Mortars (BCMs) as well as testing the mechanical properties of the five grades of developed BCMs. There are three main stages of development: preparation, prototyping and testing. The preparation includes a literature review, planning and material sourcing. The prototyping phase consists of mould fabrication, material preparation, specimen fabrication and curing. The first two stages of the research took place in the laboratory of the Institute for Advanced Architecture of Catalonia (IAAC). The final stage (testing) is carried out with the support of an industry partner: BAC Engineering Consultancy Group (BAC ECG), at their material testing lab in Barcelona.

Various mixtures are developed and tested for tensile and compressive strength. The aim is to gather sufficient data to simulate the material performance within computational design environments such as Rhino's Grasshopper and specifically the structural analysis plug-in Karamba3D.

It is proposed that at the following stage of the research various computational design and fabrication techniques for heterogeneous or graded material allocation will be explored to inform an optimal allocation of lower or higher quantities of biochar in correspondence to case-specific structural needs. The design possibilities of the material system are at an initial stage of exploration. The goal is to propose a design solution maximising the carbon sequestration potential and aiming to mitigate the environmental impact of concrete construction.

3.2 Carbon Sequestering Biochar Cementitious Composites

3.2.1 Material System: Ingredients

The research investigates the addition of biochar within a range of material mixtures of cementitious mortars. Five sets of mixtures with varying quantities of biochar are developed and standardised testing specimens are cast. The material matrix is simplified to three main ingredients: Biochar; Sand and Cement. A detailed explanation of each ingredient is provided below.

Biochar

Biochar is the main testing variable in the research. It is the material obtained when the organic matter undergoes thermal decomposition under a limited supply of oxygen at a relatively low temperature (<700 °C) (Vigneshwaran et al. 2021; Cha et al. 2016). It is a very stable carbon-rich material, which can sequester carbon for thousands of years. On average 1 kg of finished biochar can sequester up to 1.8 kg of carbon dioxide (Hettinger 2017). This value ranges from 0.5 to 2.8 kg depending on the type of feedstock and the conditions of pyrolysis (temperature and atmospheric pressure) (Peredo-Mancilla et al. 2018; Iberahim et al. 2019; Popescu et al. 2014; Byrne and Nagle 1997).

The feedstock from which biochar is produced has a great impact on the properties and carbon sequestration potential of the biochar. For instance, hardwood feedstock has a higher value of negative embodied carbon in comparison to softwood. This is due to the amount of carbon that the tree photosynthesises during the natural growth process. As biochar is commonly produced from forest management feedstock, the embodied carbon of the source material together with the processing, transport and pyrolysis conditions have to be taken into consideration when calculating the net embodied carbon of the biochar. The calculation is therefore case specific and may vary greatly. As an example, just looking at the feedstock, it is estimated that softwood biochar will have a carbon-neutral embodied carbon value whereas hardwood biochar will attain a negative embodied carbon value.

For the research described in this paper, the biochar is sourced from a Spanish company based in Catalonia: CarbonVivo (CarbonVivo 2022). The biochar they provide is derived from forest management biowaste which is composed mostly of wood feedstock but can include small quantities of leaves and other biowaste. The sourced biochar is with a particle size of 1–10 mm, which is the smallest particle size available from the producer.

The biochar is produced using Kon-Tiki Kilns. This type of open kiln is less energy-consuming and inexpensive making biochar production more easily accessible. The temperature of pyrolysis is estimated to range between 600 and 700 °C. It is estimated that the embodied carbon of the sourced biochar varies from −1 kgCO_2/kg to −2.5 kg CO_2/kg. The biochar is further processed in order to be used in the material system.

Cement

The developed cementitious material system relies on the most commonly used cement mixture which is based on Ordinary Portland Cement. This is done in order to create a fair representation of the state material composition with a standard range performance. Specifically, a type of "CEM II BL" cement is used. This is a medium-range versatile cement used commonly in building construction in the region of Catalonia. The "BL" indicates that the cement consists of 21–35% limestone (estimated from the ICE Cement, Mortar, Concrete model). The associated embodied carbon is 0.66 kg CO_2/kg and the density is 1440 kg/m^3.

Sand

Standard sand for construction with a particle size of 1 mm is used. For future research, it is recommended to use a graded particle size for improved material packing.

3.2.2 Material System: Preparation Procedure

The preparation process is described and visualised below (Fig. 1). Stages 1–3 indicate the preparation of the biochar itself. As the material does not come in an appropriate particle size it has to be processed. Stages 4 to 7 indicate the preparation of the dry-to-wet mix of the mortar.

1. The biochar is ground down from 1 to 10 mm particle size to > 1 mm particle size using an electric grain mill grinder.
2. The ground biochar is sieved through a 300 μm size sieve. All particles that do not pass through the sieve are returned to the grinder. This process repeats until all particles

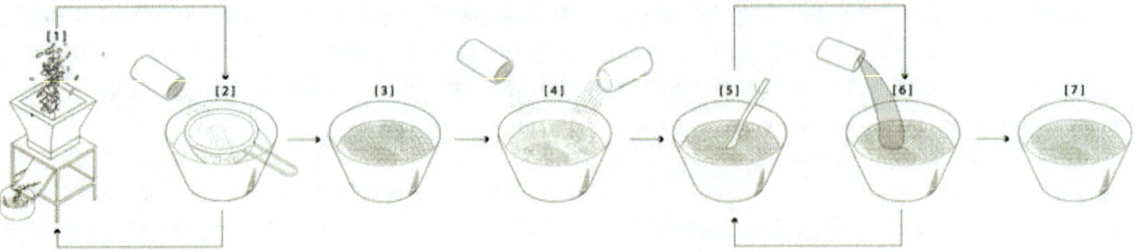

Fig. 1 Material preparation

are sieved and a maximum particle size of 300 μm is achieved.

3. The fine biochar powder is stored in a container with sodium polyacrylate pads in order to maintain low humidity as the biochar exhibited hydrophilic characteristics.
4. Based on the recipes for the five grades of BCMs designated during the planning, the appropriate quantities of biochar, sand, and cement (measured by volume) are added to a container and prepared as dry mixes.
5. After each of the five dry mixes is prepared, it is mixed using a mortar mixer and kept in a humidity-controlled environment.
6. When the moulds for the specimens are ready and covered with a release agent, the dry mix is taken out for the final mixing and water activation. In an iterative procedure, water is added to the dry mix while it is mixed with a mortar mixer. Water is added to the dried mix in steps and the quantity is recorded until workable viscosity is achieved.
7. The prepared mortar is cast in a standardised mould, covered with a damp towel for 48 h and sequentially unmoulded and left for curing for 26 days.

3.3 Material System: Grades

Five different compositions are created with a ranging proportion of biochar from 5 to 50% measured by volume. The proportion of sand is kept constant at 25%. The literature review showed that most research on BCMs is limited to investigating small quantities of biochar up to 20% by volume as partial cement replacement due to the identified decrease in strength as the quantity of biochar increases above 2% of biochar by volume. This research investigates the addition of larger quantities within a simple matrix of sand and cement. The five grades of BCMs that are developed and tested have the following quantities of biochar: 5%; 10%; 20%; 40%; 50% (Fig. 2).

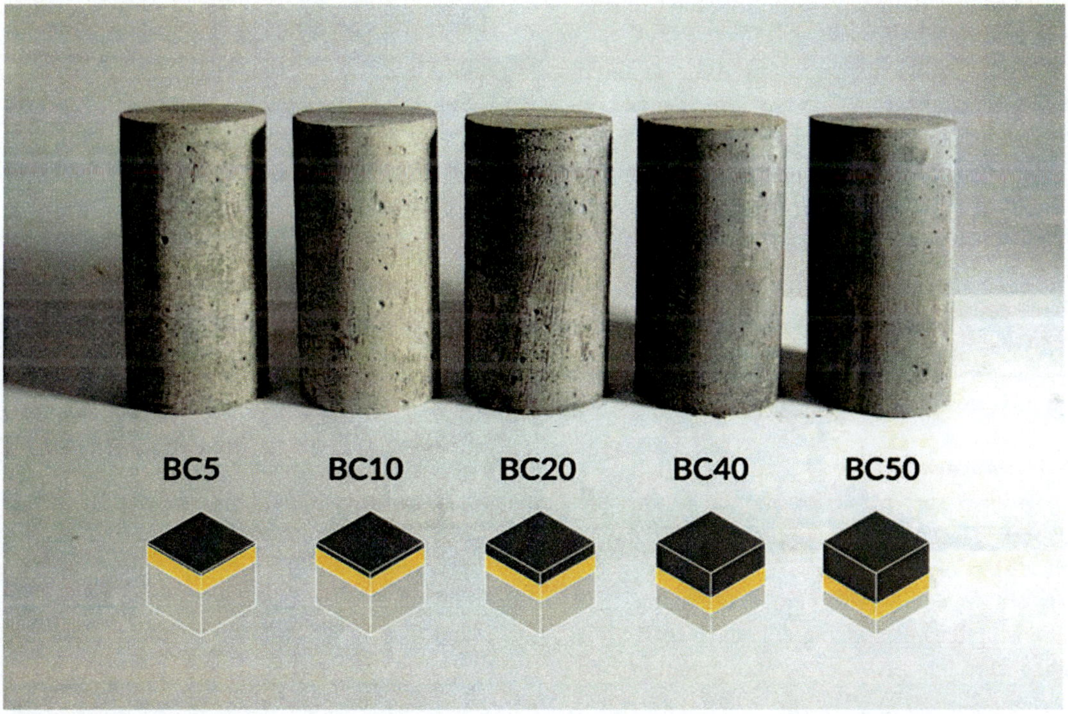

Fig. 2 BCM's developed material system

	BC 5	BC 10	BC 20	BC 40	BC 50
Cement	14	13	11	7	5
Sand	5	5	5	5	5
Biochar	1	2	4	8	10
Water per kg	0.35 l	0.35 l	0.35 l	4L	5L
Total % of Biochar	**5.0**	**10.0**	**20.0**	**40.0**	**50.0**
Embodied carbon (1kg)	**620**	**442**	**126**	**-338**	**-487**
Mass (1m3)	**1419**	**1357**	**1232.5**	**985**	**860.5**
Density (p)	p = 1.419 g/cm^3	p = 1.357 g/cm^3	p = 1.2325 g/cm^3	p = 0.985 g/cm^3	p = 0.8605 g/cm^3
Price (1ton)	**97**	**107**	**125.5**	**164**	**182.5**

Fig. 3 Developed material mixtures

As biochar is the main study variable of the material system it is crucial to investigate what is the impact of the cement substitution with biochar. The amount of biochar and cement are proportionally linked. As the quantity of biochar increases the quota of cement decreases. It is observed that samples with higher quantities of biochar require a higher amount of water during the mixing (Fig. 3).

The primary calculations showed that at about 25% of biochar within the mixture the carbon footprint of the sand and cement is completely offset, essentially creating a net embodied carbon-neutral material. Furthermore, it is calculated that the density of the material system can be programmed with the quantity of biochar ranging from 1.4 g/cm^3 at 5% biochar to 0.8 g/cm^3 at 50% biochar. The price of the material is estimated to double between the 5 and 50% mixtures (Fig. 3).

3.3.1 Material Testing

Standardised specimens for mortar testing are produced. The specimens are rectangular cuboids with a size of 40 × 40 × 160 mm. Six specimens are made for each grade of BCM. The relationship between the ingredients to strength is studied through practical experimentation and standardised testing. The tests are performed with a specialised hydraulic press for cementitious mortar testing (Fig. 4). Testing the specimens is conducted following a standardised procedure and conforming to EN 196 Methods of Testing Cement. The specimens were tested for tensile strength first and after breaking a total of twelve tests for compressive strength were conducted for each mixture.

4 Results

4.1 Observations

When the mechanical tests were performed, it was observed that BCM10 performed significantly worse in comparison to the rest of the mixtures. It is possible that this is caused by impurities and flaws in the cement used for the mixture. The tests with this composition have to be repeated and at this stage are discarded from the experiment. The rest of the mixtures showed a consistent decline in strength as the cement quantity replaced by biochar increased.

During the preparation of the specimens, it was observed that due to the high water absorption of biochar it is necessary to increase the amount of water to achieve the workable viscosity of the composite. The amount of water needed to hydrate the composite extends proportionally to the amount of biochar. The water

Fig. 4 Hydraulic press for materials testing: tensile strength set up on the left; compressive strength set up on the right

absorption leads to a strong binding behaviour of the biochar composite. After the specimens were cured for 28 days, they were weighed and compared. It is evident that the higher amount of biochar results in a lighter composite. Due to the higher water ratio, the biochar composite exhibits a slightly longer curing time.

4.2 Tensile Strength

Flexural strength is a metric that represents the tensile strength of homogeneous materials such as concrete and mortar. The rule of the thumb is that the flexural strength of concrete is about 10–20% of the compressive strength, however, this correlation was not observed with the 5 and 20% BCMs. Concrete and cementitious mortars are generally poor in tension and rely on reinforcement such as rebars or fibres to compensate for it.

There are studies examining the tensile strength of biochar-cementitious composites. Most of the studies that investigate the effect of biochar addition on the tensile strength of the biochar-cementitious composite agree that with about 1% addition of biochar the tensile properties are improved (Gupta and Kua 2020; Cosentino et al. 2019). It is suggested that the higher specific surface area of the biochar is the main reason for the increase in flexural strength, which contributed to the enhanced interaction with the matrix. The current research builds up to this by investigating the addition of higher amounts of biochar.

When biochar was introduced as a partial cement replacement it was interesting to see that there was no decrease in flexural strength until more than 20% of biochar is added. Furthermore, a slight increase was observed as the quantity of biochar increased from 5 to 20%. This indicated that the addition of biochar has a beneficial effect on the flexural strength of cementitious composites when added up to 20% by volume. With higher quantities of biochar, the performance decreased in a linear manner (Fig. 5).

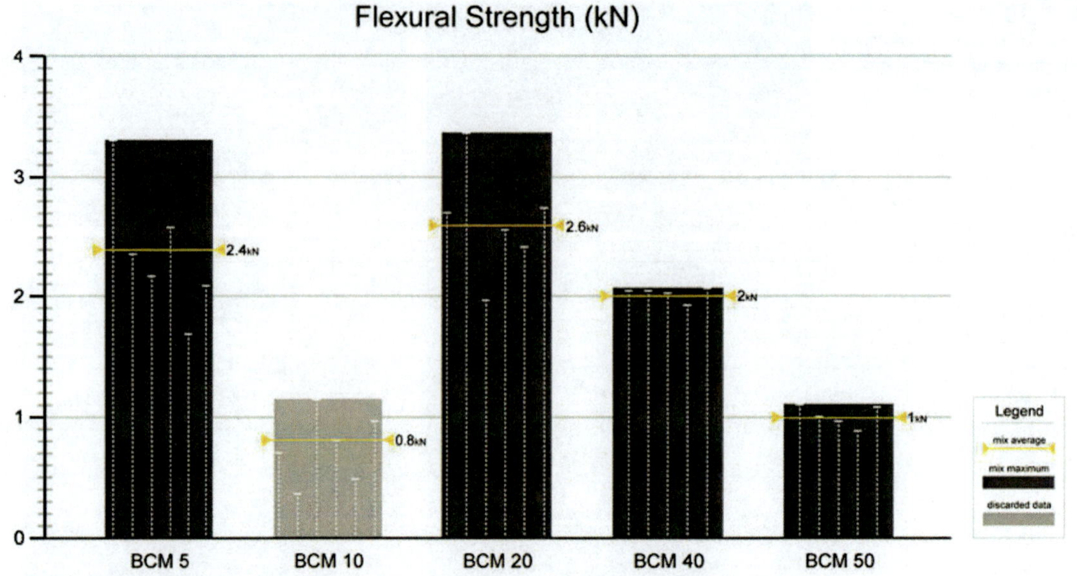

Fig. 5 Flexural strength testing data

4.3 Compressive Strength

Cementitious materials such as concrete and mortars work well in compression and can take up high loads depending on their composition. When it comes to the effect that biochar has on the compressive strength of the composites, most studies agree that a high concentration of biochar compromises the strength of the composite while low concentrations (below 2%) can improve the strength (Cuthbertson et al. 2019; Akhtar and Sarmah 2018; Gupta et al. 2018a, b, c; Gupta et al. 2020; Asadi Zeidabadi et al. 2018; Mrad and Chehab 2019).

One of the few studies where higher concentrations of biochar have been examined is that of Navaratnam et al. (2021). In comparison to the data obtained in this research, the values are slightly lower. In the current research at 5% the compressive strength is recorded at 45.7 MPa and at 20% at 33.9 MPa. This difference could be due to the different compositions of the mixtures with respect to sand quantity or the type of cement.

Similarly, to other studies, the conclusion is that the addition of biochar in higher quantities than 5% leads to a decrease in compressive strength. The same trend is recorded in this research as it is demonstrated that by increasing the amount of biochar in the mixture the compressive strength decreases (Fig. 6).

When putting the obtained data against some industry standards for concretes and mortars, it becomes clear that BCMs have multiple applications in the concrete construction sector. Standard concrete compressive strength ranges from 2500 to 4000 psi which is lower that the values of the BCM20 composite. The BCM50 mixture has the lowest compressive strength but can be used as both type N and type O mortar. The BCM40 qualifies for low-performance type S mortar. Considering both the compressive and flexural strength as well as the net embodied carbon, the most intriguing grade of BCMs is the BCM20. The mixture has a compressive strength that makes it applicable to building construction as it scores in the high range of standard performing concretes that can be used in columns, slabs, and other building elements (Fig. 7).

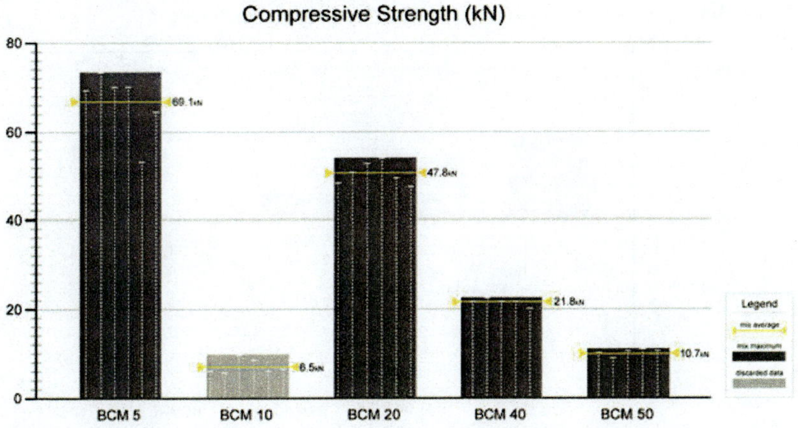

Fig. 6 Compressive strength testing data

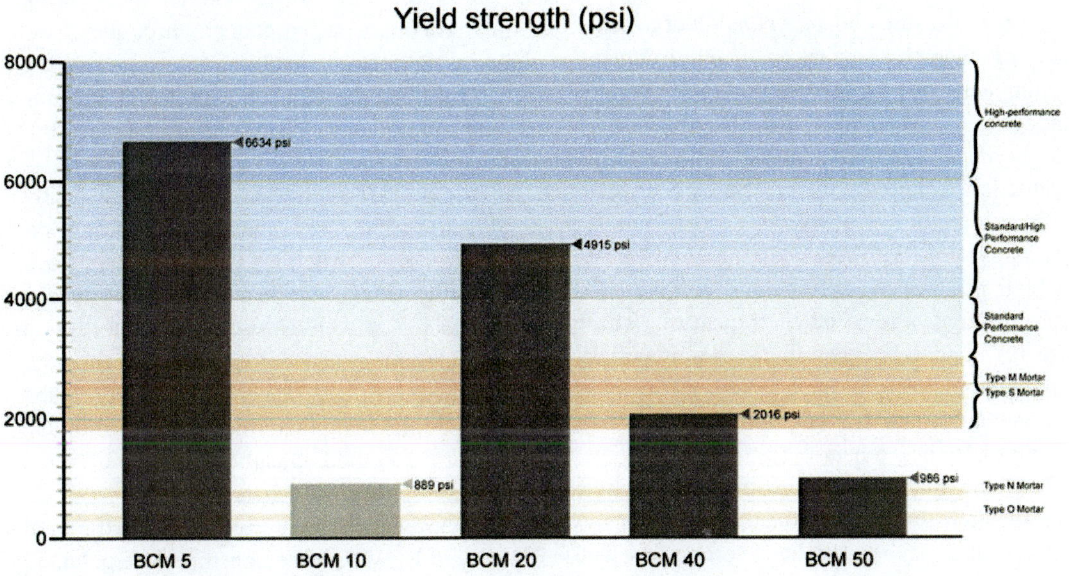

Fig. 7 Yield strength data and standards for concrete and cementitious mortars

4.4 Net Embodied Carbon

The embodied carbon of each mixture was estimated based on the values of the ingredients and their quantities. The values are taken into consideration and only the raw material is related to the embodied carbon. As demonstrated with quantities of biochar above 25% of the embodied carbon of the cement and sand is sequestered and a net carbon-negative material is created (Fig. 8).

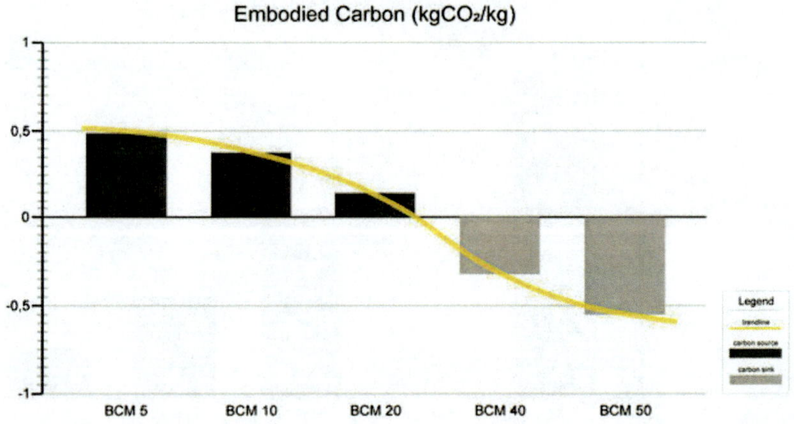

Fig. 8 Net embodied carbon of the developed mixtures

5 Discussion and Conclusions

It is evident that the concrete construction sector has a large negative impact on the environment. To mitigate and reduce this impact, various solutions are investigated and developed both from the material engineering side and from the construction technology side. The cement manufacturing sector is developing solutions for material systems with reduced embodied carbon and high performance relying on additives and waste materials from other industries. These strategies include changes in the raw materials, cement reduction, the introduction of carbon-negative amendments, and novel fabrication techniques, among others.

From the material composition perspective, there are several types of industrial waste or by-products that have pozzolanic properties and therefore can be used as either full or partial cement replacements. Some of the most used by-products are blast furnace slag which derives from the production of iron and steel; silica fume—a by-product of the manufacturing of silicone; and fly ash—a by-product of the coal industry. However, the potential of using waste resources from the agricultural industry is not widely explored, especially when it comes to the concrete construction sector.

The research proposes an alternative solution for a sustainable concrete amendment that derives from the agricultural industry. It is proposed that biochar can be used within cementitious materials to decrease their net carbon footprint through its carbon-sequestering ability. Biochars are produced throughout the globe by the local agricultural industries making them an abundant product in comparison to some of the previously discussed industrial by-products.

It is important to encourage the use of locally produced biochar. This requires material investigation prior to construction as the type of feedstock and organic matter used can affect the performance of the biochar as well as its carbon sequestration potential. Another aspect to be considered is the transportation of the material. The closer the source is to the manufacturing facility and/or site the lower net embodied carbon and therefore the higher carbon offset is generated.

The current research explores the structural capacity of BCMs and demonstrated the potential of designing cementitious materials with net neutral or negative embodied carbon with the controlled addition of biochar. The mechanical properties exhibited showed a consistent relationship between the quantity of biochar and the strength of the material. Due to the identified impact of the addition of biochar on the structural performance of the mixture as well as the carbon sequestration potential, the research proposes that a multi-material allocation would allow for an increase in carbon sequestration without compromising the structural properties. This aspect is crucial for the further development of both the material system and the fabrication process. The developed material system is deemed suitable for additive manufacturing and in the next stages of the research will be tested in a progressive-cavity pump system for concrete printing. The research explored the intersection of material, design, and fabrication, focusing on structural performance. The material system can be applied to a functionally graded design informed by computational simulation and optimisation processes. Even though the research is still in development it demonstrates a possible change in the design practice that is driven by sustainable material systems and creates symbiotic relations between the agricultural and construction industries.

References

Akhtar A, Sarmah AK (2018) Novel biochar-concrete composites: manufacturing, characterization and evaluation of the mechanical properties. Sci Total Environ 616–617:408–416. https://doi.org/10.1016/j.scitotenv.2017.10.319

Asadi Zeidabadi Z, Bakhtiari S, Abbaslou H, Ghanizadeh AR (2018) Synthesis, characterization and evaluation of biochar from agricultural waste biomass for use in building materials. Constr Build Mater 181:301–308. https://doi.org/10.1016/j.conbuildmat.2018.05.271

Birol F (2019) World energy outlook 2019. World Energy Outlook Series

Boslaugh SE (2022) Encyclopedia Britannica. https://www.britannica.com/science/pyrolysis

Byrne C, Nagle D (1997) Cellulose derived composites—a new method for materials processing. Mat Res Innovat 1:137–144. https://doi.org/10.1007/s100190050031

CarbonVivo (2022) CarbonVivo. https://carbonvivo.com/. Accessed 13 Oct 2022

Cha JS (2016) Production and utilization of biochar: a review. J Ind Eng Chem. https://doi.org/10.1016/j.jiec.2016.06.002

Cosentino I, Restuccia L, Ferro GA, Tulliani JM (2019) Type of materials, pyrolysis conditions, carbon content and size dimensions: the parameters that influence the mechanical properties of biochar cement-based composites. Theoret Appl Fract Mech 103. https://doi.org/10.1016/j.tafmec.2019.102261

Cuthbertson D, Berardi U, Briens C, Berruti F (2019) Biochar from residual biomass as a concrete filler for improved thermal and acoustic properties. Biomass Bioenerg 120:77–83. https://doi.org/10.1016/j.biombioe.2018.11.007

Delbeke J, Runge-Metzger A, Slingenberg Y, Werksman J (2019) The Paris agreement. Towards a climate-neutral Europe: curbing the trend. https://doi.org/10.4324/9789276082569-2

Estokova A, Vilcekova S, Porhincak M (2017) Analyzing embodied energy, global warming and acidification potentials of materials in residential buildings. Proc Eng 180:1675–1683. https://doi.org/10.1016/j.proeng.2017.04.330

Gupta S, Kashani A, Mahmood AH, Han T (2021a) Carbon sequestration in cementitious composites using biochar and fly ash—effect on mechanical and durability properties. Constr Build Mater. https://doi.org/10.1016/j.conbuildmat.2021.123363

Gupta S, Kua HW (2020) Combination of biochar and silica fume as partial cement replacement in mortar: performance evaluation under normal and elevated temperature. Waste Biomass Valorisation 2807–2824. https://doi.org/10.1007/s12649-018-00573-x

Gupta S, Kua HW, Tan Cynthia SY (2017) Use of biochar-coated polypropylene fibres for carbon sequestration and physical improvement of mortar. Cement Concr Compos 83:171–187. https://doi.org/10.1016/j.cemconcomp.2017.07.012

Gupta S, Kua HW, Pang SD (2018a) Biochar-mortar composite: manufacturing, evaluation of physical properties and economic viability. Constr Build Mater 874–889. https://doi.org/10.1016/j.conbuildmat.2018.02.104

Gupta, S., Kua, H. W., Pang, S. D. (2020) 'Effect of biochar on mechanical and permeability properties of concrete exposed to elevated temperature', Construction and Building Materials, p. 234. https://doi.org/10.1016/j.conbuildmat.2019.117338

Gupta S, Kua HW, Low CY (2018b) Use of biochar as carbon sequestering additive in cement mortar. Cement Concr Comp 110–129. https://doi.org/10.1016/j.cemconcomp.2017.12.009

Gupta S, Kua HW, Pang SD (2018c) Biochar-mortar composite: manufacturing, evaluation of physical properties and economic viability. Constr Build Mater. https://doi.org/10.1016/j.conbuildmat.2018.02.104

Gupta S, Muthukrishnan S, Kua HW. (2021b) Comparing influence of inert biochar and silica rich biochar on cement mortar—hydration kinetics and durability under chloride and sulfate environment. Constr Build Mater 268. https://doi.org/10.1016/j.conbuildmat.2020.121142

Hahn J (2021) Atmospheric CO^2 is "our biggest resource" says carbon-negative plastic brand made of air'. https://www.dezeen.com/2021/06/24/carbon-negative-plastic-biochar-made-of-air-interview/

Hettinger D (2017) Closing the loop—biochar as carbon negative technology. Living Web Farms. http://living.webfarms.org/biochar-as-carbon-neg-tech/

Iberahim N, Sethupathi S, Goh C, Bashir M, Ahmad W (2019) Optimization of activated palm oil sludge biochar preparation for sulphur dioxide adsorption. J Environ Manage 248. https://doi.org/10.1016/j.jenvman.2019.109302

Li H, Wang Z, Zhang X, Jia Y (2017) Biochar concrete : a new technique for carbon sequestration, pp 1–19

Matovic D (2011) Biochar as a viable carbon sequestration option: global and Canadian perspective

Mrad R, Chehab G (2019) Mechanical and microstructure properties of biochar-based mortar: an internal curing agent for PCC. Sustainability (Switzerland) 11. https://doi.org/10.3390/su11092491

Nair JJ, Shika S, Sreedharan V (2020) Biochar amended concrete for carbon sequestration. IOP Conf Ser Mater Sci Eng 936. https://doi.org/10.1088/1757-899X/936/1/012007

Navaratnam S, Wijaya H, Rajeev P, Mendis P, Nguyen K (2021) Residual stress-strain relationship for the biochar-based mortar after exposure to elevated temperature. Case Stud Constr Mater 14

Peredo-Mancilla D, Ghouma I, Hort C, Ghimbeu C, Jeguirim M, Bessieres D (2018) CO_2 and CH_4 adsorption behavior of biomass based activated carbons. Energies

Polytechnic University of Bari (2021) From farm to façade: finding new uses for agricultural waste. https://ec.europa.eu/regional_policy/en/newsroom/news/2021/04/04-12-2021-from-farm-to-facade-finding-new-uses-for-agricultural-waste

Popescu CA (2014) Equilibrium and dynamic vapour water sorption properties of biochar. Polymer degradation & stability. Elsevier

Rockwood D, Ellis M, Liu R, Zhao F, Fabbro K, He Z, Derbowka D (2020) Forest trees for biochar and carbon sequestration: production and benefits. Appl Biochar Environ Safety. https://doi.org/10.5772/intechopen.92377

Schmidt H (2014) The use of biochar as building material. www.biochar-journal.org/en/ct/3

Sizmur T, Quilliam R, Puga AP, Moreno-Jiménez E, Beesley L, Gomez-Eyles JL (2016) Application of biochar for soil remediation. Agric Environ Appl Biochar: Adv Barriers. https://doi.org/10.2136/sssaspecpub63.2014.0046.5

Sirico A, Bernardi P, Belletti B, Malcevschi A, Dalcanale E, Domenichelli I et al (2020) Mechanical characterization of cement-based materials containing biochar from gasification. Constr Build Mater 246. https://doi.org/10.1016/j.conbuildmat.2020.118490

Suarez-Riera D, Restuccia L, Ferro GA (2020) The use of biochar to reduce the carbon footprint of cement-based. Proc Struct Integr 199–210. https://doi.org/10.1016/j.prostr.2020.06.023

Tan K, Pang X, Qin Y, Wang J (2020) Properties of cement mortar containing pulverized biochar pyrolyzed at different temperatures. Constr Build Mater 263

United Nations Environment Programme (2020) 2020 Global status report for buildings and construction: towards a zero-emissions, efficient and resilient buildings and construction sector—executive summary

Vigneshwaran S (2021) Fabrication of sulfur-doped biochar derived from tapioca peel waste with superior adsorption performance for the removal of Malachite green and Rhodamine B dyes. Surf Interfaces

World Green Building Council (2019) Bringing embodied carbon upfront: Coordinated action for the building and construction sector to tackle embodied carbon. World Green Build Council 35

Irregular Architecture: The Possibility of Systems Thinking for Bamboo Architecture

Jed Long

Abstract

Bamboo is a material that has been used for thousands of years in a wide variety of cultures distributed throughout the world in different geographic and climatic settings (Clark et al (2015) Bamboo Taxonomy and Habitat). The availability of standardized, more durable industrial products has led to the displacement of bamboo within communities with a history of bamboo use (Sharma (2010) Seismic Performance of Bamboo Structures). Research has sought to recontextualize the architectural use of bamboo from vernacular material to modern construction product, citing the sustainable, structural, and social benefits of working with bamboo (Van der Lugt (2017) Booming bamboo: the (Re) discovery of a sustainable material with endless possibilities). However, it is faced by a seemingly insolvable conundrum; how do you standardize an irregular material to meet the regulatory and structural requirements of a modern building industry? The focus of this paper is to posit an alternative question; does bamboo need to be standardized? By examining the different ways bamboo has been used in vernacular cultures and in contemporary design, the paper will highlight systems and methodologies that respond to the irregularity of bamboo, focusing on both architectural systems and broader socio-technical responses to working with bamboo. It is an area of research currently unexplored and presents a novel approach to architecture where systemization allows for a flexibility that can account for non-standardized materials.

Keywords

Bamboo · Vernacular architecture · Sustainability · Material standardization · DRenewables

1 Introduction

Bamboo is a material that has been used for thousands of years in a wide variety of cultures distributed throughout the world in different geographic and climatic settings (Clark et al. 2015). It has formed a key component of different vernacular architectural traditions because of its local availability, rapid rate of growth, unique structural properties, and versatility as a material for construction and use in everyday life. In this form bamboo can be considered to have existed outside of 'modernity' (Dufrenot 2012) in a

J. Long (✉)
School of Architecture + Design, University of Tasmania, Launceston, TAS, Australia
e-mail: jed.long@utas.edu.au

M. R. Thomsen et al. (eds.), *Design for Rethinking Resources*, Sustainable Development Goals Series,
https://doi.org/10.1007/978-3-031-36554-6_34

setting where its material durability and value were dependent upon its context of use. As the various 'bamboo cultures' of use (Yu 2007) developed, their relationship with bamboo has changed, redefining its meaning within the context of a globalized and industrialized world.

The availability of standardized, more durable industrial products has led to the displacement of bamboo within informal architectural contexts (Sharma 2010). Over the course of the twentieth century, cultures with a history of bamboo use have gained access to industrialized products, enabling new forms of construction that have replaced traditional local materials such as bamboo (Yu 2007). However, over the last twenty years, a counter narrative has emerged that champions bamboo as a sustainable building product, celebrating its capacity for local harvest and construction with social and cultural benefits (Van der Lugt 2017). Through the production of landmark architectural works such as Simon Velez's ZERI Pavilion or Ibuku's Green School, bamboo has been promoted as a rapidly renewable resource with unique structural properties that offers a connection to local wisdom and an antidote to the "produced-ness" of modern industrial life (Frey 2013).

These two alternative narratives are the product of two very different contexts of use and the subsequent perceived value of bamboo. The simultaneous decline in bamboo's use within informal architectural contexts and the promotion and increased use of bamboo within an international architectural context presents an inherent contradiction. On one hand, while bamboo has proven to be highly suited to vernacular construction as demonstrated by its widespread use across multiple cultures and contexts for thousands of years, it is being displaced by industrialized building products that are standardized and more durable than bamboo. Conversely, the perceived value of bamboo by the international design community extends beyond its cost and material capacity to include intangible qualities such as its ecological, social, and cultural credentials. As such bamboo is celebrated as a sustainable building material with the potential to help address the UN'S Sustainable Development Goals at various stages of its value chain.

This paper seeks to examine the tension that exists between the desire to promote the use of bamboo at a macro scale as a sustainable building material and the challenges faced when working with bamboo at a micro scale, particularly within informal, low socio-economic settings. The apparent paradox that emerges demonstrates that there is no singular understanding of bamboo and rather it exists in a plurality of roles, dependent upon the socio-material context of its use. In other words, there are many different contexts in which bamboo may or may not be appropriate for use, and many ways of working with bamboo.

Through a review of specific case studies, common themes are established that demonstrate an alternative way of practicing architecture that is responsive to the material, cultures and context of the site and can be situated within broader architectural discourse.

2 Translating the Traditional Use of Bamboo

There are many different forms of bamboo architecture, distributed throughout a variety of different cultures spread across Asia, Africa, and the Americas. Within these cultures, bamboo formed part of the vernacular architectural response to the environmental context, that made use of available resources and technological knowledge to meet the specific needs and values of the community that built them (Oliver 2007). It was architecture of necessity, because it was reliant on locally available materials to create climatically appropriate buildings (see Fig. 1).

In most cases, bamboo was used as part of a hybrid system of construction which was able to address issues of durability and irregularity through combination with other semi-processed materials such as timber, stone, mud or thatch. For instance, the *gassho-zukuri* style *Minka* housing found in the Gifu prefecture of Japan utilizes timber as the primary structure and bamboo as a

Fig. 1 Gassho-Zukuri style Minka housing in Shirakawa-go, Japan (*Image* Jed Long)

secondary structure particularly for the battens and purlins of its roof (May 2010). These buildings are characterized by the thick straw thatch of its A-frame roof, which also ensured the bamboo was not exposed to sun or rain. Within these houses, the upper levels were used for sericulture (silk-worm farming) and the smoke from internal fires rose to warm the space. This had the added benefit of carbonizing the bamboo through the constant exposure to heat and smoke, protecting the bamboo from insect or borer infestation. This demonstrates how bamboo was only used where appropriate and in combination with other locally available materials to create a housing typology that responded directly to the needs of its occupants and the local climate (see Fig. 2).

The durability of bamboo was relative to the local context of use. Because it was often used as a semi-permanent, locally available material it was accepted that after a few years, the bamboo would need to be replaced. A typical connection methodology was to lash the bamboo structure together, as this allowed individual members to be substituted if they began to degrade (Dunkelberg 1985). It was a process governed by the climatic, economic, cultural, and material context of site. However, once these same communities gained access and the economic means to purchase imported industrialized products, bamboo came to be seen as a low quality, impermanent material and is often referred to in various contexts as 'poor man's material' (Yu 2007) (see Fig. 3).

Over the course of the twentieth century, as the vernacular use of bamboo began to be displaced through the introduction of industrialized materials, a paradox emerged with designers and engineers from outside bamboo cultures beginning to speculate on how bamboo could be utilized in contemporary design. Floyd McClure's seminal text 'Bamboo as a Building Material' (1953) formed the basis of a growing narrative promoting the use of bamboo in developing countries because of its vernacular tradition and perceived cultural and ecological credentials. McClure identified the key benefits and challenges of working with bamboo (McClure 1953):

Benefits:

- Culms are of a size and shape that allow for easy handling, storage, and processing.
- Unique structural properties—high strength to weight ratio.

Fig. 2 Traditional roof made from lashed bamboo and teak leaf, Chiang Mai, Thailand (*Image* Jed Long)

Fig. 3 Traditional Bena style housing built from a combination of locally available materials including bamboo in Wogo Village, NTT, Indonesia (*Image* Jed Long)

- Unidirectional fibers make it easy to divide the culm by hand—e.g., splitting or sawing.
- Grown at a village scale.

Constraints:

- Variable dimension—hard to mechanize the fabrication of bamboo.
- Uneven surface—culms are rarely straight, have irregular spaced internodes and rate of taper.
- Tendency to split.
- Low durability.

The research of McClure demonstrated that bamboo has unique structural, material, ecological and cultural credentials that set it apart from other materials and warranted further investigation for its architectural use. However, it also highlighted that as a natural, unprocessed material it did not have the durability and material regularity of comparable processed materials. Building on the work of McClure, subsequent research (Narayanamurti and Mohan 1972; Hidalgo López 1981; Dunkelberg 1985) continued to develop a narrative promoting bamboo as a sustainable, local, and culturally appropriate material for development programs throughout the equatorial regions of the world that was hampered by one key concern- standardization. While advances in research and design have improved the durability and understanding of how to build with bamboo, as a natural material it remains a heterogenous object whose variability in size, dimension and structural capacity make it almost impossible to standardize for modern construction (Janssen 1995).

3 Standardization

Efforts to standardize bamboo have formed a key component of research on the structural use of the material. This process has been supported through the creation of national and international standards that have helped to establish a common language for research and a reference for designers and engineers (Trujillo 2018). The desire to standardize bamboo is derived in part from the recognition that while objectively bamboo has many great ecological and structural qualities, there is a challenge in the implementation and construction of bamboo architecture due to building regulations and the requirement of specific construction knowledge. To date, research has succeeded in establishing standardized procedures for testing and processing bamboo but is unable to address the challenge that the material itself remains' non-standard and thus resistant to use within a regulated and globalized building industry. This suggests that the solution lies in either processing bamboo into a standardized material or accepting that there may be other ways of working with bamboo where standardization is not a priority (see Fig. 4).

For some, the most logical solution to the challenge of standardizing a natural material is to process it. A bamboo pole can be split into strips that are treated, and dried to a prescribed state. This process creates a dimensioned module that can be reconstructed into various laminate bamboo products. These products have homogenous physical and structural characteristics enabling use in widespread construction. There are several initiatives such as the Environmental Bamboo Foundation's Rumah Bambu Lestari program that demonstrate how this process when coupled

Fig. 4 Splitting bamboo into strips (*Image* Kai Wasikowski)

Fig. 5 Grading bamboo strips for lamination, Vietnam (*Image* Jed Long)

with consideration of the entire value chain to account for environmental, social, and cultural benefits can help to address many of the SDGs and a need for safe, durable, and standardized housing (see Fig. 5).

The transformation of bamboo into a laminate product changes its material value, altering its material, physical and cultural properties. Rather than being a round and hollow form with unidirectional fibers that is semi-processed, relatively cheap and lightweight, it has become a processed, rectilinear, solid form that is significantly heavier, more expensive, and visually very different. The processing of bamboo transforms it into a new material 'laminate bamboo' that solves some of the challenges of working with bamboo but also removes all the qualities that may have attracted someone to bamboo in the first place. While laminate bamboo creates new opportunities for the use of bamboo globally in normative construction, it does not provide a solution for those wanting to make use of bamboo in its natural state since it alters its material form (see Fig. 6).

The aspiration for a homogenous outcome for a heterogenous object creates a contradiction that requires a socio-material response as well as a structural response. This is because the choice to use bamboo as an architectural material is dependent on its perceived value, which is objective and dependent on context. As mentioned previously, the displacement of bamboo by industrialized materials within informal architectural contexts is due in part to the challenges of working with a natural material that does not have the same durability as industrialized products. Processing bamboo provides a solution to this challenge by changing the perceived value of bamboo. However, it does not provide a solution for those wanting to work with bamboo in its natural state. If bamboo is perceived to be a desirable material of choice due to its aesthetic, social, ecological, or structural qualities then there is a need for the designer to recognize that as a natural material, bamboo will remain non-standard and subsequently require alternative ways of thinking with the flexibility to allow for its irregularity.

4 Irregular Architecture

There are clear benefits that drive a desire to standardize and integrate bamboo into a homogenous global building industry. But as the building industry continues to increase its consumption of finite resources (OECD 2018) an alternative question can be posited—is the desire for homogeneity the only solution for the built environment or can we rethink the way in which we use resources and takes lessons from working with heterogenous materials such as bamboo that require design to consider social, cultural, and ecological concerns? By examining the way in

Fig. 6 Preparing laminate bamboo beams for shipment, Indonesia (*Image* Jed Long)

which informal and formal design responds to the material variability of bamboo, this paper promotes the idea of 'Irregular Architecture', a solution that requires design and structural systems to maintain a level of flexibility to allow for material variability and places greater agency upon the craftspeople that resolve these challenges in-situ.

The wide variety of vernacular architectural styles that make use of bamboo demonstrates there are many ways of working with heterogenous materials. However, through examination of the different cultures of use, common themes can be extracted that suggest an alternative approach for building with bamboo that does not rely upon standardized forms.

As discussed earlier, semi-processing bamboo into strips or flattening it into panels transforms bamboo into a different form that creates new opportunities. In traditional cultures, examples of this can be seen in the roofs of *Tongkonan* houses in South Sulawesi where layers of flattened bamboo form roofs up to 1 m thick (Widjaja 2019) or in the Dorze and Gamo communities of Ethiopia that weave walls of their houses from strips of bamboo (Asilis 2022).

A second approach is to create a structural system that contains the tolerance or flexibility to account for bamboo's irregular form. This can be achieved either through combining bamboo with complimentary materials or devising structural systems that create redundancy through multiple points of connection (see Fig. 7).

A key example of this can be found in the Caldas region of Colombia where eighteenth-century Spanish builders reimagined the local vernacular building tradition by integrating European construction knowledge to form a new hybrid architectural style (Hidalgo López 2003). The irregularity of the bamboo was overcome by encasing it in a mud or plastered finish like wattle and daub that was known as *bahareque*. When doors or openings were required, timber could be used to form the frames and lintels and over time this technique became highly sophisticated. The city of Manizales offers some of the best examples of this with multistorey buildings imitating European styles seemingly appear to be crafted from stone, when in fact they are made from bamboo and earth. Bahareque has the added benefit on encasing the bamboo which preserves it from degradation and insect infestation. This case study demonstrates that bamboo's non-uniformity can be accommodated through hybrid building systems made from complimentary materials (see Fig. 8).

Fig. 7 A wall in Manizales, Colombia made from bahareque. The concrete render is applied to panels of split bamboo attached to whole bamboo poles. By rendering the bamboo buildings can take on rectilinear forms, giving the impression of masonry construction. (*Image* Jed Long)

Fig. 8 The *Ede House* presents a different iteration of bamboo being used as part of a hybrid structural system. Hanoi Museum of Ethnology, Vietnam (*Image* Jed Long)

A different example of hybrid structural system that accommodates bamboo's variability is the *Ede* houses of Vietnam. Timber forms the primary structure and bamboo is used as a secondary structure reducing the need for the bamboo to have a consistent dimension. This idea can be seen repeated in many different contexts such as the Sumbanese *Uma Mbatangu* or the *Kampung Bena* in East Nusa Tengarra (Achmad 2019) (see Fig. 9).

These common themes carry through to contemporary construction with bamboo and are perhaps most evident in the work of Simon Velez, who employed a hybrid structural system of timber and bamboo for the creation of the ZERI Pavilion at the Hannover World Expo in 2000.

Fig. 9 The 1:1 prototype of the ZERI Pavilion, by Simon Velez. Built using a hybrid structural system of timber and bamboo, Manizales, Colombia (*Image* Jed Long)

While Simon Velez built upon traditional Colombian methods of construction to create new innovative forms.

Alternatively, bamboo's irregularity could be accounted for through structural systems that create redundancy through multiple points of connection. Frei Otto and his team at the University of Stuttgart helped to define this novel approach to the design of bamboo structures through the publication of 'IL31: Bambus-Bamboo' in 1985. Drawing upon Klaus Dunkelberg's research of vernacular architecture (Dunkelberg 1985) and extensive analysis of traditional ways of working with bamboo, Otto and his colleagues speculated on new structural solutions that took advantage of the unique structural and material properties of bamboo (Gaß 1985) (see Fig. 10).

The outcome of this research is best demonstrated through the work of Indonesian design firm Ibuku, and the construction of the Green School in Bali. Working in collaboration with Colombian based bamboo expert Jorg Stamm, the buildings were designed to take full advantage of bamboo's physical properties, utilizing structural systems first discussed in IL31 (Gaß 1985). The flexibility and round form of bamboo were seen to suit grid shells and parabolic forms, leading to the key innovation of building hyperbolic columns from bamboo. They are built by setting out rings at the top and bottom of the column that are divided into a set number of

Fig. 10 Reciprocal bamboo tower holding the roof of Sharma Springs by Ibuku, Bali, Indonesia (*Image* Ibuku)

Fig. 11 Bamboo model making at Bamboo U (*Image* Soben Giordan)

points. The bamboo connects the two points making a lattice that can be joined at each intersection. This system allows for the variability of bamboo as it has a large degree of tolerance and does not require identical units of bamboo for its construction. Rather than mimicking traditional vernacular forms, the buildings of the Green School demonstrate a hybridization of knowledge systems to champion new ways of working with bamboo based upon structural systems with enough flexibility to accommodate the irregular nature of bamboo (see Fig. 11).

The design process for bamboo buildings offers an alternative to normative construction practices. Rather than extensively detailing the building through two dimensional orthographic drawings, Ibuku designs and documents its buildings through models, which allows form and construction technics to evolve. This process responds to the irregular nature of bamboo and places added agency with the skilled craftspeople who translate the model into a building. Because no two pieces of bamboo are the same, each connection presents an individual problem that must be resolved in-situ (Arce-Villalobos 1993). Ibuku can take advantage of Indonesia's long history of skilled craft with bamboo to champion a different form of design and construction that allows for the heterogenous nature of bamboo by making use of local renewable resources and production and combines high- and low-tech building practices with local craft and vernacular wisdom (see Fig. 12).

By considering the implications of designing for material variability, an alternative approach to design emerges that is contingent upon social inputs, since design decisions cannot be conveyed through prescriptive orthographic details and must be resolved through the course of building. This process speaks to Latour's notion of the hybrid, which Jeremy Till describes as a social construction that rubs together things and people, architecture and life (Till 2009). Bruno Latour's (1993) suggestion that production should be at once both intentional and participative describes a condition that can be understood through the lens of working with bamboo. The design of bamboo architecture such as the Green School, requires the guiding hand of a design team and structural expertise of engineers but is only achieved through the embodied knowledge of skilled craftspeople who can resolve the design in-situ. Once completed the building then continues to evolve in response to the programmatic requirements of its users as well as the material upkeep of its form (see Fig. 13).

By reviewing the process of building with bamboo a connection can be drawn to a field of architectural research that is defined through

Fig. 12 Constructing a parabolic column, Bali, Indonesia (*Image* Soben Giordan)

Fig. 13 Setting out the form of Woven Sky, Woodford, Australia (*Image* Kai Wasikowski)

Jeremy Till's concept of Lo-Fi architecture (Till 2009), a theory that built upon the work of Latour and Henri Lefebvre (1991) to promote architecture, as something more than an aesthetic consideration, that also includes the social and the ethical. The idea of architecture as object, is displaced by a process that is collective and requires design and production to be a negotiation rather than an imposition, alert to the physical and social context in which it is situated (Till 2009). The heterogenous and unprocessed nature of bamboo, situates bamboo in alignment with these concepts, as it requires a response that is highly contingent upon context and the interrelation between craft and design. The most successful bamboo projects tend to demonstrate an interplay between the agency of those involved and an overarching structure that provides sufficient constraint and guidance (see Fig. 14).

Industrialization and globalization have promoted the processing of resources into plastic forms that are homogenous, separated from local identity, upon which design is imposed. Working with bamboo challenges this idea by requiring

Fig. 14 Selecting bamboo poles for harvest, Crystal Waters, Australia (*Image* Kai Wasikowski)

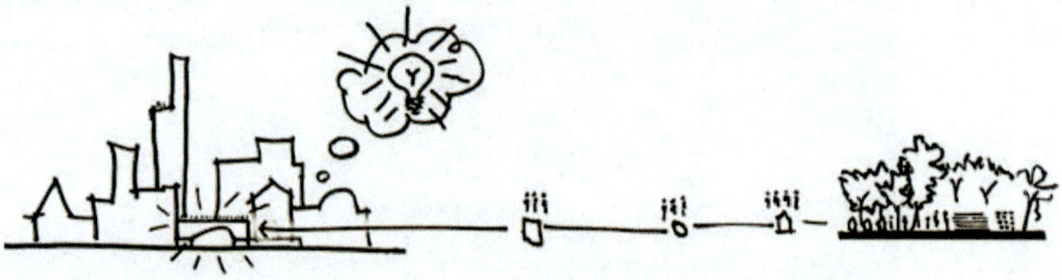

Fig. 15 Simplicity of value chain allows for the social, cultural, economic and ecological impacts of a material to be assessed. (*Image* Scott Francisco)

the design to be contingent and aware of material variability, responding to the unique identity of each member. Bamboo architecture expands agency beyond design professionals, requiring the input of multiple actors at multiple stages, in turn setting itself apart from other forms of architectural production when considering how architecture can play an active role in achieving the UN's Sustainable Development Goals (SDGs) (see Fig. 15).

As a natural material, the social, ecological, and economic impacts of bamboo can be assessed through a simple value chain that connects the cultivation of bamboo to the construction of architecture. In making this assessment the SDGs help to define the different impacts at each stage of the process, beginning with the cultivation and management of bamboo forests which aligns with 'SDG 15—Life' on Land. The harvest of bamboo which is most often used close to the point of harvest is an example of 'SDG 12—Responsible Consumption and Production'. While the final built makes use of bamboo that is most likely carbon negative, aligns with 'SDG 11—Sustainable Cities and Communities'.

The simplicity of bamboo's value chain provides an example of how architecture can play a role in achieving the UN SDGs, extending the value of a material beyond physical and economic attributes to include social and ecological concerns. This is significant when rethinking the way in which we use resources, because, unlike finite fossil-based resources with complex value chains that are extractive and incur significant ecological costs through production, bamboo is a rapidly renewable resource that can be used locally in a semi-processed state and has net positive ecological and social impacts.

By discussing the role bamboo can play in achieving the SDGs, this paper is setting out the perceived macro benefits of bamboo. The work of designers such as Ibuku or Simon Velez demonstrates that bamboo's heterogenous form offers an alternative approach to practicing architecture where flexible structural systems and contingent design can account for material irregularity. However, these projects are able to do this because they have the budget and technical knowledge to do so. When discussing the implementation of these ideas at a micro scale within informal architectural settings, this becomes much more challenging as the sociocultural context is vastly different.

Most of the construction involving bamboo occurs informally in low socio-economic communities with a history of vernacular use (Dunkelberg 1985). In these settings, the low resilience of communities, renders the broader ecological and social benefits of bamboo largely irrelevant, with users prioritizing materials that are economical, durable and aspirational. As such, a different approach is required to reframe how bamboo is perceived and offer simple methods of working with bamboo that is not dependent on technical skill and provide durable outcomes (see Fig. 16).

One example that showcases the successful delivery of bamboo housing for low-income communities, is the work of RAW Impact, an NGO based in Cambodia and founded by Australian aid workers Troy and Nicole Roberts. RAW Impact has been working with vulnerable communities on the outskirts of Phnom Penh, providing a case study of how the SDGs can be implemented at a local scale within marginalized communities. Through the construction of bamboo housing, RAW Impact looks to increase economic and ecological resilience, while also empowering the community through new livelihoods and skills development (see Fig. 17).

The design for the community has evolved out of an ongoing dialogue between RAW Impact, the local community and design studios run at the University of New South Wales and Sydney, facilitated by Casey Brown and Cave Urban. This provides an example of how north–south partnerships can build capacity and deliver local solutions that make use of local resources and fabrication (SDG 17—Partnerships for the Goals). Through a process of learning by making and community feedback, the designs have evolved to suit the local climatic, social, and economic context (SDG 11—Sustainable Cities and Communities). The housing modules are built from locally sourced bamboo which has led to RAW Impact planting their own bamboo plantation to source material directly (SDG 15—Life on Land and SDG 12—Responsible Consumption and Production). Construction is undertaken by international volunteers working

Fig. 16 Bamboo houses built by RAW Impact, Cambodia. (*Image* Nina Annand)

Fig. 17 Khmer workers prefabricating a bamboo roof for RAW Impact, Cambodia. (*Image* Nina Annand)

with local workers who are largely young Khmer men. The Khmer builders were initially unskilled in bamboo construction and so their employment doubles as an opportunity to develop new skills in bamboo construction and create new livelihood opportunities for them (SDG 4—Quality Education and SDG 8—Decent Work and Economic Growth). To ensure a consistent standard of construction, RAW Impact has innovated a construction methodology that accommodates the irregular nature of bamboo and the use of unskilled workers (SDG 9—Industry, Innovation, and Infrastructure). Quality control is achieved through metal frames that are used as a jig to create a standardized form. This reduces the complexity of the construction process by creating a series of repeatable tasks that can be taught to workers (see Fig. 18).

An integral component of this process is the continual reflection and revision involved in improving systems and designs. This forms a feedback loop giving agency to the local community to help determine what should be constructed and how. The design and build

Fig. 18 The lower level of one of the houses built by RAW Impact, Cambodia. (*Image* Nina Annand)

methodology provide a system for construction, but the individual selection and connection of members as well as the final details and finishes are determined by those that build and occupy the houses. While there has been some design input and construction advice from outside designers, most decisions have been made by RAW Impact and those building and living in the structures. This offers a different paradigm for architecture, where the architect and engineer sit adjacent to the construction process, relinquishing individual agency in favor of mutual input by a wide range of actors.

The self-build process that is fostered by RAW Impact is not a new concept, rather it builds upon the legacy of vernacular architecture and more recent architectural movements such as Walter Siegel's self-build housing (Grahame 2015). Making use of lightweight, locally available, and inexpensive materials reduces the need for specialized construction knowledge and creates an opportunity for self-determination and empowerment for those living in low-income housing (Grahame 2015). It provides a different approach to architecture that can be situated with a broader field of architectural theory through correlation with ideas such as Jeremey Till's notion of Lo-Fi Architecture (2009) or the work of Latour (1993) and Lefebvre (1991).

5 Conclusion

The process of working with a heterogenous material defies conformity to the standardized and homogenous nature of contemporary design and construction, requiring a different approach to architecture that assimilates high- and low-tech building practice with local vernacular knowledge and global visions of sustainability and resource use. Responding to the socio-material context of a specific project, the concept of irregular architecture recognizes that bamboo's role as an object occupies a plurality of possibilities, and its relevance is dependent upon the perceived value of those that use it. While there is a tension that exists between the desire to promote bamboo for its macro social, cultural, and ecological qualities and the challenges of building with bamboo, it can be addressed in part through the adoption of design practice that accounts for variation and includes local knowledge as a mechanism of resolving challenges in-situ. By emphasizing flexibility and systemization over standardization, designers can rethink how natural resources such as bamboo are utilized to allow for the involvement of multiple actors and celebrate irregularity as a social condition.

References

Achmad ZH (2019) Typology of Bena traditional architecture, Flores. Local Wisdom: Jurnal Ilmiah Kajian Kearifan Lokal 11(2)

Arce-Villalobos OA (1993) Fundamentals of the design of bamboo structures. Eindhoven University of Technology

Asilis L (2022) Vernacular bamboo structures around the world. https://bamboou.com/design-vernacular-bamboo-structures-around-the-world/. Accessed 10 Jan 2023

Clark L, Ruiz-Sanchez E, Londono X (2015) Bamboo taxonomy and habitat

Dufrenot X (2012) A sociological approach of bamboo exploitation within development-how the space of opinion reveals the 'Real-Development' and its contradictions. In: 9th world bamboo congress, pp 801–7

Dunkelberg K (1985) Bamboo as a building material. In: IL 31: Bambus-Bamboo. University of Stuttgart, pp 38–263

Frey P (2013) Simon Velez: architect mastering bamboo. Actes Sud

Gaß (1985) Bambus = Bamboo. IL: 31. Institut fur Leichte Flachentragwerke, Universitat Stuttgart

Grahame A (2015) 'This isn't at all like London': life in Walter Segal's self-build 'anarchist' estate. https://www.theguardian.com/cities/2015/sep/16/anarchism-community-walter-segal-self-build-south-london-estate. Accessed 10 Jan 2023

Hidalgo López O (1981) Manual de Construcción Con Bambú, Estudios Técnicos Colombianos

Hidalgo López O (2003) Bamboo: the gift of the gods

Janssen JJA (1995) Building with bamboo

Latour B (1993) We have never been modern. Harvester Wheatsheaf, New York

Lefebvre H (1991) Critique of everyday life. Verso, London

May J (2010) Handmade houses & other buildings: the world of vernacular architecture

McClure FA (1953) Bamboo as a building material

Narayanamurti D, Mohan D (eds) (1972) The use of bamboo and reed in building construction. United Nations, New York

OECD (2018) Global material resources outlook to 2060. In: Global material resources outlook to 2060. https://doi.org/10.1787/9789264307452-en
Oliver P (2007) Built to meet needs: cultural issues in vernacular architecture. Built to meet needs: cultural issues in vernacular architecture. Routledge
Sharma B (2010) Seismic performance of bamboo structures
Till J (2009) Architecture depends. MIT Press
Trujillo D (2018) Developments in structural design standards with bamboo. In: 11th world bamboo congress
Van der Lugt P (2017) Booming bamboo: the (Re) discovery of a sustainable material with endless possibilities
Widjaja EA, Indonesia LIP (2019 The spectacular indonesian bamboos. PT. Gudang Garam
Yu X (2007) Bamboo: structure and culture: utilizing bamboo in the industrial context with reference to its structure and cultural dimensions

Fermented Weaves—A Visual Record of Design Enquiry

Phil Ayres, Adrien Rigobello, Claudia Colmo, You-Wen Ji, Jack Young, and Karl-Johan Sørensen

Abstract

The technologies of weaving and fermentation represent two of humanity's most enduring technologies. In the Fungal Architectures project, we investigate their combination in the development of a novel architectural construction concept that utilises mycelium-based composites in conjunction with Kagome weaves. Underlying this investigation is an effort to rethink resources in the making of architecture. This is considered in terms of raw material sourcing with a focus on locality, alternative materialities, material forming processes and the capital infrastructures required for production. Our intention is to develop an architecture that is not reliant upon advanced industrial technologies for its production. Far from being anachronistic, our aim is to reveal fermentation and weaving as deep reservoirs of innovative, disruptive potential, leading to new architectural expressions and production techniques that create a platform for promoting inclusion, equitability, knowledge sharing, community practice and empowerment through accessible craft mastery. In this visual essay, we outline our contributions in relation to the state of the art and offer insight into our practice-based research methodology. This is visually presented as a journey through the project development. Our practice-based approach combines physical and digital experimentation with the objective of developing both the construction concept and the requisite design tools and workflows.

Keywords

Fermentation · Mycelium-based materials · Bio-fabrication · Kagome weaving · Computational design

1 Introduction

Evidence of the use of microbially mediated methods to effect desirable transformations of matter through the technique of fermentation, permeates human history. Across time and across cultures, fermentation has provided accessible means to produce, preserve and enhance food and beverages, extract valued material components through selective decomposition and to produce a wide variety of organic compounds with broad utility.

In the research presented here, we focus on a relatively novel use of fermentation, mediated by filamentous fungi, to bind substrates and produce materials with varying mechanical, functional

P. Ayres (✉) · A. Rigobello · C. Colmo · Y.-W. Ji · J. Young · K.-J. Sørensen
Royal Danish Academy, IBT, CITA, Chair for Biohybrid Architecture, Copenhagen, Denmark
e-mail: phil.ayres@kglakademi.dk

M. R. Thomsen et al. (eds.), *Design for Rethinking Resources*, Sustainable Development Goals Series,
https://doi.org/10.1007/978-3-031-36554-6_35

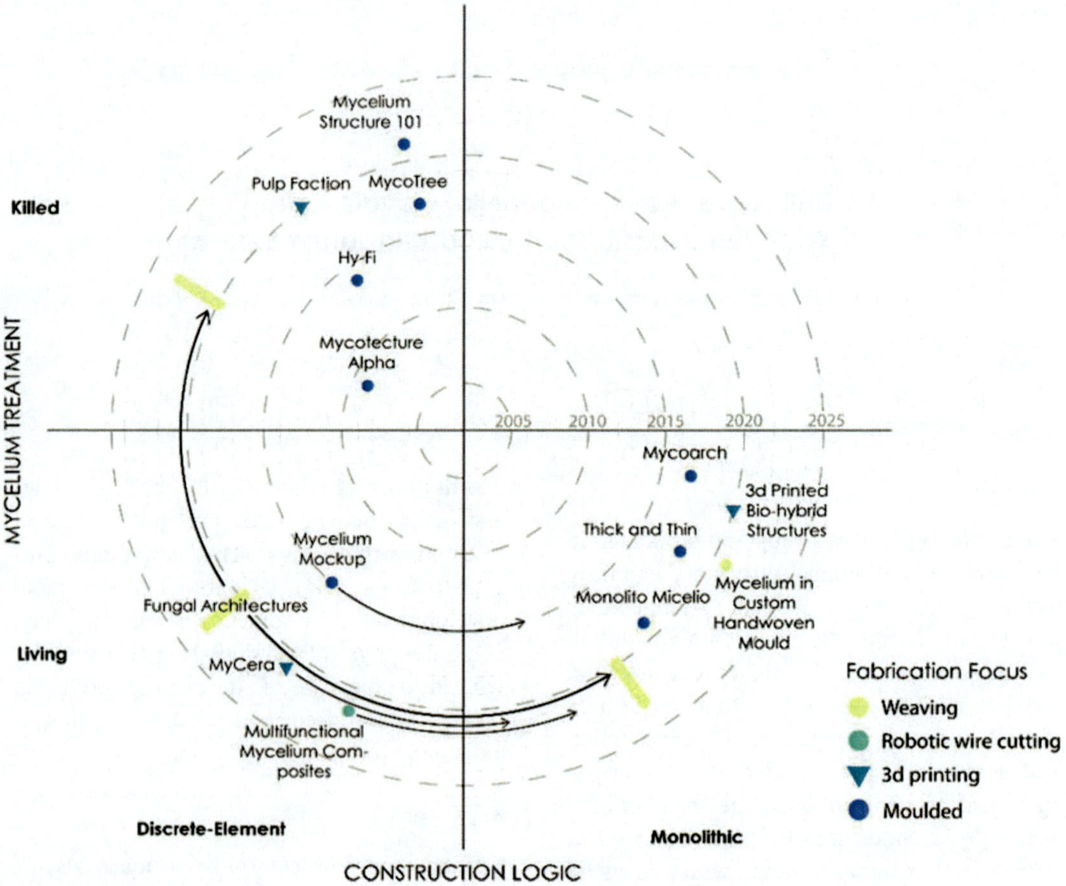

Fig. 1 Mapping of the rapidly expanding field of architecturally focused research projects exploring the use of mycelium-based composite materials through speculative constructions and various fabrication processes. The mapping reveals a sub-set of projects with a vector of investigation that spans the categories of discrete-element and monolithic construction logics. In this hybrid methodology, monolithic structures are targeted through the assembly of living discrete elements that continue to grow and bind themselves together. We investigate a novel approach to this methodology in the full-scale studies presented in this paper

and aesthetic properties. Over the last two decades, research into mycelium-based materials has rapidly accelerated, largely driven by the prospect of creating viable, sustainable and biodegradable alternatives to materials with negative environmental impacts across a wide range of sectors (Fig. 1).

Sapotrophic basiodiomycetes are the most commonly used class of fungi in the production of mycelium-based composites (MBC). This class of fungi are specialised lignocellulosic decomposers and perform cornerstone roles in a variety of ecological cycles. The appropriation of fungi to grow materials necessitates the sourcing and preparation of lignocellulosic substrates. These can be readily found in abundance as agricultural and land-management residues and waste streams (Monforti et al. 2015). When MBC reach their end-of-life, they can be recycled or biodegraded. MBC production can therefore act as a viable component of a circular economy (Grimm and Wösten 2018).

MBC exhibit many properties suited for application within building construction. They can perform well thermally (Xing et al. 2018) and acoustically (Pelletier et al. 2013), are lightweight, fast growing, have low embodied energy and can act as a carbon sink if produced using carefully considered production protocols (Livne 2022). Where these materials underperform mechanically in comparison to common structural materials, the diligent design of structural form can provide pathways to exploitation (Heisel et al. 2017). Within the Fungal Architectures project, our MBC research has extended the state-of-the-art by considering the bulk volume of the substrate as a design space that can be composed with orientated fibres to differentiate and improve mechanical characteristics. This has been explored in the context of compressive behaviour (Rigobello and Ayres 2022) and bending behaviour (Rigobello et al. 2022). In the case of bending behaviour, results showed a significant increase in performance in comparison to the literature and demonstrate clear vectors for further improvement.

Another dimension of innovation is in the combination of MBC with Kagome weaving (Fig. 2). This concept aims to address the need to contain inoculated substrates to target shape whilst the mycelium colonises and fully binds the substrate. Typically, this is achieved through moulds which are often fossil fuel based and disposable. The Kagome weave can be made of lignocellulosic material and therefore acts as a stay-in-place mould, surface reinforcement and nutrition source, as it is colonised by the mycelium (Figs. 3, 4, 5, 6 and 7). Kagome is a triaxial weaving technique that embodies simple principles for generating double curvature (Ayres et al. 2018). As such, this craft can be easily learned and, with time and practice, mastered by anyone with the motivation. Within our research, the principles underlying Kagome have been instrumentalised in a digital workflow for generating principled pattern topologies of arbitrary design targets (Ayres et al. 2021) (Figs. 8, 9, 10 and 11). The weave method allows complex geometries, including high genus morphologies, to be approximated using straight strips of material. This is of particular interest in the context of architectural fabrication as it provides a means of rationalising production by relieving the need to cut, or otherwise shape, components to the target geometry. Geometry emerges through the interaction of material properties and the arrangement of material in a principled topology. However, to achieve this, the material must have local pliability to achieve interlacing, whilst having sufficient stiffness to generate shell-action across the lattice. This material consideration presents the key challenge to the transfer of weave principles to architectural scale, leading to sacrifices in the direct transfer of weaving principles and properties, and therefore limiting the application of craft know-how (Ayres et al. 2020).

By using a proprietary method of timber steaming and compression, recent investigations in the Fungal Architectures project have achieved the direct transfer of weaving methods using solid timber sections up to 60 × 30 mm. Interlacing is manually achieved using basic and widely available tools—clamps and lever arms (Figs. 12 and 13).

As per their basket-scale counterparts, these architecturally scaled elastic grid-shells do not require mechanical or chemical fixings during or post-assembly—the lattice is held together through local jamming action resulting from the network of reciprocal triangles that are the fundamental tectonic motif of this weave method (Figs. 14, 15, 16, 17, 18, 19, 20, 21 and 22).

We believe that the direct transfer of principles across scales in weaving, and the production of materials through local sourcing and fermentation techniques, can help facilitate greater empowerment, inclusivity, equitability, knowledge sharing and community practice within the built environment. The first evidence of this, resulting directly from applying our research within the Fungal Architectures project, is through engagement with the Tasatrupgaard resident's association, Denmark. Here, we have helped to establish an independent field lab, run by Anders Hansen, for the production of mycelium composite materials by local residents. The composite panels will be used to clad a newly constructed community oven building, designed

Fig. 2 Proof-of-concept study showing the constituents of the construction concept—substrate material (centre-left), mycelium spawn (bottom-centre) and a Kagome weave which acts as a stay-in-place mould and reinforcement. The Kagome weave technique allows complex morphologies to be realised using straight strips of lignocellulosic material. The use of straight material strips rationalises fabrication, and their lignocellulosic composition provides a pathway for full colonisation and integration during the mycelium growth phase

by architect Gitte Juul (Hoff Sonne 2022). Such initiatives represent promising socially driven efforts in material production, with positively entangled consequences on the local built environment—across resource sourcing and place making. Here, the industrial imperative of 'scaling up', with its attendant concern to achieve production efficiency and quantity to satisfy market needs, is challenged. These socially concentrated efforts are better described through the notion of 'scaling deep'—creating impactful changes in behaviours and norms through cultural innovation (Moore et al. 2015).

Fig. 3 Proof-of-concept study showing the inoculated substrate packed into the Kagome weave. Analogous to other casting processes, a form is required to hold the inoculated substrate to shape whilst the mycelium colonises and binds. This process can take a few days to a few weeks depending on the mycelium strain, substrates and production protocol used. By conceptualising the formwork as a lignocellulosic stay-in-place mould and reinforcement, we relieve the use of plastics and/or disposable forms common to many production approaches described in the literature

Fig. 4 The first colonised proof-of-concept study after approx. 25 days growth. The rattan weave has been colonised and bound to the contained substrate in most areas. Delamination of the weave from the substrate is visible on the foreground branch. This is largely due to shrinkage from de-naturing using a heat-based process that kills the organism and reduces moisture content within the bulk volume

Fig. 5 A second proof-of-concept component. Greater control of growth parameters, in combination with a modified production protocol employing a substrate pre-inoculation stage prior to weave packing, has resulted in a well ripened and consistent skin encapsulating the weave

Fig. 6 Two studies of doubly-curved geometries—synclastic (top), anti-clastic (bottom)—investigating the integration and bio-compatibility of structurally performing woven carbon-fibre lathes within the mycelium material. The different geometries are generated by altering the regular topology of the weave through the introduction of a non-hexagonal polygon into the weave pattern. The synclastic geometry (top) is produced by introducing a pentagon in the centre. The anti-clastic geometry (bottom) is produced by introducing a heptagon in the centre. Rattan weaves, woven at a smaller density to hold and contain inoculated substrate, sandwich the carbon-fibre and provide a conformal surface for the mycelial growth

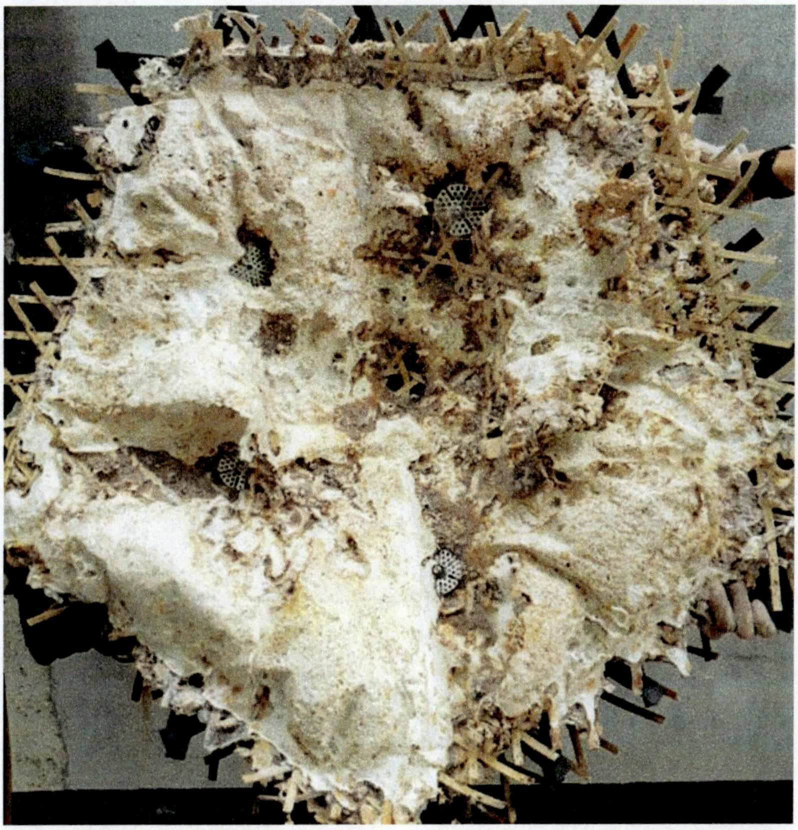

Fig. 7 Result of the synclastic prototype after approx. 25 days growth. Scaling up the component size beyond that of controlled lab conditions is challenging. Witnesses to these challenges include contamination on the rattan ends and inconsistent growth. However, whilst initially aiming for consistency of colonisation and surface conformity, the rich landscape of pockets and differentiated material becomes suggestive of spatial qualities to be articulated

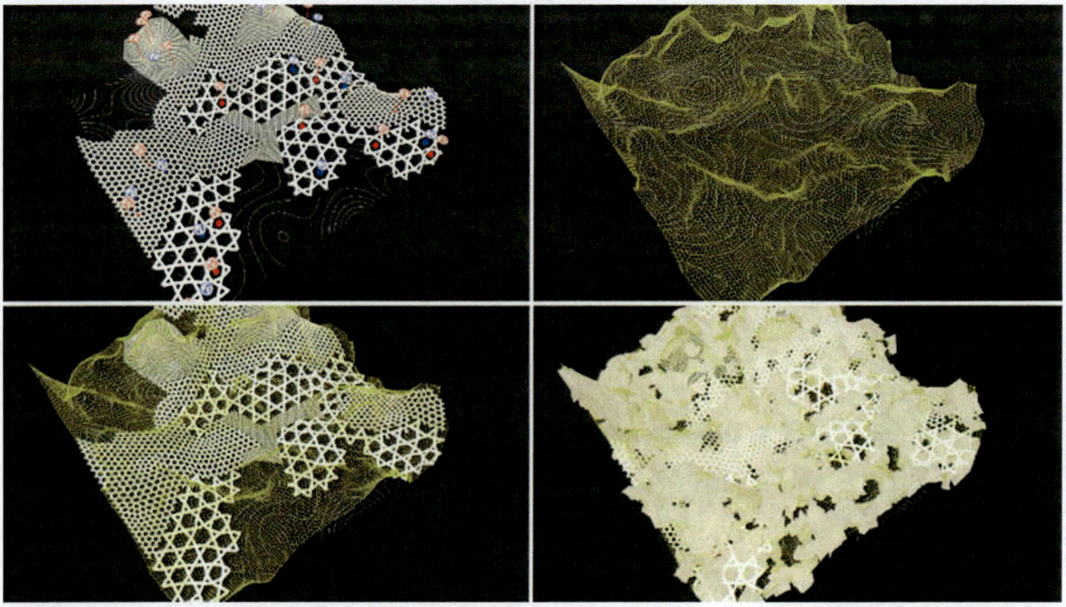

Fig. 8 A digital articulation of the suggestive spatial conditions provided by the growth prototype shown in Fig. 7. Here, we explore a double weave configuration that separates the roles of structure, shown as a white weave, and skin, shown as a yellow weave. This separation creates an additional design vector for creating pockets and varied spatial conditions that nuance distinctions between internal and external. The frame in the bottom right speculates on the partial colonisation of the weaves. We also explore the possibility of the structural weave being woven at different densities, generated through a sub-division strategy in the digital model

Fig. 9 Investigating an alternative weave density strategy compared to the discrete approach shown in Fig. 8. Here, we explore continuous density change and how this might be driven in the digital model through different design parameters. On the left, the weave density is driven by proximity to an arbitrary feature. On the right, the weave density is driven by surface curvature

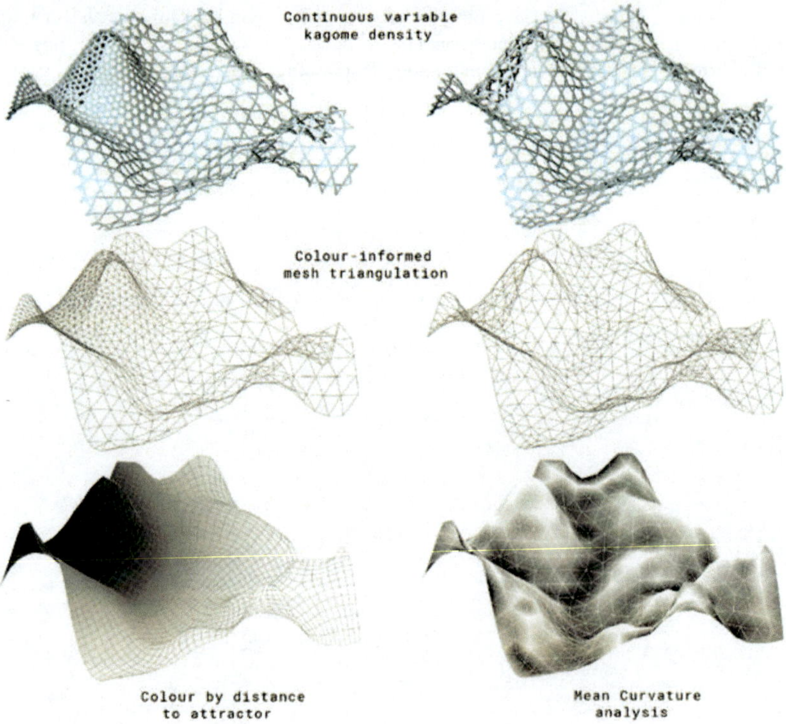

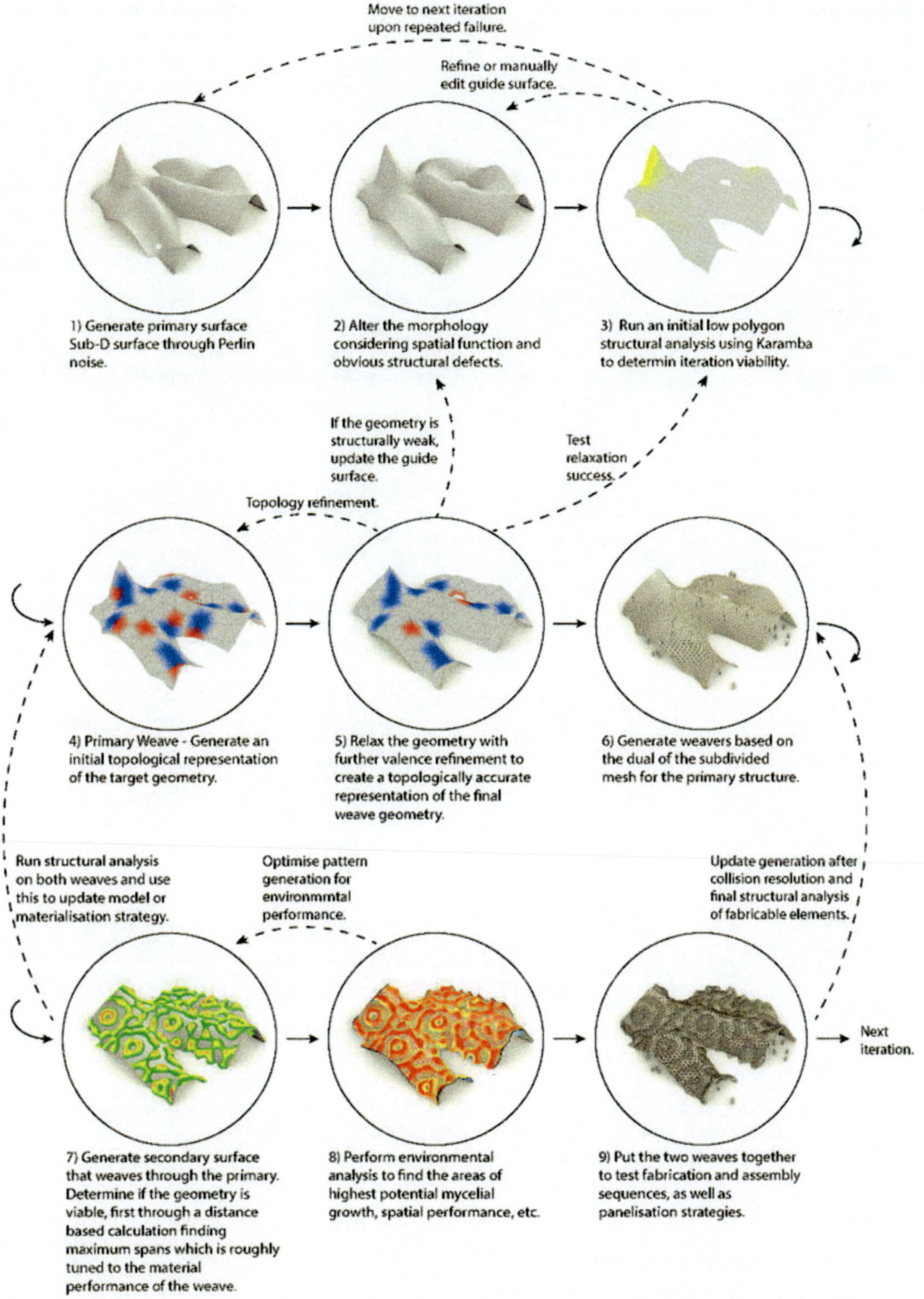

Fig. 10 Preliminary workflow definition for weave generation and design refinement. Design refinement is informed through structural form, spatial organsiation and environmental analysis. Design refinement leads to geometric alteration. Geometric alterations must be fed back to determine if there is an impact on the weave topology. This is due to the geometry being an emergent property of weave pattern topology in interaction with material properties. In the digital model, a significant enough change in geometry results in topological adjustment of the base design mesh

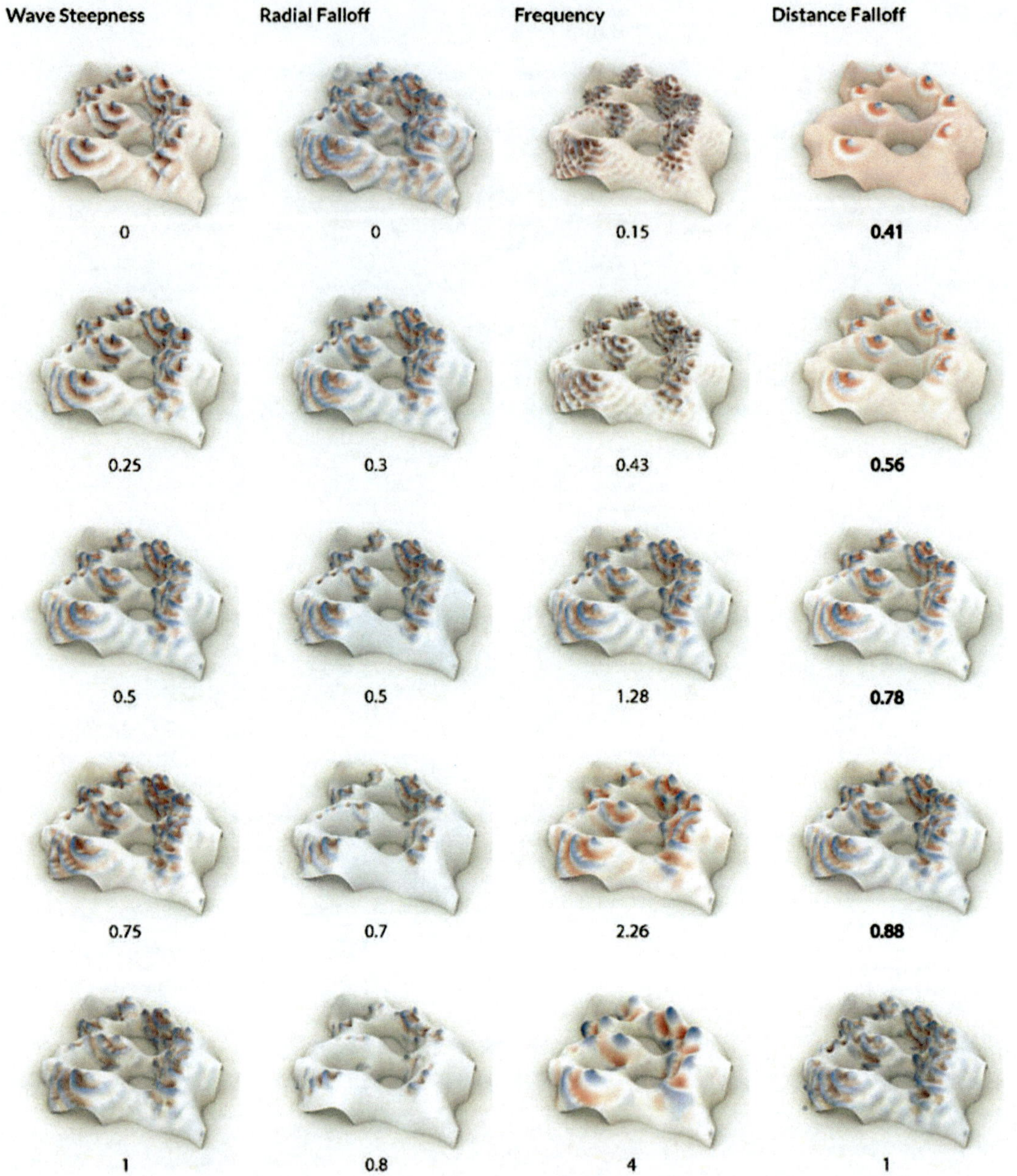

Fig. 11 Surface geometry of the secondary weave is adjusted through parameterised feature control. This matrix defines the geometric features that are controlled (columns) and illustrates the morphological outcomes across changes in the parameter (rows)

Fig. 12 Structural Kagome weave study using 40 × 20 mm section solid beech members. The timber is steamed and compressed prior to weaving. This treatment allows the timber to be bent to significantly smaller radii of curvature than untreated timber. This plasticity remains whilst the timber has a high moisture content. As the moisture content decreases, the timber re-stiffens. This period of plasticity facilitates the direct transfer of weaving principles to large-scale timber sections with architecturally relevant load-bearing capacity. No mechanical or chemical fixings are required during or post-assembly

Fig. 13 Evaluating the geometric deviation between the model and the physical weave. In weaving, topological accuracy is of primary concern as this directly impacts macro-scale geometry. Having established the 'geometric basin' defined by the specific topology, local geometry can be adjusted to achieve a closer net approximation of the design target

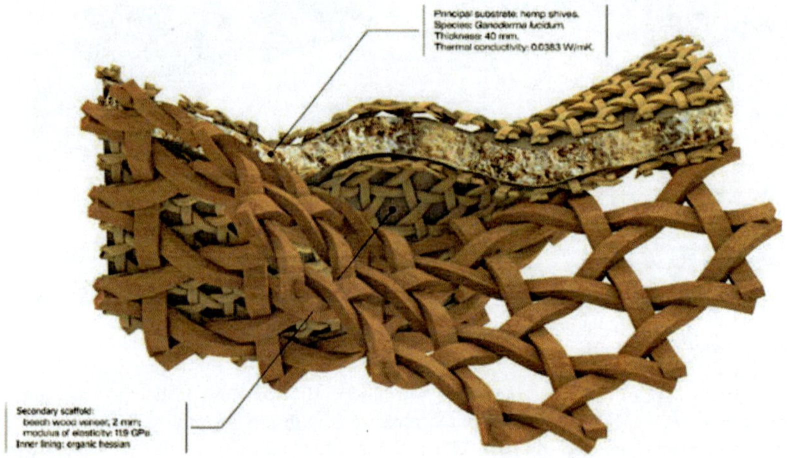

Fig. 14 Detail design study exploring the various layers of the construction concept and indicating notional performance requirements

Fig. 15 Speculative design study of a Fungal Architecture. The architectural motif of the ruin is re-interpreted through a fermented and woven tectonic. Common to the motif are indeterminate qualities poised between building and landscape, culture and wilderness, completeness and fragmentation, decay and renewal. Through the active use of fermentation, the notion of decay, in relation to the ruin, takes on new agency and scales of operation by acting as a process of material construction in the context of mycelium-based composites

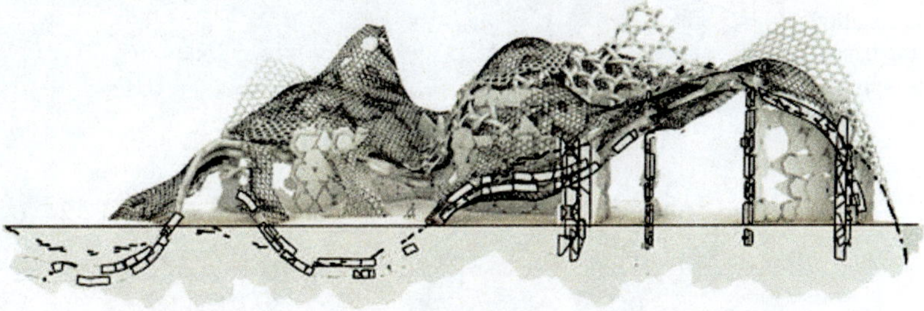

Fig. 16 Sectional study revealing the use of panel elements to create spatial sub-divisions within the grid-shell structure. Additional archaeological qualities are revealed in relation to the ground, adding to the ruinous spatial and tectonic vocabulary being developed

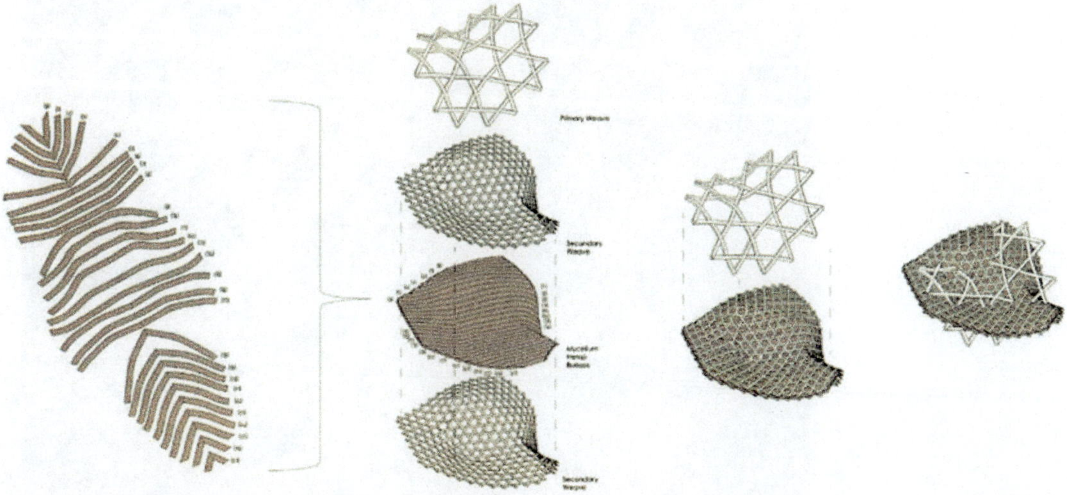

Fig. 17 Study of constituent layers for a full-scale prototype fragment of the Fungal Architecture. On the left are the cutting patterns for hessian 'burlap sausages' to be pre-inoculated and grown in controlled lab conditions. A secondary consolidating growth phase was planned once fully assembled

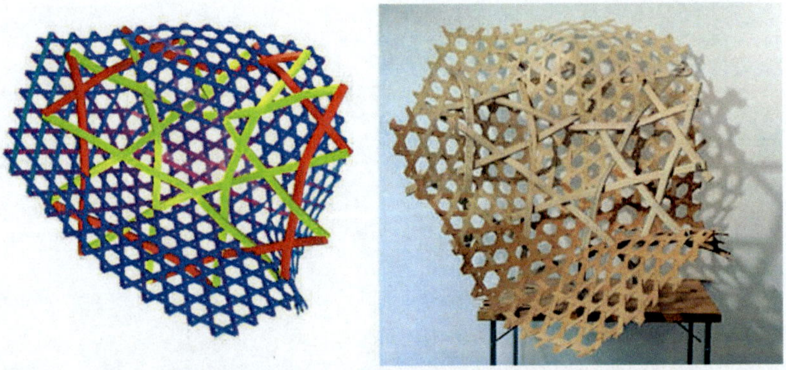

Fig. 18 Digital model (left) and preliminary assembly of the full-scale prototype fragment showing the relationship between the primary weave and the first layer of the secondary weave (right)

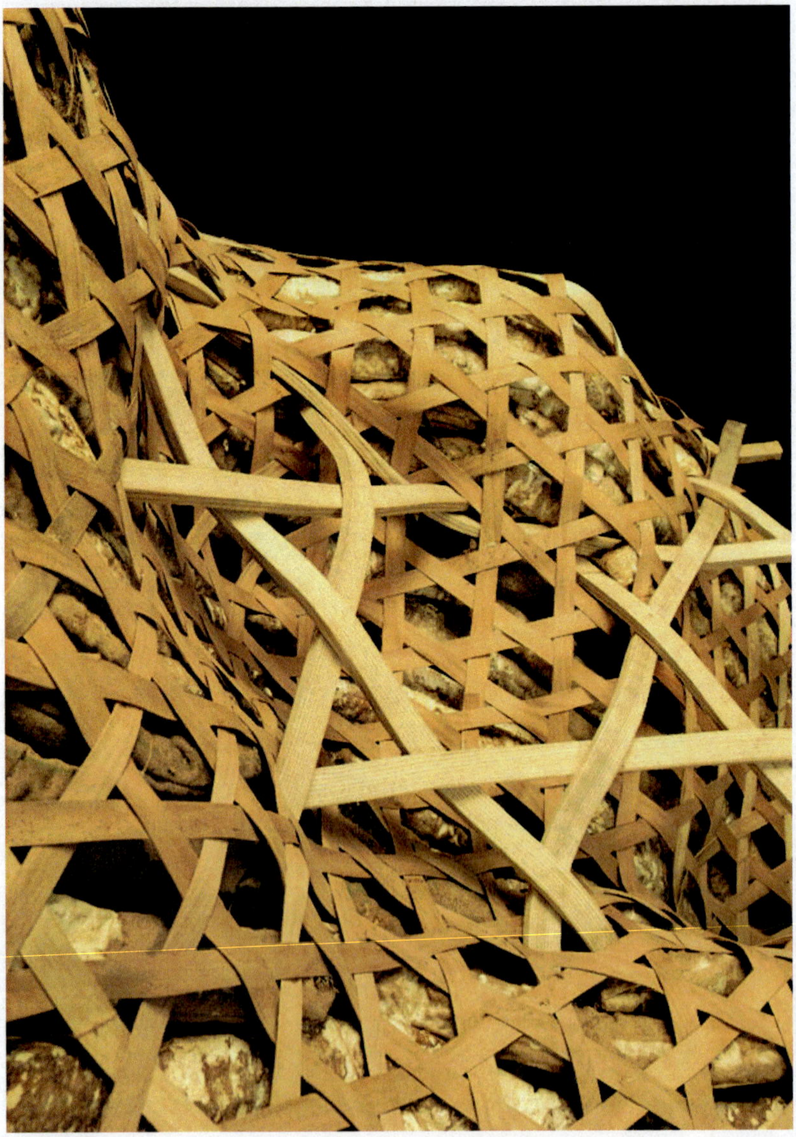

Fig. 19 Detail of the completed prototype. The design intention was to have the secondary weave fully colonised by mycelial growth, encapsulating and binding all elements. In this case, mycelium growth has stalled as seen by the fact the hessian jackets are not fully colonised

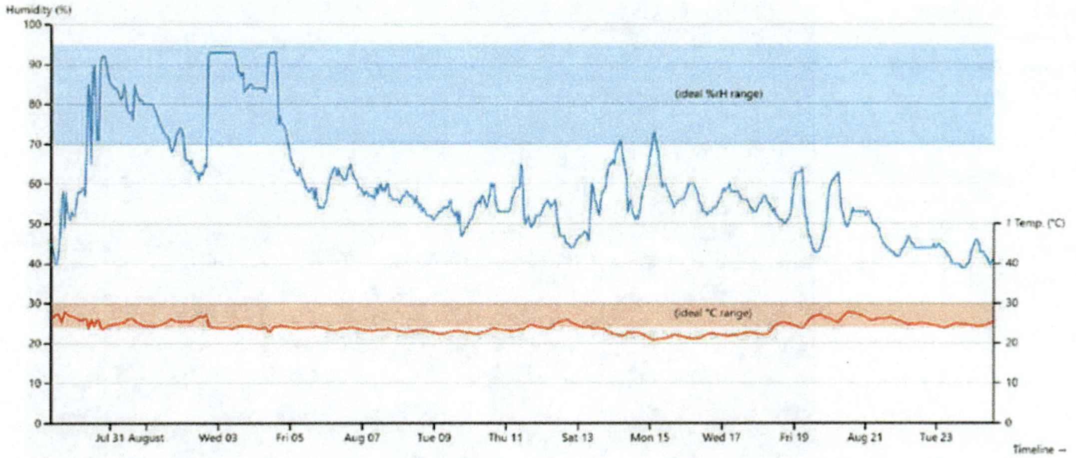

Fig. 20 Environmental record of the growth conditions for the prototype. The data provides an indicator for why the mycelium growth did not colonise as expected—humidity levels were not adequately maintained during the growth period

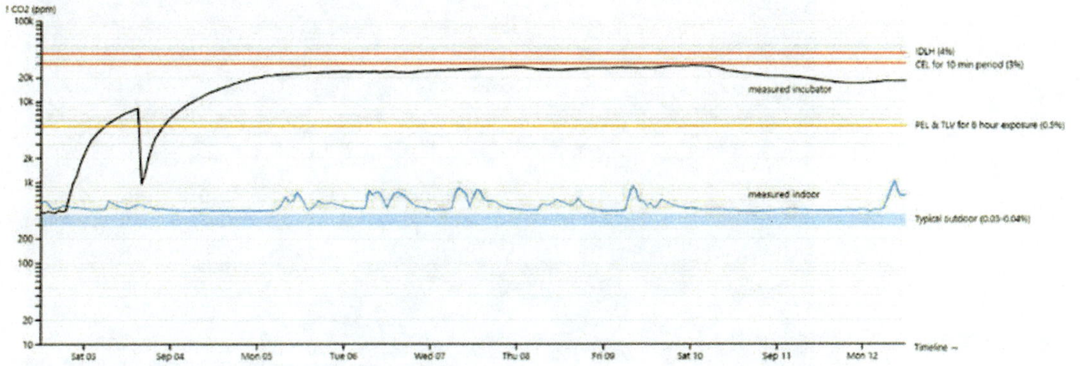

Fig. 21 Environmental record of ambient CO_2 levels v. Incubator levels where good mycelium growth is achieved. This data illustrates another vector of environmental control missing in the prototype growth. As the prototype was exposed to ambient atmospheric CO_2 levels, we can observe that these are far below the concentrations required to promote mycelial growth. The drop in CO_2 levels generally acts as a trigger for a phase shift in growth, from vegetative to fruit forming. This correlated with what was observed on closer inspection of the prototype, where many primordia could be seen forming. Maintaining the required environmental conditions for desired growth is a key challenge of working in the field at larger length scales. The concentrations of CO_2 produced as a metabolic product of the fermentation process should also be noted against health risk levels

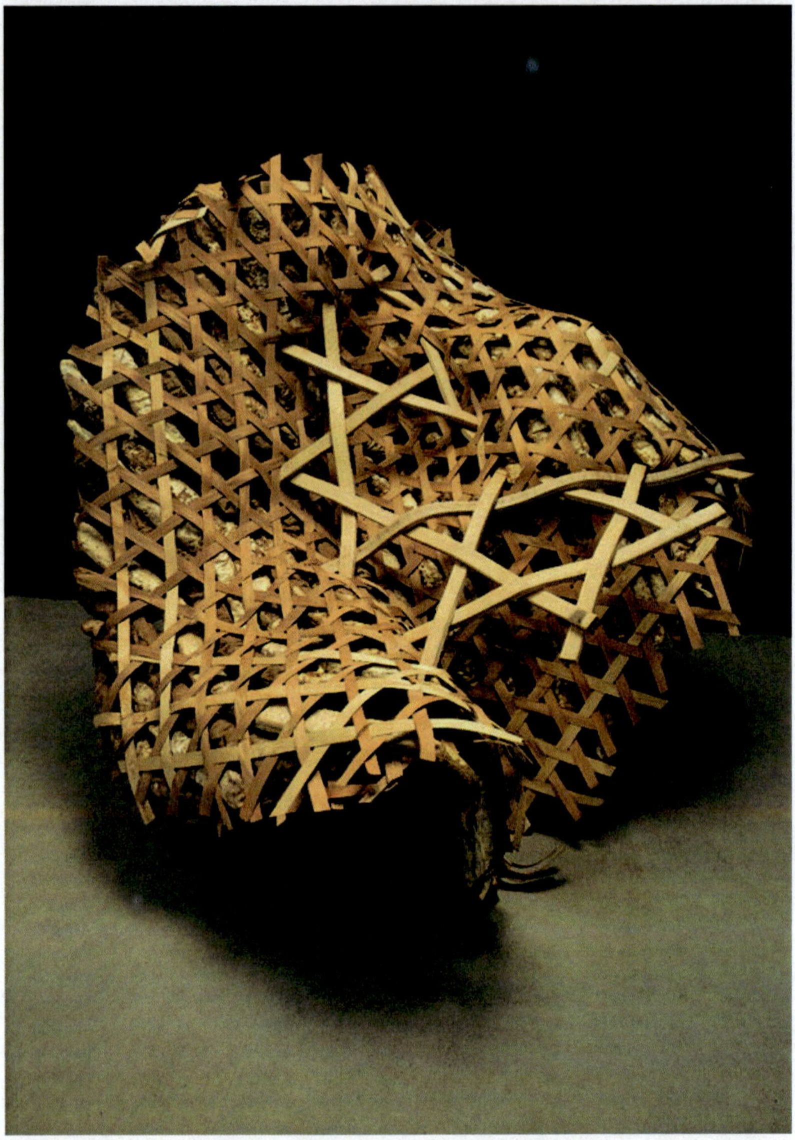

Fig. 22 Completed prototype fragment. Where the objectives related to mycelial growth have not been satisfied, the objectives related to testing the interaction and effect of the two weave densities are successful and lays the foundations for the next research iterations

Acknowledgements This work contributes to the Fungal Architectures project (FUNGAR), which is funded by the European Union's Horizon 2020 research and innovation programme FET OPEN "Challenging current thinking" under grant agreement No. 858132.

References

Ayres P, Martin AG, Zwierzycki M (2018) Beyond the basket case: a principled approach to the modelling of Kagome weave patterns for the fabrication of interlaced lattice structures using straight strips. In: Advances in architectural geometry 2018. Chalmers University of Technology, pp 72–93

Ayres P, Bornaz S, Orlinski A, Heimrath M, Martin AG (2020) Architectural scale Kagome weaving: design methods and fabrication concepts. In: Fabricate: design & making. UCL Press, pp 178–185

Ayres P, You-Wen J, Young J, Martin AG (2021) Meshing with Kagome singularities: topology adjustment for representing weaves with double curvature. In: Advances in architectural geometry. Ponts Chaussées, pp 188–207

Grimm D, Wösten HA (2018) Mushroom cultivation in the circular economy. Appl Microbiol Biotechnol 102:7795–7803

Heisel F, Schlesier K, Lee J, Rippmann M, Saeidi N, Javadian A, Block P et al (2017) Design of a load-bearing mycelium structure through informed structural engineering. In: Proceedings of the world congress on sustainable technologies, pp 1–5

Hoff Sonne FG (2022) Forskere 'tryller' svampe om til mursten og pølser i jagten på grønne byggematerialer. https://videnskab.dk/kultur-samfund/forskere-tryller-svampe-om-til-mursten-og-poelser-i-jagten-paa-groenne. Accessed 12 Jan 2023

Livne A, Wösten HA, Pearlmutter D, Gal E (2022) Fungal mycelium bio-composite acts as a CO_2-sink building material with low embodied energy. ACS Sustain Chem Eng 10(37):12099–12106

Monforti F, Lugato E, Motola V, Bodis K, Scarlat N, Dallemand JF (2015) Optimal energy use of agricultural crop residues preserving soil organic carbon stocks in Europe. Renew Sustain Energy Rev 44:519–529

Moore ML, Riddell D, Vocisano D (2015) Scaling out, scaling up, scaling deep: strategies of non-profits in advancing systemic social innovation. J Corp Citizsh 58:67–84

Pelletier MG, Holt GA, Wanjura JD, Bayer E, McIntyre G (2013) An evaluation study of mycelium based acoustic absorbers grown on agricultural by-product substrates. Ind Crops Prod 51:480–485

Rigobello A, Ayres P (2022) Compressive behaviour of anisotropic mycelium-based composites. Sci Rep 12 (1):6846

Rigobello A, Colmo C, Ayres P (2022) Effect of composition strategies on mycelium-based composites flexural behaviour. Biomimetics 7(2):53

Xing Y, Brewer M, El-Gharabawy H, Griffith G, Jones P (2018) Growing and testing mycelium bricks as building insulation materials. In IOP conference series: earth and environmental science, vol 121. IOP Publishing, p 022032

Regenerative Material-Human Ecologies: Investigating Mycelium for Living and Decentralized Architectures in Rwanda

Nina Sharifi, Yutaka Sho, Daekwon Park, Morgan Noone, and Kiana Memarandadgar

Abstract

The perpetuation of highly carbon-intensive construction practices by wealthy industrialized nations in non-industrialized contexts, in conjunction with the globally asymmetrical effects of climate change, has brought about long-lasting ecological and economic issues. The present work aims to embody an alternative approach to reconsidering resources, defined here as the material, knowledge, and economic flows that constitute an ecology of building, for non-industrialized contexts through the application of regenerative design principles. Sited in a semi-rural site in Kigali, Rwanda, this project investigates the potential for the co-development of living biocomposite construction through cooperative knowledge and supply ecosystems that enable decentralized architectures through local design agency. The unique properties of mycelium, the filamentous networks of fungal organisms, are utilized to test the adaptability of fabrication and construction models through local flows of material and information. The work aims to (1) test, prototype, fabricate, and install mycelium-based modular systems onsite, and (2) focus existing local knowledge on the effort of building a sustainable framework for design-build that enables continuity-in-place. By engaging with local stakeholders, including farmers, craftspeople, mushroom growers, organizations, and builders, the project envisions the reconsideration of resources through the culmination of a self-supporting material-human ecosystem.

Keywords

Regenerative design · Non-industrialized contexts · Mycelium · Local knowledge · Sustainable framework

1 Introduction

Construction and demolition waste account for over 35% of landfills (Papargyropoulou et al. 2011) and 27% of global greenhouse gas (GHG) emissions (IEA 2019). Carbon emissions initiatives range from international to local scales (Behuria 2021), but high-emission material life cycles continue to be found in design and construction practices (Hertwich 2021). Sustainable building certification systems such as LEED,

N. Sharifi (✉) · Y. Sho · D. Park · M. Noone · K. Memarandadgar
School of Architecture, Syracuse University, Syracuse, USA
e-mail: nmsharif@syr.edu

Y. Sho
e-mail: ysho@syr.edu

D. Park
e-mail: dpark103@syr.edu

M. R. Thomsen et al. (eds.), *Design for Rethinking Resources*, Sustainable Development Goals Series,
https://doi.org/10.1007/978-3-031-36554-6_36

BREEAM, and others (Olanrewaju et al. 2022) have been developed with consideration toward embodied carbon life cycle of materials and environmental metrics (Hafez et al. 2023), measures which may play a role in advancing the integration of novel material approaches in industrialized contexts. There is an urgent need, however, for material innovation that responds to energy, environmental, and cost criteria in non-industrialized contexts where waste and emissions-intensive construction prevails without access to wealthy industrial countries' problem-solving technologies and other resources (COP27 2022). In response to increasing demand for affordable housing and other building types in many African countries, strategies have been developed for design and construction that improve energy efficiency and life cycle performance while keeping costs low (Purchase et al. 2021). Development in non-industrialized regions, however, is often inundated by foreign interests (Mahmood et al. 2020) and relies on imported concrete and other emissions-intensive materials for construction (Bah et al. 2018). In Africa, the circular economy concept (Bilal et al. 2020) is being promoted to foster innovation that embodies principles of regenerative design (Attia 2018). Organizations such as the African Circular Economy Alliance promote the elimination of waste and pollution, using and reusing local materials, and regenerating the natural ecology. The major role that buildings play in the generation of waste, pollution, and resource extraction calls for the development of architectural materials that embody regenerative principles.

As designers seek materials with lower carbon emissions, fungal mycelium has emerged as a basis for biocomposite modular systems in buildings. Mycelium, the filamentous network of hyphal elongated cells that make up the majority of fungal organisms, binds organic substrates that can be sourced from abundant agricultural and construction waste such as bran (Sisti et al. 2021), grain husks, or waste wood material (Jones et al. 2020). As mycelium secretes enzymes to partly digest organic substrates, its hyphal network grows and becomes denser in response to environmental factors including temperature, substrate type, and growth time, which can be controlled in fabrication (Pelletier et al. 2013). It is low-cost and biodegradable, and can be recycled upon mechanical disaggregation (Holt et al. 2012).

Mycelium biocomposites with solid lignocellulosic substrates, such as wood fiber, have been studied for their structural properties under varied growing conditions (Soh and Ferrand 2023; Elsacker et al. 2019) and for linkages between cell growth behavior and material characteristics such as density (Sun et al. 2022), which can affect weight and strength. Mycelium biocomposites have also been tested for thermal and acoustic properties (Schritt et al. 2021; Jiang et al. 2017; Jones et al. 2020). While examples of built projects have been recorded, such as the mycelium pavilion at the 2019 Dutch Design Week by artist Pascal Leboucq, full-scale exhibit structures have not been accompanied by documentation of the methods used to produce the materials, construction details, or specific design strategies that enabled integration of the biocomposite. Full-scale testing of a living material system for long-term construction requires investigation of growth factors as they relate to material properties. Finally, mycelium biocomposites can be fabricated with little energy and widely available organic waste materials, which may assist the development of local, decentralized manufacturing networks that align with regenerative and circular economic principles. Such networks could provide a bulwark against dependency on imported construction materials and labor while fostering a local product and design network. However, most mycelium explorations target the global North and applications where technology, materials, expertise, and controlled lab spaces are readily available. Documentation of how to implement co-development of mycelium-based material systems and architecture in non-industrialized regions is needed to realize the benefits of a circular material design, manufacture, and reuse ecosystem.

The development of mycelium-based biocomposites is aided by their geometric plasticity and modifiable growth behavior (Appels et al.

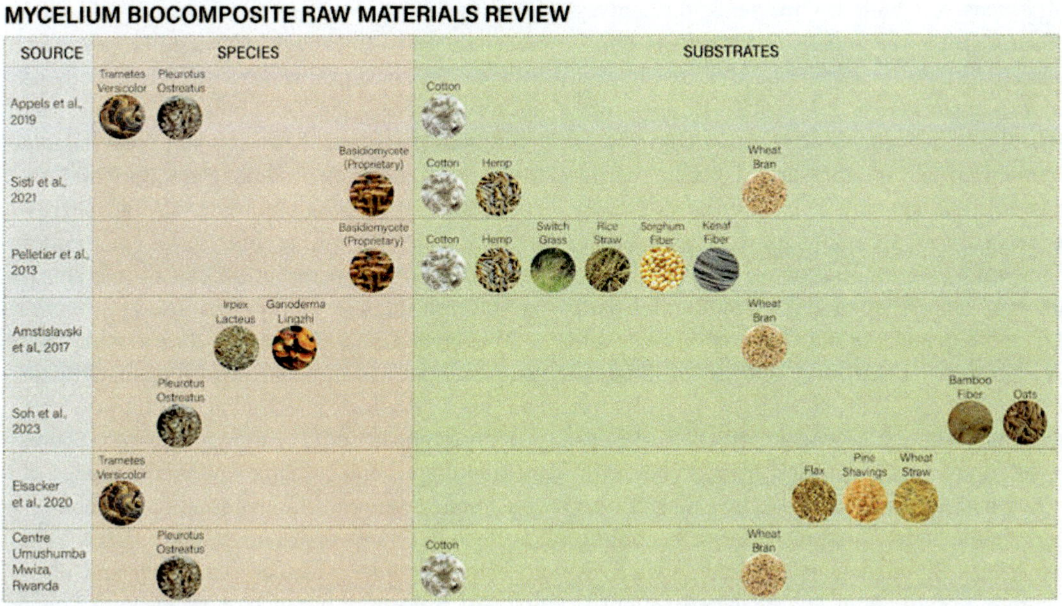

Fig. 1 Summary chart of an array of raw materials for mycelium biocomposites found in prior work

2019). The potential for interior finish applications, including ceiling panels or partitions, acoustic panels, and insulation, has been documented (IEA 2018; Pelletier et al. 2013). A wide range of mycelium and organic substrates have been used to produce biocomposite variations (Girometta et al. 2019) as seen in Fig. 1.

The ability to control certain physical properties of a mycelium biocomposite module, such as a block or panel, is useful in considering possible applications in the built environment. Whether developing structural, insulation, or finish materials, regenerative design principles prompt consideration toward emissions, waste, and human and ecological health criteria. A regenerative paradigm as a point of reference suggests the need to reevaluate the limited ability of conventional material components to address such criteria. Polystyrene foams and concrete, for example, while lauded for their material properties, remain part of the fossil fuel-based linear consumption-and-waste paradigm. For certain applications in which insulation, structure, freedom of form, and versatility are needed, mycelium biocomposites may provide viable alternatives, particularly where industrialized supply chains are not present in many regions in the global South.

A case study is presented in a semi-rural community in Rwanda, where the team has practiced and conducted research since 2008. Industrial building materials such as concrete, steel, and finish materials are imported from Uganda, Kenya, Tanzania, and China, and tariffs and transport costs are high. Landlocked and steep terrains of the country add cost, and the construction is said to be 30–40% more costly than neighboring Kenya. The industrial materials are out of reach for low-income Rwandans, 60% of the population living on less than $1.25 a day (Weatherspoon et al. 2021). Research on 370 rural homes and their built environment conducted in 2019 showed only 3% of the household had access to water within their properties, and 11% had access to electricity (Sho Forthcoming). With few paved roads and limited vehicular access, to build affordably and sustainably in Rwanda is a Herculean task.

Industrially produced construction materials may be inaccessible for most in Rwanda, but soil, labor, and knowledge of self-building processes are abundantly available (Sho 2014a). Typically, rural homes are built by residents themselves with sun-dried adobe blocks or wattle-and-daub. Rwandan agriculture supports a diet of cassava, potatoes, amaranth, and bananas. Entities like Kigali Farms have introduced mushroom growing entrepreneurship, for sale of edible mushrooms at urban supermarkets. In a news article, Audace Hirwa, Director of Documentation, Publication, Communication and Technology Promotion Unit at Rwanda Agricultural Board (RAB) commented on the country's focus on mushrooms as an effective means of income generation and cheap source of protein for Rwandans. "The edible mushrooms reduce poverty, stunting and malnutrition among children," he explained, adding that medicinal mushrooms invigorate the body's immune system. According to the article, 32 companies and 20 cooperatives are currently cultivating around 629,200 mushroom growing kits involving 123 individual farmers, harvesting around 523,600 tons per year (Nkurunziza 2021). With evolving practices and government policies, there may be opportunity in the complementarity between existing agricultural and construction knowledge and the development of mycelium-based biocomposites.

Developing alternative modes of construction in non-industrialized communities requires collaboration to avoid the historically common inculcation of foreign culture into indigenous design identities through architectural artifacts. Global development entities may excel at solving immediate problems or beautifying urban built environments, yet they have missed qualitative factors such as locally acceptable aesthetics and the economic and knowledge gaps the development projects may exacerbate (Sho 2014a). In Rwanda's capital city Kigali, soaring office towers, shopping malls, and luxury housing projects stand as architectural evidence of the miraculous economic comeback from the 1994 genocide (Effective Logistics and Consultancy Group 2020). These urban infrastructural and construction projects are made with imported cement, steel, and glass, and many are contracted by Chinese companies (Chinese Loans to Africa Database 2022; Malik et al. 2021). While their Euromodernist aesthetics may signify progress, little of the industrialized construction methods are transferred to the local building sectors, and the actual spaces and their programs are available for a small fledging middle class (Goodfellow 2017; Sho 2014a; Goodfellow and Smith 2013). In contrast, in rural regions, 95% of homes necessitated repairs, made difficult by the shortage of cash income (Sho Forthcoming).

Where earth-based brick or block structures are commonly found, but pose certain climatic and practical issues in Rwanda, the introduction of mycelium-based design and fabrication methodologies could enable alternative modes of construction through the material's affordability, low technical and skill barriers in construction, and circular life cycle. The scalability and plasticity of mycelium-based biocomposites present opportunities for expression of construction logics and aesthetic identities. There is potential to complement existing knowledge, skills, and material networks, introduce labor opportunities, and foster mutual augmentation with growing economic activity, as outlined below in Fig. 2.

Agricultural waste, or agriwaste, was identified as an affordable and locally abundant source of raw materials to work with in Kigali. The Rwanda Agriculture Board (RAB) uses cottonseed hulls as a substrate in their mushroom center in Southern Province, as do Kigali Farms (Umubyeyi 2021a), who also suggested bean straw, wheat straw, and maize residual materials. Other substrates used in Rwanda are rice bran, wheat bran, maize husks, and bean and soybean residual materials. These substrates are mixed with water and limestone powder or ash to produce the biocomposite (Umubyeyi 2021b).

In addition to organic agriwaste, Rwanda produces an abundance of inorganic waste. In Rwanda, manufacturing, use, importation, or sale of single-use plastic bags has been prohibited since 2008, but plastic water bottles and takeout containers are regularly used. As seen in Fig. 3, plastic and other unrecycled material waste flows

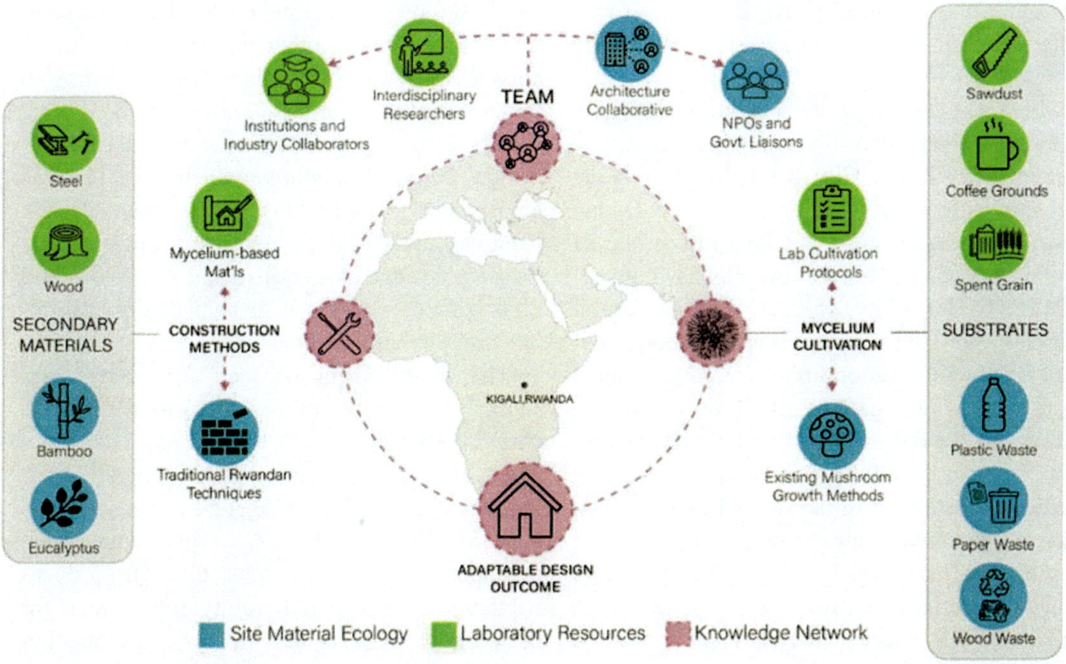

Fig. 2 Outline of teams, materials, and information flows that foster the development and integration of mycelium biocomposites

Fig. 3 Flows of waste matter in Rwandan districts (Effective Logistics and Consultancy Group 2020)

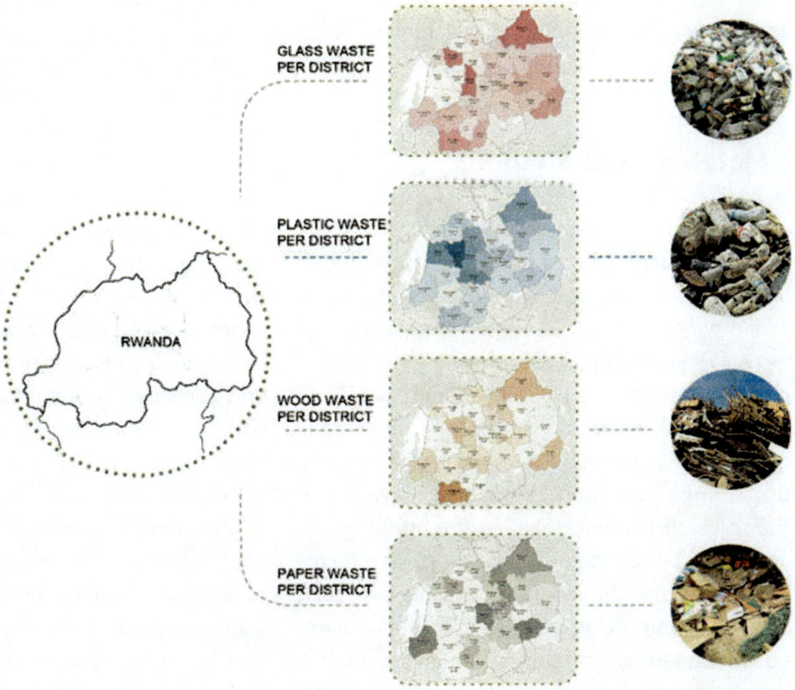

continue to be produced in Kigali City and outlying areas (Effective Logistics and Consultancy Group 2020).

Mitali Diogene represents Agruni Company Group, a private waste management company contracted by Kigali City for 12 of its sectors. The 350–450 tons of waste they collect each day are dumped in the open Nduba landfill, one of the primary dumpsites for the city (Umubyeyi 2021a). The majority of rural and semi-rural households discard their waste in the woods until there is enough to set on fire. Integrating cleaned waste such as plastics into mycelium-based biocomposites may present an opportunity to prevent combustion and other emissions from waste practices. Plastics are of particular interest in this work because many are nonbiodegradable, and plastic waste is a widely acknowledged global problem. The present work asks the question of whether plastic waste materials can be used to test, through design, the concept of a nonbiodegradable material occupying a space within the circular life cycle of a biomaterial-based composite. Given the ubiquity of plastic waste, such an exercise may produce information useful to the integration of nonbiodegradables in composites in different contexts.

2 Materials and Methods

This work aims to evaluate the potential for a mycelium biocomposite in a material-human ecology that draws together agricultural, building, and waste systems. Modular, unit-scale, and full-scale development were necessary to observe and understand factors such as climate, available materials, and physical parameters that influenced design outcomes. Prototyping, fabrication, and building integration approaches were tested at the research laboratory and at the building site in Rwanda. To leverage the learning advantages of both locations, the laboratory setting provided a controlled interior climate that could be sanitized, which allowed mixing of raw materials in a contamination-free space. It also provided the advantage of instrumentation to record temperature, humidity, and other parameters that influence the growth of mycelium and its binding to the substrate materials. The Rwanda setting provided the dynamic complexity endemic to design testing, including a fluctuating climatic environment, locally sourced raw materials, the local workers that made up a labor force with its own knowledge and skills, and economic and organizational leaders who helped to forge collaborative relationships.

The development of local knowledge and supply chain networks may help to address the chronic shortage of low-cost and self-build architecture and support income generation for the rural poor using the agencies of design and building. This project seeks to demonstrate such an approach on a small scale through (i) documentation of growth and prototyping methods for the purpose of repeatability, (ii) documentation of visual observations of mycelium behavior under changing conditions; and (iii) design and construction of a full-scale installation to evaluate constructability, test configurations of modular units, and consider the potential for further design applications in the built environment.

The work of prototyping was conducted in two environments where the team was co-located: a controlled research laboratory, and a kitchen space in Rwanda. The laboratory setup consisted of a fabrication chamber for the mixing of raw materials and mycelial growth phase of the biocomposites. The chamber, shown in Fig. 4, was constructed in an interior laboratory space using a simple wood frame structure sealed over with 3 mil clear top sheeting to prevent dust and microbes from contaminating the workspace. Cut-outs were made for a filtered forced-air supply fan and exhaust vent, with the chamber positively (outward) pressurized while in use. Within the chamber were a steel worktable and shelving units, the latter of which were wrapped in another layer of sheeting to protect the prototypes during their growth period. Finally, the biocomposite prototypes once pressed into formwork were themselves wrapped with air

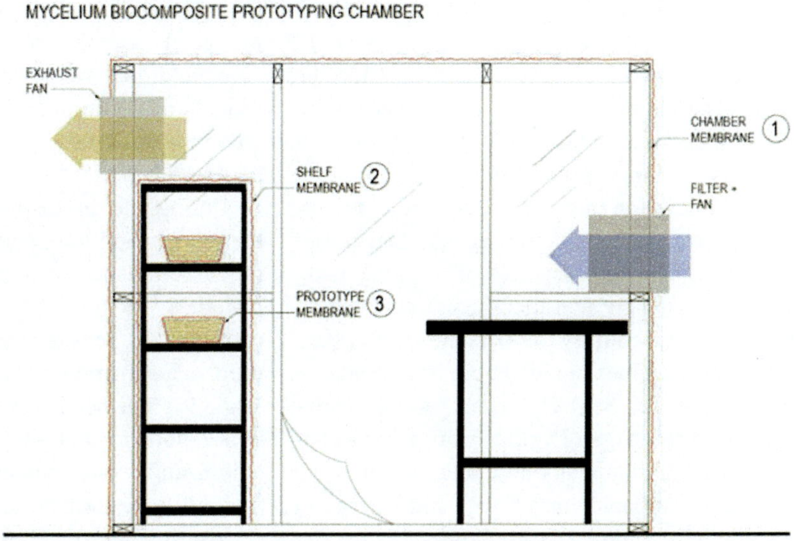

Fig. 4 Section drawing of laboratory-based mycelium biocomposite prototyping chamber with three layers of dust membranes to prevent contamination during the growth process

exchange via filter patches, which is needed for the mycelium to grow aerobically and colonize the substrate.

The fabrication of prototypes between the two setup locations was marked by important differences. The laboratory setting could be sanitized easily, including the steel work surface, and airflow mechanically controlled. Despite differences in the prototyping spaces between the laboratory and Rwanda, the process for fabricating prototypes was similar, and was carried out in a series of steps over a period of 5–10 days. The raw substrates, including cleaned agriwaste, were inoculated with mycelium, enough water to achieve a consistency similar to loose soil (ratios were modified throughout the process), and in some cases, additional pulverized nutrients such as flour to augment the mycelium's enzymatic "digestive" process and increase the rate and extent of mycelial binding and colonization. The mixes were then allowed to colonize in bags for later use, and when a certain amount of growth had occurred, the mixes were lightly pressed into thermoformed plastic formwork and covered, with controlled fresh air exchange to allow aerobic colonization. The modules were removed and dried, in some cases using additional heat from a low-temperature oven (between 90 and 100 °C) to accelerate drying without cooking the material. Sun-drying and forced-air drying can also be implemented, as long as debris and dust are controlled.

While the laboratory-based fabrication process enabled a degree of control over environmental conditions, the process in Rwanda presented unique discoveries and challenges. Firstly, Rwanda has two annual rainy seasons bringing torrential rains in short periods of time. Because most roads are unpaved, soil baking and compaction during the dry seasons prevent water absorption, creating trenches, gullies, and waterfalls everywhere in rural areas. While mycelium grows a protective hydrophobic skin naturally, it is not waterproof on its own; therefore, applications with exposure to rain would need to be finished with coatings such as beeswax, which is available through local honey farms. The second challenge is the prevention of contamination in the lab. Unpaved roads in combination with particularity of the Rwandan soil cause fine dust to circulate during dry seasons, so a contaminant-free environment must be maintained while providing a small amount of air exchange for the mycelium to fully colonize the cleaned substrate. The prototyping work the team conducted in Rwanda took place during the dry season, so coverings such as clean fabrics and sheet materials were used during the growth phase. The third challenge is the fabrication of formwork necessary to shape and contain the loose mix of mycelium inoculate and raw materials while the growth and binding process takes place. The thermoformed plastic that was used in the original laboratory setting was not available to construct formwork in which to grow the mycelium mixes. In the absence of a functioning national postal service, and the exorbitant cost of shipping using private carriers, any non-local materials must be brought to the country in person. Few digital fabrication machines are available in Rwanda, at the Westerwelle Start-Up Haus Kigali Makerspace and the Kigali Digital Fabrication Laboratory, at a high monetary and energy cost.

Locally available materials were used in keeping with the project's goal of evaluating the potential for a system for growth, production, and self-building. Three materials were used to fabricate the formwork for panels to be installed in an office interior where acoustic soundproofing was requested. The first formwork was constructed with paper, which proved to be the least robust, and gradually disintegrated in the presence of moisture, but was fast and low-cost to construct. The second formwork was made of ceramic by local artisans and provided a viable inexpensive alternative that could be reused many times. Ceramics are also available in rural areas, where it is a priority for this work to test the potential of a biocomposite growth and construction ecosystem. Finally, steel sheet forms were used, and provided a smooth, non-porous surface against which the biocomposites grew successfully. The steel sheet forms required materials and tools not readily available in rural areas, however, and would add a nontrivial cost to construction. Other materials that were considered, and are planned for inclusion in future testing, included wood, packed earth, and locally available textiles.

Finally, the question of whether the locally produced nonbiodegradable waste plastics could be integrated into the biocomposite was investigated. In both the laboratory-based setting and the Rwanda-based setting, the organic raw material substrates were combined with clean, ground waste plastic aggregate (approximately 1 mm diameter) during the mixing phase and pressed into formwork. Several prototypes were produced using the mix, and revealed variation in color and texture. A summary diagram of the process of mixing the raw materials, growth (or colonization), and drying of the prototypes can be seen in Fig. 5.

During the prototyping process in Rwanda, the organic and inorganic substrates varied based on an abundance of plastic waste and agriwaste materials. In the production of flat panel modules, banana leaves were coarsely ground and used as a substrate, with the addition of between 7 and 9 g per panel of finely ground sorghum or wheat flour to increase the rate of growth. Full colonization was achieved in five days with multiple mixes where airflow was controlled, as shown in Fig. 6. These modules were light, yellow-to-white in color, foamlike, rigid, and had enough compressive strength to be stood on by a grown person without visibly modifying the original geometry. Two prototypes that were not adequately covered during the growth process were not fully colonized, and were possibly contaminated by airborne microbes. After 16 days they were removed, and were dark brown in color, which was close to the original color of the raw materials mix prior to colonization. They held together with a weak, weblike mycelium network when picked up, but disintegrated under minimal stress. Due to the risk that they were contaminated, these panels were not crumbled and remixed into new modules, but isolated and disposed of separately.

3 Results: Translating Prototyping Outcomes to Full-Scale Building Investigations

The first set of prototypes and formwork-making processes had been developed for the intended end-use of an interior acoustic soundproofing panel in accordance with the needs of an office; however, there was an opportunity to engage with agricultural activity in a direct way through design and fabrication. Small mushroom growing huts, domed-shaped structures typically less than 3 m at ground-level diameter, had sprung up in

Fig. 5 Top, diagram of fabrication process for mycelium-based biocomposite prototypes; Bottom, materials-gathering and finished prototypes with integrated organic and inorganic (plastic) waste materials

MATRIX OF RAW MATERIALS USED IN RWANDA-BASED PROTOTYPE FABRICATION

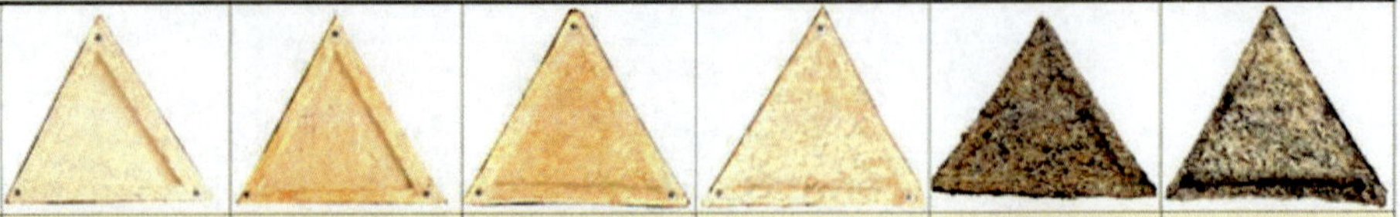

MYCELIUM	*Pleurotus Ostrealus*	*Pleurotus Ostrealus*	*Pleurotus Ostrealus*	*Pleurotus Ostrealus*	*Pleurotus Ostrealus*	*Pleurotus Ostrealus*
SUBSTRATE	Banana leaves + Shredded paper	Banana leaves	Shredded paper	Shredded paper	Shredded paper	Shredded paper
ADDED NUTRIENT	Wheat flour	Sorghum	Sorghum + Wheat flour	Wheat flour	Sorghum	Cassava
GROWTH TIME	7 days	5 days	5 days	5 days	16 days	16 days
FORMWORK	Cardboard with vinyl lining	Cardboard with vinyl lining	Steel lined with plastic	Cardboard with vinyl lining	Cardboard with vinyl lining	Cardboard with vinyl lining

Fig. 6 Matrix of Rwanda-based fabrication modifying growth variables to produce changes in material appearance, structure, and density

this landscape due to a recent boom in edible mushroom cultivation by local small-holding farmers. Mushrooms were grown on the ground of these huts, which consisted of lightweight sheet materials and a mix of bamboo and other framing materials. The use of a single horizontal ground surface limited total growing area to less than 4–5 m^2. The research team, presented with the design problem of constructing the growing huts with mycelium-based biocomposites, produced additional formwork with the goal of developing a block module that could be stacked.

In the mushroom hut prototype, one design criterion was that the processes of fabricating modules and full-scale construction had to be accessible to non-designers, and lightweight enough for women and elderly, who already participate in self-building in Rwanda (Sho 2014b, c). Biowelding, a term used to describe a process in which natural growth methods or organic materials are used to bind objects together, was introduced into the fabrication process as a way to simplify construction using the same principles already present in the existing growth method. Before the mycelium modules are dried, and thus still living, additional fresh mycelium mix can be used between modules in a manner similar to mortar, binding blocks together vertically and horizontally. This small addition of wet mix then dries and desiccates quickly, killing the mycelium and preventing further unwanted growth. With the inherently stable geometry of the dome-shaped hut, the biowelding was successful in attaching the block modules when the walls of the hut were constructed, as shown in Fig. 5. The bond was strongest when two modules were joined just prior to the completion of the growing phase. The pieces continued growing while attaching to each other, and afterward dried out together. Future fabrication must also take desiccation into account, as the extent of drying affects the final size of each module, and therefore the organization and fit of the overall structure. For an exterior standalone hut application, the exposed surfaces would require a coating to waterproof the structure; a beeswax, biopolymer, shellac, or other locally sourced coating could be used in such an application (Fig. 7).

The mushroom growing hut project remains in development, and full-scale mockups are planned in the future for construction onsite, where they can be tested in the extremes of the rainy and dry seasons in Rwanda. In addition, secure funding and a reliable supply chain will need to be established in conjunction with the construction process in order to be self-sustaining and locally owned. In the near term, the focus of

Fig. 7 Top, section drawing of possible configuration in Rwanda; Bottom, agricultural mushroom hut mock-up

the work remains on the interior office applications, where ongoing projects in Kigali were ready to integrate the panel modules.

A shortage of interior architectural elements in Rwanda presents many opportunities. Interior construction materials including gypsum boards, drop-ceiling and soundproofing panels, partitions and screens, tiles, insulations, and paint are mostly imported from China. They are limited in selection, inconsistent in supply, of lower quality, and high in life cycle carbon emissions. Locally sourced and produced mycelium panels could offer replacements to such imported materials, while adding new knowledge to the

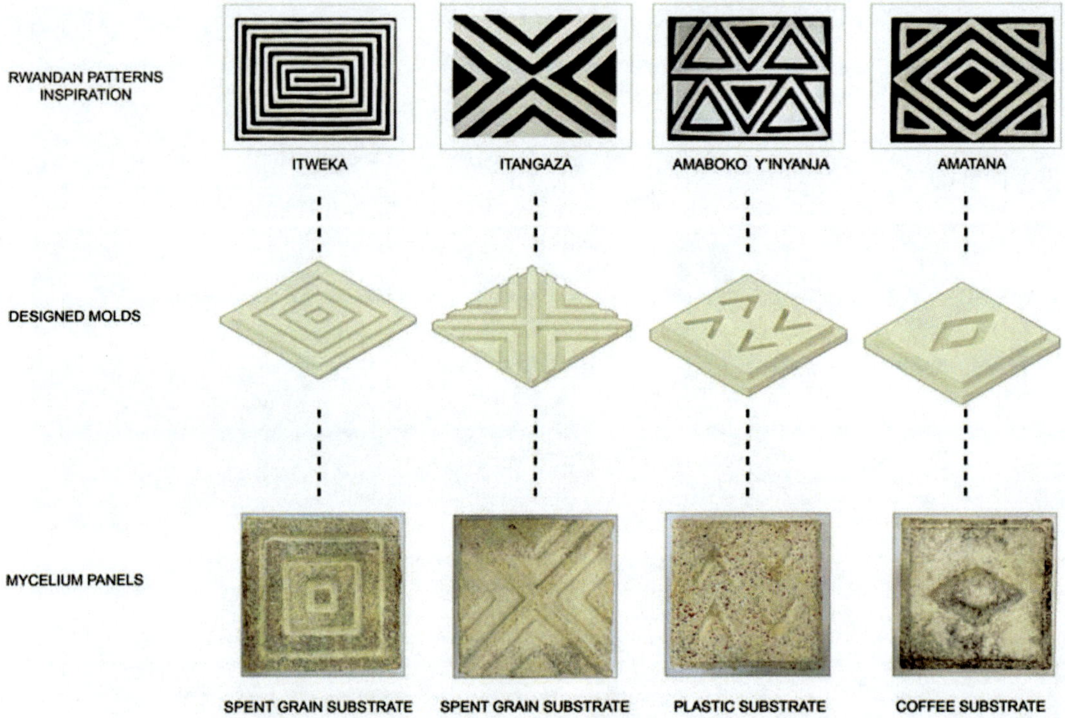

Fig. 8 Adapting Rwandan textile patterns to interior panel geometries

local construction industry. The research team worked closely with the local staff in the project, who were instrumental in piecing together the knowledge network related to the production of mycelium-based materials in Rwanda. Rwandan architects, crafts artisans, mushroom farmers, business leaders, and builders were engaged to find resources and create prototypes for the project. Many were surprised, intrigued, and impressed by the resultant panels. We received many positive responses and enthusiasm for the project, especially from mushroom farmer Sister Pascasie at Centre Umushumba Mwiza who, very much familiar with mycelium herself, expressed shock when she found that architectural materials could be made from it. The only concern noted by Rwandans was the use of a food product in the production process. Food scarcity and hunger are urgent problems throughout Africa and certainly in Rwanda. We found a sensitivity toward using edible materials in the creation of a product other than food, which should be addressed in future work.

The development of interior acoustic panels continued with design testing of panel size, geometry, and method of integration. Traditional pattern motifs from woven partitions and wall decorations found in indigenous architecture (Kanimba and Pee 2008) became an origin point for exploring geometric pattern expression, as seen in Fig. 8. The application of simple patterns demonstrated the ability of the biocomposite to articulate surface geometry while remaining lightweight and allowing multiple options for mounting or framing.

For the interior office application, the team aimed to work only with materials and tools that could be locally and easily sourced in Rwanda. These included slender wood pieces for framing, joinery that would work with hand tools and available hardware such as nails rather than screws, to eliminate the need for a drill. Slotted frame systems were developed to hold panels in such a way as to enable changing or replacement over time, as shown in Fig. 9. Also requested was a lightweight dropped ceiling system, which

Fig. 9 Framed interior partition structure designed for ease of assembly/disassembly

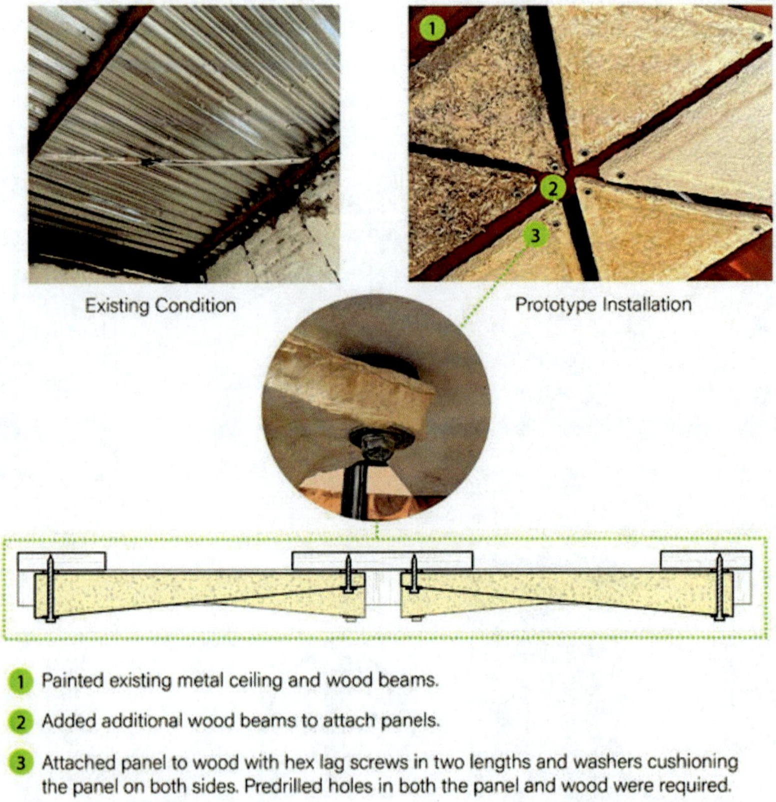

Fig. 10 Detail and installation of a biocomposite interior ceiling-mounted acoustic panel system

was designed with a similar frame that could be constructed with hand tools and existing skills (Fig. 10). Due to the slight shrinkage (between 8 and 12% in all directions, in our observation) of panels, the team has designed both panels and the frame to accommodate for shrinkages.

The products of research that are exported to Rwanda from the laboratory included the knowledge and techniques required to begin production onsite in Rwanda, as well as an illustrated manual created in the laboratory describing the fabrication process. The team in Rwanda acted as a catalyst to cultivate a network of knowledge exchange and also provided access to the multiple sites and projects with which the whole team ultimately engaged. The acoustic panels are planned for vertical-surface integration and ceiling mounting in offices in an existing workspace in the semi-rural neighborhood in which the Rwanda-based team has been working. In addition, the team was given access to the Masoro Learning and Sports Center, a recently completed project by General Architecture Collaborative, and the team has been invited to design a hanging acoustic surface installation there. A concept rendering can be seen below, in Fig. 11. This continuation of this work is planned.

Fig. 11 Rendered view of planned suspended acoustic surface in the Masoro Learning and Sports Center, Rwanda

4 Discussion and Conclusions

The outcomes of investigating mycelium-based hybrid materials for the Rwandan semi-rural site must incorporate the local construction culture and knowledge, without relying on industrialized supply chains and infrastructure. Next steps include providing further directions in best manufacturing process, evidence of structural and acoustic performance of mycelium panels, and implementation of mushroom growing structures. Further developments must embody low-carbon principles of ecosystem complementarity, including the analysis of organic and inorganic waste as substrate, and assessment for sequestering high-carbon materials, streamlining the reuse of agricultural byproducts, matching of material production with the existing mushroom economy, and skills and knowledge creation. Designing mycelium-based materials and cooperative models for living architectures in Rwanda revealed previously untapped resources such as agriwaste and plastics, farming knowledge applied to construction, and new collaborations. The learnings gained from reconsidering resources by embarking on a co-development of waste-based mycelium biocomposites with the larger, dynamic ecosystem of materials and knowledge that enable material innovation, have fostered the interdisciplinary and cross-cultural framework between the global South and North, science and design, and living and static systems upon which the human-material ecology could be developed for semi-rural and semi-urban environments. Operating within Rwandan design constraints has enabled participatory modes of making involving mushroom farmers and distributors, waste management industry, local and expatriate designers, rural self-builders, and urban professional contractors—a construction mode decidedly distinct from those existing in the South or North. Because of its carbon-negative performance, available low- and high-end market niches, and the relative ease of knowledge transfer, the mycelium building component research continues to provide opportunities for decentralized regenerative architectures. In recent years, the world has witnessed the environmental degradation in which industrialized economies have been the major culprit, while non-industrialized economies have been disproportionately impacted. For this reason, in addition to designing sustainable and marketable building materials and methods, the building disciplines are called to design new engagements between the global South and North. The mycelium research team believes that this project contributes a small and a tangible step toward such engagement.

Acknowledgement The authors would like to acknowledge the support of Syracuse University and collaborations with our teams in the laboratories and in Rwanda, including Cynthia Twagirayezu, Carene Umubyeyi, Sister Pascase, James Setzler, Ivania Rivera, and Helna Zhen.

References

Appels FVW, Camere S, Montalti M et al (2019) Fabrication factors influencing mechanical, moisture- and water-related properties of mycelium-based composites. Mater Des 161:64–71. https://doi.org/10.1016/j.matdes.2018.11.027

Attia S (2018) Design principles of regenerative design. In: Attia S (ed) Regenerative and positive impact architecture: learning from case studies. Springer International Publishing, Cham, pp 19–32

Bah EM, Faye I, Geh ZF (2018) The construction cost conundrum in Africa. Housing market dynamics in Africa. Palgrave Macmillan UK, London, pp 159–214

Behuria P (2021) Ban the (plastic) bag? Explaining variation in the implementation of plastic bag bans in Rwanda, Kenya and Uganda. Environ Plann C: Polit Space 39:1791–1808. https://doi.org/10.1177/2399654421994836

Bilal M, Khan KIA, Thaheem MJ, Nasir AR (2020) Current state and barriers to the circular economy in the building sector: towards a mitigation framework. J Clean Prod 276:123250. https://doi.org/10.1016/j.jclepro.2020.123250

Chinese Loans to Africa Database (2022) The Boston university global development policy center, Boston. https://www.bu.edu/gdp/chinese-loans-to-africa-database/. Accessed 11 Jan 2023

COP27 (2022) Historic climate deal sealed at COP27 as climate conference takes a leap to save lives and livelihoods. https://cop27.eg/#/news/259/Historic%20Climate%20Deal%20Sealed%20a. Accessed 20 Nov 2022

Effective Logistics and Consultancy Group (2020) Baseline study on waste collection and recycling countrywide. Ministry of Infrastructure (MININFRA), Kigali, Rwanda, p 59

Elsacker E, Vandelook S, Brancart J et al (2019) Mechanical, physical and chemical characterisation of mycelium-based composites with different types of lignocellulosic substrates. PLoS ONE 14:e0213954. https://doi.org/10.1371/journal.pone.0213954

Girometta C, Picco AM, Baiguera RM et al (2019) Physico-mechanical and thermodynamic properties of mycelium-based biocomposites: a review. Sustainability 11.https://doi.org/10.3390/su11010281

Goodfellow T (2017) Urban fortunes and skeleton cityscapes: real estate and late urbanization in Kigali and Addis Ababa: urban fortunes and skeleton cityscapes. Int J Urban Reg Res 41:786–803. https://doi.org/10.1111/1468-2427.12550

Goodfellow T, Smith A (2013) From urban catastrophe to 'model' city? Politics, security and development in post-conflict Kigali. Urban Stud 50:3185–3202. https://doi.org/10.1177/0042098013487776

Hafez FS, Sa'di B, Safa-Gamal M et al (2023) Energy efficiency in sustainable buildings: a systematic review with taxonomy, challenges, motivations, methodological aspects, recommendations, and pathways for future research. Energy Strategy Rev 45:101013https://doi.org/10.1016/j.esr.2022.101013

Hertwich EG (2021) Increased carbon footprint of materials production driven by rise in investments. Nat Geosci 14:151–155. https://doi.org/10.1038/s41561-021-00690-8

Holt GA, Mcintyre G, Flagg D et al (2012) Fungal mycelium and cotton plant materials in the manufacture of biodegradable molded packaging material: evaluation study of select blends of cotton byproducts. J Biobased Mat Bioenergy 6:431–439. https://doi.org/10.1166/jbmb.2012.1241

IEA (2018) Global status report. https://www.iea.org/reports/2018-global-status-report. Accessed 12 Jan 2023

IEA (2019) Global status report for buildings and construction 2019, IEA, Paris https://www.iea.org/reports/global-status-report-for-buildings-and-construction-2019. Accessed 12 Jan 2023

Jiang L, Walczyk D, McIntyre G et al (2017) Manufacturing of biocomposite sandwich structures using mycelium-bound cores and preforms. J Manuf Process 28:50–59. https://doi.org/10.1016/j.jmapro.2017.04.029

Jones M, Mautner A, Luenco S et al (2020) Engineered mycelium composite construction materials from fungal biorefineries: a critical review. Mater Des 187:108397. https://doi.org/10.1016/j.matdes.2019.108397

Kanimba MC, Pee L van (2008) Rwanda: its cultural heritage, past and present. Institute of National Museums of Rwanda, Kigali

Mahmood H, Alkhateeb TTY, Furqan M (2020) Exports, imports, foreign direct investment and CO_2 emissions in North Africa: spatial analysis. Energy Rep 6:2403–2409. https://doi.org/10.1016/j.egyr.2020.08.038

Malik A, Parks B, Russell B et al (2021) Banking on the belt and road: insights from a new global dataset of 13,427 Chinese development projects. AidData at William & Mary, Williamsburg, VA, pp 23–36

Nkurunziza M (2021) Rwanda banks on innovation for mushroom production. https://www.newtimes.co.rw/

article/189538/News/rwanda-banks-on-innovation-for-mushroom-production. Accessed 11 Jan 2023

Olanrewaju OI, Enegbuma WI, Donn M, Chileshe N (2022) Building information modelling and green building certification systems: a systematic literature review and gap spotting. Sustain Cities Soc 81:103865. https://doi.org/10.1016/j.scs.2022.103865

Papargyropoulou E, Preece C, Padfield R, Abdullah AA (2011) Sustainable construction waste management in Malaysia: a contractor's perspective. In: Proceedings of the management and innovation for a sustainable built environment, The Netherlands

Pelletier MG, Holt GA, Wanjura JD et al (2013) An evaluation study of mycelium based acoustic absorbers grown on agricultural by-product substrates. Ind Crops Prod 51:480–485. https://doi.org/10.1016/j.indcrop.2013.09.008

Purchase CK, Al Zulayq DM, O'Brien BT et al (2021) Circular economy of construction and demolition waste: a literature review on lessons, challenges, and benefits. Materials 15:76.https://doi.org/10.3390/ma15010076

Schritt H, Vidi S, Pleissner D (2021) Spent mushroom substrate and sawdust to produce mycelium-based thermal insulation composites. J Clean Prod 313:127910. https://doi.org/10.1016/j.jclepro.2021.127910

Sho Y (2014a) Northern designers in Africa: a political diary of building a house in Rwanda. In: 102nd ACSA annual meeting proceedings, globalizing architecture/flows and disruptions, pp 186–193

Sho Y (2014b) Home experiments: earthbag construction as a teaching tool in Rwanda. pp 24.670.1–24.670.15

Sho Y (2014c) Looking like developed: aesthetics and ethics in Rwandan housing projects. J Architect Educ 68:199–208. https://doi.org/10.1080/10464883.2014.937237

Sho Y (Forthcoming) Five aesthetics of the global development industry: building low-cost housing in Rwanda. Plan J 7:2

Sisti L, Gioia C, Totaro G et al (2021) Valorization of wheat bran agro-industrial byproduct as an upgrading filler for mycelium-based composite materials. Ind Crops Prod 170:113742. https://doi.org/10.1016/j.indcrop.2021.113742

Soh E, Le Ferrand H (2023) Woodpile structural designs to increase the stiffness of mycelium-bound composites. Mater Des 225:111530. https://doi.org/10.1016/j.matdes.2022.111530

Sun W, Tajvidi M, Howell C, Hunt CG (2022) Insight into mycelium-lignocellulosic bio-composites: essential factors and properties. Compos A Appl Sci Manuf 161:107125. https://doi.org/10.1016/j.compositesa.2022.107125

Umubyeyi C (2021a) Telephone interview with Rwanda Agriculture and Animal Resources Development Board (RAB) official, 8 Jan 2021

Umubyeyi C (2021b) Telephone interview with Laurent Demuynck, director of Kigali Farms, 15 Jan 2021

Weatherspoon DD, Miller SR, Niyitanga F et al (2021) Rwanda's commercialization of smallholder agriculture: implications for rural food production and household food choices. J Agric Food Ind Organ 19:51–62. https://doi.org/10.1515/jafio-2021-0011

Production of Thermoplastic Starch Pellets and Their Robotic Deposition for Biodegradable Non-standard Formworks

Benjamin Kemper

Abstract

The climate crisis challenges architects, designers, and engineers to explore alternative opportunities for more sustainable fabrication processes. Biopolymers have emerged as a potential material to replace petroleum-based plastics used in building and construction processes. This research aims to re-evaluate the production of non-standard building elements and introduce bio-based and biodegradable materials for formworks in architecture. This research paper investigates the production of thermoplastic starch (TPS) pellets and connected digital fabrication techniques. It studies the effects of varying ratios of the plasticizer on the behavior and properties of the material. TPS pellets are further processed using a large-scale robotic 3D printing setup, utilizing the Fused Deposition Modeling (FDM) method. The initial printed results using a robotic pellet extrusion system are presented, analyzed, and evaluated. The advantages and challenges of this approach are discussed within the scope of the architectural research field. This paper focuses on the production and digital fabrication techniques of TPS pellets, with the primary goal of developing a sustainable, bio-based, and bio-compostable system for concrete formwork in architecture.

Keywords

Biopolymers · Sustainable fabrication · Thermoplastic starch · Concrete formwork · Robotic 3D printing

1 Introduction

In 2015, the United Nations established 17 shared goals for peace and prosperity for humanity and the planet (UN General Assembly 2015). The present research endeavors to adopt a sustainable approach to utilizing the planet's resources in the building industry, an issue that the discipline of architecture must address with immediacy. These goals may be deemed an opportunity to re-examine building and construction processes through innovation (goals 9, 11) and responsibly transition to bio-based and compostable materials instead of relying on conventional material systems (goals 12, 13).

B. Kemper (✉)
Digital Building Technologies, ETH Zurich, Zurich, Switzerland
e-mail: benjamin.kemper@hs-anhalt.de

B. Kemper
Materiability Research Group, Anhalt University of Applied Sciences, Dessau, Germany

M. R. Thomsen et al. (eds.), *Design for Rethinking Resources*, Sustainable Development Goals Series,
https://doi.org/10.1007/978-3-031-36554-6_37

The early results of this research were briefly discussed in "Bio-Formwork" (Kemper 2022). The construction industry is a significant consumer of plastics, comprising 19% of total consumption (Geyer et al. 2017). The majority of buildings are constructed using traditional materials, each with distinct requirements. These materials often rely on finite resources. Due to their affordability and advantageous characteristics such as robustness and lightweight, petroleum-based polymers are frequently employed in building materials, including insulation, window frames, piping systems, packaging (Häkkinen et al. 2019, p. 5), and formworks (Kemper 2022, p. 65).

However, the continued use of plastics in the building industry is problematic due to their harmful environmental footprint caused by incorrect recycling, high energy consumption, pollution of the oceans, destruction of marine life, micro debris, and harmful additives (Kretzer 2017, p. 102). This research addresses this issue by exploring opportunities to replace petroleum-based plastics with bio-based and biodegradable polymers in the built environment.

Non-standard formwork is one application of engineering materials in architecture and construction that aims to decrease the consumption of resources. By fabricating unique molds using digital fabrication techniques, such as robotic milling (Mostafavi et al. 2019, p. 427) and additively printed formworks (Peters 2014; Jipa et al. 2018; Leschok and Dillenburger 2019), it is possible to reduce the amount of material needed for traditional formworks and the consumption of concrete. These processes are efficient in resource reduction but still depend on petroleum-based polymers.

Despite the reduction in material used for concrete components achieved through the utilization of complex, single-use formworks (Meibodi et al. 2017, p. 213), printed plastic formwork contributes to a greater environmental impact of the element. As such, this research aims to investigate bio-based and biodegradable materials as an alternative solution to address the challenges of disposing of valuable resources.

2 Biopolymers

Biopolymers have a similar chemical structure to traditional plastics, making them a viable alternative. Unlike conventional plastics derived from petroleum, biopolymers are composed of polymers from biological matter such as sugar cane, starch, or cellulose found in trees, straw, and cotton. These biomaterials can be engineered to exhibit biodegradability, compostability, robustness, and durability properties (Goodall 2011, pp. 2–6; Kemper 2022, p. 65).

Biopolymers can be classified into four main groups (Chandra and Rustgi 1998; Clarinval and Halleux 2005; Jamshidian et al. 2010, pp. 552–553):

1. Produced by microorganisms or genetically modified bacteria
2. Produced by chemical synthesis from bio-derived monomers
3. Derived from fossil resources
4. Derived from natural raw resources.

These types of biopolymers can be further divided into two categories: bio-polyesters and agro-polymers (Avérous and Pollet 2012, p. 16). Agro-polymers originate from natural resources and can be converted into a biopolymer through simple procedures, which differ significantly from the other groups. On the other hand, bio-polyesters are transformed through biotechnology and chemical synthesis into a biopolymer. The following subchapter will investigate a biopolymer from the bio-polyesters (PLA) group and a polymer from the agro-polymers (starch) group, highlighting each material's benefits and disadvantages.

It is essential to consider that using corn starch for non-food applications may pose a potential conflict as corn is a valuable food crop. This can affect food security and may also compete with food uses for the available corn supply, potentially driving up food prices and presenting a higher risk for developing nations. However, corn starch for non-food applications can also create a revenue stream for farmers. In 2020, approximately 57.14% of the industrial starch market was used for food, with the remaining being divided among uses such as feed (3.57%), the paper industry (28.57%), pharmaceuticals (5.36%), and other applications (5.36%). This market's compound annual growth rate is projected to be 5.75% by 2025 (Adewale et al. 2022, pp. 2–3).

In this research context, it was found that multiple types of starches, including rice and wheat starch, can be used for printing with similar results. This allows for broader distribution and reduces reliance on a single material. Locally sourced materials can be used, eliminating the need to import specific starch types to fabricate 3D-printed TPS structures. Still, this raises ethical concerns as food materials are being used for non-food purposes. It is acceptable on a small scale, but it becomes a pressing issue as usage increases. Politics, industry, and research must work together to ensure the development of diverse techniques for sustainable processes and the use of various bio-based material resources. Ultimately, finding all possible solutions to break free from the dependency on petroleum-based plastics is crucial.

2.1 Polylactic Acid (PLA)

A notable instance of bio-polyesters is Polylactic Acid (PLA), which is generated through biotechnology by converting lactic acid into a thermoplastic polyester. The production of PLA offers several benefits, including the use of lactide monomers from lactic acid extracted by the fermentation of corn, the fixation of carbon dioxide by the corn plant, energy conservation, the enhancement of the agricultural economy, and the ability to modify the material's physical properties (Dorgan et al. 2000, pp. 1–2).

The production of PLA combines agricultural, biological, and chemical technologies. Frequent utilization of PLA is found in food packaging, specifically for fresh produce such as fruits and vegetables, accounting for approximately 70% of its usage (Jamshidian et al. 2010, p. 561). Another typical application of PLA is as a filament for 3D printing.

Bioplastics, as commonly referred to, have a range of applications in everyday life, such as in the form of bioplastic cutlery and cups, as well as in architecture. While PLA is often perceived as an ecologically friendly material, it only breaks down effectively under specific industrial settings not commonly encountered in nature. It generally takes two to three months to degrade in composting plants with warm temperatures (55 °C) and a moisture-rich environment (Kale et al. 2007, p. 260; Madhavan Nampoothiri et al. 2010, p. 8498; Teixeira et al. 2021, p. 27). It can take years for the material to degrade when exposed to natural conditions such as soil and oceans (Lamberti et al. 2020, p. 2556; Kemper 2022, p. 66).

2.2 Thermoplastic Starch (TPS)

Given the challenges in the biodegradability of Polylactic Acid (PLA), research has been directed toward investigating alternative materials, such as polysaccharides, specifically thermoplastic starch (TPS). TPS has been widely employed in (food) packaging and biomedical applications (de Carvalho and Trovatti 2016, p. 5; Khan et al. 2016, p. 6). TPS is obtained from biomass sources, for example, starch from maize, potatoes, and wheat (Khan et al. 2016, p. 4), and can be disposed of through composting. The degradation of starch-based polymers is seen as a potential solution for issues related to petroleum-based polymers (Saiah et al. 2012, p. 75; Kemper 2022, p. 66).

Compared to PLA, TPS has a higher potential for sustainable use in architectural elements. Furthermore, TPS's greenhouse gas emissions and cradle-to-gate non-renewable energy usage are significantly lower than conventional plastics, making it a more environmentally friendly option (Shen and Patel 2008, p. 163). According to the material database website materialarchive.ch, TPS is considered carbon–neutral, meaning that the material decomposes the same amount of CO_2 sequestered during plant growth. However, it is important to consider the CO_2 production of the entire process, including cultivation, transport, and production, in a life cycle assessment (de Pietro and Winterthur 2022).

TPS is described as a modified natural material. It is chemically and physically altered from the biopolymer starch found in nature (de Pietro and Winterthur 2022). Starch granules swell and form a gel if their outer layer is broken through mechanical work and flow (Saiah et al. 2012, p. 58). When exposed to warm water, the soluble component of the starch evaporates through the granule wall, causing swelling and bursting of the grains (Andersen et al. 2000, p. 16; Khan et al. 2016, p. 3). Native starch granules can be transformed into a thermoplastic using a plasticizer in combination with temperature and shear (Wang et al. 2008, p. 110). A suitable ratio of plasticizers in the formulation leads to a starchy material consisting of entangled polysaccharide chains, referred to as thermoplastic starch (Saiah et al. 2012, p. 58). TPS is typically fabricated through extrusion or solvent casting (Saiah et al. 2012, p. 58). It is essential to note that TPS exhibits severe shrinkage and deformation during the setting phase, resulting in deformation and cracks (Shanks and Kong 2012, p. 113).

3 Materials and Methods

3.1 State of the Research

This research paper evaluates two scientific fields. The first field examines the 3D deposition of bio-based and biodegradable materials, while the second field focuses on the latest research on freeform formworks for optimized structures, specifically 3D printing with plastics for 3D formworks and the possibility of replacing them with biopolymers.

3.1.1 Bio-based Materials for 3D Printing

The field of 3D printing using bio-based materials is multifaceted, with ongoing research in sustainable and ecological design. However, there needs to be more advancement in applying bio-based and biodegradable materials for large-scale architectural projects.

The work of Eric Klarenbeek & Maartje Dros and Atelier Luma highlights the potential of using algae as a substitute for petrochemical plastics by converting it into a biopolymer, creating innovative and accessible products. The studio focuses on aquafarming seaweed, which absorbs carbon dioxide during growth and binds carbon to biomass, with the resulting starch used for biopolymer formulation (Klarenbeek et al. 2017).

In 2014, the Mediated Matter Group (MIT) led by Neri Oxman published the project "Water-Based Robotic Fabrication," which examined the additive manufacturing of regenerated, biodegradable hydrogel composites such as chitosan and sodium alginate using a multi-chamber extrusion system to achieve a hybrid biomaterial with varying properties in specific areas (Mogas-Soldevila et al. 2014).

More recently, CITA/Royal Danish Academy has explored cellulose-based robotic printing and its impact on material properties and geometry during the extended curing phase of the bio-based material. Rather than altering the material formulation to decrease shrinkage, the research team aims to computationally monitor and evaluate the curing behavior to anticipate specific results and integrate this knowledge into the design process (Rossi et al. 2022).

3.1.2 3D Formwork

In architecture, the use of non-standard building elements is becoming increasingly relevant. Although the research in this area is still developing, advancements in technology and material science make non-standard elements a progressively crucial part of the built environment. The benefits of individualized building components include design flexibility through innovative shapes, increased efficiency through additive manufacturing, and reduced waste. Using unique formworks for concrete structures can improve structural integrity and reduce material usage.

Formworks play an important role in ensuring the geometry of concrete elements during the setting and curing phases. Research has shown that the cost of formworks for standard concrete elements can account for up to 50% of the total cost (IfB, D-BAUG, ETHZ), while for individualized elements, the cost increases between 80 and 90% (Burger et al. 2020, p. 49). Therefore, multiple research projects are focusing on the optimization of concrete structures through the use of 3D-printed plastic formworks.

A common challenge with these individualized 3D-printed formworks is the presence of two forces during and after casting: hydrostatic and granular pressure (Leschok and Dillenburger 2019, p. 191). The "Dissolvable 3DP Formwork" project addresses this issue through design adjustments such as incorporating of a hexagonal rib structure, which provides additional support in areas where it is needed (Leschok and Dillenburger 2019, pp. 191–193).

Research on Big Area Additive Manufacturing (BAAM) has gained significant attention in architecture. BAAM is a large-scale Fused Filament Fabrication (FFF) system that creates large-scale, additively manufactured parts (Roschli et al. 2019, p. 275; Paolini et al. 2019, pp. 5–6). One of the critical applications of BAAM is in the design of formworks for concrete, which allows for the creation of large-scale, unique building components. Studies have highlighted the cost-effectiveness of using BAAM for formworks, as it can significantly reduce (up to 50%) labor and material costs compared to traditional wooden molds (Love et al. 2019, pp. 2, 9).

3.2 Preliminary Studies

Preliminary studies have been conducted to examine the suitability of TPS for 3D printing and formworks. The material properties evaluated included strength, shrinkage, durability, curing time, viscosity, stickiness, processability, and cost. Encouraging results have been obtained by incorporating hemp fibers into the TPS composition prior to heating. During the setting stage, TPS tends to shrink due to water evaporation. Incorporating hemp fibers helps mitigate this shrinkage and improves the material's tensile strength (Fig. 1).

Fig. 1 Effect of hemp fiber percentages on TPS shrinkage using LDM (150 × 225 mm); moist TPS (top) and cured geometry (bottom); 0.00%, 0.79%, 3.10% hemp fibers (left to right)

At the laboratory scale, various experiments were undertaken to alter the properties of the TPS. Due to its simple production process, using natural additives is a feasible option to extend the material palette. By adding bio-additives to the mixture, the results demonstrated the suitability of these additions and the evident modifications in material properties.

An initial liquid deposition modeling (LDM) approach was utilized to investigate the extrusion of TPS in its moist form. A Delta Wasp 60,100 printer was adapted to deliver liquid TPS to the printing head. The aim was to examine the printability of TPS in combination with various additives. However, this method was considered inappropriate due to the dynamics during the setting process (Fig. 2).

The first studies on the use of TPS in 3D printing and formworks revealed that TPS has the potential to meet the demands of these applications. Using a pellet FDM heat extrusion approach, it was possible to overcome the limitations of the material and provide a suitable fabrication strategy. Due to the thermal stability of TPS, it can be extruded multiple times (Saiah et al. 2012, p. 64), allowing for the conversion of cured TPS back to the plastic state and vice versa. The melting point of the TPS used in the experiment was between 200 and 230 °C. Testing on a German RepRap 3D printer with a modified Mahor.xyz small-scale pellet extruder confirmed the feasibility of FDM 3D printing with TPS (Fig. 3).

3.3 TPS Pellet Production Process

Preliminary testing has demonstrated that pellet extrusion is a viable method for 3D depositing TPS. By altering the base formulation to improve material properties and incorporating additives

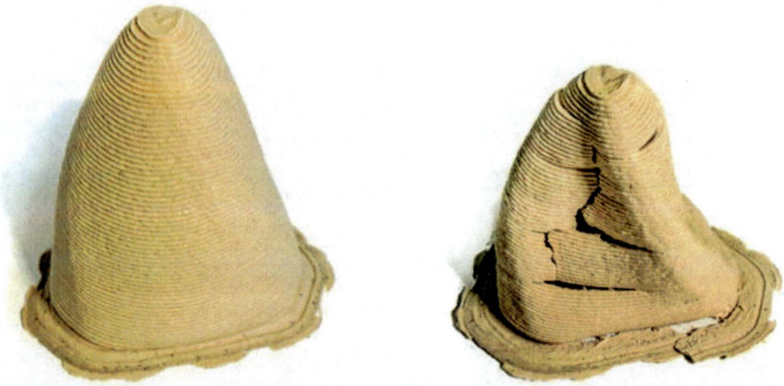

Fig. 2 LDM 3D printing of TPS (left); dried LDM print and resulting deformation caused by high shrinkage due to water evaporation (right)

Fig. 3 Initial small-scale TPS pellet extrusion test prints (120 × 80 mm)

Fig. 4 TPS pellets with various bio-additives (left to right: corn starch, raw cellulose fibers; corn starch, hemp fibers; corn starch, chitosan; corn starch, coconut fibers; rice starch, Indigofera; rice starch, spirulina)

that provide additional material characteristics, it is possible to produce usable pellets on a laboratory scale (Fig. 4).

In order to ensure optimal performance with the specific setup, it is crucial to adhere to the recommended pellet size and form as outlined by the manufacturer. Dyze Design advises that the ideal pellet should be spherical, with a diameter of 3–5 mm and minimal variations in size. Additionally, they recommend other pellet types such as repro/flat disks, considered an overall suitable format, and chopped 2.85 mm cylinders, considered to be overall acceptable (Dyze Design 2022). Given the experimental nature of the material and the laboratory setup, various strategies were implemented to research the production of cylindrical pellets.

3.3.1 Casting and Blending

In order to evaluate the suitability of TPS for 3D deposition using pellet extrusion, two liquid solutions of TPS were cast in large containers (Table 1).

After a two-week drying period at room temperature, the resulting sheet material was manually cut into small pieces (max. 30 mm) and shredded in a blender. This strategy led to the initial 3D deposition tests with a small-scale pellet extruder (Mahor.xyz) using TPS, demonstrating the suitability of TPS pellet heat extrusion. This process has the disadvantage of a prolonged drying time and increased susceptibility to contamination from microorganisms. Additionally, the resulting TPS flakes are of varying sizes, which do not meet the requirements for pellet extrusion and require extensive sorting and sifting.

3.3.2 Extrusion and Direct Slicing

The next series of tests utilized a method inspired by industrial pasta production. The base recipe was modified by reducing the amount of water to a minimum (Table 2). The mixture was heated and stirred using an automatic and powerful heating and stirring device. The heating process caused the starch to swell, forming a solid dough. This dough was then extruded into filament elements with a 500–1,000 mm length using a regular meat grinder. Special attachments were constructed to slice the extruded material in its

Table 1 Two formulations for casting TPS for subsequent blending

	Formulation A (%)	Formulation B (%)
Distilled water	72.60	68.20
Glycerol 99.50%	3.20	9.10
Corn starch	14.50	13.60
White vinegar 5.00%	9.70	9.10

Table 2 Two groups of basic TPS formulations for pellet production

	TPS A (brittle)		TPS B (elastic)	
Distilled water	250 g	25.00%	250 g	22.22%
Glycerol 99.50%	125 g	12.50%	250 g	22.22%
Corn starch	500 g	50.00%	500 g	44.44%
White vinegar 5.00%	125 g	12.50%	125 g	11.11%

moist state directly at the end of the machine. The slicing system featured rotating blades directly in front of six outlets with a diameter of 3.5 mm at the end of the meat grinder. However, due to the sticky nature of the material, this method was found unsuitable for creating pellets directly from the moist dough, as it often led to clogging in the setup.

3.3.3 Extrusion and Filament Cutting

In the subsequent stages of the process, various post-production strategies were implemented. Four methods were evaluated for the production of TPS pellets:

1. The extruded TPS filament was allowed to dry for two to three days at room temperature, after which it was cut into small pieces (3-5 mm in length) using an all-purpose slicer. This method yielded pellets with a suitable size and shape.
2. The extruded TPS filament was allowed to dry for a maximum of one day at room temperature, after which it was cut into small pieces using an all-purpose slicer. This method resulted in TPS elements that were softer and easier to cut.
3. The extruded TPS filament was allowed to dry for a maximum of one day at room temperature, after which it was cut into small pieces (3 mm in length) using a large-scale paper cutter. This method allowed for more precise control over the length of the pellets.
4. The TPS filament was extruded and dried in a drying cabinet at 105 °C for four hours. After drying, the TPS filament was cut into small pieces (approximately 4 mm long) using a filament granulator setup designed by a DIY enthusiast and hobbyist (JohnSL - Random Products 2020). The TPS filament was fed into a chamber using a standard filament feeder (powered by a stepper motor), where a modified Forstner bit chopped the dried filament into small pellets.

Two basic TPS formulations (A: brittle and B: elastic) were examined (Table 2).

3.4 Print Setup

The initial experiments on 3D deposition using TPS were carried out at the Lab of the Materiability Research Group within the design faculty at the Anhalt University of Applied Sciences. The equipment used for these experiments can be broadly categorized into the following subgroups:

3.4.1 Robotic Setup

An HRC-capable robot (KUKA lbr iiwa 14 R820) mounted on a mobile platform (KUKA flexFELLOW H900 Standard), which operates the KUKA Sunrise.OS. The toolpath for the robot was generated within the Sunrise Workbench in JAVA and the Grasshopper 3D plugin KUKA|prc (Fig. 5).

3.4.2 End Effector

A large-scale pellet extruder (Dyze Design Pulsar™) was used, which featured three individual heat zones (top, bottom, nozzle), a 348 mm extrusion screw with three zones (feed, melting, and metering zone) (d = 16 mm), and heating power of 1,100 W. The nozzle had a diameter of 5 mm, intended for large-scale manufacturing objects. The system was liquid-cooled (Dyze Design 2022). Pellets were manually inserted into the end effector via gravity feeding. The extruder controller was separated from the robot

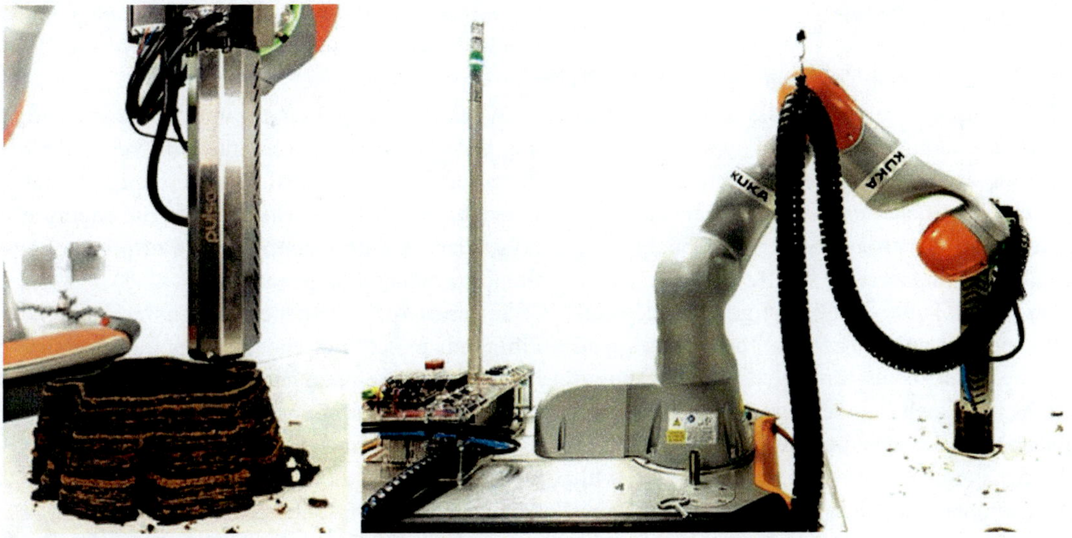

Fig. 5 TPS extrusion process (left) and 3D printing setup (right) showing control unit, liquid cooling, Kuka iiwa, and large-scale pellet extruder

Table 3 Printing parameters for Dyze design pulsar and Kuka LBR iiwa

	Formulation A (brittle)	Formulation B (elastic)
Extruder temperature Top Bottom Nozzle	190 °C 200 °C 220 °C	185 °C 200 °C 215 °C
Feed rate	3.24 mm/s 50.91 mm^3/s	8.07 mm/s 126.78 mm^3/s
Speed toolpath	4.50 mm/s	7.50 mm/s
Cylinder diameter	80.00 mm	80.00 mm
Walls	1	1
Layer height	2.00 mm	2.00 mm
Air cooling	No	No

controller at this experimentation stage, allowing for the investigation of the actual temperature values of the different heat zones and feed rate of new materials. The hardware used to control the extruder was a simple 3D printer controller board (RAMPS 1.4+ Arduino Mega), running Marlin firmware and connected with a computer executing the open-source application OctoPrint (Fig. 5). The parameters used for the robotic and extruder setup are presented in Table 3.

4 Results

With the previously described equipment and parameters, standard cylinder shapes were printed. Each cylinder had a diameter of 80 mm. This object geometry allows for comparability between various TPS formulations and facilitates comparison with the digital representation, enabling the detection of any potential anomalies.

4.1 Printing Results

Analyzing the initial TPS prints (Figs. 6 and 7) produced using the large-scale pellet extruder reveals several key findings and hypotheses. One important aspect is the temperature–time ratio (TTR) of TPS pellet extrusion. The results indicate that the glycerol concentration of the TPS formulation impacts the TTR. Specifically, a higher glycerol concentration slightly reduces the TPS's melting point. The time the pellets spend in the heat zone is also crucial. The melt time, as well as the melt temperature, decreases with higher glycerol concentrations. It is important to note that compensating a higher feed rate with higher temperatures does not result in the same material properties; instead, it can lead to the burning or melting away of the material. Therefore, it is essential to ensure that the material is kept in the heated zone for an appropriate time.

When the fitting TTR parameters are reached, the material flow is constant, with no notable flaws or errors detected. Furthermore, it was noted that when TPS melts, the material begins to crystallize. A higher concentration of glycerol is found to reduce this phenomenon.

In terms of layer adhesion, it is essential to differentiate between the adhesion of the first layer to the base plate and the adhesion between subsequent layers. Currently, it is necessary to use double-sided adhesive tape to help the first layer adhere to the base plate. However, the adhesion between subsequent layers did not present any difficulties. Additionally, it was found that the nozzle temperature must be

Fig. 6 3D printed cylinders of formulation A (left) and formulation B (right)

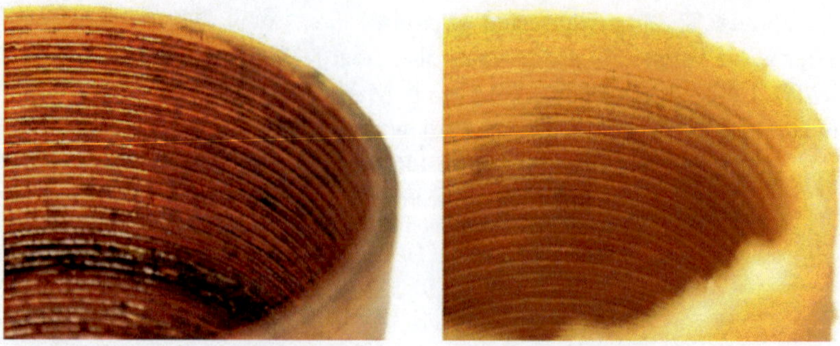

Fig. 7 Detail of 3D printed cylinder of formulation A (left) and formulation B (right)

Table 4 Print results for two different TPS formulations

	Formulation A (brittle)	Formulation B (elastic)
Wall thickness	6.00 mm	8.50 mm
Cylinder diameter	77.00 mm	76.00 mm
Shrinkage	3.75%	5.00%
Height	98.00 mm	66.00 mm
Color pellets	Matt white/light yellow	Matt white/light yellow
Color print	Glossy amber	Matt golden/brown honey
Further print prop	Glassy, crystalline, brittle, strong	Elastic, little crystallization

slightly higher than the melting point of TPS to create a smooth and sticky material flow at the end of the extruder. Lower temperatures result in less layer adhesion.

The maximum height was not reached due to a limited quantity of TPS pellets. Object A, with a height of 98 mm, was successfully printed, and the material distribution remained stable. The cylindrical shape helped to stabilize the print, and it is speculated that larger prints will not pose a problem. The wall thickness can be increased to create a more robust object, and structural additives or compressed air cooling can be utilized to enhance the structural properties. Compared to the LDM approach, the shrinkage observed in TPS pellet printing is low, between 3 and 5%. This shrinkage should be considered and taken into account during the design and material formulation processes by incorporating additives, fibers, and fillers. The pellets must be dried at 105 °C before being used in the large-scale extruder. If not appropriately dried, embedded water in the material will evaporate and accumulate at the top of the extruder, causing small explosions and steam and leading to extruded material with air bubbles similar to what is seen in conventional plastic pellet printing (Table 4).

4.2 Material Results

Analyzing the material properties of the first printed TPS pellets revealed several insights (Fig. 8):

- The material properties, such as brittleness and elasticity, were present in the printed objects. However, it was observed that the higher glycerol concentration in the TPS formulation had a more significant impact on the printing parameters than on the actual elasticity of the printed material. The produced pellets were found to have less ductility compared to the flexibility.
- When the temperature of the heat zones of the extruder was slightly higher than the material's melting point, the TPS began to crystallize.
- It was found that the extruded material undergoes slight changes after deposition, and it takes approximately one day for the final material properties to develop. For example, formulation A becomes more robust and less brittle, while formulation B regains some elasticity. Two key findings were made:

  - The most significant finding was the impact of the glycerol concentration in the formulation on the material properties. A higher concentration reduces the time needed for heating at a slightly lower melting point (5–10 °C), resulting in less crystallization, a higher feed rate, and a higher print speed.
  - A simple test series placing material fragments of the extruded TPS in water for several days, and documenting the results every 12 h, showed that the material dissolves quickly in water (Fig. 9). After one day, the material becomes soft and shows signs of water absorption. After three days, the material can be easily squished by hand into a watery

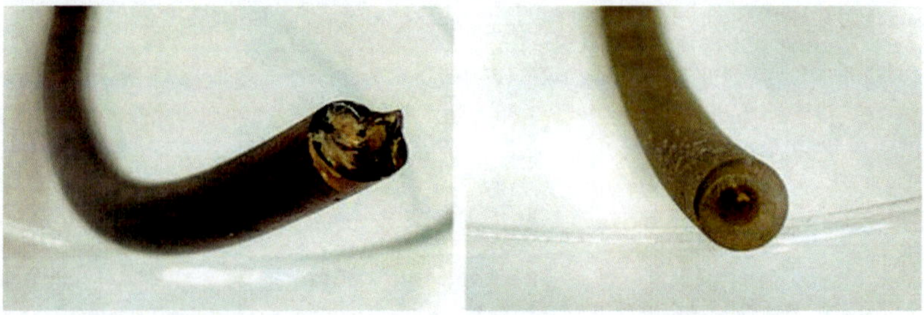

Fig. 8 Detailed images of the crystalline structure of extruded TPS material; left: Sample A (brittle), right: Sample B (elastic)

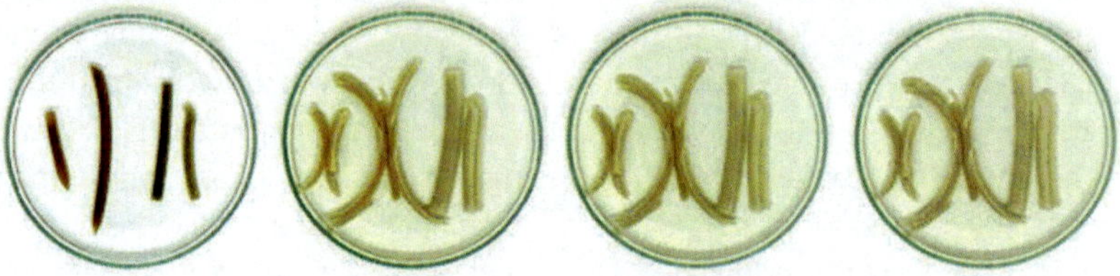

Fig. 9 The dissolution of the material in water over time; left: Sample A (brittle), right: Sample B (elastic) at 0, 12, 24, and 36 h

starchy paste. This test series indicates that the extruded material retained the parent material's hydrophilic properties, suggesting that the produced objects are bio-based, easily compostable.

4.3 Formwork Tests

This study evaluated 3D-printed test cylinders for a basic concrete test. The experiment involved casting both TPS objects with a standard concrete mixture consisting of CEM I 42.5 R cement, 0/2 mm aggregate size sand, and a mixing ratio of 1:3.5:0.4 (cement:sand:water). The cylinders were observed for five days before the formwork was removed.

Upon examination, significant differences were noted between the two TPS formulations. Cylinder A (Formulation A, brittle) exhibited minimal visible effects on the concrete, with only minor cracks observed. This was attributed to using less plasticizer, resulting in a denser, crystalline, almost glass-like material structure. In this small-scale experiment, the material could resist the water in the concrete for a sufficient period until the concrete began to cure. However, the material became softer due to the hydrophilic nature of TPS.

In contrast, cylinder B (Formulation B, elastic) was significantly more susceptible to water. The printed elastic formwork displayed less crystallization and a less closed material surface, allowing water to penetrate easily. As a result of its strong hydrophilic properties, large cracks, separation of single layers, and slight expansion of the formwork were observed (Fig. 10).

The process of removing the formwork was straightforward. Cylinder A absorbed a small amount of water from the concrete and humidity. As a result, the formwork was softer than the initial printed output and could be removed without needing external mechanical or thermal energy. The individual layers were observed to stick together as blocks that could be easily pried

Fig. 10 Comparison of concrete cylinders cast with TPS formulations after five days of curing; left: Sample A, right: Sample B

Fig. 11 Comparison of formwork remnants after removing TPS cylinders; left: Sample A, right: Sample B

out by hand. The only visible cement remnants were found on the inner surface of the TPS formwork.

Cylinder B absorbed a significant amount of water, making the removal of the formwork effortless. The adhesion between the layers was dissolved, allowing the formwork to be removed layer by layer. Cement remnants were found only on the surface (Fig. 11).

As a result, the outcomes of the two distinct TPS formulations differed significantly. While cylinder A exhibited good results and a detailed surface structure in the simple concrete casting process, cylinder B was rendered unusable. The substantial loss of water evaporating into the TPS formwork B led to two notable effects: Firstly, the formwork was unable to maintain the shape of the cylinder, resulting in an irregular shape. Secondly, due to the distorted water-cement ratio, the concrete could not cure correctly, leading to the appearance of cracks in the material (Fig. 12).

Fig. 12 Comparison of concrete cylinders produced using TPS formwork after removal; left: Sample A, right: Sample B

5 Discussion and Conclusion

5.1 Architectural Application and Use Case

In conclusion, the potential architectural applications for the described material and technique have been thoroughly examined. Considering the superior properties of conventional fossil fuel-based and non-renewable polymers is essential. The hydrophilic nature of pure TPS may be considered a disadvantage, but its dissolvability and bio-receptivity can also be considered beneficial. Possible applications include formworks for optimized concrete structures, as long as the material can withstand the water content of the cement until its reaction. However, the potential contamination of bio-based TPS through cement is a concern that should be addressed. It should be noted that the TPS and concrete experiments presented here are still on a small scale, and particular dynamics, such as the hydrostatic pressure of the concrete, will emerge when scaled to architectural size. Therefore, an approach to tackle these issues is threefold:

1. Incorporating additional external structures (exoskeleton) that follow the toolpath and material logic. TPS offers an environmentally friendly and cost-effective option, allowing for increased usage. The expected pressure and weak spots can be calculated and incorporated into the design and printing process.
2. Building an endoskeleton and voids into the formwork to reduce tensile forces from the inside. The dissolvable nature of TPS would support the ease of removal and could create a unique design aesthetic led by the material and its constraints.
3. Enhancing the reinforcement properties and counteracting tensile forces by incorporating natural fibers, such as long hemp fibers and short cellulose fibers, into the material.

5.2 Further Research

The material experiments described in this paper were among the first of their kind, providing crucial indications and raising numerous follow-up questions and topics for further investigation. In particular, further research in this area will focus on the following areas:

- Optimization of printing and material properties: This paper demonstrates a sensitive connection between printing and material properties, and further work is needed to optimize these factors. Key areas to be improved include the material feed rate, resulting print speed, and heat reduction.

More data must be generated to optimize material properties and printing speed by defining the optimal feed rate-temperature ratio.

- Introduction of bio-additives: Adding bio-additives can direct the material properties toward desired application properties. One promising additive that could be explored is chitosan, which reduces the hydrophilic behavior and increases the material's resistance to water exposure. Bio-based fibers can be implemented to increase the material's tensile strength. Bio-based filler materials can also be introduced to reduce shrinkage and ensure higher shape accuracy.
- Evaluation and classification of 3D-printed TPS biopolymers: It is necessary to evaluate and classify the materials according to biodegradability and polymer characteristics to contextualize this research fully. This includes testing and evaluating the materials to place them into context.

These research areas will provide valuable insights and pave the way for future advancements of 3D-printed TPS biopolymers.

References

Adewale P, Yancheshmeh MS, Lam E (2022) Starch modification for non-food, industrial applications: market intelligence and critical review. Carbohydr Polym 291:119590–119612. https://doi.org/10.1016/j.carbpol.2022.119590

Andersen PJ, Ong S, Christensen BJ (2000) Sheets having a starch-based binding matrix. United States Patent 6083586, 4 Jul 2000

Avérous L, Pollet E (2012) Biodegradable polymers. In: Avérous L, Pollet E (eds) Environmental silicate nano-biocomposites. Springer, London, UK, pp 13–39. https://doi.org/10.1007/978-1-4471-4108-2

Burger J, Lloret-Fritschi E, Scotto F et al (2020) Eggshell: ultra-thin three-dimensional printed formwork for concrete structures. 3D Print Addit Manuf 7:49–59. https://doi.org/10.1089/3dp.2019.0197

Chandra R, Rustgi R (1998) Biodegradable polymers. Prog Polym Sci 23:1273–1335. https://doi.org/10.1016/S0079-6700(97)00039-7

Clarinval A-M, Halleux J (2005) Classification of biodegradable polymers. In: Smith R (ed) Biodegradable polymers for industrial applications. Woodhead Publishing, pp 3–31. https://doi.org/10.1533/9781845690762.1.3

de Carvalho AJF, Trovatti E (2016) Biomedical applications for thermoplastic starch. In: Susheel K, Avérous L (eds) Biodegradable and biobased polymers for environmental and biomedical applications. Scrivener Publishing LLC., Hoboken, NJ, USA, pp 1–23. https://doi.org/10.1002/9781119117360.ch1

de Pietro J, Winterthur G (2022) Material-Archiv—Thermoplastische stärke. In: Material-Archive. https://materialarchiv.ch/de/ma:material_1790?q=tps. Accessed 13 Jan 2023

Dorgan JR, Lehermeier H, Mang M (2000) Thermal and rheological properties of commercial-grade poly(lactic acid)s. J Polym Environ 8:1–9. https://doi.org/10.1023/A:1010185910301

Dyze Design (2022) Pulsar™ pellet extruder | dyze design docs. In: Dyze design docs. https://docs.dyzedesign.com/pulsar.html#_3d-printer-guideline. Accessed 19 Aug 2022

Goodall C (2011) Bioplastics: an important component of global sustainability. Biome bioplastics

Geyer R, Jambeck JR, Law KL (2017) Production, use, and fate of all plastics ever made. Sci Adv 3:25–29. https://doi.org/10.1126/sciadv.1700782

Häkkinen T, Kuittinen M, Vares S (2019) Plastics in buildings: A study of Finnish blocks of flats and daycare centres. Ministry of the Environment, Helsinki, Finland

Jamshidian M, Tehrany EA, Imran M et al (2010) Polylactic acid: production, applications, nanocomposites, and release studies. Compr Rev Food Sci Food Saf 9:552–571. https://doi.org/10.1111/j.1541-4337.2010.00126.x

Jipa A, Bernhard M, Dillenburger B (2018) Submillimetre formwork: 3D-printed plastic formwork for concrete elements. In: Beig K (ed) 2017 TxA Emerging Design + Technology Conference Proceedings. Austin, TX, USA, pp 70–79

JohnSL - Random Products (2020) Plastic Pellet Maker Part 7. Online video. https://www.youtube.com/watch?v=gGjAa8ds8_k. Accessed 15 Jan 2023

Kale G, Kijchavengkul T, Auras R et al (2007) Compostability of bioplastic packaging materials: an overview. Macromol Biosci 7:255–277. https://doi.org/10.1002/mabi.200600168

Kemper BN (2022) Bio-formwork: large scale 3d deposition of thermoplastic starch in architecture. In: Contributions to the 22. Nachwuchswissenschaftler*innenkonferenz (NWK). Open Conference Proceedings, Brandenburg an der Havel, Germany, pp 65–70

Khan B, Bilal Khan Niazi M, Samin G, Jahan Z (2016) Thermoplastic starch: a possible biodegradable food packaging material—A review. J Food Process Eng 40:1–18. https://doi.org/10.1111/jfpe.12447

Klarenbeek E, Changemakers, Museum Boijmans van Beuningen (2017) Algae lab—Klarenbeek & Dros with Atelier Luma, Luma Arles. In: Klarenbeek &

Dros—Designers of the unusual. http://ericklarenbeek.com/. Accessed 13 Jan 2023

Kretzer M (2017) Information materials: smart materials for adaptive architecture. Dissertation, ETH Zurich

Lamberti FM, Román-Ramírez LA, Wood J (2020) Recycling of bioplastics: routes and benefits. J Polym Environ 28:2551–2571. https://doi.org/10.1007/s10924-020-01795-8

Leschok M, Dillenburger B (2019) Dissolvable 3dp formwork: water-dissolvable 3d printed thin-shell formwork for complex concrete components. In: Bieg K, Briscoe D, Odom C (eds) ACADIA 19: ubiquity and autonomy. Proceedings of the 39th annual conference of the association for computer aided design in architecture. association for computer aided design in architecture (ACADIA), Austin, TX, pp 188–197

Love L, Post B, Roschli A et al (2019) Feasibility of using BAAM for mold inserts for the precast concrete industry. Oak Ridge National Laboratory, Tennessee, USA

Madhavan Nampoothiri K, Nair NR, John RP (2010) An overview of the recent developments in polylactide (PLA) research. Bioresour Technol 101:8493–8501. https://doi.org/10.1016/j.biortech.2010.05.092

Meibodi A, Bernhard M, Jipa A, Dillenburger B (2017) The smart takes from the strong: 3d printing stay-in-place formwork for concrete slab construction. In: Menges A, Sheil B, Glynn R, Skavara M (eds) Fabricate 2017. UCL Press, Stuttgart, Germany, pp 210–217. https://doi.org/10.2307/j.ctt1n7qkg7.33

Mogas-Soldevila L, Duro-Royo J, Oxman N (2014) Water-based robotic fabrication: large-scale additive manufacturing of functionally graded hydrogel composites via multichamber extrusion. 3D Print Addit Manuf 1:141–151. https://doi.org/10.1089/3dp.2014.0014

Mostafavi S, Kemper BN, Du C (2019) Materializing hybridity in architecture: design to robotic production of multi-materiality in multiple scales. Archit Sci Rev 62:424–437. https://doi.org/10.1080/00038628.2019.1653819

Paolini A, Kollmannsberger S, Rank E (2019) Additive manufacturing in construction: a review on processes, applications, and digital planning methods. Addit Manuf 30:100894. https://doi.org/10.1016/j.addma.2019.100894

Peters B (2014) Additive formwork: 3D printed flexible formwork. In: Gerber D, Huang A, Sanchez J (eds) Design agency: Proceedings of the 34th annual conference of the association for computer-aided design in architecture (ACADIA 2014). Riverside Architectural Press, Los Angeles, CA, USA, pp 517–522

Roschli A, Gaul KT, Boulger AM et al (2019) Designing for big area additive manufacturing. Addit Manuf 25:275–285. https://doi.org/10.1016/j.addma.2018.11.006

Rossi G, Chiujdea R, Colmo C et al (2022) A material monitoring framework: tracking the curing of 3d printed cellulose-based biopolymers. In: Dörfler K, Parascho S, Scott J (eds) Realignments: toward critical computation: proceedings of the 41st annual conference of the association for computer aided design in architecture. Acadia Publishing Company, Online, pp 308–317

Saiah R, Gattin R, Sreekumar PA (2012) Properties and biodegradation nature of thermoplastic starch. In: El-Sonbati AZ (ed) Thermoplastic elastomers. IntechOpen, London, UK, pp 57–78

Shanks R, Kong I (2012) Thermoplastic starch. In: El-Sonbati AZ (ed) Thermoplastic elastomers. IntechOpen, London, UK, pp 95–116

Shen L, Patel MK (2008) Life cycle assessment of polysaccharide materials: a review. J Polym Environ 16:154–167. https://doi.org/10.1007/s10924-008-0092-9

Teixeira S, Eblagon KM, Miranda F et al (2021) Towards controlled degradation of poly(lactic) acid in technical applications. C 7:1–43. https://doi.org/10.3390/c7020042

UN General Assembly (2015) Transforming our world: the 2030 agenda for sustainable development. UN General Assembly

Wang N, Yu J, Chang PR, Ma X (2008) Influence of formamide and water on the properties of thermoplastic starch/poly(lactic acid) blends. Carbohydr Polym 71:109–118. https://doi.org/10.1016/j.carbpol.2007.05.025

MYCOlullose: Fabricating Biohybrid Material System with Mycelium-Based Composites and Bacterial Cellulose

Natalia B. Piórecka, Peter Scully, Anete K. Salmane, Brenda Parker, and Marcos Cruz

Abstract

In an attempt to become more sustainable, architects have increasingly begun employing biomaterials in design practices. The current ethos merely considers biomaterials as an eco-friendly substitute for conventional solutions. This paper attempts to elevate the benefit of using biomaterials by creating and evaluating a hybrid material formed as a result of collaboration between living organisms of mycelium, and by-product of kombucha brewing, SCOBY bacterial cellulose. The fabricated artefact provides an example whereby a combination of distinct living organisms in a complimentary fashion holds potential for the discovery of previously unimagined goals, new value systems, qualities and speculative applications that can embrace this discovered affordance. The hybrid assemblage, explored through a material-driven approach, demonstrates that mycelium can effectively grow within the cellulose produced by bacteria, where mycelium is a structural, bulk element and bacterial cellulose is a sacrificial mold hosting the growth and gradually being integrated within the fungal skin. The manufacturing methodology used principles of fabric forming fabrication as a primary technique to create geometrical expressions, by translating 2D sewn bacterial cellulose membranes, into 3D structures with mycelium composite filling. The resulting hybridised biomaterial system with augmented characteristics combines complementary properties such as mechanical strength of BC performing well in tension and the mycelium-based composites in compression.

Keywords

Mycelium · Bacterial cellulose · Hybrid biomaterials · Material system · Multi-materiality · Fabrication · Rethinking resources · Living organisms

1 Introduction

The global demand for environmentally friendly materials resulted in new research studies seeking novel biomaterials aiming to address United

N. B. Piórecka (✉) · P. Scully · A. K. Salmane · B. Parker · M. Cruz
Bio-Integrated Design Lab, Bartlett School of Architecture, University College London (UCL), London, United Kingdom
e-mail: natalia.piorecka.21@ucl.ac.uk

P. Scully
e-mail: p.scully@ucl.ac.uk

A. K. Salmane
e-mail: a.salmane@ucl.ac.uk

B. Parker
e-mail: brenda.parker@ucl.ac.uk

M. Cruz
e-mail: m.cruz@ucl.ac.uk

M. R. Thomsen et al. (eds.), *Design for Rethinking Resources*, Sustainable Development Goals Series,
https://doi.org/10.1007/978-3-031-36554-6_38

Nations Sustainable Development Goals and strategies. While that emerging practice can provide innovative solutions and applications, biomaterials are currently often considered through the lens of established norms as surrogates or alternatives to conventional materials (Karana et al. 2018). We valorise their characteristics in so far as they may offer up sustainable options for existing ambitions. However, they may offer pathways towards previously unimagined goals, opening the horizon for potentially new value systems, qualities and speculative applications that can embrace this discovered affordance.

Throughout the design industry, the shift of the emphasis from geometrical expression to material-driven approaches and production modes has been observed (Oxman 2010; Gazit 2016). Within the scope of abilities to program functionality, biomaterial systems along with the use of digital and computational tools, result in unique fabrication strategies, which could become more accessible to architects and designers, therefore, enabling biomaterial applications with a positive, environmental impact on a much greater scale (Gazit 2016).

At present, biomaterial development focuses on research, analysis and tuning of optimal growth conditions to generate homogenous, mono-biomaterials. There are a limited number of studies touching on one or more materials working together. Looking at biomaterial systems where the rules within are organically related, intimate and adaptable may be the key (Oxman 2010).

Drawing from the architectural and construction industries, the multi-functionality of the building system is achieved with a combination of various materials' assemblage and varied functionality, rather than with the usage of a single homogeneous material (Mostafavi et al. 2019). The approach of multi-materiality and material assembly applied to the living components raises the question if we could achieve a strong material bond resulting from the intergrowth as opposed to typical non-active material layering (Oxman 2010).

This paper will elaborate on to what extent, does the collaboration between mycelium and bacterial cellulose host the potential for sustainable fabrication methods for both existing and future architectural goals and structures. It will investigate what may be the opportunities of the hybrid material in the current and future architectural context and how the future development towards digitalised workflow can leverage regional abundances of resources through scientific and manufacturing understandings towards delivering contemporary and architectural responsibilities.

Both mycelium-based composites and bacterial cellulose are broadly researched materials and their intersection has already raised curiosity among scientists (Hoenerloh et al. 2022). The study will primarily explore the promising organic link between the materials, where the recognised limitations can be mitigated through the interaction between them. The collected data will inform the qualitative compatibility judgement and examine what opportunities and benefits it can introduce as a hybrid material system. Secondly, the study is looking at the material characteristics at the interfaces of scaled-up production and fabrication of the two materials, utilising unique, individual properties, to assess the material behaviours and the potential relationship between the two. Lastly, the study aims to develop a methodology and workflow of material assembly combining living materials synergistically through material-driven curiosity, where both materials complement each other, creating unique design opportunities.

1.1 What are Mycelium-Based Composites?

Mycelium-based composite (referred to from now on as: MBC) has been well studied over the last few years in order to understand its possibilities and limitations (Appels et al. 2019; Alemu et al. 2022). It has been explored as an opportunistic composite that could be tuned to specific needs depending on a range of factors including feed substrate, mycelium species and growth conditions (Jones et al. 2020; Manan et al. 2021). Bulk MBC materials, are created with the mycelium's ability to bind particles through the process of decay. Moreover, with the decomposition processes, the MBCs offer the opportunity of the so-called "bio-welding"—binding between two individual units of the living material, as an aspiring method for fabrication and assembly (Saporta et al. 2019; Modanloo et al. 2021). By utilising locally resourced organic waste in the form of natural fibres, the material offers an economically viable, lightweight composite with low cost and minimal energy production (Jones et al. 2017). The resulting MBCs are known to perform well in compression, however, they are not durable in tension. Due to that fact, various experimental studies aimed to further synthesise MBCs strength with the appropriate use of the geometry (Goidea et al. 2020).

The feasibility of current fabrication techniques for cultivating 3D mycelium forms are mostly limited to moulding with an impermeable mould or a more recently developed alternative—extrusion based 3D printing. Therefore, ongoing research among designers and scientists aims to develop effective ways to fabricate MBCs, in order to capitalise on the unique properties that the material displays (Ghazvinian and Gursoy 2021).

1.2 What is Bacterial Cellulose?

SCOBY (Symbiotic Culture of Bacteria and Yeast)—a gelatinous pellicle, composed of bacterial cellulose (referred to from now on as BC)—is formed on the air–liquid interface as a by-product of kombucha tea fermentation (Gazit 2016). The layer of BC, supports the microbial culture in kombucha and protects the fermenting liquid (Römling and Galperin 2015). The pellicle, also commonly known as a tea mushroom, is mainly built with high purity cellulose, with no hemicellulose, pectin or lignin, which makes it a potential alternative to the high energy and cost demanding cellulose obtained from plants (Esa et al. 2014; Yu et al. 2021; Derme et al. 2022).

With low waste and energy input, similar to MBC, BC is a lightweight and fully biodegradable material which shape, thickness and strength can be manipulated with the vessel, media composition and growth time (Gazit 2016). Despite this, BC alone has still limited capabilities to fulfil the current demand for high-performance biomaterials within the field of architecture (Esa et al. 2014; Yu et al. 2021). Previous studies have been investigating potential means of translating the harvested in static conditions, flat biofilm into 3D structures on both micro and macro scale. With the production process not being efficient and consistent for large quantities of production (Gazit 2016; Fernandes et al. 2020; Derme et al. 2022), various techniques and post-treatment methods were developed, but are yet not applicable within macro scale contexts (Stoica-Guzun et al. 2012; Arevalo Galllegos et al. 2016; Patchan et al. 2016; Greca et al. 2018; Laromaine et al. 2018; Derme et al. 2022). Consequently, with the need of the design industry for transferable solutions across several scales continuous research is necessary to elaborate on the scaled 3D biofabrication processes of BC, currently not fully resolved in scales applicable to architecture.

1.3 Mycelium-Bacterial Cellulose Biohybrid

Few studies have considered the juxtaposition of the two living materials together. A study by Elsacker et al. (2021) used the bacterial cellulose as reinforcement for the growth of the mycelium within a MBC composite. Another study conducted by Hoenerloh et al. (2022) demonstrated that mycelium and BC can be effectively grown in the company of one another, focusing on utilising the growth of BC along the MBC (Hoenerloh et al. 2022). This study elaborates further on the multi-organism composites experimentation by undertaking an opposite approach to material assembly. Inspired by the findings of Elsacker et al. (2021), observing strengthening of internal bonding of the mycelium material after adding BC, this paper considers cellulose as a feed for mycelium aiming for organic connection and gradual intergrowth within materials.

1.4 Challenges and Opportunities for Biohybridity

To achieve sustainable development objectives, outlined in Sect. 4, multiple studies consider MBC and BC as renewable materials with the focus to maximise the performance of each as a homogenous material. In contrast, this experimentation takes a bottom-up approach towards the unexplored territory of unexpected goals and potential developments of a new class of functionalised biomaterials (Fig. 1). This paper will examine the potential of sustainable fabrication of the investigated, novel Mycelium-BC synthesis, challenging limitations of components by turning individual material weaknesses into a collective advantageous solution, fulfilling each other's needs outperforming the materials individually.

Based on the material properties and the prospective, intimate relationship of the hybridised bio-materials, a novel fabrication technique is proposed. In order to obtain the desired form MBC, the bulk material considered here needs a host for the growth, whilst for the same purpose BC when considered as a membrane, needs support to maintain the 3D geometrical form. To achieve a complementary system with the two materials, the fabrication strategy is drawn from principles of fabric forming, in a similar fashion to the strategy applied by Yogiaman et al. (2020) with cotton knitting. This experimentation is using sewing as the main manner to create the geometrical form from the BC, permitting the augmentation and control. The MBC functions as a filling and reinforcement for the sewn BC. Simultaneously, the BC provides an envelope for mycelium composite. Acting as a sacrificial mould built with the cellulose the sewn BC envelope will be integrated into the growing geometry by being fused though the growth of mycelium into the fungal skin. As defined by Drisko and Sanchez (2012), the two dissimilar components would be hybridised, meaning that they would blend together, making a single entity with either enhanced or completely new characteristics.

The significant advantage of the two discussed biomaterials is their ability to utilise waste through growth. Once applied on a larger scale, it could have a significant impact on how we manage regional abundancies of resources. In short, BC can be cultivated in the growth media that use the waste from glucose-rich sources as a nutrient. Similarly, mycelium can biodegrade the cellulose-based agricultural waste and transform it into a viable material. Once the fabrication strategy is effective and well defined it could

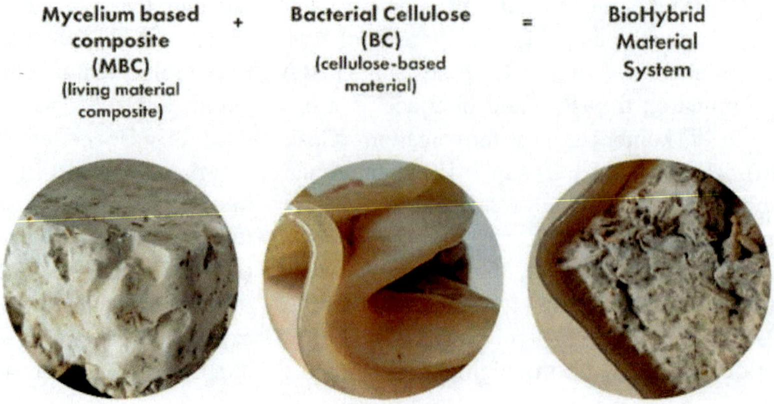

Fig. 1 Material concept diagram towards biohybrid material. *Image credits* N. B. Piórecka

provide a subsequent sustainable, attractive and accessible moulding alternative to the current manufacturing MBC methods (Colmo and Ayres 2020; Soh et al. 2020). That might broaden the design perspective, utilising local resources and addressing the economic and social viability (Jones et al. 2020).

2 Materials and Methods

The experimentational testing aimed to investigate the individual and collaborative affordance of the Mycelium-BC hybrid material fabrication, in the context of leveraging regional resources and exploring future architectural goals that the material offers. The success was measured and assessed based on multiple criteria centred around the material relations. These included biological compatibility between the mycelium and BC and their uninterrupted growth, species selection, material preparation, optimal growth conditions for both materials individually and collaboratively, minimised cross-contamination, state of each materials in order to achieve their interaction, potential for scaled-up material production, and biomaterial system assemblage through fabrication. Additionally, those criteria were used to determine optimal species selection and fabrication strategy in order to maximise the probability of success, as discussed below.

2.1 Biological Compatibility and Species Selection

Considering the biological compatibility of fungi and cellulose, the former can use *cellulase* enzymes to degrade the latter into smaller compounds such as glucose and cellobiose that could subsequently be metabolised (Lynd et al. 2002; McNamara et al. 2015; da Silva et al. 2021). The consumed part of glucose is used by fungi for biomass formation (anabolic process) and the rest for energy generation (catabolic process). The study selected a cellulose-degrading species, white-rot fungi (Floudas et al. 2012), *Ganoderma lucidum* as a competent mushroom species from the division of Basidiomycota, which are known for being cellulose degraders (Edwards et al. 2008) with the highest *cellulase* activity once compared with *Pleurotus ostreatus, Trametes versicolor* and *Phanerochaete chrysosporium* (Fauzi et al. 2018).

For the secretion of the cellulose used in the experimentation, the Gram-negative bacteria *Gluconacetobacter xylinus* (formerly known as *Acetobacter xylinum*) was used, as a model organism for the commercial fermentation and BC production, due to the high yield of cellulose (Of et al. 1954; Lahiri et al. 2021).

2.2 Production of Mycelium-Based Composite

The mycelium-based composite was cultivated using hemp waste from East Yorkshire (UK) farm, sieved to approximately 1.5 cm. Two strategies of sterilisation of the waste fibres were compared in the context of scaled-up substrate preparation and energy efficiency: (1) pasteurised—temperature maintained above 70 °C for 2 h and (2) sterilised—in Autoclave at 121 °C for 20 min. Since both methods were comparable in terms of cultivation success rate, the study continued with the pasteurisation as a less energy-intensive process, able to accommodate larger amounts of the bulk substrate within a similar timeframe, as opposed to volume restrictions of the autoclave.

For the initial cultivation, the *Ganoderma lucidum* strain (OSMOSE_GL/PE-P22-G2), was purchased from the Osmose studio (London, UK). The *Ganoderma lucidum* used is a local wild strain, isolated in South Yorkshire (UK) ensuring that the local biodiversity is not impacted by the distribution of non-regional organic matter. The following material cultivation used the overgrown hemp substrate from the prior batch to cultivate further the bulk material, in a proportion of 2:5, reducing the number of steps in the material production workflow. Optimal growth of the MBC has been achieved in the air temperature of 25 °C and air humidity of 70%.

2.3 Production of SCOBY Aka BC

Using the traditional technique of kombucha brewing to cultivate the BC, the study cultivated the cellulose in static conditions testing various recipe iterations aiming for the uniform growth of the pellicle. The most stable and unified growth of the BC pellicle used the recipe (per 1 L): 3 black tea bags, 100 ml of a starter tea and 80 g of cane sugar. The initial certified SCOBY were ordered from Kombuchaorganic, UK. The BC was grown at room temperature of 22 °C, up to 5 mm thickness, which through a series of conducted tests proved to be most suitable for the following experimentation in terms of strength and flexibility both in wet and dry states.

Combining SCOBY and mycelium requires careful considerations of pH levels. The balanced cellulose secretion is achieved at pH: 2.5–3.5, where the acidity limits the contamination, hosting safe fermentation. In comparison, the most favourable for mycelium growth of *Ganoderma lucidum* is pH 5 (Jayasinghe et al. 2008; Subedi et al. 2021). To avoid the possible growth constraint of mycelium tissues due to pH difference, the produced bacterial cellulose was post-processed to raise the cellulose pH level with cleaning and purification methods, such as washing and bleaching applied in the later stages of experimentation (Han et al. 2018).

2.4 Biomaterial System Assemblage: Experiment 1 and 2: Materials State and Assembly

The key link between the two organisms is cellulose; the experimentation goal is to encourage mycelium to partially degrade cellulose within the BC and establish the material relation as presented in the Experiment 1 (Fig. 2). Evaluating layering choreography along with the state of the materials used within the organism's life cycle will determine conditions for the most effective growth.

In the following Experiment 2, material was assembled testing various states of the material: (a) Living Mycelium with dried BC, (b) Living mycelium with wet BC, (c) Dry mycelium with wet BC. The experiment aimed to test mycelium's ability to attach to wet or dry BC as well as testing if the BC itself is able to mechanically attach to the mycelium sample (Fig. 3).

To increase production yield same material recipes were used with 25 kg of hemp as a substrate for mycelium cultivation. Expansion of BC membranes size required increased growth area. Six 12L (550 × 360 × 90 mm) Allstore containers were prepared to produce workable A3 sheets of BC, cultivated in 2L growth media.

2.5 Experiment 3A: Cotton Fabric Fabrication

Prior to applying fabrication methods of sewing to BC, as a control it was first tested with the cotton fabric (nearly 90% cellulose), similar to

Fig. 2 Experiment 1—Close up of the Petri-dish test of the MBC growing in the presence of pre-grown BC. After the mycelium growth developed within the substrate, the continuous growth of mycelium over the cellulose pellicle was observed. *Image credits* N. B. Piórecka

Fig. 3 Experiment 2—Interaction between various states of biomaterials: **a** Living mycelium with dried BC; **b** Living mycelium with wet BC and resulting interaction (right image); **c** Dry mycelium with wet BC. *Image credits* N. B. Piórecka

successful application by Yogiaman et al. (2020). The experiment assessed method viability for iteration of geometries, seeking opportunity in both sewing technique and ability of mycelium to bind and grow though organic fibres. As shown in Fig. 4, cotton fabric experiment looked at various bespoke geometries from base tube geometry (a), to geometries testing potential creation of textures (ridges) (b), self-binding abilities though the sewn material (c), integration of the openings (d), creation of unified panels (e) and modular geometries for potential assembly through strategies like interlocking or aggregation (f), using binding properties of mycelium through the so called "bio-welding" to connect modules. The two panel prototypes (d, e) introduced wet BC sheet within the created opening with no fixing.

In this instance, fabric was sewn into shape using industrial sewing machine, in order to create a 2D envelope, cleaned by soaking in ethanol and water and then filled with pre-grown MBC, converting flat sheets into a 3D formations as displayed in Fig. 5. Packed with MBC units were protected by wrapping in perforated cellophane foil for the initial growth period of 7 days in the low-tech growth chamber. The growth was then continued for another 7 days without the foil, allowing for further growth over the cotton fabric surface, then samples were dried. Developed fabrication techniques were subsequently implemented in the experiment 3B with the BC.

2.6 Experiment 3B: Bacterial Cellulose Fabrication

The experimentation was continued with the harvest of uniformly grown 5 mm thick, A3 BC sheets. Tests included post-treatment of SCOBY, as a measure for eliminating potential cross-contamination between the organisms after material setup and assembly. Two cleaning methods were tested: (1) SCOBY washed with deionised water; (2) SCOBY purified with two successive 1.0 M NaOH washings at 90 °C, followed by the treatment with 1.5% (w/w) aq. NaOCl for 2 h, tested by Amarasekara et al. (2020) as an effective method for purification. The SCOBY only washed with deionised water still had the presence of bacteria within its three-dimensional fibre structure, therefore the matter of potential cross-contamination in later stages of the growth was present. Therefore, the

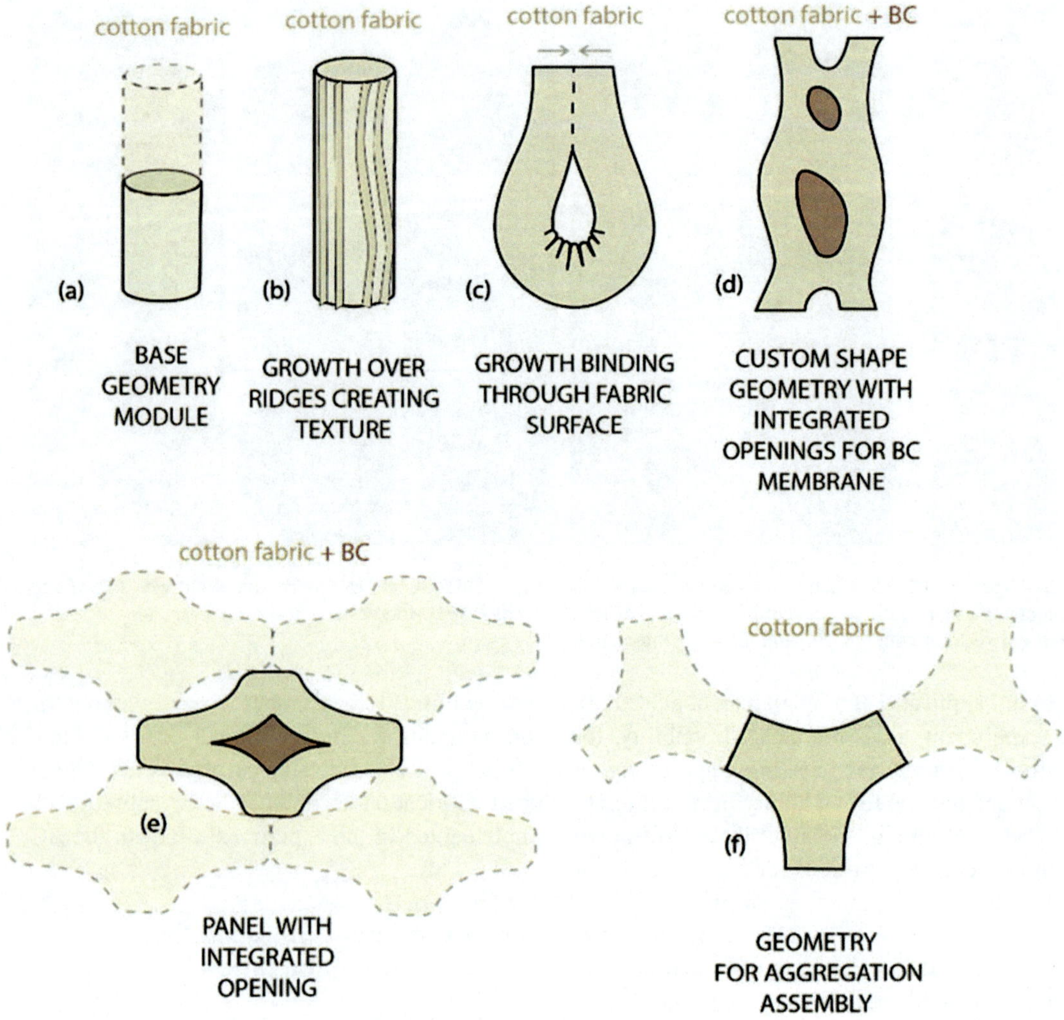

Fig. 4 Experiment 3A: cotton fabric geometries. *Image credits* N. B. Piórecka

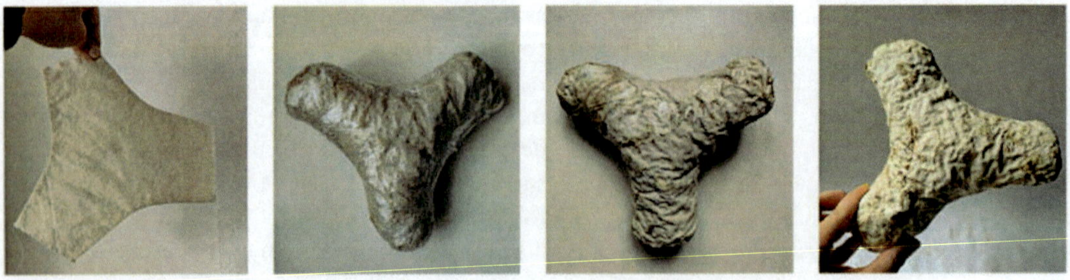

Fig. 5 Converting flat cotton fabric sheets into 3D formations. *Image credits* N. B. Piórecka By author

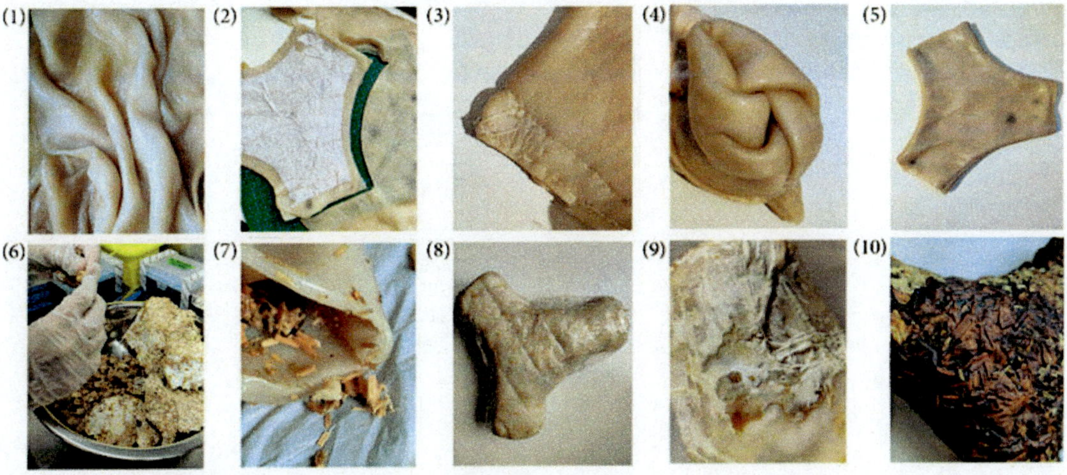

Fig. 6 Biohybrid material making workflow (1) harvesting the BC, (2) cutting desired shape, (3) hand sewing, (4) flipping the BC inside out, (5) cleaning, (6) preparing mycelium substrate, (7) filling with MBC, (8) wrapping in perforated cellophane foil, (9) growing without foil for surface growth and (10) drying. *Image credits* N. B. Piórecka

methodology implemented the cleaning process of immersing BC in a 0.5 M NaOH and soaking in water to eliminate NaOH solution leftovers before placing the MBC. Cleaning was applied after sewing to minimise steps of chemical cleaning treatment.

Both BC and MBC were grown simultaneously. While BC cellulose was cultivated, the substrate for mycelium growth was pasteurised and pre-cultivated with mycelium following the procedures described before. Once the cellulose was harvested, a similar process sequence was used as described in experiment 3A (Fig. 6).

Building on the geometries in the experiment 3A, experiment 3B applied the hand sewing fabrication method within the geometries presented in Fig. 7. It focused on modular geometries (a, b) and bespoke geometries with openings of the continuous BC membrane (c). Along with the sewn BC envelopes filled with MBC, the modular geometry was also created with an additional scaffold (d, e). The role of the scaffold was to lock the sheet of BC in place allowing for more precise control over the geometry, accommodating unified connection joints between the units, applicable when considering the bio-welding as a binding technique for the aggregation assembly. The scaffolded unit was made of wood and metal elements, soaked in bleach and water for cleaning, wrapped with cellulose sheets and filled with MBC. The cellulose sheets used were thinner (2.5 mm) than in the other experiments to be lighter and adhere to the scaffold more easily. The experiment aimed to assess the material hybridisation implemented into a 3D shape, its opportunity for achieving free standing prototypes as well as considering potential ways of increasing the precision of the fabrication, aiming for a seamless translation into digital fabrication between the physical prototype and designed digital form. The steps of the production can be generalised according to the diagram in Fig. 8.

3 Results

All experiments conducted assessed the combined potential of bacterial cellulose and mycelium in a hybridised material system and tested the developed fabrication methodology.

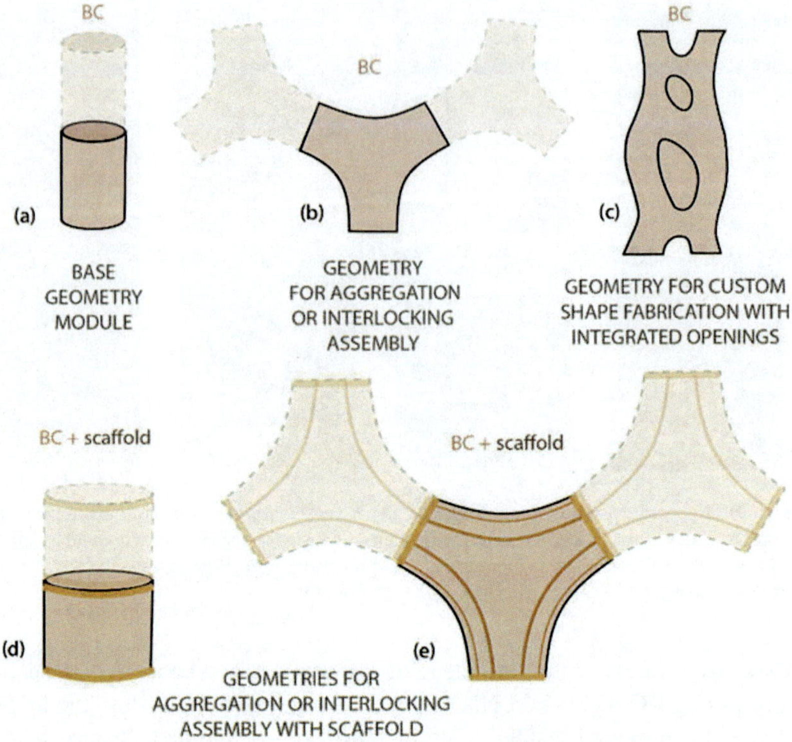

Fig. 7 Experiment 3B: bacterial cellulose geometries. *Image credits* N. B. Piórecka

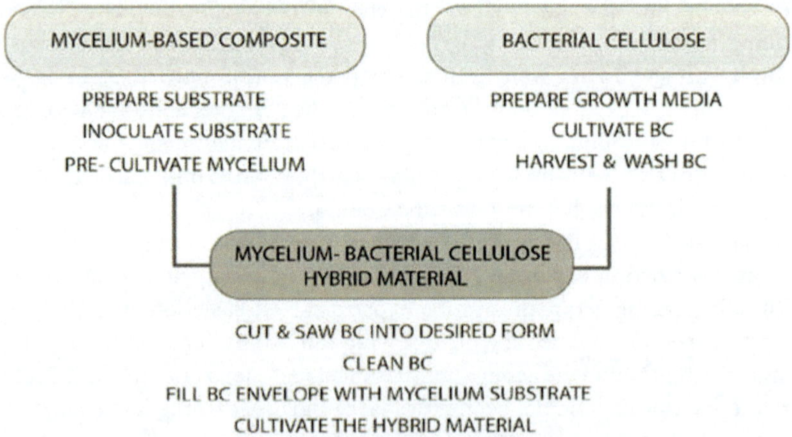

Fig. 8 Summary of the material preparation, method and production. *Image credits* N. B. Piórecka

3.1 Experiment 1 and 2: Materials State and Assembly

The first experiment proved that mycelium can grow in the presence of the BC without disturbance and is able to fully overgrow the container with the BC, displaying signs of biological compatibility between the living materials.

The most favourable growth of mycelium in Experiment 2 was observed when the BC was in the wet state in the material setup (b). The growth of the mycelium hyphae towards BC was clearly visible at the interface of the materials after two days of growth. Subsequently, the BC membrane was gradually consumed by the living mycelium that continued to grow up to 14 days, and created a fungal skin over the BC surface. The attachment was still strong while the samples dried.

In contrast, in the material setup (a) no strong attachment between the materials was observed —the layers could be easily separated. The wet BC membrane in the material setup (c) placed on the dried mycelium adjusted its shape to the textured MBC surface and physically stuck to the surface while drying, maintaining the dried shape. However, there was no attachment and the materials could be easily separated mechanically, similar to the setup (a). The material stages explored in the setup (b) were the most successful in terms of observing an interaction between materials and for that reason were used in the following experiments. It is hypothesised that the moisture present within the pellicle might have been an important factor in the hybridisation of the two, additionally supporting the mycelium growth.

3.2 Experiment 3A: Cotton Fabric Fabrication

The results of the experiment using cotton fabric have shown that mycelium can grow through a cellulose rich envelope of the cotton fibres and confirmed that the method of sewing and filling is viable. The cotton fabric samples filled with MBC grew without interruptions, gained strength with the mycelium growth and became rigid once the samples were dried. The mycelium growth was visible on the external surface where it started to form the fungal skin growing through the cotton fabric, strengthening the bond between the materials. The panels with the introduced wet BC sheet in the openings, displayed strong attachment between the BC and cotton fabric without any fixing applied. The attachment between the cellulose fibres themselves could be a further step to explore in future iterations. The described results of the experiment are presented below on the Fig. 9.

3.3 Experiment 3B: Bacterial Cellulose Fabrication

The results of the mycelium and bacterial cellulose hybrid prototype were even more effective in terms of the growth rate once compared to the cotton fabric. The juxtapositioned dried results are presented in the Fig. 10.

The growth of mycelium within the wet BC was already visible after 3 days of incubation and on day 7 also on the external surface of the cellulose. As in experiment 2 the use of the wet cellulose matrix allowed to capture and retain moisture supporting the growth of the MBC within the BC with quicker and more intensified mycelium overgrowth, as opposed to the relatively dried control cotton samples. The air permeability of the BC pellicle (Yang et al. 2016) was sufficient for the uninterrupted mycelium growth inside the BC.

Over the process of drying the prototypes stiffened and gained rigidity, resulting in the formation of free-standing structures, captured in Fig. 11. The surface of the fresh, thick and hydrated cellulose pellicle was gradually covered and partially degraded by the mycelium, integrating the BC within the fungal skin and fusing it with the rest of the material. Moreover, the integrated openings within the sewn BC membrane, host further design opportunities of creating semi-enclosed, translucent zones, that due to the properties of BC could partially allow the light to pass though as shown in the Fig. 11 (middle right) The overall shrinkage of circa 10%, did not disrupt the overall biohybrid system integrity, however the shrinkage in the thickness of the BC influenced the aesthetics of the external surface, displaying visible texture, corresponding to the filling fibres of the mycelium substrate used inside (Fig. 10, top right). That suggests that the fibres used as substrate for mycelium growth could dictate the final texture, giving an interesting design opportunities.

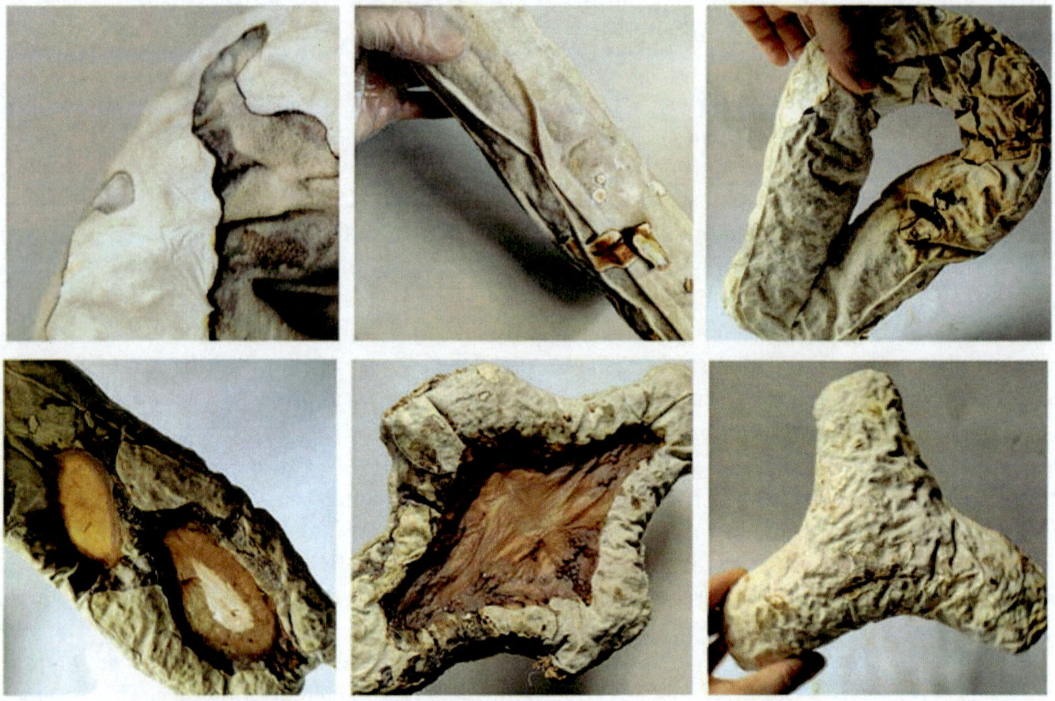

Fig. 9 Results of the experiment 3A. *Image credits* N. B. Piórecka

Fig. 10 Dried cotton fabric (top) and BC (bottom) prototypes. *Image credits* N. B. Piórecka

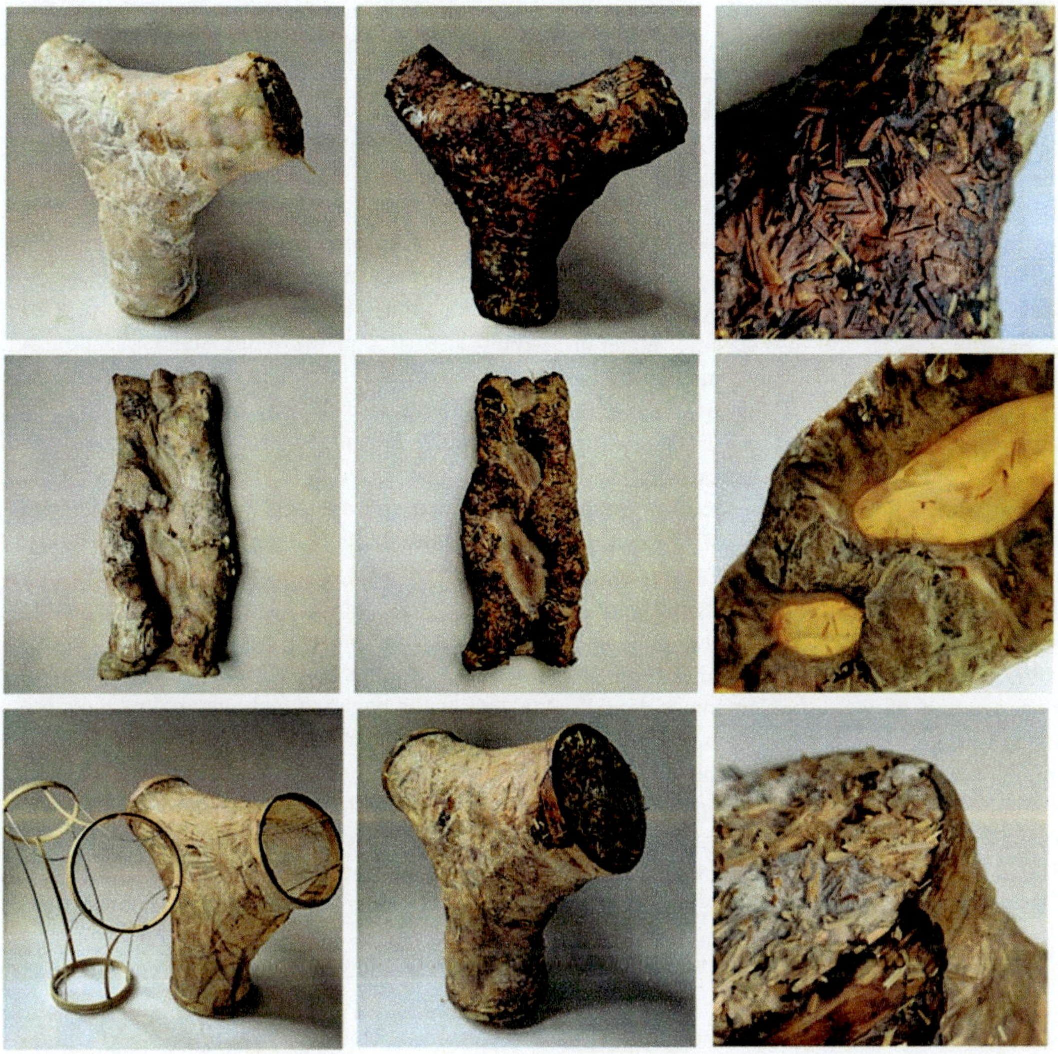

Fig. 11 Results of the experiment 3B. *Image credits* N. B. Piórecka

The free standing BC scaffold unit, filled with mycelium, also achieved the material interaction. However the growth quality was inferior compared with the previous prototypes due to the premature moisture loss. In the larger scaffold unit the BC dried before the mycelium could fully grow through it. The thickness of the BC sheet storing moisture has an influence on the growth of the mycelium and therefore should be considered in future scaffold iterations. The resulting prototypes displayed unusual aesthetic and tactile value changing their colour, texture, transparency and structure along the growth and decay processes. A range of various stages were observed: wet, shiny, flexible and lightly coloured surfaces of BC and white, soft and fluffy mycelium filling that resulted in robust elements with a mat, dark and tensioned surface with a texture imprinted by the firm mycelium composite merged into one.

4 Discussion

The results of this iterative experimental approach displayed an intriguing relation between the materials produced by living organisms through processes of fermentation, growth and decay. The creation of three-dimensional forms of BC and MBC though various methodologies and strategies is an ongoing attempt in the field mostly considering the materials separately. Proposing an assembly driven by the specific material properties, behaviour and characteristics, translating the flat membrane of the BC and bulk mycelium based composite into a free standing form hosts meaningful opportunities supporting design and fabrication feasibility of both. Further material tests of the Mycelium-BC biohybrid should be conducted to precisely evaluate the biohybrid characteristics looking at mechanical strengths, economic, social and ecological value in comparison to other materials available on the market. On top of that, supplementary experimentation would be required to assess the impact of the diversified conditions on the biohybrid elements for indoor and outdoor applications as well as its commercialisation viability and industrial scalability. Nonetheless, the resulting fabrication technique is further opening a discussion regarding application and utilisation of the two as a biologically driven material system. The relationship observed between materials with structural and functional complementary properties, hosts a superior opportunity for further exploration within the architectural industry. One is performing well in tension providing a growth vessel and a sacrificial mould and the other is performing well in compression, providing the bulk filling binding fibres as a "natural glue".

Through the iterative design process the conventional production method, derived from fabric forming, usually using concrete, was applied to the unconventional material system explored in this study, proving to be a suitable method for a BC and MBC hybrid fabrication. Looking at the resemblance of the production technique with concrete fabric forming, the future development and prototype scale up may follow similar strategies as employed in projects as "Single Cast Fabric Formwork" (Richard et al. 2010) or "Fabric Forms" (Culver et al. 2016). That could allow to transform the protypes into big scale mock up models, testing the hybrid system feasibility in 1:1 scale. The fabrication processes developed, despite currently focusing on a hand craft, are designed to be transferable towards further automation and scaled up material production. The automation of the processes would also support the future standardisation of the created biohybrid material system with the aim to offer new value system with nature as its inseparable element. Moreover, it is important to mention distinctive aesthetic values the material combination uncovers. Considering that both biomaterials of BC skin and mycelium composites are not yet well established individually within general public, the aesthetic value of the hybridised solution may have its enthusiasts and critiques. Evoking both admiration or fascination as well as the negative feelings, undoubtedly the biohybrid offers a unique set of aesthetic and tactile characteristics, opening new design opportunities with the beauty of undiscovered acts of nature, where the nature undertakes the role of the designer itself.

Moreover, uncovering a potential novelty and promoting interdisciplinary workflow of blurring the lines between science, design and fabrication is challenging the common understanding of how we currently source materials. Careful and respectful collaboration with nature by applying biomaterials through a system approach may lead us to truly sustainable solutions and alternatives.

The fabrication process implemented sought optimisation and simplification during each step of the fabrication working with low tech strategies, considering what impact the Mycelium-BC biohybrid may have throughout its lifecycle. Minimising potential liabilities and enhancing the restorative approach towards resources works towards the UN Sustainability Goals, namely Industry, Innovation and Infrastructure, Sustainable Cities and Communities, Climate Action and Life on Land.

The proposed system as a fully biodegradable material may contribute towards reducing the negative impact on our ecosystem and human health, air and land pollutions, prevent soil degradation, biodiversity loss, global warming and diminishing of planetary resources. It aimed to underlay the need to reduce energy consumption and processing applied, reducing the chemical usage to necessary minimum and replacing the energy intensive process of sterilisation to more sustainable, low tech, replacement. A quantitative evaluation using Life Cycle Assessment will further identify areas for improvement.

By rethinking the resources procurement with the application of living materials created with mycelium and bacteria we are able to utilise local agricultural waste streams, and ensure the biodiversity is not disrupted, by using locally available species. Once waste disposal is considered, the system should not affect the environment as the untreated biohybrid material would simply decay. For this reason, the system could be applied and adjusted to various bioclimate zones and regions, towards achieving regional affordance of the local resources, with the appropriate management ensuring ecological safety. On the production scale any cut-out pieces of cellulose during the process could be utilised as additional reinforcement for the MBC to further enhance the material's mechanical properties (Elsacker et al. 2021). Using locally available resources would also minimise the need for transportation, enhancing the economic viability. With the standardisation of the fabrication technique and material production it would most likely first become available to architects and designers to then reach the broader public and make it accessible to local people, even where specialised equipment is not available.

A fully biodegradable hybrid material and the combined potential of both the BC and MBC is a promise promoting a new life cycle of material systems focusing on rethinking the use of resources with the circular economy approaches of waste utilisation and integration of biology within the design process. Introducing a concept of growth and decay as tools opened up a multiple opportunities for an efficient and sustainable production. The emerging notion of collaborating with nature may lead us to revaluate our design approaches leading towards temporary elements, objects and architecture of a limited life span. Conception of "destined to decay" building elements, could become a strong proposal for augmenting current architectural industry standards and perceptions.

5 Conclusions

The research in this investigation focused on the organic compatibility between the mycelium—based composite and bacterial cellulose as well as the potential influence of fabrication strategies within the broader architectural context. Implementation of an iterative, experimental notion of material-driven thinking through making allowed to create a base, organic material interaction. It explored the biological compatibility of mycelium and bacteria-produced cellulose, as well as the efficient fabrication sequence of assembly to create a biohybrid material system. This contribution towards fabrication logic, with further research, could result in a novel fabrication intelligence.

The results demonstrated that the mycelium can effectively grow within the cellulose produced by bacteria, where the mycelium is a structural, bulk element and the bacterial cellulose is a sacrificial mould and a feed envelope hosting the mycelium growth. The manufacturing methodology used principles of fabric forming fabrication as a primary technique to create geometrical expression, by translating 2D sewn bacterial cellulose membranes, into 3D structures with mycelium composite filling. The fabrication used harvested and cleaned bacterial cellulose and living mycelium to create a hybridised biomaterial system with augmented characteristics, combining complementary properties such as mechanical strengths of BC performing well in tension and MBC in compression. Comparing various states of bacterial cellulose, the best growth rate of the mycelium-based composite was observed within the wet pellicle, cleaned with NaOH after the sewing process, which with the proceeding growth became gradually integrated within the fungal skin.

Grounding this research in the material characteristics, in and of themselves, but also at their interface with each other, will support research outcomes that are rationalised towards the 'wants' of the materials and their hybrids, allowing this to drive and shape goals. Doing so through all the attributes that these 'new' goals may have, factors such as regional resources, waste utilization, scaleability, low energy and cost processes, mechanical and structural performance are to service social and planetary goals. These all should be structured in the trades within the research towards feasible delivery of contemporary architectural responsibility.

Authors Contributions The main author N. B. Piórecka was responsible for the conceptualization and design of the study, as well as the analysis and interpretation of the data. She contributed to the writing and revising the manuscript.

P. Scully was the main thesis supervisor, provided critical and insightful feedback regarding the research methodology, production and fabrication processes.

A. K. Salmane provided important insights that shaped the direction of laboratory work, experiments design and protocols.

B. Parker and M. Cruz contributed constructive criticisms to improve the study and ensure accuracy. All should be structured in the trades within the research towards feasible delivery of contemporary architectural responsibility.

References

Alemu D, Tafesse M, Mondal AK (2022) Mycelium-based composite: the future sustainable biomaterial. Int J Biomater. https://doi.org/10.1155/2022/8401528

Amarasekara AS, Wang D, Grady TL (2020) A comparison of kombucha SCOBY bacterial cellulose purification methods. SN Appl Sci 2. https://doi.org/10.1007/S42452-020-1982-2

Appels FVW, Camere S, Montalti M, Karana E, Jansen KMB, Dijksterhuis J, Krijgsheld P, Wösten HAB (2019) Fabrication factors influencing mechanical, moisture- and water-related properties of mycelium-based composites. Mater Des 161:64–71. https://doi.org/10.1016/J.MATDES.2018.11.027

Arevalo Galllegos AM, Herrera Carrera S, Parra R, Keshavarz T, M.N. Iqbal H (2016) Bacterial cellulose: a sustainable source to develop value-added products—A review | Gallegos | BioResources

Bush R, Chan T, Powell A (2010) Single cast fabric formwork | Concrete fabrication. https://richardbush.wordpress.com/2010/12/03/single-cast-fabric-formwork/. Accessed 14 Jan 2023

Colmo C, Ayres P (2020) 3d printed bio-hybrid structures investigating the architectural potentials of mycoremediation

Culver R, Koerner J, Sarafian J (2016) Fabric forms: the robotic positioning of fabric formwork. Robot Fabric Architect, Art Des 2016:106–121. https://doi.org/10.1007/978-3-319-26378-6_8

da Silva CJG, de Medeiros ADM, de Amorim JDP, do Nascimento HA, Converti A, Costa AFS, Sarubbo LA (2021) Bacterial cellulose biotextiles for the future of sustainable fashion: a review. Environ Chem Lett 19:2967–2980. https://doi.org/10.1007/S10311-021-01214-X/TABLES/6

Derme T, Mitterberger D, di Tanna U (2022) Growth based fabrication techniques for bacterial cellulose: three-dimensional grown membranes and scaffolding design for biological polymers. In: Proceedings of the 36th annual conference of the association for computer aided design in architecture (ACADIA), pp 488–495. https://doi.org/10.52842/CONF.ACADIA.2016.488

Drisko GL, Sanchez C (2012) Hybridization in materials science—Evolution, current state, and future aspirations. Eur J Inorg Chem 5097–5105. https://doi.org/10.1002/EJIC.201201216

Edwards IP, Upchurch RA, Zak DR (2008) Isolation of fungal cellobiohydrolase I genes from sporocarps and forest soils by PCR. Appl Environ Microbiol 74:3481–3489. https://doi.org/10.1128/AEM.02893-07

Elsacker E, Vandelook S, Damsin B, van Wylick A, Peeters E, de Laet L (2021) Mechanical characteristics of bacterial cellulose-reinforced mycelium composite materials. Fungal Biol Biotechnol 8:1–14. https://doi.org/10.1186/S40694-021-00125-4/FIGURES/8

Esa F, Tasirin SM, Rahman NA (2014) Overview of bacterial cellulose production and application. Agric Agric Sci Procedia 2:113–119. https://doi.org/10.1016/J.AASPRO.2014.11.017

Fauzi M, Mubarik NR, Jayanegara A (2018) Screening of cellulose- and lignin-degrading fungi for improving nutritive quality of ruminant feed. MATEC Web of Conferences 197:06001. https://doi.org/10.1051/MATECCONF/201819706001

Fernandes I de AA, Pedro AC, Ribeiro VR, Bortolini DG, Ozaki MSC, Maciel GM, Haminiuk CWI (2020) Bacterial cellulose: from production optimization to new applications. Int J Biol Macromol 164:2598–2611. https://doi.org/10.1016/J.IJBIOMAC.2020.07.255

Floudas D, Binder M, Riley R, Barry K, Blanchette RA, Henrissat B, Martínez AT, Otillar R, Spatafora JW, Yadav JS, Aerts A, Benoit I, Boyd A, Carlson A, Copeland A, Coutinho PM, de Vries RP, Ferreira P, Findley K, Foster B, Gaskell J, Glotzer D, Górecki P, Heitman J, Hesse C, Hori C, Igarashi K, Jurgens JA,

Kallen N, Kersten P, Kohler A, Kües U, Kumar TKA, Kuo A, LaButti K, Larrondo LF, Lindquist E, Ling A, Lombard V, Lucas S, Lundell T, Martin R, McLaughlin DJ, Morgenstern I, Morin E, Murat C, Nagy LG, Nolan M, Ohm RA, Patyshakuliyeva A, Rokas A, Ruiz-Dueñas FJ, Sabat G, Salamov A, Samejima M, Schmutz J, Slot JC, John FS, Stenlid J, Sun H, Sun S, Syed K, Tsang A, Wiebenga A, Young D, Pisabarro A, Eastwood DC, Martin F, Cullen D, Grigoriev IV, Hibbett DS, (2012) The paleozoic origin of enzymatic lignin decomposition reconstructed from 31 fungal genomes. Science 1979(336):1715–1719. https://doi.org/10.1126/SCIENCE.1221748/SUPPL_FILE/FLOUDAS.SM.PDF

Gazit M (2016) Living matter : biomaterials for design and architecture. Massachusetts Institute of Technology

Ghazvinian A, Gursoy B (2021) Challenges and advantages of building with mycelium-based composites: a review of growth factors that affect the material properties

Goidea A, Floudas D, Andreen D (2020) PULP FACTION: 3D printed material assemblies through microbial biotransformation. Fabricate 2020:42–49. https://doi.org/10.2307/J.CTV13XPSVW.10

Greca LG, Lehtonen J, Tardy BL, Guo J, Rojas OJ (2018) Biofabrication of multifunctional nanocellulosic 3D structures: a facile and customizable route. Mater Horiz 5:408–415. https://doi.org/10.1039/C7MH01139C

Han J, Shim E, Kim HR (2018) Effects of cultivation, washing, and bleaching conditions on bacterial cellulose fabric production 89:1094–1104. https://doi.org/10.1177/0040517518763989

Hoenerloh A, Ozkan D, Scott J (2022) Multi-organism composites: combined growth potential of mycelium and bacterial cellulose. Biomimetics (Basel) 7:55. https://doi.org/10.3390/BIOMIMETICS7020055

Jayasinghe C, Imtiaj A, Hur H, Lee GW, Lee TS, Lee UY (2008) Favorable culture conditions for mycelial growth of Korean wild strains in Ganoderma lucidum. Mycobiology 36:28. https://doi.org/10.4489/MYCO.2008.36.1.028

Jones M, Huynh T, Dekiwadia C, Daver F, John S (2017) Mycelium composites: a review of engineering characteristics and growth kinetics. J Bionanosci 11:241–257. https://doi.org/10.1166/JBNS.2017.1440

Jones M, Mautner A, Luenco S, Bismarck A, John S (2020) Engineered mycelium composite construction materials from fungal biorefineries: a critical review. Mater Des 187:108397. https://doi.org/10.1016/J.MATDES.2019.108397

Karana E, Blauwhoff D, Hultink E, Camere S (2018) When the material grows: a case study on designing (with) mycelium-based materials. Undefined

Lahiri D, Nag M, Dutta B, Dey A, Sarkar T, Pati S, Edinur HA, Kari ZA, Noor NHM, Ray RR (2021) Bacterial cellulose: production, characterization, and application as antimicrobial agent. Int J Mol Sci 22. https://doi.org/10.3390/IJMS222312984

Laromaine A, Tronser T, Pini I, Parets S, Levkin PA, Roig A (2018) Free-standing three-dimensional hollow bacterial cellulose structures with controlled geometry via patterned superhydrophobic–hydrophilic surfaces. Soft Matter 14:3955–3962. https://doi.org/10.1039/C8SM00112J

Lynd LR, Weimer PJ, van Zyl WH, Pretorius IS (2002) Microbial cellulose utilization: fundamentals and biotechnology. https://doi.org/10.1128/MMBR.66.3.506-577.2002

Manan S, Ullah MW, Ul-Islam M, Atta OM, Yang G (2021) Synthesis and applications of fungal mycelium-based advanced functional materials. J Bioresour Bioprod 6:1–10. https://doi.org/10.1016/J.JOBAB.2021.01.001

McNamara JT, Morgan JLW, Zimmer J (2015) A molecular description of cellulose biosynthesis. Annu Rev Biochem 84:895–921. https://doi.org/10.1146/ANNUREV-BIOCHEM-060614-033930

Modanloo B, Ghazvinian A, Matini M, Andaroodi E (2021) Tilted arch; implementation of additive manufacturing and bio-welding of mycelium-based composites. Biomimetics 6. https://doi.org/10.3390/BIOMIMETICS6040068

Mostafavi S, Kemper BN, Du C (2019) Materializing hybridity in architecture: design to robotic production of multi-materiality in multiple scales. 62:424–437. https://doi.org/10.1080/00038628.2019.1653819

Of S, Hift H, Ouellet L, Littlefield JW, Sanadi DR, Hunter FE, Kaufman S, Gilvarg C, Cori O, Ochoa S, Kielley WW, Kielley RK, Krebs HA, Cohen PP, Lardy HA, Wellman H, Lee KH, Eiler JJ, Slater EC, Cleland KW (1954) Synthesis of cellulose by Acetobacter xylinum. 2. Preparation of freeze-dried cells capable of polymerizing glucose to cellulose. Biochem J 58:345. https://doi.org/10.1042/BJ0580345

Oxman N (2010) Material-based design computation. Massachusetts Institute of Technology.

Patchan MW, Chae JJ, Lee JD, Calderon-Colon X, Maranchi JP, McCally RL, Schein OD, Elisseeff JH, Trexler MM (2016) Evaluation of the biocompatibility of regenerated cellulose hydrogels with high strength and transparency for ocular applications. J Biomater Appl 30:1049–1059. https://doi.org/10.1177/0885328215616273

Römling U, Galperin MY (2015) Bacterial cellulose biosynthesis: diversity of operons, subunits, products and functions. Trends Microbiol 23:545. https://doi.org/10.1016/J.TIM.2015.05.005

Saporta S, Clark M, Associate P (2019) "Bio-Welding" of mycelium-based materials references

Soh E, Chew ZY, Saeidi N, Javadian A, Hebel D, le Ferrand H (2020) Development of an extrudable paste to build mycelium-bound composites. Mater Des 195:109058. https://doi.org/10.1016/J.MATDES.2020.109058

Stoica-Guzun A, Stroescu M, Jinga S, Jipa I, Dobre T, Dobre L (2012) Ultrasound influence upon calcium carbonate precipitation on bacterial cellulose membranes. Ultrason Sonochem 19:909–915. https://doi.org/10.1016/J.ULTSONCH.2011.12.002

Subedi K, Basnet BB, Panday R, Neupane M, Tripathi GR (2021) Optimization of growth conditions and biological activities of Nepalese Ganoderma lucidum strain Philippine. Adv Pharmacol Pharm Sci 2021. https://doi.org/10.1155/2021/4888979

Yang J, Chuanshan Zhao, Jiang Y, Han W (2016) The research of adding bacterial cellulose to improve the strength of long-fiber paper. In: 4th international conference on machinery, materials and computing technology (ICMMCT 2016)

Yogiaman C, P. Pambudi C, Kumar Jayashankar D, Chia P, Quek Y, Tracy K (2020) Knitted bio-material assembly: cultivating building materials for sustainable urban development

Yu K, Spiesz EM, Balasubramanian S, Schmieden DT, Meyer AS, Aubin-Tam ME (2021) Scalable bacterial production of moldable and recyclable biomineralized cellulose with tunable mechanical properties. Cell Rep Phys Sci 2. https://doi.org/10.1016/J.XCRP.2021.100464

Water Resources Management in a Regenerative Design Approach

Alessandro Stracqualursi
and Maria Beatrice Andreucci

Abstract

Increasingly scarce and less affordable, water is taking on the characteristics of a precious resource able to increase marked inequalities. The issue of water resources management at the micro-scale (building) stands as a priority field of investigation, and as a central approach in the transition to sustainable architecture, recognizable as 'Regenerative Design'. Water scarcity represents a pending risk, derived from multilayer interferences that threaten consequences on the availability and affordability of resources, especially in the urban built environment. The proposed contribution offers an original method for building-scale water management, developed as a strategic framework encompassing mitigation and adaptation strategies. The research adopts a mixed-methodology for data collection and analysis, articulated in three phases, and widens the scope of analysis to all the components that contribute, even indirectly, to water consumption along all phases of the building's life-cycle. A set of 11 key performance indicators is proposed, as instruments for the model-based testing as well as for the design process benchmarking. The outcome is a replicable design support tool to come up with sensible functional choices thus preventing water scarcity risk, as a catalyst for progressing UN SDG 6 in synergy with SDGs 11, 12, 13 and while reducing trade-offs with SDG 3.

Keywords

Water management · Water scarcity · Circular building · Climate adaptation · Recycle and reuse

1 Introduction

Many questions arise in front of the quality of the buildings we live in and in which we roughly spend 90% of our time (Kelly and Fussell 2019), resulting in high demand for resources, including water. The ecosystem in which these structures are placed, currently the urban built environment, inevitably makes them an integral part of the dynamics that govern it: a system mostly dependent on unidirectional flows of resources (water, energy, waste, etc.) depleted in human society, figuratively expressed as linear metabolism (Girardet 2019). This anthropogenic influence on the water resources is also exacerbated by the effects of Climate Change, which determine a progressive alteration of natural water cycle, compromising accessibility and safety (IPCC 2021).

A. Stracqualursi (✉) · M. B. Andreucci
PDTA, Sapienza University of Rome, Rome, Italy
e-mail: alessandro.stracqualursi@uniroma1.it

M. R. Thomsen et al. (eds.), *Design for Rethinking Resources*, Sustainable Development Goals Series,
https://doi.org/10.1007/978-3-031-36554-6_39

Relevant problems loom and require to face considerable challenges. The key issues are related to the exploitation of water inside buildings, directly and indirectly, which exacerbates the macro risk of water scarcity (World Economic Forum 2022). Currently in Europe, water withdrawal is destined to five main sectors: domestic (9.6%); agriculture, forestry and fisheries (58.3%); mining, manufacturing and construction (10.6%); energy production and air conditioning (18.2%); and tertiary (3.3%) (EEA 2022). Among the most demanding challenges is that of transforming the metabolic flow of buildings (Pearlmutter et al. 2021) and making them an integral part of a system that does not limit the pressure on the environment but transforms it into positivity. The importance of water management is taking on an increasingly interdisciplinary and multi-scale interest in the approach to problems. However, many studies have drawn attention on water governance in cities at the macro-scale (urban water management) with less attention at the micro-scale of buildings (Fletcher et al. 2015; Romano and Akhmouch 2019; Langergraber et al. 2020).

The factors that contribute to fuelling this critical condition are attributable to 5 main dimensions—environmental, technological, economic, social, political—and 3 key factors—climate change, population growth, urbanization—that will radically mark the evolution of urban systems in the coming years due to the alteration of natural hydrological cycles (IPCC 2022); overpopulation (UN/DESA 2022); and expansion of vulnerable environments (UN/DESA 2019). Building design and urban planning generally limit the goal of reducing consumption to a field of direct use of resources, offering partial solutions to containing demand in the building operational stage. Weighting each single water consumptive process of the building system in its entire life-cycle without neglecting the indirect contribution (i.e., in a regenerative approach), allows instead to define a more effective long-term sustainable intervention scenario.

We refer to indirect causes, as integrative parameters of water demand, that establish a correlation with water consumption. Demand is held by water use in the construction and performing water services, requiring the resource and altering its quality, subtracting it from (scarce) availability. The dynamics of Embodied Water (EW), Water-Energy Nexus (WEN), and Non-Revenue Water (NRW) can be related to these variables. Water is a renewable resource that is widely used in the construction sector, as an embodied 'invisible' element. EW expresses the water volume of a product, associated with water consumption in the processes of extraction of raw materials, production of materials, processing, and construction of the building (Fay et al. 2000). This value is generally quantified through the Water Footprint Indicator (Hoekstra et al. 2011) or as the result of a Life Cycle Assessment (LCA). WEN establishes the connection between energy production and water consumption. Although fully interrelated, water and energy resources have historically been investigated and regulated independently, and by different bodies. The resources consumption does not take place unidirectionally: WEN takes into account the consumption of water resources for energy production, but also of energy consumption in the treatment, distribution, disposal, heating, and cooling of water (Magagna et al. 2019). Further relationships are investigated in multidimensional relationships with Food, Soil, and Waste. Additional threatening effects are recorded in water dispersed for system inefficiencies. This water is defined as NRW. The losses in the water supply alternate between real losses, when the water is supplied but not consumed, and apparent losses, when the water is consumed but not recorded by the meter (unbilled authorized consumption or measurement errors). NRW has been globally estimated at 126 km^3/year, which has a financial cost/value of USD 39 billion/year (Liemberger and Wyatt 2019).

On these premises, the research has consequently been conducted aiming at three main general objectives: (i) the identification of key issues and challenges as far as 'water sustainability' in building and construction is concerned; (ii) the formulation of possible strategies and solutions for sustainable water management in buildings following a regenerative approach; and (iii) the development and dissemination of innovative integrated frameworks and tools able to support the building 'water sustainability' assessment.

2 Materials and Methods

The study proposes a synthesis of the progress of research on the water resources management in buildings, the elaboration of Key Performance Indicators (KPIs) for the evaluation of water management within them and a regenerative strategic implementation model for the integrated intervention to prevent the water scarcity through design. The research adopted a mixed-methodology which incorporates both quantitative and qualitative methods of data collection and analysis (Creswell 1999). The methodology was articulated in three phases:

- In the first, analytical-descriptive phase, together with an analysis of the regulatory framework and certification systems for green buildings (GB), the investigation conducted through a literature review aimed at interpreting and condensing all the elements that collaborate in defining the context, such as the stress factors to the natural and urban hydrological system, through the identification of specific metrics and performance indicators.
- In the second, elaborative-experimental phase, intervention scenarios for the prevention of the risk of scarcity in buildings were developed, proposing a regenerative strategic implementation model. The UN's Sustainable Development Goals (SDGs) were taken as a reference. In particular, in order to identify the main design goals, the SDG 6—*Clean Water and Sanitation* was analyzed in all its targets, highlighting synergies and trade-offs with the other SDGs.
- In the third, model testing phase, the strategies were assessed. In this process, the established KPIs are used as evaluation metrics. The output is a semi-qualitative impact assessment, with benefits and barriers for the three dimensions of Environment-Society-Economy, placing the environmental field as a priority area of interest. The scope is to evaluate the effectiveness of the model and establish a basis for future advancements.

Figure 1 illustrates the research's workflow: the study begins with the outcomes of literature review carried out with the web-based tools of Scopus, Web of Science, and Google Scholar. The main search terms included: 'Sustainable Water Management', 'Efficiency', and 'Reuse'. The review is aimed at the identification of current trends (2017–2022), with specific reference to older studies. The result is that water management is linked to many kinds of interdisciplinary and multi-scale approaches, but this work limits the focus on the topics of regenerative design and circularity, as themes closer to architecture. The most relevant issues are deduced, and relevant studies are reported. From the analysis of the SDGs, the main objectives to strive for are highlighted. From the comparative evaluation of GB protocols and integrated assessment tools, the KPIs are then proposed. Subsequently, a novel strategic framework is proposed. Finally, the results of the model testing phase are presented.

3 Results

3.1 Regenerative Design and Circularity

The conventional limiting approach to resource management will have to transition to a regenerative model. Regenerative Design stands as a positive approach to the interaction between artificial and natural systems. Design can no longer be just about reducing negative impacts but needs to adopt a net-positive thinking and regenerative sustainability (Lyle 1996). To have a positive impact, a holistic conception of environmental, economic, and social systems is needed, and that they are aligned with planetary limits (Rockström et al. 2009). Above all, architecture can no longer act as a structure

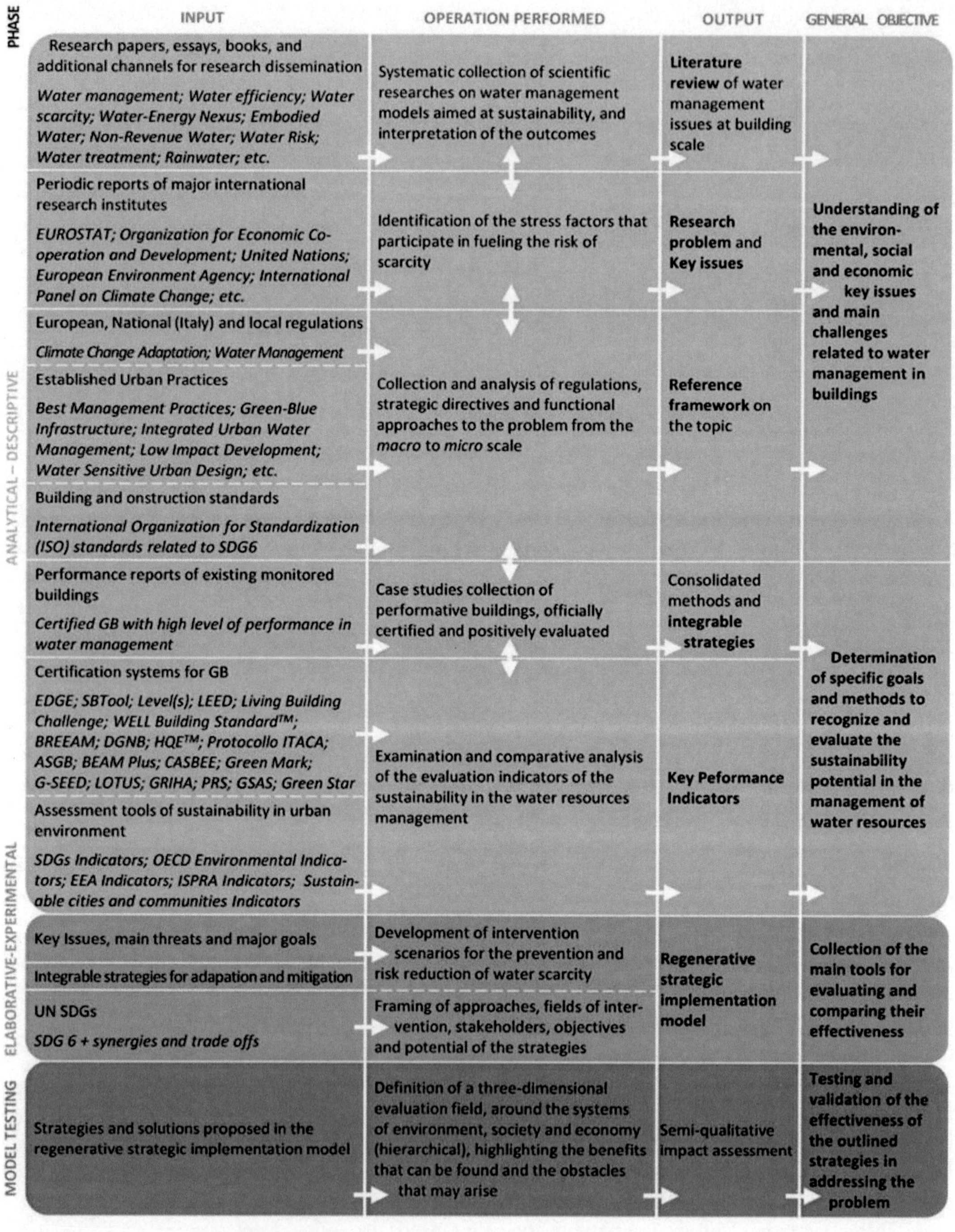

Fig. 1 Workflow of the research mixed-methodology

detached from the natural cycles of the environment and design cannot ignore a concept of multi-scalar approach that takes into consideration the complex relationships.

To regenerate the built environment is necessary to seize the opportunities that arise from the introduction of the principle of circularity in the economic system. There are many current initiatives to implement this model (Kalmykova et al. 2018), so much so as to place the circular economy as a structural component for the progress of the entire Europe (European Commission - DG COMM 2020). To achieve a paradigm shift, cyclical and regenerative innovations are required at every design scale (Prieto-Sandoval et al. 2018). The restoration and maintenance of the water cycle, as well as the treatment, recovery, and reuse, are among the greatest challenges of a circular city (Atanasova et al. 2021).

3.2 UN SDGs and Specific Goals

Sustainability is the center of an ever-changing debate, commonly defined as a holistic approach that considers the environment, society, and the economy, recognizing them as pillars for lasting prosperity (Brundtland 1987). It underlies more of an objective than of an intrinsic quality. The *United Nations 2030 Agenda* transformed this aim in 17 SDGs and 169 related targets. In this framework, environmental sustainability is one of the main themes. In particular, SDG 6 strives for equal access to water and hygiene, with efficiency in use and promoting an integrated and in-depth approach—up to the basin scale—in management. Synergies are substantial with SDG 15, while many trade-offs occur with SDG 3. But it is in comparison with SDG 11, SDG 12, and SDG 13 that we can understand the relevance for the environment, especially urban, of water management at the micro-scale. By relating the targets of these SDGs, the path to take becomes clearer (Fig. 2). In this framework, the role of regenerative design will be prominent in bringing our society ever closer to achieving the SDGs, tackling with determination effects of Climate Change and with a potentially very positive impact on people and environment (Andreucci and Marvuglia 2021).

Sustainability in water management requires to follow four main goals:

1. Preserve the quantity of accessible water resources;
2. Improve the quality of waters and wastewater;
3. Increase the resilience of the building and community water system;
4. Implement circularity in the management of water resources flows.

The above goals refer to the purpose of conserving available resources (quantity), of controlling their chemical-physical-biological state (quality), of limiting risk factors (resilience), and of transforming flow dynamics (circularity). At the building-scale, these translate into three specific goals:

1. Reduce water consumption;
2. Minimize the impact of building on water resources;
3. Balance the mix of water resources to cover the needs.

3.3 KPIs

Multiple assessment criteria—relating to the Water category—offered by the major international GB Certification Systems (20 Standards and 112 criteria) supported by methods and tools for integrated sustainability assessment (8 sets of indicators and 93 criteria), have been reinterpreted to overcome an evaluation framework limited to the site boundaries and to some phases of the building life-cycle. At the base lie the metrics, quantitative or qualitative, absolute or relative. Having set the goals, the references are then compared. The indicators are broken down by affinity to the general objectives (Figs. 3 and 4). For the integrated sustainability assessment indicators, two intermediate categories were considered: the fields of assessment within water management and shared specific goals. For each set, the specific number of useful indicators

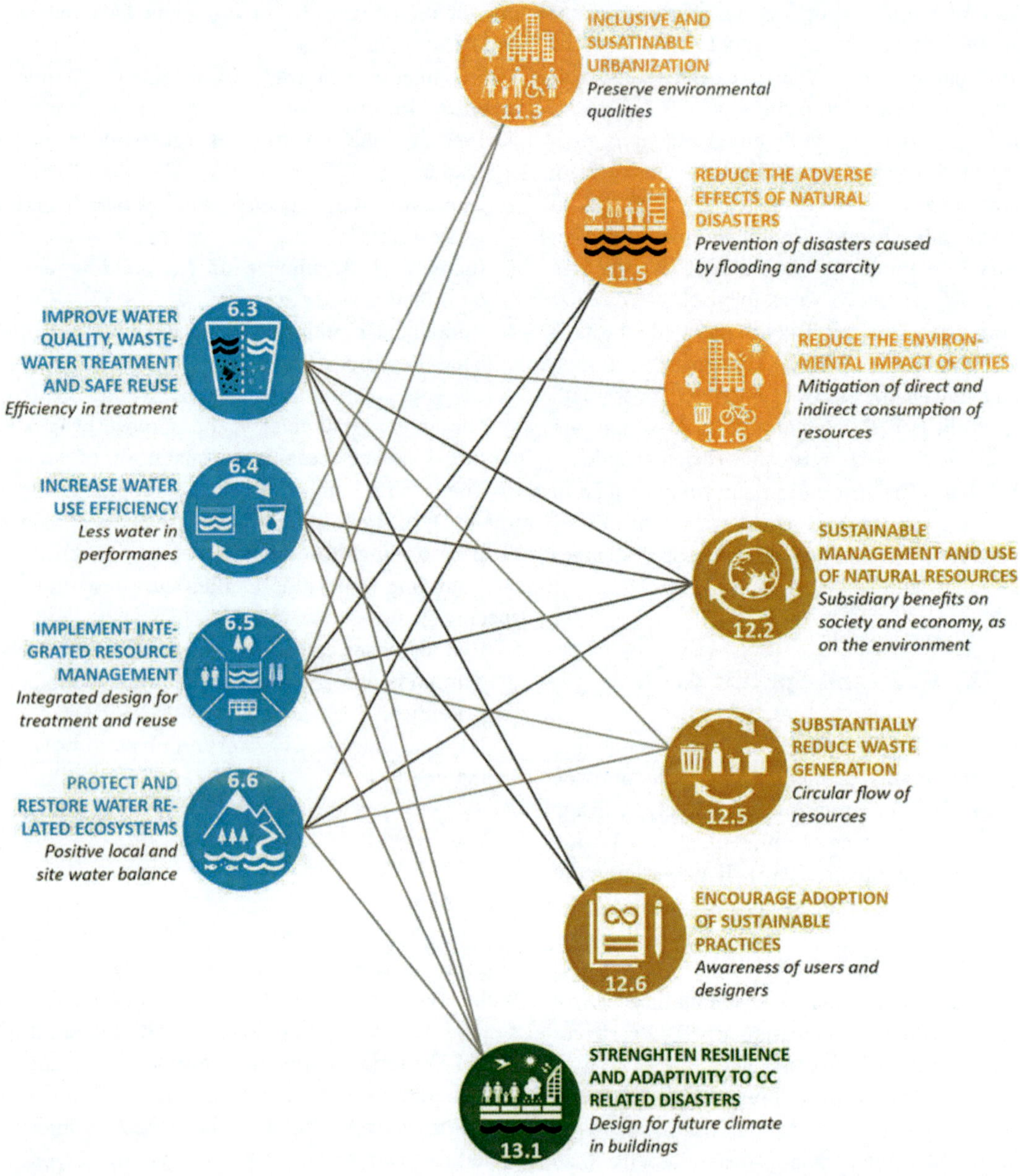

Fig. 2 Main targets for a regenerative approach in water management and interlinkages among SDGs

(step 1) is distributed among thirteen fields of evaluation (step 2), conformable to thirteen specific goals (step 3) related to the four general goals (step 4).

Filtering and implementation processes led to the elaboration of 11 KPIs, structured on 15 sub-indicators, for the evaluation of the sustainability of water management in buildings (Tables 1 and 2). Each indicator was divided into 5 performance levels, the thresholds of which were established through benchmarking on statistical data.

Indicators '1' and '2' collaborate to define a given quantity of water that is used in the building use phase, in restricted survey periods, by reading the water efficiency, but also energy

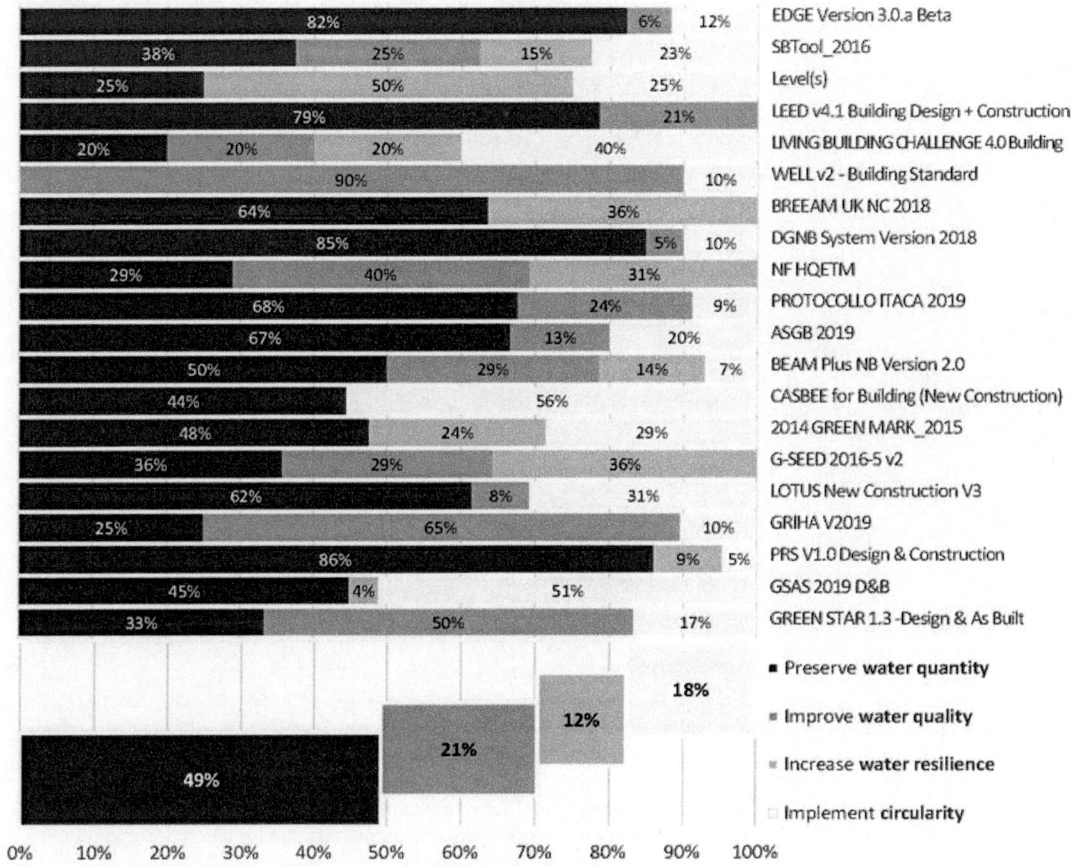

Fig. 3 GB certification systems analyzed and distribution of credits, within the *Water* category, according to the four general goals identified (above), and average percentage value of points distributed among all 20 systems (below)

efficiency, of the appliances and user demand. The following '3', '4', '5', and '6', monitor the progress toward the goal of reducing the water impact of the building system, evaluating the volume of water incorporated in all life stages, the apparent uses due to losses, and the effect of construction on the soil and on the quality of the water inside the site, from rain to waste. The KPIs '7' and '8' represent aggregate indicators that help to understand the balancing capacity between the components of the demand coverage mix and the level of pressure on the available resources, to establish the water autonomy capacity of the system from the city network and the environment. The field of observation of sustainability is then extended to social dynamics '9' and '10', and economic-financial '11'.

In '1', there are different units of measurement according to the different objects analyzed and their functionality. In '2' and '3' a measure of the demand between residences and offices is proposed. In '4', '5', '6', '7', and '8', there is a relative value of the effectiveness of the interventions and resource pressure through a percentage measurement. '9' rates through a Boolean variable no/yes, which indicates the presence or absence of solutions for this purpose, while '10' considers a reference scale where the minimum is the absence of involvement and the maximum is the proactive condition or empowerment level. For '10', four synthetic indicators of period and impact of investments were selected.

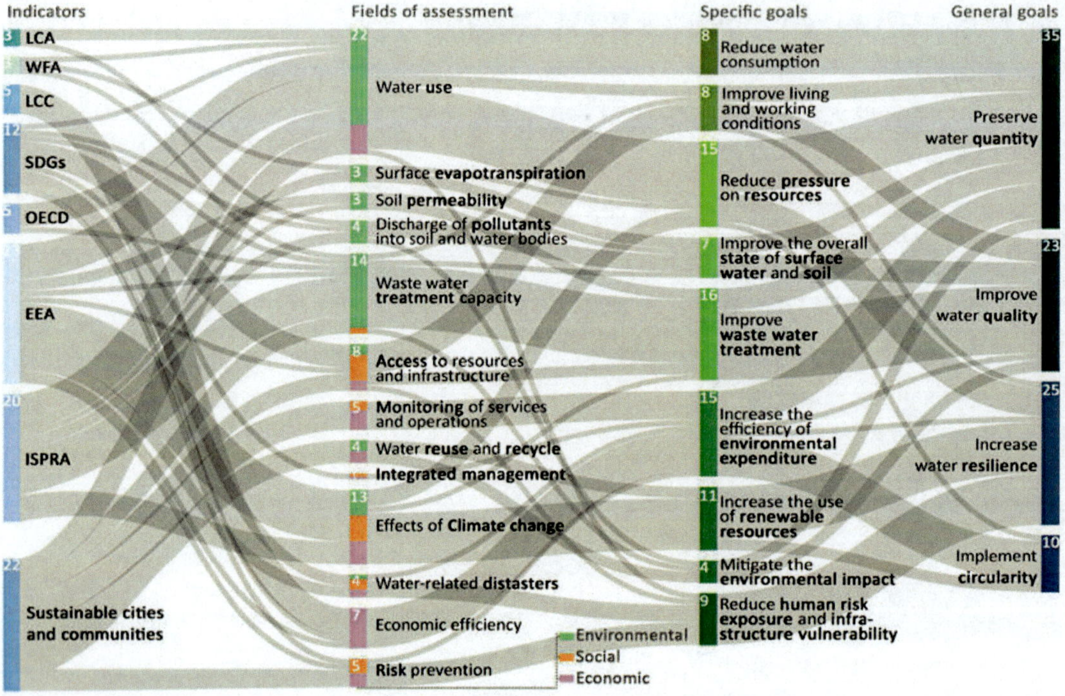

Fig. 4 Sankey chart with the connections between set of indicators for sustainability assessment. The numbers indicate the quantity of criteria for single indicator and their distribution among common fields of assessment, specific and general goals. The colours in the second column indicate affinity with the environmental, social, or economic sphere. Acronyms: *LCA* Life Cycle Assessment, *WFA* Water Footprint Assessment, *LCC* Life Cycle Cost, *OECD* Organization for Economic Co-operation and Development, *EEA* European Environment Agency, *ISPRA* Italian national institute for protection and research on environment

3.4 The Regenerative Strategic Implementation Model

The model was developed as a strategic framework, distinguishing between 'impact mitigation' and 'behavior adaptation' measures, aligned with the specific challenges to face and sustainable water management goals (Table 3).

The efficiency of indoor and outdoor water systems (EI and EO) is a dual strategy on a product scale that is intended to all water appliances, facing the challenge of technological alignment with the aim of achieving high performance in provision of controlled quantities with limited losses. Awareness raising (AU for user awareness and AD for design awareness) is an adaptive strategy aimed at regulating demand among buildings users and their designers. Its application involves community choices and administrative guidelines, clashing with the social challenges of coping and empowerment. The containment of the water impact is a mitigative strategy that aims to reduce the indirect depletion of resources. EW must be regulated above all in the building envelope (EW-O), involving manufacturers of building materials and designers who choose them, to mitigate the impact of the pre and post-use phases of the building. The Smart Water Management (SWM) means putting all water appliances in communication in a coordination and optimization structure, which minimizes structural inefficiencies, such as unaware use and losses. Quality conservation is an adaptive strategy, aimed at active and passive systems of drainage and rainwater runoff, and wastewater treatment, with the aim of collecting, storing, and reusing these resources (OSM and DWT). Finally, diversification is a

Table 1 List of the KPIs, with the description of indicators and sub-indicators

Nr	Name	Definition	Time and frequency
1	Resources for appliances	Potable water used in water appliances	Building operation phase
		Electricity used in water appliances	Building operation phase
2	Water for occupants needs	Potable water required for daily activities	Daily
3	Embodied water	Embodied (or virtual) water in Pre-use (Production + Construction) and End-of-use phases of Building life-cycle	Occasional
4	Coverage of the monitoring system	Main water use monitored with smart meters and sub-meters	Occasional
5	Stormwater Runoff Management in the site	Permeability coefficient of the site in relation to the predevelopment hydrological conditions	Hourly
		Rate of rainwater temporarily stored, treated, and released	Occasional
6	Effectiveness of water purification	Rate of stormwater treated on site[a]	Daily
		Rate of wastewater (gray, black, and process) treated on site[a]	Daily
7	Reuse potential	Coverage of non-potable water needs using local resources from alternative sources	Annual
8	Water Exploitation Index plus (WEI+)	Level of pressure on local water resources exerted by human activity	Seasonal (summer)
9	Wellbeing	Presence of solutions that through the integrated use of water contribute to aesthetic quality, recreational pleasure, and comfort	Occasional
10	Participation	Participation level of the occupants in city programs or urban policies of empowerment and advocacy for water conservation	Annual
11	Life cycle cost	Payback period of the cumulative strategies adopted in a water resources management plan	Occasional
		Incidence of implementation costs for strategies on the water supply and sanitation systems, in relation to the total cost of the building over the entire life-cycle	Occasional

[a]According to the qualitative values of Biochemical Oxygen Demand (BOD5), Chemical Oxygen Demand (COD), Total suspended solids (TSS), and other compounds, as per the limits set for reuse and discharge into water bodies (Water Framework Directive (2000/60/EC) and Successive amendments)

strategy that takes on a double effect: mitigative by application in the use of water (DA), thanks to the resource allocation, which aims to expand the benefits of a regenerative management of the water cycle of the building; and adaptive by source in water use (DS) in order to expand access to supply sources, for an incremental exploitation of resources.

Point to efficiency means bringing indoor consumption—of taps, sanitary ware, appliances, and systems—as close as possible to an amount that is essential to perform its functions. The flow rate and unit consumption values of all the possible appliances involved in the domestic use of water in residences and offices are then traced, and which collaborate to achieve a daily operating capacity of 130 l/ca in the former and of 50 l/ca in seconds. In a complementary way, the outdoor consumption of taps and external systems should reach a stable consumption of 30 l/ca/d, even if in this context the efficiency variables of the systems are very large.

Containing the demand is one of the key strategies for reducing consumption. In raising

Table 2 List of the KPIs with the object of the evaluation, level of rating (from 1 = min. to 5 = max.)

KPI	Object/Intended use	Level					Unit of measure
		1	2	3	4	5	
1	WC	>6.0	≤ 6.0	≤ 5.5	≤ 4.5	≤ 3.5	l/min
	Urinal	>4.0	≤ 4.0	≤ 3.0	≤ 2.0	≤ 1.0	
	Faucet	>13.0	≤ 13.0	≤ 10.0	≤ 8.0	≤ 6.0	
	Shower	>13.0	≤ 13.0	≤ 10.0	≤ 8.0	≤ 6.0	
	Mixed products	>13.0	≤ 13.0	≤ 10.0	≤ 8.0	≤ 6.0	
	Bathtub	>200	≤ 200	≤ 185	≤ 170	≤ 155	
	Dishwasher	>12.0	≤ 12.0	≤ 10.0	≤ 9.0	≤ 7.0	
	Washing machine	>41.0	≤ 43.0	≤ 45.0	≤ 48.0	≤ 50.0	
	Washer-dryer	>41.0	≤ 43.0	≤ 45.0	≤ 48.0	≤ 50.0	
	Bathtub	>200	≤ 200	≤ 185	≤ 170	≤ 155	l/cycle
	Dishwasher	>12.0	≤ 12.0	≤ 10.0	≤ 9.0	≤ 7.0	
	Washing machine	>41.0	≤ 43.0	≤ 45.0	≤ 48.0	≤ 50.0	
	Washer-dryer	>41.0	≤ 43.0	≤ 45.0	≤ 48.0	≤ 50.0	
	(I) Sprinkler	>1.00	≤ 0.80	≤ 0.60	≤ 0.40	≤ 0.20	kWh/m^3
	(D) Circulator pump	>0.40	≤ 0.35	≤ 0.30	≤ 0.25	≤ 0.20	
	(D) Waste water	>3.00	≤ 2.40	≤ 1.80	≤ 1.20	≤ 0.60	
	(D) Stormwater	>1.50	≤ 1.50	≤ 1.30	≤ 1.15	≤ 1.00	
	(DHW) Generic faucet	>0.50	≤ 0.50	≤ 0.40	≤ 0.30	≤ 0.25	kWh/use
	(DHW) Kitchen faucet	>0.45	≤ 0.45	≤ 0.35	≤ 0.25	≤ 0.20	
	(DHW) Shower	>0.35	≤ 0.35	≤ 0.25	≤ 0.20	≤ 0.15	
	(DHW) Dishwasher (14p.)	>0.80	≤ 0.80	≤ 0.70	≤ 0.60	≤ 0.49	kWh/cycle
	(DHW) Washing machine (7 kg)	>0.64	≤ 0.64	≤ 0.56	≤ 0.48	≤ 0.41	
	(DHW) Washer-dryer (7 kg)	>3.76	≤ 3.76	≤ 3.11	≤ 2.58	≤ 2.10	
2	Residential	>200	≤ 200	≤ 165	≤ 145	≤ 120	l/ca/d
	Office	>120	≤ 120	≤ 85	≤ 65	≤ 40	
3	Residential	>30	≤ 30	≤ 20	≤ 15	≤ 10	kl/m^2
	Office	>35	≤ 35	≤ 25	≤ 20	≤ 15	
4	Coverage	≤ 50	≤ 70	≤ 85	≤ 99	>99	%
5	Soil permeability	≤ 30	≤ 40	≤ 50	≤ 75	>75	%
	Rainwater drained	≤ 50	≤ 70	≤ 85	≤ 95	>95	%
6	Stormwater	≤ 50	≤ 70	≤ 85	≤ 99	>99	%
	Waste water	≤ 50	≤ 70	≤ 85	≤ 99	>99	
7	Water for reuse	≤ 45	≤ 60	≤ 80	≤ 99	>99	
8	Pressure on local water resources	>40	≤ 40	≤ 25	≤ 15	≤ 5	%
9	Technological solutions available	No	–	–	–	Yes	No/Yes
10	Participation level	No involvement	Passive	Active	Reactive	Proactive	

(continued)

Table 2 (continued)

KPI	Object/Intended use	Level					Unit of measure
		1	2	3	4	5	
11	Payback period	>10	≤ 10	≤ 8	≤ 6	≤ 5	Years
	Acquisition cost	>25	≤ 25	≤ 15	≤ 8	≤ 5	%
	Operating and maintenance cost	>20	≤ 20	≤ 10	≤ 8	≤ 5	
	Demolition and disposal cost	>5	≤ 5	≤ 3	≤ 2	≤ 1	

I Irrigation, *D* Distribution, *DHW* Domestic Hot Water

Table 3 The regenerative strategic implementation strategic model

Goals	1—Reduce water consumption		2—Minimize the impact of buildings on water resources		3—Balance the mix of water resources to cover the needs
Threats	Demand growth due to technological inadequacy and social dynamics		Erosion of available and renewable water resources		Variability of meteorological and hydrological models
Causes	Systems obsolescence Consumption patterns		Speculative construction Soil sealing Pollution Infrastructure leakage		Water Stress Rainfall reduction Alteration of the water cycle
Mitigation strategies	Efficiency		Impact reduction		Diversification

	EI Efficiency of indoor water systems	EO Efficiency of outdoor water systems	EW-O EW optimization in the building envelope	SWM Smart Water Management	DA Diversification by application in water use
Adaptation strategies	Awareness		Quality conservation		Diversification
		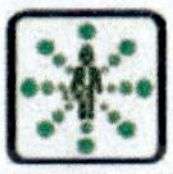			
	AU User awareness	AD Designer awareness	OSM On-site stormwater management	DWT Decentralized wastewater treatment	DS Diversification by source in water use

awareness of users, administrative, economic, and regulatory measures are relevant, starting with public information, which moves between different broadcasters as well as endogenous and exogenous communication channels. Financial incentives are economic tools that offer direct or indirect stimuli. Positive direct incentives collect environmental subsidies and tax deductions, or loans to incentivize innovation and direct consumption choices, while negative direct incentives aim to correct inappropriate behavior, through penalties and fines. Indirect incentives make users aware by putting them in front of variable prices for consumption thresholds.

Finally, obligations and limitations are intended to regulate the water consumption in a compulsory manner, in times of need, while the rules aim to regulate it.

AD is influenced by education and training received by designers through a cycle of studies or received sporadically with events such as conferences and workshops, or constantly through professional training courses. A great input comes from subsides: facilitated taxation on professional income; start-up funds for innovative businesses; professional incentives to participate in action programs, to be able to deal with a flexible administration, and to receive professional recognition from external bodies, such as for ISO 14001 or LEED, BREEAM certifications, etc.

EW-O caters to the building envelope as it is the class of technological unit with the greatest impact on the resource consumption and one of the pivots of energy flows in the building, becoming part of the dynamics of WEN. It also represents a variable of awareness in design choices, depending on its water impact. A comparison between the elements is necessary: for this purpose a quantitative analysis of EW per unit of area was previously conducted on 24 different envelope packages, including vertical and horizontal upper and lower elements (Stracqualursi 2022). For each solution, its thermal transmittance (U) is also measured in order to establish the correlation between water demand and thermal insulation capacity (Fig. 5). The water demand for production and construction of

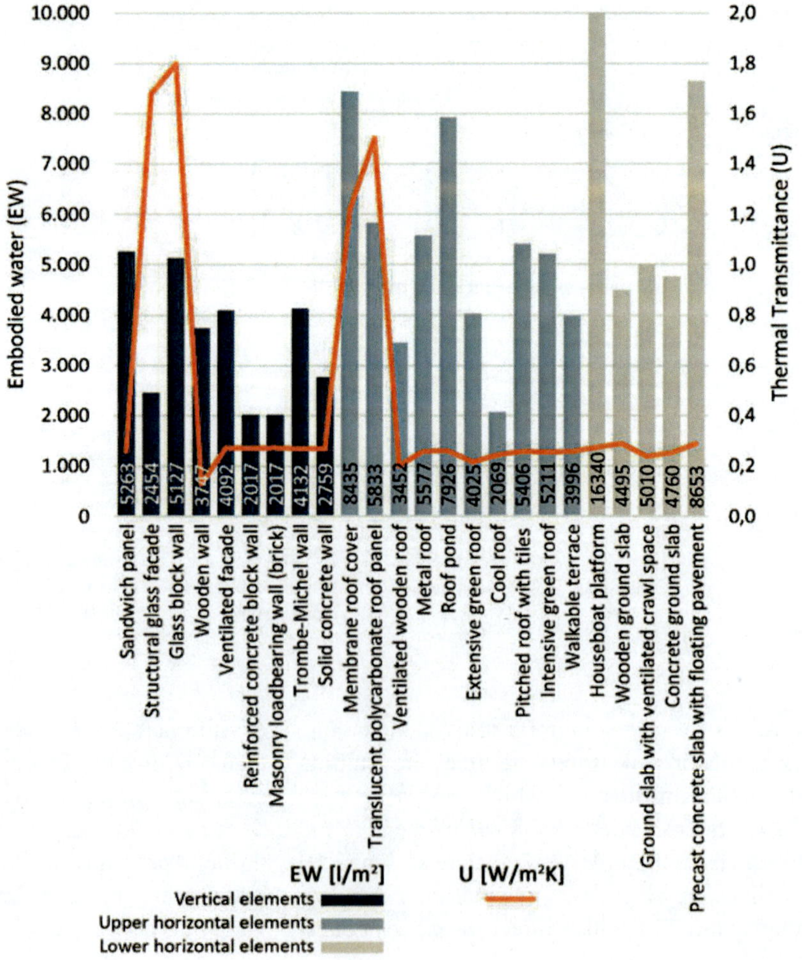

Fig. 5 EW and U for the different envelope elements analyzed. The components are ordered, from left to right, according to the increasing surface mass

1 m^2 of building envelope is on an average of 5.111 l. Implementation and type of materials are the main variables in determining these values. The use of synthetic insulating and waterproofing materials, combined with a high number of layers of the package, and metallic finishes make the envelope packages heavier, compared to traditional masonry or wood solutions and dry systems.

SWM in building is divided into 5 system levels (Li et al. 2020), alternating ACT automated control technologies and ICT data exchange, in phases of monitoring and processing of water consumption data, with the triple purpose of: making the system more efficient by detecting leaks; inform users; and increase management flexibility. The first physical layer, which monitors pipes and systems, provides a reading of the operational status of the infrastructure; the detection and control level works to establish the automation, responsiveness, connectivity with network and software, and interoperability with external users capabilities; the communication layer establishes a two-way connection between users and managers; the visualization of the data informs users and managers of the trends in progress; the processing and analysis of data makes the system a valid tool to support operational choices.

OSM is the containment of stormwater flow within the site, through drainage and collection into tanks, and slowing the extension of impervious surfaces to avoid run-off and lowering of the water table. The priority scale should place the collection at the top for reuse and place the discharge into the combined sewer as a last resort. Adaptation is implemented through the collection, drainage, and treatment tools derived from nature-based solutions. Storage tanks with induced flow can be placed at the top of the building, using gravity for water distribution, or at the base, flanked by a booster pump. The drainage can be supported by green roofs and all the vertical green system, which offer great potential in terms of implementing technologies for the treatment, including gray water. Waterproofing is a process to be contrasted with hard permeable flooring for the paths and soft for the vegetated spaces, implementing fountains and pools with induced flow flanked by trees as integrated elements of the drainage system.

In DSW, decentralized wastewater treatment systems (DEWATS) perform the functions of separating internal discharges and local treatment of all polluted waters—black water, gray water, and other processed water—so as to recycle. The treatment is associated with the contaminants typology, the pollution load, and the intended end use of the effluent, through three treatment levels: primary/physical, for the removal of suspended solids with sedimentation, through degreaser, IMHOFF tank or septic tank; secondary/biological for the abatement of biodegradable organic matter and removal of non-sedimentable solids, with activated sludge or membrane systems, attached biomass, moving bed, biodisc, and discontinuous cycle reactors; tertiary/physical–chemical, which disinfects the outgoing water reducing the microbial load, through pressure filters with activated carbon or sand, UV sterilizers, or natural phytoremediation systems.

DA exploits water for collateral functions, which can be associated with three categories: regulation, which uses water as a mitigating tool for indoor and outdoor microclimate; recreation, as a characterizing element of collective spaces; and symbolism, as an element of the definition of the contemplative space. For the first category, two functions are available: evaporative and radiative cooling. The evaporative cooling employs direct, indirect, and combined systems. Experimentation in the building sector occurs in applications of water reuse for direct evaporative cooling of interior spaces by exploiting the vertical flow of water and providing thermal capacity to the envelope—within integrated systems, such as water wall and roof pond. The radiative cooling is exploited by active and passive chilled beam systems or water to air exchangers. The second category is linked to ecosystem services pertaining to health and wellbeing. This is through the intervention on aesthetics, acoustics—especially with soothing sounds in workstations—and wellness, with a therapeutic effect through SPA, hydromassage,

KPIs		ADVANTAGES: ENVIRONMENTAL	ADVANTAGES: SOCIAL	ADVANTAGES: ECONOMIC	BARRIERS: ENVIRONMENTAL	BARRIERS: SOCIAL	BARRIERS: ECONOMIC
EI	1, 11	Water and energy saving, with reduced CO_2 emissions	Effectiveness of the system without adaptation of habits	Reduced operating costs and initial investments briefly recoverable	Greater energy impact and polluting load of some appliances in the production stage	Inaccessibility for disadvantaged population groups	Initial costs above average
EO	1, 11	Wide margin for reducing irrigation demand	Effectiveness of the system without adaptation of habits	Reduced operating costs and initial investments briefly recoverable	Outdoor use more consumptive than indoor use	Possible dynamics of conspicuous consumption for the care of outdoor areas	Initial and maintenance costs for automated irrigation systems and swimming pool covers
AU	2, 10, 11	Stabilization of water demand at low levels, with savings potential independent of the nominal life of the appliances	Participation increases knowledge and stimulates sensitivity, with positive knock-on effects	High long-term economic savings	Inadequate information and narrowness of channels can lead to zero water savings	Acceptance of new policies or voluntary programs often has low consensus	Water pricing confronts effects that are difficult to predict and many variables
AD	2, 10, 11	Capacity to address the issue of water efficiency even in the pre and post operational phases	Propulsive thrust to improvement thanks to motivational systems of gratification and competitiveness	Growth in the real estate value of certified buildings and in job demand in the professional sector of GB	The effects of unawareness of designers are reflected in the building efficiency	Accessibility to eco-oriented design is restricted to high-end designers	Joining GB certification programs depends on the client's economic capabilities
EW-O	3, 11	Holistic view of environmental impacts over the entire life cycle to optimize the use of water resources	Availability of clear information and support for evidence-based design	Decision support for achieving the economic efficiency of the project	Greater energy impact and polluting load of some appliances in the production stage	Disorganized definition of social impacts that are strictly contextualized and not generalizable	Lack of integration with complementary economic methods
SWM	4, 11	Diversified and efficient management of water demand, with reduced apparent consumption and waste	Greater awareness offered to users	More accurate and effective pricing potential. Reduced operation and maintenance costs	Infrastructure compatibility and operativity issues	Uncertainty in the correspondence between information and user adaptation	High costs for hardware and software instrumentation. Custom pricing limited by the high incidence of fixed costs
OST	5, 6, 11	Ecosystem services with environmental benefits on multiple levels	Amenity and well-being in the areas involved	Reduction of operation costs related to the supply of potable water	Complexity of execution and implementation, with possible side effects of alteration of the local ecosystem	Demand for wide spaces, subtracted from other activities. Possible conflict conditions with the normative	High construction and maintenance costs, also associated with land use
DWT	6, 11	Reduction of the building's impact on the quality and availability. Recovery of substances and by-products	Landscape integration and potential aesthetic value of certain natural systems	Savings on fixed and variable operation costs of supplied water	Constant threat of possible damage to the local environment and occupants' health	Possible discomfort of various kinds, for biological treatment systems	High construction and maintenance costs with long payback times
DA	7, 9, 11	Greater ductility and usability of the resource which makes the system water resilient	Favourable microclimatic and well-being conditions with low energy consumption	Greater investment attractiveness and increase in the value of the site	Variable effects on comfort and well-being depending on the environmental conditions	Water use in unconventional functions can be perceived as a deleterious application	Experimental nature of many passive solutions which makes them costly
DS	7, 8, 11	Increased water resilience and implemented circularity	Greater security and perception of system stability	Positive influence on land ownership values and occupant image	Management risks for people and the environment	Variable perception and acceptance of use from unconventional sources	High investment and reuse costs, with long payback periods

Fig. 6 Graphic summary of advantages and barriers for the strategy framework

and bio-pools. The last category creates a symbolic relationship between the water element and the distribution of interior spaces and paths.

DS is the final process placed at the top of regeneration, which after the reduction of demand and impact, culminates in the integration of alternative sources. The water resources available within the entire building site are superficial, underground, wastewater, rain, and air humidity. The period and area of availability, the possible applications, and the variables that characterize their usability must be investigated specifically.

3.5 Model Testing Phase

The application of the model offers benefits and encounters barriers in the environmental, social and economic spheres. The assessment for individual strategies is summarized in Fig. 6. It refers to a present scenario, taking into consideration existing technologies. The recorded effects relate to the present and the immediate future. The focus is set on building design and context is limited to urban environment. Even if water is a natural resource and its availability is strongly linked to local circumstances, it is assumed that consumption in new buildings is linked to system efficiency and to the daily demand of residents, which makes it a generalizable picture and not strictly localized. Overall, the environmental benefits derive from: indoor and outdoor water, with associated reductions in energy and Average Incremental Carbon Costs for Water Efficiency labeled products; stabilization and improved water demand management capacity; ecosystem services that preserve the quantity and quality of rainwater, also making it available for recycle; increased resilience and flexibility of water as a resource. The main social benefit is produced by the proactive involvement of all users, for daily use, and for personal and common well-being. At the same time, the effectiveness of the savings is mainly delegated to appliances, not involving the user habits. Inconveniences caused by detection and maintenance are frequently reduced. In the economic dimension, affordable operating costs are the benefit with the highest value. Savings are found both on the variable component of the tariff (quantity of use), and on the fixed costs (distribution and treatment). Overall, water-efficient buildings with a certification acquire a relatively higher real estate value.

There are, obviously, also obstacles. Firstly, the environmental barriers are derived from the global impact of water efficiency appliances. Water is mainly saved but the consumption of other resources is considerable. Recycling/reuse, especially on site, is a risky process in terms of hygiene and health. The social barriers are linked to willingness of users to accept compromises and to overcome cultural barriers related to the fear of using recycled water. Furthermore, general information on water saving is often inadequate and risks becoming counterproductive and leading more to consumerism than to preservation. This empties the importance of the concept, which from driver turns into distrust. Finally, the economic obstacles linked to initial investments must be considered, with higher acquisition costs and increased maintenance costs in relation to additional components. Recycling/reuse systems require investments for infrastructure upgrading and maintenance, with a low return in micro-scale systems such as buildings.

4 Concluding Remarks

We have the right to a clean, healthy, and sustainable environment (UN General Assembly 2021) and to an equal society (European Commission 2021). The natural environment is a capital that involves us globally. A healthy economic system should be planned to prosper, respecting these margins with a constant commitment that must lead us to "Create to regenerate". Moving from actual sustainability to regeneration (Reed 2007), translates into progressing from 'green' to 'regenerative' design (Cole 2012; Mang and Reed 2020). The transition toward a Circular Built Environment (Ellen MacArthur Foundation and ARUP 2018) is therefore essential.

To close the circle, the outcome—albeit preliminary—is a prototype of an operative design supporting tool. It is clear to think that solving an extended risk such as water scarcity through the building design is impossible. This is why a method, or rather, replicable design guidelines have been developed. The proposed KPIs are diversified parameters useful not only for the testing phase. They are also designed to facilitate the benchmarking process in comparing distinct buildings or different scenarios for the same building, as analysis tools integrable into, as an example, a digital twin. The strategies are multifaceted because they are oriented toward different solutions, since to regenerate complex systems must be managed simultaneously. The effort is to make the building process, in all phases, integrated into the economic, social, and natural cycles of the environment, minimizing conflicts and discontinuity. In conclusion, simplify, recycle, and create synergies.

The research shows limits from which development perspectives have been identified. First: the absence of direct application of the model, for example through a case study. Second: the limitation of effective applicability in the reality of the professional world and not just the academic one. The next steps to be taken, from a research advancement perspective, are mainly three:

1. Modeling, simulation, and testing of the proposed system to validate its effectiveness and provide a series of intervention alternatives for designers;
2. Synthesis of the results for encouraging the adoption of a flexible tool for concrete usability in the professional world;

3. Proposal of solutions integrated with the building envelope, as interactive components with indoor and outdoor water resources.

References

Andreucci MB, Marvuglia A (2021) Investigating, implementing and funding regenerative urban design in a post-COVID-19 pandemic built environment: a reading through selected UN sustainable development goals and the European green deal. In: Andreucci MB, Marvuglia A, Baltov M, Hansen P (eds) Rethinking sustainability towards a regenerative economy. Springer International Publishing, Cham, pp 395–413

Atanasova N, Castellar JAC, Pineda-Martos R et al (2021) Nature-based solutions and circularity in cities. Circular Econ Sustain 1:319–332. https://doi.org/10.1007/s43615-021-00024-1

Brundtland GH (1987) Our common future—Call for action. Environ Conserv 14:291–294

Cole RJ (2012) Transitioning from green to regenerative design. Build Res Inf 40:39–53. https://doi.org/10.1080/09613218.2011.610608

Creswell JW (1999) Mixed-method research: introduction and application. In: Handbook of educational policy. Elsevier, pp 455–472

EEA (2022) Use of freshwater resources in Europe. https://www.eea.europa.eu/ims/use-of-freshwater-resources-in-europe-1

Ellen MacArthur Foundation, ARUP (2018) From principles to practices: first steps towards a circular built environment. https://emf.thirdlight.com/link/ufe6ol7qbkm-a9mzju/@/preview/1?o

European Commission (2021) European pillar of social rights—Action plan. European Commission, Bruxelles, BE

European Commission - DG COMM (2020) Circular economy action plan: for a cleaner and more competitive Europe. Publications Office of the European Union

Fay R, Treloar G, Iyer-Raniga U (2000) Life-cycle energy analysis of buildings: a case study. Build Res Inf 28:31–41. https://doi.org/10.1080/096132100369073

Fletcher TD, Shuster W, Hunt WF et al (2015) SUDS, LID, BMPs, WSUD and more—The evolution and application of terminology surrounding urban drainage 12:525–542. https://doi.org/10.1080/1573062X.2014.916314

Girardet H (2019) The metabolism of cities. In: The living city. Routledge, pp 170–180

Hoekstra AY, Chapagain AK, Mekonnen MM, Aldaya MM (2011) The water footprint assessment manual: setting the global standard. Routledge

IPCC (2021) AR6 climate change 2021: the physical science basis. Cambridge University Press

IPCC (2022) AR6 climate change 2022: impacts, adaptation and vulnerability. Cambridge University Press

Kalmykova Y, Sadagopan M, Rosado L (2018) Circular economy—From review of theories and practices to development of implementation tools. Resour Conserv Recycl 135:190–201. https://doi.org/10.1016/j.resconrec.2017.10.034

Kelly FJ, Fussell JC (2019) Improving indoor air quality, health and performance within environments where people live, travel, learn and work. Atmos Environ 200:90–109. https://doi.org/10.1016/j.atmosenv.2018.11.058

Langergraber G, Pucher B, Simperler L et al (2020) Implementing nature-based solutions for creating a resourceful circular city. Blue-Green Syst 2:173–185. https://doi.org/10.2166/bgs.2020.933

Li J, Yang X, Sitzenfrei R (2020) Rethinking the framework of smart water system: a review. Water 12:412. https://doi.org/10.3390/w12020412

Liemberger R, Wyatt A (2019) Quantifying the global non-revenue water problem. Water Supply 19:831–837. https://doi.org/10.2166/ws.2018.129

Lyle JT (1996) Regenerative design for sustainable development. John Wiley & Sons

Magagna D, Hidalgo I, Bidoglio G, et al (2019) Water: energy nexus in Europe. Publications Office of the European Union Luxembourg

Mang P, Reed B (2020) Regenerative development and design. Sustain Built Environ 115–141. https://doi.org/10.1007/978-1-4614-5828-9_303

Pearlmutter D, Pucher B, Calheiros CSC et al (2021) Closing water cycles in the built environment through nature-based solutions: the contribution of vertical greening systems and green roofs. Water 13. https://doi.org/10.3390/w13162165

Prieto-Sandoval V, Jaca C, Ormazabal M (2018) Towards a consensus on the circular economy. J Clean Prod 179:605–615. https://doi.org/10.1016/j.jclepro.2017.12.224

Reed B (2007) Shifting from 'sustainability' to regeneration. Build Res Inf 35:674–680. https://doi.org/10.1080/09613210701475753

Rockström J, Steffen W, Noone K et al (2009) Planetary boundaries: exploring the safe operating space for humanity. Ecol Soc 14

Romano O, Akhmouch A (2019) Water governance in cities: current trends and future challenges. Water 11. https://doi.org/10.3390/w11030500

Stracqualursi A (2022) Un modello operativo di gestione delle risorse idriche nel Regenerative Design (An Operational Model for Water Management in Regenerative Design). Sapienza University of Rome

UN General Assembly (2021) The human right to a clean, healthy and sustainable environment (A/HRC/RES/48/13)

UN/DESA (2022) World Population Prospects 2022: Summary of Results. UN DESA/POP/2022/TR/NO. 3. United Nations, Department of Economic and Social Affairs, Population Division, New York, USA

UN/DESA (2019) World Urbanisation Prospects 2019: Volume I: Comprehensive Tables. The United Nations Department of Economic and Social Affairs (UN/DESA), New York

World Economic Forum (2022) Global risk report 2022. WEF, Cologny/Geneva, CH

The Value of Waste

Extending the Circular Design Framework for Bio-Based Materials: Reconsidering Cascading and Agency Through the Case of Biopolymer Composites

Mette Ramsgaard Thomsen, Gabriella Rossi, Anders Egede Daugaard, Arianna Rech, and Paul Nicholas

Abstract

Pressing issues of sustainability drive the pursuit of new circular bio-based design frameworks in order to limit the use of finite resources and fossil fuel energy consumption. In architecture and the built environment these efforts have fostered a renewed interest in bio-based materials as the only truly sustainable resource for building materials. However, bio-based materials are different to the resources that cascade through technological cycles. With the increased interest and application of bio-based material in the building sectors comes the realisation that they do not map directly on to existing models of circular design for the built environment. In this paper we present an effort to extend the circular design framework for biopolymer composites in architectural applications. The paper takes point of departure in a mapping of existing strategies for cascading in the built environment. By presenting two biopolymer composites as case studies, we examine how introducing this class of materials into the circular design framework challenges the actions and implication of cascading and reconfigures the weighting of agency and control. The conceptual, technical and practical implications of the usage of such materials in future building practice are speculated on through the creation of six design scenarios that expand the role of cascading creating a new connectivity between material practices.

Keywords

Circular design · Cascading · Bioeconomy · Waste streams · Sustainable materials · Biopolymer composites

1 Introduction

Our societies are under great pressure to rethink the way we employ our resources. The climate crises and the need to decrease CO_2 has led to a renewed focus on transitioning to materials arriving from the biosphere. Frameworks such as the European Green Deal (European Commission. Directorate General for Research and Innovation 2021) foreground bio-based materials as solutions for carbon neutrality, renewability and embedded circularity through biodegradability and composting. In architecture,

M. Ramsgaard Thomsen · G. Rossi (✉) · P. Nicholas
CITA / Royal Danish Academy, Copenhagen, Denmark
e-mail: garo@kglakademi.dk

A. E. Daugaard · A. Rech
Danish Polymer Center, Danmarks Tekniske Universitet, Kogens Lyngby, Denmark

M. R. Thomsen et al. (eds.), *Design for Rethinking Resources*, Sustainable Development Goals Series,
https://doi.org/10.1007/978-3-031-36554-6_40

engineering and construction the transition to a bio-based material paradigm holds potential of significant impact. Globally, the built environment represents 40% (United Nations Environment Programme 2020) of all materials and 50% of energy consumed (Waste Generation 2020, 2022). A transition into a more intense use of bio-based building materials based on grown resources and upcycled agricultural waste is argued to empower a step change in resource deployment. However, bio-based materials are fundamentally different to traditional geo-sphere driven resources. Shaped by growth cycles and formed by their environment, bio-based materials are easy to manufacture and modify, but also malleable, prone to weathering and characterised by shorter life spans. These characteristics shape their implementation in the built environment.

This paper asks how a transition to bio-based material extends circular design thinking in architecture and construction. By taking the point of departure as the key concept of cascading, the paper identifies fundamental changes to how building materials can be fabricated, deployed, maintained, reused and recycled. The paper first contextualises the idea of cascades in existing architecture circular design strategies. Through a mapping of circular design architectural strategies to the Circular Design Systems Diagram (McArthur 2019) we present an overview of emerging cascade thinking in architecture which belong mainly to finite resources of technological cycles. This mapping is then extended to bio-based materials. Finally, we use the case-study of the 3d-printing of biopolymer composites and the creation of six future use scenarios to exemplify these changes and discuss the conceptual, technical and practical implications on future building practice.

2 Circular Design in Architecture and the Built Environment

Circular Design aims to shift from a 'take-make-waste' linear economic model to one that preserves and enhances natural capital, optimises resource yields and minimises system risks by managing finite stocks and renewable flows (Stahel 2016). Based on three principles—designing out waste and pollution, keeping products and materials in use and the regeneration of natural systems (McArthur 2013)—it is a driver of new business models that foster reuse and extend service life through repair, refitting and refurbishing, which turn old goods into as new resources. As such, Circular Design points to systemic changes in which multiple actors across the production and consumption value chain act together to retain as much value as possible. In architecture, as in other sectors, there are multiple levels of engagement: from a governmental-institutional position driving new legislation, from industry position building new business models for value retention and through design reshaping products enabling materials to have longer life spans across multiple cascaded functionalities.

In architecture and the built environment Circular Design is seen as a step change in intensifying the use of resource across its life span. While Design for Sustainability has many layers of resource driven, industrial and societal implication, the most popular frameworks in contemporary practice, such as Green Design, Cradle-to-Cradle or biomimicry is concerned with circularity at the "Product design innovation level" (Ceschin and Gaziulusoy 2016). Most contemporary architecture and construction adoptions of Circular Design approaches revolve around two of six strategic Rs: Recycling and Reuse (Dokter et al. 2021; Yang et al. 2022). This is suggested either through circularity at the product and component scale through material passports, components banks and complex data pipelines which allows to disassemble, retrieve and reuse materials across buildings, or through adaptability, redesign and reuse at the larger building scale (Hamida et al. 2022). From a lifecycle perspective, Circular Design-in-construction strategies tackle all phases of the value chain, from design and manufacturing, through installation, maintenance and operation, all the way to end of life and disposal via the established infrastructure of BIM and LCA software. They aim to improve the value chain through "slowing resource use by prolonging building lifetime, closing the resource loop through by diverting End-of-Life materials from landfills and narrowing resource use through design efficiency" (Chen et al. 2022).

2.1 Understanding Practice Through the Lens of the Circular Design Cascades

When interpreting architectural practice through the Butterfly Diagram (Fig. 1), cascading can be thought of in respect to six central measures: ***retain, refit, refurbish, reclaim & reuse, remanufacture*** and ***recycle*** (Cheshire 2021). Each of the cascade levels retains different degrees of the value of the designed artefact. In the following we outline these emphasising the different actors and processes that they involve across the value chain.

Cascading starts with the effort to ***retain*** building occupation through building for longevity and by practices of repair. In private homes these cycles of repair might be undertaken by the inhabitant, while in other building stocks these are undertaken by skilled labour managed and scheduled by a building manager teams. The second cascade level points to ***refitting*** as a practice of continually adapting building as (Askar et al. 2022). Refitting might adopt practices of maintenance and repair while intensifying these to transform building programme and performance. These processes similarly address an overlap of private and skilled labour and can engage architects to develop a design strategy. The third cascade practice of ***refurbishing*** intensifies these processes further. Primarily planned by architects and undertaken by contractors, processes of refurbishing rethinks the building stock by adapting the existing structure into new structures. Adaptation is seen as the central proponent of circular design in architecture allowing buildings to attain longer life spans while simultaneously answering changing needs (Cheshire 2021). Adaptation can renew strategic

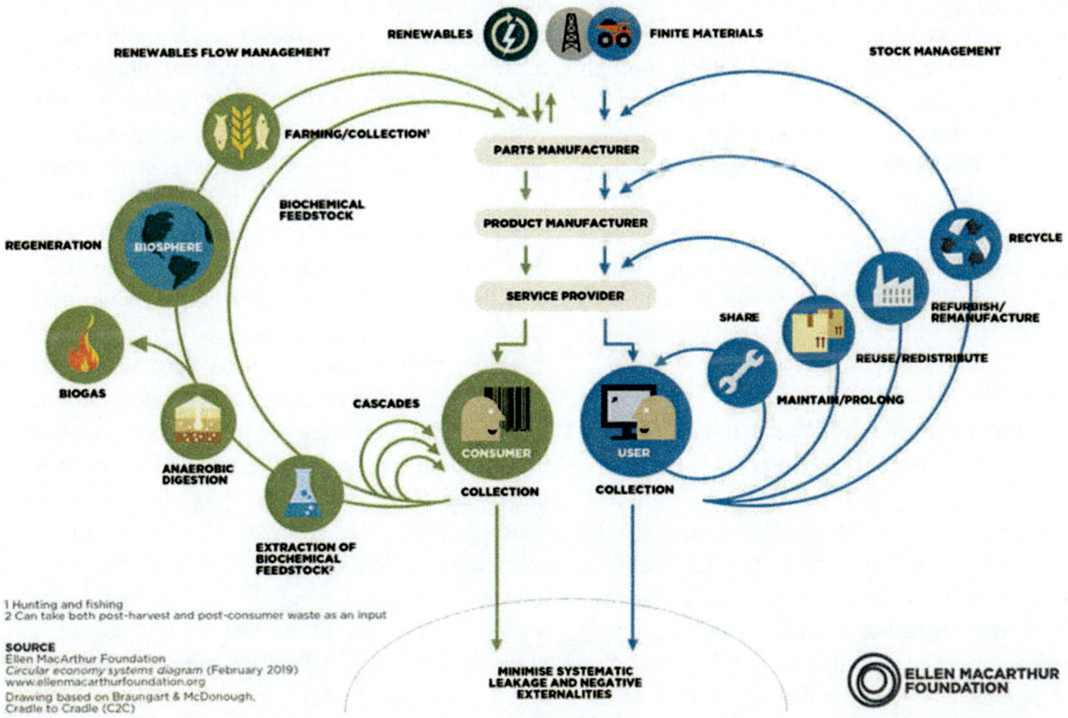

Fig. 1 Circular economy system diagram (Ellen McArthur Foundation) reproduced from https://ellenmacarthurfoundation.org/circular-economy-diagram

layers of the building (Habraken 2005) while retaining others. The fourth cascade level of ***reclaim & reuse*** defines processes of using the building stock as a resource of building elements. Through processes of deconstruction, building elements are harvested and reused in new build or refitting processes. These processes are mainly enacted by deconstruction companies together with architects who play a central role in creating design solutions that enable disassembly and reuse (Manelius 2017). The fifth cascade level, ***remanufacture,*** downcycles elements further into new elements and components, executed by building deconstruction companies (O'Grady et al. 2021). These might downcycle structural materials such as structural timber or concrete elements as tertiary building elements such as flooring or wall finishing. The material loses part of its initial value while gaining new value through design and remanufacture. Processes of remanufacture necessitate industrial actors able to build adequate infrastructures and processes of refabrication. The sixth cascade level ends with material ***recycling***. As highlighted by circular design principles generally, recycling is the least valuable practice of cascading as its focus on the material alone undoes the value added through design and manufacture. This is undertaken by demolition and waste management companies who collect material for recycling. In building processes of recycling channel back materials to fabrication in which they are mixed varying degree of intensity with virgin materials. Recycling includes links between waste management processes and fabrication.

2.2 Arguing for the "Biosphere"-Side of the Butterfly Diagram

The cascade structure in circular design for the built environment has a very clear limitation: it only addresses the finite *"technological cycles"* side of the McArthur circular design butterfly model (Leipold and Petit-Boix 2018). They also have a tendency towards the upper part of the diagram with more emphasis on value retention in already existing products and building elements, which have already extracted resources and utilized energy to be shaped. Additionally, that agency is mainly placed within skilled labour actors thereby negating the direct agency of the occupant. This approach, while offering immediate areas of action for designers, industries, and policy makers alike, disregards the complementary side of the McArthur circular design butterfly model. That of the *"biological cycles"*. To enter into bio-based material practices for architecture we need to expand our understanding of cascading and engage the fundamental differences in the network of actors and processes that bio-based material practice include.

3 Bio-Based Materials for Architecture

In architecture, bio-based materials can be described across three material perspectives: harvested materials, designed materials and living materials (Ramsgaard Thomsen and Tamke 2022). *Harvested materials*, such as timber, bamboo, hemp or straw are increasingly used in architecture, as processes of design and manufacture are maturing (Mahapatra and Gustavsson 2008). At the same time, new classes of harvested materials such as Acetobacter Xylinum producing cellulose sheets (Derme et al. 2016) and Mycelium elements (Ayres et al. 2023; Bitting et al. 2022) are currently being investigated. At the other end of the spectrum lies the functionalisation of *living materials* utilising their ability to metabolise and transform agents of material performance or the production of energy. The aspiration of integrating living microbial fuel cells into the built environment; to use building facades hosting mosses or algae to purify air (Cruz et al. 2019) or dedicated bio factories producing heat or nutrition (Peruccio and Vrenna 2019; Wurm and Pauli 2016), define new ways of understanding the building as the host of a new set of organisms that extend a building's performances (Ramsgaard Thomsen and Tamke 2022). While promising, these techniques are still at their infancy for large scale production and require adequate lab settings for

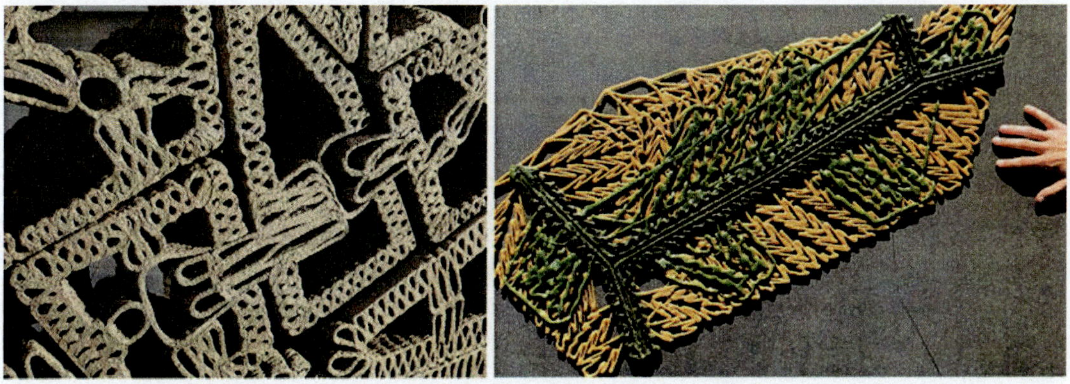

Fig. 2 The two cases of biopolymer developed at CITA, in collaboration with DTU, using respectively Xanthan Gum and Collagen Glue as binders

their success (Bitting et al. 2022). The attractiveness of the third category: *designed materials* such as biopolymers and bio-composites, lies in their capacity to be produced at large scale, and flexibility in absorbing waste streams into their material composition as such enabling cascading at a material level. Our research sits within a growing interest in additive manufacturing of biopolymer composites. While casting might provide a high-volume production method of modular elements, 3d-printing allows for a larger design space not only through geometrical freedom which has performance and formal drivers (Goidea et al. 2020), but also allow for water content grading (Mogas-Soldevila et al. 2015) and possibilities for integrating data-driven shrinkage-compensation models (Dritsas et al. 2018).

The argument presented in this paper uses the case of 3d-printed biodegradable biopolymer composites as a particular way of viewing the potential differences that bio-based materials offer to architectural production. Biopolymer composites are typically composed of a binder—a biopolymer- mixed into a solvent and reinforced by fibres—the filler. Their attractiveness lies in their designability: the choice of materials and proportions; and how these considerations coupled with fabrication systems give the material its key properties. Our research explores the potential of recipe grading as an enabler for interface between fabrication, design and material properties and performance of the printed artefacts. In this paper we present two material systems which are suitable for 3d-printing and fitting to a digital fabrication pipeline while also being truly biodegradable and entirely based on waste stream materials.

In the following we will firstly define the two materials systems based on Xanthan Gum and Collagen respectively (Fig. 2), then outline a strategy for understanding the cascading of biopolymers at both the material and architectural level. This will lead to the rethinking of a cascading framework for the bio-based material in architecture.

3.1 Xanthan Gum Bio-Composites for Self-Supporting Building Blocks

Our Xanthan Gum bio-composite is entirely bio-based constituted mainly of waste-stream materials. Xanthan Gum is a by-product waste of cheese production and is a common emulsifier and stabiliser used in the food industry. In our material recipe, we use Xanthan Gum as a binding biopolymer for the 3d printed slurry. When dispersed in water it forms a gel, to which filling and reinforcing agents can be added. Our recipe uses recycled paper flock as a fibre element, again a waste stream material being upcycled into a structural reinforcer of the

Fig. 3 **a** Cellulose Enclosures: 3d-printed blocks with the Xanthan Gum composite material. The blocks present a tectonic language which integrates aeration, structure and joinery. The blocks have a size of 50 × 70 cm on average. **b** Cellulose Screens: scaling up the prints employing a large-scale gantry printer allows us to test the performance of self-supporting block measuring 2 m height. The material performance and print pattern scales seamlessly

material mix. The slurry is then shaped through the 3d printing nozzle and will cure into a solid material as water content slowly evaporates.

We use this material to make self-supporting interior partition blocks. Printing toolpath design needs to consider the drying process. The porous patterns enable the exposure of the material to air allowing water content to evaporate. Once dry, the blocks can be shaped with conventional woodworking tool (drills, saw sanding) and off-cuts can be rehydrated and reused to print more components. Our demonstrators showcase the capacity of this material to be stacked vertically in an assembly Cellulose Enclosure (Fig. 3a) and to carry their own self-weight (Fig. 3b). In large scale tests (Fig. 3b) we have found that material behaviour scales proportionally, with 2 m tall pieces printed on a large scale 3d-printing gantry being structurally sound. The material exhibits further interesting properties relevant to the architectural quality such as thermal insulation, acoustic absorption and hygroscopic regulation.

3.2 Collagen Glue Bio-Composites for Building Panels

Our collagen glue bio-composite is also a bio-based employing waste-stream materials. The collagen glue binder itself is a by-product of the treatment of bones and tissues from the meat industry. It exhibits two main advantageous behaviors: it is thermally responsive, which means that its setting is reversible, and it is water soluble. The collagen glue is used as a binding matrix for a mix of waste-stream fibres including bark, cotton, wood flour and seagrass. The choice of fibers affects not only the structural properties of the mix, but also its texture, color and smell. The main difference to the Xanthan Gum-based slurry is that, rather than requiring a high volume of water to achieve a printable viscosity, the collagen glue requires heat. Extrusion temperatures are between 50° and 60°. At that temperature range the collagen glue is entirely recyclable and can be remelted and reused without any loss in its structural properties. It has mechanical properties similar to nylons, acetals and epoxy resins, without the environmental impact.

In the demonstrator Radicant (Fig. 4), we test the printing of large panels with this material. Again, we develop intricate print patterns to steer the distribution of the material in a series of bespoke interior architectural panels. Each panel is composed of multiple variations of the base recipe by grading the fibre content. The composition is consolidated during the 3D printing process as each new print layer seamlessly merges with the subsequent layer as soon as the

Fig. 4 Radicant: 3d-printed blocks with the bone glue bio composite material. The panels are composed of interlacing print beads made through material grading, with different fibers and concentrations. The panels size average is 30 × 50 cm

temperature of the print drops below 30°. The resulting print is stable but remains malleable as it further cools and water evaporates. In experiments we have found that this allows further forming. By laying prints on the formwork, we can achieve double curved panels. End of life possibilities vary from reheating and reprinting, to grinding down the material to pellets for recycling to biodegradation through composting.

4 Circularity with Architectural Biopolymer Composites

The two cases-studies described above demonstrate different potentials and use-cases for the application of biopolymers as a means of upcycling waste stream materials into building materials. In the following we will introduce cascading logics for biodegradable biopolymers first at a material processing level and then at an architectural component and use scenario level.

4.1 Biopolymer Composites: Specific Cascade Thinking at Material Scale

At a material level, a transition to biopolymer materials, presents a range of new possibilities for reusing and recycling (Fig. 5). Like most other materials, biopolymer materials go through different stages of reuse, recycling, and reprocessing, where each additional level of handling and processing adds to the environmental impact of recycling and revalorising the materials.

Direct reuse is preferable (1), and, for these materials, it is the most benign and environmentally friendly valorisation. Waste resulting from initial processing steps can typically be valorised by direct feeding into the process, either in its pure form or, as in the cases above, by using minor amounts of water (2). This stage corresponds directly to the mechanical recycling of classical plastic materials, effectively reducing the amount of in-process and fabrication waste materials. Considering a lower quality material, aged or substantially polluted over a longer lifetime, these types of fractions can be disassembled entirely into their separate components through dissolution techniques (3). Through disassembly, rinsing and purification processes can be applied to bring the quality and purity of the material up, and new materials can be reformulated from the base components to enter the cycle at the very highest level again (4). At the fifth level, an attractive pathway is possible for bio-based materials that can act as feedstock for use in the rapidly growing bioeconomy. Materials of poor quality due to degradation or pollution can still be valuable feedstock for fermentation processes, leading to the preparation of a broad range of products or energy recovery through incineration (5). Finally, the very last stage of the process can be either returning the materials to nature by composting (6).

The material cascading is enabled by its sensibility to water. Where this limits the outdoor applications of biopolymers, it simultaneously enables valorisation and revitalisation of waste materials at across multiple cascade at all stages of the process.

4.2 Biopolymer Elements: Specific Cascade Thinking at Architectural Scale

Existing methods for understanding cascading in bio-based materials at the level of architecture aims to preserve absorbed carbon and increase total quantity of material through processes of downgrading (European Commission. Directorate General for Internal Market, Industry, Entrepreneurship and SMEs 2018). This is in line with the circular thinking for materials of the "technological cycles" and is in part driven by the long time and high processing effort that is needed to transform the raw resource into a

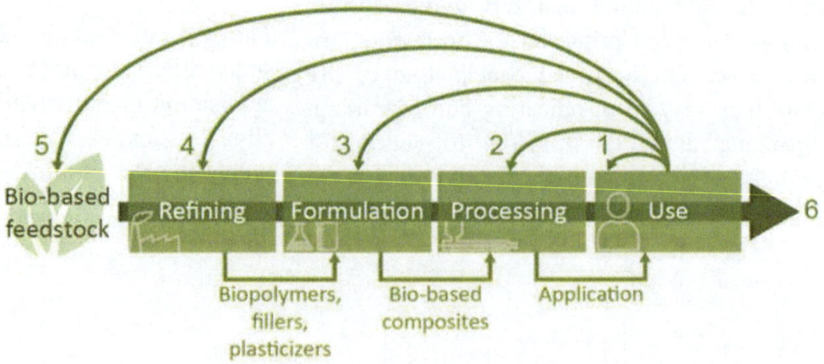

Fig. 5 Valorisation of biopolymer materials after end-of-life and how they enter a recycling cascade

usable element for construction in traditional bio-based materials such as timber. However, as explained in the previous section, the biopolymer composites we describe above are biodegradable, based on waste-stream materials and relatively easy to process and quick to remanufacture. These characteristics uncover opportunities to create new architectural cascade thinking specific to their biological cycles. Rather than emphasise design dismantlement and value retention of individual building elements, which concentrates activity with larger and more systemic actors, these new cascades activate the more local agencies of the occupant, maintenance team, contractor and architect.

In the following, we propose a new interpretation of the butterfly diagram specific to architectural circular design thinking which also integrates the emerging considerations from the usage of biopolymer composites in construction (Fig. 6). In order to further unpack the relations captured by this diagram, these are described in the form of everyday scenarios to highlight how actors, actions and implications intersect within these new biological cascades.

4.2.1 Retain Scenario: Thicken

The retain scenario takes point of departure in maintenance and repair. The wall in the entry hall has recently been damaged and received a large dent. Localised heating applied with a simple hair dryer makes the material malleable, allowing the inhabitant to push the displaced material back with a small trowel. DIY hand work, simple tooling and a bit of time is enough for the result to be as good as new. Later that Autumn, the inhabitant recoats a degraded part of the south façade with a new layer which contains a higher ratio of seagrass to better resist the rains of the winter. Over 20 years of living and maintaining the building, this yearly cycle has resulted in the southern walls becoming 10 cm thicker.

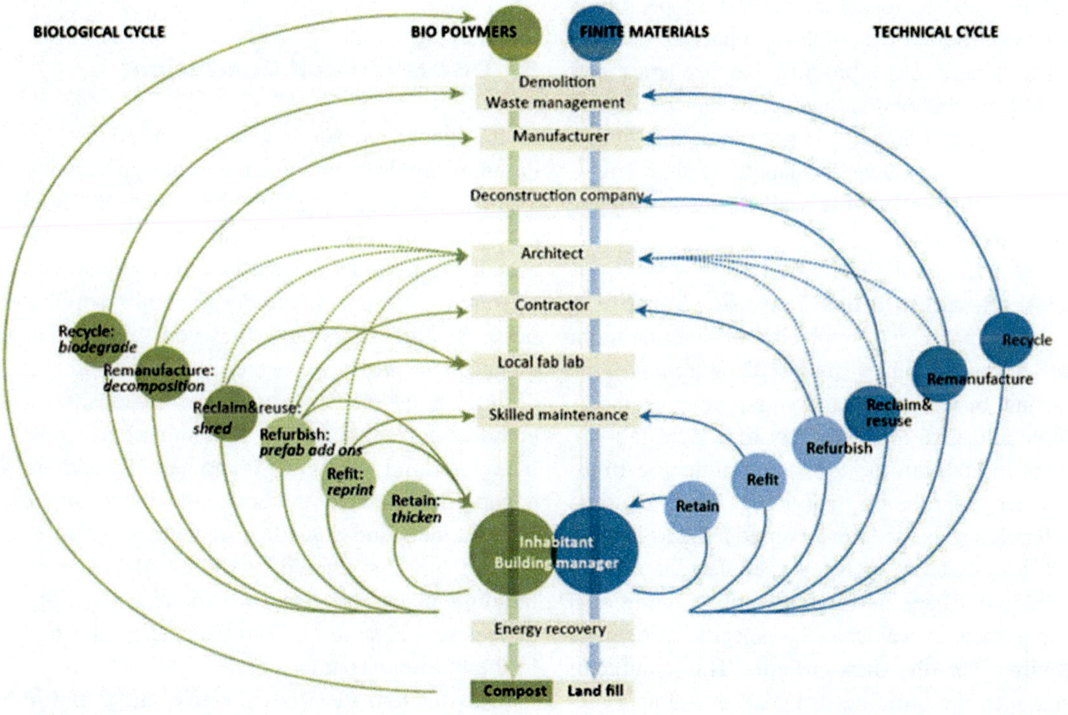

Fig. 6 In our redrawing of the right side of the diagram we emphasise the correlation between the different levels of cascading and with which actor agency is placed

4.2.2 Refit Scenario: Reprint

The refit scenario intensifies these practices to transform an aspect of building performance. When the inhabitants moved in, the same entry hall was not fitted with storage, benches or coat hooks. They contact a local contractor who operates a room-scale 3d-bio-printer and is able to print these new elements directly onto the existing floor and walls of the house. The builder uses a bark material that is sourced from a local sawmill waste and gives the newly fitted interior a warm timber look. Waste spilled during fabrication is immediately sieved and reused in the fabrication process.

4.2.3 Refurbish Scenario: Prefabricate Add-Ons

An old timber building is being refurbished by a contractor. After purchasing pre-fabricated standard size bio-panels from a local building market, they use heat to alter the edge of each panel so that they fit into the non-standard gaps between the structural timber. A new cellulose wall is installed to subdivide one of bedrooms into two: here the cellulosic material has the added benefit of improving the acoustics and modulating internal humidity levels. Where the ceiling is not straight—the floor to ceiling height varies by 15 cm over the length of this wall—they are able to cut the wall to shape using a simple hand saw.

4.2.4 Reclaim/Reuse Scenario: Shred

A local contractor is replacing an interior fitout into a rented office space. They remove the existing biopolymer panels and other elements whole and take them directly to a local fab-lab, which in addition to 3d-printing offers local biopolymer circulation services. The returned materials are mechanically shredded and heated, and the material mass is supplemented with additional fillers or diluted with an additional biopolymer to achieve the specific properties required for the new fit-out. The contractor returns to the building to install the same materials as they had removed, now in a new form that has been designed by an architect.

4.2.5 Remanufacture Scenario: Decomposition

After 10 years of heavy use, the same interior fitout has become substantially run down and damaged. Small areas have become contaminated. The manufacturer sends out a recovery team to break it up and return it to the manufacturer's facilities. Here the pieces are put directly into a dissolution process to separate the biopolymer from the fillers. After subsequent rinsing and purification processes, the manufacturer is able to use the same base components within a new material, and on a different product line.

4.2.6 Return Scenario: Biodegrade

In the return scenario, elements can be composted locally by the inhabitant in small scale residential cases, or taken by a contractor to a biomass collection point for larger quantities of material. The material is rinsed and purified and then used as foodstuff for the growing of mushrooms entering the food chain.

5 Discussion and Conclusions

In this paper we address the potential for bio-based materials to extend circular design thinking in architecture and construction, where the field recognises the need to intensify the use of bio-based resource, the current lack of a framework for circular design in bio-design limits application. By focusing our study on biopolymer composites, we have reinterpreted the circular design framework for biopolymer composites in architectural applications. The aim of our new interpretation is to highlight and partially address a gap in architectural knowledge between the better understood cascading of technical materials and the less defined cascading of bio-based building materials. Here, we identify the links between six R-strategies and the agency of actors in the built environment.

Our diagram identifies the relevant actors in the architectural value chain, starting with the inhabitant as the first level. It traces the cascades specific to biopolymer composites, based on the properties of two described biopolymer composites which are suitable for 3d-printing at architectural component scales, biodegradable and entirely based on waste stream materials. The relations captured in the diagram are unpacked in everyday scenarios of future building practice to highlight how the actors, actions and implications intersect within these new biological cascades.

In contrast to the technical side of the butterfly diagram, this mapping reveals that on the biological side new opportunities for maintenance and fabrication lead to a more manifold agency between actors within the inner cascade loops. This is linked to the relative malleability and processability of the bio-polymers composites, which lower the requirements for skill and machinery when inserting, adapting or removing components made from this material. It also impacts the outer cascades, where water rather than more chemically or energy intensive processes can be used to reclaim the ingredients of the composite material. As showcased in the scenarios, cascades within the biological cycles present opportunities to go beyond simply replicate the emphasis on design dismantlement and the value retention of individual building elements. Instead of concentrating activity with larger and more systemic actors, this paper implies new conceptual and technical approaches based in the local agencies of the occupant, maintenance team, contractor and architect.

References

Askar R, Bragança L, Gervásio H (2022) Design for Adaptability (DfA)—Frameworks and assessment models for enhanced circularity in buildings. Appl Syst Innov 5(1):24. https://doi.org/10.3390/asi5010024

Ayres P, Rigobello A, You-Wen J, Colmo C, Young J, Sørensen K-J (2023) Investigating a design and construction approach for fungal architectures. In: Gengnagel C, Baverel O, Betti G, Popescu M, Thomsen MR, Wurm J (eds) Towards radical regeneration. Springer International Publishing, pp 571–583. https://doi.org/10.1007/978-3-031-13249-0_45

Bitting S, Derme T, Lee J, Van Mele T, Dillenburger B, Block P (2022) Challenges and opportunities in scaling up architectural applications of mycelium-based materials with digital fabrication. Biomimetics 7 (2), Article 2. https://doi.org/10.3390/biomimetics7020044

Ceschin F, Gaziulusoy I (2016) Evolution of design for sustainability: from product design to design for system innovations and transitions. Des Stud 47:118–163. https://doi.org/10.1016/j.destud.2016.09.002

Chen Q, Feng H, Garcia de Soto B (2022) Revamping construction supply chain processes with circular economy strategies: a systematic literature review. J Clean Prod 335:130240. https://doi.org/10.1016/j.jclepro.2021.130240

Cheshire D (2021) The handbook to building a circular economy. RIBA Publications. http://ebookcentral.proquest.com/lib/kadk/detail.action?docID=6796468

Cruz M, Beckett R, Ruiz J (2019) Bioreceptive concrete wall. In Brayer M-A, Zeitoun O, Papapetros S, Bianchini S, Quinz E, Collet C (eds) Mutations/Créations: La fabrique du vivant | Catalogue Exposition (p. Chapter 7). Editions HYX / Editions du Centre Pompidou. https://editions.centrepompidou.fr/en/exhibition-catalogues/exhibition-catalog-mutations-creations-la-fabrique-du-vivant/1506.html

Derme T, Mitterberger D, Di Tanna U (2016) Growth based fabrication techniques for bacterial cellulose: three-dimensional grown membranes and scaffolding design for biological polymers, pp 488–495. https://doi.org/10.52842/conf.acadia.2016.488

Dokter G, Thuvander L, Rahe U (2021) How circular is current design practice? Investigating perspectives across industrial design and architecture in the transition towards a circular economy. Sustain Prod Consum 26:692–708. https://doi.org/10.1016/j.spc.2020.12.032

Dritsas S, Halim SEP, Vijay Y, Sanandiya NG, Fernandez JG (2018) Digital fabrication with natural composites: design and development towards sustainable manufacturing. Constr Robot 2(1–4):41–51. https://doi.org/10.1007/s41693-018-0011-0

European Commission. Directorate General for Internal Market, Industry, Entrepreneurship and SMEs (2018) Guidance on cascading use of biomass with selected good practice examples on woody biomass. Publications Office. https://data.europa.eu/doi/10.2873/68553

European Commission. Directorate General for Research and Innovation (2021) European green deal: research & innovation call. Publications Office. https://data.europa.eu/doi/10.2777/33415

Goidea A, Floudas D, Andréen D (2020) Pulp Faction: 3d printed material assemblies through microbial biotransformation. In: Burry J, Sabin J, Sheil B, Skavara M (eds) Fabricate 2020: making resilient architecture. UCL Press, pp 42–49

Habraken NJ (2005) Change and the distribution of design. Open House Int 30(1):6–12. https://doi.org/10.1108/OHI-01-2005-B0003

Hamida MB, Jylhä T, Remøy H, Gruis V (2022) Circular building adaptability and its determinants—A literature review. Int J Build Pathol Adapt. https://doi.org/10.1108/IJBPA-11-2021-0150

Leipold S, Petit-Boix A (2018) The circular economy and the bio-based sector—Perspectives of European and German stakeholders. J Clean Prod 201:1125–1137. https://doi.org/10.1016/j.jclepro.2018.08.019

Mahapatra K, Gustavsson L (2008) Multi-storey timber buildings: breaking industry path dependency. Build Res Inf 36(6):638–648. https://doi.org/10.1080/09613210802386123

Manelius AM (2017) Rebeauty—Nordic built component reuse. Vandkunsten Architects. https://vandkunsten.com/content//2019/01/NBCR-20170201-web.pdf

McArthur E (2013) Towards the circular economy Vol. 1: an economic and business rationale for an accelerated transition. https://www.ellenmacarthurfoundation.org/assets/downloads/publications/Ellen-MacArthur-Foundation-Towards-the-Circular-Economy-vol.1.pdf

McArthur E (2019) Circular economy systems diagram: the butterfly diagram. https://ellenmacarthurfoundation.org/circular-economy-diagram

Mogas-Soldevila L, Duro-Royo J, Lizardo D, Kayser M, Sharma S, Keating S, Klein J, Inamura C, Oxman N (2015) Designing the ocean pavilion: biomaterial templating of structural, manufacturing, and environmental performance 9

O'Grady T, Minunno R, Chong H-Y, Morrison GM (2021) Design for disassembly, deconstruction and resilience: a circular economy index for the built environment. Resour Conserv Recycl 175:105847. https://doi.org/10.1016/j.resconrec.2021.105847

Peruccio PP, Vrenna M (2019) Design and microalgae. Sustainable systems for cities. AGATHÓN | Int J Archit, Art Des 6:218–227. https://doi.org/10.19229/2464-9309/6212019

Ramsgaard Thomsen M, Tamke M (2022) Towards a transformational eco-metabolistic bio-based design framework in architecture. Bioinspir Biomim 17 (4):045005. https://doi.org/10.1088/1748-3190/ac62e2

Stahel WR (2016) The circular economy. Nature 531 (7595). Article 7595. https://doi.org/10.1038/531435a

United Nations Environment Programme (2020) 2020 global status report for buildings and construction: towards a zero emission, efficient and resilient buildings and construction sector. https://globalabc.org/sites/default/files/inline-files/2020%20Buildings%20GSR_FULL%20REPORT.pdf

Waste generation (2020) (2022) Statistsches bundesamt (Destatis). https://www.destatis.de/EN/Themes/Society-Environment/Environment/_Graphic/_Interactive/waste-management-quantity.html

Wurm J, Pauli M (2016) SolarLeaf: the world's first bioreactive façade. Arq: Archit Res Q 20(1):73–79. https://doi.org/10.1017/S1359135516000245

Yang Y, Guan J, Nwaogu JM, Chan APC, Chi H, Luk CWH (2022) Attaining higher levels of circularity in construction: scientometric review and cross-industry exploration.J Clean Prod 133934. https://doi.org/10.1016/j.jclepro.2022.133934

Sustainable (Re)Development in Post Industrial City Regions Centering Circular Systems of Food, Energy, Water, and Waste: A Case for Detroit

Geoffrey Thun, Tithi Sanyal, and Kathy Velikov

Abstract

For architects and urban designers working on sustainable urban projects and aiming to integrate UN SDG frameworks, a broader set of methods and practices is required than those that have traditionally been part of the discipline. To apprehend and evaluate sustainable development project goals and impacts, a collaborative and systems-oriented design approach is required to situate design within its broader complex of networks and processes and to expand project priorities and interventions to operate in multiscalar and integrated ways. This paper describes a framework that evaluates the Food Energy Water (FEW) Nexus across scales from Region (Great Lakes Megaregion), to State (Michigan), to the City of Detroit to illuminate issues of equity and access across FEW domains while aiming to leverage circular systems towards more just and sustainable urban futures. Working with stakeholders in Detroit, the project illustrates a design-based methodology and strategies to enhance circular synergies, reduce waste flows, and evaluate proposal impacts based on Co_2e per capita in the resulting urban design schemes. Quantification of systemic impacts alongside visualizations of resulting building and urban configurations, engineering components, and public space design are presented. Lessons from Detroit offer portable strategies for other post-industrial cities and urban centers where issues of equity and access to food systems are pressurized.

Keywords

Sustainable urban design · FEW nexus · SDG integration · Systems-oriented design · Detroit

1 Introduction

The UN Sustainable Development Goals (SDGs) aim to address complex and interconnected problems, which require integrated and system-based and cross-sectional approaches for architects and urban designers. As recognized by the United Nations Development Programme (UNDP), challenges to sustainable development "cannot be dealt with in isolation" and solutions "must be integrated" (UNDP SDG Integration 2022). Toward this "nexus approach," frameworks such as the Food Energy Water (FEW) Nexus, have the capacity to integrate across silos

G. Thun · K. Velikov (✉)
Taubman College of Architecture and Urban Planning, University of Michigan, Ann Arbor, USA
e-mail: kvelikov@umich.edu

T. Sanyal
School of Architecture, University of Virginia, Charlottesville, USA

M. R. Thomsen et al. (eds.), *Design for Rethinking Resources*, Sustainable Development Goals Series,
https://doi.org/10.1007/978-3-031-36554-6_41

of both research and implementation, and simultaneously address multiple SDGs (Liu et al. 2018). It has also been argued that the FEW-Nexus could also strengthen synergies across SDGs (*Ibid*). However, FEW-Nexus approaches need to expand from privileging environmental science and engineering principles that look to optimize stocks and flows across food, water, and energy systems and leverage efficiency and circularity, toward "design-led" approaches that can address multiple stakeholders in broad and integrated ways (Roggema et al. 2021). Design-led approaches, where architectural and urban design thinking plays a central role, are more able to balance and consider synergies and trade-offs across needs of social sustainability, equity, and community in relation to socio-technical system optimization.

This project paper describes a design-led approach developed by the University of Michigan team for the city of Detroit within the context of a global multi-institutional project. The work was funded by The Sustainable Urbanisation Global Initiative (SUGI)/Food-Water-Energy Nexus and focused on six cities/bioregions around the world: Amsterdam, Belfast, Detroit, Doha, Sydney, and Tokyo. The research and design framework developed by the authors' team evaluates the FEW Nexus across scales to illuminate issues of equity and access across domains while aiming to leverage circular systems towards the design of more just and sustainable urban futures in alignment with the SDG's.

Detroit is a city that has been a central actor in the history of American industrialization, race relations, and economic decline. It is a case study in shrinking post-industrial cities, and most recently, in regeneration and resizing of a post-industrial legacy city. Several broader patterns exist today within which the FEW Nexus relative to future sustainable urban development is considered. These include spatial fragmentation due to industrial infrastructure and patterns of escalating vacancy amplified by forces that include systematic redlining, disinvestment, failure of service provision, poverty, and corruption (Dewar and Thomas 2012; Detroit Future City 2017). Detroit's highly distributed population is for the most part underserved by physical and social infrastructures, the product of an ongoing history of racial segregation and tensions (Thomas and Bekkering 2015). The contemporary development context is limited in part by the capital demands of environmental site remediation, parcel assemblage, and the spatial sprawl of Detroit's urban fabric, coupled with its current levels of vacancy that together produce a context where service provision in untenable, and for decades underdelivered (Surgrue 2005). This has contributed to a concentration of conditions referred to as 'food deserts' and 'food swamps', which define a lack of access to healthy food options (Pothukuchi 2011, 2015; Detroit Food Policy Council 2016). Documentation of the linkages between these contexts and chronic health disparities are extensive (Budzynska et al. 2013). Yet this context has produced one of the most uniquely cultural entrepreneurial environments in the nation with respect to the ways in which grassroots organizations and community activists related to social justice and food security have begun to mobilize to rebuild the city towards more just futures with specific focus on the production of local food in residential and mixed-use contexts (Allnutt 2019; Detroit Food Policy Council 2017; Hoey and Sponseller 2018).

2 Methods

Through this project we have developed a transferrable framework and workflow for designers and architects to work with other stakeholders in support of sustainable development and evaluation of FEW Nexus-based urban design proposals (Fig. 1). This process includes (i) documentation and visualization of existing FEW system analysis and spatialization at the regional, city and neighbourhood scale, (ii) identification of specific local issues and contexts (cultural, social environmental, infrastructural, policy, and systemic) that shape the deployment of FEW principles in the urban development, and (iii) development of urban design prototypes that integrate these priorities into engineered components and design, and (iv) quantification of the

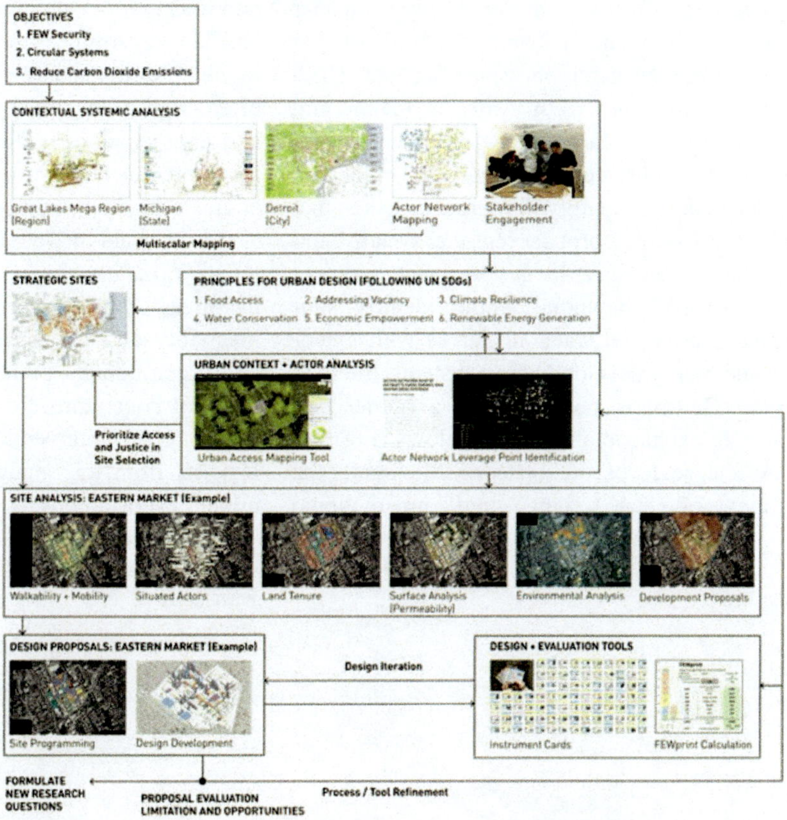

Fig. 1 Design research process framework and workflow to integrate FEW-nexus principles into specific local urban design proposals aimed to lower CO_2 emissions per capita

environmental impacts of the design proposals' carbon dioxide emissions using the FEWprint tool developed with our collaborators from TU Delft (Ten Caat et al. 2022). In addition to the FEWprint tool our team has also developed several processes and tools to assist design teams in identifying priorities responding to local contexts to address FEW systems in the development of urban design proposals:

(i) Multiscalar Mapping techniques for visualization of FEW systems contexts
(ii) FEW Material Flows and Interactions Visualizations
(iii) FEW Actor-Network Mapping Process and application to online tools
(iv) MultiModal Access Mapping Tools for application to issues of food access and site identification
(v) FEW Design Toolkit and stakeholder engagement processes.

2.1 Multiscalar FEW System Mapping

Multiscalar mapping engages cross-scalar thinking in imagining FEW futures for designers by describing the nature of FEW systems as they manifest across cascading scales in a specific geography. We unpack the contextual narratives of FEW systems and their interactions across three scales provisionally defined through *ecosystems* scale (regional), *jurisdictional* scale (state), and *operational* scale (city). Recognizing the global nature and extents of supply chains and extraterritorial urban impacts these techniques reflect the unique locational contexts (social, cultural, environmental, infrastructural, climatic, economic) of specific urban locations. A detailed description of the multiscalar mapping process and related insights has been published in (Thün et al. 2021).

Each scale of consideration brings into view alternating narratives and priorities for apprehending FEW systems locally. We consider here the ecosystems networks defined by the Great Lakes Megaregion, spanning the U.S.-Canada border in North America (Thün and Velikov 2017; Thün et al. 2015). Each cartography depicts the broader urban, environmental, commercial, and energy resource contexts that undergird Detroit's broader regional systems and network connections. In this territory, where a resource-based viewpoint would identify substantial energy and water resources with extensive and highly developed food systems, the U.S. city of Detroit is positioned as a border city central to exchange and material flows (Fig. 2).

At the scale of the State of Michigan, the realities of state rights and jurisdictional dominance reveals high differentiation in policies and law regarding systemic governance relative to neighboring states and provinces producing often highly constrained contexts for its cities and municipalities. The cartographies describe the interconnected food, energy, water, and waste contexts beyond the city (Fig. 3).

At the scale of Detroit, we see the operational legacies of its shaped industrial and cultural histories, contexts of racial and economic inequality and the legacies of a post-industrial city. Each context points to different opportunities and constraints for developing FEW-nexus based urban interventions in Detroit. (Fig. 4) Insights from this analysis inform priorities for urban design proposals (see Fig. 1).

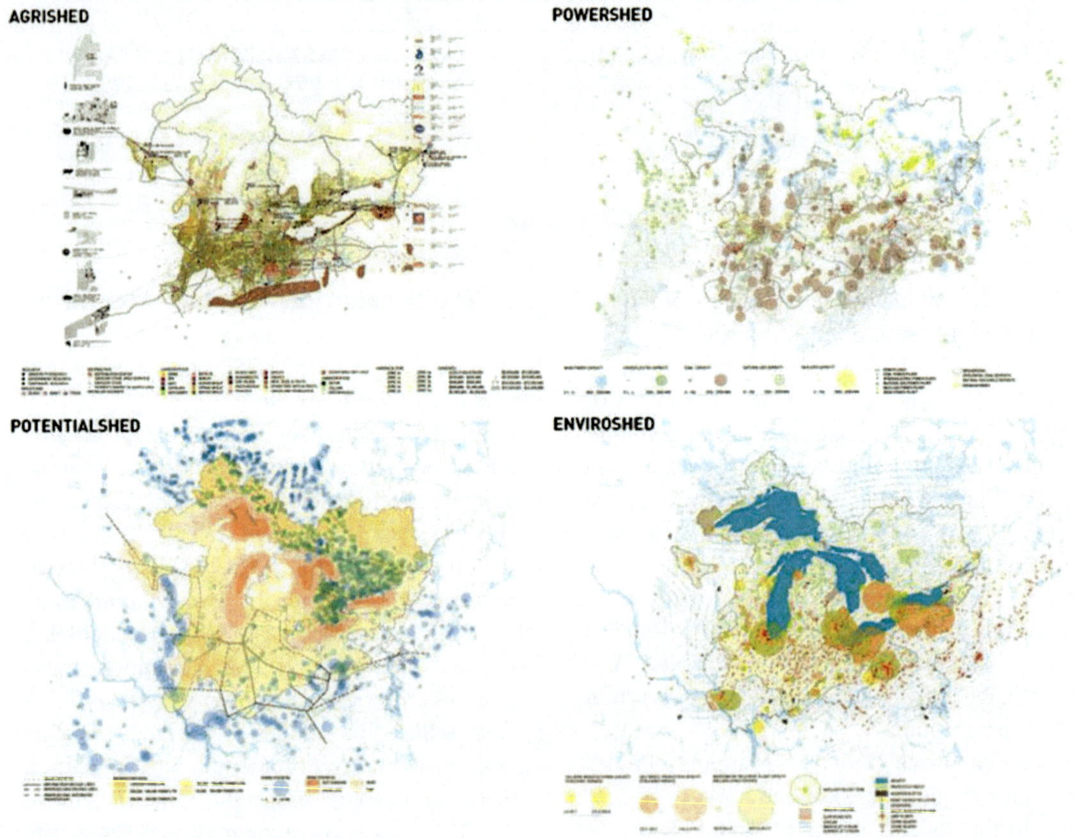

Fig. 2 Thematically differentiated 'Sheds' of the Great Lakes Megaregion: (from Top Left) The Agrished, the PowerShed, the potentialshed and the enviroshed

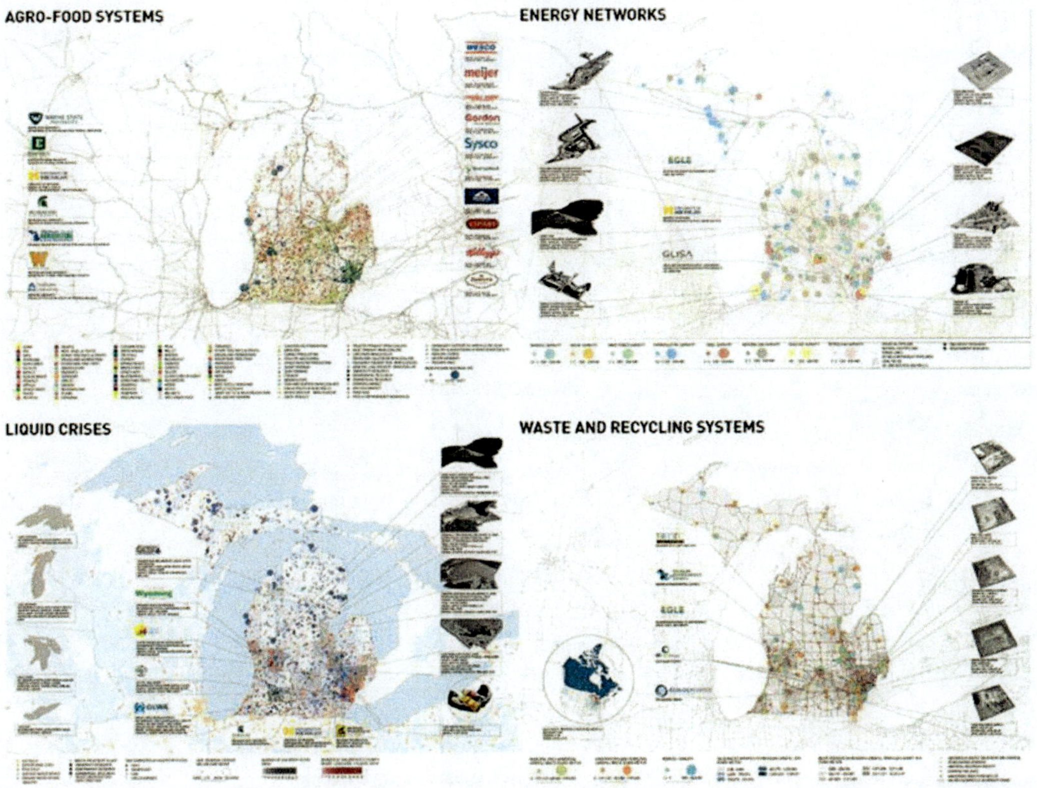

Fig. 3 Thematically differentiated Cartographies of Michigan: (from Top Left) Agro-food system, energy networks, liquid crises waste and recycling systems

2.2 FEW Material Flows and Interactions: Visualizations in the Detroit Context

Concurrent with the mappings, the sustainability systems experts on our team undertook a "material and energy flow analysis (MEFA)" to quantify the stocks and flows within the FEW systems of the Detroit Metropolitan Area (Liang et al. 2019). The method calculates flows using nitrogen, phosphorus, energy, and water as currencies for measurement. This research points out that a standard household and its associated services are the main drivers of the FEW Nexus; there is a high per capita intake of phosphorus and output of nitrogen into the water bodies, and the electricity sector is the largest consumer of water especially to generate thermoelectric power. This research identifies three distinct policy-based areas of concern that shape design considerations to reduce the environmental impacts of FEW systems and strengthen resource efficiency including (i) "optimizing dietary habits of households to improve phosphorus use efficiency"; (ii) "improving effluent-disposal standards..., promoting adequate fertilization, and enhancing the maintenance of wastewater collection pipelines" to regulate nitrogen emission levels; (iii) "improving water use efficiency of thermoelectric power plants" to reduce water withdrawal (Ibid).

2.3 FEW Actor-Network Mapping

Complimenting the multiscalar mapping and material energy flow and analysis, actor-network

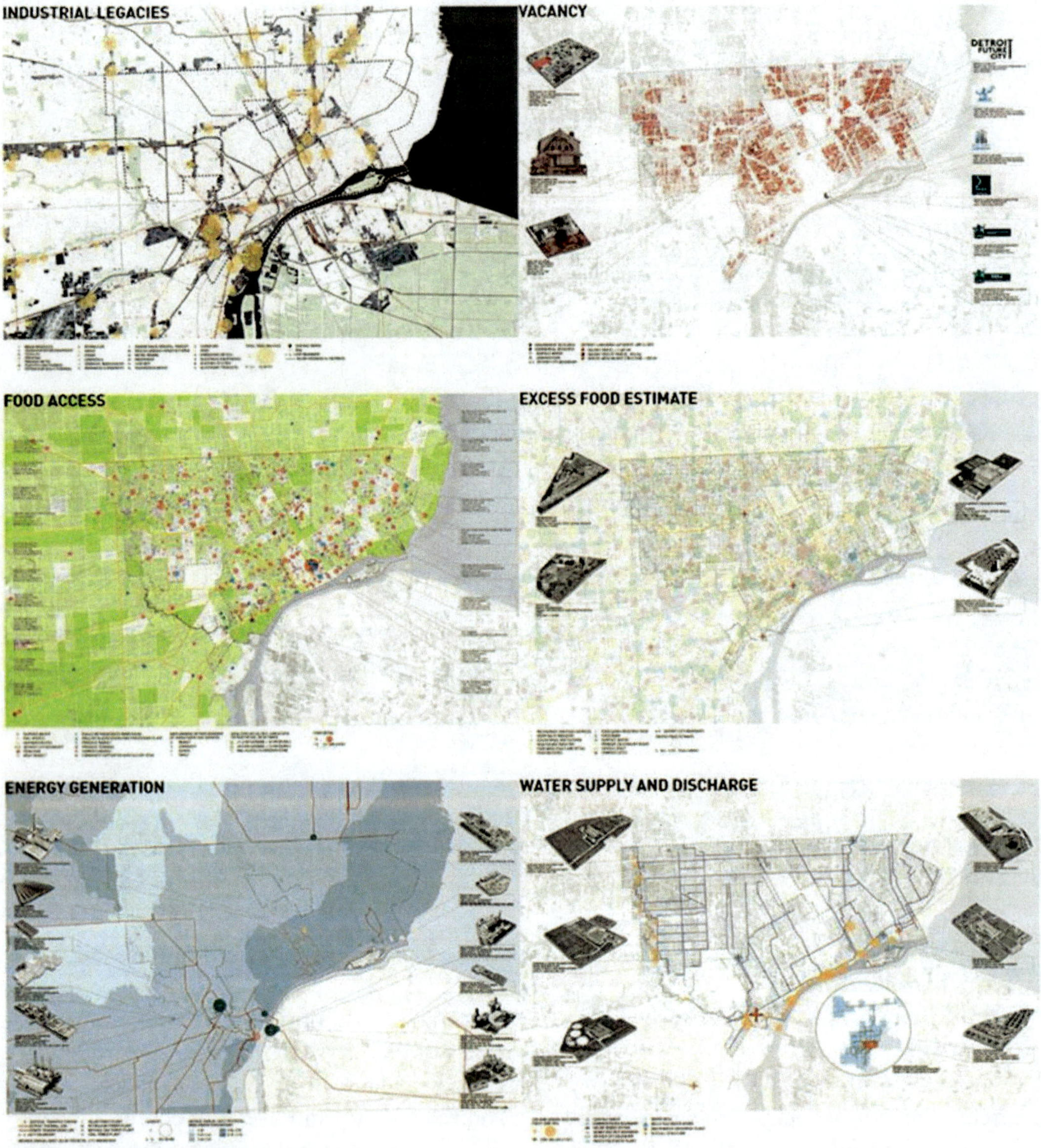

Fig. 4 Thematically differentiated cartographies of Detroit: (from Top Left). Industrial legacies, vacancy, food access, excess food estimates, energy generation, water supply and discharge

mapping further illuminates the FEW context in urban areas. The Actor-Network map is a technique utilized to define relationships between different entities (actors), networked relationships and interrelationships within the FEW nexus within a given urban context. Developed in the web-based relationship mapping tool KUMU (https://kumu.io/), the mapping identifies national, regional and local programs, initiatives, policies, funding, and stakeholders within the system and assists in identifying points of intervention. These groups, financing sources, laws and policies assist in identifying sources of support and stakeholders for engagement to be leveraged in project design development and delivery.

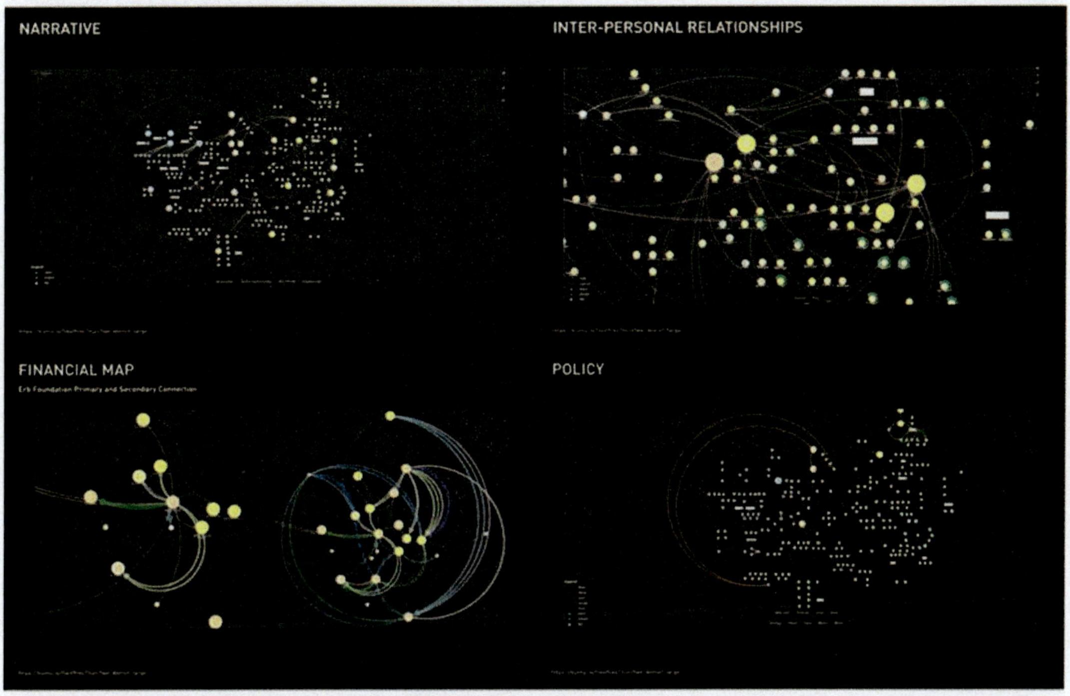

Fig. 5 FEW actor-network visualizations for detroit food networks

FEW Actor-Network mapping reveals key actors in both the policy and implementation space while exposing gaps within the network that assist in identifying spaces of design intervention (Fig. 5). Considering food access as a priority in Detroit, we see organic growth of food production and distribution network that caters to the residents in the city. Focusing on specific instances within the actor-network reveals a series of socio-political influences especially with respect to grassroots initiatives working towards food justice in Detroit.

While the Actor-Network describes existing networks, it can also point to initiatives and entities in need of greater connectivity to increase impact. For instance, a crucial aspect of the FEW Nexus is determining ways of achieving circularity, thus limiting the use of virgin materials, diverting waste from landfills, and reducing environmental pollution. In Detroit, few organizations are working in smaller networks to manage and repurpose waste. Here, the Actor-Network reveals an opportunity for synergetic development between policymakers, small-businesses and urban-agricultural practices, nonprofits, and corporate entities through urban design and planning initiatives such as curbside design strategies, organic waste centers within cities, and developing training and research centers to innovate methods of generating circular products following cradle to cradle principles.

2.4 FEW Design Toolkit and Techniques for Design Engagement

There is a history of urban designers working on complex and systemic questions to develop game-based tools (computational and analog) for analysis and stakeholder engagement and to enable shared visualization and design iteration and co-creation (Brković Dodig and Groat 2019). Our project has developed the (i) Urban Access Mapping Tool, (ii) a set of FEW Instrument Cards, and a (iii) participatory mapping process model for community design engagement.

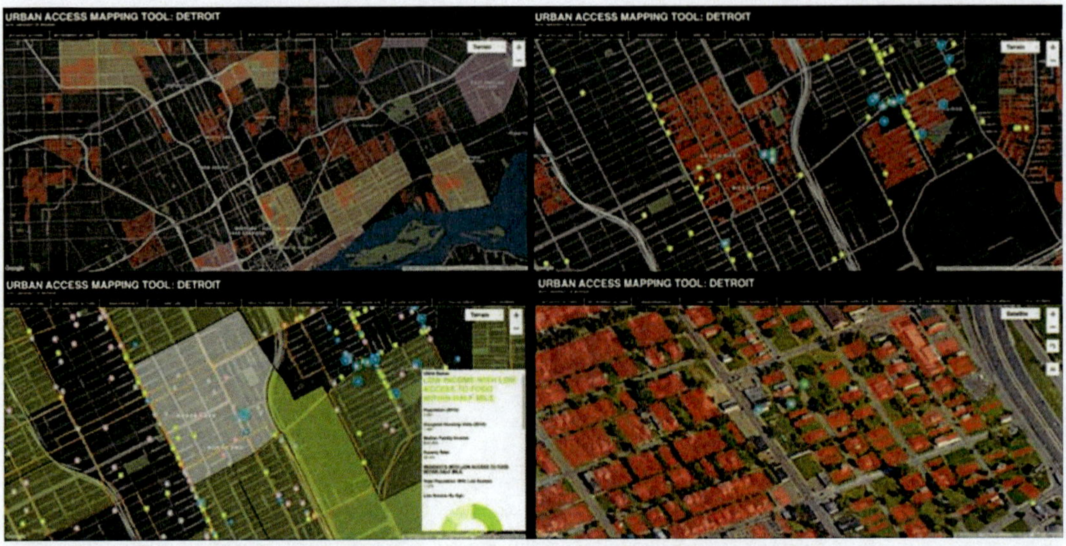

Fig. 6 Urban access mapping tool Detroit: overlays of current development projects proposed by the DPDD, residential 'inaccess hotspots', food access metrics, and access enabling actors in focus area

The Urban Access Mapping Tool (Figs. 6 and 7) aggregates data across a range of urban accessibility metrics (Food, Health, Learning and Mobility) as essential urban rights that should form a central dimension of any new urban design proposal aiming to address social justice and principles of equitable development (Independent Group of Scientists 2019; Detroit Future City 2017; EPA 2020). The tool is built on the GoogleMap API and aggregates metrics by census tract from a range of US Governmental data sources that can be accessed through a set of pull-down menus. Information regarding for profit, non-profit and NGO agencies working to deliver access across these thematic areas, as access *enabling* actors, are also spatialized within the tool (Thün and Velikov 2020). The team has prioritized issues of food access and advancement of urban agriculture and circular food production and waste systems within project site determination.

While the priority of the mapping tool is to provide rapid visually comprehensible access to urban statistics, data, and their existing spatialization in the city, FEW Instrument Cards focus on defining the programs, technologies, and products that can be incorporated into urban design proposals through multistakeholder engagement (Fig. 8). The cards depict specific system components, their relative costs, performance, configurations, spatial requirements, operational logics and systemic impacts. Cards are thematically organized across themes of *Access*: Food, Health, Learning, Mobility, and FEW systems: Food production, Energy savings and generation, and Water processing, storage, retention and Waste circularities, reduction, and processes. The Instrument Cards are at once a digital artifact recording the data gathered as part of systems research undertaken within the project and a physical design assist tool. The cards facilitate groups working on the proposed urban design projects to discuss, assess and debate and co-design project components and technologies within moments of project development and group meetings across a diverse range of consultants and stakeholders within interdisciplinary and public contexts.

As a means to prototype stakeholder engagement for FEW-based urban design proposals we undertook participatory mapping workshops that utilize both the Urban Access Mapping Tool and the FEW Instrument Cards in an iterative design exercise (Fig. 9). Teams utilized the mapping

Fig. 7 Urban access mapping tool: 2015 USDA food environment Atlas data for food accessibility by census tract; food actors, service areas within a 5 or 10 min walk of perceived high quality food retailers; details of a particular food actor

Fig. 8 FEW instrument cards

Fig. 9 Mapping process and stakeholder engagement workshops

tool to explore, assess and understand site related access issues. New insights and 'user identified' sources of data are manually added to the mapping tool to enhance the limitations of official datasets (Thün and Velikov 2020).

3 Results and Discussion

3.1 Detroit as a Living Lab: Local Stakeholders

In parallel with the multiscalar mapping and analysis process, we engaged with Detroit-based community partners to understand how their respective efforts within current developments in Detroit might serve as local test beds for urban design engaging the FEW Nexus. Each of these partners constitutes an important actor within the city that is working to advance dimensions of equitable development and socially engaged change in Detroit.

- **Oakland Avenue Urban Farm (OAUF)** identifies as the "nation's first "Agri-Cultural" urban landscape" and works toward producing "healthy foods, sustainable economies, and an active cultural environment" (Oakland Avenue Urban Farm 2020). OAUF's mission is to support sustainable cultural and ecological systems in the community to catalyze equitable redevelopment.
- **RecoveryPark**, a non-profit organization that aims to "leverage the city's underutilized assets: a large available workforce, abundant open space, access to fresh water, extensive infrastructure and a wide array of manufacturing and technology companies" towards the production of an innovative locally produced food enterprise that creates job opportunities for Detroit's residents returning to the work force (Recovery Park 2019).
- **Detroit Planning and Development Department (DPDD)** has been rethinking models of development in the city, including innovative zoning, establishment of the Detroit Land Bank Authority (2019) as a powerful tool that enables the city to be proactive in defining how development occurs, and careful community engagement processes to enable development while limiting gentrification. The team engaged with staff in discussions to identify strategic neighborhoods in which to develop the FEW Nexus urban design prototypes.

3.2 Design Proposal Development, Evaluation, and Implementation of FEW Principles

Four sites were selected that embodied project priorities: Oakland Avenue Urban Farm, RecoveryPark, Eastern Market, and the Riverbend neighborhood, the first two of which are elaborated on in this paper (Fig. 10). For each site, the team developed a speculative design proposal for urban form, density, and mixed-use infill addressing existing social models, urban farming, and secondary and tertiary food processing, through feedback with FEWprint tool analysis.

Key priorities advanced across the proposals can be summarized as:

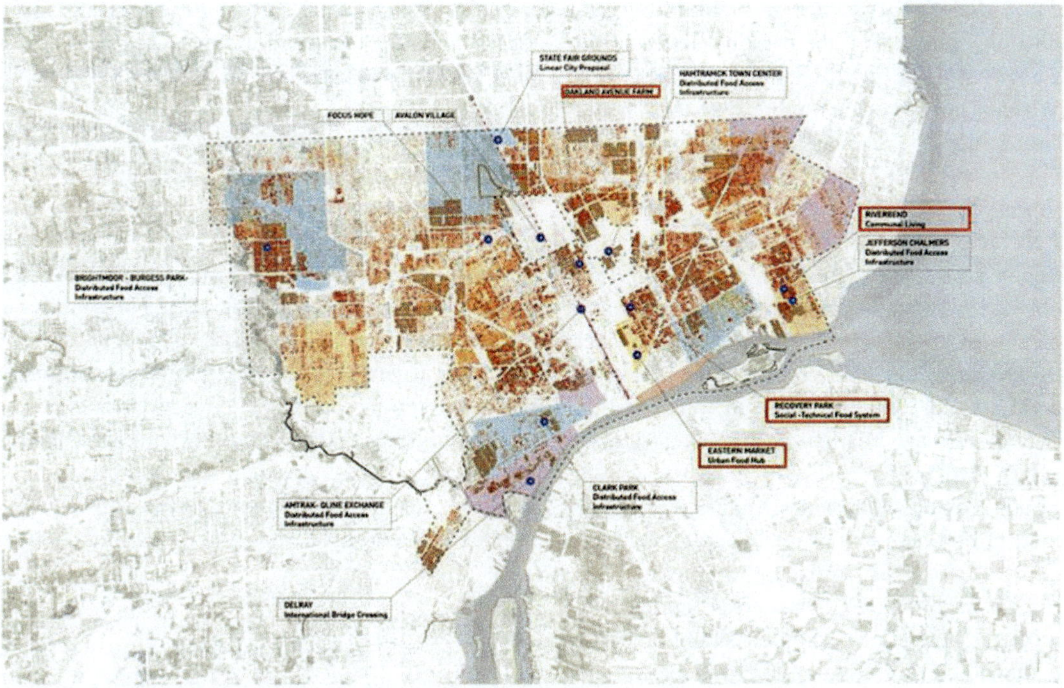

Fig. 10 Detroit design sites and proposal development distribution

- **Leveraging Vacancy and Sustainable Site Planning Practices**. Given the role of DLBA, a primary assumption is that aggressive FEW system planning guidelines can be introduced at each site. These principles also align with and advance core goals and actions in the Detroit Sustainability Action Agenda (City of Detroit 2019).
- **Equitable Food Access**. Each scheme addresses questions of access to healthy food through measures that include local production to provide a portion of the diet of residents. In addition, commercial programs are included to offer a location where healthy food can be purchased, community or commercial kitchens to enable upcycling of food products and food entrepreneurship, and spaces for locally operated educational programs that encourage healthy food practices.
- **Dietary Transformation.** Food consumption accounts for 10–30% of a household's carbon footprint and can typically be a higher portion in lower-income households (Center for Sustainable Systems 2020). Anticipating a transition toward more healthy eating practices, we assume the current U.S. national diet to describe the existing condition and project a transition toward a "universal healthy reference diet" developed by the EAT-Lancet Commission (Willett et al. 2019).
- **Domestic Energy Demand and CO_2 Emissions Reduction**. Heating and Cooling in residential buildings are estimated to account for 44% of energy consumption in U.S. homes in 2020 (US EPA 2020). Residential electricity use in 2018 emitted 666.5 mmt CO_2e, 10% of the U.S. total (US EPA 2020). All new residential buildings in the proposed scenarios have been designed to meet Passive House standards (PHIUS+ 2018), with appliance and lighting performance targets aligned with targets based on LEED v 4.1 (USGBC 2021).
- **Renewable Energy**. Each proposed scenario incorporates onsite renewable energy production with targets to offset the energy

demands associated with site-wide food production, and where there is possible, offset energy demand for residential buildings.

- **Water Conservation and Diversion**. Water conserving appliances that meet targets outlined in the LEED v4.1 standard are utilized for all new proposed residential buildings. Rainwater recovery for use in on site irrigation systems and other water demands are coupled with on-site stormwater retention and bioswales to minimize projected stormwater to district CSOs. The proposed scenario also accounts for 4% increase in precipitation due to climate change in the Great Lakes (ELPC 2019).
- **Food Waste Utilization**. Organic waste, primarily food and animal waste associated with each proposal is utilized for biodigestion (scaled according to proposed scenario) with biogas generation supplementing energy demand and residual digestate utilized for soil production and upcycling. An increase in domestic recycling fraction from 15 to 30% is assumed, based on Michigan target (MDEQ 2017).
- **Mobility CO_2 Demand Reduction**. Cars and light trucks emitted 1.1 billion metric tons CO_2e or 17% of the total U.S. GHG emissions in 2018 (US EPA 2020). Current patterns of mobility via car ownership, vehicular miles traveled, and fuel used are offset through neighborhood electric roundtrip car sharing programs, multimodal transit hubs, and reduced parking accommodation within each site plan area.

Evaluation of each scheme is undertaken through the FEWprint tool (8.5 v). The tool developed by TU Delft aims to account for the per capita CO_2e of a mixed-use development, and models future proposals based on a priority of meeting the food demands for a given neighborhood through local production (Ten Cat et al. 2022). While not all schemes developed have the capacity to meet this requirement, each is evaluated through the tool to produce a comparison of per capita and total CO_2e of the existing site condition (2021), and for the proposed development scheme (2030).

3.3 Oakland Avenue Urban Farm Proposal

Oakland Avenue Urban Farm Context: OAUF represents an emerging model where community-led activism and land assembly strategies have produced a centrally located public space coupled with food production, education, and cultural programs (Fig. 11) (Oakland Avenue Urban Farm 2020). As such it is a compelling precedent for exploring urban design and development that could be represented in other precincts within the city.

OAUF Design: This urban design scenario builds on the masterplan developed by Akoaki (https://www.akoaki.com/) and develops a set of strategies for producing new single-family detached homes and townhouses that densifies development surrounding the 'farm' as a form of open cultural park focused on sustainable food and cultural production and a community-managed argi/cultural public space (Fig. 12). This prototype is aimed for potential replication around similar high-vacancy contexts anchored by local community non-profit organizations as a new model of 'platting plan' within Detroit's former residential areas and focusses on incremental infill of vacant lands coupled with mixed use reinforcement of existing arteries. The precinct incorporates innovations in water resource management, micro-livestock rearing facilities, and renewable distributed energy generation. New housing typologies are incorporated for indoor and outdoor growing and low-cost high-performance building envelope systems. Within the site boundary, the existing population is estimated at 427, with an average household size of 2.2. The projected population residing within the site is envisioned to be 2,676, with an average household size of 3.3.

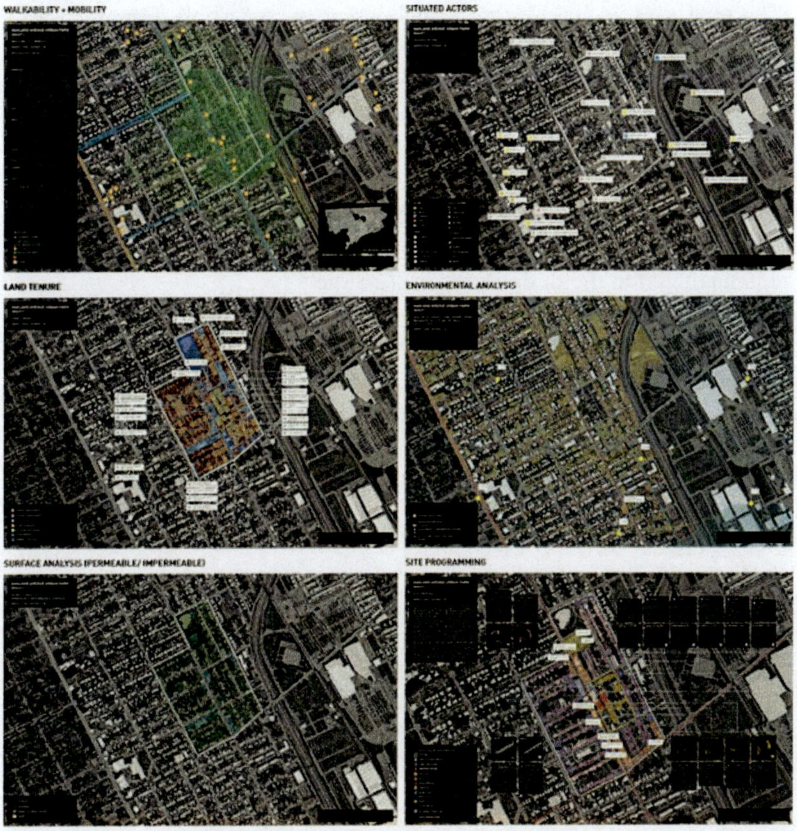

Fig. 11 Site Analysis: Oakland Avenue Urban Farm. (from Top Left) Walkability and mobility, land tenure. Environmental analysis, surface analysis (Permeable/Impermeable), and Site Programing

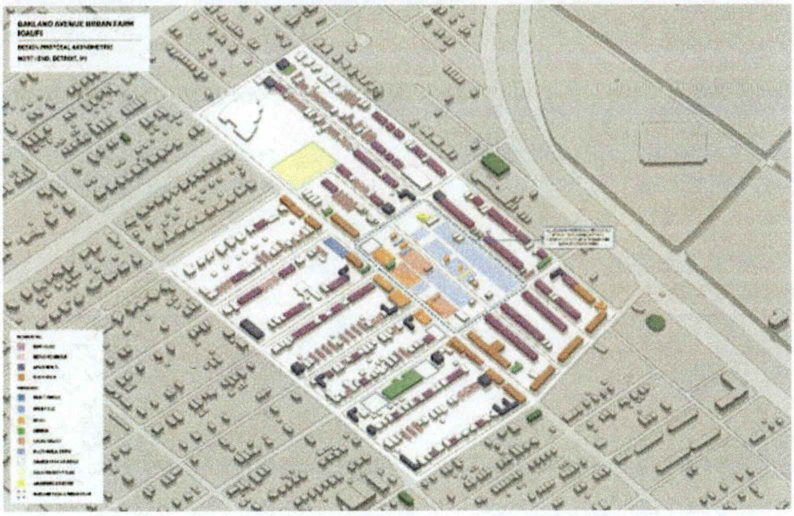

Fig. 12 Design proposal: Oakland Avenue urban farm urban design massing

The total area for the design development is 36.7 ha. The project includes a series of new housing developments including single detached homes, townhouses, apartments, and mixed-use housing accounting for 86,937 m^2. The new housing is designed to passive house standards. The neighborhood also includes existing single detached homes amounting to 20,370 m^2 which remain unaltered in the design proposal. All new residences have access to either backyard farming or greenhouse cultivation. The greenhouse structures are attached to single family homes and on rooftops of apartments and mixed-use buildings. The total urban farming area of 68,938 m^2 includes 56,966 m^2 of open-field farming, 11,919 m^2 of greenhouse cultivation, and 54 m^2 of chicken production. The commercial corridors along Oakland Avenue and Owen streets contain retail space accounting for 3626 m^2. The project proposes 225 round-trip carshare parking spots as an attempt to reduce the need for personal vehicle ownership within the site boundary (Fig. 13).

OAUF Discussion: FEW Systems Overview: FEWprint tool (v8.5) analysis indicates the proposed design can reduce the per capita carbon emission from 10,975 (existing scenario) to 2,975 kg/year. The total food produced in this neighborhood is 233 tonnes/year. The scheme fulfils 39% of vegetables, 38% of fruits, 31% of pulses and legumes, 32% of starchy roots, 4% of

Fig. 13 Oakland Avenue Urban Farm: FEW urban design visualizations. (Clockwise from Top) streetscape infill, urban agriculture as cultural space, OAUF social infrastructure

eggs, and 8% of nuts and seeds of the Lancet diet for residents within the project boundary. In terms of local energy generation, the project includes (i) a small solar farm at the north of the site, and (ii) rooftop photovoltaic on new single detached homes and townhouse typologies, generating 3,991 MWh/year of electricity. A small-scale floating drum anaerobic digester of 464.4 m^2 with containers holding 87.56 m^3 is proposed. The anaerobic digester generates 52,284 m^3/year of biogas energy and 38 tonnes/year of digestate as a nutrient-rich fertilizer for urban farming. Through rooftop rainwater collection, the project collects 25,181 m^3/year of rainwater which is directed towards urban farming. The project envisions a change in the domestic recycling fraction from 15% (current MI standards) to 30% in the design proposal. Further, 100% of organic domestic waste is processed by the anaerobic digester (Fig. 14).

3.4 RecoveryPark Proposal

RecoveryPark Context: RecoveryPark Farms is a for-profit urban farming initiative in Detroit that integrates technical farming systems with large-scale greenhouse production (Fig. 15). RecoveryPark Farms has commissioned Hamilton Anderson Associates (https://www.hamilton-anderson.com/), to masterplan the expansion of its food production facility. The master plan of 123 ha between Canfield and Harper Street has been prepared to attract future investment and philanthropic support for its efforts which focus on community job creation within a FEW-nexus paradigm.

RecoveryPark Design: The team has modified the design proposal by Hamilton Anderson to advance FEW-nexus considerations. The design proposal includes the masterplan area and proposes an additional residential zone and a solar farm on the former Detroit Incinerator facility, increasing the area to 135 ha. This prototype takes the form of a large-scale urban agriculture campus including housing for transitional agricultural workers in recovery. The core components of the design strategy include high-performance greenhouses, open field farming, rooftop rainwater collection systems, bioswales connecting segments of existing blue-green infrastructure, an autonomous geothermal system to heat greenhouses, and solar power generation. Within the site boundary, the existing population is estimated at 338, with an average household size of 2.2. The projected population for this design scheme is 2,209, with an average household size of 3.88. The project proposes 72

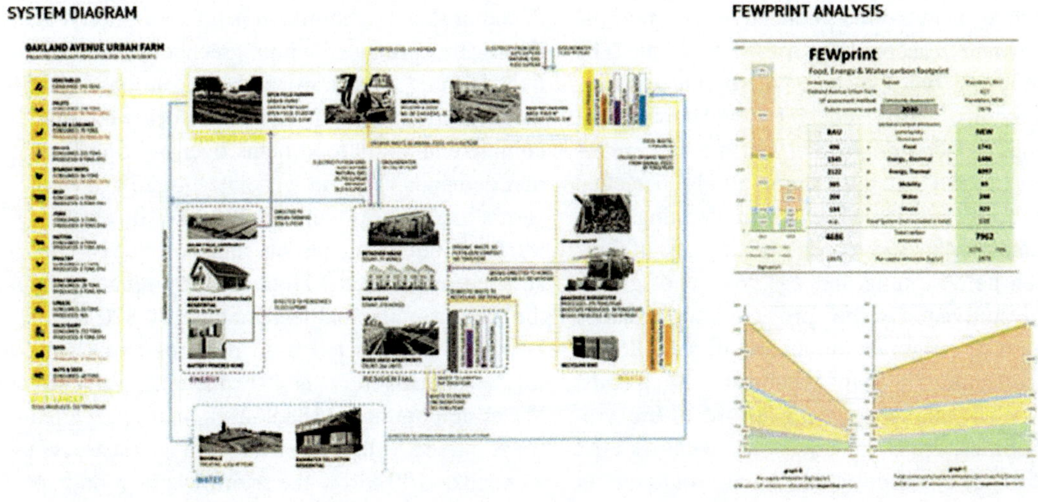

Fig. 14 FEW system diagram and FEWprint analysis: OAUF

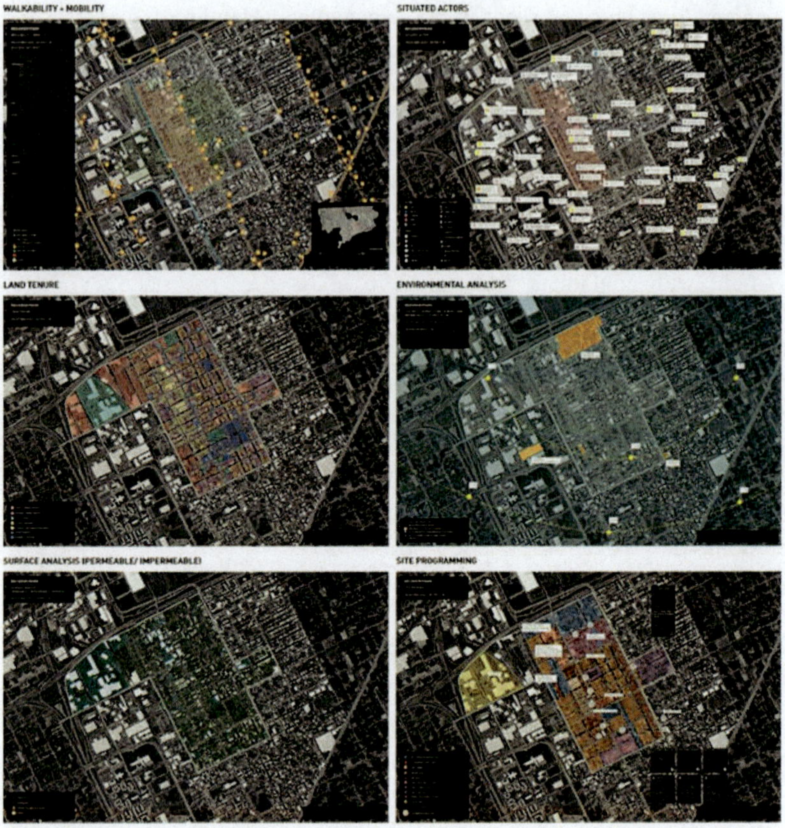

Fig. 15 Site analysis: RecoveryPark. (from Top Left) walkability and mobility, situated actors, land tenure, environmental analysis, surface analysis (Permeable/Impermeable), and Site Programing

round-trip carshare parking spots as an attempt to eliminate personal vehicle ownership within the site boundary.

The project includes a series of housing types designed to Passive House standards that are supportive of individual, collective, and familial arrangements, accounting for 85,610 m^2. The new housing types include micro-unit housing, single-family infill housing, and multi-unit housing, in addition to high-performance renovations to the existing single-family detached homes on the site. The urban farming area accounts for 484,368 m^2, comprising 190,405 m^2 of open field farming and 293,964 m^2 of greenhouse cultivation. The project also includes commercial programs comprised of a 3,136 m^2 salvage center, 7,888 m^2 administration campus, 4,607 m^2 of market space, 6,410 m^2 of processing facilities, 2,857 m^2 equestrian facilities, and 2,880 m^2 of community education programming (Figs. 16 and 17).

Recovery Park Discussion: FEW Systems Overview: FEWprint tool (v 8.5) analysis indicates the proposed design can reduce the per capita carbon emission from 53,947 (existing scenario) to 4,384 kg/year. The total food produced in this neighborhood is 4,656 tonnes/year. Food items produced within greenhouse or open-field farming are prioritized based on the market value and productivity, with a focus on high commercial value food items over meeting local diet demands. The site generates 46,252 MW of geothermal energy, and 17,286 MW of electricity from photovoltaics on roofs of newly constructed or renovated homes and by repurposing the Detroit municipal incinerator. 160,820 m^3 of rainwater is collected from rooftops and is utilized for irrigation and residential purposes. Water amounting to 188,962 m^3/year is diverted from streets into bioswales and constructed wetlands. 100% of the stormwater runoff is diverted from the central wastewater treatment

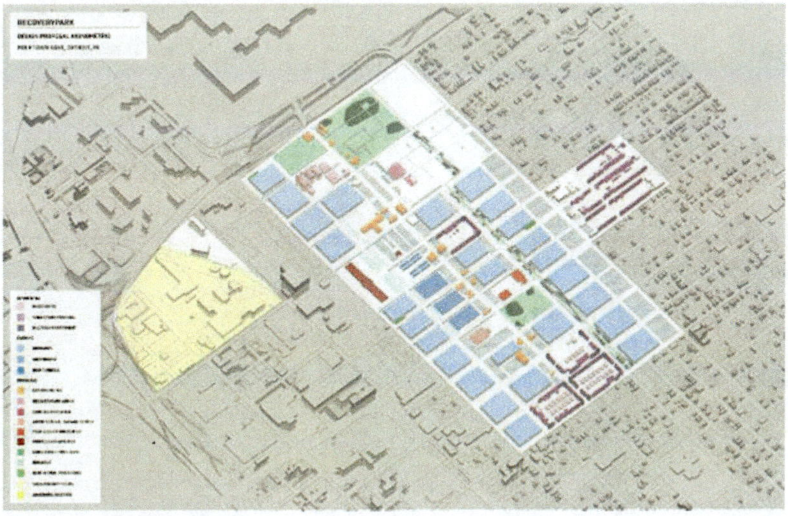

Fig. 16 Design proposal: RecoveryPark

Fig. 17 RecoveryPark: FEW urban design visualizations. (Clockwise from top) Constructed Wetlands as Public Space, multi-unit and accessory housing, Greenhouse Urbanism

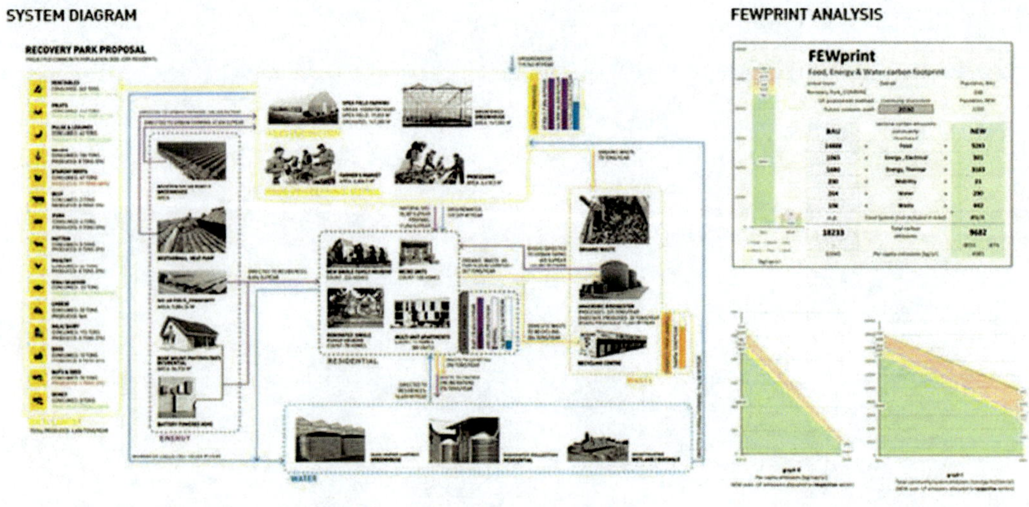

Fig. 18 FEW System diagram and FEWprint analysis: RecoveryPark

plant. 100% of organic domestic waste is processed by the anaerobic digester. An anaerobic digester on a campus of 580 m^2 utilizing containers holding 71.9 m^3 is proposed to generate 42,880 m^3/year of biogas energy and 32 tonnes/year of digestate as fertilizer for urban farming (Fig. 18).

4 Conclusion

“As a result of these cybernetic, sub-theoretical developments, many architects wanted to design systems but, on the whole, they were expected to design buildings…there is a sense in which the brief given to an architect has widened during the last decades,” wrote Gordon Pask (Pask 1969). Architectural designers, whose training has traditionally focused on the design and crafting of physical objects and spaces, even today have few shared toolsets and frames of reference with which to approach design within entangled and multiscalar situations such as the UN SDGs implicate, not to mention the need for data-driven impact assessment that informs design. And yet, it is precisely the architectural designer’s ability to think iteratively, systemically, and to shape and structure complex and competing entities into synthetic propositions with larger teams that is necessary for the task of working on sustainable development and urban design today.

This paper outlines a collaborative design process that centers how FEW systems-based urban design benefits from a design led process that includes a multi-perspectival lens, stakeholder engagement, and specific computational and tools that evaluate and integrate context specific issues beyond those afforded by stock and flow analysis of current engineering-led methods. Tools such as the FEW print evaluation framework offer models for ways in which design-led solutions can be quantified in terms of CO_2e outcomes to inform decision making and scenario evaluation. Key strategic considerations such as the challenges of food access and policy-based limitations of urban composting within the case of Detroit point to ways in which urban design efforts can center and surface policy questions in the development of FEW-Nexus based schema using design scenario possibilities. Through the methods and approach outlined in this paper, the authors are hopeful that a broader community of designers may engage with aspects of FEW Nexus urban design strategies to explore the ways in which SDGs might be leveraged to shape local change through a systems-based approach to equitable design transformation in post-industrial city contexts.

Acknowledgements Project: Moveable Nexus (M-Nex): Design-Led Urban Food, Water And Energy Management Innovation In New Boundary Conditions Of Change.

International Institutional Partners and PIs: University of Michigan (U-M): Geoffrey Thün, Delft University of Technology (DUT): Prof. Andy van den Dobbelsteen, KEIO University (KU): Prof. Wanglin Yan, Queens University Belfast (QUB): Prof. Greg Keeffe, Qatar University (QU): Prof. Sami Sayady.

U-M/RVTR Design Research Team: Kathy Velikov (CoPI), Tithi Sanyal (Research Associate), Ife Owolabi, Linda Lee, Lucas Denit, Andrew Kremers, Jiushuai Zhang Dan McTavish, Mary O'Malley.

Funding: M-Nex is a granted project of the collaborative research area Belmont Forum (1314551). We appreciate and acknowledge the financial support of the U.S. National Science Foundation (1832214) and the University of Michigan. Any opinions, findings, and conclusions or recommendations expressed are those of the authors and do not necessarily reflect the views of sponsoring agencies.

References

Allnutt B (2019) A black-led food co-op grows in Detroit. City Lab. https://www.citylab.com/equity/2019/01/black-owned-food-coop-detroit-dbcfsn-d-town-farm-yakini/580819/. Accessed 21 Feb 2019

Brković Dodig M, Groat LN (eds) (2019) The Routledge companion to games in architecture and urban planning: tools for design, teaching, and research, 1st edn. Routledge. https://doi.org/10.4324/9780429441325

Budzynska K, West P, Savoy-Moore RT, Lindsey D, Winter M, Newby PK (2013) A food desert in Detroit: associations with food shopping and eating behaviours, dietary intakes and obesity. Public Health Nutr 16(12):2114–2123

Center for Sustainable Systems, University of Michigan (2020) Carbon footprint factsheet. Pub. No. CSS09-05. http://css.umich.edu/sites/default/files/Carbon%20Footprint_CSS09-05_e2020_0.pdf. Accessed 20 Feb 2020

City of Detroit (2019) Detroit sustainability action agenda. https://detroitmi.gov/sites/detroitmi.localhost/files/2019-06/Detroit-Sustainability-Action-Agenda-Web.pdf. Accessed 10 June 2021

Detroit Food Policy Council (2016) Creating a food secure Detroit: policy review and update. Urban Press Detroit, Detroit, MI. https://detroitfoodpolicycouncil.net/knowledge-center/reports. Accessed 20 Feb 2019

Detroit Food Policy Council (2017) Detroit food metrics Report 2017. Detroit Food Policy Council and Detroit Health Department, Detroit, MI. https://detroitfoodpolicycouncil.net/knowledge-center/reports. Accessed 21 Feb 2019

Detroit Future City (2017) 139 square miles (20–71). Island Press, Detroit, MI

Dewar M, Thomas JM (eds) (2012) The city after abandonment. University of Pennsylvania Press, Philadelphia, PA

DLBA (2019) Policies & procedures. https://buildingdetroit.org/our-policies/. Accessed 31 Jan 2020

EPA (2020) Smart growth and equitable development. https://www.epa.gov/smartgrowth/smart-growth-and-equitable-development#others. Accessed 29 Apr 2019

Hoey L, Sponseller A (2018) "It's hard to be strategic when your hair is on fire": alternative food movement leaders' motivation and capacity to act. Agric Hum Values 35(3):595–609

Independent Group of Scientists appointed by the Secretary-General (2019) Global Sustainable Development Report 2019: the future is now—Science for achieving sustainable development. United Nations, New York, NY. https://sustainabledevelopment.un.org/globalsdreport/2019. Accessed 27 Apr 2020

Liang S, Qu S, Zhao Q, Zhang X, Diagger G, Newell J, Miller S, Johnson J, Love N, Zhang L, Yang Z, Xu M (2019) Quantifying the urban food-energy-water nexus: the case of the Detroit metropolitan area. Environ Sci Technol 53(2):779–788. https://doi.org/10.1021/acs.est.8b06240

Liu J, Hull V, Godfray HCJ et al (2018) Nexus approaches to global sustainable development. Nat Sustain 1:466–476. https://doi.org/10.1038/s41893-018-0135-8

Michigan Department of Environmental Quality (MDEQ) (2017) State of Michigan's Environment. https://www.michigan.gov/-/media/Project/Websites/egle/Documents/Reports/Department/2017-triennial-report.pdf?rev=be9d5a2ba80846299170f1f302ba4c63. Accessed 17 Feb 2019

Oakland Avenue Urban Farm (2020) About. http://www.oaklandurbanfarm.org/about.html. Accessed 20 Feb 2019

Pask G (1969) The architectural relevance of cybernetics. Archit Des, 494–496

PHIUS+ Multifamily Quality Assurance Workbook (v2.2) (2018). https://www.phius.org/phius-certification-for-buildings-products/project-certification/documents-for-download. Accessed 5 June 2020

Pothukuchi K (2011) The Detroit food system report. Detroit Food Policy Council, Detroit, MI. https://detroitfoodpolicycouncil.net/knowledge-center/reports. Accessed 20 Feb 2019

Pothukuchi K (2015) Five decades of community food planning in Detroit: city and grassroots growth and equity. J Plan Educ Res 35(4):419–434

Recovery Park (2019) About. https://www.recoverypark.org/. Accessed 20 Feb 2019

Roggema R, Yan W, Keeffe G (2021) A moveable nexus: framework for FEW-design and planning. In: TransFEWmation: towards design-led food-energy-water systems for future urbanization. Springer, pp 9–37. https://doi.org/10.1007/978-3-030-61977-0_2

Sugrue TJ (2005) The origins of the urban crisis: race and inequality in Postwar Detroit. Princeton University Press, Princeton, NJ

Ten Caat PN, Tenpierik MJ, Sanyal T, Tilie N, van den Dobbelsteen A, Thün G, Cullen S, Nakayama S, Karanisa T, Monti S (2022) Towards fossil free cities–Emission assessment of food and resources consumption with the FEWprint carbon accounting platform. Clean Environ Syst 4:1–26. https://doi.org/10.1016/j.cesys.2022.100074

Thomas JM, Bekkering H (2015) Introduction: land and change in Detroit. In: Thomas JM, Bekkering H (eds) Mapping Detroit, land, community and shaping a city. Wayne State University Press, Detroit, MI, pp 1–6

Thün G, Sanyal T, Velikov K (2021) Mapping the food energy water nexus across cascading scales: contexts for Detroit from region to city. In: Roggema R (ed) TransFEWmation: towards design-led food-energy-water systems for future urbanization. Springer, London, pp 171–207

Thün G, Velikov K (2020) Rethinking the design and delivery of social infrastructure-A systems based methodology and proposition. In: Lastman R (ed) AMPS proceedings series 19.1. The city and complexity—Life, design and commerce in the built environment. AMPs, London, UK, pp 388–399

Thün G, Velikov K (2017) Shed cartographies: a methodology for regional research. In: Waldheim C, Lyster C, White M, Ibáñez D (eds) Third Coast Atlas: Prelude to a plan of the Great Lakes Region. ACTAR, Barcelona, ES, pp 84–95

Thün G, Velikov K, Ripley C, McTavish D (2015) Infra Eco Logi Urbanism: a project for the Great Lakes Megaregion. Park Books, Zurich CH

UNDP SDG Integration (2022) About. https://sdgintegration.undp.org/about. Accessed 5 OCT 2022

U.S. Environmental Protection Agency (EPA) (2020) Inventory of U.S. Greenhouse Gas Emissions and Sinks 1990–2018

USGBC LEED v4.1 Building Design and Construction (2021). https://www.usgbc.org/leed/v41. Accessed 10 May 2021

Willett W, Rockström J, Loken B, Springmann M, Lang T, Vermeulen S, Garnett T et al(2019) Food in the anthropocene: the EAT–Lancet Commission on healthy diets from sustainable food systems. Lancet (British edition) 393(10170):447–492 (Research Support, Non-U.S. Gov't. Elsevier Ltd., New York)

Yan W, Roggema R (2019) Developing a in design-led approach for the food–energy–water nexus cities. Urban Plan 4:123

Can Digital Matchmaking Boost Circular Construction? Lessons from Reusing the Glass of Centre Pompidou

Catherine De Wolf, Sultan Cetin, and Nancy Bocken

Abstract

Digitalization is driving innovation towards a circular economy in various industries—but the construction industry is lagging behind. The building industry, a growth sector due to increasing urbanization, is at the same time actively depleting our resources, generating waste, and emitting greenhouse gases at a tremendous scale and speed. This chapter argues that we must urgently shift from a linear take–make–waste model to a circular one whereby we utilize our resources wisely and keep them from becoming waste. The experience of reusing the glass from the Centre Pompidou in Paris, France, confronted the architects with the many challenges we face when renovating a building with circular principles. Finding architects to use the iconic bent glass instead of crushing it for recycling (or worse, for disposing it in a landfill) turned out to be a time-consuming task. Adopting artificial intelligence and digital information sharing to match materials for reuse with people who can reuse them is exactly what the construction industry needs for a paradigm shift towards circularity.

Keywords

Reuse · Centre Pompidou · Circular construction · Digital matchmaking

C. De Wolf (✉)
ETH Zurich, Department of Civil, Environmental and Geomatic Engineering, Zurich, Switzerland
e-mail: cdewolf@ethz.ch

S. Cetin
TU Delft, Faculty of Architecture and the Built Environment, Delft, Netherlands

N. Bocken
Maastricht University, Maastricht Sustainability Institute (MSI), Maastricht, Netherlands

1 Introduction

It is evident that the building industry is a growth sector due to increasing urbanization and yet, at the same time, one that is actively depleting our resources, generating waste, and emitting greenhouse gases at a tremendous scale and speed. This essay argues that we[1] must urgently shift from a linear take–make–waste model to a circular one whereby we rethink our resources wisely and keep them from becoming waste. Adopting digital innovation is exactly what the construction industry needs in order to achieve this shift (World Economic Forum 2022). This also responds to Sustainable Development Goals such as 'Industry, innovation and infrastructure', 'Sustainable cities and communities',

[1] By "we," I am referring to key actors of the construction sector: designers, architects, engineers, contractors, etc.

M. R. Thomsen et al. (eds.), *Design for Rethinking Resources*, Sustainable Development Goals Series,
https://doi.org/10.1007/978-3-031-36554-6_42

'Responsible consumption and production', and 'Climate action'.

Today, we construct buildings using "linear economy" principles: we extract resources from the earth, produce construction materials, use them, and then dispose of them when we no longer need them. In the process, we deplete global resources such as sand (United Nations Environment Programme 2019). What's more, we produce a huge amount of waste: in 2018 and in the United States alone, we generated 600 million tons of construction and demolition waste (United States Environmental Protection Agency 2016). We also intensify the climate crisis: the building sector accounts for around 38% of our global greenhouse gas (GHG) emissions (United Nations Environment Programme 2020). Of course, we know this results in a loss of value and resources, but we clearly haven't sufficiently considered how to fix this costly problem. And it is getting worse. By 2060, we expect to add 2.4 trillion ft^2 (230 billion m^2) of floor area due to the growing population, urbanization, and higher demands—this is equivalent to constructing one New York City every month for 40 years (Architecture 2030). The question is, how do we break the pattern of resource depletion, waste generation, and pollution in the construction sector?

A shift to a "circular economy" model would yield maximum value from building materials by extending their service life through maintenance and repair and reusing them as new resources when they reach the end of their life. Yet, while pioneering circular projects have been promising, a static industry with little cross-sector communication impedes the practical implementation of circular construction on a large scale (Nußholz et al. 2020). This was not always the case: reuse was common until about a century ago, when, due to the high price of land in dense urban areas like Manhattan, we began demolishing buildings rather than disassembling them. Because of the time required to carefully disassemble buildings and reuse their materials, we have moved away from the circular construction model. But it doesn't have to be that way.

Digital technologies can help us disassemble and reuse much quicker, cheaper, and smarter than before. What if we used Building Information Modeling (BIM), to store information about materials available for reuse? What if we used artificial intelligence (AI) to match people wanting to design with reused building materials with people who have materials from demolition sites? How about we apply algorithms to find the right match between supply and demand? If we had better management tools, we could use *buildings* instead of the *earth* as a mine: the buildings would become depots and banks of materials.

The construction industry has been slow in adopting digital innovation, even though it is everywhere and growing: blockchain technology is now used by over 20% of traditional banks (Graffeo 2021), the Internet of Things is shaping smart cities in the making, and AI can be found everywhere from online marketplaces like Amazon to matchmaking platforms like Tinder. Though designers in these fields are taking considerable steps forward, the construction industry has not explored these technologies enough to solve its biggest current challenge: transitioning the built environment to a circular economy. The Excess Materials Exchange (EME) platform is an example of a digital marketplace that uses resource passports so companies can exchange excess materials with each other (Excess Materials Exchange 2022). This concept of digital matchmaking, plus exploring technologies such as AI and blockchain, needs to be extrapolated to the building industry.

The construction sector could use emerging digital technologies to make this transition happen. In academia and practice, new digital tools are being developed or used to manage circular construction. Take the Centre Pompidou[2] in Paris, France, for example (Fig. 1). The architects faced many challenges when renovating the structure's façade with circular principles. They needed to replace the bent glass due to changing

[2] The author's own experience is described in the documentary Samsara (Circular Engineering for Architecture and DiCE 2021).

Fig. 1 Renovation of façade and reuse of its glass from the centre Pompidou by maximum, Paris, France. ©Alexandre Attias and maximum architecture

security norms. The value chain in today's construction industry does not make it easy to reuse this glass. Often, it gets thrown away or, in the best-case scenario, recycled. Surely, there must be architects who dream of reusing such iconic bent glass instead of crushing it for recycling or, worse, disposing it in a landfill.

Spoiler: the architects were able to save the glass (as you can see on the right side of Fig. 1), but not without time, passion, and dogged persistence. Could digital technologies have helped these architects navigate this process more easily? This essay argues that we need digital innovation to match materials with users to make circular construction possible.

2 Transitioning to Circular Construction

Narrow, slow, close, regenerate. Let us start with the basics. What is a "circular economy"? It is a system that supports sustainable development to secure the resources to provide for current and future generations. This is achieved by applying four distinct strategies: narrowing, closing, slowing, and regenerating resource loops (Bocken et al. 2021). **Narrowing** the loop uses fewer resources per product, e.g., through structural optimization, so that fewer materials are needed. An example is the Guastavino vaulting system found in hundreds of buildings across the United States, which uses a uniquely thin layer of tiles (Ochsendorf 2010). **Slowing** the loop is about using buildings for longer, e.g., through adaptive reuse of building spaces. An example is Apple's reuse of a historic theatre building in Los Angeles, United States (Foster + Partners 2021). **Closing** the loop ensures that materials are used again, e.g., through reusing materials after demolition. An example is the reuse of window frames from all EU nations in Samyn and Partners' European Council building in Brussels, Belgium (Samyn and Partners 2016). **Regenerating** improves natural ecosystems in which materials exist, e.g., through green facades that help purify the air. An example is the Torre de Especialidades in Mexico City, Mexico, built with Made of Air technology sucking up air pollution (Made of Air 2022). Most of these strategies have been applied only in niche projects, but a collaboration, information sharing, and automation would enable building designers to take a more holistic, circular approach. In this essay, the case of Centre Pompidou's renovation shows how we can 'close' the loop by reusing the bent glass in Maximum Architecture's new project (Fig. 1), but we could also note how its structure already 'narrows' the loop by optimizing water-filled (for avoiding fire-protection) steel columns, how its renovation 'slows' the loop by keeping the building in use longer, and how its presence 'regenerates' the neighbourhood from a social perspective by hosting culture.

Monitor, Manage, Match. If we would like to automate circular construction, or automate the reuse of the bent glass from the Centre Pompidou, we need to: (a) collect and **monitor** data—we need to know what the materials in our buildings are, what resources we use, and their condition; (b) **manage** data—we need to know when these materials and resources become available and where we can store and distribute them; and (c) find the right **match** between supply and demand—we need to know who has materials to offer and who wants to turn them into resources. These monitoring, managing, and matching strategies are already being applied at the scale of an individual building in the Swiss Federal Laboratories for Materials Science and Technology's (EMPA) experimental NEST building, for example. This building is a full-scale experiment, showing the latest innovation in Swiss research and industry. Two particular NEST units demonstrate how circular strategies can be managed in construction: in the Urban Mining & Recycling (UMAR) unit, all resources are reusable, recyclable, or compostable; and in the Sprint unit, Covid-19 flexible offices were designed with largely reused materials (Swiss Federal Laboratories for Materials Science and Technology (EMPA) and Swiss Federal Institute of Aquatic Science and Technology (Eawag) 2018, 2021).

Today, the work of predicting recoverable materials is performed by a limited pool of experts who measure and collate data manually. In this field, efficiency is key, so it is important to understand the available components as quickly as possible in order to connect with potential end users. Today's matchmaking methods are manual and very time-consuming. For example, pioneering reuse practitioners in Switzerland such as Baubüro In Situ hire 'material hunters' who go around the country looking for poles indicating demolition sites[3] and then contact the building owners (De Wolf et al. 2020). There is not yet a feasible circular matchmaking strategy that can be broadly applied in construction practice, but digital technology could improve efficiency of the process.

3 Use Digital Technologies for More Efficient Matchmaking

Digital technologies could have better streamlined the reuse of the glass from the Centre Pompidou, especially for digital matchmaking, digital tracking, and creating online connections. Numerous digital technologies already exist that could support those endeavours. Learning from this reuse experience, laboratories[4] have been created to explore how digital innovation can foster a circular economy.

We need to connect the end-of-life of buildings with the start-of-life of other buildings making the reuse of building materials more effective, user-friendly, and widespread. Existing platforms list materials available for reuse, but in practice, the major challenge is finding the right match. Companies struggle to find architects who design with reused materials, construction sites in which the materials fit, contractors with the skills needed to reuse them, and so on. To solve this, we need to create a **'Tinder' for materials, buildings, and people**. Utilizing digital tools that have proven their feasibility and effectiveness in other sectors, we need to develop algorithms adapted to the construction sector to tackle the challenge of **matching supply and demand**,

[3] In Switzerland, construction sites must be staked out with poles to show how high a new construction will become.

[4] Examples are the Circular Engineering for Architecture (CEA—www.cea.ibi.ethz.ch) lab and the Digital Circular Economy (DiCE—www.dice-lab.com) lab.

acting as 'matchmakers' for owners, users, designers, suppliers, and workers across the entire value chain.

In the last several years, **Industry 4.0**, also known as the "fourth industrial revolution", has triggered a paradigm shift, particularly in the manufacturing sector, and has emerged as a promising technology framework used for integrating and extending manufacturing processes at both intra- and inter-organizational levels through the use of cyber-physical systems (Xu et al. 2018). Tracking devices even enable us to use the **Internet of Things (IoT)** to gather, store, and transmit data on the condition and availability of materials such as Centre Pompidou's glass using cloud computing and wireless sensor networks. The application of IoT in a circular economy is dispersed across various fields, covering topics from smart cities (Larson 2022) to sustainable product lifecycle management (Ingemarsdotter et al. 2019).

If **online platforms**[5] that list all the materials locally available for reuse were well known by all stakeholders in the supply chain, architects such as those at Centre Pompidou could list their glass on such a platform. By connecting supply and demand, digital sharing platforms and marketplaces create circular markets, thus facilitating communication and collaboration between value chain actors. Ideally, we need access to digital representations of a built asset, what we often call **Building Information Modeling (BIM)**. However, the use of BIM is not as widespread as the sector would have hoped today. Its use would help to create what we call a "**material passport**" that includes information about the components in a building such as material type and properties, dimensions and geometry, quantity, condition, location, etc. A material passport with digitally registered data describing a component's characteristics, location, history, and ownership status would serve to facilitate the sharing or selling of materials such as the glass from Centre Pompidou on a digital platform. The Madaster platform, which provides owners incentives to record their data and store the passports of buildings, is an example of such an initiative (Madaster 2022). Material passports can be connected to a digital twin, a virtual replica of the physical world, which is already commonly used in the automobile, aerospace, and process industries to simulate performance (Çetin et al. 2021). Such a connection can potentially extend the service life of building elements through predictive maintenance. Other platforms, such as Material Mapper, provide insight into all buildings about to be demolished, rebuilt, or newly built so that the type and quantity of materials available for reuse can be estimated (Material Mapper 2022). Many other platforms exist on the market, but we lack standardization on how to map materials and globally measure circularity so that these platforms may truly thrive.

To help facilitate resource and waste optimization by matchmaking materials available for reuse with designs from reused materials, we can use AI, and in particular **Machine Learning (ML)**, in combination with **Big Data Analytics**. ML trains algorithms to learn from data and identify patterns for decision-making. More specifically, deep learning, a subset of ML, is based on neural networks. While used regularly in other sectors, it has been recently applied to building demolition records' datasets (Akanbi et al. 2020). Though not a common use case in the field of **computer vision**, which trains computers to interpret and understand the visual world, some research has brought computer vision methodology to the construction site too. We can learn a lot from the many images that exist of the Centre Pompidou. For any building, even things like simple Google Street View images, social media images, historical archives, or national demolition records can enable us to map and match materials generated from deconstruction and demolition projects (Structural Xploration Lab 2018). Algorithms can indeed identify patterns in the data for building components' end-of-life treatments in order to make the right match (Raghu 2021). Companies such as FaSA (Facade Service Applicatie 2022) and Sportr.ai (Spotr 2022) use **AI** to predict the

[5] For example, www.useagain.ch/links/, www.rotordc.com/store/, or www.cycle-up.fr.

Fig. 2 QR code engraving in reused beams for tracking and tracing material passports for matchmaking. ©Anna Buser

need for façade maintenance. ZenRobotics uses AI to build smart sorting robots for recycling (ZenRobotics 2022). **Geographical Information Systems (GIS)** are often used as a data set to identify, map, and manage resources in building stocks for future reuse or recycling. This system is also used to support urban mining in the built environment. For example, GIS data of the city of Zurich is currently being used to identify the materials available in buildings (Raghu 2022). Using AI techniques, scenario modelling regarding future demolitions can help predict which materials will become available for reuse. For a further level of detail, drone imagery can be used in combination with AI; this is what businesses such as AeroScan offer to assess the condition of building assets (Aeroscan 2022).

Another challenge in reusing materials such as the glass from the Centre Pompidou is the transaction itself. The construction industry's supply chain is highly fragmented, and designers are reluctant to share information about their projects. Researchers are exploring **blockchain technology** (Tapscott and Vargas 2019; Hunhevicz et al. 2020; Li and Kassem 2021), a secure distributed peer-to-peer system that enables transparent value transactions without the need for central authorities and intermediaries, as a way to trace materials. Securely sharing data on where products have been and where they are shipped to along the supply chain is also what TradeLens, powered by IBM, is doing in the global trade and shipping industry (TradeLens 2022). Circulor also enables the traceability of supply chains through blockchain and tagging (Circulor 2022). Indeed, tagging technologies such as Quick Response (QR) codes, Near-Field Communication (NFC), and Radio Frequency Identification System (RFID) chips enable this transparent material tracking. In pilot projects that reuse building materials, such as Buildings As Material Banks (BAMB) (2022) or the ETH Reuse Dome (Gordon et al. 2022), QR codes are used to track and trace materials through different building life cycles (Fig. 2).

4 Collaborate and Be Critical

Complexity & interdependencies. Buildings are unique in the sense that they have a long lifespan, are composed of many parts (site, structure, façade, finishes, furniture), and use diverse resources (materials, land, energy, water, nutrients). Consequently, many stakeholders are involved, including architects, developers, designers, engineers, governments, occupants, clients, contractors, material suppliers, and demolishers. As a result, the industry is extremely compartmentalized. If it is going to transition to a circular model, construction designers need to collaborate with computer scientists, lawmakers, engineers, business developers, and others. Digital tools, such as BIM and IoT, could make it easier for designers to communicate, coordinate, fabricate, and visualize, but not in isolation. One digital technology is rarely used without another: e.g., Material Passports are created using BIM and can be connected to the IoT to sell reused elements on digital platforms.

New business models. Visionary actors in the construction sector know they must reinvent their business models to transition to a circular economy. European research project Circular X presents a series of business models as an inspiration for circular experimentation in construction, including buildings-as-a-service, movable housing solutions, and temporary material banks (Circular 2022). Buildings can be seen as trees by "supplying energy and raw materials for reuse" (Kraaijvanger 2022). The digital technologies discussed here enable these new circular business models. The Material@Hand service from construction materials multinational Saint-Gobain, for example, uses robotic measurements, BIM, and a tracking app to determine the exact material quantities needed for partitioning walls—avoiding the 15–25% waste created by the current industry's bulk package delivery of gypsum boards (Saint-Gobain 2022). The global design, engineering, and architecture firm Arup, the Ellen MacArthur Foundation, and the World Green Building Council recently introduced a Circular Building Design Toolkit at Glasgow's COP26 climate conference, to bring together tools for designing more circular buildings (España 2021).

Rebound effect. As we have seen many times in the past, technological innovations often unwittingly create one new problem while solving another. Designers should become aware of the actual net benefits for environmental, economic, and social sustainability, as well as potential trade-offs and rebound effects of implementing digital technologies. For example, we can't use blockchain technology to facilitate transactions of reused materials without thinking about the energy consumption of this Industry 4.0 technology.

As we have seen, digital technologies can help transition from a linear to a circular model in the built environment. Due to rapid urbanization, powerful driving forces will be needed to generate a fundamental paradigm shift from a linear to a circular model in construction. The digital technologies described here will help tackle urgent societal urbanization needs to ensure that buildings, not raw materials, are the resources for construction in the future. Architects of Centre Pompidou's renovation site were able to make the iconic glass available for reuse in a beautiful project by Maximum Architecture (2022). But digital technologies for matchmaking reused materials are essential for a happy—circular—ending.

Acknowledgements The authors would like to thank Pavillon de l'Arsenal (https://www.pavillon-arsenal.com/fr/edition-e-boutique/collections/recherches-et-experimentations/11032-faire-et-refaire-du-verre.html), Raphaël Ménard, 169 Architecture, Elioth, by Egis (elioth.com/en/beaubourg-caterpillar/), and Maximum Architecture (www.maximumarchitecture.fr) for the case study as well as Alexandre Attias and Anna Buser for the images. This research was funded by Vlaanderen Circulair—Call CirculaireBouweconomie—Vlaams Agentschap Innoveren & Ondernemen (VLAIO).

References

Aeroscan (2022) Vastgoed volledig digitaal vastgelegd. https://www.aeroscan.nl/. Accessed 13 Oct 2022

Akanbi LA, Oyedele AO, Oyedele LO, Salami RO (2020) Deep learning model for Demolition Waste Prediction in a circular economy. J Clean Prod 274:122843. https://doi.org/10.1016/j.jclepro.2020.122843

Architecture 2030 Why The Building Sector?—Architecture 2030. https://architecture2030.org/why-the-building-sector/. Accessed 13 Oct 2022

Bocken N, Stahel W, Dobrauz G et al (2021) Circularity as the new normal: future fitting Swiss business strategies. PwC Switzerland

Buildings As Material Banks (BAMB) (2022) Reversible experience modules (REMs). https://www.bamb2020.eu/topics/pilot-cases-in-bamb/rem/. Accessed 13 Oct 2022

Çetin S, De Wolf C, Bocken N (2021) Circular digital built environment: an emerging framework. Sustainability 13:6348. https://doi.org/10.3390/su13116348

Circular Engineering for Architecture, DiCE (2021) Samsara, the story of reusing the glass from the Centre Pompidou

Circular X (2022) Cases. https://www.circularx.eu/en/cases/construction. Accessed 13 Oct 2022

Circulor (2022) Industrial supply chain traceability. In: Circulor. https://www.circulor.com. Accessed 13 Oct 2022

De Wolf C, Hoxha E, Fivet C (2020) Comparison of environmental assessment methods when reusing building components: a case study. Sustain Cities Soc 61:102322. https://doi.org/10.1016/j.scs.2020.102322

España Z (2021) Arup and EMF introduce circular building design toolkit at COP26. https://www.arup.com/news-and-events/arup-and-emf-introduce-circular-building-design-toolkit. Accessed 13 Oct 2022

Excess Materials Exchange (2022). https://excessmaterialsexchange.com/en_us/. Accessed 13 Oct 2022

Facade Service Applicatie (2022) Platform voor digitale informatie in de vastgoedmarkt. https://facadeserviceapplicatie.nl/. Accessed 13 Oct 2022

Foster + Partners (2021) Apple Tower Theatre. In: ArchDaily. https://www.archdaily.com/965106/apple-tower-theatre-foster-plus-partners. Accessed 13 Oct 2022

Gordon M, Batallé A, De Wolf C et al (2022) Automating detection for reuse of building materials: a case study

Graffeo E (2021) JPMorgan and Citi are using blockchain technology, and other banks are considering allowing clients to hold crypto in bank accounts. In: Business insider. https://web.archive.org/web/20210810125833/, https://markets.businessinsider.com/news/currencies/blockchain-technology-financial-institutions-jpmorgan-bitcoin-citi-cryptocurrency-transactions-btc-2021-2. Accessed 13 Oct 2022

Hunhevicz JJ, Brasey P-A, Bonanomi MM, Hall D (2020) Blockchain and smart contracts for integrated project delivery: inspiration from the commons. In: EPOC 2020 working paper proceedings. Engineering project organization society (EPOS)

Ingemarsdotter E, Jamsin E, Kortuem G, Balkenende R (2019) Circular strategies enabled by the internet of things—A framework and analysis of current practice. Sustainability 11:5689. https://doi.org/10.3390/su11205689

Kraaijvanger (2022) Circular building: a second life is always a new beginning. In: Kraaijvanger. https://kraaijvanger.nl/en/expertise/circular-building-design-and-architecture/. Accessed 13 Oct 2022

Larson K (2022) Beyond smart cities: emerging design and technology. https://mit-online.getsmarter.com/presentations/lp/mit-beyond-smart-cities-online-short-course/. Accessed 13 Oct 2022

Li J, Kassem M (2021) Applications of distributed ledger technology (DLT) and blockchain-enabled smart contracts in construction. Autom Constr 132:103955. https://doi.org/10.1016/j.autcon.2021.103955

Madaster (2022) The cadastre for materials and products. In: Madaster Glob. https://madaster.com/. Accessed 13 Oct 2022

Made of Air (2022) Carbon-negative materials. https://www.madeofair.com. Accessed 13 Oct 2022

Material Mapper (2022) Reusable building materials in your area. https://materialmapper.com/. Accessed 13 Oct 2022

Maximum Architecture (2022) Papillon. https://www.maximumarchitecture.fr/projets/papillon. Accessed 13 Oct 2022

Nußholz JLK, Rasmussen FN, Whalen K, Plepys A (2020) Material reuse in buildings: implications of a circular business model for sustainable value creation. J Clean Prod 245:118546. https://doi.org/10.1016/j.jclepro.2019.118546

Ochsendorf J (2010) Guastavino vaulting: the art of structural tile. Princeton Architectural Press, New York

Raghu D (2021) Wasteways: a strategic framework for existing buildings as material banks to enable component reuse. In: IAAC blog. https://www.iaacblog.com/programs/wasteways-strategic-framework-existing-buildings-material-banks-enable-component-reuse/. Accessed 13 Oct 2022

Raghu D (2022) Enabling component reuse from existing buildings through machine learning, using google street view to enhance building databases. In: van Ameijde J, Gardner N, Hyun KH, Luo D, Sheth U (eds) Post-carbon: proceedings of the 27th CAADRIA conference, Sydney, pp 577–586

Saint-Gobain (2022) Material@Hand Service. https://www.circularx.eu/en/cases/62/saint-gobain-material-hand-service. Accessed 13 Oct 2022

Samyn and Partners (2016) EUROPA: new headquarters of the Council of the European Union

Spotr (2022) Inspect millions of buildings in seconds. https://www.spotr.ai/. Accessed 13 Oct 2022

Structural Xploration Lab (2018) Habitats in time: mapping Geneva's embodied carbon legacy. Doctoral Seminar

Swiss Federal Laboratories for Materials Science and Technology (EMPA), Swiss Federal Institute of Aquatic Science and Technology (Eawag) (2018) Urban mining & recycling. https://www.empa.ch/web/nest/urban-mining. Accessed 13 Oct 2022

Swiss Federal Laboratories for Materials Science and Technology (EMPA), Swiss Federal Institute of Aquatic Science and Technology (Eawag) (2021) Sprint: from dismantling to re-use as fast as possible. https://www.empa.ch/web/nest/sprint. Accessed 13 Oct 2022

Tapscott D, Vargas RV (2019) How blockchain will change construction. Harv. Bus. Rev.

TradeLens (2022) Supply chain data and docs. https://www.tradelens.com/. Accessed 13 Oct 2022

United Nations Environment Programme (2019) Rising demand for sand calls for resource governance. In: UN Environ. http://www.unep.org/news-and-stories/press-release/rising-demand-sand-calls-resource-governance. Accessed 13 Oct 2022

United Nations Environment Programme, Global Alliance for Buildings and Construction (2020) 2020 Global status report for buildings and construction: towards a zero-emission, Efficient and resilient buildings and construction sector. Nairobi

United States Environmental Protection Agency, Office of Land and Emergency Management (2016) Sustainable management of construction and demolition materials. https://www.epa.gov/smm/sustainable-management-construction-and-demolition-materials. Accessed 13 Oct 2022

World Economic Forum (2022) 4 promising digital technologies for circular construction. In: World Economic Forum. https://www.weforum.org/agenda/2022/09/4-promising-digital-technologies-for-circular-construction/. Accessed 13 Oct 2022

Xu LD, Xu EL, Li L (2018) Industry 4.0: state of the art and future trends. Int J Prod Res 56:2941–2962. https://doi.org/10.1080/00207543.2018.1444806

ZenRobotics (2022) Heavy Picker—World's strongest recycling robot. https://zenrobotics.com/en/heavy-picker. Accessed 13 Oct 2022

Tak for Sidst: A Field Study of Demolition in Denmark

Tom Buckland

Abstract

Construction and demolition waste contributes huge amount of emissions to the climate crisis. The EU has asserted that moving toward a circular economy is one strategy to address climate change, of which resolving issues with construction and demolition is an important aspect. Denmark is leading the way in terms of construction and demolition waste recycling. However, the Danish Ministry of Interior and Housing has suggested that the current 85% construction and demolition waste recycling rate can be improved on. Through a hands-on field study of a state-of-the-art sustainable demolition project in Copenhagen, this paper investigates crucial collaborators for architects when circular economy principals are applied and practiced. Strategies and important stakeholders are detailed.

Keywords

Sustainability · Construction · Demolition · Recycling

T. Buckland (✉)
Det Kongelige Akademi, Frederiksberg, Denmark
e-mail: info@tombuckland.site

1 Introduction

The planet is in a climate crisis. In summer of 2021 UN Secretary-General António Guterres (2021) declared that "the evidence is irrefutable: greenhouse gas emissions are choking our planet and placing billions of people in danger" (Guterres 2021). The contribution of the construction industry to greenhouse gas production is significant. In the EU, the European Commission (2020) asserts that the construction and use of buildings accounts for 50% of all extracted resource and energy consumption. According to the European Commission (2015), waste generated from construction and demolition such as concrete, bricks, glass, wood, metals, and plastic accounts for more than a third of all waste generated in the EU. United Nations Sustainable Development Goal 12 (SDG 12), Responsible Consumption and Production, intends to "ensure sustainable consumption and production patterns" (UN n.d.). In reference to SDG 12, The Sustainable Development Report (UN 2021), asserts that "a path for sustainable consumption and production requires circular economy approaches, designed to reduce or eliminate waste and pollution, keep products and materials in use, and regenerate natural systems" (p. 50). The current global economy is mostly linear rather than circular. Materials are collected, processed into products, and eventually disposed of in landfills or incinerated. According to the

M. R. Thomsen et al. (eds.), *Design for Rethinking Resources*, Sustainable Development Goals Series,
https://doi.org/10.1007/978-3-031-36554-6_43

European Commission (2021) in the Circular Economy Action Plan, revisions of material recovery targets for construction and demolition waste (CDW) are necessary in order to maximize economic, social, and environmental sustainability. Since 2008, the EU has recommended construction and demolition waste be separated at source as part of its ongoing transition to a circular economy. This recommendation underscores the economic and environmental potential of utilizing recycled building materials, thereby preserving embedded carbon. The European Commission (2021) asserts that few enterprises comply with the rules regarding separation of CDW, something that needs to be remedied if the European Union is to move toward a circular economy. In Denmark, building and construction waste accounts for 39% of all waste at 5,023,253 tons. Wood waste represents 2.5% of that, at 154,893 tons, metal, 444,046 tons, while 3,910,264 tons are categorized as mixed building waste (Danmarks Statistik 2019). Out of 5,023,253 tons of CDW, 4,381,182 tons are recycled (Danmarks Statistik 2019). The Danish Ministry of Interior and Housing (2021), suggest that although 85% of CDW, which accounts for more than a third of total waste in Denmark, is recycled and utilized, there is potential for more appropriate reuse and recycling of materials.

This paper asks, if Denmark is to improve on an 85% CDW recycling rate who are important collaborators for architects when circular economy principals are applied and practiced in the field?

2 Method

The idea of reusing building materials is not a new one. Throughout history scarce or expensive materials have motivated deconstruction and recycling. The ancient Romans used the word "spolia" to refer to reused stone. During the urban renewal of Paris coordinated by Baron Haussman in the mid-1800s, large swaths of tenement buildings were demolished and Haussman ordered reusable materials to be sold at public auctions (Rotor, para 4). Korhonen et al. (2018) define a circular economy as one characterized by cyclical material flows, renewable energy sources, and cascading 1-type energy flows. They assert that the circular economy "limits the throughput flow to a level that nature tolerates and utilizes ecosystem cycles in economic cycles by respecting their natural reproduction rates" (Korhonen et al. 2018, p. 39). The logic of circular economy suggests that rather than ending up in landfills, waste should be integrated into new products that can be recycled again when no longer useful. Koutamanis (2018) indicates that urban mining has been discussed in relation to the possibilities offered by circular economy. Rotor (2021), explains:

> In its purest, conceptual sense the circular economy promises an endless cycle of reusable resources. With regard to the construction industry, the concept often revolves around the central metaphor of the urban mine, which understands the city as an accumulation of valuable materials and reusable elements. (para. 1)

Bachmann (2019) claims that "construction in the Anthropocene era will rely on the recovery and recycling of building materials from the urban ecosystem" (p. 7), harkening back to earlier times when materials were reused many times over. As Bachmann (2019) explains, urban mining regards the built environment as a repository of raw materials that can be collected or extracted under appropriate circumstances to be used again in new constructions. Jarre et al. (2020) assert that cascading use is another aspect of a circular economy, and literature often reveals overlaps in ideas between the two theories. Generally, cascading use refers to "a river over a sequence of plateaus" (Sirkin and ten Houten 1994, p. 10) that describes a process wherein biomass material (forest products) are efficiently utilized by cycles of recycling and then ultimately incinerated as fuel for energy generation (Jarre et al. 2020, p. 2). Conceptual considerations and practical applications of cascading use are widespread in the forestry industry (Jarre et al.).

These interrelated fields, concepts, principals, and practices have major implications for the way architects and the construction industries function. In order to investigate the practical application of circular economy principals this paper employs the demolition of a building at Jagtvej 169B and the recycling of the building's constituent materials as a case for field study. Jagtvej 169B was built in 1860 as the home of Aldersro Brewery in the Østerbro neighborhood of Copenhagen. In 1870 the building switched to production of cigars and cigarettes which continued until the 1970s when it was converted to offices. The building was demolished in late 2021. The Jagtvej 169B project is unusual in that the developer had a goal to recycle a large amount, about 95%, of the components and materials from the building and sought out a demolition company that would be able to undertake such a task efficiently.

This field study was conducted in a way in the style of what landscape architect Kahn (2011) describes. The fieldwork took place both *in* the professional field of architecture and disciplines that intersect with it and *on* a physical field at the demolition site. Data was collected through field research consisting of participant observation as a demolition laborer for Søndergaard Nedrivning A/S. Participant observation, defined by Dewalt et al. (1998), is a method of research in which information is gained by participating and observing, while recording, in field notes. Field notes consisted of observations and recordings in the form of writing, photographs, videos, document gathering, conversations with coworkers and community members, and interviews with stakeholders. Interviews were conducted with Kasper Sørensen, Produktionsdiretør at Søndergaard, Jens Thamdrup, Construktion Arkitekt and Senior Projektchef at ATP, and Thomas Bo Nielson, Innovation Engineer at RGS Nordic.

3 Results

Developer ATP Ejendomme began acquiring properties adjacent to Jagtvej 169 in 2011 with successive purchases accumulating into a large corner lot. In the intervening years one major Copenhagen architect put forth a proposal. The 11-storey design would have allowed the preservation of the original building, surrounding it on three sides while several stories would cantilever over it (J. Thamdrup, personal communication, December 3, 2021). The proposal allowed for a significant rental area, about 15,000 m^2, or 375% on the property. However, the design faced considerable objection from the community due to its height and concerns about infringement on daylight access to neighboring properties. Since then, a different architect, Dissing + Weitling, was brought on who also attempted to design around the existing building. Since the existing business case was for 15,000 m^2 it was clear to ATP Ejendomme that the new building should achieve the same area. However, Dissing + Weitling determined it was not possible to design a project of that scale while preserving both existing structures. ATP negotiated with the community and municipality to establish a "lokaleplan," the permit procedure required in Denmark, and a permit was granted to demolish Jagtvej 169B while leaving 169A, the slightly newer building to the west, standing (J. Thamdrup, personal communication, December 3, 2021).

The project has faced some criticism from the community (K. Sørensen, personal communication, December 2, 2021). One neighbor and local architect said that the demolition is a "tragedy and should be a scandal" due to what he considered the wasteful demolition of a historic building in the interest of a few DGNB points (community member, personal communication, December 10, 2021). Sørensen himself conceded that the "most sustainable thing to do would have been to leave the structure and facade of the building standing" in a "soft-strip" style, typical of extensive renovations, rather than demolish the entire building, including the foundation (K. Sørensen, personal communication, December 2, 2021). Another neighbor said he was "worried about what is coming next," but the knowledge that materials are being recycled was some consolation (community member, personal communication, 10/11/21).

Kasper Sørensen, Produktionsdiretør at Søndergaard (2021), explains that typically, demolition of buildings is carried out as quickly as possible. Most wood materials are sent to the incinerator, bricks are crushed for fill under roads, and metal is sent to be melted down. The ambition of the developer was to reuse materials from the building and dedicate funds to reach their goal. To achieve it ATP included in the tender for the demolition contractor the requirement to complete an assessment of resources, "resource-mapping," which would consist of a meticulous mapping and cataloging of all the materials and components in the building (J. Thamdrup, personal communication, December 3, 2021). Thorough environmental assessments are required and typical for demolition to identify the location and quantity of materials like PCB and asbestos. However, the requirement for resource-mapping is unusual (K. Sørensen, personal communication, December 2, 2021). ATP hired the engineering firm EKJ for the exhaustive resource-mapping process. Søndergaard was selected as the demolition contractor as they were demonstrably the company most well equipped to take on such a challenge. At Jagtvej 169B, materials and components were assessed and mapped. J. Thamdrup (2021) explains that "sanitary installations were mapped, lamps, suspended ceiling, doors, door knobs, radiators, windows, everything" (personal communication, 2021). Many components like windows, doors, and radiators were sold through an online auction, while other materials, such as approximately 1200 square meters of Oregon pine floorboards, were donated locally to Rasmus Greve, a carpenter (K. Sørensen, personal communication, December 2, 2021). His company, Loud Living, will be producing furniture for the new building at Jagtvej 169 (R. Greve, personal communication, 1/2/21). Removing the floorboards was one of the first tasks I was assigned while working in the field. The floorboards were covered with carpet that had to be removed before prying them up from an additional two layers of lumber, with power and network cables running between them. It was a difficult and time-consuming process to remove the boards carefully and efficiently, in a way that they retained their value.

Timber in pre-1950 Copenhagen buildings is typically old growth, meaning that the tree lived a long, natural life before being cut down. Jagtvej 169 was no exception, making wood a primary concern on the project and one of the most challenging. Having been built in 1860, the structural timbers had ring counts that suggested the trees were 200–300 years old when they were cut down. Old growth timber is characterized by tightly spaced growth rings as a result of slow growth over a long period making it desirable for its density and perceived beauty, in contrast to widely spaced growth rings of second growth timber as a result of fertilization and watering making it less dense, less visually appealing, and structurally weaker than old growth timber. Costs usually prohibit time-consuming assessment of wood and whether it could serve a better purpose than being incinerated for energy generation. However, increasingly, regulation and market changes are pushing more meticulous separation and consideration of construction and demolition waste (K. Sørensen, personal communication, December 2, 2021). Thomas Bo Nielson, innovation engineer at RGS Nordic (2021), one of the waste disposal contractors at Jagtevej 169 explains, while wood cannot be melted down and reconstituted like metal it can be broken down to smaller fragments and bound together with adhesives to form building materials like Oriented-Strand Board (OSB), chip board, and medium-density fibreboard (MDF) (T. Bo Nielson, personal communication, November 26, 2021). Because of the complexity of the supply chain, it can be difficult to determine whether cascading used of recovered wood from buildings has environmental benefits. Niu et al. (2021) have concluded that "reusing structural timber components is beneficial regarding combating global warming" (p. 7). A study by Weber-Blaschke and Richter (2013) "revealed a considerable amount of recovered wood from the building sector available in conditions which allow a high-quality material cascading and thus a secondary utilization" (p. 313). But they conceded that "to apply the concept of cascading of recovered wood in the most beneficial way, further research regarding the ecological and economic benefits, the necessary logistics, and the technical processing is necessary" (p. 313).

Knauf (2015) noted that "previous environmental comparisons of waste wood management (waste wood recycling or its use for bioenergy production) do not take into account the alternative use of the fresh wood saved by recycling" (p. 3). Knauf (2015) concludes that "material use as a first use is far superior to initial direct use for energy production" (p. 3). Wood recovery and recycling is therefore a way to recapture value and carbon in order to combat climate change. The urgent need for sustainability in construction and demolition is obvious and well documented. Sustainable demolition practices and recycling of materials can cut emissions through reduction of extraction, production, and transportation. "Generally speaking," says Sørensen, "reusing a piece of wood is better than burning it" (K. Sørensen, personal communication, December 2, 2021). Wood has a lot of potential for reuse, in a wide spectrum of applications from practical applications on construction sites to high-end counter tops and tables. If wood recycling is to be effective, it has to span across that spectrum of use (K. Sørensen, personal communication, December 2, 2021). The challenges to wood recycling are very large in scope. At a high level, inefficient recycling of wood is a common practice worldwide, and improving it requires cooperation between many actors. On the other hand, wood recycling at a large scale requires thousands of minute, time-consuming tasks like taking nails out of boards (T. Bo Nielson, personal communication, November 26, 2021).

RGS Nordic was responsible for management of wood waste that was not already bought or donated for a specific purpose. There were three dumpsters at the site that represented three streams of waste. Generally, pieces of solid wood that were two meters or longer went in the first dumpster. Søndergaard machine operators were careful not to damage large beams and floorboards in order to allow direct recycling. Pieces of solid wood that were less than 2 m that were free of paint and other contaminants were sent to chip board production, while small pieces of plywood and anything with glue went to the incinerator (K. Sørensen, personal communication, December 2, 2021). This careful work of removing, sorting, and storing took considerably longer than the rapid, albeit wasteful, demolition more typical in the industry. Furthermore, it relied on the demolition worker to make dozens, perhaps hundreds, of small decisions every day about where pieces of waste should be put. This is ensured only by having systems in place that make it possible and easy as possible to do so and supervisors that were able to motivate and instruct workers in the appropriate way.

Despite the extra length of the contract ATP is paying only 5–8% extra for demolition, excluding external expenses like scaffolding and crane rentals (K. Sørensen, personal communication, December 2, 2021). J. Thamdrup (2021) says the project is DGNB certified, a process conducted by Sveco as consultant. Jagtvej 169B is the first project in Denmark that has DGNB points awarded for demolition of a previous construction. DGNB points can only be awarded if work is planned, so therefore the resource-mapping provides an estimate for potential recycled materials and proves it is intentional. To complete this process Søndergaard documented all materials recycled or sold and surpassed estimates by 2%, resulting in a 97% recycling of materials (J. Thamdrup, personal communication, December 3, 2021). Furthermore, as J. Thamdrup (2021) explains that "even if there was no value according to environmental things, there would be a value just according to the building process" (personal communication, December 3, 2021). That is to say, if no carbon emissions were spared by mapping resources and carefully deconstructing and separating recoverable parts, the resource-mapping would provide value through savings of time and effort in negotiating unforeseen circumstances that typically arise during demolition and preparation for new constructions.

4 Discussion

At Jagtvej 169B the demolition phase of the project was mostly guided by the ambitions of the developer. The primary stakeholders from an architectural perspective were the customer, ATP

Ejendomme, who presented the tender including goals for the demolition, the demolition contractor Søndergaard, engineer, EKJ, who conducted the resource-mapping, and the demolition workers who carried out the demolition. Two documents emerged as crucial tools. In the Jagtvej 169B project, Søndergaard workers sought "to harvest resources to the highest value" (K. Sørensen, personal communication, December 2, 2021). It is an ambition that is instructed by the tender and facilitated by the resource-mapping. Through the tender and resource-mapping, stakeholders were able to align on goals and collaborate effectively. In order for improvement of recycling, it has to be planned for and implemented from the outset by motivated stakeholders. Resource-mapping provides environmental benefits along with cost savings and improvements in efficiency. There is a benefit to the customer, as J. Thamdrup says, the resource-mapping allowed the demolition process to go more smoothly with less surprises as well as facilitating DGNB points to be applied to the new construction. If material can be diverted from incineration in order to be directly recycled there is a benefit of saving carbon emissions. The resource-mapping is a crucial tool that described the building and allowed the demolition contractor and their employees to conduct operations smoothly, dismantling the building and distributing its composite components and materials for reuse in other applications. The resource-mapping makes the internals of the building visible and manageable to site supervisors and in turn, demolition workers, laying out a framework for how to treat building materials and where to put them.

Furthermore, this analysis has revealed the importance of the demolition worker as a collaborator in the sustainable demolition landscape. Demolition workers and the decisions they make on a task-by-task basis hold the balance of whether a project will succeed or fail in terms of sustainable ambitions. They are the front line of the circular economy in the built environment. The skills one needs to hone to remove fixtures, appliances, trim, floorboards, windows, and myriad other building materials intact and efficiently should not be underestimated. The presence of demolition workers as more than an abstract labor unit in discussions about the circular economy is a glaring omission.

The circular economy and practical applications of circular principals in cases such as Jagtvej 169B presents the possibility of having a market of previously used building materials where instead of a designer dictating what building materials will be used, reused building materials instruct what can be designed. The quality of building materials and reliability of supply relies on the workers handling them. If demolition workers are to be relied on for collaboration on they should be included in discussions about the value of recycling and the ways in which it is conducted.

This paper has investigated the practical applications of circular economy principals and the stakeholders involved through a field study of Jagtvej 169B. The case demonstrates concrete, replicable ways to reduce waste, keep building materials in use, thereby complying with UN Sustainable Development Goal 12, reducing CDW from 39%, and improving upon Denmark's 85% CDW recycling rate by up to 12%. The question of whether this historical building should have been torn down is an important one, but one that is beyond the scope of this paper. From the perspective of the developer, it was necessary. The case of Jagtvej 169B provides a model for a more sustainable demolition process that can be applied to less contentious developments.

References

Bachmann G (2019) Urban resource exploration—producing structures in closed-loop materials cycles. In: Hillebrandt A, Riegler-Floors P, Rosen A, Seggewies J-K (eds) Manual of recycling: buildings as sources of materials. Detail, Wuppertal, pp 10–16

Danmarks Statistik (2019) Affaldsproduktion efter branche, tid og behandlingsform. http://www.statistikbanken.dk/10294. Accessed 3 Feb 2022

DeWalt KM, DeWalt BR, Wayland CB (1998) Participant observation. In: Hand book of methods in cultural anthropology, pp 259–299

European Union: European Commission (2015) Communication from the commission to the European parliament, the council, the European economic and social committee and the committee of the regions closing the loop—an eu action plan for the circular economy. https://eur-lex.europa.eu/legal-content/EN/TXT/?uri=CELEX:52015DC0614

European Union: European Commission (2020) Communication from the commission to the European parliament, the council, the European economic and social committee and the committee of the regions. A new Circular Economy Action Plan for a cleaner and more competitive Europe. https://eur-lex.europa.eu/legal-content/EN/TXT/qid=1583933814386&uri=COM:2020:98:FIN

Guterres A (2021) The evidence is irrefutable: greenhouse gas emissions are choking our planet & placing billions of people in danger [Tweet]. Twitter. https://twitter.com/antonioguterres/status1424649118312435714

Jarre M, Petit-Boix A, Priefer C, Meyer R, Leipold S (2020) Transforming the bio-based sector towards a circular economy—what can we learn from wood cascading? Forest Policy Econ 110:101872. https://doi.org/10.1016/j.forpol.2019.01.017

Kahn A (2011) Field note 1. In: Ewing S (ed) Architecture and field/work. Routledge

Knauf M (2015) Waste hierarchy revisited—an evaluation of waste wood recycling in the context of EU energy policy and the European market. Forest Policy Econ 54:58–60. https://doi.org/10.1016/j.forpol.2014.12.003

Korhonen J, Honkasalo A, Seppälä J (2018) Circular economy: the concept and its limitations. Ecol Econ 143:37–46. https://doi.org/10.1016/j.ecolecon.2017.06.041

Koutamanis A, van Reijn B, van Bueren E (2018) Urban mining and buildings: a review of possibilities and limitations. Res Cons Recycl 138:32–39. https://doi.org/10.1016/j.resconrec.2018.06.024

Ministry of Interior and Housing (2021) National Strategy For Sustainable Construction

Niu Y, Rasi K, Hughes M, Halme M, Fink G (2021) Prolonging life cycles of construction materials and combating climate change by cascading: the case of reusing timber in Finland. Resour, Conserv Recycl 170:105555. https://doi.org/10.1016/j.resconrec.2021.105555

Rotor (2021) Urban Mine Inc. Rotor DB. https://rotordb.org/en/stories/urban-mine-inc

Sirkin T, ten Houten M (1994) The cascade chain: a theory and tool for achieving resource sustainability with applications for product design. Resour, Conserv Recycl 10(3):213–276. https://doi.org/10.1016/09213449(94)90016-7

United Nations Department of Economic and Social Affairs (n.d.) The 17 goals. Sustainable Development. https://sdgs.un.org/goals

United Nations Department of Economic and Social Affairs (2021) The sustainable development goals report. https://doi.org/10.18356/9789210056083

Weber-Blaschke G, Höglmeier K, Richter K (2013) Potentials for cascading of recovered wood from building deconstruction—a case study for south-east Germany. Res Cons Recycl 78:81–91. https://doi.org/10.1016/j.resconrec.2013.07.004

Enhanced Databases on City's Building Material Stock. An Urban Mining Method Based on Machine Learning for Enabling Building's Materials Reuse Strategies

Areti Markopoulou, Oana Taut, and Hesham Shawqy

Abstract

The climate crisis and the growing urbanization needs urge design and construction practices to shift their focus to the anthroposphere as a source of, rather than just a destination for, building materials. The concept of *urban mining* is revisited by many to manage the existing building material stock exploring the potential for reuse in new constructions. By combining image-based segmentation with cadastral data in a GIS database, this paper proposes an end to end process for an integrated web-based application enclosing and delivering cross-referenced data about a city's material stock. The study uses unstructured data from Open Street View to identify relevant patterns for making estimations of the quantity, state, and projected availability cycle of concrete, brick, stone, metal, timber, and glass in building facades. By applying predictive modelling at the city scale, the algorithm can identify, geolocate and quantify façade materials with a present accuracy of 87%. The developed method proposes an alternative automated way to physical inspections which makes it applicable to different cities that lack registers of building data. A data repository map is developed in the format of an interface, so that it can be used by different stakeholders including decision-makers in the formulation and planning of urban material reuse strategies, as well as designers in the early stage of circular design processes. Tackling the accuracy limitations of machine learning, the paper concludes with studying the potential of combining the use of image unstructured data with statistical data of existing building registers including age of buildings and preservation state.

Keywords

Machine learning · Urban mining · Mapping resources · Material reuse · Circular construction

1 Introduction

Buildings are responsible for half of the global primary material consumption while they are accountable for almost 40% of global climate emissions, of which 50% are produced during construction. In recent years the increased demand of resources required, the amount of waste related with the built environment and the reduced building lifetimes have led to a growing

A. Markopoulou (✉)
IAAC, Institute for Advanced Architecture of Catalonia, Barcelona, Spain
e-mail: areti@iaac.net

O. Taut · H. Shawqy
IAAC, Institute for Advanced Architecture of Catalonia, Barcelona, Spain

M. R. Thomsen et al. (eds.), *Design for Rethinking Resources*, Sustainable Development Goals Series,
https://doi.org/10.1007/978-3-031-36554-6_44

interest in urban mining (UM) in buildings (Koutamanis et al. 2018; Arora et al. 2020). Current urban mining practices and policies, though, are primarily focusing on reusing construction and demolition waste (CDW) by downcycling them (Pomponi and Moncaster 2017). The majority of demolition waste, for instance, is used as filler when building roadways, which prevents its further reuse or recycling. There is very little research on how CDW can be reused as primary materials in new building construction, and this is due to the lack of building information on the amount, state, and reuse possibility of the existing building material stock. Current urban mining initiatives for building materials and components reuse rely on the expert survey and analysis of field data, in conjunction with other relevant census or cadastral data sources from public building registers (Kumar and Ai Lin Teo 2019). However, on the one hand, public building data lack key information on materials and on the other, the need for extensive prior field research results in a process that requires high costs and sophisticated equipment, making it difficult to implement in low-budgeted administrations or in countries where extensive public open data is not available (Raghu et al. 2022). Developing, therefore, quick, feasible, and replicable methods to create inventory of existing building stock reuse through automating the expert field audit and inspections is a key research field that requires further exploration.

In the field of urban policies there is an increasing number of urban administrations that in their effort to minimize the negative impacts of the built environment and to achieve their goals for reducing the consumption of primary materials, they invest both economically and intellectually to set up policies and the necessary infrastructure that enhance Circular Economy (CE). The principles of CE lie in limiting material and resource loss, minimizing waste, and enabling design for reuse (Ellen MacArthur Foundation 2012). Most of these administration efforts, however, focus on matching demolition and construction material flows in cities, which means finding a balance between yearly demolition and construction material provision and demand (Arora et al. 2021). Although such scenarios could be ideal for cities and while recycling of demolition waste has a sizeable potential, the reality is that in many urban environments—including cities like Amsterdam that has provided high investment in CE—this potential falls short on achieving the policy goals required for reducing primary material consumption set forth by the 2016 Paris Agreement. The current study contributes to creating material databases of the entire stock of existing buildings, beyond, therefore, the analysis of buildings planned for demolition. Such building information data becomes significant for contributing to the formulation of long-term urban policies and planning strategies for building reuse that tackles issues of mismatch between annual demolition and construction material provision and demand.

2 Material and Methods

As a growing field related to translating unstructured visual data into meaningful semantic by using AI, Computer Vision captures multiple tasks related to automatic identification of features in images. Due to the ease to source image data regarding urban environments from open sources, this direction is particularly relevant for the current study and has been explored with different end goals in several previous research, forming the field of Visual Facade Classification. Depending on the goal of the classification, two approaches are mainly observed:

- multiple object detection frameworks, specifically R-CNN (Lee et al. 2020); or
- pixelwise labelling commonly called semantic segmentation (Fröhlich et al. 2010) (Fig. 1).

The scope of material reuse, however, which is the scope of this study, goes beyond identifiable building features (windows, railings, decoration) and differentiation between different facade finishes requires pixelwise comprehension. In this regard, semantic segmentation method which has

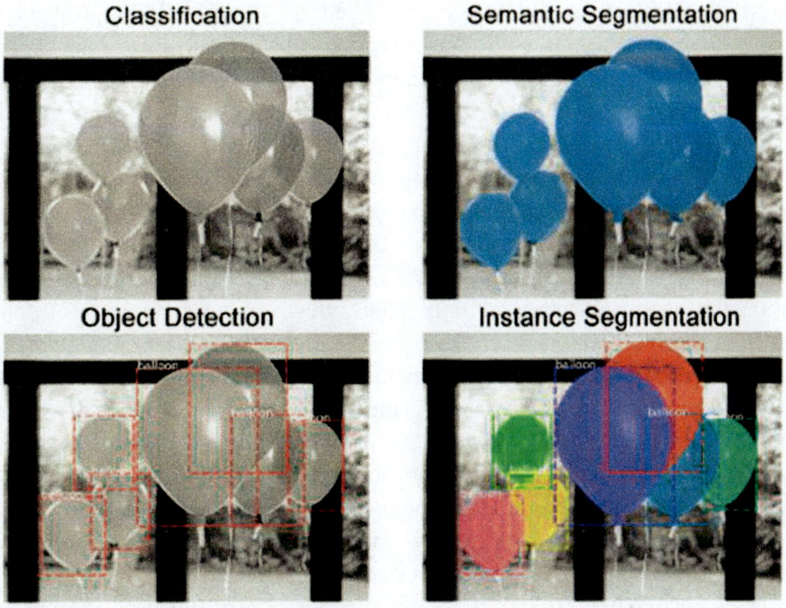

Fig. 1 Image showing different computer vision tasks (Abdulla 2018)

been previously used for facade material recognition, and material segmentation (Bell et al. 2015), has been selected as most relevant for this study. While pixelwise image segmentation can predict the material mix of a given building facade, many applications related to urban-scale material reuse strategies, such as material density mapping, require relating the segmented image to a geographic information system (GIS) (Kang et al. 2018).

The need for material aware segmentation of thousands of images, essential to make city scale estimations, called for an artificially cognizant network, in other words a deep neural network, capable of pixelwise image classification. Several viable options were considered to respond to the need stated as follows: starting from a ground-truth image, generating an equivalent image mask, and grouping in different color patches the pixels of high probability of representing the same material. After evaluating relevant work on similar data types, a type of generative adversarial neural network (GAN) (Goodfellow et al. 2020), specifically a conditional GAN based on the Pix-to-Pix framework (Isola et al. 2017) was selected. To train this model, a sizable dataset of pair images, containing ground-truth and equivalent labelling mask, is necessary, and its collection and manual labelling was a necessary first step.

2.1 Collection

Although existing urban facades sets follow a labelling strategy suited to identify building elements, indiscriminate of their materials, and are therefore unsuitable for the training task, best practices and characteristics observed in the popular CMP dataset were followed as closely as possible (Tylecek 2012). In accordance with these, images included in our dataset were required to be orthogonal, removing lens distortion insofar as possible. The overall strategy and the intended application also dictate the data collection source and parameters. Both for the initial training dataset and for the prediction database, images must be retrieved from an openly available and extensive street level imagery source. Google StreetView API was found to satisfy both requirements. Apart from the requirements of a suitable GAN training dataset, the need for quantitative evaluation at a global level as well as geo-referencing of this data, calls for standardization of the images collected to obtain orthogonal views with a consistent focal length, and with a viewpoint and angle that can be related back to the building ID in GIS mapping.

The first step was therefore finding a suitable method to extract orthogonal images using GoogleStreetViews API. This interface allows to extract views following a schema of user parameters such as location lat/long values, view direction, view horizontal and vertical angle. Considering the image collection methodology of this service, by Google car, it is understood that the camera position is always fixed to the street and 360° angles are available at each point. Due to this fact, an uncontrolled amount of street view images could be extracted from one street, showing the same buildings from different angles. This wouldn't be suitable as data pertaining to the same building in multiple pictures cannot be easily referenced back to a single building ID, therefore a strategy was necessary to define a single "main facade" image for each building. To collect the best orthogonal image for each facade, geodata was first extracted from an open-source online mapping service, specifically OpenStreetMap, which retrieves building footprints and street centerlines by their geographic locations (latitude, longitude). The centroid of each building polygon was projected to the street centerline segments, and the intersection point was used as camera position. The camera orientation was then calculated based on the line deviation from true North. With these parameters, an algorithm was created to automatically write the API requests. This process was used first to collect the training dataset, and then to collect one relevant facade for each building in the Barcelona Metropolitan area in preparation for the prediction step. The image below shows the camera viewpoint in relation to the building centre point. This process proves successful in retrieving the best orthogonal view of the façade in approximately 60% of the cases but requires further review to account for corner buildings and larger buildings, especially when placed on narrow streets. Possible solutions are to apply the centroid projection algorithm to building facade segments and find the furthest available view within a set range (Fig. 2).

Training Dataset Collection/Variation/Criteria

Based on initial tests using the Pix-to-Pix model selected, a dataset containing 800 images was the minimum requirement for suitable training of the model. Dealing with a relatively small dataset, the selection strategy is important to ensure the model can generalize across features bracing the

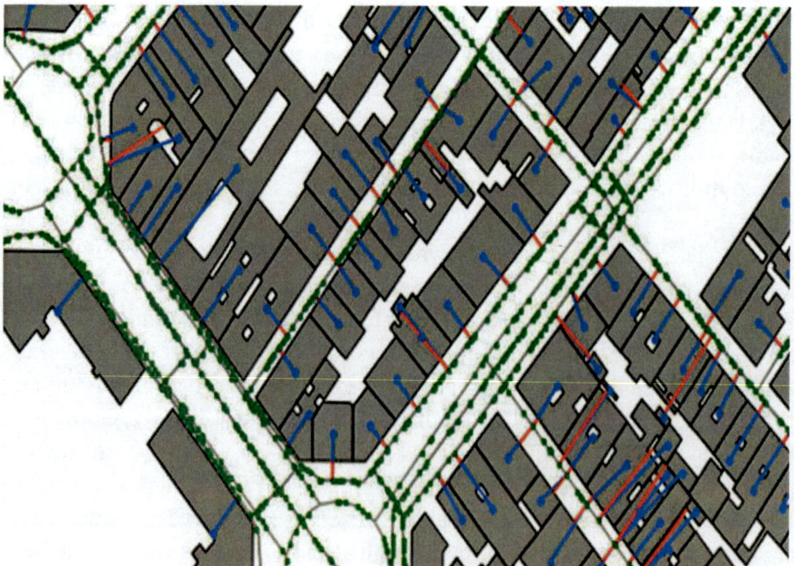

Fig. 2 Image standardization technique showing the centroid projection to street, in blue, and the adjusted camera position and angle, in red

full variation of the city in study. Using information from the open city data, selection criteria were embedded considering buildings from different neighbourhoods, heights, and typologies. A manual screening is also necessary to ensure variation across architectural styles, dominant finishing materials such as concrete, bricks, stone, as well as sufficient representation of special use materials such as wood, metal, and glass. The 800 images so collected were distributed in sets of 60 to a group of researchers at the Institute for Advanced Architecture of Catalonia, with clear labelling instructions and a one-hour long labelling tutorial. For each image, 6 labels were used to identify different material classes: concrete, bricks, stone, glass, wood, and metal were each assigned their colour code to be easily distinguished by human and computer alike. To label different materials for individual images, an image processing software was used for hand-labelling. For each material region, a new label was added associated with the colour code. Consistent with similar projects, the researchers reported spending 6–8 min per image, which amounts to a total labelling time of approximately 70% of the total spent on the Neural Network training aspect of this project.

2.2 Training

Pix-to-Pix Architecture

To go from 800 to 3200 labelled images (the number of images required to estimate materials for the metropolitan area of Barcelona), a Pix2-Pix deep neural network was employed. Developed originally by Phillip Isola et al. this model is a particular type of generative adversarial neural network that uses semantic labels in addition to random noise to generate a new image starting from a given condition image. This task is commonly known as image-to-image translation. A particularity of GAN, the model consists of two neural nets, a generator, and a discriminator, working in adversarial relationship to each other. In the case of Pix2Pix, the generator is a U-net, a type of architecture already used in state-of-the-art image segmentation research, while the discriminator is a patch GAN network, recognized to produce sharper images, by judging between real and fake at the scale of the image patches rather than averaging over the entire image at once.

Training Runs and Changes

While training this model, different methods of labelling the dataset were tested, starting from images with 10 labels that included a wider domain of mapping IDs. This method was quite challenging as the model complexity was getting higher and the output mapping images were distorted. It was noticed that model performance was degrading along the training process. Finally, images with only 6 labels were used for the training as it was indicating a good model performance.

The model has been trained using Pix2Pix GAN model on Google Colab, this online platform allows writing python code and utilizing Google's great computing resources. The model performance has been monitored using Wandb.ai platform, which affords analytics tools to compare between different models' architecture performance and accuracy.

For each model, different hyper parameters were tested against the accuracy to validate the model performance. The network's performance was evaluated using the metric of both generator and discriminator loss. It's a standard metrics used in GAN models. It indicates the difference between the predicted labelled image and the ground truth. The final model was trained for 1500 epochs only as the network performance stopped evolving at this point. That model was tested with new images that were not used in the training dataset and it was able to correctly generate labelled images in 65% of testing cases. More details about the final model performance will be provided later in the results section (Figs. 3, 4, 5, and 6).

Fig. 3 Google street imagery and ground-truth labels versus the predicted labelled images comparison to evaluate different models performance

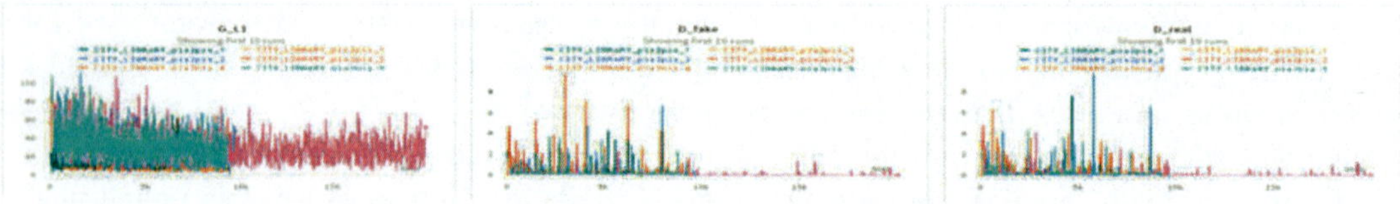

Fig. 4 Generator and discriminator loss comparison to evaluate different models' performance

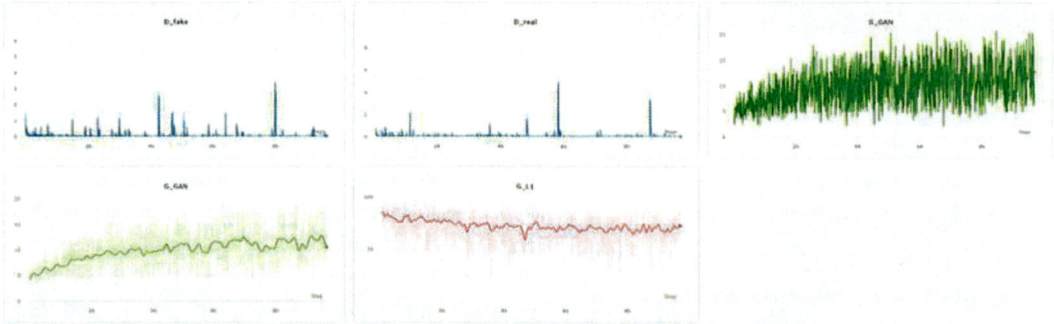

Fig. 5 Final model evaluation graphs using the metric of both generator and discriminator loss

Merging the Existing Geodata with the Material Data (CSV)

As per our research statement, automatic image comprehension achieved by training the deep generative neural network model resolves only part of the problem statement. Enabling inference in relation to additional cadastral data about existing buildings is key to accomplishing the project scope. To this end, with an acceptable level of accuracy, the model trained as above is used to make a prediction for one image of each building facade and reference it back to open urban data through the building ID. Such data, available for the city of Barcelona includes year of construction, building use, building area, historical protection status, building occupancy, etc.

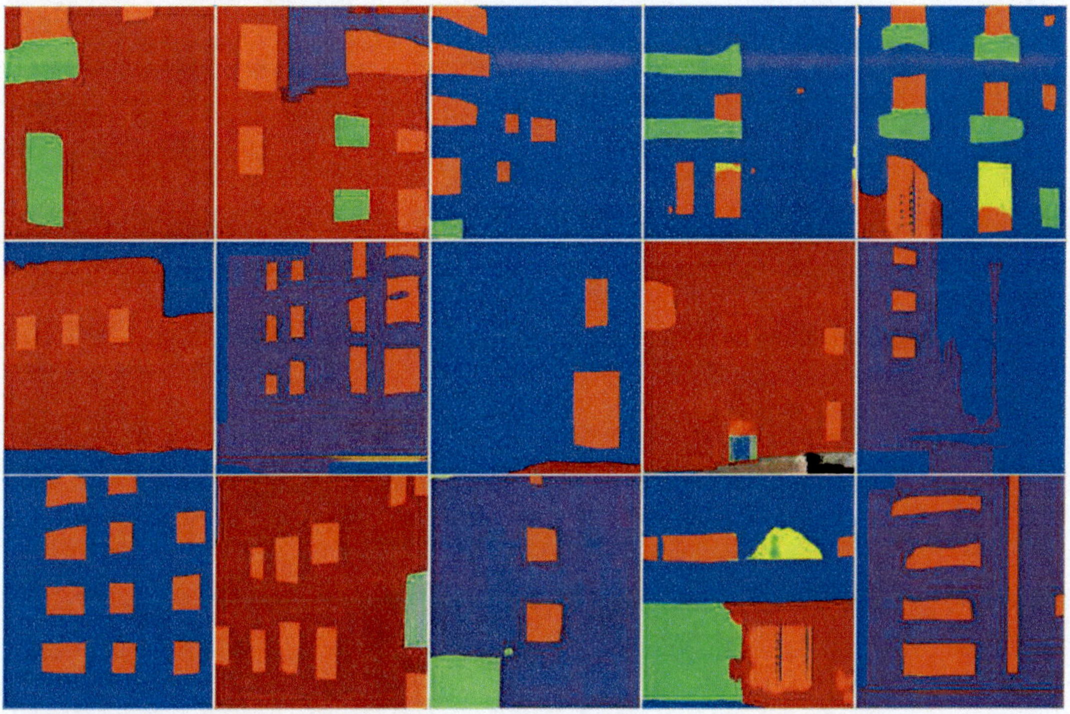

Fig. 6 Final model generates fake images sample

Additional columns containing facade level predictions (facade percentage and facade area of plaster, brick, stone, metal, timber, and glass) were added to the database, which was then exported in CSV and .shp format. The suitability of the workflow was then tested by asking the research study group of Institute for Advanced Architecture of Catalonia to use the resulting database to drive insights on Urban Mining potential in their research design studio.

2.3 Interface

Having the compiled database in a CSV file format, allowed the flexibility of exchanging the data with different platforms. During our research both offline and online user-friendly tools were needed to allow users to interact with the materials dataset. The two frameworks attached below have been proposed as preliminary prototypes which involves the development virtual environment for the analysis and evaluation of the building's material recognition model. Both frameworks have the flexibility to be used in all levels of urban analysis and materials extraction (Fig. 7).

Integration with Rhino-GH

The first setup, the integration of the Geojson and csv files with Grasshopper3d (node-based algorithmic environment within Rhinoceros3D modelling software) allows users to interact with the data locally on their machines. A Grasshopper3d script was developed to visualize three-dimensional geodata within the Rhino software associated with the metadata extracted from the complied buildings materials database.

That script afforded 3D data enquiry on multiple levels; visualizing building materials for the whole city or smaller neighbourhoods. The end user would start using this script by picking the desired Geojson and csv files for a city, then define the search parameters such as search domain, radius, material filter, buildings age

Fig. 7 Materials prediction sample

Fig. 8 The interface of the local model using Rhinoceros and Grasshopper3d script

domain, single or multiple regions, and finally would end this process by extracting both 3D geometries and material metadata (Fig. 8).

Deploying the Model on the Web

The second setup, the web deployment of the model provides the general framework for the collaborative mapping environment. This lightweight online environment contains a variety of functionalities including a user interface for data visualization, search boxes, media library, material charts, building material vs Google street imagery visualization, and data export (Fig. 9).

This framework has a wide domain of potentials, allowing users to explore data in three different levels.

1. Explore data in a map scale, showing the building's materials for a complete city, filter data using a layering system, and extract a csv table of an average materials percentages.
2. Explore data in a region scale, having the ability to select several buildings using a region, and extract a csv table of an average materials percentages plus buildings count.
3. Explore data for individual buildings, where the user can get the information for the

Fig. 9 The landing page of the web interface

Fig. 10 Map view of the interface, a legend key for material percentages on the left side, and search parameters on the right side

building's age, typology, google street view image, and material percentage radar Chart (Figs. 10, 11, and 12).

The collaborative platform has been developed using a client–server web application architecture to facilitate communication between the compiled material dataset and the web platform front-end interface which visualize data on the browser and receive search parameters from users. This workflow was developed using Mapbox, Openstreetmaps, Javascript, Python, and the prototype was finally deployed using a Github server.

The geometrical representation of analytical data can vary and be shaped in multiple formats such as vector layers, images, 3D interactive

Fig. 11 Region views of the interface and materials charts on the right side for the selected region

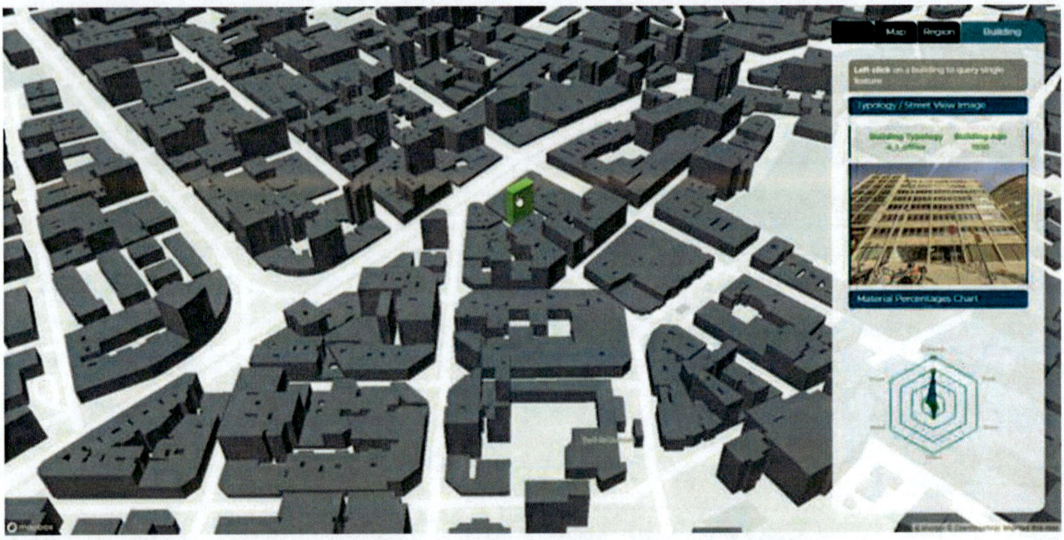

Fig. 12 Building view of the interface, blocks information, GSV image, and materials radar chart

model, charts, and statistical data for the material percentages as a result from the GAN model. These different layers and data are being stored on the server and could be requested using Mapbox and Javascript.

The user interaction takes place on the platform interface that consists of two main parts corresponding to the associated activities: A main wide layout that shows the map plus the Geojson file, and a side bar that shows statistical data plus raster images. This simple interface affords a higher level of communication between the user and the prototype (Fig. 13).

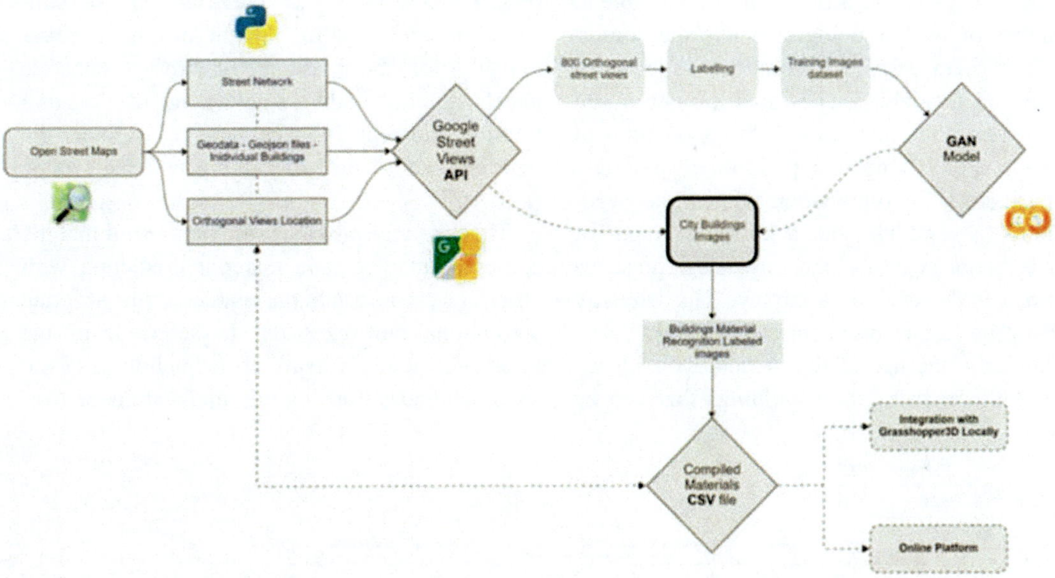

Fig. 13 Technical workflow diagram

3 Results

Observing the results of the use of both the database and interface for detecting and clustering different materials, provides critical feedback regarding the value, suitability, and necessary improvements of the proposed process. When tasked with mapping different material availability at the urban scale, it is immediately apparent that facade information alone surfers from a very narrow understanding of the quantities and types of material. For example, wood is not a predominant facade finishing material, but it does appear in structural and non-structural components throughout the building. An example of how to face this limitation is the formulation of a filtering strategy that uses multiple database fields in an established sequence to screen for the probability that different wooden components are present. By combining the filters in different orders, and applying different thresholds, a workflow allows to map the confidence level that timber floors or wooden trusses are present in buildings without the need to perform an on-site survey (Fig. 14).

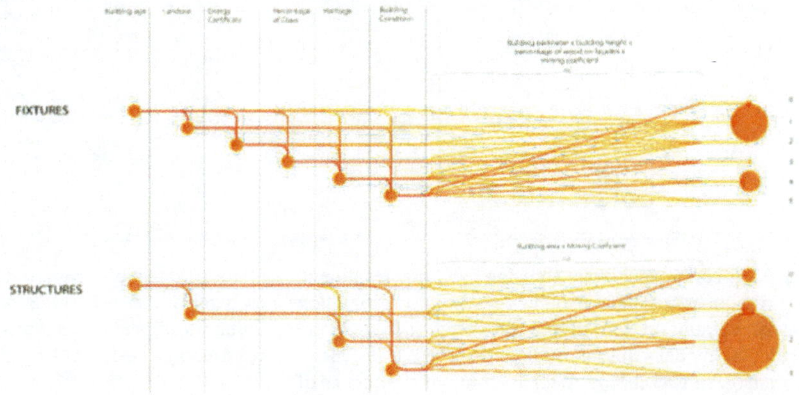

Fig. 14 Decision tree showing the connection between combinations of data filters to probability of finding wood in the skin or the structure of buildings

The strategy above proves that it's possible to formulate building typologies where the combination of data extracted from the street view façade and available cadastral information can estimate material quantities within a set limit of accuracy. For example: a building age of less than 25 years, in combination with a glass estimation of over 50%, and a building height of over 7 floors suggests structural steel presence within a 70% level of accuracy. This strategy needs to be further researched (Fig. 15).

Similarly, the use of the database for identifying concrete in existing buildings that can be reused, showcased that building year of construction, and building height can be a better indicator of the presence of precast concrete panels, and that adding a percentage of glass as a filtering element is necessary to exclude the buildings most probably built on steel frame (Fig. 16).

This exercise provides sufficient proof that the combination of facade material prediction with other census information enables more accurate predictions and expands its scope from the facade to other elements of the building. Given the relational nature of this application of ML,

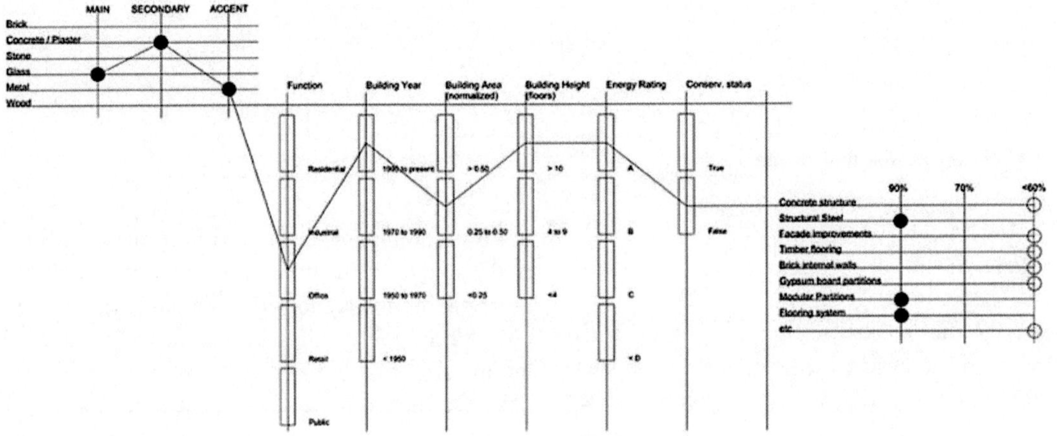

(a)

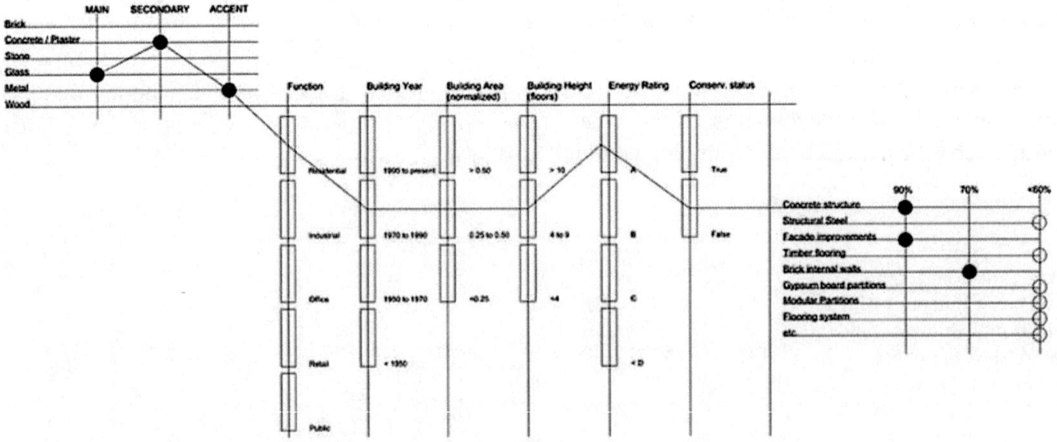

(b)

Fig. 15 Workflow diagram showing the variables, filters, and results of a potential building typology estimation workflow built on top of the proposed fade material prediction model. Buildings classified based on main, secondary, and accent material pass through a sequence of filters using public cadastral information to estimate the probability that a building scale condition is met. **a** and **b** show different estimations based on the building use, construction year, and height

Fig. 16 The mapping of cast in situ concrete considers buildings constructed after 1960 with a proportion of a minimum 40% of the facade predicted as plaster, and a facade area in the 90 percentiles. The result is presented as a geolocated gradient map as well as a bar chart computing neighbourhood totals

the process can be expanded to include comprehension of different types of image-based datasets. Satellite view for example will provide important information about the building state (if roof is in bad state, the building is likely unused, while a newly repaired roof is likely a sign of major renovation), and about the roof finishing material, thus augmenting the scope and accuracy of the material estimation. Similarly, different sources of information from city initiatives and public data can be added to narrow the scope of material identification where necessary. For example, when working on identifying reuse ready brick in the former industrial area of Barcelona, data can be sourced about future rezoning proposals and combined it with existing databases to reveal the brick buildings that will be proposed for demolition in the next ten years.

Addressing the machine learning model used, specific topics affecting the prediction accuracy must be addressed. Finetuning and obtaining highly accurate GAN models is a very wide and technical topic. The selected architecture and the reference model used come pre-set with the most recent best practices, and these hyperparameters were used for training on the dataset. The model trained for 1500 iterations, and the accuracy plots appear to be stabilized by this point. While continuing training may provide an increase in accuracy, it is likely that other aspects must be addressed first including the size and quality of the labelled dataset and the consistency of the test data presented to the trained model.

Deep learning is known to require large amounts of training data, and popular benchmarking datasets for the state-of-the-art models contain upwards of a few thousands images, reaching to hundreds of thousands for the more general use models. The AI used in this study is trained on 800 labelled images, which passed data augmentation techniques, like rotation and mirroring, to artificially increase the number of samples. Therefore, increasing the quantity of original samples can have a significant effect on the model performance. However, the quality and consistency of the manual labelling as performed by the research group that has participated in the study is an equally interesting pound to address (Fig. 17).

More comprehensive documentation for the labelling standard is required as well as quality assurance and filtering of unsuitable labelled samples prior to training. Another possible improvement can be observed when the retrieved building database in more detail. While approximately 60% of the images achieve the desired standard—meaning the main building image-centred, its entire facade visible and its position being orthogonal—there is a large percentage of outliers presenting important distortions and

Fig. 17 From top to bottom: **a** mislabelling of stone as bricks. Bias to labelling metal; **b** bias to labelling metal. Dismisses stone features on facades **c** bias to labelling wood. Mislabelling **d** labelling key

Fig. 18 Top left to bottom right: **a**, **b** samples of the correctly standardized images; **c**, **d** samples of aggressively cropped images; **e**, **f** samples of buildings hidden behind tall fences; **g** sample of non-orthogonal image; **h** sample of complex urban layering, making human and computer comprehension difficult

making a coherent prediction either difficult or impossible; for instance, more than one buildings are featured at different angles or depths, only a small percentage of the building ground floor is visible, building is covered by tall monotonous fence, trees cover most building features, fully orthogonal view is not achieved, and others (Fig. 18).

4 Conclusions

This study explores the methods and techniques for creating an open repository to map the material availability in the existing building stock of a city. The study enhances building public datasets that can contribute to the circular economy in construction including datasets of material quantification in relation to the state of components, age of building, or renovation/ demolition states, among others. The contribution of this paper is the creation of an integrated web-based application that could be used by decision-makers during early planning stages or to be tailored so that it can be accessed by architects for implementing the data in early stage design of circular buildings that use existing building stock as new materials for construction. Observing the lack of public data of buildings related to materials and components, the presented techniques for developing material libraries at urban or regional scale can become crucial datasets for supporting the development of Circular Economy policies in cities, including for instance, decisions on the site, size, or nature of circular hubs for storing and processing the components and materials that can be reused.

Furthermore, the technique can be upscaled and applied in different territories where data on the built environment is scarce, and allow, therefore, different stakeholders such as planning experts or construction companies to make more informed decisions towards the reuse of BDW. The technique and results of the study can also become data sources for existing digital reused material marketplaces (RMMs) contributing to match offer and demand within proximity geolocations, maximizing the values of the circular (and local) economy.

The value of the process of this study lies not in achieving near 100% accuracy but in the use of open unstructured data as a basis, making it a highly repeatable and expandable application, and attempting to propose a solution that crosses the boundaries of the original case study and can democratize access to data for local decision-makers in countries where open public information is less accessible. Furthermore, the purpose of this study is to identify specific buildings characteristics that limit the accuracy of the proposed technique for material detection and therefore, identify areas of action where the use of additional techniques should be considered, such as Laser Scanning Technologies, detailed site surveys by aerial and terrestrial robots or physical inspections and on-site photogrammetry.

References

Abdulla W (2018) Object detection tasks. https://engineering.matterport.com/splash-of-color-instance-segmentation-with-mask-r-cnn-and-tensorflow-7c761e238b46. Accessed 13 Jan 2023

Arora M, Raspall F, Cheah L, Silva A (2020) Buildings and the circular economy: estimating urban mining, recovery and reuse potential of building components. Resour, Conserv Recycl 154

Arora M, Raspall F, Fearnley L, Silva A (2021) Urban mining in buildings for a circular economy: planning, process and feasibility prospects. Resour, Conserv Recycl 174

Bell S, Upchurch P, Snavely N, Bala K (2015) Material recognition in the wild with the materials in context database. Department of Computer Science, Cornell University

Ellen MacArthur Foundation (2012) Towards the circular economy: an economic and business rationale for an accelerated transition. J Ind Ecol. Isle of Wight. https://www.werktrends.nl/app/uploads/2015/06/Rapport_McKinsey-Towards_A_Circular_Economy.pdf. Accessed 04 Oct 2022

Fröhlich B, Rodner E, Denzler J (2010) A fast approach for pixelwise labeling of facade images. In: 2010 20th international conference on pattern recognition, August, pp 3029–3032

Goodfellow I, Pouget-Abadie J, Mirza M, Xu B, Warde-Farley D, Ozair S, Courville A, Bengio Y (2020) Generative adversarial networks. Commun ACM 63 (11):139–144

Isola P, Zhu JY, Zhou T, Efros AA (2017) Image-to-image translation with conditional adversarial networks. In: Proceedings of the IEEE conference on computer vision and pattern recognition, pp 1125–1134

Kang J, Körner M, Wang Y, Taubenböck H, Zhu X (2018) Building instance classification using street view images. ISPRS J Photogramm Remote Sens 145 (Part A):44–59. https://doi.org/10.1016/j.isprsjprs.2018.02.006

Kumar V, Ai Lin Teo E (2019) Towards a more circular construction model: conceptualizing an open-BIM based estimation framework for urban mining. In: CIB world building congress

Koutamanis A, Van Reijn B, Van Bueren E (2018) Urban mining and buildings: a review of possibilities and limitations. Resour Conserv Recycl 138:32–39

Lee K, Hong G, Sael L, Lee S, Kim HY (2020) MultiDefectNet: multi-class defect detection of building façade based on deep convolutional neural network. Sustainability 12(22):9785

Pomponi F, Moncaster A (2017) Circular economy for the built environment: a research framework. J Clean Prod 143:710–718

Raghu D, Markopoulou A, Marengo M, Neri I, Chronis A, De Wolf C (2022) Enabling component reuse from existing buildings through machine learning—using google street view to enhance building databases. In: van Ameijde J, Gardner N, Hyun KH, Luo D, Sheth U (eds) POST-CARBON—proceedings of the 27th CAADRIA conference, Sydney, 9–15 Apr 2022, pp 577–586

Tylecek R (2012) The CMP facade database. Research Report CTU–CMP–2012–24, Czech Technical University

Post Rock: From Designing a Building Material to Designing a Business Ecosystem

Meredith Miller, Thom Moran, and Christopher Humphrey

Abstract

Post Rock is an ongoing research initiative developing new architectural materials made from waste plastics. The authors, academic researchers in architecture and advanced fabrication, have developed novel processes for thermoforming that yield variable, stone-like patterning. This paper discusses a recent reframing of the technical and design research through an immersion in evidence-based entrepreneurship. Lessons from over 100 interviews with building industry professionals not only provided key input at an early stage of commercialization, but also shaped the team's thinking around circular strategies for architecture. The paper will demonstrate the feedback loops between insights shared from professionals "in the field" and refinements to Post Rock materials and fabrication processes "in the lab." Four key hurdles to commercializing this building technology will be discussed: identifying a customer segment, defining sustainability and climate change goals, ensuring fire safety, and connecting cross-sector regional resources. Not only are the findings relevant to this project's advancement, but they also point to broader challenges to circular design models. The work-in-progress shared here illustrates the importance of expanding design efforts beyond a singular focus on the product itself and toward co-designing a regional ecosystem of stakeholders across multiple sectors for waste products, particularly plastics, to become a climate-positive component of future buildings.

Keywords

Post rock · Waste plastics · Circular design · Immersion entrepreneurship · Business ecosystem · Materials

1 Introduction: Building Material Paradigms

The climate emergency has prompted sweeping audits of global energy use and greenhouse gas emissions sources, leading to an increased awareness of the embodied energy of construction materials within the allied fields of architecture, engineering and construction. With this growing concern for material life cycles, the practice of categorizing materials as mineral or bio-based has emerged as a way of distinguishing their relative global warming potential (Churkina 2020). Derived from inorganic primary resources extracted from the earth, mineral-based materials

M. Miller (✉) · T. Moran · C. Humphrey
University of Michigan, Taubman College of Architecture and Urban Planning, Architecture, Ann Arbor, USA
e-mail: mlmmlm@umich.edu

M. R. Thomsen et al. (eds.), *Design for Rethinking Resources*, Sustainable Development Goals Series,
https://doi.org/10.1007/978-3-031-36554-6_45

such as steel and concrete are the obvious target for embodied carbon reduction, though their strength and durability make them challenging to replace. Wood is commonly known as the most extensively used bio-based material in construction, but other bio-based building products are in wide use, such as bamboo or plant-fiber insulation. Where the energy demands of processing or transporting these materials are not excessive, wood and bio-based materials can yield a positive carbon balance, essentially storing the atmospheric carbon absorbed and chemically altered during the plant's lifetime.

Synthetic polymers confound the mineral versus bio-based paradigm. Industrially produced plastics are "organic," in the chemical sense that hydrocarbons are central to their molecular structure. Plastics derived from petrochemicals contain carbon traces of the prehistoric life that was pressed into fossil fuels over hundreds of millions of years. Thus, in addition to the energy required across the oil extraction, chemical processing, and manufacture of plastic products, the carbon trapped within the polymer chains can release greenhouse gasses into the atmosphere if the plastic is allowed to decay or burn. Waste-to-energy facilities in the U.S. were responsible for 12 million tons CO^2e in 2016, more than half of which resulted from incinerating plastics (Royte 2019). Improving recycling rates, cutting down single-use plastics and decoupling plastic production from fossil fuel resources are all important strategies for mitigating the climate impacts of the plastics industry. However, it is also important to think about the staggering amounts of plastic that have already entered the world's systems–over 60 million tons of polymers were produced in North America just last year (Beyond Plastics 2021). The global warming potential locked within all that synthetic plastic in circulation would have devastating impacts on CO_2 reduction goals if allowed to escape.

Buildings offer long-term storage for carbon-sequestering materials such as mass timber and other bio-based products. When designed for reuse and reassembly, recycled plastic components could offer a smaller but still significant piece of the puzzle, especially where plant-based products cannot be used in construction application due to their performance properties. The material composition of building assemblies that require water-resistance such as foundations, envelopes, and roofing must be considered within the framework of embodied carbon. Reusing resources that have already been extracted and processed can help reduce the embodied energy intensity of new building materials. Could plastic waste, especially streams of plastic waste that are currently untapped, become resources for future building?

This paper will discuss lessons on circularity learned from pursuing commercial applications for Post Rock, an experimental architectural material made from waste plastics and other materials. Addressing the UN Sustainability Goal #12, Post Rock research and development rethinks the value chains of building materials. While this process of translating architectural research is still at an early phase, the ongoing efforts have revealed a range of insights relevant to the UIA conference's themes, including regional resources, rethinking waste, material life cycles, and the role of design and architects within these cross-sector domains.

2 Design Experiment to Designing a Product

Post Rock is a material research initiative developing new building materials from waste plastics and aggregates. Since 2015, a team of architects and advanced fabrication specialists within a North American university's architecture department have been exploring methods for converting plastic material from problematic waste streams into a durable and aesthetic composite. This work began as an architectural design experiment inspired by a recently named geological phenomenon, plastiglomerate. Formed when organic and inorganic materials such as stone, sand, and seashells fuse with polymer plastics, plastiglomerates are the result of plastics accumulating in marine and intertidal ecosystems (Corcoran 2014). Given the inherent durability of both stone and plastic, plastiglomerates are likely to last for a very long time, leaving behind a layer in the geological record indicating a human-dominated period of earth's history. While plastiglomerate rock specimens must meet a threshold of hardness and mineral composition to be included in this classification, their appearance varies with the different component parts of human and natural origins. Partially melted fragments of shampoo bottles, braided fishing rope, seashells, and other objects agglomerate within a medium of rock swirled with unnatural colors. Smooth and rough, synthetic and natural, banal and uncanny, this hybrid geologic material is being called the stone of the anthropocene (Corcoran et al. 2014).

The ambition behind the initial Post Rock research effort was to speculate on the architectural potential of this hybrid, durable material. Rather than deriving plastiglomerate from the polluting processes that led to its formation, however, the team aimed to divert plastic waste from damaging waste streams by finding new end-of-life solutions. Initial prototyping simulated the geological processes that form plastiglomerates at a small scale in the team's architectural fabrication lab. A wide range of thermoplastics (PP, LDPE, PE) and mineral aggregates (sand, reclaimed concrete) were combined and heated in molds. The process design evolved to ensure the smaller plastic fragments melted enough to bind the mixture while larger fragments (plastic and non-plastic) remained at least partially intact, acting as the aggregate within the fluid medium of melted plastic. This approach created a heterogenous surface patterning similar to the unique, hybrid appearance of plastiglomerates, where component parts such as the curved rim of a flowerpot or a broken piece of concrete remained legible within smooth blends of plastic and sand (Fig. 1).

From a design point of view, the team embraced the way in which different "waste" components remained visible, rather than blending into a homogenous new whole. As an authored version of plastiglomerate, Post Rock had the potential to not only absorb plastic waste, but also to visually communicate the dispersed geography of its component parts. Used in architectural applications, Post Rock could define an aesthetics of waste, giving new value to an otherwise undesirable class of materials through design.

Fig. 1 Early post rock test sample, 2016

From a technical perspective, the team aimed to develop thermoforming processes that were low-energy, choosing polymers with a lower flow-temperature and using gravity to move the melted plastics in the mold and create various patterns. Moving on from the initial small-scale prototypes, the team has developed thermoforming methods[1] that apply robotically controlled movements to produce unique patterns and varied textures at the surface of the material while maintaining a modular shape and size. This technical progression of the research has been driven in large part by a pivot toward thinking of Post Rock as a viable commercial building product. Having shared early prototypes in architecture exhibitions and design publications, the team sought to expand beyond the limited impact of small-scale, one-off projects. The key question became: how can Post Rock have a measurable impact on the enormous amount of plastic waste? According to the Environmental Protection Agency (EPA), only 8.7% of the approximately 36,000 tons of plastic discarded in 2018 in the United States was recycled (EPA 2022). The global construction market is a $12.5

[1] The team has patented its novel method of applying heated molds and controlled gravity-assisted movements during the casting period. Moran et al. Monolithic thermocasting of polymer mixtures for architectural applications. United States Letters Patent No. 11,623,372. Issued April 11, 2023.

trillion industry (Oxford Economics 2021)—perhaps the only industry large enough to absorb all this waste and put it to use. The challenge, then, was to change the team's approach in order to chart a path toward commercialization. What began as a series of design-driven projects and installations within the limited realm of architecture academia evolved into a business-informed design of a circular building product and, more importantly, the ecosystem of particular stakeholders necessary to support a circular building product. This paper will describe the team's application of evidence-based entrepreneurship as a methodology for gaining insights from industry to shape research toward greater impact.

3 Evidence-Based Entrepreneurship

Architects often express the ambition for our work to reach beyond an esoteric audience and to solve problems in the world. Less commonly do architects describe precise methods for identifying where and how architecture can be effective. Evidence-based entrepreneurship (EBE), also known as the lean start-up methodology, is a systematic approach to business development that emphasizes learning at the early stages. Popularized by Eric Ries, author of *The Lean Startup* (Ries 2011), this methodology has gained a following among tech industry executives and aspiring entrepreneurs. A central principle of this approach is for the entrepreneur/inventor to "get out of the building," or seek input from people who represent the end-user of the proposed business or innovation early in the process. The idea is not to pitch the innovation but to listen to people's experiences within the specific domain that the entrepreneur/inventor hopes to enter. This process of "getting out of the building" and "talking to people" is referred to as Customer Discovery, an early step in learning and validating the core assumptions behind a business idea before heavy investment is made (Constable et al. 2014). The U.S. National Science Foundation (NSF) has adopted the EBE methodology in its Innovation Corps (ICorps) program, which annually runs several cohorts, each comprising about 24 academic teams seeking to commercialize their technology or innovation (National Science Foundation Innovation Corps 2022). Similar to its application in the private sector, the NSF trains researchers in this approach to learn early on what problems their innovation might solve in the "real world." The intensive seven-week course introduces academic researchers to EBE tools with a focus on Customer Discovery, encouraging them to get "out of the lab" in order to talk with people whose experiences can provide insights into the field their research is aimed to impact.

As participants in the National ICORPS Summer Cohort #2 (online, 6/28/2021–8/12/2021), the authors conducted interviews with professionals across the building design, construction, and product manufacturing industries to more narrowly define a user group, their needs, and the specific hurdles to producing a Post Rock product that could meet these needs. To qualify for this national program, the team participated in the Midwest ICORPS Node in July 2018. At this earlier date, the building application was still undetermined. The fundamental assumption that needed testing was more broadly the idea that architects would choose to specify a recycled plastic product in their building out of a combination of sustainability goals and aesthetic preferences.

Given these questions, the team drilled down to the decision-making factors that informed a particular material choice. EBE literature encourages asking questions that prompt the interview subject to reflect on recent behaviors and experiences, rather than speculate on what he or she might do in a hypothetical future scenario. Rather than ask: "Would you, (Project Architect at leading American design firm), specify a high-recycled-content rainscreen product in a future mid-rise building?", the team would inquire about past experiences: "Can you tell me about a recent time you've selected a new rainscreen product? What was the team looking for and how did you rule out other options?" For an architect, the response may focus on a specific project where various pressures of cost, design ambition, and approval processes steered the cladding decision toward certain products and finishes. For a builder or installer, the questions may prompt stories about installation issues, pricing challenges, or back-and-forths between the builders and the design team to reduce risks posed by a new or untested facade assembly. In these conversations, the purpose was to gather anecdotal evidence from experienced professionals that could confirm or challenge the assumed role Post Rock could fill within the building industry. In aggregate, the interviews revealed patterns and tendencies that pointed to the values, pressures, and pain points affecting material selection from different stakeholder perspectives.

Through these two programs, the team has conducted 150 interviews, 84 of which have engaged industry professionals believed to represent the target customer (architects, developers/owners, and builders), with the remainder representing the broader product ecosystem (competitors, manufacturers, engineers, material suppliers, building code officials, regulating bodies). Interviews were conducted primarily over video conference and were not recorded to promote candidness and protect privacy. The researchers took thorough notes in a spreadsheet that enabled key-word coding of responses and categorizing the interview subjects by stakeholder category. The following sections will summarize the findings of these two rounds of customer discovery interviews and describe the impact of these findings on the research trajectory. Finally, a few key interactions with relevant stakeholders will be shared in greater detail to show specific models of circular and/or plastic building material ecosystems that provide valuable insights into the challenges and opportunities of pursuing a commercial product in an industry that is notoriously resistant to change.

4 Applying Feedback Between Field and Lab

While other applications for Post Rock thermoforming technology remain on the horizon, the first stage of customer discovery interviews focused the efforts toward rainscreen facade panels as the initial target application. A rainscreen facade assembly is an opaque, exterior wall assembly in which an outer layer of finish or cladding materials are held off the primary weather barrier of the wall with an air gap of at least 3/8″ (10 mm) between. Thus, the outermost layer of rainscreen panels is not load-bearing but provides the primary aesthetic component of the building facade. Half the architects and builders the team interviewed in the first round referenced the widespread adoption of rainscreen technology in certain regions of the United States, despite it being a more costly wall assembly than more typical wall assembly methods. While rainscreens have been prevalent in European markets for decades, it has experienced recent growth in the U.S., in part due to new energy standards for building envelope design. One key new standard for non-residential buildings over four stories high, ANSI/ASHRAE/IES Standard 90.1, Energy Standard for Buildings Except Low-Rise Residential Buildings (ANSI/ASHRAE/IES 2019), require a continuous layer of insulation which is more easily and affordably achieved with a rainscreen assembly. Given the visibility of the rainscreen facade panels, many of the architects interviewed by our team cited the need for a greater number of options on the market in order to differentiate their projects from that of others while meeting their project goals (cost, sustainability, and durability being the most common). In terms of performance, rainscreen facade panels must withstand loads from wind and self-weight (a significant criteria for the design of attachment systems and substructures); the rainscreen material must hold up to UV radiation; and depending on the building type and location of installation, the panels must meet certain tiers of fire safety and health regulations.

Of the thirty-two professionals interviewed in the first round in 2018, 18 or more than half, were with architects and interior designers, including architects at large international design firms known for innovative buildings, mid-size design-build firms, small boutique practices, LEED-driven corporate firms, and technology-driven companies with alternative delivery

Fig. 2 Column prototype installation, in group exhibition (name withheld for anonymity) 2017

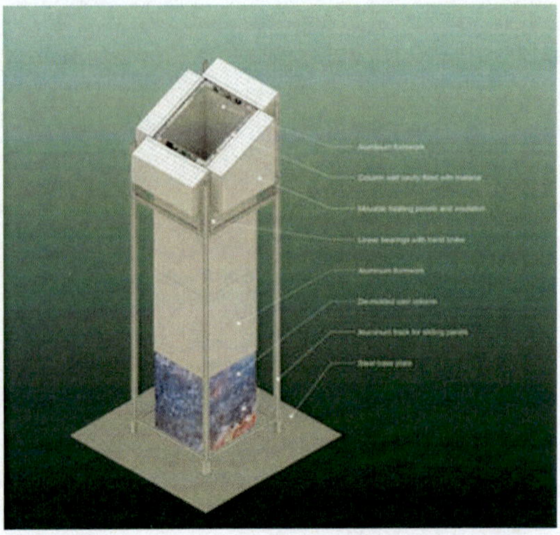

Fig. 3 Heated mold in "slip-forming" configuration to produce hollow, monolithic components

models. Six interview subjects represented manufacturers of rainscreen or cladding panels and three came from construction firms. Others included professional construction and reuse associations, architecture journalists, and material distributors. In addition to confirming the rising use of rainscreen construction in the U.S., these "user groups" offered different points of view on rainscreen products including finish preferences, sustainability concerns, lead time, compatibility with existing back-up systems, ease of installation.

These insights informed the next phase of prototyping, which shifted away from the prior focus on monolithic casts and toward a modular panel system that would integrate with conventional facade construction. The largest Post Rock prototypes at that time were large (48″–96″ or 122–244 cm in height), hollow components that could make up a primary building element, such as a wall or column (Fig. 2). The fabrication method used heated molds in an incremental manner similar to slip forming monolithic concrete structures (Fig. 3).[2] One successful example produced through this early phase of prototyping is the "Tripod," a freestanding assembly consisting of (3) 30″ × 30″ × 50″ hollow, monolithic components connected by hardware. Its stone-like surface patterning resulted from sand (40%) and a mixture of heterogeneous post-consumer polymer mix sorted for color (Fig. 4). While the appearance and durability proved promising for architectural applications, the process offered little control over important qualities including consistency and dimension of wall thickness, predictability of surface appearance and mixture distribution.

Fig. 4 Tests of modular mold configuration, 2019

With the new goal to produce panels that could be easily lifted and installed by one person and mechanically fastened to a third-party rainscreen back-up system, the research team made significant modifications to the fabrication process. Modular, reusable molds constructed from sheet aluminum and aluminum T-slot framing were developed to minimize material waste in the production of panels. The smaller mold size allowed the team to utilize the department's

[2] Fabrication process patented in the U.S, Mexico, and Canada.

six-axis Kuka robot, affixing the heated molds to the arm so that the gravity-assisted flow of materials could be controlled during heating and cooling. Successful prototypes (12″ × 12″ × 4″ hollow blocks with ½″ walls) have been produced using this process of widely varying mixtures (Fig. 4), including.

i. post-commercial regrind mixed polyethylene and polypropylene and sand (varying percentages from as little as 5% sand up to 60% sand)
ii. 50% sand, 40% post-consumer polypropylene, 10% virgin polypropylene
iii. 100% industrial shredded TPU (Thermoplastic Polyurethane)

The molds currently being developed will produce square panels with dimensions more closely approaching that of rainscreen cladding, at 20″ × 20″ (50 × 50 cm) with a thickness of 2″ (50 mm).

The pivot from many molds with unique geometry to a limited set of reusable molds with modular dimensions also put emphasis on the patterning of different colors and textures of the plastic ingredients as the primary options that could be customized. Architects spoke about the importance of a major finish material as having a "story" or "narrative" that aligned with a project's overall sustainability messaging. With the possibility of aligning the feedstocks to different geographies of waste, Post Rock could not only perform sustainably but also communicate a story of material reuse through its heterogenous aesthetic.

5 Lessons on Building Circular Ecosystems

Having narrowed the target application to rainscreen panels (Fig. 5), the team joined the 2021 NSF National ICorps Cohort with a new set of hypotheses to test. The seven-week course required a minimum 100 interviews. Primary among these hypotheses was the belief that architects seeking a lightweight, customizable rainscreen panel with a sustainability narrative, would choose Post Rock over other options such as UHPC, GFRC, or terracotta. Early on it became clear that this statement includes a number of untested assumptions. First, what tier or cost range of rainscreen product would a Post Rock panel compare to? Would a Post Rock rainscreen panel measure up to existing products of a similar cost range in terms of safety and durability requirements? What metrics of sustainability are architects using to evaluate products, particularly those belonging to the facade assembly? Given the varying exposures to risk among project team members, are architects even the primary "customer" or are owners and builders key decision makers to convey the merits of a recycled-content rainscreen product?

This section will address findings from the second stage of customer discovery interviews and the shift from designing a product and process to designing an ecosystem.[3] The table below breaks down the stakeholder categories of the 118 interviews completed between June 28 and August 12, 2021. (Note that this category tallies add up to 127 because some interviews included more than one interview subject and some interview subjects represented more than one category, for example, an architect who also acts as a zoning commissioner).

Stakeholder category	Interview count
Architects	35
Competitors	22
Building owners/Developers	15
Contractors	13
Key partners	13
Engineers	10
Advisors	8
Material suppliers	6
Zoning commissioners	5

[3] Interviews were conducted under the condition that proprietary information would not be published. This paper shares generalized conclusions derived from patterns in interview responses. Any information linked to a person's identity here is publicly available elsewhere.

Fig. 5 Post rock rainscreen panel designed to integrate with existing wall assembly systems

Relevant insights from the interviews fall within three related areas of further study: **sustainability, fire safety, regional resources**.

Sustainability and Climate Change Mitigation: The Post Rock team went into customer discovery with the belief that transforming plastic waste into a recycled product with a durable application would undeniably be (and be perceived as) sustainable. Not only did material reuse divert waste from landfills, incineration, or more problematic disposal scenarios, the use of non-virgin polymers would have a smaller carbon footprint compared to newly processed and manufactured polymers. The majority of people interviewed cited sustainability as a driving concern for their work. Speaking to people across architecture firms, construction companies, engineering firms, and sustainable design consultants, the team heard a range of priorities from focusing on natural ecosystems to energy efficiency to human health. A number of certification platforms, such as U.S. Green Building Council's LEED score or Living Building Institute's Passive House Standards played a role in how professionals measured sustainable impact. However, between interviews conducted in the summer of 2018 and those conducted summer of 2021, one focus area of sustainable design became quite clear: there was a dramatic increase in mentions of carbon footprint. (Two out of 32 interviewees brought up building's carbon footprint in 2018 versus 42 out of 118 interviewees in 2021.) This change reflects a more general awareness of embodied energy in architecture and other sectors. While improving the operational energy efficiency of buildings has been a particular focus of sustainable design and construction for over two decades, accounting for the carbon impacts across the full life cycle of building materials (resource extraction, processing, manufacturing, transportation, and installation) has revealed some hidden environmental costs of energy-efficient construction. A recent study reports: "While the average share of embodied GHG emissions from buildings following current energy performance regulations is approximately 20–25% of life cycle GHG emissions, this figure escalates to 45–50% for highly energy-efficient buildings and surpasses 90% in extreme cases" (Rock 2019).

Increasingly, life-cycle assessments (LCAs) and Environmental Product Declarations (EPDs) serve as tools for quantifying the whole-life impact of a building product and making this data accessible to decision makers across the design and construction timeline. Material products that support a "circular economy" are one way to reduce the negative impacts of construction and the manufacturing sector that supports construction but reporting frameworks that rely on "cradle-to-gate" LCAs do not always account for the costs and benefits of using recycled content in place of virgin equivalents. The authors' interviews with design and construction professionals revealed an uneven landscape in terms of how embodied energy is influencing building product selections and construction approaches. With these methods still emerging, major industry players with larger economic capacity than smaller producers have an edge to perform the necessary studies and obtain third-party verification. Aware of these issues, the Post Rock team has been working with potential suppliers of waste plastics to understand the availability and quality of data that would plug into transparent reporting of a Post Rock's building product. Understanding upstream factors such as energy type and use, material composition, and processing and transport needs at this early phase of commercialization can help the team weigh decisions about supply that would later impact the product's global warming potential values and other key impacts.

Fire Safety Certification: While waste plastic is seemingly an ideal construction material to reach carbon sequestration goals, the flammability of petroleum-based polymers raised concern throughout the customer discovery process. Polymer products are no stranger to construction assemblies with flooring, roofing, siding, decking and other products commonly composed of thermoplastics, thermosets and composites. Within facade assemblies, laminar products such as vapor barriers, insulation, weather-proof membranes, and multi-material finish composites are an industry standard that obtain code compliance through the reliance on a non-flammable surface material. Recent building fire catastrophes, such as the Grenfell Tower in London and Torch Tower in Dubai, cite facade cladding as the main catalyst for flame spread, putting facade materials under increased scrutiny for their ability to propagate flame spread (Messerschmidt 2020). Compliance with National Fire Protection Association (NFPA) 285 and American Society for Testing and Materials (ASTM) E84 was often cited as a major barrier to commercializing a polymer composite as a facade cladding material. Along with that warning, several interview subjects offered the example of Bill Kreysler as proof that it could be done. His fiber-reinforced polymer panels recently obtained the rigorous NFPA 285 certification for use on the facade of the San Francisco Museum of Modern Art extension. Having started his career fabricating boats, Kreysler evolved his business model and material processes to meet the needs of architects designing complex geometries. Snohetta's design for the SF MoMA extension relied on Kreysler's FRP panels to create the rippling white surfaces of its enclosure while keeping the overall weight of the building under a limit set by existing foundations. The authors conducted a number of interviews with various stakeholders involved in the SFMoMA design and construction, including Kreysler. As a "competitor" and "advisor," his experience offered insights from his process of scaling up an experimental polymer material to meet the needs of a project of this scale.

Two key insights from this conversation include, first, the importance of building a team with experts who can fill gaps in the R&D process, particularly when it comes to material engineering and fire certifications and secondly, the perceptions of plastic as a hurdle to widespread adoption among architects and building owners. Following Kreysler's model, the authors are establishing relationships crossing industry and academia to conduct incremental studies of fire propagation behavior at a bench scale, well before attempting NFPA 285 which is specific to an assembly, not a material. However, by using waste thermoplastics, Post Rock differs from the virgin thermosets used in Kreysler's panels. The SFMoMA FRP panels included certain industrial byproducts as flame retardants, in addition to the presence of glass fibers, to slow down combustion (Kreysler 2021). Our team aims to avoid incorporating substances that undermine the carbon benefits of sequestering waste plastics. Another "competitor" working through similar commercialization issues, Netherlands-based start-up Pretty Plastics was founded by architect Peter van Assche and partners. Having first tested their 100% recycled plastic cladding tiles on interiors and a temporary pavilion, they have also had to reformulate their feedstocks to address fire safety standards without having to use retardants that would add too much weight, diminishing the knock-on benefits of using a lightweight cladding (Assche 2022).

Reusing waste materials over virgin feedstocks presents challenges with contamination and consistency, but it also presents the opportunity of finding thermoplastic feedstocks that already have flame resistant properties. The following section describes the team's process of forging regional connections with potential suppliers of waste feedstocks.

Regional Resources: The previous categories addressed key features or properties to incorporate in a market-ready Post Rock product to come. Anticipating embodied energy certifications as well as fire safety certifications emphasizes the importance of transparency throughout the product development. At this early stage designing a recycled-content product, it became clear that these "downstream" features or properties (how the product would perform) will rely on the "upstream" characteristics of the material supply chain (how the product's feedstocks would be sourced and processed).

Following the intensive seven-week period of customer discovery interviews, which gave insights into stakeholder needs at the user end of the life cycle (architects, builders, building owners), the team shifted to focus on stakeholders representing the supply side of the value chain (manufacturers, producers of waste plastic). Located in the Great Lakes region, where the automotive industry has and continues to loom large in the regional economy, the research team has begun building relationships with potential suppliers of waste plastics. Automotive plastics, especially those used within the interior of the car, have performance requirements driven by similar concerns as building codes. Like small building enclosures, motor vehicles contain materials that come into close contact with occupants and harsh exterior conditions. The minimum thresholds of health and human safety, UV resistance, and smoke and flame resistance for automotive plastics are parallel to those of the building industry. However, the metrics and standards for automotive components are specific to that industry, primarily the National Highway Traffic Safety Administration (NHTSA). To predict whether a product made from plastic recycled from, for example, a dashboard, would meet the fire-resistance requirements for a building cladding product (Standards ISO 5660–1:2015 and NFPA 271) would require translating between two separate industry-specific codes and standards. Most experts in fire-safety engineering are familiar with one particular industry, and one consulted by the team advised that testing. This example sheds light on a larger barrier to more widespread reuse and recycling of materials across different sectors—a lack of coordination among regulating standards, placing the burden of testing and proof-of-concepts on innovators (often small businesses and researchers).

One major automaker headquartered in southeast Michigan has expressed interest in Post Rock research as a potential end-of-life alternative for waste plastics from their two production facilities in the state. Like many major corporations in the U.S., they have committed to a clean energy transition and to finding circular solutions to reduce waste. One of the company's chief executives stated: "[Our company's] zero-waste initiative aims to divert more than 90% of its manufacturing waste from landfills and incineration globally by 2025" (Silverstein 2022). Under these initiatives, this company created a circular solutions department which is actively seeking creative solutions for a number of challenging waste streams. This department provided the research team with a list of over 100 different plastic waste categories generated by its regional manufacturing activities, totaling 7000 metric tons per month. Of this monthly amount, most is either reused internally or mechanically recycled via third-party recyclers. Approximately 10% is incinerated for energy recovery. 200 metric tons of various plastic materials per month are landfilled. With input from engineers the Post Rock team reviewed and evaluated this list of potential feedstocks to determine whether any of these post-industrial polymer waste streams would meet the performance criteria needed for a building cladding. The top five materials by mass include: PET, PPE, LDPE, HDPE, and PS. Based on preliminary analysis, the team has narrowed in on a few candidates of automotive polymer waste for further evaluation. By producing small scale samples, preliminary fire-behavior data can be obtained through cone calorimeter testing at third-party lab. This data will inform the final feedstock selection and mix in conjunction with other performance criteria for the panel.

While more research and development are needed to refine the product, the work to date has revealed a need (lower carbon façade materials) and an underutilized regional resource (automotive plastic waste) that could be addressed through the creation of a new value chain. Understanding stakeholder's needs, both downstream and upstream, at this early phase of commercialization allows the team to continue to adapt and refine its business model. Given the material characteristics of the recycled Post Rock product itself, changes in the value chain are changes to its visual and material qualities. The colors and textures would derive from the automotive plastic feedstocks. Reflecting current trends in car design and marketing, this edition of Post Rock will revive some of the early design goals, where the legible traces of various inputs would visually communicate a particular geography of waste. Moving upstream, engineers from the automotive partner have suggested altering the material composition of their plastic components if it improves its compatibility with Post Rock manufacturing needs. This idea speaks to the need to coordinate with stakeholders across value chain to improve rates of material recovery by designing products and materials in ways that anticipate future reuse. Design-for-disassembly principles have informed the development of a more modular Post Rock panel, rather than custom shapes and components the team initially prototyped, so that the life of the product could be extended before recycling would be needed.

6 Conclusion

In order to reuse one class of plastics in another industry or application, a number of hurdles must be confronted–hurdles that are not only technical but also logistical and cultural. Extending the life cycle of a particular material through its incorporation within a new durable product can require forming new relationships among committed stakeholders across different sectors. While prototyping continues to be a valuable technique for advancing the commercial potential of Post Rock, evidence-based entrepreneurship has become an important feedback channel to test the team's core assumptions. Customer discovery interviews have also helped the team navigate the complex ecosystems around building material manufacturing, from supply chains to regulatory bodies to end-users. Talking to people about their various experiences–whether evaluating, specifying, manufacturing, installing, or ultimately living with cladding materials–has not only shaped the team's research agenda but also provided a broad view into the level of cross-sector coordination needed for circular strategies to transform the design and construction industry.

Business development tools such as EBE and Customer Discovery were developed for entrepreneurs, who by definition take on financial risk in search of high profit margins. However, placing these methods in the context of a mission-driven design innovation empowered the Post Rock team to imagine other forms of risk and gain: environmental, social, aesthetic. Within research institutions, the process of identifying a market fit for academic innovations is referred to as "commercialization." For the authors of this paper, whose approach is more informed by design practice than scientific models of research, this process has become an important extension of architectural practice–defining ways for our work to be of service, and for whom. Eschewing the paradigm of designer-authored singularity more common to architectural practice, the Post Rock research initiative aims to work with regional stakeholders to co-design an upcycling ecosystem. In this way, the work can contribute to a larger effort changing building practice at a scale commensurate with that of plastic waste.

References

ANSI/ASHRAE/IES (2019) Standard 90.1 Energy standard for buildings except low-rise residential buildings. pp 45–53

Assche P (2022) Personal interview with peter von Assche and Reinder Bakker by the authors

Beyond Plastics (2021) The new coal: Plastics & climate change

Churkina G et al (2020) Buildings as a global carbon sink. Nat Sustain 3:269–276. https://doi.org/10.1038/s41893-019-0462-4

Constable G et al (2014) Talking to humans: Success starts with understanding your customers

Corcoran P et al (2014) An anthropogenic marker horizon in the future rock record geological society of America today, vol 24, no 6. pp 4–8

Environmental protection agency, Plastics: Material Specific Data. https://www.epa.gov/facts-and-figures-about-materials-waste-and-recycling/plastics-material-specific-data

Kreysler W (2021) Personal interview with William Kreysler by the authors

Messerschmidt B (2020) NFPA Journal: Data void. https://www.nfpa.org/News-and-Research/Publications-and-media/NFPA-Journal/2020/May-June-2020/Features/Grenfell. Accessed 3 Oct 2022

National Science Foundation—Innovation Corps (2022). https://beta.nsf.gov/funding/initiatives/i-corps. Accessed 8 Oct 2022

Oxford Economics (2021) Future of construction: A global forecast for construction to 2030.

Ries E (2011) The lean startup

Rock M et al (2019) Embodied GHG emissions of buildings–The hidden challenge for effective climate change mitigation. Applied Energy

Royte E (2019) Is burning plastics a good idea? National Geographic. https://www.nationalgeographic.com/environment/article/should-we-burn-plastic-waste

Silverstein K (2022) Environment and energy leader: GM has become an 'energy star' with efficiency and circular programs. https://www.environmentalleader.com/2022/05/gm-has-become-an-energy-star-with-efficiency-and-circular-programs/. Accessed 7 Oct 2022

Circular Economy Principles as Obstacles to Creativity?—A Study of Architects' Expectations of Challenges and Opportunities

Mia B. Münster and Marie-Jo Gutenkauf

Abstract

In the design phase, 80% of the environmental impact of products and services is determined. The implementation of a circular economy (CE) has the potential to reduce waste and encourage the reuse of resources. A CE enables firms to reduce pressure on the environment and stimulates the fight against climate change by adjusting the take-make-use-dispose paradigm to an approach that focuses on keeping materials in closed loops. A significant portion of the environmental impact of products and services is determined in the design phase, which potentially gives architects, who are involved in the early stages of the creative process, responsibility for meeting sustainability requirements. It has been claimed that sustainability requirements may reduce creativity. Therefore, it is relevant to investigate whether and how considering circular economy principles in the design phase will influence creativity. To investigate and comprehend the influence that principles of CE may have on creative outcomes, 21 Danish architects whose job duties particularly concentrate on artistic or creative work were interviewed. This study presents architects' thoughts on how CE might impact their creativity and discusses expected opportunities and challenges related to considering CE in the creative process. By contributing to research on architects and the transition towards a CE, the project touches on two of the 17 United Nations Sustainable Development Goals: Goal 11, which pertains to sustainable cities and consumption, and Goal 12, which concerns responsible consumption and production (United Nations in The 17 Goals. Department of Economic and Social Affairs, 2015).

Keywords

Circular economy · Creativity · Design process · Sustainability · Architects

1 Introduction

According to The European Commission (2018), 80% of the environmental impact of products and services is determined in the design phase. A circular economy (CE) enables firms to reduce pressure on the environment and stimulates the fight against climate change by adjusting the take-make-use-dispose paradigm to an approach focusing on keeping materials in closed loops,

Mia B.Münster (✉)
School of Design, Polyu Design, Hong Kong Polytechnic University, Kowloon, Hong Kong
e-mail: mbmuens@polyu.edu.hk

M.-J. Gutenkauf
Copenhagen Business School, Frederiksberg, Denmark

M. R. Thomsen et al. (eds.), *Design for Rethinking Resources*, Sustainable Development Goals Series,
https://doi.org/10.1007/978-3-031-36554-6_46

ultimately reducing waste. The concept of a CE is therefore seen as a new way of thinking about business (Ellen MacArthur Foundation 2013). Both academics and policymakers agree that design plays a crucial role in the transition toward a CE (Moreno et al. 2016; The European Commission 2018; Dokter et al. 2021). Since architects are involved in the initial phase of any design process, it has been argued that they are in a unique position to influence and implement new ways of building sustainable practices (Ruiz-Pastor et al. 2021; Sumter et al. 2021).

Architecture shapes the everyday lives of people, keeping them comfortable, healthy, and safe, but it also makes up a large part of the world's resources. As one of the most resource-intensive industries, the field of architecture has enormous potential to impact a CE transition (Creative Denmark 2021). According to the Circular Economy Action Plan, the built environment requires vast amounts of resources. The construction sector accounts for about 50% of all extracted material and is responsible for over 35% of the EU's total waste generation (The European Commission 2020). The designs of buildings are key to how they are used, their impact on their surroundings, and how long they remain fit for their purpose. Therefore, designing new buildings in line with CE principles is an important way of implementing CE in the built environment (Arup and Ellen MacArthur Foundation 2022). However, many architects, particularly those who are more experienced and have been trained in the past, have not been educated on the application of CE principles.

Creativity is required to meet the demands of clients or society in terms of both functional and aesthetic design solutions. Charter (2018) states that designing for a CE requires thinking about how to enable CE principles in the early creative stages of the design process. Previous research suggests that the demands of clients or society can be interpreted as restrictions by designers and obstacles to creativity (Ruiz-Pastor et al. 2021). Several studies also claim that sustainability requirements may reduce creativity (Collado-Ruiz and Ostad-Ahmad-Ghorabi 2010; Mohanani et al. 2014).

Implementing a CE in built environments is, in many regions, no longer just an option. From January 2023 onwards, it is expected that life cycle assessments (LCA) will have to be carried out for all new buildings over 1000 m^2 in Denmark (Indenrigs–og Boligministeriet 2021; Arkitektforeningen 2022). Therefore, it is relevant to study how and whether architects can think and work in a more circular manner without compromising creativity and artistic design. This study explores the potential impact of CE principles on creativity, as revealed through interviews with 21 experienced architects whose job roles primarily focus on artistic or creative work.

The paper explores architects' experiences with and expectations of including CE principles in their creative process, focusing especially on revealing whether and how the CE is seen as an obstacle to creativity. We do this by asking two questions: "How are CE principles expected to impact creativity in the design process?" and "What challenges and opportunities do architects expect when implementing CE principles in the design process?" By answering these questions, we aim to contribute to the research concerning the transition toward a CE in the field of architecture. Moreover, we expect the findings to be useful for management in architectural studios. The paper is structured into the sections: an introduction, the background, methods, results, and a discussion with concluding remarks.

2 Background

2.1 The Circular Economy and Architecture

Geissdoerfer et al. (2017) define the concept of a circular economy as follows: "A regenerative system in which resource input and waste, emission, and energy leakage are minimized by slowing, closing, and narrowing material and energy loops. This can be achieved through long-lasting design, maintenance, repair, reuse, remanufacturing, refurbishing, and recycling."

Design certainly rests at the heart of the circular economy. It requires redesigning, among other things, products, business models, cities, and the linear systems that have lasted for the past centuries (Ellen MacArthur Foundation 2015). To support designing for a CE, several methods, tools, and frameworks have been proposed. For instance, Bocken et al. (2016) developed a framework of strategies that aims to facilitate the shift from a linear economy to a circular economy, defining terminologies of slowing, closing, and narrowing resource loops. Within the field of architecture, CE principles have been described as focusing on waste management and the reuse of construction and demolition waste (Joensuu et al. 2020; Munaro et al. 2020). According to previous studies, practices such as adaptive reuse, design for disassembly, design for repair, and remanufacturing have been introduced to encourage CE principles within the field of architecture (ibid.).

2.2 Creativity

With creativity playing a crucial role in the design process, theories on creative work and processes are relevant to this study and will help to understand the design process. Mumford et al. (2002) identified two different processes associated with creative work: activities leading to idea generation and activities needed to implement ideas. Furthermore, Csikszentmihalyi (1988) argues that conceptualization, meaning the identification of an unsolved problem, requires creativity and can be acknowledged as a creative task itself. Regarding the creative process, Amabile (1983) introduced the componential model of creativity, which describes the stages of the creative process. As visualized in Fig. 1, the first stage of the process is called "task presentation" and involves identifying the goal or problem. This stage can be kicked off either by an individual's strong intrinsic motivation or by an external source, such as an assignment from a project manager. The next stage, "preparation," includes the preparation of a successful process, meaning a phase for building up knowledge, skills, and other relevant information. The third phase is called "idea generation," followed by "idea validation," which involves analyzing ideas against criteria for a task to ensure the usefulness or appropriateness of the novel ideas emerging from the idea generation stage. Finally, the creative process ends with the "outcome assessment," where decisions are made based on the results of the previous phase. The creative process applies to all degrees of creativity, from low to high. The final level of creativity depends on the levels of these three factors of individual creativity and the extent to which each stage of the process is completely realized.

2.3 Creativity Within a Circular Economy

Several studies have been performed on the role of designers in a CE. Sumter et al. (2021) identified key competencies for designers who specialize in design for CEs: circular systems thinking, design for recovery, design for multiple use cycles, circular business propositions, circular user engagement, circular materials and manufacturing, design impact assessment, circular economy collaboration, and circular economy storytelling. Although creativity is implicit in these competencies, the study does not specifically mention creativity or examine the relationship between CE and creativity.

Munster et al. (2022) show how a specific group of designers and architects, those involved in retail and hospitality design, are motivated to engage in the transition toward a CE but express, among other things, a lack of knowledge and systems and do not trust themselves to take a leading role in the transition toward a CE. Without studying the creative process in detail, their findings indicate that designers see their creative abilities as strengths that could be utilized.

Some scholars have claimed that sustainability requirements reduce creativity (Collado-Ruiz and Ostad-Ahmad-Ghorabi 2010; Mohanani et al. 2014). Collado-Ruiz and Ostad-Ahmad Ghorabi (2010) show that exposure to examples of environmental information caused fixation and reduced overall creativity in the idea-generation

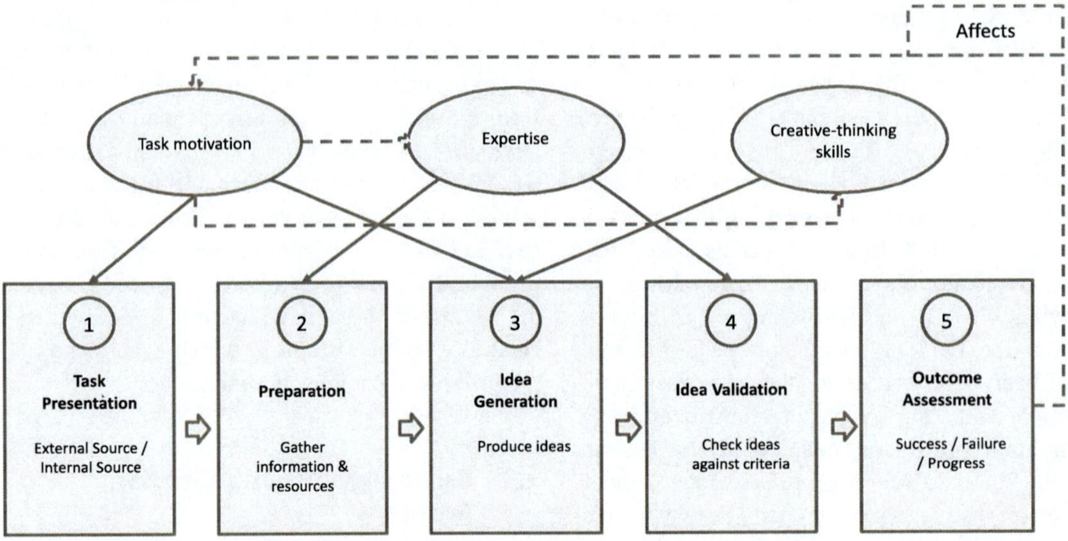

Fig. 1 Componential model of creativity. Author's creation, adapted from Amabile (1983)

process among students. The results of their experiment prove that detailed information significantly reduces the creativity of design ideas, but that soft information, on the other hand, does not. Mohanani et al (2014) shows that framing something as a "requirement" may cause a fixation on requirements, where designers' preoccupations with satisfying explicit requirements inhibit their creativity.

A study of creativity and CE conducted by Ruiz-Pastor et al. (2021) shows that the introduction of guided questions on CE principles during the design process leads to no significant differences in creativity or circularity. The indefinite results in these studies support the idea that there is still a need for research on how to foster creativity when designers are required to be creative while considering CE.

3 Method

3.1 Research Design

Given the explorative nature of this study, a qualitative approach was deemed appropriate (Crotty 1998). Allowing architects to express thoughts and experiences in their own words provides insight into possible ways of dealing with issues pertaining to sustainability and creativity. Our focus is therefore on understanding designers' expectations within the context of their practice as opposed to instances and frequencies of a particular response (Gioia et al. 2013).

3.2 Participant Selection

An interview guide was designed and tested prior to the interviews to enhance reliability (Yin, 2003) and ensure that respondents would be able to both understand the topics in focus and express their thoughts and experiences in their own words. Topics in focus ranged from former experiences with sustainability initiatives in a professional context to knowledge of CEs in general, and finally to experiences and considerations of involving CE in the creative process.

Respondents met all the following criteria: they (1) had earned a degree in architecture from an accredited institution, (2) had at least five years of professional experience, and (3) worked in full-time positions as architects, either in a design studio as in-house designers or as independent architects, (4) their job particularly focused on the creative process or on creating artistic solutions, such as retail design, which is constraints," indicating that being an architect is not only about being creative. This was supported by others, who said, "Architecture is a constrained form of art, in that we do have regulations, and we do have a budget most of the time."

Although implementing CE principles in the design process might be challenging, the architects still viewed this task positively:

"There are always new regulations coming along [that we must follow], and we are always being challenged by clients. So, having to take one more thing in as one of the parameters for how something is going to look, I don't think there are many architects who would see that as a problem. Of course, it will present some challenges, but probably mostly in terms of price rather than the creative side, I think."

The seriousness of the reason behind the challenges was also clear to the architects, who mentioned that they, like everyone else, are obliged to fight climate change: "Everybody needs to accept these constraints and take them in a positive way."

In general, the architects acknowledged the challenges and always mentioned opportunities as well: "You can also take the constraints as something positive." Another offered the following thought: "It would be hard for me not to say that it [considering CE principles] is a limitation, but it could also be motivating. Isn't it always like that? Other possibilities open up."

4.2 Opportunities Arising from Implementing CE Principles in Designs

CE as Fostering Curiosity

The participants expressed that CE principles go hand in hand with their artistic design approaches, as they develop the participants' mindsets and the way they work, reinforcing the curiosity needed for the artistic design. For example, one respondent said the following: "I think that [including CE principles] truly goes hand in hand with this curiosity [that architects need], this way of developing our mindset, and the way we work."

Another architect put it this way and emphasized that design is always about creating something new; therefore, considering new requirements can be seen as inspiring rather than limiting: "I think that design benefits from the idea of circularity and sustainability. It feeds curiosity [...], and, I mean, it underlines the idea that there is never a fixed solution. We cannot just go with what we did before."

Such expressions indicate that architectural design might benefit from CE principles, as new challenges stimulate architects' curiosity: "We are used to handling lots and lots of requirements and restrictions, so I see it purely as a possibility for development. It could be really exciting."

As such, the constraints arising from circular economy principles were, in most cases, perceived as positive: "There are constraints, and everybody needs to accept these constraints and take them in a positive way."

CE as a Driver of Creativity

The participants mentioned that considering CE principles can enhance creativity and that they feel stimulated by the principles. One architect said, "I think it could be a good chance for development. It would be inspiring," indicating that the discussion of CE had triggered her creativity, and she was already inspired to grapple with it.

Some mentioned that CE principles would allow them to think in a different way and that it might be stimulating to change their focus: "It is possible that you can think of it in another way [...] I think you can get more creative." Another respondent stated, "I think you need to embrace the circularity [...] you should be happy about it and not feeling constrained, but actually feeling that it's a driver. A positive drive."

Constraints as Idea Generators

Architects acknowledged that there might be challenges in the process, but they highlighted that they believed that the consideration of CE principles would generate new ideas: "It will challenge you to change your focus a little bit. So, in that respect, you can feel challenged, and maybe you're going to get a new idea."

Others described the link between constraints and new ideas in the following ways: "It would probably cause some problems, but it doesn't need to. Of course, it might be that a lot of ideas get shot down because, oh no, we can't do that—but it might also be that that leads us to other things that we might not otherwise have thought about."

> I see it as a possibility for development. With design, you are always working with limitations in one way or another, so I see this clearly as a possibility to develop in the right direction—but also in the correct societal direction with respect to this. I see that as a huge benefit and also as a responsibility.

Easing Consciences Through a CE

Another opportunity mentioned by the architects is that a CE could help ease their consciences about the effect that architecture has on the environment. Several architects expressed a guilty conscience over the high level of resource consumption and waste production associated with their practice. For example, one respondent said that she still felt like a part of the problem: "I would much rather like to be, you know, part of the solution than part of the problem. And I feel that I'm still a little bit too much part of the problem by building new [structures]."

Others directly mentioned that they had a bad conscience about their work: "I feel bad about working in a profession that consumes so many resources. When we use natural stone, for example, we use resources that will never be returned to the environment from where we took them."

The hope that CE could change that was expressed by several architects. For example, "I would really like it if my work didn't leave me with a guilty conscience. That would make me feel better about what I do." And as another one said, "I definitely see it [CE] as a possibility for development. Personally, I think I would be happier about my product if circularity was built in from the start."

Architects as Advocates for a CE

The architects acknowledged that they could potentially have a central role in the transition toward a CE due to their involvement in the early stages of the design process and their close collaborations with clients, contractors, and suppliers.

> Architects become more, in my opinion, the puzzle collectors; they collect all different industries together, and it's important for the architects to have a very clear vision of what we are creating.

As such, architects could take upon themselves the role of educating and consulting other stakeholders about including CE principles. Some mentioned that their clients would not have the time to look into the sustainability issues of projects and that they would expect architects to provide information: "I know that they [the clients] won't look into it, and they want me to look into it."

Another architect suggested that architects who have worked with CE for a long time could eventually become advocates for CE and guide others, either as consultants or in the companies where they work:

> Actually, I could almost see myself as a salesperson [for a circular economy], selling the idea to the brand I work for, because there definitely isn't anyone else in the company who is thinking about these things.

Such advocates might be able to evaluate design proposals and come up with suggestions for alternative materials that would be more sustainable. Or, as one respondent suggested, they could build up a library of materials from buildings that have to be torn down: "A place where materials could be stored so that others could use them—I would like that."

Architects would need knowledge and practical experience with CE to educate other stakeholders, but the role of architects could involve convincing stakeholders of the value of the CE in the future and engaging with that possibility.

5 Discussion and Conclusion

The findings suggest that the participants were motivated, curious, and had a strong will to play a role in the transition toward a CE. Our findings confirm former studies from scholars such as Dokter et al. (2021) and Münster et al. (2022) suggesting that more knowledge, systems, and legislation are needed, and that the market might not always be ready for the transition to a CE. However, our findings also indicate that creativity is not likely to suffer from considering CE principles. On the contrary, the interviewed architects described their practice as "a constrained form of art," and their responses indicated that they were trained to handle problems, deal with limitations, and still come up with creative solutions. Thus, even without having much experience with handling the limitations that CE principles might cause, it was obvious that the new constraints triggered the architects' creative minds during the conversations. It is therefore likely that considerations of CE principles will not have a negative impact on creativity; rather, it is likely that CE considerations will be seen as solutions that architects would like to be part of. Besides the environmental benefits, the interviewed architects suggested that considering CE principles in the design process might (1) foster curiosity, (2) drive creativity, and (3) generate new ideas. The architects also saw CE as a way to (4) ease guilty consciences caused by working in a field that intensely exploits natural resources. Finally, some architects suggested that knowledge of CE principles might (5) provide them with the opportunity to become advocates for CE, thereby suggesting a new role for architects in organizations and in society.

The interviews provide examples both in terms of architects using creativity to solve problems in the design process, and in terms of identifying unsolved problems, such as a lack of knowledge and a local materials library. Our findings suggest that CE considerations should be involved from the beginning of the design process, which would be in the "task presentation" and "preparation" stages (Amabile, 1983), but more studies are needed to identify the possible impacts on creativity in the different stages. This could be done by studying architects as they work through the different stages of the design process. The "task presentation" and "preparation" stages are also stages in which external factors, such as legislation, project management, and clients, are involved. This indicates that the transition to a CE will not be solved solely by the architects but requires a mindset shift among all stakeholders involved in innovative projects. Further studies on collaboration between stakeholders are needed to understand the impact of external resources on creativity.

In conclusion, our study has provided insights into the potential impact of CE principles on creativity in the field of architecture, as revealed through interviews with a small sample of architects. Our findings suggest that CE considerations may not have a negative impact on creativity, but rather may drive curiosity, creativity, and generate new ideas. While the study supports previous research on the need for more knowledge, systems, and legislation to support the transition to CE, it also highlights the need for further studies to fully understand the impact of CE on creativity in the different stages of the design process. Additionally, it is important to note that the transition to CE will require a mindset shift among all stakeholders involved in innovative projects, thus further studies on collaboration between stakeholders are needed to fully understand the impact of external resources on creativity.

References

Amabile TM (1983) The social psychology of creativity: A componential conceptualization. J Pers Soc Psychol 45(2):357

Arkitektforeningen (2022) Når LCA bliver et fælles ansvar. https://arkitektforeningen.dk/kalender/open-source-naar-lca-bliver-et-faelles-ansvar/. Accessed 25 Sep 2022

Arup, Ellen MacArthur Foundation (2022) From Principles to Practices: Realising the value of the circular economy in real estate

Bocken N, de Pauw I, Bakker C, van der Grinten B (2016) Product design and business model strategies for a circular economy. J Ind Prod Eng 33:308–320. https://doi.org/10.1080/21681015.2016.1172124

Charter M (ed) (2018) Designing for the circular economy. Routledge

Collado-Ruiz D, Ostad-Ahmad-Ghorabi H (2010) Influence of environmental information on creativity. Design Studies 31

Crotty M (1998) The foundations of social research – Meaning and perspective in the research process. SAGE Publications, Thousand Oaks, London, UK

Csikszentmihalyi M (1988) Motivation and creativity: Toward a synthesis of structural and energistic approaches to cognition. New Ideas Psychol 6:159–176

Creative Denmark (2021) Designing the irresistible circular society. https://ddc.dk/wp-content/uploads/2021/10/creative_denmark_white_paper_designing_the_irresistible_circular_society_small.pdf. Accessed 25 Sep 2022

Dokter G, Thuvander L, Rahe U (2021) How circular is current design practice ? Investigating perspectives across industrial design and architecture in the transition towards a circular economy. Sustain Prod Consum 26:692–708. https://doi.org/10.1016/j.spc.2020.12.032

Ellen MacArthur Foundation (2013) Towards the circular economy—Economic and business rationale for an accelerated transition

Ellen MacArthur Foundation (2015) Towards a Circular Economy: Business Rationale for an Accelerated Transition. Ellen MacArthur Foundation 20

Ellen MacArthur Foundation (2016) What is the Circular Economy. In: Video. https://youtu.be/jbBTwKOllrg

Geissdoerfer M, Saverget P, Bocken N, Hultink EJ (2017) The circular economy—A new sustainability paradigm? J Clean Prod 143:757–768

Gioia DA, Corley KG, Hamilton AL (2013) Seeking qualitative rigor in inductive research. Organ Res Methods 16:15–31. https://doi.org/10.1177/1094428112452151

Indenrigs- og Boligministeriet (2021) LOV nr 2156 af 27/11/2021, Lov om ændring af byggeloven. https://www.retsinformation.dk/eli/lta/2021/2156. Accessed 21 Sep 2022

Joensuu T, Edelman H, Saari A (2020) Circular economy practices in the built environment. J Clean Prod 276:124215. https://doi.org/10.1016/j.jclepro.2020.124215

Mohanani R, Ralph P, Shreeve B (2014) Requirements fixation. In: Proceedings—International Conference on Software Engineering, pp 895–906. https://doi.org/10.1145/2568225.2568235

Moreno M, De los Rios C, Rowe Z, Charnley F (2016) A conceptual framework for circular design. Sustainability (Switzerland) 8. https://doi.org/10.3390/su8090937

Mumford MD, Scott GM, Gaddis B, Strange JM (2002) Leading creative people: Orchestrating expertise and relationships. Leadersh Q 13:705–750

Munaro MR, Tavares SF, Bragança L (2020) Towards circular and more sustainable buildings: A systematic literature review on the circular economy in the built environment. J Clean Prod 260

Münster MB, Sönnichsen SD, Clement J (2022) Retail design in the transition to circular economy : A study of barriers and drivers. J Clean Prod 362. https://doi.org/10.1016/j.jclepro.2022.132310

Ruiz-Pastor L, Mulet E, Chulvi V, Royo M (2021) Effect of the application of circularity requirements as guided questions on the creativity and the circularity of the design outcomes. Journal of Cleaner Production 124758. https://doi.org/10.1016/j.jclepro.2020.124758

Sumter D, de Koning J, Bakker C, Balkenende R (2021) Key competencies for design in a circular economy: Exploring gaps in design knowledge and skills for a circular economy. Sustainability (switzerland) 13:1–15. https://doi.org/10.3390/su13020776

The European Commission (2018) Sustainable Product Policy. https://ec.europa.eu/jrc/en/research-topic/sustainable-product-policy. Accessed 23 May 2021

The European Commission (2020) Circular economy action plan. In: European Union. https://environment.ec.europa.eu/strategy/circular-economy-action-plan_en. Accessed 28 Sep 2022

United Nations (2015) The 17 goals. Department of economic and social affairs. https://sdgs.un.org/goals. Accessed 25 Sep 2022